CAMBRIDGE LIBRARY COLLECTION

Books of enduring scholarly value

Botany and Horticulture

Until the nineteenth century, the investigation of natural phenomena, plants and animals was considered either the preserve of elite scholars or a pastime for the leisured upper classes. As increasing academic rigour and systematisation was brought to the study of 'natural history', its subdisciplines were adopted into university curricula, and learned societies (such as the Royal Horticultural Society, founded in 1804) were established to support research in these areas. A related development was strong enthusiasm for exotic garden plants, which resulted in plant collecting expeditions to every corner of the globe, sometimes with tragic consequences. This series includes accounts of some of those expeditions, detailed reference works on the flora of different regions, and practical advice for amateur and professional gardeners.

The Student's Flora of the British Islands

This textbook was originally published in 1870, but is here reissued in the third edition of 1884. Its object was 'to supply students and field-botanists with a fuller account of the Flowering Plants and Vascular Cryptograms of the British Islands than the manuals hitherto in use aim at giving'. Sir Joseph Dalton Hooker (1817–1911), one of the most eminent botanists of the later nineteenth century, was educated at Glasgow, and developed his studies of plant life through expeditions all over the world. (Several of his other works are also reissued in the Cambridge Library Collection.) A close friend and supporter of Charles Darwin, he was appointed to succeed his father as Director of the Botanical Gardens at Kew in 1865. The flora is followed in this reissue by an 1879 catalogue of British plants compiled by the botanist George Henslow (1835–1925), intended as a companion volume.

Cambridge University Press has long been a pioneer in the reissuing of out-of-print titles from its own backlist, producing digital reprints of books that are still sought after by scholars and students but could not be reprinted economically using traditional technology. The Cambridge Library Collection extends this activity to a wider range of books which are still of importance to researchers and professionals, either for the source material they contain, or as landmarks in the history of their academic discipline.

Drawing from the world-renowned collections in the Cambridge University Library and other partner libraries, and guided by the advice of experts in each subject area, Cambridge University Press is using state-of-the-art scanning machines in its own Printing House to capture the content of each book selected for inclusion. The files are processed to give a consistently clear, crisp image, and the books finished to the high quality standard for which the Press is recognised around the world. The latest print-on-demand technology ensures that the books will remain available indefinitely, and that orders for single or multiple copies can quickly be supplied.

The Cambridge Library Collection brings back to life books of enduring scholarly value (including out-of-copyright works originally issued by other publishers) across a wide range of disciplines in the humanities and social sciences and in science and technology.

The Student's Flora
of the British Islands

JOSEPH DALTON HOOKER

CAMBRIDGE
UNIVERSITY PRESS

CAMBRIDGE
UNIVERSITY PRESS

University Printing House, Cambridge, CB2 8BS, United Kingdom

Cambridge University Press is part of the University of Cambridge.
It furthers the University's mission by disseminating knowledge in the pursuit of
education, learning and research at the highest international levels of excellence.

www.cambridge.org
Information on this title: www.cambridge.org/9781108069663

© in this compilation Cambridge University Press 2014

This edition first published 1884
This digitally printed version 2014

ISBN 978-1-108-06966-3 Paperback

Selected botanical reference works available in the
CAMBRIDGE LIBRARY COLLECTION

al-Shirazi, Noureddeen Mohammed Abdullah (compiler), translated by Francis Gladwin: *Ulfáz Udwiyeh, or the Materia Medica* (1793) [ISBN 9781108056090]

Arber, Agnes: *Herbals: Their Origin and Evolution* (1938) [ISBN 9781108016711]

Arber, Agnes: *Monocotyledons* (1925) [ISBN 9781108013208]

Arber, Agnes: *The Gramineae* (1934) [ISBN 9781108017312]

Arber, Agnes: *Water Plants* (1920) [ISBN 9781108017329]

Bower, F.O.: *The Ferns (Filicales)* (3 vols., 1923–8) [ISBN 9781108013192]

Candolle, Augustin Pyramus de, and Sprengel, Kurt: *Elements of the Philosophy of Plants* (1821) [ISBN 9781108037464]

Cheeseman, Thomas Frederick: *Manual of the New Zealand Flora* (2 vols., 1906) [ISBN 9781108037525]

Cockayne, Leonard: *The Vegetation of New Zealand* (1928) [ISBN 9781108032384]

Cunningham, Robert O.: *Notes on the Natural History of the Strait of Magellan and West Coast of Patagonia* (1871) [ISBN 9781108041850]

Gwynne-Vaughan, Helen: *Fungi* (1922) [ISBN 9781108013215]

Henslow, John Stevens: *A Catalogue of British Plants Arranged According to the Natural System* (1829) [ISBN 9781108061728]

Henslow, John Stevens: *A Dictionary of Botanical Terms* (1856) [ISBN 9781108001311]

Henslow, John Stevens: *Flora of Suffolk* (1860) [ISBN 9781108055673]

Henslow, John Stevens: *The Principles of Descriptive and Physiological Botany* (1835) [ISBN 9781108001861]

Hogg, Robert: *The British Pomology* (1851) [ISBN 9781108039444]

Hooker, Joseph Dalton, and Thomson, Thomas: *Flora Indica* (1855) [ISBN 9781108037495]

Hooker, Joseph Dalton: *Handbook of the New Zealand Flora* (2 vols., 1864–7) [ISBN 9781108030410]

Hooker, William Jackson: *Icones Plantarum* (10 vols., 1837–54) [ISBN 9781108039314]

Hooker, William Jackson: *Kew Gardens* (1858) [ISBN 9781108065450]

Jussieu, Adrien de, edited by J.H. Wilson: *The Elements of Botany* (1849) [ISBN 9781108037310]

Lindley, John: *Flora Medica* (1838) [ISBN 9781108038454]

Müller, Ferdinand von, edited by William Woolls: *Plants of New South Wales* (1885) [ISBN 9781108021050]

Oliver, Daniel: *First Book of Indian Botany* (1869) [ISBN 9781108055628]

Pearson, H.H.W., edited by A.C. Seward: *Gnetales* (1929) [ISBN 9781108013987]

Perring, Franklyn Hugh et al.: *A Flora of Cambridgeshire* (1964) [ISBN 9781108002400]

Sachs, Julius, edited and translated by Alfred Bennett, assisted by W.T. Thiselton Dyer: *A Text-Book of Botany* (1875) [ISBN 9781108038324]

Seward, A.C.: *Fossil Plants* (4 vols., 1898–1919) [ISBN 9781108015998]

Tansley, A.G.: *Types of British Vegetation* (1911) [ISBN 9781108045063]

Traill, Catherine Parr Strickland, illustrated by Agnes FitzGibbon Chamberlin: *Studies of Plant Life in Canada* (1885) [ISBN 9781108033756]

Tristram, Henry Baker: *The Fauna and Flora of Palestine* (1884) [ISBN 9781108042048]

Vogel, Theodore, edited by William Jackson Hooker: *Niger Flora* (1849) [ISBN 9781108030380]

West, G.S.: *Algae* (1916) [ISBN 9781108013222]

Woods, Joseph: *The Tourist's Flora* (1850) [ISBN 9781108062466]

For a complete list of titles in the Cambridge Library Collection please visit:
www.cambridge.org/features/CambridgeLibraryCollection/books.htm

THE STUDENT'S FLORA

OF THE

BRITISH ISLANDS.

THE STUDENT'S FLORA

OF THE

𝕭𝖗𝖎𝖙𝖎𝖘𝖍 𝕴𝖘𝖑𝖆𝖓𝖉𝖘

BY

Sɪʀ J. D. HOOKER, K.C.S.I., C.B.,

M.D.; D.C.L. OXON.; LL.D. CANTAB., GLOTT., ET DUBL.; F.R.S., L.S., AND G.S.;
DIRECTOR OF THE ROYAL GARDENS, KEW.

THIRD EDITION

London:

MACMILLAN & CO.

1884.

LONDON:

R. Clay, Sons, and Taylor,

BREAD STREET HILL.

PREFACE.

THE object of this work is to supply students and field-botanists with a fuller account of the Flowering Plants and Vascular Cryptogams of the British Islands than the manuals hitherto in use aim at giving.

For the plants regarded as composing the British Flora proper, I have mainly followed the *London Catalogue of British Plants*, 7th ed., 1874 ; being fully satisfied that I should thus best serve the interests of British Botany. The difficult task of determining which of the many doubtfully indigenous or naturalized plants should be regarded as British by adoption or otherwise has in the successive editions of this Catalogue been settled by the two botanists most competent to form an opinion by many years of research and by matured judgment—Messrs. H. C. Watson and J. Boswell. It is true, I may think that some of the Species they have introduced have less claims than some they have rejected, but this applies to very few cases indeed.

The Ordinal, Generic, and Specific characters are to a great extent original, and drawn from living or dried specimens or both. After working them out, I have consulted the usual British and Continental Floras, and collated the descriptions throughout with Mr. Boswell's (an author usually quoted by his earlier name of Syme) edition of *English Botany*, of the descriptions in which work I cannot speak in terms of too high praise. By this method of re-description, whilst I believe I have avoided some errors of my predecessors, I have no doubt made others of my own ; such creep into all endeavours to describe most or all of the organs of many Species : and if I have made many such blunders, a part may be attributable to the fact that various Genera were described amidst constant interruptions, and all under pressure of official duties.

The terminology employed is as simple as is attainable with a due regard to precision of language.   In the choice of terms I have followed Oliver's *Lessons in Elementary Botany;* usually avoiding such as are used in single Orders only, or are of special signification in single Orders or Genera.   For modifications of the fruit the choice of terms presents great difficulty ; and I have therefore very much confined myself to such as are required to avoid periphrasis, as capsule, drupe, berry, utricle, follicle, pod, &c. (about which there is no ambiguity), and to *achene* for the dry indehiscent 1-seeded carpels of apocarpous fruits.   For Grasses, Compositæ, &c., the term *fruit* is itself sufficiently explicit, its nature being explained in the Ordinal description.   The term *nutlet* for the parts of the fruit of *Boragineæ* and *Labiatæ* I have borrowed from Asa Gray.

The Keys to the Genera are naturally arranged, but in *Umbelliferæ* I have added an artificial key, as being useful for the determination of a Genus before the whole Order has been studied.   I have given no keys to the Species, preferring curt diagnoses which embrace the more important characters of the plant ; finding, moreover, from experience, that such keys promote very superficial habits among students.

For the areas and elevations inhabited by the plants of the British Isles I am mainly indebted to Mr. Watson's admirable works.   The areas occupied more or less continuously by the Species are here defined by the counties, which thus indicate their limits.   Where the words "northwards" and "southwards" are used it implies that the plant ranges to Shetland in the former case, and to both Cornwall and Kent in the latter.   In this Edition I have in all cases mentioned Ireland when the Species inhabits that country ; and when rare or local in Ireland, its limits are taken from the *Cybele Hibernica* of More and Moore, a standard work.   I have in like manner definitely ·mentioned the Channel Islands.   I have been urged by very competent botanists to include the Faroe Islands, as really more British geographically than are the Channel Islands ; but, if I did so, Iceland should also be included, and on the whole I have thought it best to retain the old limits of the British Flora. The extra-British distributions I worked out myself for most of the British plants, making large use of Nyman's *Sylloge* (ed. 2).

Of the altitudes, I have chosen the highest the species attains, and indicated the region where this is attained ; when no elevation is given, the Species is not known to ascend to 1,000 feet, and may be assumed to be a "low-ground" plant. To the doubtfully indigenous Species I have often added Watson's opinion as to whether they are "aliens," "colonists," or "denizens," &c. It may be well to repeat here his definitions of these terms, premising that by "native" is meant that the Species has not been introduced by human agency :— *

"A denizen is a Species suspected to have been introduced by man, and which maintains its habitat. A colonist is one found only in ground adapted by man for its growth and continuous maintenance. An alien has presumably been introduced by human agency."

The estimates of the numbers of Genera in the Orders, and of Species in the Genera, are taken from the *Genera Plantarum ;* they serve to indicate to the student the relative extent of these groups. The indications of their affinities and properties are necessarily extremely brief. The etymologies of the generic names I have endeavoured to reduce to really useful limits. Only such English names as are pretty well known are given, and for these I have in many cases been guided by Dr. Alexander Prior's *Popular Names of British Plants,* a very good book.

In the First and Second Editions I recorded my obligations to Professor Oliver, Mr. Baker, Professor Dickson, Mr. G. Griffiths, and the Rev. E. J. Linton, for valuable observations and suggestions ; to Mr. Baker especially for aid in classifying the critical forms of *Rubus, Rosa,* and *Hieracium.*

In this, the Third Edition, I have introduced many improvements in the classification and characters of the Orders, Genera, and Tribes, adopted in

* The vagueness of these definitions is unavoidable ; and their correct application in many cases is exceedingly difficult. Few who have not gone into the subject have an idea of how many plants would disappear from our Flora were the soil left undisturbed by man and the lower animals which he rears. I think it probable that both the Shepherd's-Purse and the common form of the Dandelion would be amongst the first to be suppressed.

Bentham's and my *Genera Plantarum*. I have also made changes in the limits of the Species of certain Genera, and of their subordinate forms, in which matter I have often had regard·to suggestions and materials laid before me by Mr. Baker (who has again revised the sheets as they passed through the press), and Mr. Nicholson ; and for the first forty-one Orders to notes made for me by Mr. Ball, F.R.S. These last have a special value, due to Mr. Ball's critical knowledge of so many European Floras, and his excellent judgment. I have further profited by the last edition (8th) of Professor Babington's accurate and critical Manual, and have collated the whole with the second edition of Nyman's *Sylloge Floræ Europææ*, and of Newbould's and Baker's edition of Watson's *Cybele*. To Mr. Arthur Bennett, F.L.S., of Croydon, I am indebted for revising the Genus *Potamogeton*, and for notes upon *Carices*.

The collation of the British Flora with Nyman's *Sylloge* has not been satisfactory throughout, because of the wide divergence of the views there upheld regarding the Species of such Genera as *Rubus*, *Rosa*, &c., from those held by English botanists. This is doubtless due to the fact that characters which are constant and strong in one country become vague and even evanescent in others ; insomuch that I am led from examination and study to believe that, in respect of the subdivision of the European forms of such Genera into Species, Sub-species, and Varieties, the materials in Britain may give one result, those in France another, in Scandinavia a third, and in Germany a fourth.

I am disposed to think that the term Sub-species (which represents a stage of evolution between Species and Variety) should be given to many forms considered by some as Species and as Varieties by others ; and that this would facilitate the better understanding especially of the larger critical Genera. The various forms of fruticose *Rubi*, for example, whether all treated as Species, or all as Varieties, present to me a mere chaos ; whereas, when treated as Sub-species and Varieties, however imperfectly, they fall into comprehensible groups, whose cross affinities may thus be more clearly enunciated.

Lastly, I have ventured to introduce into this Edition, under the description of the flowers of various Genera, characters concerned in the process

of fertilization,—as, whether wind-fertilized (anemophilous), insect-fertil-
ized (entomophilous), or self-fertilized ; also whether honey is secreted
in the flower ; and whether the stamens and stigma ripen together (homogamous), or the anthers first (proterandrous), or the stigma first (proterogynous). For most of the information under these heads I am indebted
especially to the observations of Hermann Müller, supplemented by those
of Sir John Lubbock and Mr. Alfred Bennett. Our knowledge of these
subjects is incomplete and rudimentary : any student may add to it ; but
great caution is required, for I suspect that individual Species are subject
to considerable variation in these respects.

ROYAL GARDENS, KEW,
June 1, 1884.

SYNOPSIS OF THE NATURAL ORDERS

(ADAPTED TO THE BRITISH GENERA).

THE arrangement of Dicotyledons here adopted adheres very closely to the Jussieuan as modified by De Candolle, which, notwithstanding its many defects (inseparable from a linear arrangement), is, I think, as good as any of those subsequently proposed,* and has the great advantage of being that most generally adopted in the Universities and Schools of Great Britain and America, and in systematic works everywhere. Its great defect is the necessity of an Apetalous division, embracing a hetero-geneous mass of Orders, which are incapable of being naturally grouped. Some of these are obviously allied to Polypetalous or Monopetalous Orders, but cannot be placed in contiguity with them without interfering with their other and closer alliances ; some again present cross affinities with two or more distant Orders ; and the greater proportion have no recog-nized near affinities. Under these circumstances, and seeing how much the retention of the Apetalous division facilitates the often difficult task of finding the Natural Order of a plant, it appears to be premature to depart from the Jussieuan system.

SUB-KINGDOM I. **Phænogamous** or **Flowering** plants. Plants provided with stamens, and ovules which after fertilization become seeds containing an embryo.

CLASS I. **Dicotyledonous** or **Exogenous** plants. *Stem* with bark, pith, and interposed wood ; when perennial increasing in diameter annu-ally by a layer of wood added to the outside of the old wood, and another of bark added to the inside of the old bark. *Leaves* with usually netted

---

* Of these the principal are : that of Brongniart, adopted in the Paris Schools ; of Endlicher, in many of the German Schools ; of Fries, by various botanists in Scandi-navia ; and of Lindley ("The Vegetable Kingdom"), which has been partially followed in England and Ind.a alone.

veins. *Flowers* with the organs mostly in fours or fives. *Embryo* with opposite or whorled cotyledons.

SUB-CLASS I. **Angiospermous Dicotyledons.** *Flowers* usually provided with a distinct perianth. *Ovules* contained in closed carpels, through the tissues of which the pollen-tube passes to effect fertilization. *Embryo* with 2 cotyledons

DIVISION I.   **Polypet'alæ.**

Flowers with both calyx and corolla (dichlamydeous). *Petals* free.—See also the exceptional *Monopetalæ*.

Exceptions. Flowers wanting either calyx or corolla occur in: 1 RANUNCU-LACEÆ; 6 CRUCIFERÆ (5, *Cardamine*, 16 *Senebiera*, and 17 *Lepidium*); 9 VIOLACEÆ (apetalous forms of *Viola*); 12 CARYOPHYLLEÆ (8 *Arenaria* § *Cherleria*, 9 *Sagina*); 26 ROSACEÆ (8 *Alchemilla*, 10 *Poterium*); 27 SAXIFRAGEÆ (2 *Chrysosplenium*); 30 HALORAGEÆ; 31 LYTHRACEÆ (2 *Peplis*); 32 ONAGRARIEÆ (2 *Ludwigia*).

Petals more or less connate or coherent occur in: 5 FUMARIACEÆ; 10 POLYGALEÆ; 13 PORTULACEÆ; 13* TAMARISCINEÆ; 16 MALVACEÆ; 20 ILICINEÆ; 28 CRASSULACEÆ (2 *Cotyledon*); 33 CUCURBITACEÆ.

SUB-DIVISION I. **Thalamiflo'ræ.** *Stamens* inserted on the receptacle (hypogynous), free from the calyx, or on a disk that terminates the pedicel. *Ovary* superior.

Exceptions. Stamens apparently perigynous or epigynous in 3 NYMPHÆACEÆ (1 *Nymphæa*) and in some 12 CARYOPHYLLEÆ.

* *Ovary apocarpous, carpels 1 or more ; ovules sutural or basal. (See also 16* Malvaceæ *and 19* Geraniaceæ.)

1. RANUNCULACEÆ. *Flowers* regular or irregular. *Stamens* indefinite ; anthers basifixed, opening by slits. *Seeds* albuminous.—Herbs with alternate leaves (except *Clematis*). (p. 1.)
2. BERBERIDEÆ. *Flowers* regular, 3-merous. *Stamens* definite, opposite the petals ; anthers basifixed, opening by recurved valves. *Seeds* albuminous.—Shrubs ; leaves alternate ; flowers often showy. (p. 14.)

** *Ovary syncarpous, 1-celled (except 3* Nymphæaceæ), *or 2-celled by a membranous septum ; ovules parietal, rarely basal.*

3. NYMPHÆACEÆ. *Flowers* regular. *Stamens* indefinite ; anthers basifixed. *Ovary* many-celled ; *ovules* scattered over the walls of the cells ; stigmas sessile. *Seeds* albuminous.—Water-herbs ; flowers showy. (p. 15.)
4. PAPAVERACEÆ. *Flowers* regular, 2-merous. *Stamens* indefinite ; anthers basifixed. *Ovules* parietal or on the surfaces of partial dissepiments ; style 1 or stigmas sessile. *Seeds* albuminous.—Herbs ; juice milky ; leaves alternate, exstipulate ; flowers usually showy. (p. 16.)

5. FUMARIACEÆ. *Flowers* irregular. *Sepals* 2. *Petals* 4. *Stamens* 6 in 2 bundles. *Ovary* 1-celled ; ovules many parietal, or 1 (by suppression) basal ; style 1 or 0. *Seeds* albuminous.—Weak herbs with exstipulate alternate leaves ; flowers usually small. (p. 19.)

6. CRUCIFERÆ. *Flowers* usually regular. *Sepals* 4. *Petals* 4. *Stamens* usually 6, 4 longer than the others. *Ovary* 1-2-celled, of 2 carpels ; ovules parietal ; style 1 or 0. *Seeds* exalbuminous.—Herbs ; leaves exstipulate, alternate ; flowers usually small and ebracteate. (p. 22.)

7. RESEDACEÆ. *Flowers* irregular. *Sepals* and *petals* 4–7 each. *Stamens* indefinite. *Ovary* 1-celled, of 2–6 carpels, at length open at the top ; ovules parietal ; stigma sessile. *Seeds* exalbuminous.— Herbs ; leaves alternate, stipules glandular or 0 ; flowers small, greenish. (p. 44.)

8. CISTINEÆ. *Flowers* regular. *Sepals* 3–5. *Petals* 5. *Stamens* indefinite. *Ovary* 1-celled, of 3 carpels ; ovules parietal ; styles 3. *Seeds* albuminous.—Shrubs ; leaves usually stipulate ; flowers yellow or red, showy ; petals fugaceous. (p. 45.)

9. VIOLACEÆ. *Flowers* irregular. *Sepals, petals,* and *stamens* 5 each. *Ovary* 1-celled ; ovules parietal ; style 1. *Capsule* 3-valved, loculicidal. *Seeds* albuminous.—Herbs ; leaves alternate, stipulate ; flowers often showy. (p. 47.)

11. FRANKENIACEÆ. *Flowers* regular. *Sepals, petals,* and *stamens* 4–6 each. *Ovary* 1-celled, of 2–5 carpels ; ovules parietal ; style 1.—A littoral herb ; leaves opposite, exstipulate ; flowers small. (p. 51.)

*** *Ovary syncarpous, 1-celled ; placenta free-central or basal.*

12. CARYOPHYLLEÆ. *Flowers* regular. *Sepals* and *petals* 4 or 5 each. *Stamens* 8 or 10. *Ovules* many ; styles 2–5. *Seeds* albuminous ; embryo curved.—Herbs ; leaves opposite, stipulate or not ; flowers usually small and pink or white. (p. 52.)

13. PORTULACEÆ. *Flowers* regular. *Sepals* 2. *Petals* 4 or more. *Stamens* 3 or more. *Ovules* 2 or more ; style 1, 2–3-fid. *Seeds* albuminous ; embryo curved.—Herbs ; leaves opposite or alternate, quite entire ; flowers small. (p. 69.)

13*. TAMARISCINEÆ. *Flowers* regular. *Sepals* and *petals* 4–5 each. *Stamens* 4 or more. *Ovules* 2 or more ; styles 3–4.—Shrubs ; leaves minute, exstipulate ; flowers small. (p. 70.)

**** *Ovary syncarpous, 2- or more-celled ; placentas axile.*

10. POLYGALEÆ. *Flowers* irregular. *Inner sepals* petaloid. *Petals* adnate to the staminal sheath. *Stamens* 8 ; anthers 1-celled. *Ovary* 2-celled, 2-ovuled ; style 1.—Herbs ; leaves alternate or subopposite, exstipulate ; flowers small, usually blue. (p. 50.)

14. ELATINEÆ. *Flowers* regular. *Sepals* and *petals* 3–4 each. *Stamens* 3–4 or twice as many, free. *Ovary* 2-5-celled ; styles 2–5. *Seeds* ribbed ; albumen scanty or 0.—Water-herbs ; leaves opposite, stipulate ; flowers minute. (p. 71.)

15. HYPERICINEÆ. *Flowers* regular. *Sepals* and *petals* 5 each. *Stamens* many, united in bundles. *Ovary* more or less completely 3–5-celled, cells many-ovuled ; styles 3–5. *Seeds* exalbuminous.—Herbs or shrubs ; leaves opposite, often gland-dotted, exstipulate ; flowers often showy, yellow. (p. 71.)

16. MALVACEÆ. *Flowers* regular. *Sepals* 5, valvate, persistent. *Petals* 5, twisted in bud, adnate to the staminal tube. *Stamens* monadelphous ; anthers 1-celled. *Ovary* many-celled, cells 1-ovuled (in British genera). *Albumen* scanty or 0; embryo crumpled.—Herbs or shrubs ; leaves alternate, stipulate ; flowers often showy. (p. 74.)

17. TILIACEÆ. *Flowers* regular. *Sepals* 5, valvate, deciduous. *Petals* 5. *Stamens* indefinite ; anthers 2-celled. *Ovary* 2–10-celled ; cells 2-ovuled ; style 1. *Seeds* albuminous.—Trees ; leaves alternate, stipulate ; flowers not showy. (p. 76.)

18. LINEÆ. *Flowers* regular. *Sepals* 4–5. *Petals* 4–5, convolute in bud. *Stamens* usually 4–5. *Ovary* 3–5- (-10-) celled, cells 1–2-ovuled ; styles 3–5. *Seeds* albuminous.—Herbs ; leaves opposite or alternate, narrow, quite entire, exstipulate ; flowers usually showy. (p. 77.)

19. GERANIACEÆ. *Flowers* regular or not. *Sepals* 3–5. *Petals* 3–5, imbricate in bud. *Stamens* definite. *Ovary* 3–5-lobed and -celled ; cells 1–many-ovuled ; styles 1 or more. *Albumen* scanty or 0 ; cotyledons plaited or convolute.—Herbs ; leaves opposite or alternate, usually stipulate ; flowers often showy. (p. 79.)

20. ILICINEÆ. *Flowers* regular. *Sepals* 4–5. *Petals* 4–5, often connate, imbricate in bud. *Stamens* 4–5. *Ovary* 3–5-celled, cells 1–2-ovuled. *Seeds* albuminous.—Shrubs ; leaves evergreen, alternate, exstipulate ; flowers small. (p. 85.)

21. EMPETRACEÆ. *Flowers* regular, diœcious. *Sepals* 3. *Petals* 3, imbricate in bud. *Stamens* 3. *Ovary* 3–9-celled, cells 1-ovuled. *Seeds* albuminous.—Small shrubs ; leaves evergreen, alternate, exstipulate ; flowers inconspicuous. (p. 86.)

SUB-DIVISION II. **Calyciflorae.** *Stamens* inserted on the calyx or disk (perigynous or epigynous).—See also the exceptional *Thalamifloræ.*

Exceptions. Stamens hypogynous in 27 SAXIFRAGEÆ (3 *Parnassia*), and in 29 DROSERACEÆ; epipetalous in some 28 CRASSULACEÆ ; almost hypogynous in some 25 LEGUMINOSÆ.

* *Ovary superior (except some 26 Rosaceæ and 27 Saxifrageæ). Stamens perigynous.*

22. CELASTRINEÆ. *Flowers* regular. *Calyx* 4–5-lobed, and *petals* 4–5, both imbricate in bud. *Stamens* 4–5, inserted on the disk. *Ovary* 3–5-celled, cells with 2 erect ovules. *Seeds* arillate ; cotyledons foliaceous.—Trees or shrubs ; leaves various ; flowers small. (p. 87.)

23. RHAMNEÆ. *Flowers* regular. *Calyx* 4–5-lobed, valvate in bud. *Petals* 4–5, minute. *Stamens* 1 opposite each petal, inserted on the calyx-tube at the edge of the disk. *Ovary* 3-celled ; ovule 1, erect in each cell.

—Shrubs; leaves alternate or opposite, stipules small; flowers inconspicuous. (p. 87.)

24. SAPINDACEÆ, *Tribe* ACERINEÆ. *Flowers* regular. *Calyx* 4–9-lobed, and *petals* 4–9, both imbricate in bud. *Stamens* 8–12, inserted on the disk. *Ovary* 2-lobed and -celled; cells 2-ovuled. *Fruit* a samara; cotyledons plaited.—Trees or shrubs; leaves opposite; flowers rather small, green. (p. 88.)

25. LEGUMINOSÆ. *Flowers* irregular, papilionaceous. *Stamens* 10, subhypogynous or inserted on the calyx-tube, all or 9 of them combined. *Ovary* of 1 carpel. *Fruit* a legume. *Albumen* 0.—Herbs or shrubs; leaves usually alternate compound and stipulate.   (p. 89.)

26. ROSACEÆ. *Flowers* regular. *Calyx* 4–5- (rarely 8–9-) lobed, imbricate or valvate in bud. *Petals* 4–5 (rarely 8–9 or 0), imbricate in bud. *Stamens* usually indefinite, inserted on the calyx-tube or disk, incurved in bud. *Ovary* of 1 or more free or connate 1- or more-ovuled carpels. *Fruit* various. *Albumen* 0.—Herbs or shrubs; leaves usually alternate, stipulate; flowers often showy. (p. 113.)

27. SAXIFRAGEÆ. *Flowers* regular. *Calyx* 4-5-lobed. *Petals* 4–5, rarely 0, imbricate in bud. *Stamens* definite. *Carpels* fewer than the petals, usually 2 connate; placentas axile, rarely parietal. *Fruit* various. *Seeds* albuminous.—Herbs or shrubs; leaves opposite or alternate, stipulate or not; flowers small. (p. 138.)

28. CRASSULACEÆ. *Flowers* regular. *Calyx* 4–12-lobed. *Petals* 4-12. *Stamens* twice as many as the petals (except 1 *Tillæa*). *Carpels* follicular, usually 5, separate.—Herbs; leaves succulent, exstipulate; flowers small. (p. 145.)

29. DROSERACEÆ. *Flowers* regular *Sepals* and *petals* 5, imbricate in bud. *Stamens* as many, hypogynous or perigynous. *Ovary* 1-celled; ovules many, parietal. *Fruit* capsular. *Seeds* albuminous.—Glandular herbs; leaves radical; flowers small, white or pink. (p. 149.)

31. LYTHRACEÆ. *Flowers* regular. *Calyx-lobes* 3–6, valvate in bud. *Petals* 3–6, crumpled in bud. *Stamens* definite. *Ovary* 2-6-celled, cells many-ovuled. *Capsule* many-seeded. *Seeds* exalbuminous.—Herbs; leaves opposite or whorled, quite entire, exstipulate; flowers often showy. (p. 153.)

** *Ovary inferior. Stamens epigynous.*

30. HALORAGEÆ. *Flowers* usually apetalous and 1-sexual. *Calyx-lobes* 2–4, valvate in bud, or 0. *Stamens* 1 or more, definite. *Ovary* 1-4-celled, cells 1-ovuled. *Seeds* albuminous.—Herbs, often marsh or aquatic; leaves opposite alternate or whorled, exstipulate; flowers very inconspicuous. (p. 151.)

32. ONAGRARIEÆ. *Flowers* usually regular. *Calyx-lobes* 2 or 4, valvate n bud. *Petals* 2 or 4, twisted in bud. *Stamens* definite. *Ovary* 1-4-elled, cells 1–many-seeded. *Seeds* exalbuminous.—Herbs; leaves oppoite or alternate, exstipulate; flowers often showy. (p. 155.)

33. CUCURBITACEÆ.   *Flowers* regular, 1-sexual.   *Calyx* 5-toothed. *Corolla* 5-lobed.   *Stamens* 3.   *Ovary* 3-celled, many-ovuled.   *Fruit* a berry.   *Seeds* exalbuminous.—Herbs with tendrils; leaves alternate, exstipulate; flowers showy or not.   (p. 160.)

34. UMBELLIFERÆ.   *Flowers* usually regular.   *Calyx-lobes* 5 or 0. *Petals* 5.   *Stamens* 5, incurved in bud.   *Ovary* 2-celled; styles 2; ovules solitary.   *Fruit* of 2 separable indehiscent dry carpels.   *Seeds* albuminous.—Herbs; leaves alternate; flowers usually umbelled, small. (p. 161.)

35. ARALIACEÆ.   *Flowers* of *Umbelliferæ*, but shrubs or trees; ovary of often more than 2 carpels.   *Fruit* of inseparable usually fleshy carpels. —Leaves alternate; flowers usually green.   (p. 186.)

36. CORNACEÆ.   *Flowers* regular.   *Calyx-lobes* 4–5 or 0.   *Petals* 4–5. *Stamens* 4–5.   *Ovary* 2-celled, cells 1-ovuled; style simple.   *Drupe* 1–2-celled.   *Seeds* albuminous.—Herbs, shrubs, or trees; leaves opposite; flowers usually small.   (p. 187.)

DIVISION II.   **Monopet'alæ** or **Gamopet'alæ.**

Flowers with both calyx and corolla (dichlamydeous).   *Petals* more or less connate into a 2- or more-lobed corolla.—See also various monopetalous genera under the exceptional *Polypetalæ.*

Exceptions.   Petals free in 43 ERICACEÆ (11 *Pyrola* and 12 *Monotropa*) and 45 PLUMBAGINEÆ.   Corolla absent in 47 OLEACEÆ (2 *Fraxinus*), and 46 PRIMULACEÆ (4 *Glaux*).

1. *Ovary* inferior.—See also 46 PRIMULACEÆ (8 *Samolus*).

* *Stamens epipetalous; see also* 42 CAMPANULACEÆ (1 Lobelia).

37. CAPRIFOLIACEÆ.   *Flowers* regular or not.   *Corolla-lobes* valvate or imbricate in bud.   *Ovary* 1–5-celled, cells 1- or more-ovuled.   *Seeds* albuminous.—Shrubs, rarely herbs; leaves opposite, exstipulate; flowers usually showy.   (p. 188.)

38. RUBIACEÆ, *Tribe* STELLATÆ.   *Flowers* regular.   *Corolla-lobes* valvate in bud.   *Ovary* 2-celled; cells 1-ovuled.   *Seeds* albuminous.— Herbs; leaves whorled or opposite, exstipulate; flowers small or minute. (p. 191.)

39. VALERIANEÆ.   *Flowers* irregular.   *Corolla-lobes* imbricate.   *Stamens* 1–3 or 5, free.   *Ovary* 1–3-celled, one cell 1-ovuled; ovule pendulous. *Seeds* exalbuminous.—Herbs; leaves opposite; flowers small.   (p. 196.)

40. DIPSACEÆ.   *Flowers* regular or not, in involucrate heads.   *Corolla-lobes* imbricate.   *Stamens* 4.   *Ovary* 1-celled; ovule 1, pendulous.   *Seeds* albuminous.—Herbs; leaves opposite, exstipulate; flowers small.   (p. 198.)

41. COMPOSITÆ.   *Flowers* in involucrate heads.   *Corolla-lobes* valvate. *Stamens* 4–5; anthers usually connate.   *Ovary* 1-celled; ovule 1, erect. *Seeds* exalbuminous.—Herbs, rarely shrubs; leaves various, exstipulate; flowers small or minute.   (p. 200.)

** *Stamens inserted on the top of the ovary.*

42. CAMPANULACEÆ   *Flowers* regular or irregular.   *Stamens* 5, separate or connate.   *Ovary* 2–8-celled, cells many-ovuled.—Herbs ; juice milky ; leaves alternate, exstipulate ; flowers usually showy.   (p. 243.)

43. ERICACEÆ, *Suborder* VACCINIEÆ.   *Flowers* regular.   *Stamens* 8 or 10.   *Ovary* 4–5-celled.—Small shrubs ; leaves alternate, exstipulate. (p. 249.)

2. *Ovary* superior.   *Stamens* epipetalous (hypogynous in 43 ERICACEÆ, 47 OLEINEÆ (2 *Fraxinus*), 54 PLANTAGINEÆ (2 *Littorella*), and 45 PLUMBAGINEÆ).

* *Corolla regular.   Stamens 8 or 10, rarely 5 or 6 ; anthers usually opening by pores.   Ovary 4–6-celled.*

43. ERICACEÆ, *Suborder* ERICEÆ.   Leafy shrubs or trees, rarely herbs. (p. 248.)

44. MONOTROPEÆ.   Leafless parasitic herbs.   (p. 257.)

** *Corolla regular.   Stamens 4–5, opposite the corolla-lobes.   Ovary 1-celled ; placenta central.*

45. PLUMBAGINEÆ.   *Styles* or *style-arms* 5.   *Utricle* 1-seeded.—Maritime, rarely alpine, scapigerous herbs ; flowers small.   (p. 257.)

46. PRIMULACEÆ.   *Style* 1.   *Stigma* capitate.   *Capsule* 5–10-valved or circumsciss, many-seeded.—Herbs ; flowers often showy.   (p. 260.)

*** *Corolla regular.   Stamens 2, 4, or 5, alternate with the corolla-lobes.* *Ovary 2-celled.   Leaves opposite (except Menyanthes).*

47. OLEACEÆ.   *Calyx* 4-fid or 0.   *Corolla* 4-lobed or 0.   *Stamens* 2. *Ovary* 2-celled, cells 2–3-ovuled.   *Fruit* a drupe or samara.—Trees or shrubs ; leaves opposite, exstipulate.   (p. 267.)

48. APOCYNACEÆ.   *Calyx* 4–5 fid.   *Corolla* 4–5-lobed, twisted in bud. *Stamens* 4–5 ; anthers basifixed.   *Carpels* 2, free below.   *Fruit* of 2 follicles.—Shrubs ; leaves opposite, quite entire ; flowers often showy. (p. 268.)

49. GENTIANEÆ.   *Calyx* 4–8-fid.   *Corolla* 4–8-lobed, twisted in bud. *Stamens* 4–8 ; anthers versatile.   *Ovary* 1-celled ; ovules many, parietal. *Fruit* usually capsular.—Herbs ; leaves opposite, quite entire (alternate, 3-foliolate in *Menyanthes*) ; flowers often showy.   (p. 269.)

**** *Corolla regular or subregular.   Stamens 4–5, alternate with the corolla-lobes.   Ovary 2–4-celled.   Leaves alternate or radical.*

50. POLEMONIACEÆ.   *Calyx* 5-lobed.   *Corolla* 5-lobed, twisted in bud. *Stamens* 5.   *Ovary* 3-celled ; stigma 3-fid.   *Fruit* capsular.—Herbs ; leaves pinnate, exstipulate ; flowers showy.   (p. 274.)

51. BORAGINEÆ.   *Calyx* 5-lobed, valvate in bud.   *Corolla* 5-lobed, imbricate in bud.   *Stamens* 5.   *Ovary* of 2 2-lobed 2-celled 2-ovuled carpels.

Fruit of 4 nutlets.—Hispid or scabrid herbs ; leaves alternate, quite entire, exstipulate ; flowers often showy. (p. 275.)

52. CONVOLVULACEÆ. *Sepals* 5. *Corolla* 5-lobed, plaited and twisted in bud. *Stamens* 5. *Ovary* 2-celled, cells 2-ovuled ; stigmas 2-fid or styles 2.—Herbs ; leaves alternate, simple (0 in *Cuscuta*) ; flowers often showy. (p. 283.)

53. SOLANACEÆ. *Calyx* 5-fid. *Corolla* 5-lobed, imbricate, plaited or valvate in bud. *Stamens* 5, often cohering. *Ovary* 2-celled ; ovules many, axile. *Fruit* a capsule or berry.—Herbs ; leaves alternate or in pairs, exstipulate ; flowers small or large. (p. 286.)

54. PLANTAGINEÆ. *Sepals* 4. *Corolla* scarious, 4-lobed, imbricate in bud. *Stamens* 4 ; anthers pendulous. *Ovary* 2-4-celled ; style and stigma filiform. *Capsule* 1-4-celled.—Herbs ; leaves alternate or radical ; flowers inconspicuous. (*Littorella* is altogether anomalous.) (p. 288.)

***** *Corolla irregular, rarely subregular. Stamens 2 or 4, rarely 5. Ovary 1-2-celled, cells many ovuled. Leaves opposite or alternate. (See also 53 Solanaceæ.)*

55. SCROPHULARINEÆ. *Calyx* 4-5-merous. *Corolla* often 2-lipped, 4-5-lobed. *Stamens* 4, didynamous, rarely 2 or 5. *Ovary* 2-celled ; ovules many, axile.—Herbs ; leaves various ; flowers often showy. (p. 290.)

56. OROBANCHACEÆ. *Sepals* 4 or 5, free or connate. *Corolla* gaping. *Stamens* 4, didynamous. *Ovary* 1-celled ; ovules many, parietal.—Herbs, with alternate scales · instead of leaves ; flowers rather large, brown or coloured. (p. 308.)

57. LENTIBULARINEÆ. *Calyx* 2-5-partite. *Corolla* 2-labiate. *Stamens* 2. *Capsule* 2-valved, many-seeded.—Marsh or water-plants ; flowers rather large for the plant. (p. 310.)

****** *Corolla irregular. Stamens 2 or 4. Ovary 2- or 4-celled, cells 1-ovuled. —Herbs or shrubs ; leaves opposite or whorled, exstipulate.*

58. VERBENACEÆ. *Calyx* cleft or toothed. *Corolla* tubular, often 2-lipped. *Stamens* 4. *Ovary* not lobed, 2-4-celled ; cells 1-ovuled. *Fruit* a drupe, berry, or of 1-4 nutlets.—Flowers small or showy. (p. 313.)

59. LABIATÆ. *Calyx* 5-cleft or 2-lipped. *Corolla* usually 2-lipped. *Stamens* 2 or 4, didynamous. *Ovary* of 2 2-lobed 2-celled 2-ovuled carpels. *Fruit* of 1-4 1-seeded nutlets.—Flowers in opposite cymes forming false whorls. (p. 313.)

DIVISION III. **Incompletæ.**

(**Monochlamydeæ and Achlamydeæ.**)

Corolla and often calyx absent.—(Petals present in some 60 *Illecebraceæ*. For· various apetalous genera see Exceptions to the *Polypetalæ* and *Mono-petalæ*.

* *Flowers not in catkins. Perianth single, inferior* (0 *in* Euphorbia).

60. ILLECEBRACEÆ. *Flowers* 2-sexual. *Calyx* herbaceous or coriaceous, persistent round the fruit. *Stamens* perigynous, opposite the sepals. *Ovary* 1-celled ; styles 2–3 ; ovules 1–2. *Utricle* 1-seeded. *Albumen* floury, embryo various.—Herbs ; leaves opposite, stipulate (except *Scleranthus*) ; flowers minute. (p. 333.)

61. CHENOPODIACEÆ. *Flowers* 1–2-sexual. *Calyx* 3–5-lobed, herbaceous, persistent round the fruit. *Stamens* 1–5, opposite the sepals. *Ovary* 1-celled ; ovule amphitropous. *Utricle* 1-seeded, indehiscent. *Albumen* floury or fleshy ; embryo annular or spiral.—Herbs ; leaves opposite or alternate, exstipulate, or stems leafless and jointed ; flowers green, inconspicuous. (p. 335.)

62. POLYGONACEÆ. *Flowers* usually 2-sexual. *Sepals* 3–6, green or coloured. *Stamens* 5–8, perigynous or hypogynous. *Fruit* usually enclosed in the sepals. *Ovules* erect, orthotropous. *Albumen* floury ; embryo curved.—Herbs ; leaves alternate ; stipules sheathing ; flowers small. (p. 343.)

64. THYMELÆACEÆ. *Flowers* 2-sexual. *Calyx* tubular ; lobes 4–5. *Stamens* definite, inserted in the tube. *Ovules* pendulous, anatropous. *Albumen* 0 or scanty ; embryo straight.—Shrubs ; leaves quite entire, exstipulate ; bark tenacious ; flowers conspicuous, sweet-scented. (p. 353.)

65. ELÆAGNACEÆ. *Calyx*, in male fl. 3–4-sepalous ; in female or 2-sexual fl. tubular. *Stamens* 4–8 at the base of the sepals in the male fl. *Ovule* erect, anatropous. *Albumen* 0 or scanty ; embryo straight.—Shrubs with silvery scales ; leaves quite entire, exstipulate ; flowers inconspicuous. (p. 354.)

68. EUPHORBIACEÆ. *Flowers* 1-sexual. *Calyx* 0 or *sepals* 2 or more. *Male : Stamens* 1 or more ; anthers didymous. *Female Ovary* 2-3-lobed and -celled ; ovules 1–2 in each cell, pendulous, anatropous ; styles 2–3. *Albumen* copious, fleshy.—Herbs or shrubs ; leaves various ; inflorescence often of many stamens and 1 pistil collected in a small calyx-like involucre. (p. 356.)

69. URTICACEÆ. *Flowers* 1–2-sexual. *Perianth* of male 3–8-lobed or -partite ; of female tubular, or 3–5-cleft, or a scale. *Stamens* opposite the perianth-lobes. *Ovary* 1-celled ; styles 1–2 or 0 ; ovule solitary, pendulous and anatropous, or erect and orthotropous. *Albumen* fleshy or 0.—Herbs or shrubs ; leaves various, stipulate ; flowers minute, green. (p. 361.)

73. CERATOPHYLLEÆ. *Flowers* 1-sexual. *Perianth* 8–12-partite, segments subulate.—*Male*, of many anthers. *Ovary* 1-celled ; style subulate, persistent ; ovule 1, pendulous, anatropous. *Albumen* 0.—A submerged herb, with whorled multifid leaves ; flowers very inconspicuous. (p. 378.)

** *Flowers not in catkins. Perianth single, superior.*

66. LORANTHACEÆ. *Calyx* 4-cleft, valvate in bud. *Stamens* one

adnate to each calyx-lobe. *Ovary* 1-celled ; ovule 1, adnate to the ovary. *Seed* erect, radicle superior ; albumen fleshy.—Parasitic shrubs ; leaves quite entire, exstipulate ; flowers inconspicuous. (p. 354.)

67. SANTALACEÆ. *Calyx* 3–5-lobed, valvate in bud. *Stamens* one adnate to each calyx-lobe. *Ovary* 1-celled ; ovules several, pendulous from a free central placenta. *Albumen* fleshy ; radicle superior.—Shrubs or herbs, often root-parasites ; leaves usually alternate, quite entire, exstipulate ; flowers inconspicuous. (p. 355.)

63. ARISTOLOCHIACEÆ. *Calyx* 3-lobed, or 1–2-lipped, valvate in bud. *Stamens* 6–12, epigynous or gynandrous. *Ovary* 4–6-celled ; ovules many. *Albumen* fleshy ; embryo minute.—Herbs or shrubs ; leaves alternate, exstipulate. (p. 351.)

*** *Flowers 1-sexual ; males in catkins, females in spikes or catkins.* *Perianth present or absent.*

70. MYRICACEÆ. *Flowers* of both sexes in the axils of imbricating bracts ; perianth 0.—*Male* of 2–16 stamens ; anthers basifixed, bursting outwards. —*Female : Ovary* 1-celled ; styles 2, filiform ; ovule 1, basal, orthotropous. *Fruit* a drupe. *Albumen* 0.—A glandular shrub ; leaves alternate, exstipulate ; flowers very inconspicuous. (p. 364.)

71. CUPULIFERÆ. *Flowers* mono-dioecious. *Males* in catkins. *Sepals* 0 or 5 or more. *Stamens* 5–20.—*Females,* sessile in an involucre of free or connate bracts. *Calyx* superior, 5–6-toothed or 0. *Ovary* 2–3-celled ; styles 2–3 ; cells 1–2-ovuled. *Fruit* 1-celled, 1-seeded, dry, indehiscent. *Albumen* 0.—Trees or shrubs ; leaves alternate, stipulate ; flowers small, green. (p. 364.)

72. SALICINEÆ. *Flowers* dioecious, without perianth, both sexes in catkins.—*Male : Stamens* 1 or more.—*Female : Ovary* 1-celled ; stigmas 2 ; ovules many, parietal, anatropous. *Capsule* 2-valved. *Albumen* 0.— Trees, leaves alternate, stipulate. (p. 369.)

SUB-CLASS II. **Gymnospermous Dicotyledons.** *Perianth* usually 0. *Ovules* not contained in close carpels, fertilized by the direct application of the pollen. *Embryo* with often whorled cotyledons.

74. CONIFERÆ. *Perianth* 0. *Male flowers* of 2–8-celled anthers, usually forming a deciduous catkin. *Female fl.* of one or more naked ovules (ovaries of some) on the scales of a cone or head, or of a solitary ovule (*Taxus*). *Albumen* fleshy ; embryo straight,—Trees or shrubs ; leaves alternate opposite or fascicled ; flowers very inconspicuous. (p. 379.)

CLASS II. **Monocotyledonous** or **Endogenous** plants. *Stem* with the wood forming longitudinal bundles irregularly disposed in the stem, not in concentric layers, and having no defined central pith. *Leaves* with usually parallel veins. *Flowers* with the organs mostly in threes or fours, never in fives. *Embryo* with a single cotyledon ; first formed leaves alternate ; radicle not branching, but throwing out adventitious roots.

Exceptions. Leaves net-veined in 79 Dioscoreæ, 80 Liliaceæ (1 *Paris*), and 84 Aroideæ (1 *Arum*). Flowers 4-merous in 86 Naiadaceæ, and 2-3-merous in 89 Gramineæ.

1. Microspermeæ. *Perianth* 2-seriate, coloured. *Ovary* inferior, syncarpous, 1- rarely 3-6-celled, placentas 3 parietal. *Seeds* minute, exalbuminous.

75. Hydrocharideæ. *Flowers* regular, 1-sexual. *Perianth* 6-partite, outer segments herbaceous, inner petaloid (except in *Elodea*). *Stamens* 3 or more. *Ovary* 1- or 3-6-celled. *Fruit* a berry.—Water plants ; leaves erect or floating, flowers usually conspicuous. (p. 381.)

76. Orchideæ. *Flowers* irregular, 2-sexual. *Stamens* 1 or 2, adnate to the style. *Ovary* 1-celled. *Fruit* capsular.—Herbs of various habit. (p. 383.)

2. Epigyneæ. *Perianth* 2-seriate, coloured (except Dioscoreæ). *Ovary* inferior, syncarpous, 3-celled. *Seeds* large, albuminous.

77. Irideæ. *Flowers* 2-sexual. *Perianth* 6-partite, petaloid. *Stamens* 3, separate ; anthers bursting outwards. *Ovary* 3-celled. *Capsule* 3-valved.—Herbs ; roots tuberous, or rootstock creeping ; leaves narrow ; flowers usually handsome. (p. 395.)

78. Amaryllideæ. *Flowers* 2-sexual. *Perianth* 6-partite, petaloid. *Stamens* 6, separate ; anthers bursting inwards. *Ovary* 3-celled. *Capsule* 3-valved.—Herbs ; leaves narrow ; flowers usually handsome. (p. 398.)

79. Dioscoreæ. *Flowers* 1-sexual. *Perianth* small, 6-partite, herbaceous. *Stamens* 6 ; anthers bursting inwards. *Ovary* 3-celled. *Berry* few-seeded.—Climbing herbs ; leaves broad, with netted veins ; flowers inconspicuous. (p. 400.)

3. Coronarieæ. *Perianth* 2-seriate, usually coloured. *Ovary* superior, syncarpous. *Seeds* albuminous.

80. Liliaceæ. *Flowers* usually 2-sexual. *Perianth* usually 6-cleft or of 6 segments, petaloid. *Stamens* 6, opposite the perianth-segments. *Ovary* 3-celled. *Fruit* various.—Herbs (except *Ruscus*) of various habit ; flowers usually showy. (p. 401.)

81. Junceæ. *Flowers* 2-sexual. *Perianth* of 6 green or brown segments. *Stamens* usually 6. *Ovary* 1-3-celled with 3 basilar, or many parietal or axile ovules. *Capsule* 3-valved.—Rushy herbs ; leaves very narrow ; flowers brown, small. (p. 413.)

82. Eriocauloneæ. *Flowers* monœcious, in involucrate heads. *Perianth* membranous or scarious ; outer of 2-3 sepals ; inner 3-lobed or of 3 scales. *Stamens* 2-3 on the inner perianth-segments. *Ovary* 2-3-celled ; ovule 1, pendulous in each cell. *Capsule* 2-3-valved.—Usually scapigerous, cellular, marsh or water herbs ; flowers small, dull-coloured. (p. 420.)

4. NUDIFLORÆ. *Perianth* 0, or rudimentary. *Ovary* superior, syncarpous, or monocarpellary.

83. TYPHACEÆ. *Flowers* monœcious, in catkins or heads. *Perianth* 0, or of scales or hairs. *Stamens* many; anthers basifixed. *Ovary* 1-2-celled; style persistent; ovule 1, pendulous. *Fruit* a drupe or utricle.— Erect marsh or water plants; leaves linear; flowers small or minute, in conspicuous spiked heads. (p. 421.)

84. AROIDEÆ. *Flowers* sessile on a spadix, enclosed in a spathe when young, 1-2-sexual. *Perianth* 0, or of scale-like sepals. *Stamens* few or many. *Ovary* 1- or more-celled. *Berry* few- or many-seeded. *Albumen* mealy.—Herbs; leaves various, often broad, net-veined; flowers with often conspicuous spathes or spadixes. (p. 423.)

85. LEMNACEÆ. Minute floating cellular green fronds. *Flowers* embedded in slits or cavities of the frond, most minute, 1-3 in a spathe. *Stamens* 1-2. *Ovary* 1-celled, 1-7-ovuled.—Fronds covering ponds; flowers very rare and inconspicuous. (p. 424.)

5. APOCARPEÆ. *Perianth* coloured and 2-seriate, or green 1-seriate, or imperfect, or 0. *Ovary* superior, apocarpous or monocarpellary. *Seeds* exalbuminous, cotyledonary end usually contracted hooked or coiled, rarely straight.

86. ALISMACEÆ. *Flowers* usually 2-sexual. *Perianth* 6-partite; inner segments or all petaloid. *Stamens* 6 or more. *Carpels* many. *Fruit* of many achenes; albumen 0; radicle very large.—Marsh or water herbs; flowers usually conspicuous. (p. 426.)

87. NAIADACEÆ. *Flowers* 1-2-sexual. *Perianth* of 4 valvate sepals, or imperfect, or 0. *Stamens* as many as the sepals, or fewer. *Carpels* 1-4, 1-ovuled. *Albumen* 0; radicle very large.—Marsh or water plants; flowers inconspicuous, green. (p. 428.)

6. GLUMACEÆ. *Perianth* 0, or of bristles or very minute scales. *Ovary* 1-celled, 1-ovuled; styles or stigmas 2-3. *Seeds* albuminous, embryo small. *Flowers* spicate, solitary in the axils of imbricating bracts (*glumes*).

88. CYPERACEÆ. *Flowers* 1-2-sexual. *Perianth* 0 or of bristles, rarely of scales. *Stamens* 1-3; anthers basifixed. *Ovary* 1-celled; style 1, stigmas 2-3 papillose; ovule 1, erect. *Fruit* compressed or 3-gonous. *Embryo* at the base of the albumen.—*Stem* usually solid, 3-gonous; leaves often grass-like, but with entire sheaths. (p. 439.)

89. GRAMINEÆ. *Flowers* usually 2-sexual. *Perianth* usually of 2 very minute scales. *Stamens* usually 3; anthers versatile. *Ovary* 1-celled, stigmas 1-2, hairy or feathery. *Fruit* terete, or grooved on one side. *Embryo* on one side of the base of the albumen.—*Stem* cylindrical, usually hollow, except at the joints; leaves with sheaths split to the base. (p. 466.)

SUB-KINGDOM II. **Cryptogams,** or **Acotyledons,** or **Flowerless** plants. *Plants* not provided with stamens and ovules as in Phænogams. *Seeds* represented by minute spores which contain no embryo.

CLASS I. **Acrogens.** Plants with a distinct stem.

SUB-CLASS **Vasculares.** *Stem* with vascular tissue. *Spores* contained in a spore case (*sporangium*), and developing a prothallus in germination.

* *Spores of one kind.*

90. FILICES. Sporangia usually very minute, situated on the margin or under surface of the leaf (frond) ; rarely larger, in separate spikes or panicles.—*Fronds* usually circinate in vernation. (p. 507.)

91. EQUISETACEÆ. Sporangia 2-valved, on the under side of peltate scales that are arranged in terminal cones. *Spores* with 4 filiform clubbed appendages rolled round them.—*Stems* erect from a creeping rootstock, cylindric, hollow, grooved, septate, simple or with whorled branches and with toothed sheaths at the joints. (p. 521.)

92. LYCOPODIACEÆ. Sporangia not very minute, situated in the axils of the leaves, or of the scales of a cone.—*Fronds* usually circinate in vernation. (p. 523.)

* *Spores of two kinds.*

93. SELAGINELLACEÆ. Sporangia not very minute, situated in the axils of the scales of a cone or at the bases of subulate leaves. Spores of 2 kinds; the larger developing a prothallus within its coat ; the smaller containing antherozoids. Decumbent or prostrate plants with small imbricating leaves of 2 forms ; or stemless water plants with subulate leaves. (p. 525.)

94. MARSILEACEÆ. Sporangia (membranous sacs) very minute, enclosed in the cells of a globose receptacle near the base of the frond. *Spores* of 2 kinds ; the larger developing a prothallus ; the smaller containing antherozoids.—Marsh or water plants, rarer on dry soils. (p. 526.)

The Student's Flora of the British Isles.

CLASS I. DICOTYLE'DONES.

SUB-CLASS I. ANGIOSPER'MÆ.

ORDER I. **RANUNCULA'CEÆ.**

HERBS rarely shrubs. *Leaves* radical or alternate, opposite in *Clematis ;* stipules 0, or adnate to the petiole. *Flowers* regular or irregular, 1–2-sexual. *Sepals* 5 or more, rarely 2–4, deciduous, often petaloid, usually imbricate in bud. *Petals* 0, or 5 or more, rarely 3, imbricate in bud, often minute or deformed. *Stamens* many, hypogynous ; anthers basifixed, dehiscence subdorsal. *Disk* 0. *Carpels* many, rarely 1, usually free, 1-celled ; stigma simple ; ovules 1 or more on the ventral suture, anatropous, erect with a ventral or pendulous with a dorsal raphe. *Fruit* of 1-seeded achenes, or many-seeded follicles. *Seed* small, albumen copious ; embryo minute.—DISTRIB. Abundant in temp. and cold regions ; genera 30 ; species 503.—AFFINITIES. With *Berberideæ* and *Papaveraceæ ;* analogies with *Alismaceæ* and *Rosaceæ.*—PROPERTIES. Usually acrid.—EXCEPTIONAL FORMS (British). Stem woody in *Clematis ;* flowers polygamous in *Thalictrum ;* sepals persistent, carpels connate in *Helleborus ;* stamens few in *Myosurus ;* follicle sometimes solitary in *Delphinium ;* berry solitary, and stigma dilated in *Actæa.*

TRIBE I. **CLEMATI'DEÆ.** *Sepals* valvate. Shrubs with opposite leaves.
1. Clematis

TRIBE II. **ANEMO'NEÆ.** *Sepals* imbricate. *Achenes* with 1 pendulous seed.
Involucre 0. Sepals 4–5, petaloid. Petals 0......................2. Thalictrum.
Involucre of 3 leaves. Sepals 4–20, petaloid. Petals 0..........3. Anemone.
Involucre 0. Sepals 5–8, petaloid. Petals 5–16, conspicuous......3* Adonis.
Involucre 0. Sepals 5, spurred. Petals small, tubular............4. Myosurus.

TRIBE III. **RANUN'CULEÆ.** *Sepals* imbricate. *Achenes* with 1 ascending seed ....5. Ranunculus.

TRIBE IV. **HELLEBO'REÆ.** *Sepals* imbricate. *Follicles* many-seeded, except in *Actæa.*

B

* *Flowers regular. Follicles many-seeded.*

Sepals petaloid. Petals 0...6. Caltha.
Sepals petaloid, deciduous. Petals small, entire......................7. Trollius.
Sepals herbaceous, persistent. Petals small, 2-lipped...........8. Helleborus.
Sepals petaloid, deciduous. Petals small, 2-lipped ....8*. Eranthis.
Sepals 5-6, petaloid. Petals large, spurred...........................9. Aquilegia.

** *Flowers irregular. Follicles many-seeded.*

Sepals many, the dorsal spurred ....9*. Delphinium.
Sepals many, the dorsal arched and hooded........................10. Aconitum.

*** *Flowers nearly regular. Fruit a berry....... ....11. Actæa.*

1. CLE'MATIS, *L.* TRAVELLER'S JOY.

Usually climbing under-shrubs. *Leaves* opposite, usually compound, exstipulate; petiole often twining. *Inflorescence* axillary or terminal; flowers proterandrous, honeyless. *Sepals* usually 4, petaloid, imbricate or valvate. *Petals* 0. *Stamens* many. *Carpels* many; ovule 1, pendulous. *Fruit* a head of sessile or stalked achenes, with long bearded styles.— DISTRIB. All temp. climates, rarer in the tropics; species 100.—ETYM. κληματίς, the Greek name for this or a plant of similar habit.

C. Vital'ba, *L.*; leaflets 3–5 remote. *Old Man's Beard.*

Hedges and thickets, from Stafford and Denbigh southd.; most common on chalky soil; not a native of Scotland or (?) Ireland; fl. July–Aug.—A climbing under-shrub. *Leaflets* 2–3 in., ovate-cordate, entire toothed or lobed; petiole persistent when twining. *Flowers* 1 in. diam., odorous, greenish-white. *Sepals* 4, pubescent. *Achenes* hairy; awns 1 in., feathery. —DISTRIB. Europe, from Holland southd., N. Africa, W. Asia.

2. THALIC'TRUM, *L.* MEADOW-RUE.

Erect perennial herbs. *Leaves* compound, stipulate. *Flowers* panicled or racemed, often polygamous, honeyless, proterogynous, anemophilous. *Sepals* 4–5, petaloid, imbricate in bud. *Petals* 0. *Stamens* many. *Carpels* few or many; ovule 1, pendulous. *Fruit* a small head of sessile or stalked achenes; style persistent or deciduous.—DISTRIB. Temp. and colder regions of the N. hemisphere; species 50.—ETYM. Probably the *Thalictrum* of Pliny.

1. **T. alpi'num,** *L.*; raceme simple, flowers few drooping, anthers linear apiculate.

Alpine and sub-alpine bogs from Shetland to York and Carnarvon; rare in Ireland; ascends to 4,000 ft.; fl. July–Aug.—*Stem* 4–10 in., wiry, simple, often stoloniferous. *Leaves* 2-ternate; leaflets ¼ to ½ in., suborbicular, glaucous beneath, obtusely lobulate. *Raceme* drooping, then erect; pedicels recurved in fruit. *Sepals* 4, purplish. *Stamens* 8–20, pendulous. *Achenes* 2–3, stipitate, curved, ribbed.—DISTRIB. N. and Arctic Europe, N. and W. Asia to Himalaya, N. America.

2. **T. mi'nus,** *L.*; stem more or less striate, panicle lax, flowers drooping, anthers apiculate, fruit erect.

Dry places from Orkney southd.; ascends to 1,800 ft. in the Lake District; Ireland; fl. July–Aug.—*Stem* ½–4 ft., stout, rigid, often zigzag, striate throughout or towards the nodes only, usually furrowed when dry. *Leaves* triangular, 3–4-pinnate; leaflets variable, ¼–1 in., acutely or obtusely lobed, sometimes stipellate; stipules adnate to the petiole, auricles spreading or reflexed. *Sepals* 4, yellow-green. *Achenes* 3–5, sessile, elliptic-oblong, straight or gibbous, 8–10-ribbed.—DISTRIB. Europe (Arctic), N. Africa, N. and W. Asia to the Himalaya, Greenland.

T. MI'NUS proper; often glaucous and glandular; stem 6–18 in., usually naked at the base.—VAR. *dunense*, Dumort. (maritimum, *Ed.* 2); branches of broad panicle spreading. Sandy coasts, Orkney to Norfolk and S. Wales.—VAR. *T. monta'num*, Wallr. (calca'reum, *Jord.*); branches of deltoid panicle erecto-patent. Dry hills, Argyll to Somerset.—The Cambridge var. *saxa'tile*, Bab. of *ma'jus* seems rather referable here.

Sub-sp. MA'JUS, Sm. (not *Jacq.*); stem 2–4 ft.,more leafy below, leaflets usually much larger. From Perth southd.; most common in the north, in copses, &c.; Ireland. The vars. *T. Ko'chii*, Fries, with spreading stipules and ovoid achenes, and *T. flexuo'sum*, Reichb. (? of Bernhardi), with reflexed stipules and larger gibbous achenes, are with difficulty distinguishable.

3. **T. fla'vum,** *L.* ; stem furrowed, panicle compound, flowers erect crowded, anthers not apiculate.

Wet places, from Fife and Argyll southd., rare in Scotland; local in Ireland; fl. July–Aug.—*Rootstock* yellow, creeping, stoloniferous. *Stem* 2–4 ft., stout. *Leaves* 3-nately 2–3-pinnate; leaflets 1–1½ in., 3-lobed. *Panicle* sub-corymbose or pyramidal; flowers pale yellow, often umbelled. *Sepals* small. *Anthers* bright yellow. *Achenes* 6–10, small, dark, 8-ribbed. —DISTRIB. Europe (Arctic), N. Asia.

VAR. *sphærocar'pum*, Lej.; panicle usually contracted, achenes broadly oblong. —VAR. *ripa'rium*, Jord.; panicle usually lax, achenes oblong.—VAR. *T. Moriso'nii*, Gmel.; panicle usually interrupted, fascicles of flowers small, achenes narrow oblong.

3. ANEMO'NE, *L.*

Perennial very acrid herbs. *Leaves* radical, lobed or divided. *Flowers* on 1- or more-fld. scapes, rarely yellow; invol. leaves 3-partite. *Sepals* 4–20, petaloid, imbricate in bud. *Petals* 0. *Stamens* many, outer sometimes imperfect or petaloid. *Carpels* many; ovule 1, pendulous. *Fruit* a head of sessile achenes, with naked or bearded styles.—DISTRIB. Cold and temp. regions; species 70.—ETYM. ἄνεμος, *the wind*, of obscure application.

1. **A. Pulsatil'la,** *L.* ; sepals 6 erect silky, outer stamens reduced to glands, achenes with long feathery styles. *Pasque-flower.*

Chalk downs and limestone pastures; York to Norfolk, Essex, and Gloucester; fl. May–June.—Silky, 4–10 in. *Rootstock* stout, woody. *Leaves* maturing after flowering, 3-pinnatifid, segments linear; involucral sessile, divided to the base into long linear segments. *Flower* 1½ in., solitary, proterandrous, inclined in bud, dull purple; peduncle lengthening after flowering. Imperfect *stamens* honeyed. *Styles* of silky achenes 1½ in.—DISTRIB. Europe N. Asia to Dahuria.

2. **A. nemoro'sa,** *L.*; sepals 6 (rarely 5-9) oblong glabrous spreading, stamens all perfect, achenes with short straight styles. *Wood Anemone.*

Woods and copses fron Sutherland southd.; ascends to 2,800 ft. in the Highlands; Ireland; fl. April-May.—Nearly glabrous, slender. *Rootstock* horizontal, woody. *Scape* 4-8 in. *Leaves* few, usually remote from the scape, petioled, 3-foliolate; leaflets narrow, subsessile, cut lobed or pinnatifid; involucral like the radical, petioled. *Flower* solitary, 1-1½ in. diam., homogamous, honeyless. *Sepals* oblong, white, rarely purple. *Achenes* downy, as long as the style.—DISTRIB. Europe (Arctic), W. Siberia, N. America.

3*. *ADO'NIS, L.* PHEASANT'S-EYE.

Herbs, annual or perennial. *Leaves* much divided. *Sepals* 5-8, petaloid, imbricate in bud. *Petals* 5-16, yellow or red, eglandular. *Carpels* many; ovule 1, pendulous. *Fruit* a spike or head of many achenes; style short, persistent.—DISTRIB. Temp. Europe and Asia; species 3-4. —ETYM. classical.

A. AUTUMNA'LIS, *L.*; annual, flowers globose, petals broad concave.

Naturalized in Suffolk, and S. counties, sporadic elsewhere, and in Scotland and Ireland; (alien or colonist, *Wats.*); fl. May-Sept.—*Stem* 8-10 in., erect, branched, very leafy. *Leaves* decompound; segments small, linear. *Sepals* greenish. *Petals* scarlet, with a dark basal spot, suberect, rather longer than the sepals. *Head* of reticulated achenes sometimes elongate.— DISTRIB. Europe, W. Asia, N. Africa; introd. in America.

4. MYOSU'RUS, *L.* MOUSE-TAIL.

Small annual herbs. *Leaves* narrow, all radical. *Scapes* 1-fld. *Sepals* 5, rarely 6-7, with a small basal spur. *Petals* 5, rarely 6-7, or 0, small, narrow, tubular. *Stamens* few. *Carpels* many; ovule 1, pendulous. *Fruit* a long spike of densely packed achenes; style short, persistent.— DISTRIB. Europe, N. Asia, S. America, Australasia; species 2.—ETYM. μῦς and οὐρά, *mouse-tail.*

M. **min'imus,** *L.*; spike slender, style very short.

Cornfields, &c., from Northumberland to Kent and Devon; Channel Islands; fl. April-June.—Glabrous, 2-6 in. *Leaves* erect, many, linear, rather fleshy. *Scapes* many, slender. *Flowers* minute, yellow-green, proterandrous. *Sepals* 5, narrow-oblong; spur appressed to the scape. *Petals* 5; limb short, ligulate. *Spike* of achenes 1-3 in.; receptacle filiform; achenes attached ventrally, minute, keeled, back mucronate.—DISTRIB. Europe, W. Asia, N. Africa; introd. in America, &c.

5. RANUN'CULUS, *L.* BUTTERCUP, CROWFOOT.

Annual or perennial usually acrid herbs. *Leaves* entire lobed or compound; stipules membranous or 0. *Flowers* usually panicled, white or yellow (the British species). *Sepals* 3-5, caducous. *Petals* usually 5, rarely 0, glandular near or above the base. *Stamens* many. *Carpels* many; style short; ovule 1, ascending. *Fruit* a head or spike of apiculate or beaked achenes.—DISTRIB. All temp. regions; species about 160.—ETYM. *rana, a frog.*

SECTION 1. **Batra′chium.** Water- or marsh-plants. *Leaves* often submerged and multifid; stipules membranous. *Peduncles* usually leaf-opposed, 1-fld. *Flowers* proterandrous. *Petals* white; gland naked, yellow, basal. *Achenes* transversely wrinkled.

The following is an attempt to group naturally the British Batrachian *Ranunculi*, after a protracted study of the large collection at Kew (Herb. Kew, H. C. Watson, Borrer, Bot. Exch. Club, G. Nicholson, &c.). The result accords in a measure with the early views of H. C. Watson (Suppl. to Cybele, 1860; and Companion to ditto, 1868). Opinious vary as to whether the 8 forms or even segregates of them should be ranked as one or more species, subspecies, or varieties; I regard them as approximately equivalent to the species I have retained under *Rubus, Rosa,* &c. Of the characters attributed to these and their subordinate forms by critical authors, I find some variable, others valueless, and still others deceptive; such especially as concern the tapering, &c., of the peduncle, the comparative length of stamens and petals, number of stamens, and especially the forms of the receptacle, achenes, and stigma.

* *Aquatic. Floating leaves usually present; submerged numerous, multifid. Petals 5-9-nerved. Receptacle hispid.*

1. **R. heterophyl′lus,** *Fries;* segments of submerged leaves spreading in all directions, peduncles hardly exceeding the leaves, flowers ½-1 in. diam., petals broadly obovate, stamens numerous. *R. aqua′tilis,* Sm.

Streams and ponds from Orkney southd., ascending to 1,050 ft. in Scotland; Ireland; Channel Islands; fl. May-June.—*Floating leaves* (rarely 0), ½-1½ in. diam., from orbicular to reniform, 3-5-lobed or -partite or 3-foliolate; basal sinus broad or narrow, segments broadly cuneate, toothed, lobulate, or laciniate; stipules broad, rounded. *Petals* much longer than the sepals. *Stamens* longer than the pistil. *Achenes* glabrous or hairy or hispid; stigma short, obtuse.—DISTRIB. Europe (Arctic), N. America, N. Asia.—*R. hetero-phyl′lus* proper; segments of submerged leaves collapsing into a tassel when removed from the water, flowers about ½ in. diam., achenes usually glabrous.—*R. pelta′tus,* Fries (*R. floribun′dus,* Bab., *R. trunc′atus,* Dumort.), has segments of submerged leaves more rigid, flowers ½-1 in. diam., petals broader, achenes usually hairy or hispid.—*R. fissifo′lius,* Schrank, is a form with laciniate leaf-lobes (Loch Maben, &c.).—*R. penicilla′tus,* Dumort. (*R. pseudo-fluitans,* Bab.), is a remarkable form with the habit, long robust stem and long leaves and peduncles, and large flowers of *fluitans,* but the hirsute receptacle of *heterophyl′lus;* it forms the passage between the two. From Derby and Warwick, to Wilts and Surrey; Ireland.—*R. triphyl′los,* Wallr., from Guernsey, an imperfectly known plant, may be a form of *heterophyl′lus.*

2. **R. mari′nus,** *Fries;* segments of shortly petioled submerged leaves spreading all round, peduncles much longer than the leaves, flowers ¼-¾ in. diam., petals narrowly obovate, stamens few or many. *R. aquatilis,* var. *Symei,* Hook. and Arn.

Brackish waters near the sea, from Caithness southd.; Ireland; fl. June–
Sept.—*Floating leaves* reniform or broader than long, basal sinus broad,
3-lobed or -partite, segments sessile or petiolulate, cuneate, crenate or lobed;
submerged not collapsing when removed from the water; stipules broad,
rounded. *Petals* not touching, much longer than the sepals. *Stamens*
shorter or longer than the pistil. *Achenes* very many, small, glabrous or
hairy; stigma usually hooked.—DISTRIB. W. Europe.
The *R. mari'nus* proper (*i.e.* of Fries) has no floating leaves (like *R.
salsugino'sus*, Hiern), and few stamens.—*R. confu'sus*, Godr., is characterized
by its more slender tapering peduncles, stamens many exceeding the
pistil, ovoid-conic receptacles and ½-ovate compressed achenes narrowed
upwards, and *R. Baudo'tii*, Godr., as having stout pedicels, stamens many
not exceeding the pistil, long conic receptacles and ½-obovate achenes with
inflated tops:—characters which I cannot verify as constant in either case.

** *Aquatic. Floating leaves rarely present; submerged numerous, multifid.
Petals 5-9-nerved. Receptacles glabrous or hairy.*

3. **R. flu'itans,** *Lamk.* ; stems long robust, submerged leaves with few
long narrow rigid tassel-like segments, peduncles much longer than the
leaves, flowers ¾–1 in. diam.. petals broadly obovate, receptacle
glabrous, achenes few large turgid.   *R. peucedanifo'lius,* Schrank.
Rivers and running streams from the Clyde southd.; Ireland; fl. June–
Aug.—*Stem* several feet long and usually stout. *Leaves* 3–9-in. long, long
petioled, black, forming flaccid or rather rigid tassels; floating leaves very
rare, 3-lobed or -partite or -foliolate, segments sometimes petiolulate;
stipules broad, rounded. *Peduncles* very long and robust. *Petals* often
more than 5, and 2-seriate. *Stamens* many, short or long. *Achenes* compressed, glabrous; stigma short, thick.—DISTRIB. Europe.—*R. Ba'chii,*
Wirtg., is a small form, more slender, with subsessile more divided leaves
and narrower petals.

4. **R. trichophyl'lus,** *Chaix ;* submerged leaves usually subsessile,
black and rigid, not collapsing when removed from the water, peduncles
stout, shorter than the leaves, flowers ⅓–½ in. diam., petals small
narrow distant, receptacle glabrous, achenes few.   *R. pantothrix,* Brot.
Water-fennel.
Still waters from Orkney southd.; Ireland; fl. May–June.—*Floating
leaves,* if present, 3-lobed, -partite, or sometimes 3-foliolate, submerged;
2-3-chotomously multifid; stipules large, rounded. *Peduncles* about
equalling the leaves, or shorter, not tapering. *Stamens* few, longer than the
pistil. *Achenes* glabrous or hairy; stigma short, thick.—DISTRIB. Europe,
W. Asia, Himalaya, N. America.—*R. ' Droue'tii,* F. Schultz, has paler,
more flaccid submerged leaves, the mid segment of the floating ones when
present often deflexed.—*R. radians,* Rev., and *R. Godronii,* Gren., and *R. diversifolius,* H. Wats., are forms with floating leaves.—VAR. *confervoides* is a
depauperated northern form from Rescobie Loch in Forfarshire, which is
the original *R. aquatilis* of Linnæus's Flora Lapponica. It is probably not
rare in the north.

5. **R. circina'tus,** *Sibth.* ; floating leaves 0, submerged small sessile
orbicular, segments in one plane rigid, peduncles much longer than the

leaves, flowers $\frac{3}{4}$ in. diam., petals broadly obovate, receptacle hispid, achenes glabrate or hispid acute. *R. diva'ricatus*, Schrank.

Still and slowly-flowing water from Forfar southd., not common; Ireland; fl. June–Aug.—Much the most distinct species of this section, very uniform in size, habit, and character. *Leaves* $\frac{1}{2}$–$\frac{3}{4}$ in. diam.; stipules wholly adnate, like leaf-sheaths. *Peduncles* tapering. *Petals* twice as long as the sepals, many-veined. *Stamens* many, longer than the pistil. *Achenes* compressed; style slender deciduous.—Distrib. Europe (local), N. America.

*** *Marsh- or mud-plants, creeping, rarely floating. Submerged leaves 0 (very rare in R. tripartitus). Petals 3–5-nerved. Receptacle glabrous or nearly so.*

6. **R. triparti'tus,** *DC. ;* leaves $\frac{1}{2}$-orbicular or reniform 3-lobed or partite, segments cuneate spreading, tips crenate, submerged when present very few and flaccid, peduncles shorter than the leaves, flowers $\frac{1}{4}$ in. diam., petals narrow, achenes few glabrous. *R. interme'dius*, Hiern.

Marshes and ditches in S. and W. England; fl. May–July.—*Stem* aerial, or floating with emerged tips. *Leaves* $\frac{1}{2}$–$\frac{3}{4}$ in. diam.; stipules broad, upper rounded free. *Peduncles* equalling the leaves or shorter. *Petals* 3-nerved, about twice as long as the calyx, pinkish. *Stamens* few, longer than the pistil. *Receptacle* slightly hairy. *Achenes* turgid; style slender, deciduous. —Distrib. W. Europe.—With difficulty distinguished from forms of *heterophyl'lus*.

7. **R. Lenorman'di,** *Schultz ;* leaves all reniform or orbicular, lobes shallow, bases contracted, peduncles equalling the petioles, flowers $\frac{1}{4}$–$\frac{1}{2}$ in. diam., petals remote oblong, receptacles glabrous, achenes many glabrous.

Marshes and ditches from the Clyde southd.; ascends to 1,600 ft. in Yorkshire; S. Ireland; fl. June–Aug.—*Stem* stout, branched, 2–8 in. long. *Leaves* $\frac{1}{2}$–1 in. diam., often opposite, rounded and more crenate than *R. hederaceus*, never spotted; stipules large, broad. *Petals* 5-nerved, twice as long as the calyx. *Stamens* few, about equalling the pistil. *Achenes* with deciduous subterminal slender styles.—Distrib. N.W. Europe.

8. **R. hedera'ceus,** *L. ;* leaves reniform angularly 5–7-lobed, lobes broadest at base, peduncles usually shorter than the leaves, flowers $\frac{1}{6}$–$\frac{1}{3}$ in. diam., petals very narrow, distant, receptacle glabrous, achenes few obtuse. *Ivy-leaved Crowfoot.*

Shallow ponds and ditches from Shetland southd., ascends to 2,200 ft. in Wales; Ireland; Channel Islands; fl. May–Aug.—Habit of *R. Lenorman'di*. *Leaves* $\frac{1}{2}$–$1\frac{1}{2}$ in. diam., usually opposite, with a $\frac{1}{2}$-lunar black patch, lobes broader than long, rarely notched; stipules various. *Petals* 3-nerved, sometimes hardly exceeding the calyx. *Stamens* few, about equalling the pistil. *Achenes* small.—Distrib. W. Europe.—*R. homoiophyl'lus*, Tenore (*R. cœno'sus*, Guss.), is a floating form.

Section 2. **Hecato'nia.** Perennial, rarely annual. *Leaves* mostly radical, stipules obscure or 0. *Stems* 2- or more-flowered. *Sepals* 5. *Petals* 5, yellow. *Achenes* not tubercled (granulate in *R. ophioglossifolius*).

* *Leaves all undivided. Gland of petals with a small scale.*

9. R. Lin'gua, *L.* ; perennial, erect, leaves sessile ½-amplexicaul lanceolate entire or toothed, achenes pitted, style broad. *Great Spear-wort.*

Marshes and ditches, from Aberdeen southd.; local in Ireland; Channel Islands ; fl. July–Sept.—Glabrous. *Root* densely fibrous. *Stem* 2–3 ft., hollow; lower nodes rooting. *Leaves* 6–10 in. ¾–1 in. broad, veins parallel and reticulated. *Flowers* 2 in. diam., handsome, sub-panicled.—DISTRIB. Temp. Europe, N. and W. Asia to the Himalaya.

10. R. Flam'mula, *L.* ; perennial, suberect creeping or ascending, leaves petioled linear- or ovate-lanceolate nearly entire, achenes minutely pitted, style minute subulate. *Lesser Spear-wort.*

Wet places ; ascends to 2,700 ft. in the Highlands ; Ireland ; Channel Islands fl. June–Aug.—Very variable, glabrous or slightly hairy, 4–12 in. *Lowest leaves* petioled, ovate, upper more lanceolate and sessile. *Flowers* yellow, rarely ¾-in. diam., proterandrous. *Head* of achenes small.—DISTRIB. Europe (Arctic), N. Asia, Africa, and America.
R. FLAM'MULA proper ; prostrate or erect, internodes straight, style of achenes short obtuse.
Sub-sp. R. REP'TANS, *L.* ; creeping, very slender, internodes arching, style of minute achenes subcylindric, style recurved. Sandy shores of Loch Leven. (N.W. Europe, Canada.)

11. R. ophioglossifo'lius, *Villars ;* annual, erect, lower leaves long-petioled broadly ovate or cordate, petals scarcely longer than the sepals, achenes small hairy minutely granulate, style minute.

Marshes, S. Hants; Jersey (extinct); fl. June–Aug.—Glabrous or slightly hairy upwards. *Root* fibrous. *Stem* 6–10 in. or more, slender, decumbent at the base, branched, hollow, furrowed. *Peduncles* furrowed. *Flowers* many, ¼ in. diam.—DISTRIB. W. and S. Europe.

** *Radical leaves divided, upper cauline entire. Gland of petals without a scale.*

12. R. auri'comus, *L.* ; perennial, leaves orbicular 3-lobed or -partite, segments of lower obtuse cuneate cut, of upper linear spreading, sepals spreading pubescent, head of downy achenes globose. *Goldielocks.*

Woods and copses, from Aberdeen southd.; ascends to 1,600 ft. in the Highlands; S. and W. Ireland rare; Jersey; fl. April–May.—Erect, 6–10 in., branched, slender, glabrous or slightly hairy upwards. *Root* fibrous. *Radical leaves* long-petioled. *Peduncles* not furrowed, pubescent. *Flowers* ¾ in. diam., seldom regular. *Petals* larger than the downy sepals, bright yellow, often imperfect (var. *depaupera'ta*). *Achenes* on tubercles of the receptacle, compressed; style slender, subulate, curved.—DISTRIB. Europe (Arctic), N. and W. Asia to the Himalaya.—Not acrid.

13. R. scelera'tus, *L.* ; annual, erect, leaves glabrous 3-lobed or -partite, segments of lower lobed obtuse, of upper linear subentire, sepals reflexed hairy, head of small glabrous achenes oblong.

Ditches, &c., from Ross southd.; Ireland; Channel Islands; fl. May–Sept. —*Root* fibrous. *Stem* 8–24 in., subcorymbose above, hollow. *Leaves* variable in lobing, upper a little hairy. *Flowers* ¼ in. diam., proterogynous. *Achenes* many, small, faces a little wrinkled, dorsal edge furrowed; style minute.— Distrib. Europe (Arctic), N. Asia, N. India to Bengal; introd. in America, &c.—Very acrid.

*** *Perennials. Radical leaves divided, upper cauline entire. Glands of petals with a small scale.*

14. **R. a′cris,** *L.* ; hairy, erect, without runners, leaves 3–7-partite, segments of lower cuneate deeply cut and lobed, peduncles not furrowed, sepals spreading pubescent, receptacle glabrous, achenes compressed margined glabrous, style hooked.

Meadows, &c., N. to Shetland; ascends to nearly 4,000 ft. in the Highlands; Ireland; Channel Islands; fl. April–Sept.—*Rootstock* straight. *Stem* 8 in.- 3 ft. *Leaves* usually all petioled, orbicular or 5-angled in outline, uppermost sessile. *Flowers* 1 in. diam., proterandrous, spreading.—Distrib. Europe (Arctic), N. Asia; introd. in America.

Var. *R. vulga′tus,* Jord.; rootstock creeping horizontal or slightly inclined.
Var. *R. Borœa′nus,* Jord.; stem glabrous below, leaf-segments very narrow.
Var. *R. tomophyl′lus,* Jord.; rootstock nearly erect, leaf-segments very narrow

15. **R. re′pens,** *L.* ; hairy, stem decumbent below with long runners, leaves 3-foliolate or 3-nately pinnatisect, segments cuneate lobed and toothed, peduncles furrowed, sepals spreading hairy, receptacle slightly hairy, achenes compressed margined glabrous, style hooked.

Waste ground from Sutherland southd.; ascends to 2,700 ft. in the Highlands; Ireland; Channel Islands; fl. May–Aug.—*Rootstock* stout, short. *Stem* 8 in.-2 ft. *Leaves* petioled, triangular or ovate; segments variable, middle usually longest. *Flowers* 1 in. diam. *Petals* generally suberect.— Distrib. Europe (Arctic), N. and W. Asia, N. Africa; introd. in America.

16. **R. bulbo′sus,** *L.* ; erect, hairy, stem swollen at the base without runners, leaves 3-foliolate or ternatisect, segments lobed, peduncles furrowed, sepals reflexed and receptacle hairy, achenes compressed margined glabrous, style short hooked.

Meadows, &c., from Caithness southd.; ascends to 1,500 ft. in the Highlands; Ireland; Channel Islands; fl. May–July.—*Stem* 6–12 in., base often as big as a walnut, sometimes corymbose above. *Leaves* variable in form and lobing. *Flowers* ½ to 1 in. diam.—Distrib. Europe, Asia, N. Africa; introd. in America.

17. **R. chærophyl′lus,** *L.* ; erect, silkily hairy, stem swollen at the base with tuberous offsets, leaves (of young plant entire) 3-foliolate or ternatisect, peduncles not furrowed, sepals spreading, receptacle glabrous, hairy head of compressed glabrous acute dotted achenes cylindric-oblong.

Jersey, St. Aubin's Bay; fl. May.—*Root-fibres* stout. *Stem* 6–12 in., usually simple, slender, 1-fld.; neck clothed with dry matted fibres. *Leaves* of young plant orbicular or broadly cuneate, toothed or lobed; later formed leaves very variable in lobing or cutting. *Flowers* ¾–1½ in. diam., bright yellow, proterandrous. *Achenes* very numerous, small, simply acute.—Distrib. France, Mediterranean region to Syria.

SECTION 3. **Echinel'la.** Annual, rarely biennial. *Leaves* rádical and cauline, divided ; stipules inconspicuous. *Sepals 5. Petals 5,* yellow. *Achenes* tubercled or spinose, compressed, margin thickened.

18. **R. hirsu'tus,** *Curtis ;* erect, leaves 3-lobed or -partite, segments obtuse cut, peduncles furrowed, sepals reflexed hairy, petals with a scale over the gland, receptacle hairy, achenes tubercled towards the margin, style straight. *R. Philono'tis,* Ehrht.

Damp ground from Argyll and Forfar southd. ; rare in Scotland ; not in Ireland; Channel Islands ; fl. June–Oct.—Hairy. *Stems* many, 6–18 in. *Leaves* variable in lobing. *Flowers* about 1 in. diam., peduncles with spreading or reflexed hairs. *Achenes* broad, much flattened.—DISTRIB. Europe, W. Asia, N. Africa.—The earlier names of *R. par'vulus,* L., and *R. Sardo'us,* Crantz, are superseded for being too inappropriate.

19. **R. arven'sis,** *L.* ; erect, lowest leaves obovate or cuneate toothed, upper 3-partite or -foliolate, segments narrow cut, peduncles not furrowed, sepals spreading, petals gland with a scale, receptacle hairy, achenes usually covered with hooked spines, style stout hooked.

Cornfields, from Perth southwd.; in Ireland near Dublin only ; (a colonist, *Wats.*) ; fl. May–July.—Nearly glabrous. *Stem* 6–24 in., solitary. *Leaves* variable. *Flowers* ½ in. diam., pale. *Petals* suberect. *Achenes* few, large. —DISTRIB. Europe, temp. Asia to India, N. Africa.

20. **R. parviflo'rus,** *L.* ; slender, decumbent, leaves orbicular or reniform 3-lobed, segments toothed, peduncles furrowed, sepals reflexed, petals 3–5 small oblong, gland with an obscure scale, receptacle glabrous, achenes faced with hooked tubercles, style short nearly straight.

Dry banks, &c., from Durham southwd.; Ireland, rare ; Channel Islands; fl. May–Aug.—Hairy. *Stems* and branches spreading, 6–8 in. *Leaves* divided to the middle or less, lowest often entire, uppermost more deeply cut into linear lobes. *Peduncles* leaf-opposed or in the forks. *Flowers* ⅜–½ in. diam. *Achenes* small.—DISTRIB. Europe from Denmark southwd., W. Asia, N. Africa; introd. in America.

SECTION 4. **Fica'ria,** *DC.* (gen.). Perennial. *Leaves* opposite, chiefly radical, entire. *Sepals* 3–5. *Petals* 8–12, yellow ; gland with a scale. *Achenes* small, not beaked.

21. **R. Fica'ria,** *L.* ; leaves cordate obtusely angled or crenate, achenes globose smooth, style minute. *Pilewort or Lesser Celandine.*

Pastures and waste places, N. to Shetland ; ascends to 2,400 ft. in Wales ; Ireland; Channel Islands; fl. March–May.—Glabrous. *Root-fibres,* stout, cylindric. *Stem* short, decumbent, branched at the base. *Leaves* variable ; petiole stout with a base dilated. *Peduncles* stout, axillary, 1-fld. *Flowers* about 1 in. diam., bright yellow, sometimes apetalous, proterandrous. Head of *achenes* globose; *Cotyledon* solitary (one suppressed).—DISTRIB. Europe (Arctic), W. Asia, and N. Africa.

VAR. *diver'gens,* F. Schultz; lobes of lowest leaves not overlapping at the base, lowest sheaths narrow.—VAR. *incum'bens,* F. Schultz; lobes of lowest leaves overlapping at the base, lowest sheaths amplexicaul.

6. CAL'THA, *L.* MARSH MARIGOLD.

Herbs with stout creeping rootstocks. *Leaves* chiefly radical, cordate. *Flowers* terminal, few, white or yellow, honeyed. *Sepals* 5 or more, petaloid, deciduous, imbricate in bud. *Petals* 0. *Carpels* many, sessile ; ovules numerous, 2-seriate. *Follicles* numerous, many-seeded. *Seeds* many, with a prominent raphe and thickened funicle.—DISTRIB. N. and S. temp. and cold regions ; species 5-6.—ETYM. κάλαθος, *a cup.*

1. **C. palus'tris,** *L.* ; stem not rooting at the nodes, leaves orbicular-reniform crenate-toothed.

Marshes and ditch-banks, N. to Shetland ; ascends to 3,400 ft. in the High-lands ; Ireland ; fl. March–May.—A coarse, glabrous, dark green, showy, very variable plant. *Rootstock* short, horizontal. *Stem* 8 in.–3 ft., suberect, prostrate, or procumbent and rooting from all the nodes. *Leaves* ½–2 in. diam., base deeply 2-lobed, sinus narrow. *Stipules* very large, membranous, glairy, quite entire in bud and enclosing the young leaf. *Flowers* 1–2 in. diam., golden yellow. *Sepals* unequal, obovate or oblong. *Follicles* ½–¾ in. —DISTRIB. Europe (Arctic), N. and W. Asia to the Himalaya, N. America. VAR. *C. vulga'ris,* Schott ; stem ascending, flowers many 1½–2 in. diam., sepals contiguous, follicles spreading, beak short.—VAR. *C. Guerange'rii,* Boreau ; stem ascending, flowers many smaller, sepals remote when expanded, follicles spreading, beak longer. Probably *C. ripa'ria,* Don, and the origin of the double-flowered *Caltha* of gardens.—VAR. *mi'nor,* Syme ; stem pro-cumbent, flower solitary ¾–1 in. diam., sepals remote, follicles erect, beak short. Mountainous places.

2. **C. radi'cans,** *Forster ;* rooting at the nodes, radical leaves deltoid obscurely 5-angled acutely toothed, base truncate or reniform.

Forfarshire, very rare ; fl. May–June.—This is a very remarkable species, or perhaps form of *C. palus'tris,* differing from all other forms of the latter in the deltoid sharply-toothed leaves and rooting nodes of the branches. It is said by Nyman to have been found by Th. Fries in E. Finland.

7. TROL'LIUS, *L.* GLOBE-FLOWER.

Erect perennial herbs. *Leaves* alternate, palmately lobed or cut. *Flowers* large, yellow or lilac. *Sepals* 5-15, petaloid, imbricate in bud. *Petals* 5-15, small, narrow, claw very short, blade with a glandular pit at the base. *Stamens* very many. *Carpels* 5 or more, sessile ; ovules many, 2-seriate. *Follicles* 5 or more. *Seeds* many, angled, testa coriaceous.—DISTRIB. N. temp. and arctic ; species 9.—ETYM. *Trol, a globe,* in old German.

T. europæ'us, *L.* ; flower globose, petals equalling the stamens.

Subalpine pastures and copses, from E. Cornwall, Worcester, and Wales to Shetland, ascending to 3,300 ft.; N. of Ireland very rare ; fl. June–Aug.— Glabrous. *Rootstock* short, crowned with rigid fibres. *Stem* 6-24 in., simple, leafy. *Radical leaves* petioled, suborbicular, 5-partite, segments cuneate lobed and cleft; cauline smaller, sessile. *Flowers* 1-1½ in. diam., pale yellow, homogamous. *Sepals* orbicular, concave. *Petals* oblong. *Stamens* short.

Follicles transversely wrinkled, keeled, beaked.   *Seeds* black, dotted.—
DISTRIB. Europe (Arctic) to the Caucasus.

8. HELLEB'ORUS, *L.*   HELLEBORE, BEAR'S-FOOT.

Coarse perennial herbs.   *Leaves* palmately pedately or digitately lobed,
upper bract-like.   *Flowers* corymbose, proterogynous.   *Sepals* 5, large,
petaloid or herbaceous, imbricate in bud, persistent.   *Petals* small, tubular,
2-lipped, honeyed.   *Stamens* many.   *Carpels* separate, or cohering below.
Follicles dehiscing at the top.   *Seeds* many, oblong, funicle thickened,
testa crustaceous shining.—DISTRIB. Europe, N. and W. Asia ; species
10.—ETYM. ἐλλέβορος, the Greek name.

1. **H. fœ'tidus,** *L.* ; stem many-fld. perennial, leaves pedate, sepals
erect.   *Stinking Hellebore, Setter-wort.*

Chalk-pastures and thickets S. and E. of England, rare, naturalized elsewhere ;
(a denizen, *Wats.*) ; fl. Feb.-March.—Glabrous below, glandular-pubescent
above.   *Stem* 1-2 ft., leafless, scarred below.   *Lower leaves* petioled, leaflets
5-7, nearly as in *H. viridis,* but the outer segments recurved, upper with
large sheaths.   *Flowers* drooping, 1 in. diam.   *Sepals* truncate, green,
bordered with dull-purple.   *Petals* shorter than the stamens.   *Follicles* 3,
wrinkled, glandular, style subulate.— DISTRIB. W. Europe, from Belgium
southd.—Plant fœtid and cathartic.

2. **H. vir'idis,** *L.* ; stem few-fld. annual, radical leaves digitate, cauline
sessile, sepals spreading.   *Bear's-foot.*

Woods, hedges, &c., chiefly on chalk in the S. and E. of England, naturalized
elsewhere ; (a denizen, *Wats.*) ; fl. March-April.—Glabrous, dark-green, 1-1½
ft.   *Radical leaves* fully developed after flowering ; leaflets 5-7, narrow,
serrate, lateral cleft.   *Flowers* inclined, 1½-2 in. diam.   *Sepals* green, oblong.
Petals 9-12, minute, shorter than the stamens, curved.   *Follicles* 3 ; style
straight, subulate.—DISTRIB. From Holland southd., but not well
established N. of the Mediterranean region, *Ball.*

8*. *ERAN'THIS, Salisbury.*   WINTER ACONITE.

Low herbs.   *Rootstock* stout, creeping.   *Radical leaves* palmate ; cauline
whorled and involucriform.   *Flower* solitary, pale yellow.   *Sepals* 5-8,
narrow, petaloid, deciduous, imbricate.   *Petals* small, clawed, 2-lipped.
Stamens many.   *Carpels* 5-6, stipitate.   *Follicles* separate.   *Seeds* many,
ovoid or globose, testa smooth crustaceous.—DISTRIB. Europe, N. Asia ;
species 2.—ETYM. ἔαρ, *spring,* and ἄνθος, *flower.*

E. HYEMA'LIS, *Salisb.* ; sepals 6-8 oblong.

In plantations, parks, &c., naturalized ; fl. Jan.-March.—*Stem* 4-6 in.   *Radical
leaves* orbicular, 3-5-partite, segments obtusely lobed ; petiole long.   *Involucre* of 2 sessile lobed bracts.   *Flower* cup-shaped, 1-1½ in. diam.   *Petals*
shorter than the stamens.—DISTRIB. W. Europe, from Belgium southd.

9. AQUILE'GIA, *L.*   COLUMBINE.

Erect perennial herbs.   *Leaves* 2-3-nately divided.   *Flowers* panicled or
solitary, handsome, proterandrous.   *Sepals* 5, regular, petaloid.   *Petals* 5,

concave, spurred behind, spur honeyed. *Stamens* many, inner imperfect. *Carpels* 5, many-ovuled. *Follicles* 5. *Seeds* many, testa crustaceous smooth or granulated.—DISTRIB. N. temp. zone; species 5-6.—ETYM. *aquila*, an *eagle*, from the form of the petals.

A. **vulga'ris,** *L.* ; spur hooked, follicles cylindric hairy.

Woods and thickets, England and Ireland, often naturalized, ascending to 1,000 ft. in Yorkshire; fl. May–July.—*Rootstock*'stout,'blackish. *Stem* 1-2 ft. slender. *Radical leaves* fascicled, petiole long, 2-3-ternately divided, segments stalked, lobed, glaucous, glabrous or hairy beneath. *Flowers* 1½-2 in. diam., loosely corymbose, drooping, blue or dull purple white (or red in garden varieties). *Sepals* ovate-lanceolate. *Petals* oblong; spur curved, involute at the tip. *Stamens* declinate, rising and dehiscing successively; inner reduced to broad wrinkled white filaments. — DISTRIB. Europe, N. Africa, N. and W. Asia to the W. Himalaya.

9*. *DELPHIN'IUM, L.* LARKSPUR.

Erect, annual or perennial herbs. *Leaves* alternate, lobed or cut. *Flowers* racemed or panicled, bracteate. *Sepals* 5, separate, or cohering below, dorsal spurred behind. *Petals* 2-4, small, 2 dorsal with spurs within the sepaline spur, 2 lateral spurless or 0. *Stamens* many. *Follicles* 1-5. *Seeds* many, testa coriaceous wrinkled or plaited. —DISTRIB. N. temp. zone; species about 40.—ETYM. δελφίν, a *dolphin*, from the form of the flower.

D. AJA'CIS, *Reichb.* (not *L.*) ; racemes long, lower bracts lobed, follicles solitary pubescent. *D. Consol'ida,* Brit. Fl. (not *L.*).

Cornfields, naturalized in Cambridgeshire, sporadic elsewhere; (alien or colonist, *Wats.*); fl. June–July.—Annual, pubescent. *Stem* 10-18 in., slender, sparingly branched. *Leaves* cut into many narrow linear lobes, lower petioled, upper sessile. *Flower* 1 in. diam., blue, white or pink. *Sepals* spathulate-oblong, spur ½ in. *Petals* 2. *Follicles* ¾ in., cylindric; style short. *Seeds* continuously plaited all round.—DISTRIB. Central and S. Europe, N. Africa; introd. in U. States.—*D. Consol'ida,* L., which has been occasionally found in England, has glabrous follicles, short racemes, and seeds with interrupted ridges.

10. ACONI'TUM, *L.* MONKSHOOD, WOLFSBANE.

Erect, perennial herbs. *Leaves* alternate, palmately-lobed or cut. *Flowers* panicled or racemed, bracteate, proterandrous. *Sepals* 5; dorsal large, arched, hooded; anterior narrowest. *Petals* 2-5, small; 2 dorsal with long claws, hooded at the tip, covered by the sepaline hood; 3 lateral small or 0. *Follicles* 3-5. *Seeds* many, testa spongy rugose.—DISTRIB. Mountains of the N. hemisphere.—ETYM. classical.

A. **Napel'lus,** *L.* ; leaf-lobes pinnatifid, raceme simple dense-fld.

Shady places near streams, in Wales, Hereford, Somerset, Dorset, and Denbigh, naturalized elsewhere; (a denizen? *Wats.*); fl. July–Sept.—*Rootstock* fusiform, black. *Stem* 1-2 ft., erect, slightly pubescent. *Leaves* palmately

5-7-partite, upper often sessile; petiole dilated at the base. *Flowers* bracteolate, 1-1½ in. diam., dark blue; pedicels erect, pubescent. *Upper sepal* at first concealing the others, then thrown back. *Spurs* of upper petals conical, deflexed. *Filaments* dilated below; anthers greenish-black. *Follicles* 3-5, sub-cylindric, beaked.—DISTRIB. Europe, N. and W. Asia to the Himalaya.—A deadly acrid poison.

11. ACTÆ'A, *L.* BANE-BERRY, HERB CHRISTOPHER.

Erect perennial herbs. *Leaves* alternate, 3-nately compound; stipules adnate. *Flowers* small, in short crowded racemes. *Sepals* 3-5, rather unequal, petaloid. *Petals* 4-10, small, spathulate, or 0. *Carpel* 1, many-ovuled; stigma sessile, dilated. *Berry* many-seeded. *Seeds* depressed, testa crustaceous smooth.—DISTRIB. Colder regions of the N. hemisphere; species 1 or 3.—ETYM. *ἀκτή*, the *Elder*, from a fancied likeness.

A. spica'ta, *L.* ; raceme simple, fruiting pedicels slender.

Copses on limestone, Yorks. and Westmorel., ascending to 1,000 ft.; fl. May. —*Rootstock* stout, black. *Stem* 1-2 ft., perennial, simple or sparingly branched. *Radical leaves* with long petioles, 2-3-ternately-pinnate; leaflets 1-3 in., ovate, acuminate, lobed and serrate, glabrous; auricles short, rounded. *Racemes* 1-2 in., solitary or few, oblong; peduncle and pedicels pubescent. *Flowers* ¼ in. diam., white. *Sepals* obtuse, caducous. *Petals* minute or 0. *Filaments* dilated above; anther cells dehiscing in front. *Berries* ½ in. long, ovoid, nearly black, on spreading pedicels.—DISTRIB. Temp. and Arctic Europe, Asia, and N. America (a red-berried var.). - Nauseous, poisonous.

ORDER II. **BERBERI'DEÆ.**

Herbs or shrubs; buds scaly. *Leaves* alternate, simple or compound, usually exstipulate. *Inflorescence* various; flowers often globose. *Sepals* petaloid. *Petals* hypogynous, numerous, distinct, multiples of 2, 3, or 4, never of 5, imbricate in bud. *Stamens* one opposite each petal; anthers opening by 2 ascending lids or valves. *Carpel* 1, 1-celled : stigma usually peltate; ovules 2 or more, basal or on the ventral suture, anatropous, raphe ventral. *Fruit* a berry or capsule. *Seeds* albuminous; embryo various.—DISTRIB. Most cool regions, except Australia and S. Africa; genera 20, species 100.—AFFINITIES with *Ranunculaceæ* and *Menispermaceæ;* analogy in anther with *Laurineæ* and in the 3-nary floral whorls with Monocotyledons.—PROPERTIES. Astringent, and yield a yellow dye. Berries of *Berberis* acid and eatable.

1. BER'BERIS, *L.* BARBERRY.

Spiny shrubs, wood yellow. *Leaves* spinous-toothed, jointed on the very short petiole, often reduced to 3-7-fid. spines. *Flowers* racemed solitary or fascicled, yellow, globose. *Sepals* 8-9, outer minute, imbricate. *Petals* 6, in 2 series, with 2 basal honeyed glands. *Stamens* 6. *Ovules* few, basal, erect. *Berry* 1-2-seeded. *Seeds* oblong, testa crustaceous; embryo straight.—Distrib. N. temp. regions, sub-trop. Asia, temp. S. America; species 50.—Etym. Arabic.

B. vulga'ris, *L.* ; leaves obovate spinous-serrate, stigma sessile. Copses and hedges from Caithness southd., naturalized only in Scotland and Ireland; fl. May–June.—An acid shrub, 4–6 ft. *Leaves* on the annual shoots 1–1½ in., alternate, shortly petioled on the woody shoots; reduced to 3–7-forked (rarely simple) spines jointed on to a very short sheath, and bearing fascicles of leaves (reduced branches) in their axils. *Flowers* ¼–½ in. diam., in terminal pendulous racemes, pale yellow, proterandrous; bracts short, triangular. *Stamens* irritable, springing forward when touched at the base. *Berry* ½ in. long, oblong, compressed, slightly curved, orange-red; stigma broad, black.—Distrib. Europe, temp. Asia, N. Africa; introd. in U. States.

ORDER III. **NYMPHÆA'CEÆ.**

Aquatic perennial herbs. *Leaves* usually floating, often peltate, margins involute in vernation. *Scapes* 1-fld. naked. *Floral whorls* all free and hypogynous, or adnate to a fleshy disk that envelops the carpels. *Sepals* 3–6. *Petals* 3–5, or more. *Stamens* many. *Carpels* 3 or more in one whorl, free, or adnate with the disk into a many-celled ovary; styles as many as carpels, stigma peltate or decurrent; ovules parietal, anatropous or orthotropous. *Fruit* a berry, or carpels separate and indehiscent. *Seeds* naked or arilled, albumen floury or 0; embryo enclosed in the enlarged amniotic sac.—Distrib. Temp. and trop.; genera 8; species 30–40.—Affinities. With *Papaveraceæ*, but not close.—Properties unimportant.

1. **NU'PHAR,** *Smith.*   Yellow Water-lily, Brandy-bottle.

Flowers yellow, globose. *Sepals* 5–6, concave. *Petals* many, small, hypogynous. *Stamens* many, inserted beneath the disk; filaments short, flattened. *Carpels* many, together forming a many-celled ovary; stigma peltate, rayed; ovules many. *Berry* ovoid, of separable carpels, ripening above water. *Seeds* small, not arilled.—Distrib. N. temp. hemisphere; species 3–4.—Etym. Arabic *naufar.*

1. **N. lu'teum,** *Sm.* ; leaves orbicular, base deeply 2-lobed, lobes usually contiguous, anthers linear, stigma 10–30 rayed. Still waters from the Hebrides and Aberdeen southd.; ascends to near 1,000 ft. in Yorkshire; Ireland; fl. June–Aug.—*Rootstock* creeping in mud; bud terminal. *Submerged leaves* membranous, waved; floating coriaceous; petiole obtusely 3-gonous at the top. *Flowers* fragrant, odour alcoholic. *Petals* 18–20, obovate-cuneate, thickly coriaceous, with a sub-terminal glandular pore, honeyed beneath. *Berry* beaked.—Distrib. Europe, temp. Asia, N. America.—Rootstock abounds in tannic acid.
N. lu'teum, proper; flower 2–3 in. diam., stigma generally entire 13–30-rayed. —Var. *N. interme'dium,* Ledeb.; flower 1½ in. diam., stigma waved at the margin 10–14-rayed.—Chartner's Lough, Northumb., and E. Perth.

2. **N. pu'milum,** *Smith;* leaves oblong deeply 2-lobed at the base, lobes at length spreading, anthers oblong, stigma lobed at the margin, rays 8-10 reaching the margin.

Small lakes in Scotland, Argyll to Elgin, rare ; Salop; fl. June–Aug.—Very similar to *N. lu'teum,* differing in the smaller more orbicular petals, and shorter anthers. *Petiole* 2-edged.—Distrib. Arctic and Central Europe, N. Asia.

2. NYMPHÆ'A, *L.* White Water-lily.

Flowers expanded, white blue or red. *Sepals* 4, adnate to the base of the disk. *Petals* in many series, inner successively transformed into stamens, adnate to the sides of the disk. *Carpels* many, their bases and the filaments sunk in the fleshy disk, and with it forming a many-celled ovary, crowned by the connate radiating stigmas ; ovules many, anatropous. *Fruit* a spongy berry, ripening under water. *Seeds* buried in pulp, aril fleshy.—Distrib. Most temp. and trop. regions, except N. Zealand and the Pacific Isles ; species 20.—Etym. dedicated by the Greeks to the nymphs.

N. al'ba, *L.* ; leaves floating orbicular base cordate quite entire.

Lakes and ponds, from Shetland southd., ascending to 1,000 ft. in the Lake District; Ireland ; fl. June–Aug.— *Rootstock* stout, fleshy; buds terminal. *Leaves* 5–10 in. diam., deeply 2-lobed at the base; lobes contiguous ; petiole very long. *Flowers* white. *Sepals* linear-oblong, back green. *Petals* oblong, obtuse, with no glandular pore. *Fruit* globose ; stigmatic rays 15–20. —Distrib. Europe (Arctic), N. Africa, N. and W. Asia to Kashmir, N. America.

Order IV. PAPAVERA'CEÆ.

Annual or perennial herbs ; juice milky or coloured. *Leaves* radical or alternate, exstipulate. *Flowers* regular, usually nodding in bud, envelopes and stamens caducous. *Sepals* 2, concave. *Petals* 4, crumpled. *Stamens* very many, hypogynous, filaments slender : anthers erect, insertion basal, bursting laterally. *Ovary* 1-celled, or 2-4-celled by prolonged placentas ; style short or 0, stigmas radiating in connate pairs opposite the placentas, or separate and alternating with these ; ovules in many rows, anatropous, parietal. *Capsule* dehiscing by pores or valves. *Seeds* many, small, albumen oily and fleshy ; embryo minute.—Distrib. N. temp. zone chiefly ; genera 17 ; species 65. — Affinities. With *Fumaria'ceæ* and *Crucif'eræ.*—Properties. Narcotic, emetic, purgative, or acridly poisonous.

* Capsule dehiscing by pores or very small valves.

Stigmas 4 or more, subsessile, forming a radiating disk..............1. Papaver.
Stigmas 4-5, deflexed on a conical clavate style....................2. Meconopsis.
** Capsule dehiscing to, or nearly to, the base by valves.
Ovary 1-celled. Seeds crested. Flower yellow................3. Chelidonium.
Ovary more or less completely 2-celled. Flower yellow..........4. Glaucium.
Ovary 1-celled. Seeds not crested. Flower violet.............4*. Rœmeria.

1. PAPA'VER, *L.* Poppy.

Annual erect herbs ; juice milky. *Leaves* lobed or cut. *Flowers* long-peduncled honeyless, proterandrous. *Ovary* 1-celled ; style short or 0, stigmas opposite the placentas united into a flat or pyramidal sessile or stalked 4–20-rayed disk ; placentas prominent. *Capsule* short, opening by very small valves under the lobes of the persistent stigma. *Seeds* small, pitted.—Distrib. Europe, N. Africa, N. Asia, one S. African and one Australian ; species 12.—Etym. obscure.

1. **P. hy'bridum,** *L.* ; leaves 2–3-pinnatifid sparingly hispid, filaments dilated upwards, capsule globose sessile bristly, stigma convex, rays 4–8.

Dry fields and waste places from Durham and Carnarvon southd.; rare·in Ireland ; Channel Islands ; (a colonist, *Wats.*); fl. May–July.—*Stem* 10–18 in., sparingly branched. *Leaves* with acute or awned lobes. *Flower* 1–2 in. diam., scarlet with a black disk. *Capsule* ½ in.; stigmatic rays reaching or exceeding the edge of the disk.—Distrib. Europe, N. Africa, W. Asia to the Himalaya.

2. **P. Argemo'ne,** *L.* ; leaves 2-pinnatifid, filaments dilated upwards, capsule clavate usually hispid, stigma convex, rays 4–6.

Waste dry places from Ross, southd.; rare in Ireland ; Channel Islands ; (a colonist, *Wats.*) ; fl. May–July.—Habit, &c., of *P. hyb'ridum,* but weaker, flowers smaller and paler, petals narrower, and capsule very different. The smallest British species.—Distrib. Europe, N. Africa, W. Asia ; introd. in America.

3. **P. du'bium,** *L.* ; leaves 1–2-pinnatifid, filaments filiform, capsule sessile obovoid glabrous, stigma 6–12-rayed.

Waste places N. to Shetland ; Ireland ; Channel Islands ; (a colonist, *Wats.*); fl. May–July.—Habit of succeeding species. Hairs of peduncles appressed. Pairs of petals unequal.—Distrib. Europe, N. Asia, N.W. India ; introd. in America.

P. du'bium proper; sap white, leaf-lobes shorter, capsule narrowing from just below the stigma to the base, lobes of stigmatic disk spreading. *P. Lamottei,* Bor. Abundant.

Sub-sp. P. Lecoq'ii, Lamotte ; sap yellow on exposure, leaf-lobes longer, capsule broadest at ⅓ below the stigma, lobes of stigmatic disk deflexed. England, Scotland, Ireland, rare.

4. **P. Rhœ'as,** *L.* ; leaves 1–2-pinnatifid, filaments filiform, capsule subglobose glabrous, stigma convex with overlapping lobes, rays 8–12.

Cornfields and waste places ; rare N. of the Tay ; Ireland ; Channel Islands ; (a colonist, *Wats.*) ; fl. June–Aug.—*Stem* branched, hispid. *Leaf* lobes ascending, with a bristle at the tip. *Peduncles* with spreading or appressed (*P. strigo'sum,* Boenn.) hairs. *Flowers* 3–4 in. diam. scarlet; pairs of petals unequal. *Capsule* stipitate.—Distrib. Europe, N. Africa, W. Asia to India. —A form entirely intermediate between *P. dubium* and *P. Rhœas* has been found in Surrey by Mr. G. Nicholson.

c

P. SOMNIF'ERUM, *L;* ; glaucous, glabrous or hispid, leaves amplexicaul sinuate-lobed or toothed, flowers large white or blue-purple, filaments slightly dilated upwards, capsule ovoid or globose stipitate. *Opium Poppy.* Cornfields and waste places, sporadic ; established in Kent, *Syme;* fl. July-Aug.—Variable in hispidity, in the shape of the capsule, colour of the flower, and black or white seeds.—DISTRIB. Europe, W. Africa, all Asia.

2. MECONOP'SIS, *Viguier.* WELSH POPPY.

Perennial herbs ; juice yellow. *Leaves* entire pinnate or pinnatifidly lobed. *Flowers* solitary or racemed, not honeyed, homogamous. *Ovary* 1-celled ; style distinct, stigmas 4 or more opposite the projecting placentas dilated or club-shaped. *Capsule* ovoid or elongate, with short valves below the persistent style. *Seeds* small, testa rugose.—DISTRIB. Mountain N. temp. regions ; species 9.—ETYM. μήκων, a *poppy,* and ὄψις, *resemblance.*

M. cam'brica, *Vig.* ; leaves pinnate, lobes pinnatifid.

Moist glens, Cornwall to Somerset, York, Westmoreland(?), Wales (ascending about 2,000 ft.) ; Ireland ; naturalized in Scotland ; fl. June.—Nearly glabrous. *Rootstock* stout, branched, tufted ; roots thick. *Stem* 1-2 ft., woolly at the base. *Leaves* petioled, pale green ; segments distinct or decurrent, ovate-lanceolate, lobed and toothed. *Flowers* 2-3 in. diam., pale yellow, peduncles long. *Sepals* hairy. *Petals* orbicular. *Style* short, stigma capitate, 4-6-rayed. *Capsule* 4-6-valved, ribbed.—DISTRIB. W. Europe, from Ireland to the Pyrenees.

3. CHELIDO'NIUM, *L.* CELANDINE.

Erect, branched, perennial herbs ; juice yellow. *Leaves* much divided. *Flowers* yellow. *Ovary* 1-celled ; style dilated at the top, with two adnate stigmas opposite the slender placentas. *Capsule* linear ; valves thin, separating upwards from the persistent placentas and style. *Seeds* with a shining testa and crested raphe.—DISTRIB. Europe to Japan ; species 2. —ETYM. doubtful.

C. ma'jus, *L.* ; leaves 1-2 pinnate, flowers small.

Waste places and hedgerows from Inverness southd., probably naturalized, elsewhere an escape ; Ireland ; Channel Islds. ; (a denizen, *Wats.*) ; fl. May-Aug.—*Stem* 1-2 ft., brittle, sparingly hairy, leafy. *Leaves* membranous, glabrous beneath ; segments 1-2 in., ovate, toothed, lobed or laciniate (*C. lacinia'tum,* DC.) ; petiole dilated at the base. *Flowers* ¾-1 in. diam., in loose few-fld umbels, yellow; pedicels slender ; bracts whorled. *Capsule* 1½ in., readily dehiscing, valves torulose.—DISTRIB. Europe (Arctic), W. Asia to Persia ; introd. in N. America.

4. GLAU'CIUM, *Tourn.* HORNED POPPY.

Glaucous herbs ; juice yellow. *Leaves* lobed or cut. *Flowers* large, yellow or purple. *Ovary* 2- rarely 3-celled ; style short or 0, with 2 deflexed stigmas opposite the placentas which meet in the axis of the ovary ; ovules very many. *Capsule* long, narrow, 2-valved almost to the base. *Seeds* many, sunk in the spongy septum, testa pitted.—DISTRIB. Chiefly Mediterianean ; species 5–6. The dissepiment (formed by the placentas as in *Cruciferæ*) is sometimes incomplete.—ETYM. γλαύκιον, from the *blue* hue of some species.

G. lu'teum, *Scop.*, ; leaves ½-amplexicaul, capsule tubercled. Sandy sea-shores Shetland ; and from the Forth and Clyde southd.; Ireland ; Channel Islds. ; fl. June–Oct.—Glaucous, sub-hispid, annual, sometimes perennial. *Stem* 1–2 ft., branched, erect or ascending. *Radical leaves* 2-pinnatifid, rough with stout hairs ; lobes pointing various ways. *Flower*- 2–4 in. diam., golden yellow : peduncles short, glabrous. *Petals* in opposite dissimilar pairs. *Pod* curved, a foot long, glabrous ; stigmatic lobes spreading.—DISTRIB. Europe, N. Africa, W. Asia ; introd. in U. States.

4*. RŒME'RIA, *DC.*

Annual herbs ; juice yellow. *Leaves* much cut. *Flowers* long-peduncled, violet. *Sepals* 4. *Petals* 4. *Ovary* 1-celled ; stigma sessile, lobes 2–4 deflexed opposite as many slender placentas ; ovules many. *Capsule* linear, 2–4-valved nearly to the base. *Seeds* many, testa rough.— DISTRIB. Cornfield plants of Europe and W. Asia ; species 2.—ETYM. *J. F. Rœmer*, a German botanist.

R. HYB'RIDA, *DC.* ; leaves 3-pinnatifid, capsule 3-valved. *Glau'cium viola'ceum*, Juss.

Dry soil, Cambridge and Norfolk ; (a colonist, *Wats.*); fl. May–June. —Habit of *Papaver Argemone*, glabrous or slightly hairy. *Stem* erect. *Leaves* 1- or 2-pinnatifid, segments tipped by a bristle. *Flower* 2–3 in. diam., violet-purple with a black disk. *Sepals* hairy. *Capsule* 2–3 in. cylindric, hispid above.—DISTRIB. Central and S. Europe.

ORDER V. FUMARIA'CEÆ.

Annual or perennial herbs ; juice watery. *Leaves* usually divided. *Flowers* racemose. *Sepals* 2, small, scale-like, deciduous. *Petals* 4, in 2 usually very dissimilar pairs ; 2 outer larger lateral, but becoming antero-posterior by a ¼-twist of the pedicel, one or both gibbous or spurred ; two inner smaller, erect, often coherent at the tips. *Stamens* (in the British species) 6, in 2 bundles opposite the 2 outer petals ; anther of central stamen in each bundle 2-celled, of lateral 1-celled. *Ovary* 1-celled ; style long or short, stigma obtuse or lobed ; ovules 2 or more, amphitropous ; placentas parietal. *Fruit* a 2-valved many-seeded capsule, or an indehiscent 1-seeded nut. *Seeds* albuminous, raphe sometimes appendaged ; embryo minute.—DISTRIB. Temp. and warm N. hemisphere, and S. Africa ; genera 7 ; species 100 —AFFINITIES. Between *Papavera'ceæ* and *Crucif'eræ.*—PROPERTIES. Astringent, acrid, and reputed diaphoretic.

1. FUMA'RIA, *L.* FUMITORY.

Annual, rarely perennial herbs, usually branched, often climbing. *Leaves* much divided ; segments very narrow. *Flowers* small, in terminal

or leaf-opposed racemes honeyed, homogamous. *Petals* 4, erect, conniving; the posterior gibbous or spurred at the base, the anterior flat; 2
inner narrow, cohering by their tips, winged or keeled at the back.
Filament of the stamen opposite the gibbous petal usually spurred at the
base. *Ovary* globose; style filiform, stigma entire or shortly lobed;
ovules 2, on 2 placentas. *Fruit* 1-seeded, indehiscent, globose.—DISTRIB.
Europe, Asia, following cultivation; species 6.—ETYM. doubtful.

1. **F. capreola′ta,** *L.* ; climbing by the twisting petioles, leaf-segments
flat, sepals ovate toothed below at least as broad and ⅓–⅔ as long as the
corolla-tube, lower petal gradually dilated at the tip, pedicels longer than
the bracts, fruit globose contracted into a neck at the base not retuse.

Fields and waste places from Orkney southd.; Ireland; Channel Islds.;
(a colonist, *Wats.*); fl. May–Sept.—*Stems* 1–2 ft. or more, branched. *Leaves*
2-pinnate; segments broad. *Racemes* lax-fld., not much elongated in fruit.
Flowers ⅓–½ in.—DISTRIB. Europe, N. Africa, W. Asia.

F. CAPREOLA′TA proper; sepals denticulate ½–⅔ as long as the corolla-
tube, petals cream-coloured often coloured after fertilization, fruit longer
than broad with 2 deep pits at the top, neck narrower than the dilated top
of the recurved pedicel. *F. pallidiflo′ra,* Jord.—From Roxburgh southd.;
Co. Down.

VAR. *F. Borœ′i,* Jord.; sepals smaller, petals redder, fruiting pedicels not
recurved.—Common.

Sub-sp. F. CONFU′SA, *Jord.*; sepals ½ as long and nearly as broad as the
corolla-tube, petals pink tipped with purple, fruit subrugose when dry a
little longer than broad with 2 broad shallow pits at the top, neck broader
than the dilated top of the erecto patent pedicel. *F. agra′ria,* Mitten.—
From Perth southd.

Sub-sp. F. MURA′LIS, *Sonder ;* flowers smaller and laxer, sepals as in *confu′sa,*
petals pink tipped with purple, fruit finely rugose when dry with 2 inconspicuous pits at the top, neck narrower than the dilated top of the erecto-
patent pedicel.—England chiefly, rare; Stirling; Belfast.

2. **F. officina′lis,** *L.* ; diffuse, leaf-segments flat, sepals ovate-lanceolate
⅓ as long and ½ as broad as the corolla-tube, lower petal abruptly dilated
at the tip, pedicel ascending longer than the bracts, fruit depressed-sphe-
rical rugose when dry top with a large shallow pit.

Waste places, N. to Shetland; ascends to 1,000 feet in N. England; Ireland;
Channel Islds.; (a colonist, *Wats.*); fl. May–Sept.—Smaller than *F. capreo-
la′ta,* leaves more divided, flowers smaller and raceme much elongated after
flowering. *Flowers* dark or pale rose-purple.—DISTRIB. Europe, N. Africa,
W. Asia; introd. in U. States.

3. **F. densiflo′ra,** *DC.* ; diffuse, leaf-segments narrow, sepals broadly
ovate toothed ½ as long as and broader than the corolla-tube, lower petal
abruptly dilated at the tip, pedicels erecto-patent about as long as the
bracts, fruit globose rugose when dry top with 2 shallow pits. *F. micran′-
tha,* Lagasca.

Waste places, from Elgin southd.; (a colonist, *Wats.*); fl. May–Sept.—
Habit of *F. officina′lis,* but weaker, leaf-segments smaller and narrower, flat

or slightly channelled, racemes short, much elongated after flowering, flowers smaller $\frac{1}{4}$–$\frac{1}{3}$ in. pale, bracts coloured.—Distrib. Europe, N. Africa, W. Asia to India.

4. F. parviflo'ra, *Lamk.* ; diffuse, leaf-segments narrow, sepals minute toothed $\frac{1}{10}$–$\frac{1}{8}$ as long and not $\frac{1}{2}$ as broad as the corolla-tube, lower petal abruptly dilated at the tip, pedicels erecto-patent equal to or exceeding the bracts, fruit globose rugose when dry top with 2 pits. *F. tenuisec'ta,* Syme.

Waste places, &c., from Mid. Scotland southd.; (a colonist, *Wats.*); fl. June–Sept.—Best distinguished by habit, by the narrow leaf-segments, small pale flowers and minute sepals.—Distrib. Europe, N. Africa, N. and W. Asia to India.

F. parviflo'ra proper ; leaf-segments channelled, racemes dense, sepals triangular-ovate $\frac{2}{3}$ as long and $\frac{1}{2}$ as broad as the corolla-tube, pedicels equalling the bracts, fruit pointed.—From Perth southd., rare.

Sub-sp. F. Vaillan'tii, *Loisel.* ; leaf-segments flat, racemes lax, sepals lanceolate $\frac{1}{10}$ as long and $\frac{1}{3}$ as broad as the corolla-tube, pedicels exceeding the bracts, fruit rounded at the top.—Yorkshire and S E. England.

2. CORYD'ALIS, *DC.*

Erect herbs with a tuberous rootstock, or weak and diffuse, or slender and climbing by tendrils. *Leaves* much divided, alternate or subopposite. *Racemes* terminal or leaf-opposed. *Floral* characters of *Fumaria,* but ovules numerous, and fruit an inflated 2-valved capsule. *Seeds* small, raphe often crested.—Distrib. Chiefly Mediterranean and Himalayan, a few N. American and S. African ; species 70.—Etym. Greek for a *Fumaria.*

1. **C. clavicula'ta,** *DC.* ; annual, branched, climbing by branched tendrils terminating the petioles, racemes leaf-opposed.

Copses, banks, and thatched roofs from Ross southd.; ascends to near 1,000 ft.; N.E. Ireland ; fl. June–Aug.—*Stems* 1–3 ft., brittle, slender. *Leaves* glaucous, pinnate, pinna 3- or digitately 5-foliolate ; segments small, ovate or oblong. *Pedicels* very short ; bracts cuspidate. *Flowers* $\frac{1}{4}$ in., straw-coloured ; spur very short. *Pods* $\frac{1}{4}$ in., linear-oblong. *Testa* shining, granulate.—Distrib. W. Europe, from Denmark to Spain.

C. lu'tea, *DC.* ; perennial, branched, diffuse, root fibrous, leaves 2–3-ternately pinnate, racemes leaf-opposed, flowers subsecund yellow.

Old walls; an escape from cultivation ; fl. May–Aug.—*Rootstock* branched. *Stem* 6–12 in.; angular. *Leaves* long petioled, leaflets oblong-ovate or obovate, entire or lobed. *Pedicels* long ; bracts lanceolate, erose. *Flowers* $\frac{1}{2}$–$\frac{3}{4}$ in.; spur short, thick, incurved. *Pods* oblong, compressed, acuminate ; style deciduous.—Distrib. W. Europe, from Belgium southd.

C. sol'ida, *Hook.* ; perennial, rootstock tuberous, stem simple, leaves 2–3-ternately pinnate, raceme terminal, flowers purple. *C. bulbo'sa,* DC.

Banks and cultivated ground, naturalized in England ; fl. April–May.—Very glaucous. *Rootstock* 1 in. diam. and upwards. *Stem* 6–10 in., stout, with one or two oblong scales below, and a few leaves about the middle. *Leaves* with stout petioles ; leaflets broad. *Flowers* 1 in.; bracts lobed, leafy; spur longer than the rest of the corolla. *Pods* narrow, lanceolate ; style persistent. *Cotyledons* connate.—DISTRIB. Europe, from Denmark southd.

ORDER VI. **CRUCIF'ERÆ.**

Herbs. *Leaves* radical or alternate, exstipulate. *Flowers* racemed. *Sepals* 4, 2 lateral (opposite the placentas) often larger and saccate at the base, imbricate in bud. *Petals* 4, placed crosswise, imbricate in bud. *Stamens* 6 (rarely 1, 2, or 4), in 2 series, hypogynous ; 2 outer opposite the lateral sepals ; 4 inner longer, in pairs opposite the other sepals. *Disk* honeyed, glands 2, 4, or 6, opposite the sepals. *Ovary* 2-celled by a vertical prolongation of the placentas, or 1-celled, or with superimposed cells ; style short or 0, stigma simple or 2-lobed, lobes opposite the placentas ; ovules 2-seriate on 2 parietal placentas, rarely solitary and erect, amphitropous or campylotropous, micropyle superior. *Fruit* a long or short 2-celled and 2-valved capsule (*pod*) ; valves deciduous, leaving the seeds on the persistent placentas (*replum*), rarely indehiscent, or of superposed 1-seeded joints. *Seeds* small, albumen 0 ; cotyledons large, plano-convex or longitudinally folded, foliaceous in germination, radicle turned up on the back of one cotyledon (*incumbent*), or facing their edges (*accumbent*).— DISTRIB. All temp. and cold regions, but chiefly of the Old World ; genera 172 ; species 1,200.—AFFINITIES. Between *Fumariaceæ* and *Capparideæ.* —PROPERTIES. All are nitrogenous and contain sulphur, are pungent, stimulant, anti-scorbutic, often acrid. Seeds oily. Testa of cress and others mucilaginous when moistened, owing to the swelling and bursting of superficial cells.

A. Pods *elongate* (*much longer than broad*), *dehiscing throughout their length, flat or turgid, not compressed at right angles to the septum.* (Pods *sometimes short in* Nasturtium, *the tip sometimes indehiscent in* Brassica. *See* Draba *in* B.

TRIBE I.   **ARABIDE'Æ.**   *Seeds* 1-seriate (or 2-seriate in *Arabis* and *Nasturtium*) ; radicle accumbent. (Flowers white, yellow or lilac.)
* *Stigmas erect or decurrent on the style* ....1. Matthiola.
** *Stigma small, simple, terminal.*
Lateral sepals saccate. Hairs forked...........................1*. *Cheiranthus.*
Pods terete, valves turgid. Seeds minute, 2-seriate ....2. Nasturtium.
Pods 4-angled. Seeds oblong...3. Barbarea.
Pods flat, valves not elastic 1-nerved....................................4. Arabis.
Pods flat, valves elastic. Funicle filiform ....5. Cardamine.
Pods flat, valves elastic. Funicle dilated ....6. Dentaria.
TRIBE II.   **SISYMBRIE'Æ.**   *Seeds* usually 1-seriate; radicle incumbent, straight, plano-convex. (Flowers white, yellow or lilac.)
Glabrous or hairs spreading, stigma obtuse........................7. Sisymbrium.
Hairs appressed 2–3-furcate, stigma obtuse ....8. Erysimum.
Hairs spreading, stigmas decurrent on the style............ ....8*. Hesperis.

TRIBE III. **BRASSICE'Æ.** *Seeds* 1-2-seriate ; radicle incumbent, longi-
tudinally folded or very concave. (Flowers yellow.)
Pods terete or angled. Seeds 1-seriate.............................9. Brassica.
Pods compressed. Seeds 2-seriate, compressed.................10. Diplotaxis.

B. Pods *short* (*not or not much longer than broad*), *dehiscing through their whole
length, broad, flat or turgid, not compressed at right angles to the septum.*
(Flowers *white or yellow.*) (Pod *sometimes long in* Draba ; *see* Nasturtium
in A.)

TRIBE IV. **ALYSSINE'Æ.** *Seeds* 2-seriate ; radicle accumbent.
Petals entire. Pods oblong, flat, many-seeded ....11. Draba.
Petals 2-cleft. Pods oblong, flat or turgid ....12. Erophila.
Petals entire. Pods circular, few-seeded ....12*. Alyssum.
Petals entire. Pods inflated, many-seeded.......................13. Cochlearia.

TRIBE V. **CAMELINE'Æ.** *Seeds* 2-seriate; radicle incumbent.
Tall herb, cauline leaves sessile auricled ....13*. Camelina.
Small scapigerous water-herb. Leaves subulate..................14. Subularia.

C. Pods *short, dehiscing throughout their length, much compressed at right angles
to the septum, which is hence very narrow.* (Pod *indehiscent in* Senebiera.)

TRIBE VI. **LEPIDINE'Æ.** *Cotyledons* straight incurved or longitudinally
folded, radicle incumbent. (Flowers white.)
Pods dehiscent, many-seeded ....15. Capsella.
Pods didymous, indehiscent, 2-seeded..............................16. Senebiera.
Pods dehiscent, 2-4-seeded ....17. Lepidium.

TRIBE VII. **THLASPIDE'Æ.** *Cotyledons* straight, radicle accumbent.
Pods on horizontal pedicels. (Flowers white.)
Pods notched. Petals equal. Filaments without scales ....18. Thlaspi.
Pods ovate. Petals very unequal. Filaments without scales ...19. Iberis.
Pods oblong. Petals unequal. Filaments with basal scales...20. Teesdalia.
Pods oblong. Petals equal. Filaments without scales ......21. Hutchinsia.

D. Pods *indehiscent or with very short valves which cover a few of the seeds only.*

TRIBE VIII. **ISATIDE'Æ.** *Pods* indehiscent, 1-celled, 1-seeded ...22. Isatis

TRIBE IX. **CAKILINE'Æ.** *Pods* transversely 2-jointed, lower joint indehiscent seedless or not, or 2-valved and 2- or more-seeded; upper joint
indehiscent, 1-2-celled. (Affinity with *Brassiceæ.*)
Lower joint slender, seedless ; upper globose, 1-seeded...........23. Crambe.
Lower joint 2-edged, 1-celled ; upper ensiform, 1-seeded ... ...24. Cakile.

TRIBE X. **RAPHANE'Æ.** *Pods* elongate, 1-celled, many-seeded, or indehiscent, or jointed, the 1-seeded joints indehiscent...............25. Raphanus.

1. MATTHI'OLA, *Br.* STOCK.

Herbs, sometimes shrubby, downy with stellate hairs. *Leaves* entire or
sinuate. *Flowers* large. *Sepals* erect, lateral saccate at the base. *Petals*
with long claws. *Pods* elongate, terete or compressed ; septum thick, often
2-3-nerved ; stigmatic lobes erect, conniving, often thickened or horned at

the back. *Seeds* 1-seriate, compressed, winged, or margined ; radicle accumbent.—DISTRIB. Europe, N. Africa, W. Asia, one is S. African ; species 30.—ETYM. After *Mattioli,* an Italian physician.

1. **M. inca'na,** *Br.* ; shrubby, erect, hoary, leaves oblong-lanceolate entire, pod eglandular. *Queen Stock.*

Sea-cliffs eastward of Hastings (now extinct), I. of Wight; (a denizen, *Wats.*) fl. May–June.—*Stem* 1-2 ft., branched. *Leaves* rarely obscurely toothed. *Raceme* 1–2 in. *Flowers* 1–2 in. diam., purple to violet. *Pods* 2–4 in., ½ in. broad ; seeds orbicular, winged.—DISTRIB. W. Europe, Canaries, Levant.

2. **M. sinua'ta,** *Br.*; herbaceous, diffuse, woolly or downy, leaves linear-obovate or -oblong, lower sinuate-toothed, pod muricate and glandular.

Shores of Wales, Cornwall (extinct?), Devon; S.E. and S. W. Ireland; Channel Isles; fl. May–Aug.—*Stem* 1–2 ft., branched above. *Root-leaves* petioled. *Raceme* 1–3 in. *Flowers* 1 in. diam., pale lilac, fragrant at night. *Pods* 3–4 in., ⅛ in. broad ; seeds winged.—DISTRIB. W. Europe, N. Africa, Levant. —Taste alkaline.

1*. *CHEIRAN'THUS, L.* WALLFLOWER.

Herbs or under-shrubs, pubescent with appressed 2-partite hairs. *Leaves* entire or toothed. *Flowers* large, racemed, yellow or purple. *Sepals* erect, lateral saccate at the base. *Petals* with long claws. *Pods* elongate, compressed or 4-angled ; valves 1-nerved, flat or convex ; stigma capitate or with 2 spreading lobes. *Seeds* 1-seriate, compressed ; radicle accumbent. —DISTRIB. N. temp. and cold regions ; species 12.—ETYM. doubtful.

C. CHEI'RI, *L.* ; leaves lanceolate acute entire.

Old walls ; (an alien, *Wats.*) ; fl. May–June.—Perennial. *Stem* shrubby below, branched, angled. *Leaves* 2–3 in. *Flowers* about 1 in. diam., fragrant, orange-yellow (in cultivation red, purple or brown). *Pods* 1–2½ in., 4-angled ; stigma subsessile; seeds shortly winged above.—DISTRIB. Central and N. Europe.

2. NASTUR'TIUM, *Br.*

Branched, terrestrial or aquatic glabrous herbs ; hairs if present usually simple. *Leaves* entire lobed or cut. *Flowers* small, usually yellow, sometimes bracteate. *Sepals* short, equal, spreading. *Petals* slightly clawed, or 0. *Stamens* 1–6. *Pods* short or long, often curved, terete, pedicels patent and curved ; valves not rigid, convex, obscurely 1-nerved ; style short or long, stigma simple or 2-lobed. *Seeds* 2-seriate, small, turgid ; radicle accumbent.—DISTRIB. N. temp. and warmer regions ; species about 20.—ETYM. *Nasi tortium,* from the bitterness distorting the face.— United with *Cochlearia* by C. *Armoracia* (Horse-radish) and others.

1. **N. officina'le,** *Br.* ; aquatic, leaves pinnate, leaflets subcordate sinuate-toothed, petals white twice as long as the sepals, pods linear. *Watercress.*

Watercourses, N. to Shetland; ascends above 1,000 feet in N. England ; Ireland ; Channel Islands ; fl. May–Oct.—Perennial, glabrous, green or olive-

brown. *Stem* 2–4 ft., rooting, often floating, fistular. *Leaflets* 3–6 pair. *Racemes* short, flowers ⅛–¼ in. diam. *Disk-glands* 4. *Pods* ¼ in., deflexed or horizontal, longer than the pedicels ; valves beaded ; seeds suborbicular, compressed.—DISTRIB. Europe, W. Asia, N. Africa ; introd. in N. America and the colonies, and choking some rivers of N. Zealand.
N. officina'le proper ; decumbent, terminal leaflet broadest and largest.—VAR. *siifo'lium,* Reichb. ; erect, tall, leaflets subequal, terminal oblong.—VAR. *microphyl'lum* is a starved terrestial state with small leaflets.

2. **N. sylves'tre,** *Br.* ; rootstock creeping, leaves deeply pinnatifid, leaflets many lanceolate more or less cut nearly equal, petals yellow twice as long as the sepals, pod linear.

Moist waste places from the Tay southd., rare in the north ; S. Ireland, rare ; Channel Islands ; fl. June–August.—*Stem* angular, flexuous. *Leaves* very variable. *Racemes* short ; flowers ¼ in. diam. *Disk-glands* 6. *Pods* ½–¾ in., curved, pedicel very slender longer or shorter than the pod ; seeds minute, hardly 2-seriate.—DISTRIB. Europe, N. Africa, temp. and subtrop. Asia ; introd. in America.

3. **N. palus'tre,** *DC.* ; leaves lyrate-pinnatifid, lobes few broad unequally cut, terminal of the lower leaves very large, petals yellow equalling the sepals or shorter, pods linear-oblong turgid. *N. terres'tre,* Sm.

Damp places from the Clyde southd.; Ireland ; fl. June–Oct.—*Stem* 1–2 ft., erect or inclined. *Leaves* pinnate or pinnatifid, terminal leaflet some- times 6 in. *Flower* ⅛ in. diam. *Disk-glands* 4. *Pods* equalling or exceed- ing their pedicels ; style very short ; seeds angular.—DISTRIB. Europe (Arctic), N. Africa, temp. and cold Asia and America.

4. **N. amphib'ium,** *Br.* ; rootstock short stoloniferous, leaves entire toothed or pinnatifid, petals twice as long as the sepals, pods oblong shorter than their pedicels. *Armora'cia,* Koch.

Wet places, York to Somerset and Kent ; Ireland ; Channel Islds.; fl. June– Sept.—*Stem* 2–4 ft., erect. *Leaves* pinnatifid when submerged, base often ½-amplexicaul. *Flowers* ¼ in. diam. *Disk-glands* 4. *Pods* ¼ in., pedicels spreading or deflexed ; style slender ; stigma large, capitate ; seeds small, oblong.—DISTRIB. Europe, N. Africa, temp. Asia.

3. BARBARE'A, *Br.* WINTER-CRESS.

Erect, glabrous, biennials. *Stem* angular. *Leaves* entire lobed or pinnatifid. *Flowers* yellow, sometimes bracteate. *Sepals* suberect, equal. *Petals* clawed. *Disk-glands* 6. *Pods* linear, elongate, compressed, 4-angled, acuminate ; valves keeled or ribbed ; style short, stigma capitate or 2-lobed. *Seeds* 1-seriate, oblong, not margined ; radicle accumbent.—DISTRIB. Temp. regions ; species about 6.—ETYM. Dedicated to St. Barbara.—The straight stiff pods, keeled valves, and 1-seriate seeds, separate this from *Nasturtium.*

1. **B. vulga'ris,** *Br.* ; leaves toothed or pinnatifid at the base, pods short acuminate, pedicels spreading, style distinct.

Hedgebanks, water-sides, from the Clyde and Aberdeen southd.; Ireland ; Channel Islds.; fl. May–Aug.—*Stem* rigid, erect, simple or sparingly branched

Lower leaves pinnate, rarely pinnatifid, terminal leaflet usually largest, cordate; upper subentire or pinnatifid with amplexicaul auricled bases. *Flowers* small, bright yellow. *Pods* ¾–1 in., broader than their slender pedicels; style ₁₀¹–⅓ in.—DISTRIB. Europe (Arctic), temp. Asia, Himalaya to 17,000 ft., S. Africa, Australia, and N. America.

B. VULGA′RIS proper; raceme about as long as broad, petals twice as long as the sepals, pods in a dense raceme 3–6 times as long as their pedicels, erect rarely spreading, seeds 1½ times as long as broad.—Common. VAR. *arcua′ta,* Reichb.; raceme elongate, petals rather more than twice the length of the sepals, pods in a lax raceme arched and spreading when young 5–8 times as long as their pedicels, seeds more than twice as long as broad.—Rare, Loughgall, Armagh.

Sub-sp. B. STRIC′TA, *Andrz.*; upper leaves entire, terminal lobe of lower oblong, flowers smaller, pods in a dense narrow raceme with erect pedicels. *B. parviflo′ra,* Fries.—Chester, York, S.E. counties.

Sub-sp. B. INTERME′DIA, *Boreau;* leaf-segments many, petals twice as long as the sepals, pods in a dense raceme erect 4–6 times as long as their pedicels, seeds nearly as long as broad. — Cultivated fields, rare.—Intermediate between *B. stric′ta* and *præ′cox.*

B. PRÆ′COX, *Br.*; leaves pinnatifid, segments narrow, petals 3 times as long as the sepals, pods long and distant scarcely thicker than their very stout short pedicels, style very short. *American Cress.*

Roadsides, &c., a garden escape; (an alien, *Wats.*); fl. April–Oct.—Very similar to *B. vulga′ris,* of which I suspect it is a cultivated form. *Seeds* ¼ longer than broad, twice as large and more ellipsoid than in *B. vulga′ris.*—DISTRIB. All Europe; introd. in U. States.—An excellent salad.

4. AR′ABIS, *L.* ROCK-CRESS.

Annual or perennial herbs, glabrous or with forked or stellate hairs. *Radical leaves* spathulate; cauline sessile. *Flowers* usually white. *Sepals* short, equal, or the lateral saccate at base. *Petals* entire, usually clawed. *Pods* linear, compressed; valves flat, keeled, veined or ribbed; stigma simple or 2-lobed. *Seeds* 1- rarely sub-2-seriate, compressed, often margined or winged; radicle accumbent.—DISTRIB. N. temp. zone; species 60.—ETYM. From Arabia, the native country of various species.—Differs from *Cardami′ne* in the more keeled less elastic pod-valves.

1. **A. petræ′a,** *Lamk.*; leaves petioled radical lyrate-pinnatifid, cauline subentire, petals spreading broadly clawed, pods spreading. *A. his′pida,* L. fil; *Cardami′ne hastula′ta,* Sm.

Alps of Wales and Scotland, ascending above 4,000 ft.; Glenade Mt., Leitrim; fl. June–Aug.—Glabrous or hairy, perennial. *Stem* 3–6 in., branched below. *Leaf-segments* short. *Flowers* corymbose, white or purplish. *Pods* ½–1 in; valves 3-nerved; seeds hardly winged.—DISTRIB. Alpine and Arctic Europe, N. Asia and N. America.

2. **A. stric′ta,** *Huds.*; hispid, radical leaves subpetiolate small obtuse sinuate-lobed, cauline few ½-amplexicaul, petals narrow cuneate suberect, pods suberect.

Rocks, N. Somerset and W. Gloucester, very rare; fl. March–May.—Perennial. *Stems* 5–10 in., ascending or erect. *Radical leaves* ciliate, hairs forked and simple; lobes oblong or triangular, pointing upwards. *Flowers* rather large, cream-coloured. *Pods* 1 in., 3–6 times as long as their pedicels; valves 1-nerved; style short; seeds oblong, slightly winged above.—DISTRIB. Europe, from Spain to Hungary.—Habit of *Sisymb. Thalia'na.*

3. **A. hirsu'ta,** *Br.* ; hispid, stem leafy, radical leaves subpetiolate toothed, cauline sessile or ½-amplexicaul, petals spreading white, pods many slender erect. *A. sagitta'ta,* DC. ; *Turri'tis hirsu'ta,* L.

Dry places, ascending to 2,700 ft. in Scotland; local in Ireland; Channel Islds.; fl. June–Aug.—Biennial or perennial. *Stems* 1–2 ft., many, slender. *Leaves* 1½–3 in., obtuse or acute. *Flowers* small. *Pods* 1½–2 in., very many, much narrower than in *A. ciliata;* style short; seeds distant, very narrowly winged all round.—DISTRIB. Europe, temp. Asia to the Himalaya, N. America.—VAR. *glabra'ta,* Syme; stem and leaves glabrous or ciliated.—Great Aran Is, W. Ireland, and Eastbourne, Sussex.

Sub-sp. A. CILIA'TA, *Br.*; glabrous or ciliate, cauline leaves sessile, base rounded, pods fewer broader, seeds closer.—Rocky shores, S. Wales, W. Ireland.

4. **A. perfolia'ta,** *Lamk.* ; nearly glabrous, glaucous, radical leaves obovate sinuate or lobed, cauline amplexicaul entire auricled, petals erect pale yellow, pods many crowded slender erect, seeds sub-2-seriate. *Turri'tis gla'bra,* L.

Dry rocky places, local, from Perth southd.; Ireland, Antrim only : fl. May–July.—Annual or biennial. *Stem* 2–3 ft., erect, with few spreading hairs about the early withering root-leaves, &c. *Pods* 1–2 in.; pedicel slender ; style very short; seeds minute, oblong, angled, not winged.—DISTRIB. Europe (Arctic), temp. Asia to the Himalaya, N. America.

A. TURRI'TA, *L.* ; stellately pubescent, leaves remotely toothed, radical petioled entire, cauline narrow oblong amplexicaul, flowers bracteate, petals obovate-lanceolate spreading, pods large long secund decurved.

Naturalized on castle walls, Cleish, Kinross-shire, &c.; fl. May–July.—Perennial. *Stem* 1–2 ft., rather robust, leafy. *Radical leaves* on barren branches of the rootstock. *Flowers* pale yellow. *Pods* 3–6 in. ; valves thick, veined; seeds oblong, winged.—DISTRIB. Central and S. Europe.

5. CARDAMI'NE, *L.* BITTER-CRESS.

Annual or perennial herbs, usually glabrous. *Leaves* usually pinnate. *Flowers* white cream-coloured or purple. *Sepals* equal at the base. *Petals* clawed, rarely 0. *Pods* elongate, linear, compressed ; valves flat, indistinctly nerved, elastic ; stigma small. *Seeds* compressed, not margined, funicle filiform ; radicle accumbent.—DISTRIB. Temp. and cold regions, rarely tropical ; species 50.—ETYM. κάρδαμον, a kind of cress.

1. **C. hirsu'ta,** *L.* ; radical leaves pinnate, leaflets broad petioled auricles 0, cauline narrower, petals small erect oblong-lanceolate, anthers yellow, pods erect.

Moist places, N. to Shetland; ascending to 3,000 ft. in Scotland; Ireland ;
Channel Islds.; fl. April–Sept.—Very variable, glabrous or hairy, 6 in.–2 ft.,
erect or diffuse. *Leaflets* 3–6 pairs, angled or sublobate. *Flowers* ½ in.
diam., white. *Pods* about 1 in., slender, torulose.—DISTRIB. N. temp. and
cold zones; in S. temp. the perennial varieties are very numerous and
and puzzling.

C. HIRSU'TA proper; annual, radical leaves rosulate, pedicels erect, stamens
usually 4, style short stout.—Open ground.
Sub-sp. C. FLEXUO'SA, *Withering;* perennial or biennial, radical leaves few,
leaflets lobed, pedicels spreading, stamens usually 6, style slender elongate.
C. sylvat'ica, Link.—Shaded places.

2. **C. praten'sis,** *L.* ; leaves all pinnate, leaflets of radical petioled
suborbicular, of cauline narrow subsessile, petals large spreading obovate,
anthers yellow, pods erect on slender pedicels, style short stout. *Lady's
Smock, Cuckoo-flower.*

Moist meadows, N. to Shetland; ascends to 3,200 ft. in Scotland; Ireland;
Channel Islds.; fl. April–June.—*Rootstock* short, stout, sometimes stoloniferous. *Stem* 1–2 ft. *Leaflets* of radical leaves ¼–¾ in.; of cauline usually
much longer, almost entire. *Flower* ½–¾ in. diam., lilac or almost white.
Pods 1–1½ in ; pedicel ½–1 in.—DISTRIB. N. temp. and Arctic regions,
Abyssinia, Himalaya, Chili.—A similar plant occurs in Tasmania.—VAR. *C.
denta'ta,* Schult.; taller, stronger, radical leaves erect or suberect, leaflets
fewer larger angled toothed.—VAR. *Haynea'na,* Welw.; habit of *C. hirsu'ta,*
leaflets many small round linear or lanceolate, flowers small white, petals
narrower. Thames near Mortlake.

3. **C. ama'ra,** *L.* ; leaves pinnate, radical leaflets suborbicular, cauline
narrow or deeply toothed, petals large obovate spreading, anthers purple,
pods erect on slender pedicels, style slender. *Bitter cress.*

River-sides, &c., scarce; from Aberdeen southd., ascending to 1,000 ft. in
Yorkshire; N.E. Ireland; fl. April–June.—*Rootstock* slender, stoloniferous.
Stem 1–2 ft., ascending, glabrous or hairy. *Leaves* all alternate. *Flowers*
½ in. diam., creamy white. *Pods* 1–1½ in.; pedicel ¼–¾ in.; style slender,
stigma minute.—DISTRIB. N. Europe and Asia to the Himalaya.

4. **C. impa'tiens,** *L.* ; leaves pinnate, petiole with stipuliform fringed
auricles, leaflets all narrow deeply cut, petals linear-obovate or 0, anthers
yellow, pods erect on short slender pedicels, style slender.

Shady copses, &c., local; from Westmoreland and York (ascends to 1,000 ft.)
southd.; casual in Ireland; fl. May–Aug.—*Rootstock* spindle-shaped. *Stem*
1–2 ft., stout, erect, very leafy, glabrous. *Leaflets* many, petioled. *Flowers*
¼ in. diam., often panicled, white. *Pods* ¾–1 in., very slender; pedicel ¼ in.
DISTRIB. Europe, temp. Asia to the Himalaya. –The stipuliform auricles,
reflexed on the stem, are anomalous structures.

6. DENTA'RIA, *L.* CORAL-ROOT.

Herbs with creeping, scaly rootstocks. *Radical leaves* few or 0 ; cauline
often opposite or 3-nately whorled. *Flowers* large, purple *Sepals* erect,
equal at the base. *Petals* clawed. *Pods* narrow-lanceolate ; valves flat,

nearly nerveless, elastic ; septum membranous. *Seeds* 1-ser'ate, compressed, not margined, funicle dilated ; radical accumbent, stalked.—DISTRIB. N. temp. regions ; species about 20.—ETYM. *dens*, from the scaly rootstock. —Differs from *Cardamine* chiefly in habit, and the funicles.

1. D. bulbif'era, *L.* ; lower leaves pinnate, upper simple entire. Woods and copses, rare ; Stafford to Kent and Sussex ; a doubtful native of Scotland ; fl. April–June.—*Rootstock* annual or biennial, white. *Stem* 1–2 ft., simple, leafless below. *Leaflets* 1–2 in., in few pairs, oblong, entire, or serrate ciliate; uppermost confluent at the base ; upper bulbiferous. *Flowers* ½–¾ in. diam., white or lilac ; pedicel slender. *Pods* erect, rarely ripening.—DISTRIB. Throughout Europe, rare in W. Asia.

7. SISYM'BRIUM, *L.* HEDGE-MUSTARD.

Annual or biennial herbs ; hairs simple. *Radical leaves* spreading ; cauline alternate, often auricled. *Flowers* loosely racemed, usually yellow, often bracteate. *Sepals* short or long. *Petals* often narrow and long-clawed. *Pods* narrow-linear, terete or 4–6-angled or compressed ; valves flat or convex, often 3-nerved ; septum membranous, nerveless or 2-nerved ; stigma simple 2-lobed or cup-shaped. *Seeds* many, 1-seriate, not margined ; radicle usually incumbent.—DISTRIB. N. temp. and cold regions ; rare in Southern ; species 80.—ETYM. doubtful.—Differs from *Brassica* only in the flat ocotyledons, and united to *Arabis* by *S. Thaliana.* Pods of 4 types :—1. *S. Thaliana, Irio,* and *Sophia ;* 2. *S. officinale ;* 3. *S. polyceratium ;* 4. *S. Alliaria.*

1. S. Thalia'na, *Hook.*; leaves toothed pubescent, flowers white, pods spreading or ascending obscurely 4-angled. *Thale-cress.* Dry soils, from Orkney southd., ascends to 1,500 ft. in Yorks.; Ireland ; Channel Islds.; fl. May–Sept.—Annual. *Stem* 6–10 in., slender, nearly leafless. *Radical leaves* rosulate, oblong, petioled ; cauline narrow, sessile, all entire or toothed. *Flowers* ⅛ in. diam.; pedicel slender. *Pods* ½–¾ in., slender, curved; seeds minute; radicle rarely accumbent. –DISTRIB. Europe (Arctic), to the Himalaya, N. Africa ; introd. in N. America.

2. S. Ir'io, *L.* ; leaves runcinate-toothed or pinnatifid glabrous, flowers yellow, pods terete slender suberect. *London Rocket.* Waste places, Berwick, Dublin ; Channel Islds.; sporadic elsewhere ; (a denizen, *Wats.*) ; fl. July–Aug.—Annual or biennial. *Stem* 1–2 ft., branched. *Radical leaves* petioled ; lobes irregularly toothed, terminal large often hastate. *Flowers* ⅛ in. diam. *Pods* 1½ in., very many, glabrous, erect, strict; valves beaded, 3-nerved ; style 0 ; seeds oblong.—DISTRIB. Central and S. Europe, N. Africa, W. Asia to the Himalaya.—Called "London Rocket" because it sprang up after the Fire of 1666.

3. S. Sophi'a, *L.* ; leaves 2–3-pinnatifid glabrous or downy, segments narrowly linear, flowers yellow, pods slender terete ascending curved, pedicels very slender. *Flixweed.* Waste places from Caithness southd., rarer in Scotland and Ireland ; a colonist? fl. June–Aug.—Annual. *Stem* 1–3 ft., branched above ; branches leafy. *Leaves* 2–4 in., finely divided; lobes spreading. *Flowers* ⅛ in. diam. *Pods* 1 in., in long racemes; valves beaded, 3-ribbed; style very short.— DISTRIB. Europe (Arctic), N. Africa, W. Asia to Himalaya, N. and S. America.

4. S. officina'le, *Scop.*; leaves runcinate-toothed or -lobed hairy, flowers yellow, pods in a leafless raceme subulate terete appressed to the stem. *Hedge-mustard.*

Hedgebanks and waste places from Orkney southd.; Ireland; Channel Islds., fl. June–July.—*Stem* 1–2 ft., terete, erect, with spreading or reflexed hairs; branches horizontal. *Leaves* variously cut or lobed, with a tendency to a large terminal lobe. *Flowers* ₁₆ in. diam., homogamous. *Pods* ½ in., tapered from the base into the almost pungent style; pedicel short, thick. —DISTRIB. Europe, W. Asia to the Himalaya, N. Africa; introd. in the U. States.

5. S. Allia'ria, *Scop.*; leaves all petioled deltoid or reniform-cordate coarsely toothed or crenate hairy beneath, pods stout long 4-angled, pedicels short stout. *Erysimum,* L.; *Alliaria officinalis,* Andrz. *Garlic-mustard, Sauce alone, Jack by the hedge.*

Hedgebanks, &c., from Ross southd.; ascends to near 1,000 ft. in England; rarer in Scotland and Ireland; Channel Islds.; fl. May–June.—Annual or rarely biennial, glabrous or with a few scattered simple hairs, rank scented. *Stem* 2–3 ft., decumbent at the base, then flexuous, erect, simple or sparingly branched. *Radical leaves* often 3 in. diam., with long slender petioles, smaller and more reniform than the cauline, which are cuneate at the base. *Flowers* ¼ in. diam., white; homogamous. *Pods* 2½ in., linear, slightly curved, rigid, subacute; valves keeled; style very short, stigma truncate; seeds oblong.—DISTRIB. Europe, N. Africa, temp. and W. Asia to the Himalaya.

S. POLYCERA'TIUM, *L.*; prostrate, leaves runcinate-pinnatifid glabrous, flowers yellow, pods 1–3 in the axils of leafy bracts cylindric curved spreading.

Roadside paths, Bury St. Edmunds, introduced by Dr. Goodenough; ballast-heaps, Fife; fl. July–Aug.—Annual, glabrous, very leafy. *Leaves* often reduced to the large triangular coarsely toothed terminal lobe. *Flowers* small, yellow. *Pods* ¾ in., in short leafy racemes, broad at the base, on very short thick pedicels; valves 3-nerved, very convex, beaded, obtuse; style evident, short, thick, stigma obtuse.—DISTRIB. Mediterranean to the Caucasus.

8. ERYS'IMUM, *L.* TREACLE-MUSTARD.

Annual biennial or perennial hoary herbs; hairs appressed, forked. *Leaves* narrow, entire. *Flowers* yellow, often fragrant. *Sepals* erect, equal or the lateral gibbous at the base. *Petals* clawed. *Pods* narrow, compressed, 4-angled or terete; valves linear, often keeled; replum usually prominent; septum membranous or corky; stigma 2-lobed or entire. *Seeds* many, 1-seriate, oblong, not winged or winged at the tip; radicle incumbent.—DISTRIB. Temp. and cold N. hemisphere; species about 70.—ETYM. ἐρύω, to draw blisters.—Near *Sisymbrium,* but cauline leaves never auricled, and hairs 2–3-partite ; differs from *Cheiranthus* in the cotyledons, and generally in the stigma.

E. cheiranthoi'des, *L.* ; leaves lanceolate, pods short suberect.

Waste places from Mid. England southd., a casual north of it ; rare in Ireland; (a colonist, *Wats.*) ; fl. June–Aug.—Annual. *Stem* 1–2 ft., erect leafy rigid terete. *Leaves* 3–4 in., subsessile, lanceolate or oblong-lanceolate, acute, narrowed at the base, obscurely toothed. *Flowers* ¼ in. diam. *Pods* 1 in., straight ; pedicel spreading ; valves strongly keeled, acute ; style very short stout, stigma truncate ; seeds oblong, smooth, very strong-tasted. DISTRIB. Europe (Arctic), N. Africa, N. Asia, N. America.

8*. *HES'PERIS, L.* DAME'S VIOLET.

Erect, biennial or perennial herbs. *Leaves* alternate, entire. *Flowers* large, handsome. *Sepals* erect, lateral gibbous at the base. *Petals* clawed. *Pods* elongate, terete or 4-angled ; valves flattish, keeled, 3-nerved ; septum membranous ; stigmatic lobes suberect. *Seeds* many, margined or not ; cotyledons incumbent.—DISTRIB. Europe and temp. Asia ; species 20.—ETYM. ἕσπερος, from some species being odorous in the evening.—Very near *Matthiola* and *Cheiranthus,* differing chiefly in the stigmas and embryo.

H. MATRONA'LIS, *L.* ; pubescent, leaves oblong-lanceolate acuminate.

Meadows, plantations, &c., rarely even naturalized; fl. May–July.—Perennial, hairs simple or branched. *Stems* 2–3 ft., erect, stout, leafy. *Leaves* 2–5 in., shortly petioled or sessile, finely irregularly toothed or serrate. *Flowers* ¾ in. diam., white or lilac, odorous in the evening, proterandrous ; pedicel ½ in., spreading. *Pods* 2–4 in., slender, cylindric, constricted here and there between the remote seeds ; pedicel ascending ; valves much narrowed at the tip ; style stout ; seeds linear-oblong.—DISTRIB. Europe, temp. Asia.

9. BRAS'SICA, *L.* CABBAGE, &c.

Herbs of various habit. *Leaves* entire or pinnatifid, often large. *Flowers* in corymbs or racemes, white or yellow, rather large. *Sepals* erect or spreading, equal or the lateral saccate at the base. *Pods* elongate, nearly terete, with sometimes an indehiscent 1-seeded beak ; valves convex, often 3-nerved, the lateral nerves flexuous ; septum membranous or spongy ; stigma truncate or 2-lobed. *Seeds* 1-seriate, oblong or subglobose ; radicle incumbent, cotyledons concave or conduplicate.—DISTRIB. Temp. Europe, Asia, and N. Africa ; species 100.—ETYM. The Latin name.

SECTION 1. **Bras'sica** proper. *Sepals* erect.

1. B. olera'cea, *L.* ; rootstock stout branched leafy at the top, leaves obovate lobed or sinuate below glaucous glabrous, upper sessile oblong dilated at the base, flowering racemes elongate, beak of pod seedless, valves keeled and nerved. *Wild Cabbage.*

Sea-cliffs, S.W. of England and Wales; Channel Islds.; (a denizen? *Wats.*) ; fl. May–Aug.—*Stem* 1–2 ft., biennial or perennial, very stout, tortuous, usually decumbent, scarred. *Lower leaves* often 1–1½ ft. *Flowers* 1 in.

diam., pale yellow, homogamous. *Pods* 2–3 in., spreading, slightly compressed; beak short, subulate; seeds globose.—DISTRIB. W. and S. coasts of Europe. Cultivated forms are *aceph'ala* (scotch kail, cow cabbage, borecole); *bulla'ta* and *gemmif'era* (brussels sprouts and savoys); *capita'ta* (red and white cabbage); *Caulora'pa* (cole rabi); *Botry'tis* (cauliflower and broccoli).

2. **B. campes'tris,** *L.*; erect, lower leaves lyrate-pinnate hispid, upper oblong or lanceolate amplexicaul and auricled, flowering racemes corymbiform, beak of pod seedless, valves 1-nerved. *B. polymorpha,* Syme.

Weeds of cultivated ground: a colonist? *Watson;* fl. June–Sept.

Mr. Dyer considers that only two primary forms of this species are to be found in cultivation or as escapes in Britain; excluding *B. Na'pus* and its forms (of which none of the leaves are hispid) as being almost exclusively Continental. He assumes that Var. *oleif'era* is the Linnean type, described as a troublesome weed in Sweden, which may be a starved state of the turnip escaped from cultivation. Mr. Watson, on the other hand, considers *B. Ra'pa, campes'tris* and *Na'pus* as all British, and affirms that the latter is wrongly described as glabrous. *B. Na'pus,* L. (?) being the rape; *B. Rutaba'ga,* L., the swede; and *B. Ra'pa,* L, the turnip, with 3 varieties *sati'va, sylves'tris,* and *Brigg'sii.*

B. CAMPES'TRIS proper (*Linn. Herb.*); leaves glaucous, flowers pale orange.— VAR. *oleif'era,* DC.; root slender spindle-shaped—(yields rape and colza).— VAR. *Na'po-brassica,* DC.; root tuberous, neck elongated. *B. Rutaba'ga,* DC.—(swedish turnip).

Sub-sp. B. RA'PA, *L.*; leaves not gaucous, flowers smaller bright yellow. VAR. *rapif'era,* Koch; root tuberous (turnip).—VAR. *campes'tris,* Koch; root spindle-shaped.—VAR. *sylves'tris,* Lond. Cat. (navew.) *B. Brigg'sii,* Wats., is an annual form from Cornwall.

3. **B. monen'sis,** *Huds.*; leaves petioled deeply pinnatifid, segments toothed, upper linear, beak of pod 1–3-seeded, valves 3-nerved.

Sea-shores on the west from Skye to S. Wales; Channel Islands: fl. May–June.— *Rootstock* usually stout, woody, perennial. *Stem* 6–24 in., erect or decumbent. *Radical leaves* with short broad-toothed segments. *Flowers* ½–¾ in. diam., pale yellow. *Pods* 1½–2½ in., spreading; beak thick; seeds globose, dark, punctate.—DISTRIB. Shores, W. and S Europe, N. Africa. (Subalpine in Pyrenees.)

B. MONEN'SIS proper; glabrous, stem nearly simple, leaves chiefly radical. S. Wales to Skye; inland at Merthyr Tydfil, S. Wales.

Sub-sp. B. CHEIRAN'THUS, *Villars;* hispid, stem branched leafy. Cornwall; Channel Islands; introduced elsewhere.

SECTION 2. **Sina'pis,** *L.* (Gen.). *Sepals* spreading.

   * *Pods erect, appressed to the stem; valve 1-nerved; cells few-seeded.*

4. **B. ni'gra,** *Koch;* stem-leaves petioled linear-lanceolate entire or toothed glabrous, pods subulate 4-angled glabrous, beak short seedless. *Black Mustard.*

Hedges and waste places, from Northumberland southd., common as an escape, wild on sea-cliffs (Syme); not wild in Scotland; S. of Ireland;

Channel Islands; (a native? *Wats*.); fl. June–Sept.—Annual. *Stem* 2–3 ft. rigid, branched, more or less hispid. *Leaves* 4–8 in., lower lyrate, terminal lobe much the longest. *Flowers* ⅓–½ in. diam., bright yellow. *Pod* ¼–½ in., subulate, beak slender; valves keeled, torulose; pedicel short, stout, erect; cells 3–5-seeded; seeds oblong.—DISTRIB. Europe, N. Africa, W. Asia to the Himalaya; introd. in U. States.

5. **B. adpres'sa,** *Boiss.*; uppermost stem-leaves linear or lanceolate quite entire hispid, pods subcylindric, beak clavate ribbed ½ as long as the valves often 1-seeded. *Sina'pis inca'na,* L. *Erucastrum inca'num,* Koch.

Sandy fields, Jersey and Alderney; casual in Ireland; fl. July–Aug.—Habit of *B. ni'gra,* but more branched; pod ½ in.; valves linear; beak 8-ribbed; seeds fewer, flattened.—DISTRIB. Europe, from Belgium southd.

** *Pods spreading; valves 3-nerved; cells few- or many-seeded.*

6. **B. Sina'pis,** *Visiani;* hispid, upper leaves toothed or lyrate-pinnatifid, pods linear angular longer than the elongate compressed rarely 1-seeded beak. *B. Sinapis'trum,* Boiss. *Sina'pis arven'sis,* L. *Charlock.*

Cornfields, N. to Shetland; ascending to 1,200 ft.; Ireland; Channel Islands; fl. May–Aug.—Annual. *Stems* 1–2 ft., usually branched. *Flowers* ½–¾ in., diam., subcorymbose, bright yellow, homogamous. *Pods* 1½–2 in.; pedicel slender, spreading; beak deciduous, straight, almost rigid, as broad as the hispid torulose valves; seeds subcompressed, dark brown.—DISTRIB. Europe, N. Africa, N. and W. Asia to the Himalaya; introd. in America.

7. **B. al'ba,** *Boiss.*; hispid with reflexed hairs, upper leaves pinnatifid, pods short beaded few-seeded, valves equalling the broad sometimes 1-seeded beak. *Sinapis,* L. *White Mustard.*

Cultivated ground, &c., from Ross southd.; Mid. Ireland rare; Channel Islands; (a colonist, *Wats.*); fl. June–July.—Annual. *Stem* 1–3 ft., erect, furrowed; branches ascending. *Leaves* all lyrate-pinnatifid or pinnate; segments cut and lobed. *Flowers* ½ in. diam., yellow. *Pods* 2 in., hispid; valve strongly ribbed, concave; beak ensiform, persistent, ribbed, often curved; cells 1–3-seeded; seeds subglobose, pale.—DISTRIB. Europe, N. Africa, N. and W. Asia to the Himalaya; introd. in the U. States.— Cult. as a salad.

10. DIPLOTAX'IS, *DC.* ROCKET.

Annual or perennial herbs. *Leaves* pinnatifid. *Flowers* yellow. *Sepals* spreading, equal. *Pod* narrow, elongate, compressed; valves 1-nerved; septum membranous; style stout or slender, stigma simple. *Seeds* many, sub-2-seriate, compressed; cotyledons as in *Brassica.*—DISTRIB. Temp. Europe, Asia, and N. Africa; species 20.—ETYM. διπλόος and τάξις, from the 2-seriate seeds.—Differs from *Brassica* in the flat pods, membranous valves, minute compressed 2-seriate seeds, and the flowers varying to pink or purplish, which is never the case in *Brassica.*

1. **D. mura'lis,** *DC.*; stem hispid leafy at the base, leaves sinuate or pinnatifid, scapes slender ascending, pods suberect linear. *Sisym'brium,* L. *Bras'sica brevi'pes,* Syme.

D

34   *CRUCIFERÆ.*   [Diplotaxis.

Roadsides and waste places, from Roxburgh southd.; E. Ireland; Channel
Isands; (a denizen, *Wats.*); fl. Aug.–Sept.—A small herb. *Leaves* usually
long-petioled. *Scapes* 6–8 in., few-leaved. *Flowers* ½ in. diam.
, yellow;
pedicels as long as the expanded flowers, petals 2–3 times as long as the
sepals abruptly obovate, style not narrowed below. *Pods* 1–2 in., slender,
narrowed above and below; valves flat, almost nerveless; style stout,
straight, cylindric.—Distrib. W. Europe, from Belgium southd., N.
Africa.
Var. *Babingto'nii*, Syme; biennial or perennial, stem-leaves several.—S. of
England, common; E. of Ireland, very rare; a denizen, *Watson.*

2. **D. tenuifo'lia,** *DC.* ; stem branched leafy, leaves pinnatifid, lobes
long narrow, pods suberect linear on very long slender pedicels. *Sisym'-
brium,* L. ; *Sina'pis,* Sm. ; *Bras'sica,* Boiss.

Waste places from the Cheviots southd.; Ireland; Channel Islands; (a denizen,
Wats.); fl. June–Sept.—Bushy, glabrous or hispid, glaucous, fœtid. *Root-
stock* woody; branches 1–3 ft. *Leaves* 3–5 in.; lobes distant, very unequal.
Flowers ¾ in. diam., yellow. *Pods* ⅓–½ in., distant, linear, narrowed at
both ends; valves flat; style stout, straight, cylindric.—Distrib. Europe,
N. Africa, W. Asia.

11. DRA'BA, *L.* Whitlow-grass.

Herbs, usually small, rarely annual, hoary with stellate down. *Leaves*
entire ; radical rosulate ; cauline sessile or 0. *Flowers* small, racemose or
corymbose, white or yellow. *Sepals* short, equal at the base. *Petals*
shortly clawed, entire. *Pods* oblong or linear, compressed ; valves flattish,
rarely ribbed ; septum membranous ; stigma simple. *Seeds* 2-seriate, compressed ; funicle filiform ; radicle accumbent.—Distrib. Temp. Arctic and
Alpine regions, chiefly of N. hemisphere, and Andes ; a prominent feature
in N. Polar regions ; species 80, all very variable.—Etym. δράβη, *acrid,*
in allusion to the taste.

1. **D. aizoi'des,** *L.* ; leaves rigid ciliate, scape leafless, petals yellow.

Walls and rocks, Pennard Castle, Glamorgan; (a native ? *Wats.*); fl. March–
May.—Perennial, densely tufted. *Leaves* ½ in., rosulate, glabrous, shining,
keeled, margins and tip white cartilaginous ciliate. *Scape* 1–5 in., rigid.
Racemes short, slightly lengthened in fruit. *Flowers* ⅓ in. diam., proterogynous. *Petals* hardly notched. *Pods* oblong, acute, glabrous ; style half
its length ; cells 10–12-seeded.—Distrib. Mountains of Central and S.
Europe, W. Asia.

2. **D. rupes'tris,** *Br.* ; erect, small, leaves oblong-lanceolate hairy,
scape-leaf 1 or 0, petals white, pods small straight oblong hairy, pedicels
straight erecto-patent.

Alpine rocks, alt. 3–4,000 ft., rare and local ; Ben Lawers, Cairngorm, Ben-
hope ; Benbulben in Ireland ; fl. July–Aug.—*Rootstock* slender. *Leaves* ¼–½
in., ciliate, rarely toothed. *Scapes* one or more, 1–2 in., slender, flexuous ;
their leaf sessile, stellately pubescent. *Flowers* ¼ in. diam., few, white.
Pods ¼ in., oblong, obtuse ; stigma subsessile.—Distrib. Arctic regions.

3. **D. inca'na,** *L.* ; erect, stellately hispid, often branched, leaves oblong cauline amplexicaul, petals white, pods linear or oblong-lanceolate usually twisted, pedicels erecto-patent. *D. confu'sa,* and *D. contor'ta,* Ehrh.

Alpine rocks, N. England, Wales, Scotland; ascends above 3,000 ft.; mountains and maritime sandhills of W. Ireland; fl. June–July.—*Rootstock* often woody and long.   *Stem* 6–14 in.   *Radical leaves* short, $\frac{3}{4}$–1 in., densely rosulate, usually much toothed ; cauline many, suberect.   *Racemes* much elongated after flowering, many-fld.   *Flowers* as in *D. rupestris.*   *Pods* $\frac{1}{4}$–$\frac{1}{2}$ in., variable in length breadth and twisting, glabrous or hairy, obtuse or subacute; pedicel short; stigma subsessile; seeds many.– DISTRIB. Alpine and Arctic Europe, Asia, and N. and S. America.—Very variable ; small specimens with few stem-leaves resemble *D. rupestris.*

4. **D. mura'lis,** *L.*; suberect or prostrate, slender, branched, stellately hispid, stem-leaves broadly ovate or cordate obtuse coarsely toothed, petals minute white, pods linear-oblong horizontal, pedicels spreading.

Limestone rocks and walls in W. England ; from Yorks. (ascending to 1,200 ft.) to Somerset; introd. in Scotland and N.W. Ireland; fl. April–May.—*Rootstock* annual or biennial, slender.   *Stem* 1–2 ft., flexuous.   *Radical leaves* small, $\frac{1}{2}$–1 in., obovate ; cauline few, distant, broader.   *Flowers* $\frac{1}{8}$ in. diam.   *Racemes* short, fruiting long.   *Pods* $\frac{1}{4}$ in., flat, obtuse, on pedicels longer than themselves; style 0; seeds 10–12, minute.—DISTRIB. Europe, temp. Asia to the Himalaya, N. Africa.

12. EROPH'ILA, *DC.*   VERNAL WHITLOW-GRASS.

Small annual or biennial herbs.   *Radical leaves* entire, spreading.   *Scapes* slender, leafless.   *Flowers* few, small, white.   *Sepals* spreading, equal at the base.   *Petals* obovate, 2-lobed or 2-partite.   *Pods* oblong, compressed ; valves 1-nerved, membranous, flat or convex ; septum membranous.   *Seeds* 2-seriate, very many and minute ; funicle capillary ; radicle accumbent.— DISTRIB. Europe, W. Asia, and N. India ; species 2–3.—ETYM. ἔαρ and φιλῶ, from flowering in spring.

E. vulga'ris, *DC.* ; leaves oblong-lanceolate toothed.   *Dra'ba ver'na,* L.

Abundant on walls, paths, &c., N. to Orkney ; Ireland ; Channel Islands; fl. March–June.—Subglabrous or delicately pubescent.   *Leaves* $\frac{1}{2}$–1 in., all radical, rosulate.   *Scapes* 1–16 in., flexuous.   *Flowers* $\frac{1}{8}$–$\frac{1}{4}$ in. diam., homogamous.   *Pods* $\frac{1}{4}$–$\frac{1}{3}$ in., on spreading pedicels.—DISTRIB. Europe, temp. Asia to the Himalaya, N. Africa ; N. America (perhaps introd.).—Seventy forms have been cultivated by M. Jordan as specific, with more or less constancy.

E. VULGA'RIS proper ; pods compressed obovate-oblong twice or more as long as broad, cells 20–40-seeded.—Ascends to 1,200 ft. in Yorkshire.

Sub-sp. E. BRACHYCAR'PA, *Jord.*; pods compressed orbicular-oblong 1–1$\frac{1}{2}$ as long as broad, cells 12–20-seeded.—From Fife southd., scarce.

Sub-sp. E. INFLA'TA, *Wats.* ; pods turgid ovoid-oblong twice as long as broad, cells 20–40-seeded.—*D. verna β* Hook.– Alpine rocks, Ben Lawers and Glen Shee, alt. 2,200–3,000 ft.; fl. June–July.

D 2

12*. *ALYS'SUM, L.·*

Herbs or small shrubs, often covered with stellate down. *Leaves* scattered or crowded, entire. *Flowers* small, white or yellow. *Sepals* short, equal. *Petals* short, entire or 2-fid. *Filaments* sometimes toothed or appendaged. *Pods* short, very various in form ; valves flat, concave or convex ; septum entire or perforate ; stigma simple. *Seeds* 2–10 ; radicle accumbent.—DISTRIB. N. and W. Asia and N. Africa ; species 80–90.— ETYM. obscure.—A polymorphous genus, almost every organ varying ; hence 10 genera have been made out of it.

A. CALYCI'NUM, *L.* ; pubescent with appressed stellate hairs, leaves linear-spathulate, sepals persistent, petals yellow, pods suborbicular.

Cultivated fields, rare and sporadic in England, Scotland, and Ireland ; (an alien, *Wats.*) ; fl. June–Aug.—Annual, hoary. *Stem* branched at the base ; branches 3–8 in., rigid, ascending. *Leaves* ¼–¾ in., few, scattered, obtuse. *Flowers* small ; filaments with 2 teeth at the base. *Fruiting racemes* elongate. *Pods* ₁/₁₈ in. diam., very numerous, on short stiff spreading pedicels, nearly orbicular, turgid with broad thin flat margins, notched at the tip ; style very short ; seeds 1–2 in each cell, narrowly winged.—DISTRIB. Central and S. Europe, W. Asia.

A. MARIT'IMUM, *L.* ; pubescent with appressed 2-partite hairs, leaves linear, sepals deciduous, petals obovate white, pods obovoid-orbicular. *Köni'ga,* Br. ; *Lobula'ria,* Desv. ; *Gly'ce,* Lindl.

Waste places near the sea ; in England and the Channel Islands ; (a denizen, *Wats.*) ; fl. June–Sept.—Annual or perennial. *Rootstock* prostrate. *Stem* 4–10 in., ascending, leafy. *Leaves* 1–1½ in., ₁/₁₀–⅜ in. broad, subacute. *Flowers* small, odorous ; filaments all simple. *Pods* ₁/₁₀ in. without a border ; pedicel slender, spreading ; valves convex, 1-nerved ; cells 1-seeded ; style slender.—DISTRIB. Maritime S. Europe and W. Asia.

13. COCHLEA'RIA, *L.* SCURVY-GRASS.

Perennial herbs. *Leaves* entire or pinnate. *Flowers* small, white. *Sepals* short, equal, spreading. *Petals* shortly clawed. *Pods* sessile or shortly stalked, oblong or globose ; valves turgid, reticulate ; septum often imperfect ; stigma simple or capitate. *Seeds* few or many, 2-seriate, not margined, tubercled ; cotyledons accumbent.—DISTRIB. Temp. and Arctic regions, chiefly littoral or Alpine ; species 25.—ETYM. *cochlcar,* from the *spoon*-like leaves.—The native British species form a well defined group of variable littoral plants, confined to N.W. Europe and the Arctic regions ; but some exotics have the habit of *Nasturtium.*

* *Valves with a dorsal nerve.*

1. C. officina'lis, *L.* ; radical leaves cordate, pods subglobose, valves reticulate, style very short. *C. polymor'pha,* Syme.

Sea-shores and high mountains, N. to Shetland ; Ireland ; Channel Islands ; fl. May–Aug.—*Stems* many, 4–10 in., glabrous, fleshy, ascending from the perennial rarely biennial rootstock. *Flowers* ¼–⅓ in. diam. *Pods* ⅛–¼ in. diam., cells 4–6-seeded.—DISTRIB. N.W. Europe, Polar regions.—A valuable antiscorbutic.

C. OFFICINA'LIS proper ; radical leaves deeply cordate orbicular or reniform, cauline amplexicaul angled toothed or lobed, pods nearly globose. VAR. *littora'lis*, Lond. Cat.—Muddy sea-shore.

Sub-sp. C. ALPI'NA, *Wats.*; radical and cauline leaves as in *officina'lis*, pods rhomboid-oblong narrowed at both ends. *C. grœnlan'dica*, Sm.—Mountains, ascending to near 4,000 ft.

Sub-sp. C. DAN'ICA, *L.*; radical leaves deltoid lobed, lower cauline similar petioled, upper amplexicaul also lobed, petals smaller, and pods as in *alpi'na.* —Sandy and muddy shores, rarer in Scotland and Ireland.

2. **C. ang'lica**, *L.* ; radical leaves oblong-rhomboid or ovate not cordate entire lobed or angled, cauline ½-amplexicaul, pods oblong or obovoid inflated much constricted at the sutures, valves reticulate, style slender.

Muddy shores in England and W. Scotland ; rare in Ireland ; Channel Islands ; fl. May–July.—Much larger than *C. officinalis*, branches 10–18 in., but connected with it by intermediates. *Leaves* more fleshy, narrower, and pod very different, sometimes ½ in., style longer, seeds larger.—DISTRIB. Cf. *C. officinalis.*

C. ANGLICA proper (VAR. *gem'ina*, Hort) ; radical leaves narrowed below, pod obovoid large much constricted at the suture.—Var. *Hor'tii*, Syme ; radical leaves rounded at the base, pod smaller ellipsoid.

** *Valves with no dorsal nerve.* ARMORA'CIA, Rupp.

C. ARMORA'CIA, *L.*; leaves linear-oblong obtuse deeply regularly crenate, radical long-petioled, cauline narrower sessile, racemes panicled, pods (immature) obovoid on long slender pedicels. *Horse-radish.*

Ditches, corners of fields, &c. ; (an alien or denizen (?) *Wats.*); fl. May–June. —*Rootstock* stout, long, cylindric. *Leaves* 8–12 in., radical 3–5 in. broad, on petioles 1 ft., waved with many spreading reticulate nerves, cordate cuneate or unequal at the base; cauline many, 4–8 in., ½–1 in. broad, more serrate than toothed. *Flowers* ⅛ in. diam. *Pods* never ripening in this country ; style slender ; stigma large, capitate; seeds described as 8–12 in a cell, smooth.—DISTRIB. Origin unknown ; possibly a cultivated form of *C. macrocar'pa*, W. and K., a native of Hungary.

13*. *CAMELI'NA, Crantz.* GOLD OF PLEASURE.

Annual, erect herbs. *Leaves* almost entire, cauline auricled. *Flowers* small, yellow. *Sepals* short, equal at the base. *Petals* spathulate. *Pods* obovoid ; valves turgid, keeled at the back, produced upwards along the base of the style ; margin flat ; septum membranous ; stigma simple. *Seeds* 2-seriate, not margined ; funicle slender, adnate at the base to the septum ; radicle incumbent.—DISTRIB. Europe and temp. Asia ; species 5–10 (all vars. of one ?).—ETYM. χαμαί and λίνον, *dwarf flax.*

C. SATI'VA, *Crantz ;* radical leaves petioled, cauline oblong-lanceolate.

Flax-fields, sporadic ; fl. June–July.—*Stems* 2–3 ft., branched above, slender. *Radical leaves* soon withering ; cauline 1–3 in., obtuse ; auricles pointed, entire or lobulate. *Flowers* ⅛ in. diam. *Petals* erect. *Pods* ¼–⅓ in., on slender spreading pedicels, obovoid, margins thin ; seeds few, oblong, punctulate.—Distrib. Central and S. Europe, and temp. Asia.—Seeds used for soap-makers' oil, oil-cake, and for feeding poultry.

14. SUBULA'RIA, *L.* Awl-wort.

A small submerged perennial scapigerous herb. *Leaves* all radical and subulate. *Flowers* small, white. *Sepals* spreading, equal. *Petals* small. *Pods* shortly stalked, oblong or nearly globose ; valves convex, ribbed ; septum membranous ; stigma sessile, entire. *Seeds* few, 2-seriate ; radicle incurved, narrowed into the incumbent radicle, and owing to their lengthened bases being turned up a transverse section of the embryo shows a radicle with apparently 3 cotyledons.—Distrib. Arctic, N. Europe, N. Asia, N.E. U. States.—Etym. *subula*, from the *awl*-like leaves.

S. aquat'ica, *L.* ; leaves cellular, scape naked few-flowered.
Gravelly bottoms of subalpine lakes in N. Wales, Cumberland, and N. to Sutherland ; ascends to 2,200 ft.; Ireland ; fl. June–Aug.—*Roots* of densely tufted matted white fibres from a small stock. *Leaves* 1–3 in., fascicled, terete, gradually tapering upwards. *Scapes* 1–3 in. *Flowers* $\frac{1}{10}$ in. diam., submerged. *Pods* small, $\frac{1}{8}$ in.; pedicel short, ascending ; seeds pale brown punctulate.

15. CAPSEL'LA, *Mœnch.* Shepherd's Purse.

Annual herbs. *Radical leaves* entire or lobed. *Flowers* small, white, pedicels slender. *Sepals* spreading, equal. *Pods* much laterally compressed, oblong or obcuneate or obcordate ; valves boat-shaped, keeled ; septum membranous ; stigma sessile. *Seeds* many, minute ; radicle incumbent. —Distrib. N. temp. regions ; species 6.—Etym. Diminutive *Capsula.* —Petals tend to be transformed into stamens.

C. Bur'sa-Pasto'ris, *Mœnch ;* pod triangular or obcordate.
A weed in all situations, ascending to 1,200 ft.; fl. March–Nov.—Glabrous or hairy, hairs branched. *Root* long, tapering. *Stems* 6–16 in., branched. *Leaves* rosulate, pinnatifid, rarely entire, end lobe triangular, cauline ; auricled. *Flowers* $\frac{1}{18}$ in. diam., homogamous. *Pods* $\frac{1}{4}$–$\frac{1}{3}$ in.; pedicel slender ; style short ; valves smooth ; seeds many, oblong, punctate.—Distrib. Temp. and Arctic Europe, N. Africa and Asia to the Himalaya ; introd. in all temp. climates.

16. SENEBIE'RA, *DC.* Wart-cress.

Annual or biennial, branched, prostrate herbs. *Leaves* entire or cut. *Flowers* in short leaf-opposed racemes, minute, white, sometimes apetalous. *Sepals* short, spreading. *Stamens* 2 or 6, or 4 when the shorter are absent. *Pods* small, didymous, indehiscent, laterally compressed ; lobes subglobose, rugose or crested ; style short or 0. *Seeds* 1 in each cell ; cotyledons induplicate, gradually narrowed into the incumbent radicle.—Distrib. Temp. and warm regions ; species 6.—Etym. *J. Senebier*, a Genevese physiologist.—Rapidly spreading weeds in the colonies, &c. Embryo in some species cyclical.

1. **S. did'yma,** *Persoon ;* lobes of fruit separating wrinkled, style minute. *Coro'nopus,* Sm.

Waste ground from Fife southd., and spreading ; S. and W. Ireland ; Channel Islands ; a colonist ; fl. July–Sept.—Annual or biennial, slightly hairy, diffuse, creeping, branched and leafy. *Leaves* finely cut, 1–2-pinnatifid ; lobes small, obovate, spreading. *Flowers* very minute, in leaf-opposed racemes, with a solitary one on the internode below, usually apetalous and diandrous. *Pods* $\frac{1}{12}$ in. broad, pedicels spreading, separating into 2 indehiscent hard lobes ; seeds reniform, punctate-striate.—DISTRIB. Temp. S. America ; a colonist elsewhere.

2. **S. Coro'nopus,** *Poiret ;* lobes of fruit not separating deeply wrinkled, the wrinkles forming a crest, style subulate. *Coro'nopus Ruel'lii,* Allioni.

Waste ground from Caithness southd. ; rare in Scotland ; local in Ireland ; Channel Islands ; fl. June–Sept.—Habit, &c., of *S. did'yma,* but glabrous, less branched ; foliage larger, less divided ; petals and stamens usually perfect ; pods twice the size, abruptly narrowed into the subulate style, lobes connate ; pedicels very short, thick, and seeds twice as large. Surface of pod variable as to sculpturing.—DISTRIB. Europe, N. Africa, W. Asia ; introd. in the U. States.

17. LEPID'IUM, *L.* CRESS.

Herbs, sometimes shrubby at the base, various in habit. *Leaves* entire or much divided. *Flowers* small, white, often apetalous. *Sepals* short, equal. *Petals* short or 0. *Stamens* 2, 4, or 6. *Pods* oblong ovate obovate or obcordate, much laterally compressed ; valves usually keeled, winged or not ; septum narrow ; stigma notched. *Seeds* 1 in each cell, rarely 2, pendulous from the septum ; radicle incumbent, rarely accumbent.— DISTRIB. Temp. and warmer regions ; species 60–80.—ETYM. λεπίδιον, from the *scale*-like form of the pods.—Stamens sometimes deformed. Cotyledons 3-partite in the common Cress (*L. sativum,* L.)

SECTION 1. **Nasturtias'trum,** *Gren.* and *Godr. Pods* much compressed, entire or notched ; valves keeled, not winged ; style minute.

1. **L. latifo'lium,** *L.* ; perennial, erect, radical leaves long-petioled oblong serrate, pod ovoid entire. *Dittander.*

Salt marshes, N.E. England, and from Wales round to Norfolk ; introd. in Fife and Berwick ; S. of Ireland ; Channel Islands ; fl. July–Aug.—Glabrous, rather glaucous. *Rootstock* elongate, stoloniferous. *Stem* 2–4 ft., much branched, leafy. *Radical leaves* often 1 ft. ; cauline narrower, upper sessile. *Flowers* minute, in short densely panicled bracteate corymbs. *Pod* $\frac{1}{12}$ in. ; pedicel short ; valves not winged.—DISTRIB. Mid. and S. Europe, W. Asia.

2. **L. rudera'le,** *L.* ; annual, erect or prostrate, lower leaves 2-pinnatifid, pod orbicular-oblong notched.

Waste places chiefly near the sea, rare and doubtfully native in Scotland ; Ireland ; Channel Islands ; fl. May–June.—Glabrous or slightly pubescent. *Radical leaves* much divided, segments narrow ; upper cauline linear, entire.

Flowers minute, usually apetalous and diandrous, in terminal and lateral corymbs. *Pods* $\frac{1}{4}$-$\frac{1}{2}$ in., flat; pedicel slender,' diverging; valves keeled, almost winged at the top; seeds compressed.—DISTRIB. Europe, N.W. Asia to the Himalaya; introd. in U. States.

SECTION 2. **Le'pia,** *DC.* *Pod* ovoid or oblong, much compressed, notched ; valves broadly winged ; style short.

3. **L. campes'tre,** *Br.* ; cauline leaves auricled toothed, anthers yellow, pods concave papillose. *Pepperwort.*

Fields and roadsides, from Lanark and Elgin southd.; rare in Scotland and Ireland; Channel Islands ; fl. May–Aug.—Glabrous or pubescent, annual or biennial. *Stem* 6–18 in., erect, simple or branched. *Radical leaves* pinnatifid or entire, cauline oblong-lanceolate. *Flowers* $\frac{1}{10}$ in. diam. *Pods* $\frac{1}{4}$ in., in spreading racemes, broadly ovate, shorter than the spreading pedicels ; papillæ scale-like when dry; seeds oblong, curved.—DISTRIB. Europe, N. Africa, N. and W. Asia to India; introd. in U. States.
L. CAMPES'TRE proper; stem simple below branched above, anthers yellow, pod papillose, style not longer than the notch.—Common in dry places.
Sub-sp. L. SMITH'II, *Hook.*; branched from the base, anthers violet, pod nearly smooth, style longer than the notch, seeds smaller. *L. hirtum,* Sm., in part.
Fields and banks ; ascends to 1,000 ft. in Ireland ; Channel Islands (W. Europe only). Var. *alatostyla,* pod not notched, produced into the style. Hants.

SECTION 3. **Carda'ria,** *DC.* *Pod* deltoid-cordate, constricted between the valves ; valves hardly keeled, not winged ; style distinct.

L. DRA'BA, *L.* ; stem flexuous leafy, leaves oblong amplexicaul toothed.
Fields, banks, and railway cuttings in Mid. and S. England ; Channel Islands ; rare and sporadic ; (an alien, *Wats.*) ; fl. May–June.—Perennial, hoary or downy. *Stem* 1–3 ft., branched above. *Leaves* 1–3 in., auricles converging, lower petioled. *Racemes* short, panicled. *Flowers* $\frac{1}{4}$ in. diam., white. *Pods* $\frac{1}{6}$ in. broad, deltoid with rounded angles, on slender spreading pedicels; valves slightly papillose, one often smaller or imperfect, enclosing the seed when dehiscing.—DISTRIB. S.E. Europe, W. Asia; introd. in U. States.

18. THLAS'PI, *L.* PENNY CRESS.

Annual or perennial, glabrous, often glaucous herbs. *Leaves* quite entire or toothed, radical rosulate, cauline auricled. *Flowers* white or rose-coloured. *Sepals* erect, equal at the base. *Petals* obovate. *Pods* short, laterally compressed, broader upwards, notched ; valves keeled or winged. *Seeds* few or many, not margined ; radicle accumbent.—DISTRIB. Temp., Alpine, and Arctic N. hemisphere, rare in South ; species 30.—ETYM. The old Greek name of the genus.

1. **T. arven'se,** *L.* ; annual, cauline leaves sagittate sinuate-toothed, pods large orbicular deeply notched, valves broadly winged all round, cells 5–8-seeded. *Mithridate Mustard, Penny Cress.*

Fields, &c., rather common; rarer in Scotland; E. Ireland ; Channel Islands; - (a colonist, *Wats.*) ; fl. May–July.—Stem 1–2 ft., usually simple, slender, erect. *Radical leaves* petioled, cauline with prominent auricles. *Flowers* ¼ in. diam., white, homogamous. *Pods* in long racemes ½–¾ in. diam., flat, pedicels slender spreading, marginal nerve delicate, lobes sometimes overlapping at the tip; style very short; seeds dark, oblong, ridged and punctate.—DISTRIB. Europe to N. Africa, N. and W. Asia to N.W. India ; introd. in U. States.

2. **T. perfolia'tum,** *L.* ; annual, cauline leaves cordate with converging auricles, pods small obcordate, valves winged above, cells 4–6-seeded.

On limestone ; Oxford (extinct), E. Gloster ; fl. April–May.—Paniculately branched ; branches 4–6 in., ascending, flexuous. *Leaves* ½–1 in. broad; radical spathulate, petioled. *Flowers* ₁₂′ in. diam., white. *Pods* ⅛ in., and pedicels horizontal ; valves turgid, wings short, marginal nerve stout, style very short; seeds pale.—DISTRIB. Mid. and S. Europe, N. Africa, N. and W. Asia.

3. **T. alpes'tre,** *L.* ; perennial or biennial, cauline leaves sagittate, pods obcordate retuse, valves winged above, cells 4–8-seeded.

Mountain districts, England, Wales and Scotland, ascending to 2,500 ft. in Forfarshire ; fl. June–Aug.—Stem 6–10 in. *Radical leaves* long-petioled, obovate, entire ; cauline ½ in. *Flowers* ⅛ in. diam. *Racemes* of pods variable. *Pod* ¼ in., curved upwards, on spreading pedicels, longer than in the preceding species, and more narrowed at the base ; marginal nerve obscure ; seeds red-brown.—DISTRIB. Europe, Himalaya.—I do not find that authentically named specimens of the following varieties altogether tally with the characters assigned to them.—Var. *T. sylves'tre,* Jord. ; notch of pod shallow, style as long as its lobes.—Teesdale ; Allen river, Northumbd.; Glen Isla and Glen Shee, Scotland.—Var. *T. occita'num,* Jord. ; notch of pod shallow, style slender longer than its lobes, radicle at times incumbent (Syme).—Limestone rocks, Settle, Yorkshire ; Llanrwst, N. Wales.—Var. *T. vi'rens,* Jord. ; notch of pod minute, style slender much exceeding it. *T. alpes'tre,* Sm.—Limestone rocks, Matlock.

·19. IBE'RIS, *L.* CANDY-TUFT.

Low, glabrous, branched, leafy herbs, often shrubby below. *Leaves* entire or pinnatifid, often fleshy. *Flowers* corymbose, all or the outer only with the 2 outer petals radiating. *Sepals* equal at the base. *Petals* white or lilac, the two outer much the longest. *Filaments* without appendages. *Pods* broad, much compressed, orbicular or ovate, tip entire or notched ; valves keeled or winged ; septum very narrow, of two lamellæ ; stigma notched. *Seeds* 1 in each cell, not margined ; radicle accumbent, horizontal, or ascending.—DISTRIB. Mid. and S. Europe, Asia Minor ; species about 20.—ETYM. Iberia (Spain), where many species grow.

I. ama'ra, *L.* ; leaves oblong-lanceolate, pods suborbicular.

Cornfields and cultivated ground, on a dry soil, chiefly in the centre and E. of England, rare in Scotland ; (a colonist, *Wats.*) ; fl. July–Aug.—Annual. *Stem* 6–9 in., erect, corymbosely branched, ribbed, the ribs minutely downy.

Leaves 1–3 in., sessile, scattered, sparingly toothed or pinnatifid, often minutely ciliate. *Flowers* ¼–⅓ in. diam., white or purplish. *Pods* in short racemes, ⅓ in. broad, flat; pedicels horizontal; notch triangular; valves narrowly winged, wings acute above; style exceeding the wings.—DISTRIB: W. Europe, from Belgium southd.—Very bitter.

20. TEESDA'LIA, *Br.*

Small, annual, glabrous, scapigerous herbs. *Leaves* rosulate, pinnatifid. *Flowers* minute, white. *Sepals* spreading, equal at the base. *Petals* equal, or 2 outer larger, with a basal pouch. *Stamens* 4 or 6, with a scale at the base of each filament. *Pods* broadly obovate or orbicular, notched or 2-lobed; valves boat-shaped, slightly winged; cells 2-seeded; stigma simple. *Seeds* 2 in each cell; radicle accumbent.—DISTRIB. S. and W. Europe and W. Asia; species 2.—ETYM. *Robert.Teesdale*, a Yorkshire botanist.

T. nudicau'lis, *Br.* ; two outer petals twice as long as the others. *T. Iberis*, DC.

Sandy and gravelly places; ascends to near 1,000 ft. in Yorks.; local·in Scotland; Channel Islands; fl. April–June.—*Stems* 4–18 in., usually many, slender, ascending. *Radical leaves* 1–2 in., numerous, lobes broad, spreading. *Flowers* ₁₂ in. diam., homogamous, corymbose then racemose. *Pods* ⅓ in., in racemes 2–9 in; pedicels short, slender, spreading; style very short. —DISTRIB. Europe, N. Africa, W. Asia.

21. HUTCHIN'SIA, *Br.*

A small, annual herb. *Leaves* rosulate, pinnatifid. *Flowers* minute, corymbose. *Sepals* short, equal at the base. *Petals* small, equal. *Filaments* without scales at the base. *Pods* broadly oblong, obtuse, much compressed; valves keeled; septum narrow; stigma sessile. *Seeds* 2 in each cell, compressed, not margined; funicle slender; radicle accumbent. —DISTRIB. Europe, N. Africa, W. Asia to N.W. India.—ETYM. *Miss Hutchins*, a zealous Irish botanist.

1. **H. petræ'a,** *Br.* ; sepals about equalling the petals.

Limestone rocks W. of England and Wales, from Dumfries and Yorkshire (ascending to 1,500 ft.) to Somerset; naturalized in Eltham churchyard, where planted by Dillenius? and in Mathew cemetery, Cork; fl. March– May.—Glabrous or sparingly hairy, slender, much branched from the base, 2–5 in. *Radical leaves* ½–1 in., lobes spreading, obovate, almost petioled; cauline shorter, pinnatifid. *Pods* ₁₆–₁₂ in., on horizontal pedicels, in a short raceme; style distinct; seeds pale.

22. ISA'TIS, *L.* WOAD.

Tall, erect, annual or biennial branched herbs. *Cauline leaves* sagittate. *Flowers* yellow, pedicels slender, deflexed in fruit. *Sepals* equal at the base. *Petals* equal. *Pods* indehiscent, 1-celled, oblong obovate or orbicular, thickened in the middle; wing or margin very broad; stigma sessile. *Seed* pendulous from the top of the cell; radicle incumbent.—DISTRIB.

Temp. Europe, Asia, and N. Africa ; species 25–30.—ETYM.   The Greek name of the genus.

I. **tincto'ria,** *L.* ; pods obovate-oblong, tip rounded, wing thick.

Wild on cliffs by the Severn,Tewkesbury ; naturalized near Guildford ; sporadic elsewhere ; (an alien, *Wats.*) ; fl. July–Aug.—Glaucous, glabrous or nearly so. *Stem* 1–3 ft., stout, erect, branched above. *Radical leaves* oblong-obovate or lanceolate ; petiole long ; cauline 3–5 in., sessile. *Flowers* ½ in. diam., in crowded panicled corymbs. *Pods* ½ in. in short racemes, pendulous, glabrous, brown when ripe ; stigma sessile in a minute notch ; seed linear-oblong.—DISTRIB. Europe, N. Asia.—The ancient Britons stained themselves with it ; later, the Saxons imported it ; it is still cultivated in Lincolnshire.

23. CRAM'BE, *L.* SEA-KALE.

Perennial herbs with stout branching stems. *Leaves* usually broad. *Flowers* white, in long corymbose racemes. *Sepals* spreading, equal. *Longer filaments* often with a tooth on the outside. *Pods* indehiscent, 2-jointed ; lower joint slender, seedless, forming a pedicel to the upper, which is globose 1-celled and 1-seeded ; stigma sessile. *Seed* globose, pendulous from a basal funicle ; radicle incumbent, conduplicate.—DISTRIB. Europe, W. Asia ; species 16.—ETYM. The Greek name of the plant.

C. marit'ima, *L.* ; leaves broad waved toothed or pinnatifid.

Sandy and shingly sea-coasts, rare, from Fife and Isla southd.; N. and W. Ireland ; Channel Islands ; fl. June–Aug.—*Rootstock* as thick as the thumb, fleshy, burrowing ; branches 1–2 ft., spreading. *Leaves* 6–10 in., fleshy, petioled, broadly ovate-cordate oblong or orbicular, glabrous and glaucous ; upper few and small. *Corymbs* much branched. *Flowers* ½ in. diam., white ; longer filaments toothed near the tip externally. *Pods* ¾ in., on slender ascending pedicels 1 in.—DISTRIB. Coasts from Finland to the Bay of Biscay and the Black Sea.—Formerly eaten wild, cultivated for about 200 years in England, whence it was introduced to the Continent.

24. CAKI'LE, *Gærtn.* SEA ROCKET.

Annual, large, fleshy, branched herbs. *Leaves* entire or pinnatifid. *Flowers* white or purplish. *Lateral sepals* gibbous at the base. *Pods* indehiscent 2-jointed ; joints angled, 1-celled, upper deciduous compressed, seed basal ; lower cuneate, 2-edged, seed pendulous ; stigma sessile. *Radicle* accumbent, sometimes oblique.—DISTRIB. Sea-shores of Europe and N. America ; species 2.—ETYM. An Arabic word.

C. marit'ima, *Scop.* ; suberect or decumbent, leaves entire or lobed.

Sandy and shingly shores, N. to Shetland, abundant ; Ireland ; Channel Islands ; fl. June–July.—Annual ; rather succulent, 1–2 ft., zigzag, ascending. *Leaves* 2–3 in., fleshy. *Flowers* ½ in. diam., corymbose ; pedicel stout. *Pods* on short thick pedicels loosely racemed, ribbed when dry ; lower joint ½ in., broader upwards ; upper ¾ in., base truncate.—DISTRIB. Europe, N. Africa, Iceland.—*C. america'na,* which is perhaps identical, extends from the Canadian Lakes to the W. Indies.

25. RAPH'ANUS, *L.* RADISH.

Annual or biennial herbs. *Radical leaves* lyrate. *Flowers* in long racemes, white or yellow, purple-veined. *Sepals* erect, lateral saccate at the base. *Pods* elongate, indehiscent, or separating into several superimposed 1-seeded joints, terete or moniliform, coriaceous or corky ; style or beak of the pod slender ; stigma notched. *Seeds* pendulous, globose ; cotyledons conduplicate or much folded.—DISTRIB. Europe and temp. Asia ; species 6.—ETYM. *Rapa,* the Latin name.

1. **R. Raphanis'trum,** *L.* ; leaf-segments usually few and remote, pod subulate not much constricted at the 4–8 faintly-ribbed joints, beak as long as the 2 or 3 last joints. *Wild Radish* or *White Charlock.*

Cornfields, N. to Shetland : Ireland; Channel Islands ; ascends to 1,000 feet; (a colonist, *Wats.*) ; fl. May–Sept.—Annual, stout, 1–2 ft., erect or spreading, hairy or hispid. *Leaves* 4–10 in., coarsely toothed or serrate, terminal lobe largest. *Flowers* ¾ in. diam., white or straw-coloured, homogamous. *Pods* 1–3 in., dehiscing at the base above the first segment, which is seedless and very small ; beak ½–¾ in., subulate, flattened.—DISTRIB. Europe (Arctic), N. Africa, N. and W. Asia to India; introd. in America.

2. **R. marit'imus,** *Sm.* ; leaf-segments many approximate horizontal or reversed, alternate often smaller, pod deeply constricted at the 2–4 strongly ribbed joints, beak slender subulate.

Sandy and rocky shores from the Clyde southd.; Ireland ; Channel Islands ; fl. July–Aug.—Very near *R. Raphanis'trum,* and perhaps the wild form of that plant, but biennial, more hispid, leaves with more numerous and closer set lobes; flowers smaller, darker yellow, rarely white ; pod with fewer joints, deeper intervals between these, stronger ribs, and a beak as long as the upper joint.—DISTRIB. W. Europe, from Holland to Spain.

ORDER VII. **RESEDA'CEÆ.**

Annual or perennial herbs, rarely shrubs. *Leaves* alternate, simple or pinnatisect ; stipules 0, or minute and glandular. *Flowers* racemed or spiked, bracteate. *Calyx* persistent, 4–7-partite, often irregular, imbricate in bud. *Petals* 4–7, hypogynous, entire or lobed, equal or the posticous larger, open in bud. *Disk* hypogynous, conspicuous. *Stamens* usually many, inserted on the disk, equal or unequal, free or connate. *Ovary* of 2- 6 connate carpels, lobed at the top, open between the stigmatiferous lobes ; ovules usually many, on 2–6 parietal placentas, amphitropous or campylotropous. *Fruit* usually a coriaceous capsule, open at the top. *Seeds* many, reniform, exalbuminous ; embryo curved or folded, radicle incumbent. —DISTRIB. Europe, W. Asia, N. and S. Africa ; genera 6 ; species 20. —AFFINITIES. Closely allied to *Cappari'deæ.*—PROPERTIES, unimportant.

1. **RESE'DA,** *L.* MIGNONETTE.

Herbs. *Leaves* entire lobed or pinnatifid ; stipules glandular. *Flowers* racemed. *Calyx* irregular. *Petals* unequal, 2-multifid, the posticous with a membranous appendage on its face. *Disk* broad, honeyed, dilated behind. *Stamens* 10-40.—DISTRIB. Europe, W. Asia, N. Africa ; species 26.—ETYM. *resedo,* being a supposed sedative.

1. **R. Lu'teola,** *L.* ; leaves linear-lanceolate undivided, sepals 4, stigmas 3. *Dyer's weed, Weld.*

Waste dry places, from Ross southd. ; rarer in Scotland ; common in Ireland ; Channel Islands ; fl. June-Aug.—Glabrous, 2-3 ft., annual or biennial, branched. *Racemes* long, spike-like. *Flowers* yellow-green. *Petals* 3-5, if 5 upper 3-4 cleft, two lateral 3-cleft, two lower entire. *Disk* large, crenate. *Stamens* 20-24. *Capsule* short, 3-lobed ; seeds subglobose, black.—DISTRIB. Europe, N. Africa, W. Asia introd. in U. States.—Yields a yellow dye.

2. **R. lu'tea,** *L.* ; leaves 2-3-fid or pinnate or bipinnatifid, lobes few distant linear obtuse, sepals and petals 6 very unequal, stigmas usually 3.

Waste places in England; very rare, and native? Scotland and Ireland ; Channel Islands; fl. June-Aug.—Biennial. *Stem* 1-2 ft., branched, ribbed, papillose. *Leaves* very various in lobing. *Racemes* dense-flowered, conical ; flowers pale yellow. *Sepals* linear, upper smaller. *Upper petals* with a 2-lobed claw and 3-fid limb, lateral 2-fid, lower entire. *Stamens* 16-20, deflexed. *Capsule* ½-¾ in., oblong, 3-toothed ; seeds obovoid, black.— DISTRIB. Europe, N. Africa, W. Asia.

R. AL'BA, *L.* ; leaves pinnate undulate glaucous, sepals 5-6, petals 5-6 all 3-fid, stigmas usually 4. *R. suffruticulo'sa,* L., and *R. fruticulo'sa,* L. *R. Hooke'ri,* Guss.

Waste places, chiefly near the sea, an outcast ; fl. June-Aug.—I follow Bentham in taking the name of *al'ba* for this, which most authors agree in considering the same with *fruticulo'sa* and *suffruticulo'sa. Flowers* white. *Stamens* 12-14. *Seeds* reniform, rough.—DISTRIB. S. Europe and N. Africa. —The true *R. alba* is more often 6-merous than is *fruticulo'sa.*

ORDER VIII. **CISTI'NEÆ.**

Herbs or shrubs. *Leaves* mostly opposite, entire ; stipules foliaceous, small, or 0. *Flowers* terminal and solitary, or in scorpoid cymes, not honeyed. *Sepals* 3-5, imbricate, 2 outer (bracts of some) small or 0 ; 3 inner often convolute in bud. *Petals* 5, rarely 3 or 0, fugacious, convolute in bud. *Stamens* many, rarely few, hypogynous, free. *Ovary* 1-celled, or divided by parietal septa ; style simple, stigmas 3 ; ovules 2 or more, on parietal placentas, orthotropous, funicle slender. *Capsule* 3-5 valved ; valves placentiferous. *Seeds* with mealy or firm albumen, testa crustaceous often mucilaginous ; embryo usually curved and excentric. —DISTRIB. Europe, N. Africa, and W. Asia ; rare in N. America ;

genera 4; species 60.—AFFINITIES, With *Bixineæ* and *Capparideæ.*—
PROPERTIES. A resinous balsam (Ladanum) is yielded by *Cisti.*

1. HELIAN'THEMUM, *Tourn.* ROCK-ROSE.

Herbs or under-shrubs. *Petals* 5. *Stamens* many, rarely few, diverging
when irritated. *Ovary* many-ovuled; style jointed at the base, stigma
capitate or 3-lobed. *Embryo* hooked, folded or circumflex.—DISTRIB.
Europe, W. Asia, and N. America; species about 30.—ETYM. ἥλιος and
ἄνθεμον, *sun-flower.*—Various species are dimorphic, some flowers having
no petals and few stamens.

SECTION 1. **Helian'themum** proper. *Style* elongate, bent upwards.
Funicle thickened. *Embryo* with the radicle bent upwards parallel to the
cotyledons.

1. **H. vulga're,** *Gœrtn.*; shrubby, leaves opposite stipulate hairy
above downy beneath, margins flat, pedicels bracteate. *H. surreia'num,*
Mill. (a garden variety); *Cis'tus tomento'sus,* Sm.

Dry soils, abundant, from Ross southd., rare in W. Scotland and Cornwall,
ascends to 2,000 ft.; fl. July–Sept.—*Branches* 3–10 in., procumbent. *Leaves*
oblong, variable. *Flowers* ¾–1¼ in. diam., yellow, homogamous; bracts
narrow. *Sepals* subglabrous, inner apiculate.—DISTRIB. Europe (Arctic),
N. Africa, W. Asia.

2. **H. polifo'lium,** *Pers.*; shrubby, leaves opposite hoary and downy
on both surfaces stipulate, margins recurved, pedicels bracteate.

Stony places, very rare, Brean Down Somerset, Babbicombe near Torquay;
fl. May–July.—Habit of *H. vulga're,* but more shrubby, leaves with recurved
and even revolute margins, and flowers white. *Sepals* tomentose, inner
obtuse.—DISTRIB. Mid. and S. Europe, N. Africa.

SECTION 2. **Tubera'ria.** *Style* straight or 0. *Funicle* thickened.
Embryo annular.

3. **H. gutta'tum,** *Miller;* annual, erect, hoary and hairy, lower leaves
opposite exstipulate, upper alternate stipulate.

Stony places, Anglesea, very rare; S. and W. Ireland; Channel Islands; fl.
June–Aug.— *Stem* 6–12 in., 2–3-chotomously branched. *Leaves* 1–2 in.,
linear- or obovate- or oblong-lanceolate. *Flowers* ⅓–½ in. diam., yellow with
a red spot at the base of the cuneate petals. *Capsule* smooth.—DISTRIB.
Europe, N. Africa, W. Asia.
H. GUTTA'TUM proper; pedicels ebracteate.—Cork, Jersey, Alderney.
Sub-sp. H. BREWE'RI, *Planch.*; pedicels bracteate.—Holyhead and Angleses.

SECTION 3. **Pseudo-cis'tus.** *Style* sigmoid. *Funicle* not thickened.
Embryo sigmoid.

4. **H. ca'num,** *Dunal;* shrubby, hoary, leaves opposite exstipulate,
pedicels usually bracteate. *Cis'tus marifo'lius,* Sm., and *C. angli'cus,* L.

Dry banks, rocks, &c., in W. England, from Westmoreland to Glamorgan, and
in Teesdale, ascending to 1,800 ft.; fl. May–July.—Woody, procumbent,

much branched, 6–8 in. *Leaves* ¼–¾ in., ovate or oblong. *Flowers* ¼–¾ in. diam., few, yellow.—DISTRIB. Europe, N. Africa, and W. Asia. *H. ca'num* proper ; leaves hoary on both surfaces hairy above, sepals pubescent and patently hairy.-- VAR. *H.vinea'le*, Pers. ; leaves hoary beneath nearly glabrous above, sepals hoary and shortly hairy on the ribs and margins.— W. Ireland, Aran Is.

ORDER IX. VIOLA'CEÆ.

Herbs or shrubs. *Leaves* radical or alternate, entire or pinnatisect, margins involute in vernation, stipulate. *Flowers* axillary, regular or irregular, solitary or cymose, 2-bracteolate. *Sepals* 5, persistent, equal or unequal, imbricate in bud. *Petals* 5, equal or unequal, hypogynous, imbricate or contorted in bud. *Disk* 0. *Stamens* 5, filaments short, broad, lower with honeyed spurs ; connectives broad, often connate, produced beyond the cells. *Ovary* sessile, 1-celled ; style simple, stigma entire cup-shaped or lobed ; ovules many, on 3 parietal placentas, anatropous. *Fruit* a 3-valved capsule, rarely a berry. *Seeds* many, small, funicle short, albumen fleshy ; embryo straight, cotyledons flat.—DISTRIB. Temp. and trop. regions ; genera 21, species 240.—AFFINITIES with *Bixineæ, Passifloreæ,* and *Frankeniaceæ.*—PROPERTIES. Emetic and laxative.

1. VI'OLA, *L.* VIOLET, PANSY, HEARTSEASE.

Low herbs, rarely shrubs. *Leaves* radical or alternate. *Flowers* on 1- rarely 2-fld. peduncles. *Sepals* subequal, produced at the base. *Petals* erect or spreading ; lower largest, spurred or saccate at the base. *Anthers* connate, connectives of the 2 lower often spurred at the base. *Style* swollen above, tip straight or oblique, stigma obtuse or cup-shaped. *Capsule* 3-valved ; valves elastic. *Seeds* ovoid or globose.—DISTRIB. All temp. regions ; species 100.—ETYM. The old Latin name.—Flowers often cleistogamous (except in sect. *Melanium*), the large-petalled appear early and often yield no seed ; the small-petalled or apetalous appear late, and are prolific.

SECTION 1. **Nomin'ium.** *Stipules* not leafy. *Upper petals* directed forwards. *Stigma* oblique.—Apetalous autumnal flowers chiefly fertile.

* *Stem very short. Leaves enlarging after flowering. Sepals obtuse.*

1. **V. palus'tris,** *L.* ; nearly glabrous, rootstock subterranean creeping, leaves reniform-cordate, style straight, stigma obliquely truncate, fruiting peduncle erect.

Swamps and bogs, N. to the Shetlands, ascends to 4,000 ft., rarer in S. England ; Ireland ; fl. April–July.—*Rootstock* white, scaly ; runners short, leafless. *Leaves* slightly crenate ; stipules glandular. *Flowers* ½ in. diam., white or lilac, scentless ; spur short, obtuse. *Lateral petals* almost glabrous. *Anther-spurs* short. — DISTRIB. Temp. and Arctic Europe, Asia, and America.

2. **V. odora'ta,** *L.* ; slightly hairy or downy, runners very long, leaves broadly cordate, spur nearly straight, style hooked, stigma oblique, fruiting peduncle decurved. *Sweet Violet.*

Banks and copses, wild in E. and S. England; and perhaps E. Ireland; naturalized as far N. as Forfar; Channel Islands; fl. March–May.—*Rootstock* short, scarred. *Leaves* deeply cordate at the base, sinus closed; stipules glandular; petiole with deflexed hairs. *Bracts* at or about the middle of the peduncle. *Flowers* fragrant, blue, white, or red-purple; lateral petals with or without a tuft of hairs; spur short, obtuse. *Anther-spurs* linear-oblong. *Capsule* pubescent; peduncle recurved.—DISTRIB. Europe, N. Africa, N. and W. Asia to the Himalaya. The following are probably hybrids with *V. hir'ta.*
V. permix'ta, Jord. ; runners not rooting, flowers pale scentless.—*V. sepin'cola,* Jord.; more hairy, runners rooting, flowers dark scentless.

3. **V. hir'ta,** *L.* ; pubescent, rootstock very short, runners short or 0, leaves subtriangular-cordate, spur hooked, style hooked, stigma oblique.

Dry soils, local, from Forfar southd.; ascends to 1,000 ft. in Yorks.; Ireland; fl. April–June.—Very near *V. odora'ta,* but more tufted and hairy, leaves narrower and more triangular, with deeper crenatures and shallower sinus; hairs of petiole more spreading; bracts lower on the peduncle; spur long and hooked; flowers inodorous or faintly scented; anther spurs lanceolate. —DISTRIB. Europe, N. and W. Asia to N.W. India.
V. calca'rea, Bab.; is a dwarf starved form with petals narrower. Gogmagog Hills and Portland.

** *Stem evident. Leaves not enlarging after flowering. Sepals acute or acuminate.*

4. **V. cani'na,** *L.* ; glabrous, main stem elongate and flowering, leaves ovate-cordate or oblong-lanceolate, stipules small narrow toothed and ciliate, fruiting peduncle erect. *Dog-violet.*

Pastures and banks from Caithness southd.; Ireland; Channel Islands; fl. April–Aug.—Very variable in size, habit, and colour of flower, glabrous or nearly so. *Leaves* long-petioled, crenate-serrate, from ¾ in., broadly ovate deeply cordate, to oblong-lanceolate and 3 in. *Bracts* at or above the middle of the peduncle, subulate or lanceolate, toothed or serrate. *Flower* ½–1¼ in. diam., blue, lilac, grey or white. *Sepals* narrow, acuminate. *Style* clavate, hooked; stigma oblique. *Capsule* oblong, 3-gonous.—DISTRIB. Europe (Arctic), N. and W. Asia to the Himalaya, N. Africa, N. America.
V. CANI'NA proper; rootstock short, runners 0, leaves narrow ovate-cordate, spur obtuse, anther-spurs 5 times as long as broad. *V. flavicor'nis,* Sm. *V. pu'mila,* Hook and Arn.
Sub-sp. V. LAC'TEA, *Sm.*; very slender, rootstock short, runners 0, leaves ovate-lanceolate, base rounded or cuneate, petals narrow grey, spur very short, capsule subglobose.—Heaths, York to Cornwall, W. Europe.
Sub-sp. V. PERSICÆFO'LIA, *Roth;* rootstock long with runners, leaves oblonglanceolate base truncate, upper narrower, petals pale lilac or white, spur very short, capsule 3-gonous. *V. stagnina,* Kit.—Bogs, E. of England, Galway; extends to Siberia.

5. **V. sylvat'ica,** *Fries;* glabrous, flowers on axillary branches from a radical rosette, leaves broadly ovate-cordate, stipules lanceolate acute

fimbriate or toothed, fruiting peduncle erect, capsule glabrous. *V. cani'na*, Sm. *Wood Violet.*

Copses and woods, from the Shetlands southd.; ascends to 3,000 ft.; Ireland; Channel Islands; fl. March–July.—*Rootstock* very short. *Leaves* in a rosette, which however is often deficient, when it is with difficulty distinguished from *V. cani'na*, from which I doubt its permanent distinctness.—DISTRIB. Europe, N. and W. Asia, N. America.

V. SYLVAT'ICA proper; rootstock short, spur short broad compressed furrowed, usually pale, base of sepals much produced in fruit. *V. Rivinia'no*, Reichb.

Sub-sp. V. REICHENBACHIA'NA, *Bor.*; flowers smaller paler, spur longer, fruiting sepals hardly produced. (Flowers earlier.)

6. **V. arena'ria**, *DC.*; small, tufted, pubescent, leaves orbicular-ovate obtuse, flowers on short axillary branches from a compact rosette, stipules small fimbriate, spur short, capsule oblong pubescent.

Upper Teesdale, alt. 2,000 ft., and Westmoreland, very rare; fl. May–June.— Whole plant about 2–6 in. diam., compact, hoary-pubescent. *Leaves* much rounder than in *V. sylvat'ica* and *cani'na*. *Sepals* lanceolate, acute, bases produced square in fruit. *Petals* broad, pale blue; spur short.—Europe (excl. Greece), Siberia, Labrador.

SECTION 2. **Mela'nium.** *Stipules* leafy. *Upper petals* erect. *Stigma* capitate, hollow, with a pencil of hairs on each side.—No dimorphic flowers.

7. **V. tri'color**, *L.*; leaves long-petioled ovate-oblong or lanceolate crenate, stipules pinnatifid, sepals with large auricles, style short straight, stigma capitate excavated. *Heartsease, Pansy.*

Pastures, banks and waste places, N. to Shetland; Ireland; Channel Islands; fl. May–Sept.—Very variable. *Stem* 4–18 in., branched, erect or ascending, angular, flexuous. *Leaves* 1–1½ in., lyrate, coarsely and remotely crenate-serrate; stipules ⅓–½ in. broad, very large; lobes spreading like a fan, linear or oblong, obtuse, lateral smaller, middle sometimes leafy. *Bracts* minute, high up on the peduncle. *Flowers* ¼–1¼ in. diam.; petals purple whitish or golden yellow, sometimes parti-coloured, very variable in size, sometimes 0.—DISTRIB. Europe (Arctic), N. Africa, N. and W. Asia to Siberia and N.W. India.—Sub-sp. *lu'tea* and *Curtis'ii* are confined to W. and Central Europe. *V. tri'color* proper and *arven'sis* are naturalized in America.

V. TRI'COLOR proper; rootstock 0, stem elongate branched, petals spreading usually longer than the sepals pale yellow or lilac, lip of stigma developed, capsule ovoid. — Cultivated ground; ascends to near 2,000 ft. in the Highlands.

Sub-sp. V. ARVEN'SIS, *Murr.*; rootstock 0, stem elongate branched, petals erect usually shorter than the sepals or 0 white or yellowish, capsule globose. —Cultivated ground; ascends to near 1,000 ft. in Scotland and Yorkshire.

Sub-sp. V. CURTIS'II, *Forst.*; rootstock branched stoloniferous tufted, petals spreading rather longer than the sepals blue purple or yellow, capsule 3-gonous. *V. sabulo'sa*, Boreau. VARS. *Mackai'i, Syme'i* and *Forste'ri* are

E

hardly distinguishable forms.—Sandy shores from the Clyde to Cornwall ;
Ireland.
Sub-sp. V. LU′TEA, *Huds.*; rootstock branched, branches slender with short
stems and underground runners, mid lobe of stipules entire, petals spreading much longer than the sepals blue purple (var. *amœ′na*) or yellow, capsule
oblong 3-gonous. *V. grandiflo′ra*, Huds. ed. 2.—Hilly districts from Mid.
England and Wales to Ross ; ascends to 2,800 ft.

ORDER X. POLYGAL′EÆ.

Herbs or shrubs, erect or climbing. *Leaves* alternate or subopposite,
simple, exstipulate. *Flowers* irregular. *Sepals* imbricate in bud ; 2 inner
larger, petaloid, winglike. *Petals* 3–5, hypogynous, 2 outer (lateral) separate
or united with the hooded lower one into a tube split at the base behind ;
2 inner equal to the outer, or smaller or 0. *Stamens* 8, filaments connate
in a split sheath which is usually adnate to the petals ; anthers 1- rarely
2-celled, opening by pores, rarely by valves. *Disk* small. *Ovary* free,
2-celled ; style simple, curved, stigma various ; ovules 1 in each cell,
pendulous, anatropous, raphe ventral. *Seeds* pendulous, testa often hairy,
arillate, albumen fleshy or 0 ; embryo straight.—DISTRIB. Temp. and
trop. regions ; genera 15 ; species 400.—AFFINITIES, distant with *Sapindaceæ, Violaceæ,* and *Pittosporeæ.*—PROPERTIES. Bitter, emetic, purgative,
and diuretic.

1. POLYG′ALA, *L.* MILKWORT.

Herbs or shrubs. *Leaves* alternate, rarely subopposite or whorled.
Flowers in terminal or lateral racemes or spikes ; pedicels bracteate and
2-bracteolate. *Petals* combined below with the staminal sheath, which
has reversed hairs within, and a viscid gland at the mouth. *Stamens* 8 ;
anthers 1-2-celled, opening by transverse pores. *Stigma* spathulate.
Capsule compressed, loculicidal at the margins. *Seeds* usually downy ;
aril very variable, 2-auricled. DISTRIB. Trop. and temp. regions ;
species 200.—ETYM. πόλυς and γάλα, being supposed to increase the milk
in cows. *Flowers* in some cleistogamous.

1. **P. vulga′ris,** *L.* ; stems many leafy, leaves scattered lower oblong
upper lanceolate, lateral nerves of inner sepals anastomosing copiously,
central nearly simple.
Heaths and meadows, N. to Shetland ; Ireland ; Channel Islands ; fl. June–
Aug.—A small wiry perennial, 2–10 in., glabrous or very rarely pubescent.
Rootstock short. *Leaves* ½–1½ in., rather coriaceous, quite entire. *Flowers*
⅓–½ in., white, pink, blue, lilac, or purple. *Sepals* purplish in flower, green
in fruit, inner elliptic-obovate. *Capsule* ⅓ in. diam., obcordate or nearly
orbicular and notched. *Aril* with nearly equal lobes.—DISTRIB. Europe
(Arctic), N. Africa, Siberia, and W. Asia.
P. VULGA′RIS proper, stems ascending, branches straight, leaves all linear or
lanceolate, racemes many-fld., bract as long as the flowering pedicel, pedicels

sepals petals and capsules not ciliate, large sepals oblong-obovate broader than the capsule. Common; ascends to near 3,000 ft. in the Highlands.—
VAR. *grandiflo'ra*, Bab.; upper leaves large, inner sepals oblong acute, flowers large dark blue. Benbulben, Sligo.

Sub-sp. P. OXYP'TERA, *Reichb.*; branches flexuous, leaves linear, inner sepals cuneate below shorter and narrower than the capsule, pedicels &c. glabrous. —Sandy shores, limestones and chalky soils, from Perth southd.; local.

Sub-sp. P. DEPRES'SA, *Wend.*; stems flexuous, leaves somewhat opposite and distichous, lower oblong spathulate, racemes fewer-flowered, bract shorter than the flowering pedicel. *P. serpylla'cea*, Weihe. –Common on heaths.—
VAR. *P. cilia'ta*, Lebel; branches prostrate tortuous, inner sepals broader than the capsule, pedicels bracts sepals and capsule ciliate. Gogmagog Hills.

2. **P. calca'rea,** *F. Schultz;* branches many rooting and proliferous umbellately spreading from the root, radical leaves rosulate, cauline oblong, inner sepals longer and broader than the obcordate capsule, central nerve branching above the middle. *P. ama'ra*, Don, not L.

Dry soil and rocks, S. and S.E. of England; Wiltshire to Kent, and Gloucester to Berks; Channel Islands; fl. June–July.—Perhaps only a sub-species of *P. vulga'ris*, approaching sub-sp. *depres'sa*, but the habit is entirely different, and the nerves of the sepals scarcely anastomose.—DISTRIB. Central and S. Europe.

3. **P. ama'ra,** *L.* ; leaves rosulate spathulate, flowering branches axillary, inner sepals narrower than the capsule, nerves simple or slightly branched free, capsule orbicular notched.

Very rare, margins of rills, in Teesdale, alt. 1,800 ft.; Wye Down, Kent; fl. June–July.—Much smaller in all its parts than *P. vulga'ris* or *calca'rea*, and readily distinguished by this character and the narrow inner sepals. The Teesdale form (*P. uligino'sa*, Fries) is rather more fleshy and has rosy flowers; the Kent form (*P. austri'aca*, Crantz) is blue flowered. I find no difference between their capsules. It is certainly the *P. ama'ra* of Linn. Herb.—DISTRIB. Europe (Arctic) from Sweden southd.

ORDER XI. **FRANKENIA'CEÆ.**

Perennial rarely annual herbs or small shrubs, with jointed branches. *Leaves* small, opposite, exstipulate. *Flowers* small, regular, solitary, in the forks of the branches. *Calyx* tubular, persistent; lobes 4–6, induplicate in bud. *Petals* 4–6, hypogynous, imbricate in bud, claw with an adnate scale. *Stamens* 4 or more, separate or connate at the base; anthers versatile. *Disk* 0. *Ovary* free, sessile, 1-celled; style slender, stigma 2–5-lobed; ovules many, in 2 series, on 2–5 parietal placentas, amphitropous with the micropyle below; funicle slender. *Capsule* enclosed in the persistent calyx, 3–5-valved. *Seeds* oblong, raphe linear, testa crustaceous, albumen mealy; embryo axile, straight.—DISTRIB. Temp. and warm regions, chiefly littoral; species about 12.—AFFINITIES with *Caryophyl'leæ* and *Tamariscine'æ.*—PROPERTIES none.

1 FRANKE'NIA, *L.* SEA-HEATH.

Characters of the Order.—ETYM. *J. Franken,* a Swedish botanist.

F. læ'vis, *L.* ; stem pubescent, leaves with revolute margins.
Salt marshes on S.E. coasts of England, Yarmouth to Kent; Channel Islands;
fl. July–Aug.—Perennial, procumbent; branches wiry. *Leaves* ¼–½ in.,
fascicled or whorled, oblong but linear from the reflexed margins, glabrous,
ciliate at the base. *Flowers* small, rose-coloured, dichogamous. *Capsule*
3-gonous.—DISTRIB. W. Europe and Africa to the Cape, W. Asia to India,
in salt plains.

ORDER XII. **CARYOPHYL'LEÆ.**

Herbs, sometimes woody below, nodes thickened. *Leaves* opposite,
bases usually connate, entire; stipules 0, or small and scarious. *Inflorescence* definite, centrifugal. *Sepals* 4–5, free or connate. imbricate in
bud. *Petals* 4–5 (rarely 0), hypogynous, rarely perigynous, imbricate or contorted in bud. *Stamens* 8–10, rarely fewer, inserted with the
petals. *Disk* annular or elongated, or of inter-staminal glands. *Ovary*
free, 1-celled, or 3-5-celled at the base; styles 2–5, free or connate, stigmatose on the inner surface; ovules 2 or more, funicles slender basal
often connate, amphitropal, micropyle inferior or transverse. *Fruit* capsular. *Seeds* many, small, albumen floury, rarely fleshy; embryo cylindric,
usually curved or annular, radicle incumbent.—DISTRIB. Cosmopolitan,
but chiefly Arctic, Alpine European, and W. Asiatic; genera 35, species
800.—AFFINITIES with *Illecebra'ceæ, Portula'ceæ,* and *Chenopodia'ceæ.*—
PROPERTIES unimportant.

TRIBE I. **SILE'NEÆ.** *Stipules* 0. *Calyx* 4–5-lobed or toothed. *Disk*
elongated, bearing the petals and stamens. *Styles* free.
* *Hilum on the face of the peltate seed. Embryo straight*............1. Dianthus.
** *Hilum lateral. Embryo annular.*
Styles 2. Capsule 4-valved...1*. *Saponaria.*
Styles 3. Capsule 6- rarely 3-valved.......................................2. Silene.
Styles 4–5.
Petals appendiculate..3. Lychnis.
Petals exappendiculate...4. Githago.
TRIBE II. **ALSI'NEÆ.** *Sepals* separate. *Disk* small. *Styles* free.
* *Stipules* 0.
Capsule cylindric, 6-valved. Petals jagged. Styles 3..........5. Holosteum.
Capsule cylindric, 8–10-valved. Petals notched (rarely entire).6. Cerastium.
Capsule globose, 6-10-valved; styles 3–5. Petals 2-fid............7. Stellaria.
Capsule 3- 4- 6- or 10-valved. Styles 3–4. Petals entire.........8. Arenaria.
Capsule 4–5-valved. Styles 4–5. Petals entire or 0.................9. Sagina.
** *Stipules scarious.*
Styles and valves of capsule 5..10. Spergula.
Styles and valves of capsule 3..11. Spergularia.
TRIBE III. **POLYCAR'PEÆ.** *Stipules* scarious. *Sepals* separate. *Disk* small.
Petals small. *Stamens* 5 or fewer. *Styles* connate at base...11*. *Polycarpon.*

1. DIAN'THUS, *L.* PINK and CARNATION.

Tufted herbs, often shrubby at the base. *Leaves* narrow, grass like. *Flowers* solitary, panicled or fascicled, dichogamous, proterandrous. *Calyx* tubular, 5-toothed, striate, with imbricating bracts at the base. *Petals* 5, entire or cut, claw long. *Stamens* 10, emerging and dehiscing 5 at a time. *Disk* elongated. *Ovary* 1-celled ; styles 2. *Capsule* 4-valved at the top. *Seeds* discoid, imbricate upon the columnar placenta, hilum ventral ; embryo straight.—DISTRIB. Europe, temp. Asia, N.W. America, N. and S. Africa ; species about 70.—ETYM. supposed to be Διός and ἄνθος, *flower* of *Jupiter*.

* Flowers fascicled.

1. **D. Arme'ria,** *L.* ; fascicles of flowers in loose cymes, bracts lanceolate downy as long as the calyx, tips subulate. *Deptford Pink.*

Fields and dry banks from Forfar to Cornwall and Kent ; (a doubtful native, *Wats.*) ; fl. July–Aug.—Annual. *Stems* 1–2 ft., few, strict, erect. *Leaves* 1–2 in., linear, lower obtuse, upper acute. *Calyx-tube* 2–3 in. cylindric, many-nerved. *Flowers* ½ in. diam. *Petals* distant, narrow, red with dark dots, toothed.—DISTRIB. Europe, W. Asia; introd. in the U. States.

2. **D. pro'lifer,** *L.* ; fascicles of flowers capitate, bracts ovate membranous as long as the calyx-tube, inner obtuse. *Tunica prolifer*, Hall.

Gravelly pastures from Perth southd. ; often a casual ; Channel Islands; fl. June–Oct.—Annual. *Stems* few, 6–18 in., sometimes branched above. *Leaves* short, linear-lanceolate, margins scabrid. *Heads* ¾ in., many-fld., bracts dry brown. *Flowers* ¼–½ in. diam., opening one by one, all but the uppermost 2-bracteate. *Calyx* very narrow, faintly ribbed. *Petals* contiguous, purplish-red, obovate, notched. *Capsule* ovoid, rupturing the calyx. —DISTRIB. Europe, W. Asia; introd. in the U. States, usually placed in *Tunica.*

** Flowers solitary or loosely cymose.

3. **D. deltoi'des,** *L.* ; leaves narrow-lanceolate downy and subscabrous, lower obtuse, flowers solitary, bracts ovate acuminate half as long as the calyx-tube, petals toothed. *Maiden Pink.*

Fields and banks, dry soil, from Inverness and Argyll to Devon and Kent ; fl. June–Sept. — Perennial, much branched; branches slender, 1 ft. *Leaves* of barren shoots ligulate. *Flowers* ¾ in. diam., rarely 2 together, inodorous. *Calyx* glabrous, strongly ribbed. *Petals* distant, obovate, rosy, spotted with white. *Capsule* cylindric.—DISTRIB. Europe.
D. deltoi'des proper ; faintly glaucous, bracts generally 2, flowers rosy.—VAR. *D. glau'cus*, L.; very glaucous, bracts usually 4, flowers white.—Edinburgh, Croydon.

4. **D. cæ'sius,** *Sm.* ; leaves scabrous at the margin, flowers usually solitary, bracts orbicular mucronate 4 times shorter than the calyx-tube, petals jagged and bearded. *Cheddar Pink.*

Limestone rocks, Cheddar ; fl. June–July.— Perennial, glaucous. *Rootstock* woody, branched. *Stems* 4–10 in., many. *Leaves* of barren shoots linear, obtuse, upper ones of the flowering stems acute. *Bracts* membranous.

Flower 1 in. diam., fragrant. *Petals* contiguous, obovate, rosy, teeth ⅛-⅙ the length of the blade. *Calyx-tube* faintly-ribbed.—DISTRIB. Belgium southd. to Lombardy and Hungary.

D. CARYOPHYI'LUS, *L.* ; leaves grooved above, margins smooth, cymes loosely panicled, bracts obovate mucronate 3–4 times shorter than the calyx-tube, petals toothed and crenate. *Wild Carnation, Clove Pink.*

Old castle walls, &c., naturalized; fl. July–Aug.—Perennial, glabrous, glaucous, stout, much branched and leafy below, 18–24 in. *Leaves* 4–6 in., recurved. *Bracts* membranous, tips herbaceous. *Flower* 1½ in. diam., fragrant. *Calyx* cylindric, faintly ribbed. *Petals* obovate, rosy, teeth ⅛-¼ the length of the blade. *Capsule* ovoid.—DISTRIB. Belgium and France to Italy, Hungary, and Greece.—Flowers dimorphic on the same individual; stamens in one form much longer than in the other. The origin of the garden carnation.

D. PLUMA'RIUS, *L.* ; leaves all acute 1-nerved, margins scabrous, cymes loosely panicled, bracts 4 rhomboid cuspidate equalling ¼ of the calyx-tube, petals fimbriate. *Wild Pink.*

Naturalized on old walls in England and Wales; fl. June–Aug.—Perennial, tufted, branched, 1 ft. *Flowers* as in *D. Caryophyllus,* but smaller, rose-purple, segments of petals ⅓-½ as long as the blade.—DISTRIB. Mid. Europe from Austria to Lombardy, and Mid. Russia.—The origin of the garden pinks.

1*. SAPONA'RIA, *L.* SOAPWORT, FULLER'S HERB.

Annual or perennial herbs. *Radical leaves* spathulate, cauline narrower. *Flowers* in panicled or fascicled cymes, white, lilac, red or yellow, honeyed, proterandrous. *Calyx* tubular, 5-toothed, obscurely nerved, ebracteate. *Petals* 5, clawed, entire or notched. *Stamens* 10. *Disk* small. *Styles* 2. *Capsule* oblong, 2-celled at the base, 4-valved at the top. *Seeds* reniform, tubercled, hilum marginal; embryo annular.—DISTRIB. Europe and temp. Asia ; species 30.—ETYM. *Sapo,* the plant having been used as a *soap.*

S. OFFICINA'LIS, *L.* ; glabrous, glaucous, leaves oblong-lanceolate.

Hedges, roadsides, and fields, naturalized before Gerard's time ; (a denizen, *Wats.*) ; fl. Aug.–Sept.—*Rootstock* white, creeping, fleshy, stoloniferous. *Stem* 1–3 ft., straight, ascending. *Leaves* 2–4 in., 3-ribbed. *Cymes* in panicled corymbs. *Flowers* 1 in. diam. *Petals* obcordate, lilac or white. *Capsule* ovoid, on a stout pedicel, enclosed in the fusiform calyx-tube.—DISTRIB. Europe, W. Asia; introd. in U. States.—A decoction is very saponaceous. Flowers often double.—*S. hyb'rida,* L., is a var. with connate upper leaves and monopetalous corolla.—VAR. *puber'ula,* Syme, is another with the upper part of the stem and calyx pubescent. Near Hightown, Lancashire.

2. SILE'NE, *L.* CATCHFLY.

Habit of *Saponaria.* *Calyx* inflated, 5-toothed, 10-nerved. *Petals* 5 ; claw narrow ; blade entire or divided, with usually 2 scales at its base.

Stamens 10, the 5 petaline sometimes adnate to the claw. *Disk* columnar. *Ovary* 1-3-celled below the middle ; *styles* 3, rarely 2-5, opposite the sepals ; ovules many. *Capsule* 6- rarely 3-valved at the top. *Seeds* with a marginal hilum ; embryo annular or ½-annular.—DISTRIB. N. temp. hemisphere ; species 800.—ETYM. σίαλον, *saliva,* from the viscidity of some species.

* *Calyx bladdery, nerves reticulate. Capsule incompletely septate.*

1. **S. Cucu'balus,** *Wibel.* ; erect, panicle many-fld., bracts scarious, petals deeply cloven, scales obscure. *S. infla'ta,* Sm. *Cucu'balus Behen,* L. *Bladder Campion, White bottle.*

Roadsides and waste places, N. to Caithness ; ascends to nearly 1,000 ft. in Yorkshire ; Ireland ; fl. June–Aug.—Perennial, branched, 2-3 ft., glaucous, glabrous or downy (VAR. *puber'ula*). *Leaves* 1-3 in., variable, ovate, obovate or oblong. *Flowers* ¾ in. diam., drooping, white, proterandrous, trimorphous (male, fem., and hermaphr.). *Capsule* globose, top conical.—DISTRIB. Europe (Arctic), N. Africa, Siberia, W. Asia to N.W. India ; introd. in the U. States.

2. **S. marit'ima,** *With.* ; diffuse, flowers 1-4, bracts herbaceous, petals shortly cleft, segments broad with two scales at the base.

Sea-shores, N. to Shetland ; rare by Alpine streams ; ascends to 3,000 ft. in the Highlands ; Ireland ; Channel Islands ; fl. June–Aug.—Very nearly allied to *S. Cucu'balus.*—DISTRIB. Shores of Europe (Arctic), from Italy to the Canaries, Norway and Finland.

** *Calyx cylindric, strongly many-ribbed, closing tightly over the capsule at the top. Capsule incompletely septate.*

3. **S. con'ica,** *L.* ; erect, hairy and glandular, dichotomously branched, flowers many erect.

Pastures and sandy heaths, local ; Kent, Norfolk, Suffolk, Haddington and Forfar ; Channel Islands ; fl. May–July.—Annual, 6-12 in. *Leaves* linear, upper acute and ribbed. *Calyx* ½ in., ampulliform, 30-ribbed, intruded at the base ; teeth subulate. *Petals* small, rosy or purple, cleft, with 2 scales at the base of the blade ; gynophore very short.—DISTRIB. Europe, N. Africa, Siberia, W. Asia to India.

*** *Calyx with broad nerves. Capsule incompletely septate.*

4. **S. gal'lica,** *L.* ; hairy and viscid, lower leaves spathulate, flowers in leafy racemose cymes, calyx-teeth setaceous, petals and scales small entire or slightly 2-fid.

Gravelly places from Moray southd., not rare, probably often an escape , Ireland ; Channel Islands ; fl. June–Oct.—Annual, 1-2 ft., erect or diffusely branched. *Leaves* variable. *Calyx* ½ in , membranous with green pubescent ribs. *Capsule* ovoid on a usually deflexed pedicel.—DISTRIB. Europe, N. Africa, Siberia, N. and W. Asia to India.

S. gal'lica proper ; flowers white or pink, petals large 2-fid.—VAR. *S.˙ quin-quevul'nera,* L. ; petals entire white with a red spot.—VAR. *S. ang'lica,* L. ; branches spreading, petals small white often jagged.

5. **S. acau'lis,** *L.* ; densely tufted, leaves small linear-subulate close set, flowers usually diœcious shortly peduncled solitary erect, calyx tubular teeth obtuse, petals and scales notched. *Moss Campion.*

Alpine rocks; ascends to 4,300 ft.; Donegal only in Ireland; fl. June–Aug.— Perennial, glabrous, forming bright green, moss-like cushions. *Leaves* ¼–½ in., channelled above, keeled below, ciliate. *Flowers* pink, rarely white, ½ in. diam., peduncles lengthening after flowering. *Calyx* faintly nerved; teeth with scarious margins. *Capsule* exserted, subcylindric, 6-toothed.— DISTRIB. Alps of Europe, N. Asia, N. America; all Arctic regions.

6. **S. Oti'tes,** *L.* ; flowering stems erect simple few-leaved viscid, radical leaves narrow-spathulate, cymes panicled, flowers small sub-diœcious erect whorled, calyx teeth obtuse, petals linear without scales.

Sandy fields and roadsides of the E. Counties, local; fl. June–Aug.—*Rootstock* woody, branched. *Stem* 1–3 ft. *Radical leaves* ½–3 in., many, slender, puberulous. *Panicle* narrow, interrupted. *Flowers* ⅛ in. diam., many, sub-erect, pale yellow-green ; bracts membranous. *Calyx* obovoid, membranous. *Petals* entire. *Stamens* and *styles* much exserted. *Capsule* ovoid, rupturing the calyx.—DISTRIB. Europe, Siberia, W. Asia to Persia.

7. **S. nu'tans,** *L.* ; pubescent, stem above and calyx viscid, radical leaves oblong-lanceolate, cauline linear, flowers in panicled or subracemed cymes drooping, calyx teeth acute, petals 2-partite, scales lanceolate. *S. paradoxa,* Sm., not L. *Nottingham Catchfly.*

Dry places and walls, from Forfar southd., local; Channel Islands; fl. May–July.—*Rootstock* woody, branched. *Stems* 2–3 ft. *Radical leaves* 2–5 in., tufted, petioled ; cauline small, narrow, sessile. *Flowers* dimorphic, honey-less, proterandous, opening and fragrant for 3 nights, 5 stamens ripening on each of the two first nights, the styles protruding on the third. *Calyx* ½ in., tubular, swollen in the middle, membranous, nerves purple. *Petals* white or pink ; segments diverging, narrow, incurved. *Capsule* erect, exceeding and rupturing the calyx.—DISTRIB. Europe (Arctic), Siberia, Dahuria, Canaries.

**** *Calyx cylindric or ovoid,* 10-*nerved. Capsule without septa.*

8. **S. noctiflo'ra,** *L.* ; softly pubescent, viscid above, leaves all oblong-lanceolate acute the lower petioled, flowers few, calyx-tube long, teeth slender, petals 2-fid, scales truncate.

Sandy fields, on the E. chiefly ; Forfar to Cornwall; Ireland; Channel Islands; a colonist; fl. July–Aug.—Annual, 1–2 ft., erect, simple or dichotomous. *Leaves* 3–4 in., ½–1½ in. broad. *Flowers* erect, open at night, fragrant. *Calyx* 1 in., narrow in flower; nerves green. *Petals* rosy within, yellow outside, segments incurved by day. *Capsule* as long as, and often ruptur-ing, the calyx.—DISTRIB. Europe, Siberia, W. Asia ; introd. in U. States. —Regarded by A. de Candolle as a Siberian and Caucasian plant of very ancient naturalization in W. Europe.

3. LYCH'NIS, *L.* CAMPION.

Characters of *Silene*, but styles 5. *Sepals* not foliaceous. *Petals* with a simple or 2-fid scale at the base of the blade. *Styles* and *carpels* opposite the sepals.—DISTRIB. N. temp. hemisphere ; species 30.—ETYM. λύχνος, from the *flame-like* floweis of some species.—Styles rarely 3-4, when the species may be referred with equal justice to *Silene.*

* *Petals 4-cleft. Capsule 5-toothed, without septa.*

1. L. Flos-cucu'li, *L.* ; flowers in loose dichotomous cymes, petals 4-cleft. *Ragged Robin.*

Moist meadows, copses, &c., N. to Shetlands ; ascends to near 2,000 ft. in the Highlands ; Ireland ; Channel Islands ; fl. May–June.—Glabrous. *Rootstock* slender. *Stem* 1–2 ft., roughish above. *Radical leaves* petioled, oblong-lanceolate, acuminate, cauline narrow. *Flowers* drooping, pedicels slender, honeyed, proterandrous. *Calyx* ½ in., veins purple ; teeth acuminate. *Petals* rosy, rarely white, segments linear ; scales long, 2-fid. *Capsule* broadly ovoid, very shortly pedicelled.—DISTRIB. Europe (Arctic), Siberia.

** *Petals notched or 2-fid. Capsule 5-toothed, with incomplete septa.*

2. L. Visca'ria, *L.* ; stem viscid at the nodes, petals notched.

Trap rocks local ; N. Wales ; Mid. and S. Scotland ; fl. June–Aug.—Glabrous, stout, 6–10 in. *Rootstock* perennial, woody. *Radical leaves* 3–5 in., very narrow-lanceolate ; petiole downy at the margins. *Flowers* very contracted, panicled, few-flowered. *Flowers* almost sessile. *Calyx* ½ in., membranous, purple, dilated upwards ; teeth short, acute. *Petals* obovate, red-purple ; scales short. *Capsule* broadly ovoid ; pedicel slender, ½ as. long as the capsule.—DISTRIB. Europe (excl. Spain and Greece) to the Caucasus, Siberia.

3. L. alpi'na, *L.* ; tufted, not viscid, cymes compact, petals 2-lobed.

Alpine moors and ravines ; Cumberland, Lancashire, and Clova Mts., ascending to 3,200 ft. ; fl. June–July.—Glabrous 4–8 in. *Rootstock* short, much branched. *Leaves* 1–2 in., crowded, narrow, linear-lanceolate. *Flowers* nearly ½ in. diam., shortly pedicelled, proterandrous ; bracts red. *Calyx* nerves faint, teeth rounded. *Petals* rosy. *Capsule* ovoid, pedicel half its length.—DISTRIB. Alps, Pyrenees, and Arctic regions.

*** *Petals 2-partite. Capsule 10-toothed, septa 0. Flowers subdiœcious.*

4. L. diur'na, *Sibth.* ; calyx reddish teeth triangular acute, petals red, capsule subglobose, teeth recurved. *L. dioi'ca a,* L. *Red Campion.*

Damp copses and hedgebanks, N. to Shetlands ; Ireland ; Channel Islands ; fl. June–July.—Softly hairy, rarely quite glabrous, viscid above. *Rootstock* slender, branched. *Radical leaves* 3–6 in., obovate, petioled ; cauline narrower. *Flowering stem* 1–3 ft., erect. *Flowers* in loose dichotomous cymes. *Calyx* ½ in., subcylindric, reddish, rarely green. *Petals* red, rarely white, lobes oblong ; scales lanceolate. *Capsule* mouth wide ; pedicel very short.-- DISTRIB. Europe (Arctic) to the Caucasus, Siberia to Baikal, Greenland.

5. L. vesperti'na, *Sibth.* ; calyx greenish, teeth elongate, petals white, capsule conical, teeth short linear-lanceolate erect. *L. dioi'ca β,* L. ; *L. al'ba,* Mill. (the earliest name). *White Campion.*

Fields, hedgerows, &c., to Orkneys; Ireland ; Channel Islands; fl. June–Sept.
—Very similar to *L. diur'na. Flowers* rarely reddish, open and fragrant in
the evening.—DISTRIB. Europe, N. Africa, Siberia, W. Asia; introd. in
U. States.

4. GITHA'GO, *Desfontaines.* CORN-COCKLE.

Characters of *Lychnis,* but calyx coriaceous with foliaceous teeth, and
entire petals without scales at the base of the blade. *Flowers* honeyed,
proterandrous. *Styles* and *carpels* opposite the petals.-—DISTRIB. Europe,
Siberia, W. Asia to Persia ; introd. in U. States, species 1.—ETYM.
obscure.

G. seg'etum, *Desf.* ; flowers solitary, calyx woolly segments much
longer than the petals. *Agrostem'ma Githa'go,* L.

Cornfields N. to the Orkneys; Ireland ; Channel Islands; (a colonist, *Wats.*) :
fl. June–Aug.—Annual ; clothed with dense white hairs. *Stems* 1–2 ft.
Leaves 2–5 in., linear-lanceolate. *Flowers* 1½–2 in. diam.; pedicels long.
Calyx 1–1½ in., cylindric-ovoid, ribs strong. *Petals* pale purple, limb obovate. *Capsule* ovoid.—Cosson regards this as a quasi-cultivated form, of
which the type is the Anatolian *A. gra'cilis,* Boiss.

5. HOLOS'TEUM, L.

Annual herbs, viscid and glandular. *Leaves* narrow. *Flowers* in terminal umbel-like cymes. *Sepals* 5. *Petals* 5, toothed or notched. *Stamens* 3–5, rarely 10. *Ovary* 1-celled ; styles 3, rarely 4 or 5 ; ovules
many. *Capsule* subcylindric, with twice as many short terminal valves
as there are styles. *Seeds* peltate, concavo-convex, rough ; embryo
horseshoe-shaped. — DISTRIB. Europe, W. Asia ; species 3.— ETYM.
doubtful.

H. umbella'tum, *L.* ; lower leaves petioled elliptic-oblong.

Old walls and thatched roofs, very rare, Norwich, Eye, and Bury ; (a denizen
or native ? *Wats.*) ; fl. April–May.—*Stem* 4–8 in., very slender, branched at
the base. *Radical leaves* ½–1 in.; cauline very few, sessile, ovate or linear.
Flowers few, erect ; pedicels ½ in., deflexed after flowering, erect after fruiting ; bracts small, membranous. *Sepals* white, edges scarious, obtuse. *Petals*
¼ in., a little longer than the sepals, white or pale pink. *Stamens* and styles
often 3 each. *Capsule* twice as long as the sepals. *Seeds* black.—DISTRIB.
Europe, N. Africa, W. Asia to N.W. India.

6. CERAS'TIUM, *L.* MOUSE-EAR CHICKWEED.

Pubescent rarely glabrous herbs, the hairs articulate, some glandular,
others not. *Leaves* small. *Flowers* white, in terminal dichotomous cymes,
proterandrous. *Sepals* 5, rarely 4. *Petals* as many, rarely 0, notched
or 2-fid, rarely quite entire or deeply cut. *Disk* of 5 honeyed glands.
Stamens 10, 5, or fewer. *Ovary* 1-celled ; styles usually 3, when 5,
opposite the sepals ; ovules many. *Capsules* cylindric, often incurved,
with twice as many short terminal valves as styles. *Seeds* compressed,

often tubercled ; embryo annular.—DISTRIB. All temp. and cold regions ; species 40.—ETYM. κέρας, from the *horn*-like capsule.

SECTION 1. **Mœnch'ia,** *Ehrh.* (gen.). *Sepals* acuminate, longer than the entire petals.

1. **C. quarternel'lum,** *Fenzl ;* glabrous, glaucous, stamens 4. *Mœnch'ia erec'ta,* Ehrh.

Gravelly pastures, &c., from the Cheviots southd.; ascends to 1,200 ft. in Wales ; Ireland ; Channel Islands; fl. May–June.—Annual. *Stems* 2–6 in., dichotomously branched from the base; branches slender, stiff. *Leaves* 1 in., radical sublanceolate ; cauline few, shorter, broader. *Flowers* few ; pedicels long, erect, stiff. *Sepals* 4, ¼ in.; margins broad, membranous, white. *Petals* 4,shorter than the sepals, oblong. *Styles* 4, short. *Capsule* subcylindric, as long as the sepals, 8-toothed.—DISTRIB. W. Europe from Holland to Hungary, N. Africa; introd. in the U. States.

SECTION 2. **Ceras'tium** proper. *Petals* notched 2-fid, or erose.

* *Annual rarely perennial, hairy and viscid except* C. triviale. *Sepals* 4–5, *about as long as the petals.* (Perhaps all sub-species of one.)

2. **C. tetran'drum,** *Curtis ;* sepals viscid, pedicels usually erect when fruiting 2–3 times as long as the capsule, bracts herbaceous, sepals 4 rarely 5 lanceolate, glandular margins narrowly membranous, capsule straight. *C. atrovi'rens,* Bąb.

Sandy and waste places usually near the sea, N. to the Shetlands; Ireland ; Channel Islands ; fl. April–Oct.—*Stem* 4–12 in. dichotomously branched from the base. *Radical leaves* obovate-lanceolate, cauline usually broader upwards. *Cymes* leafy. *Flowers* ¼ in. diam. *Petals* notched, veins branched. *Capsule* scarcely longer than the sepals.—DISTRIB. W. Europe from Sweden to Spain and eastwards to Hungary.

VAR. *C. pu'milum,* Curtis ; branching from above the middle, upper bracts with narrow membranous margins, petals notched, veins branched, fruiting pedicel short curved, capsule curved. *C. glutino'sum,* Fries.—Dry banks, rare, Worcester to Devon, Surrey.

3. **C. semidecan'drum,** *L.* ; pedicels a little exceeding the sepals deflexed between flowering and fruiting, bracts half-membranous, sepals usually 5 glandular acute margins broadly membranous, capsule slightly curved.

Walls and banks, N. to Shetland ; ascends to 1,000 ft. in Yorkshire ; Ireland ; Channel Islands; fl. March–May.—*Stem* 1–10 in., erect or decumbent, sometimes nearly glabrous, branched from the base. *Leaves* as in *C. tetrandrum.* *Cymes* few- or many-fld. *Petals* erose, with simple veins, shorter than the sepals. *Stamens* 4–5, or 10. *Capsule* exserted.—DISTRIB. Europe, N. Africa.—The earliest to flower.

4. **C. glomera'tum,** *Thuillier ;* fruiting pedicels suberect shorter than the sepals, bracts all herbaceous, sepals acute with few glands and narrow membranous margins, petals as long as the sepals 2-fid rarely 0, capsule twice as long as the sepals cuived. *C. vulga'tum,* L., and *visco'sum,* L., in part.

Dry places, N. to Shetland ; ascends to 1,000 ft. ; Ireland ; Channel Islands ; fl. April–Sept.—Habit of the preceding but usually larger, less glandular and cymes more fascicled, at first subcapitate.—DISTRIB. Europe (Arctic), N. Africa, W. Asia to the Himalaya, Greenland ; introd. in U. States.

5. C. trivia'le, *Link ;* pedicels longer than the sepals reflexed between flowering and fruiting, primary bracts wholly herbaceous, margins of secondary sometimes membranous, sepals obtuse margins broad membranous, petals 2-fid, capsule twice as long as the sepals curved. *C. visco'sum,* L., of Sm. and Hook. and Arn. *C. vulgatum,* Fries.

Waste places, N. to Shetland ; ascends to 3,600 ft. in Scotland ; Ireland ; Channel Islands ; fl. April–Aug.—Similar to *C. glomera'tum,* but usually perennial, often with leafy barren shoots and lax cymes. *Flowers* proterandrous.—DISTRIB. Europe from the Arctic circle southd., N. and W. Asia, the Himalaya, N. Africa ; introd. in U. States.

C. trivia'le proper ; perennial, decandrous, hairs not glandular, sepals pubescent.—VAR. *C. holosteoi'des,* Fries ; stem sparingly pubescent, leaves dark smooth shining, flowers large. Tidal rivers, Newcastle, Wigton, Perth.
—VAR. *pentandrum ;* annual, pentandrous, capsule shorter, sepals as in *trivia'le.* Sea-shores.—VAR. *alpestre,* Lond. Cat. (*alpinum,* Koch) ; dwarf, flowers much larger. Scotch Mts.

** *Perennial, downy or woolly. Petals 5, twice as long as the sepals, 2-fid.*

6. C. arven'se, *L.* ; stems hairy all round, leaves linear-lanceolate, bracts and sepals subacute, margins and tip membranous, seeds acutely tubercled.

Sandy fields and waste places, from Inverness southd. ; rarer in Scotland ; local in Ireland ; fl. April–Aug.—*Branches* 6–10 in., tufted, ascending. *Leaves* crowded on the basal shoots. *Cymes* many-fld. *Sepals* oblong-lanceolate, glandular. *Capsule* inclined, a little longer than the sepals, pedicel erect.—DISTRIB. Europe (Arctic), N. Africa, Siberia, W. Asia to the Himalaya, N. America, Fuegia, Chili.
VAR. *pubes'cens,* Syme ; leaves soft and pubescent, cymes 3–10-flowered.—
VAR. *Andrew'sii,* Syme ; leaves rigid glabrescent, midrib strong beneath, flowers subsolitary.

7. C. alpi'num, *L.* ; stems hairy all round, leaves ovate or oblong-ovate obtuse pubescent, bracts herbaceous obtuse, sepals obtuse with a membranous margin, seeds tubercled.

Alpine and subalpine rocks, Westmoreland, Wales, and Scotland ; ascends to near 4,000 ft. ; fl. June–Aug.—Habit of *C. arvense,* but leaves much broader, flowers fewer and much larger, ¾–1 in. diam., and capsule almost twice as long as the sepals on a spreading pedicel.—DISTRIB. Mountains of Europe, N. America, and all Arctic regions.
C. alpi'num proper ; villous, cymes 1–several-fld., sepals faintly margined, seeds small subacutely tubercled.—VAR. *Smith'ii ;* loosely tufted, less hairy and glandular, leaves broader, flowers subsolitary, seeds larger obtusely tubercled. *C. latifo'lium,* Sm. not L. *C. lana'tum,* Lamk.—VAR. *Edmondstone'i,* Wats. (*nigres'cens,* Syme) ; deep green, leaves broader. Unst, in Shetland.—The true *C. latifo'lium,* L., of the Alps, with globose capsules and inflated testa, is not British.

8. **C. tri'gynum,** *Villars ;* stem with alternating hairy lines, leaves small narrow oblong-lanceolate glabrous, cymes 1-3-fld., bracts glandular or glabrous, margins broadly membranous. *Stella'ria cerastoi'des,* L. Alpine and subalpine rills, rare, Mid. Scotland and Ireland ; ascends to 3,700 ft. ; fl. July-Aug.—A smaller and more delicate species than the two preceding alpine ones, nearly glabrous. *Leaves* ¼-½ in., distant, obtuse, often recurved. *Flowers* ½ in. diam., homogamous ; pedicels very slender. *Sepals* linear-oblong, spreading, 1-nerved. *Petals* deeply 2-fid. *Styles* usually 3. *Capsule* longer than the sepals.—DISTRIB. Alps of Europe, N. and W. Asia to the Himalaya, Arctic regions.—Intermediate between *Cerastium* and *Stellaria.*

7. STELLA'RIA, *L.* STITCHWORT.

Slender, usually glabrous herbs. *Leaves* narrow or broad. *Flowers* in dichotomous cymes, white, small, honeyed, proterandrous. *Sepals* 5, rarely 4. *Petals* 5, rarely 4, 2-fid or 2-partite. *Stamens* 10 (rarely 8 5 or 3) more or less perigynous, rising and dehiscing in 2 sets. *Disk* annular elongating, or of 5 interstaminal glands. *Ovary* 1-celled ; styles 3, or 5 and opposite the petals, ovules many. *Capsule* short, splitting below the middle into as many simple or 2-fid valves as there are styles. *Seeds* compressed, granulate ; embryo annular.—DISTRIB. All temp. and cold regions ; species 70.—ETYM. *Stella,* from the *star*-like flowers.

SECTION 1. **Mala'chium,** *Fries* (gen.). *Sepals* free to the base. *Styles* 5, rarely 3. *Capsule* with 5 2-fid valves.

1. **S. aquat'ica,** *Scopoli ;* stems diffuse decumbent angular slightly glandular above, leaves ovate=cordate. *Cerastium aquaticum,* L.

Borders of ditches, streams, &c., from York southd. ; fl. July-Aug.—Perennial. *Stem* 1-3 ft., brittle, branched, trailing over bushes. *Leaves* 1-1½ in., membranous, lower shortly petioled, acute, sometimes ciliate. *Flowers* ½ in. diam., axillary. *Sepals* lanceolate, enlarged in fruit. *Petals* white, lobes diverging. *Capsule* 1 in., ovoid, a little longer than the sepals ; pedicel deflexed, tip curved.—DISTRIB. Europe, N. Africa, N. and W. Asia.

SECTION 2. **Stella'ria** proper. *Sepals* free to the base. *Stamens* subperigynous. *Styles* 3. *Capsule* with 6 entire valves.

2. **S. nem'orum,** *L.* ; stem ascending glabrous or hairy all round, leaves ovate acuminate, lower subcordate long-petioled, upper cauline sessile, cymes lax, petals longer than the sepals.

Shady places from Dumbarton and Moray to S. Wales and Hereford ; ascends to 2,700 ft. ; Channel Islands ? fl. May-Aug.—Glabrous or pilose with jointed hairs and slightly glandular. *Stem* 1-2 ft., stout, terete, brittle, shining. *Leaves* 1-3 in., membranous, ciliate. *Flowers* ½-¾ in. diam. ; pedicels very slender. *Sepals* lanceolate, obtuse, margins scarious. *Capsule* ovoid, as long as the sepals ; pedicels spreading or reflexed.—DISTRIB. Europe (Arctic), excl. Greece and Turkey.

3. **S. me'dia,** *Vill.* ; stem procumbent with a line of hairs, leaves ovate acuminate, lower petioled, upper sessile, cymes many-flowered, petals shorter than the glandular sepals sometimes 0. *Chickweed.*

Cultivated and waste ground, N. to Shetland; ascends to 2,700 ft.; Ireland; Channel Islands; fl. March–Oct.—One of the commonest and most variable of plants, 6–18 in., easily recognized by the line of hairs on the stem and branches. *Stamens* 3, 5, or 10.—Distrib. All Arctic and N. temp. regions; naturalized elsewhere.

S. me'dia proper; pedicels pubescent, sepals hairy, petals 5, variable or 0, stamens 5 (3 in *S. Boræa'na*, Jord., and *Als.pal'lida*, Dum.) (10 in *S. neglec'ta*, Weihe), seeds obtusely tubercled

Sub-sp. S. umbro'sa, Opitz.; more erect, leaves more acuminate, pedicels glabrous, sepals lanceolate glabrous with raised points, seeds acutely tubercled. *S. Elizabe'thæ*, Schultz.—From Perth southd.

4. **S. Holos'tea,** *L.* ; stem suberect 4-angled, angles rough, leaves sessile connate lanceolate acuminate ciliate, petals twice as long as the almost nerveless sepals.

Copses, hedgerows, &c., from Caithness southd.; ascends to 1,900 ft. in the Highlands; Ireland; Channel Islands; fl. April–June.—Perennial. *Stem* 1–2 ft., decumbent at the base, brittle at the nodes, hairy above. *Leaves* 1–4 in., rigid. *Flowers* ½–¾ in. diam., white, pedicels slender. *Capsule* globose.—Distrib. Europe, W. Asia.—Flowers sometimes double; petals occasionally laciniate.

Section 3. **Lar'brea,** *St. Hilaire* (gen.). *Sepals* united at the base into a conical tube. *Stamens* very perigynous. *Styles* 3. *Capsule* 6-valved.

5. **S. palus'tris,** *Ehrh.* ; glaucous, glabrous, stem suberect 4-angled, leaves very narrow sessile margins even, peduncles very long axillary, petals longer than the 3-nerved sepals. *S. glau'ca*, With.

Marshy places, not uncommon from the Clyde and Forth to Surrey and Dorset; rare in Ireland; Channel Islands; fl. May–July.—Perennial. *Stems* 1–2 ft., very slender. *Leaves* 1–2 in., oblong-lanceolate, or linear-oblong. *Flowers* ½–¾ in. diam., few, distant. *Bracts* membranous. *Sepals* lanceolate, acute, margins broadly scarious. *Capsule* ovoid, as long as the sepals; pedicel spreading.—Distrib. Europe, Siberia, W. Asia to the Himalaya, Greenland.

6. **S. gramin'ea,** *L.* ; glabrous, stem suberect 4-angled, leaves very narrow sessile ciliate, cymes branched, petals equalling the 3-nerved sepals.

Dry pastures, hedgebanks, &c., N. to Shetland; ascends to 1,000 ft. in Yorks.; Ireland; Channel Islands; fl. May–Aug.—Perennial, not glaucous *Stem* 1–3 ft. *Leaves* as in *S. glauca*, but ciliate. *Flowers* ½–¾ in. diam., many. *Bracts* scarious, ciliate. *Pedicels* reflexed after flowering, then spreading. *Sepals* acute. *Capsule* ovoid, nodding, a little longer than the sepals.—Distrib. Europe (Arctic), Siberia, W. Asia to the Himalaya.—*S. scapig'era*, Willd., said to be found by Don in Perth and Inverness-shire, is a cultivated abnormal form, with short stems, imbricate leaves, long erect solitary peduncles, and small flowers.

7. S. uligino'sa, *Murr.* ; glaucous, nearly glabrous, stem 4-angled, leaves oblong or ovate-lanceolate, cymes few-fld., petals shorter than the acuminate sepals.

Wet places, N. to Shetland; ascends to 3,300 ft. in the Highlands; Ireland; Channel Islands; fl. May–July.—Perennial, 3–18 in., erect or diffuse, variable in size, habit, and breadth of leaves, 1- or more-fld., glabrous, or with a few hairs at the bases of the leaves, which are narrowed at both ends, and callous at the tip. *Bracts* scarious. *Flowers* ¼ in. diam. *Tube of calyx* funnel-shaped. *Capsule* ovoid. *Seeds* minute.—DISTRIB. Europe (Arctic), N. Africa, N. and W. Asia to the Himalaya, N. America.

8. ARENA'RIA, *L.* SANDWORT.

Annual or perennial herbs, often tufted. *Leaves* broad or narrow. *Flowers* white or pink, in dichotomous cymes. *Sepals* 5. *Petals* 5, entire or slightly notched, rarely 0. *Stamens* 10, rarely 5, inserted on the disk. *Disk* annular, or of inter-staminal honeyed glands. *Ovary* 1-celled ; styles 3-4 ; ovules many, rarely few. *Capsule* short, with as many entire or 2-fid valves as there are styles. *Seeds* compressed, smooth or tubercled ; embryo annular.—DISTRIB. all temp. and cold regions ; species 130.— ETYM. *Arena,* from many growing in *sand.*

SECTION 1. **Alsi'ne,** *Wahl.* (gen.). *Flowers* hermaphrodite. *Disk* annular. *Capsule* with 3–4 entire valves. *Seeds* many, funicle not swollen or appendaged.—Leaves linear-setaceous in all the British species.

1. A. ver'na, *L.* ; densely tufted, leaves crowded subulate, flowering branches slender few-fld., oblong petals and capsule rather longer than the lanceolate sepals.

Dry rocks, pastures and banks, N. to Shetland, local; ascends to 2,500 ft. in the Highlands; Ireland; fl. May–July.—Perennial, bright green, sparingly hairy and glandular. *Rootstock* woody; branches 2-4 in, densely tufted, forming a green cushion. *Leaves* ¼-½ in., 3-nerved. *Bracts* acute, margins scarious. *Flowering* branches strict. *Flowers* ½ in. diam., white, proterandrous ; pedicels slender, glandular. *Petals* oblong, hardly longer than the sepals.—DISTRIB. Mid. and S. Europe, N. Africa, N. America. *A. ver'na,* proper ; leaves apiculate, lower not appressed.—VAR. *Gerar'di,* Wahlb.; leaves not apiculate, lower appressed.— Cornwall.

2. A. hir'ta, *Wormsk.* ; densely tufted, leaves crowded subulate obtuse, peduncles pubescent 1-flowered, petals lanceolate and capsule shorter than the acute 3-nerved sepals. *A. rubel'la,* Hook.

Rocky tops of Breadalbane Mts. and Ben Hope, alt. 2,500 to 4,000 ft., very rare; fl. July–Aug.— General character of *A. verna,* of which it may be an Arctic sub-species, but of laxer habit, smaller in all its parts, yellow-green and purplish ; leaves more flaccid and obtuse; flowers usually solitary ; petals shorter; styles commonly 4; seeds smaller and more orbicular.— DISTRIB. Arctic regions.

3. A. uligino'sa, *Schleich.* ; stems loosely tufted ascending, peduncles filiform 1 3-flowered, leaves subulate semiterete obtuse nerveless, petals

oblong as long as the ovate acute 3-nerved sepals.   *Sper'gula stric'ta,*
Swartz ; *Alsi'ne stric'ta,* Wahl.

Banks of a rill, Widdy-bank Fell, Teesdale, alt. 1,800 ft.; fl. June–July.—
Perennial, glabrous, 2–3 in., very slender, habit of *Sagina.   Leaves* ¼–½ in.,
curved, upper pairs few and distant. *Peduncles* 1–2 in. *Flowers* ½ in. diam.,
white. *Capsule* ovoid. *Seeds* reniform, rugose on the disk.—DISTRIB. W.
Europe (Arctic), Lapland to Italy, Greenland.

4. **A. tenuifo'lia,** *L.* ; erect, very slender, leaves subulate acute 3–5-
nerved, cymes many-fld., petals oblong half as long as the lanceolate
3-nerved sepals, capsule 3-valved equalling the sepals or longer.

Sandy fields and waste places from York southd., chiefly in the E. counties;
Channel Islands; fl. June–Aug.—Annual, 2–8 in., simple or branched.
Leaves crowded below, upper pairs remote. *Flowers* ½ in. diam., white.—
DISTRIB. Europe, N. Africa, Siberia, W. Asia to India.
A. tenuifo'lia proper; glabrous, stamens 10, capsule equalling the sepals.—
VAR. *A. lax'a,* Jord.; calyx glandular, stamens 5, capsules longer than the
sepals. Great ₁Wilbraham.—VAR. *hyb'rida,* Vill. (*A. visco'sa,* Schreb.);
peduncles and sepals glandular, stamens 8–10, capsule broader at the base.
Thetford.

SECTION 2. **Arena'ria** proper. *Flowers* hermaphrodite. *Disk* annular.
Capsule with 3 2-fid valves. *Seeds* many.—Leaves broad in all the British
species.

5. **A. triner'via,** *L.* ; diffuse, pubescent, leaves petioled ovate acute
3–5-nerved ciliate, flowers solitary or cymose, sepals obscurely 3-ribbed,
seeds smoothed arilled. *Mœhrin'gia,* Clairv.

Moist copses, hedgebanks, &c , N. to Ross; Ireland ; Channel Islands; fl.
May–July.—Annual, branched, flaccid ; branches 4–18 in. *Leaves* ½–1 in.
Flowers ¼ in. diam., rarely 5-androus, proterogynous; pedicels long, slender.
Sepals lanceolate, longer than the obovate-lanceolate petals, middle nerve
hairy. *Capsule* subglobose, shorter than the sepals.—DISTRIB. Europe
(Arctic), Canaries, Siberia, W. Asia, Greenland.

6. **A. serpyllifo'lia,** *L.* ; decumbent or suberect, pubescent, leaves subsessile ovate acuminate 1–3-nerved ciliate, cymes many-flowered, bracts
foliaceous, sepals with 3–5 hairy ribs.

Wall-tops, &c., N. to Orkneys, ascends to 2,000 ft.; Ireland ; Channel Islands;
fl. June–Aug.—Annual, very variable in habit, grey-green, branched, hairs
recurved on the stem and peduncles. *Leaves* ¼–½ in., shortly petioled,
rather rigid. *Flowers* ⅛ in. diam. *Sepals* with narrow margins, longer than
the petals. *Seeds* rough, shining, not arilled.—DISTRIB. Europe (Arctic), N.
Africa, N. and W. Asia to the Himalaya; introd. in the U. States.
A. serpyllifo'lia proper (*A. sphærocar'pa,* Tenore); rigid, sepals ovate-lanceolate, capsule ovoid, its pedicel ascending.—VAR. *glutino'sa,* Koch; shorter,
stouter, more glandular, capsule more swollen below. Isle of Wight.—VAR.
A. leptocla'dos, Guss.; weak, sepals lanceolate, capsule narrower, its pedicel
spreading.

7. **A. cilia′ta,** *L.* ; procumbent or ascending, leaves oblong-spathulate obtuse 1-nerved, flowers subsolitary, bracts foliaceous, sepals oblong-lanceolate with 3 hairy ribs much shorter than the petals.

Ireland, Orkney and Shetlands; fl. June–July.—Perennial, dark green, hairs reflexed. *Stems* 3–6 in., numerous, matted, tips ascending. *Leaves* ¼–¾ in., petioled. *Flowers* nearly ½ in. diam. *Sepals* subacute, margins membranous. *Petals* spathulate. *Capsule* ovoid, as long as the sepals. *Seeds* not arilled. — DISTRIB. Arctic and Alpine Europe, E. to Crete.

A. CILIA′TA proper; pubescent, leaves ciliate, ribs of sepals hairy.—Sligo Mountains, alt. 1,000–1,700 ft.

Sub-sp. A. NORVE′GICA, *Gunn.*; almost glabrous, leaves denser shorter broader more fleshy, peduncles shorter, ribs of sepals glabrous.—Shetlands and Orkney.

SECTION 3. **Ammode′nia,** *Gmel.* (gen.). *Flowers* polygamous. *Disk* glandular, 10-lobed. *Capsule* fleshy, usually 3-valved. *Seeds* 1–2, large. *Honcke′nya,* Ehrh.

8. **A. peploi′des,** *L.* ; creeping, fleshy, leaves ovate acute recurved, flowers 1–3 together axillary subsessile, sepals obtuse. *Sea Purslane.*

Sandy and pebbly shores, N. to Shetland; Ireland; Channel Islands; fl. May-Aug.—Perennial, dark green, glabrous. *Rootstock* creeping; branches 4–8 in., ascending. *Leaves* ¼–½ in., decussate, margins cartilaginous. *Flowers* ¼ in. in diam., pedicels compressed. *Sepals* with membranous margins. *Petals* of male fl. as long as the sepals, of female shorter. *Stamens* 10, the alternate shorter. *Styles* 3–5. *Capsule* globose. *Seeds* obovoid, concavo-convex.—DISTRIB. W. Europe from the Arctic regions to Spain, Arctic America.—Used as a pickle in Yorkshire.

SECTION 4. **Cherle′ria,** *L.* (gen.). *Flowers* polygamous. *Sepals* united at the base. *Petals* 0 or minute. *Disk* with 5 large glands. *Seeds* few, minute, smooth.

9. **A. Cherle′ri,** *Benth.* ; densely tufted, leaves closely imbricate linear-subulate 3-gonous, flowers solitary, sepals obtuse. *Alsine Cherleri,* Fenzl. *Cherleria sedoides,* L. Cyphel.

Lofty Scotch mountains, alt. 2,500–3,000 ft.; fl. June–Aug.—Perennial, forming mossy, yellow-green cushions 6–12 in. diam., with a very long tap root. *Leaves* ¼–¾ in., obtuse, ciliate, grooved above. *Flowers* sessile, proterandrous. *Sepals* 3-nerved, margins membranous. *Petals* 0, or in the male fl. subulate. *Capsule* ovoid, shorter than the sepals.—DISTRIB. Alps of Central and S. Europe, Pyrenees.

9. SAGI′NA, *L.* PEARL-WORT.

Very small, tufted, annual or perennial herbs. *Leaves* subulate, connate at the base. *Flowers* small, solitary, pedicelled, proterandrous. *Sepals* 4–5. *Petals* 4–5, entire, sometimes minute or 0. *Stamens* 4, 5, 8, or 10. *Ovary* 1-celled; styles 4–5, opposite the sepals; ovules many. *Capsule* 4–5-valved to the base.—DISTRIB. Temp. and cold N. and S. hemispheres; species 8.—ETYM. doubtful.

F

* *Flowers 4- rarely 5-merous. Petals minute.*

1. **S. apet′ala,** *L.* ; annual, primary and lateral shoots all flowering, radical leaves sub-rosulate, petals minute or 0.

Dry banks, wall tops, &c.; fl. May–Aug.—A slender, wiry herb, 4–10 in. *Leaves* $\frac{1}{10}$–$\frac{1}{4}$ in. *Flowers* $\frac{1}{12}$ in. diam., green, pedicels capillary.—Distrib. Europe, N. Africa, W. Asia ; doubtfully indigenous in the U. States. S. apet′ala proper ; branches ascending, leaves ciliate at the base mucronate, pedicels erect, sepals at length spreading obtuse exceeding the capsule.— From Perth and Forfar southd.; rare in Scotland ; Ireland ; Channel Islands.
Sub-sp. S. cilia′ta, Fries ; decumbent, glandular-pubescent, leaves more or less ciliate mucronate, sepals always appressed to the capsule 2 outer mucronate. *S. ambigua*, Lloyd.—From Aberdeen southd., Ireland and Channel Islands ; rather rare.
Sub-sp. S. marit′ima, Don ; decumbent or ascending, glabrous, leaves obtuse or apiculate, sepals suberect in fruit broad obtuse.—*S. marit′ima* proper ; ascending, slender, internodes long, capsule about equal to the sepals. Sandy sea-shores, N. to Shetland, Ireland and Channel Islands.—Var. *S. deb′ilis*, Jord.; decumbent, slender, internodes long, capsule a little shorter than the sepals.—Var. *S. den′sa*, Jord.; tufted, slender, internodes short, capsule as in *deb′ilis*. Christchurch, Hants, and Wisbeach.—Var. *alpi′na*, Syme ; ascending, stoutish, internodes short, capsule shorter than the sepals. Top of Ben Nevis, *Don.*

2. **S. procum′bens,** *L.* ; perennial, stems many, primary shoot flower-less, lateral slender with fascicled branchlets usually procumbent and rooting, petals very small.

Waste places, paths, banks, &c.; ascends to 3,800 ft. in the Highlands; Ireland ; Channel Islands ; fl. May–Sept.—Branches 1–8 in. *Leaves* glabrous or ciliate (var. *spino′sa*, Bab.), obtuse, mucronate, longer than in *S. apet′ala*. *Flowers* usually solitary, sometimes 5-merous. *Sepals* spreading in fruit. *Styles* recurved during flowering. *Capsule* a little longer than the sepals, pedicels erect or curved at the tip.—Distrib. Europe (Arctic), N. Africa, N. and W. Asia to the Himalaya, Greenland, N. America, Fuegia.

** *Flowers 5- rarely 4-merous. Petals as long as the sepals or longer.* (*Perennial, with a leafy flowerless central stem, many lateral flowering branches, minute subulate leaves, and slender erect 1-flowered pedicels.* Spergella, *Reichb.*)

3. **S. Linnæ′i,** *Presl ;* glabrous, leaves mucronate, petals longer than the glabrous obtuse sepals, capsule 5-valved almost twice as long as the usually appressed sepals.

Alps of Perth and Forfar to Sutherland ; ascends to 2,700 ft.; fl. June–Aug.— Distinguishable from the 5-merous forms of *S. procum′bens* only by the longer white petals, erect styles, and usually longer capsules with appressed sepals.—Distrib. Arctic and Alpine Europe, Siberia, N. America. S. Linnæ′i proper ; branches many prostrate rooting, pedicels curved erect in fruit. *S. saxat′ilis*, Wimm. ;*Sper′gula saginoi′des*, Sm.
Sub-sp. S. niva′lis, Fries ; densely tufted, leaves broader, pedicels always erect, petals shorter.—Ben Lawers, Skye, and Clova Mountains.

4. **S. subula'ta**, *Presl ;* tufted, more or less glandular and hairy, leaves narrowed to the awned tip, petals as long as the lanceolate obtuse subglandular sepals, pedicels long curved after flowering then erect, capsule rather longer than the appressed sepals. *Sper'gula,* Swartz.

Heaths, dry pastures, &c.; ascending to 2,700 ft. in the Highlands ; N. and W. Ireland ; Channel Islands ; fl. June–Aug.—DISTRIB. Mid. and W. Europe, N. America (doubtfully indigenous, *Gray*).

5. **S. nodo'sa**, *E. Mey.* ; glabrous or glandular, leaf-buds many in the axils of the subulate acute leaves, peduncles short always erect 1–2-flowered, petals and capsules much longer than the oblong obtuse sepals. *Sper'gula,* L. *Knotted Spurrey.*

Moist heaths and sandy places, N. to Shetland ; ascends to 1,800 ft. in Yorks. ; Ireland ; Channel Islands ; fl. July–Aug.—Much the largest and largest-flowered species of the genus. *Branches* 4–10 in., decumbent, curved, wiry, rooting at the nodes. *Radical leaves* ½ in., cauline, usually ⅛–¼ in. *Flowers* ¼ in. diam., proterandrous.—DISTRIB. Europe (Arctic), N. Africa, Siberia, W. Asia to the Himalaya, N. America.

10. SPER'GULA, *L.* SPURREY.

Annual herbs, with forked or fascicled branches. *Leaves* opposite, with abbreviated leaf-buds in their axils, whence the foliage appears whorled ; stipules small, scarious. *Flowers* white, in peduncled cymes. *Sepals* 5. *Petals* 5, entire. *Stamens* 5 or 10. *Ovary* 1-celled ; styles 5, opposite the petals ; ovules many. *Capsule* with 5 entire valves. *Seeds* compressed, margined or winged ; embryo annular.—DISTRIB. Weeds of cultivation in temp. regions ; species 2–3.—ETYM. *spargo,* from *scattering* its seeds.

S. arven'sis, *L.* ; leaves linear-subulate semiterete rather fleshy.

Cornfields, &c., N. to Shetland ; Ireland ; Channel Islands ; ascends to 1,000 ft. ; fl. June–Aug.—More or less pubescent and glandular. *Stems* 5–18 in., branched from the root, geniculate. *Leaves* ½–1½ in., in distant pairs grooved beneath. *Flowers* ⅛–¼ in. diam., in terminal subumbellate cymes ; peduncles slender, spreading or reflexed. *Sepals* ovate, obtuse, rather shorter than the white petals. *Capsu'e* sub-globose.— DISTRIB. Europe (Arctic), N. Africa, W. Asia to N.W. India ; introd. in N. America.
S. arven'sis proper (*S. vulga'ris,* Bœnn.) ; seeds papillose, wing narrow or 0.—
VAR. *S. sati'va,* Bœnn. ; more viscid, seeds smooth or punctulate, winged.

11. SPERGULA'RIA, *Persoon.* SANDWORT-SPURREY.

Diffuse herbs, with the foliage and inflorescence of *Spergula. Stipules* membranous connate and surrounding the leaf-bases. *Sepals* 5. *Petals* 5, rarely 0, entire, white or red. *Stamens* 2–10. *Ovary* 1-celled ; styles 3. *Capsule* 3-valved. *Seeds* compressed, often winged ; embryo annular or hooked.—DISTRIB. Temp. and warm regions ; often littoral ; species 3–4. —ETYM. a derivative from *Sper'gula.*

The species are very variable, and may be regarded as sub-species of one.

68   *CARYOPHYLLEÆ.*   [SPERGULARIA.

1. **S. ru'bra,** *Pers.* ; annual or biennial, leaves linear flat acute, stipules cleft, capsule equalling the calyx, seeds plano-convex tubercled, margins thickened not winged. *Arenaria ru'bra,* L. *Lepig'onum ru'brum,* Fr.

Gravelly and sandy soils, from Ross southd.; rare in Ireland; Channel Islands; fl. June–Sept.—Pubescent and glandular above. *Stem* much branched from the base; branches 4–12 in., spreading, prostrate. *Leaves* ¼–½ in. *Stipules* connate, silvery, torn. *Flowers* ¼ in. diam., solitary or in subracemose cymes; pedicels short, spreading or reflexed, erect in fruit. *Petals* rosy, shorter than the obtuse lanceolate sepals. *Stamens* 5 or 10.— DISTRIB. Europe, N. and S. Africa, N. and W. Asia to India, America, Australia.

2. **S. sali'na,** *Presl;* annual or biennial, leaves ½-cylindric acuminate, stipules entire short, capsule longer than the calyx, seeds orbicular plano-concave smooth or papillar, margin thickened winged or not. *Lepig'onum sali'num,* Kindb.

Muddy and rocky places by the sea, N. to Shetland; Ireland; Channel Islands; fl. June–Aug.—More or less pubescent and glandular. *Branches* stout from a small rootstock, compressed. *Stipules* dark, deltoid. *Pedicels* equalling or longer than the capsule. *Petals* shorter than the calyx, rose. *Capsule* ¼ in. *Seeds* pale brown.—DISTRIB. Europe, N. and W. Asia, N. and S. Africa, N. America.

S. sali'na proper; seeds smooth, pedicels equalling the bracts.—VAR. *L. me'dium,* Fries; pedicels shorter than the leaf-like bracts, seeds nearly smooth.—VAR. *L. neglec'tum,* Kindb.; glandular above, upper pedicels longer than the scarious bracts, seeds papillose.

3. **S. me'dia,** *Pers.* ; perennial, usually glabrous, leaves ½-cylindric subacute, stipules usually entire, pedicels long, capsule twice as long as the calyx or less, seeds orbicular, smooth, margins thickened and broadly-winged. *S. mari'na,* Leb. *Arenaria marginata,* DC. *Lepig'onum sali'-num,* Wahl.

Muddy salt marshes, from Orkney southd.; fl. June–Aug.—*Rootstock* woody; branches stout, compressed. *Leaves* fleshy, stipules broad. *Flowers* ⅓ in. diam. *Petals* pale, as long as the sepals. *Capsule* ⅓ in. (the largest of the genus).—DISTRIB. Europe, N. and W. Asia, N. and S. Africa, America, Australia.

4. **S. rupes'tris,** *Lebel ;* perennial, glandular-pubescent, leaves ½-cylindric fleshy acute, stipules subentire, capsule equalling the calyx, seeds pyriform compressed, margin thickened not winged. *Lepig'onum rupes'tre,* Kindb. Suppl. *L. rupic'ola,* Kindb. (Bab.).

Rocky places near the sea, from Ross southd., not common; fl. June–Aug.— Habit and size of *S. me'dia,* but glandular, with very different seeds.— DISTRIB. France, Spain, Italy, and probably elsewhere in N. and S. temp. regions.

12. POLYCAR'PON, *L.*

Annual herbs. *Leaves* flat, opposite or whorled; stipules scarious. *Flowers* small, in crowded bracteate cymes. *Sepals* 5, keeled, entire.

Petals 5, small. *Stamens* 3–5. *Ovary* 1-celled ; style short, 3-fid ; ovules many. *Capsule* 3-valved.—DISTRIB. Various warm and temp. regions ; species 6.—ETYM. πόλυς and καρπός, from the *abundant capsules.*

P. tetraphyl'lum, *L.* ; lower leaves in whorls of 4, flowers 3-androus. Channel Islands, Cornwall, Devon and Dorset, in sandy and waste places ; fl. June–July.—*Stems* 3–6 in., prostrate. *Leaves* ½ in., obovate, upper opposite. *Flowers* ⅛ in. diam.—DISTRIB. Europe, Asia, Africa, &c.

ORDER XIII. PORTULA'CEÆ.

Herbs, rarely small shrubs. *Leaves* opposite or alternate, quite entire ; stipules scarious. *Inflorescence* various. *Sepals* 2, imbricate in bud. *Petals* 4 or more, distinct or united at the base, imbricate in bud. *Stamens* 4 or more, free or adnate to the petals, filaments filiform. *Disk* small or 0. *Ovary* usually free, 1-celled ; style simple or 3-fid, branches stigmatose all over ; ovules 2 or more, on long often connate basal funicles, amphitropal, ascending, micropyle inferior or transverse. *Capsule* dehiscing transversely or 2–3-valved. *Seeds* 1 or more, compressed, hilum marginal ; embryo terete, hooked or annular and coiled round the mealy albumen.—DISTRIB. Cosmopolitan, but chiefly American ; genera 15 ; species 125.—AFFINITIES. Close to *Caryophylleæ, Ficoideæ,* and *Mollugineæ.*—PROPERTIES. Purslane is a good salad and a potherb.

Petals united at the base. Stamens 3......................................1. Montia.
Petals distinct. Stamens 5...1*. Claytonia.

1. MON'TIA, *L.* BLINKS.

A small, annual, glabrous herb. *Leaves* usually opposite. *Flowers* minute, solitary or few and shortly cymose, white. *Petals* 5, hypogynous, connate at the base. *Disk* small, hypogynous. *Stamens* hypogynous, usually 3, opposite and attached to the base of the petals. *Ovary* free ; styles short, 3-fid ; ovules 3. *Capsule* globose, 3-valved. *Seeds* 1–3, compressed ; embryo annular.—DISTRIB. N. and S. temperate Arctic and cold regions.—ETYM. J. de Monti, an Italian botanist.

M. fonta'na, *L.* ; leaves spathulate, flowers drooping then erect. Brooks and marshes, N. to Shetland ; ascends to 3,200 ft. in the Highlands ; Ireland ; Channel Islands ; fl. May–Aug.—Pale-green, 1–5 in., usually flaccid, branched, tufted. *Leaves* ¼–¾ in., sub-opposite. *Flowers* ₁⁄₁₀ in. diam. *Bracts* scarious. *Petals* a little longer than the obtuse sepals. *Capsule* obovoid. *Seeds* shining.
VAR. *M. mi'nor,* Gmel. ; stem short, cymes terminal and axillary, tubercles of seed conical.—VAR. *M. rivula'ris,* Gmel. ; stem elongate flaccid, cymes all axillary, tubercles of seeds flattened.

1*. CLAYTO'NIA, *L.*

Glabrous succulent herbs. *Radical leaves* petioled, cauline alternate and opposite, exstipulate. *Flowers* in terminal cymes. *Petals* 5. *Stamens* 5,

opposite and adnate to the bases of the petals. *Ovary* free ; style entire or
3-fid at the tip ; ovules few. *Capsule* membranous, 3-valved.—DISTRIB.
America, N.W. Asia, Australia ; species 20.—ETYM. Dr. J. J. Clayton,
an American botanist.

1. C. PERFOLIA′TA, *Don ;* radical leaves rhomboid, cauline 2 connate.
A garden escape, rapidly becoming naturalized in many places ; fl. May–July.
—Annual, tufted, fleshy, 6–12 in. *Cauline leaves* connate into a suborbicular
blade. *Flowers* small, white.—DISTRIB. N.W. America.

2. C. ALSINOI′DES, *Sims ;* radical leaves ovate acuminate, cauline sessile
orbicular.
A garden escape, rapidly becoming naturalized near Glasgow and elsewhere ;
fl. May–July.—Annual. *Flowers* more numerous and much larger than in
C. perfoliata, and petals chiefly bifid.—DISTRIB. N.W. America.

ORDER XIII*. TAMARISCI′NEÆ.

Shrubs or small trees. *Leaves* very small, often scale-like, imbricate,
amplexicaul, exstipulate. *Inflorescence* of solitary or panicled axillary
spikes. *Sepals* 5, rarely 4, imbricate in bud. *Petals* 5, rarely 4, distinct
or connate below, imbricate in bud. *Stamens* 4, 5, 8, or 10 inserted on
the disk, distinct or connate below ; anthers versatile. *Disk* hypogynous
or slightly perigynous, 10-glandular. *Ovary* free, 1- or imperfectly 2–5-
celled ; styles 2–5, distinct or connate, or 2–5 sessile stigmas ; ovules 2 or
more, basal, erect, anatropous, raphe ventral, micropyle inferior. *Capsule*
2 5-valved. *Seeds* erect, usually more or less comose or winged, albumen
fleshy farinaceous or 0 ; embryo straight, cotyledons flat.—DISTRIB. Cold,
temp. and hot regions, often in sandy or saline places ; genera 5 ; species
40.—AFFINITIES. With *Caryophylleæ, Portulaceæ,* and *Frankeniaceæ.*—
PROPERTIES. *Tamarix* yields manna and galls, and its ashes soda.

TAM′ARIX, L. TAMARISK.

Sepals 4–5, distinct. *Petals* 4–5, distinct or connate at the base. *Stamens*
4, 5, 8, or 10. *Ovary* narrowed upwards ; styles 3–4, short, thick ; ovules
many. *Capsule* 3-valved. *Seeds* many, with a lateral and terminal pencil
of hairs, albumen 0 ; embryo ovoid.—DISTRIB. Of the Order ; species 20.
—ETYM. The Tamaris, a river of Spain, where Tamarisk abounds.

T. GAL′LICA, *L.* ; glabrous, disk acutely 5-angled. *T. anglica,* Webb.
S. and E. coasts of England, and Channel Islands, planted ; fl. July–Sept.—
An evergreen shrub or small tree, 5–10 ft. *Branchlets* excessively slender
and feathery. *Leaves* on the branchlets extremely minute, closely imbricate,
triangular, auricled, keeled ; on the older wood much larger, ⅛ in., subulate.
Flowers ⅛ in. diam., white or pink, in catkin-like obtuse spikes 1 in.
Sepals lanceolate. *Petals* persistent. *Anthers* apiculate. *Capsule*
3-gonous.—DISTRIB. Shores of Atlantic and Mediterranean, W. Asia to
N.W. India.

ORDER XIV. **ELAT'INEÆ.**

Herbs, often minute, or under-shrubs. *Leaves* opposite or whorled, entire or serrate, stipulate. *Flowers* small, axillary, solitary or cymose. *Sepals* and *petals* each 2–5, distinct, imbricate in bud. *Stamens* 2–5, or twice as many, hypogynous, distinct; anthers versatile. *Ovary* free, cells and styles 2–5; stigmas capitate; ovules many, on the inner angles of the cells, anatropous, raphe lateral or ventral. *Capsule* septicidal; valves flat concave or inflexed, separating from the axis and septa. *Seeds* straight or curved, raphe on the concave side, testa often rugose, albumen scanty or 0; embryo cylindric, straight or curved, cotyledons small.— DISTRIB. Scattered over the globe; genera 2; species 20.—AFFINITIES. With *Caryophylleæ* and *Hypericineæ.*—PROPERTIES. Supposed to be acrid.

1. ELAT'INE, *L.* WATERWORT.

Very small, submerged, creeping, glabrous herbs. *Leaves* spathulate. *Flowers* minute, axillary. *Sepals* 2–4, membranous. *Petals* 2–4. *Ovary* globose. *Capsule* membranous; septa evanescent after bursting, or adhering to the axis. *Seeds* cylindric, straight or curved, ridged and pitted. —DISTRIB. Temp. and sub-trop. regions; species 6.—ETYM. obscure.

1. **E. hexan'dra,** *DC.* ; flowers pedicelled 3-merous, capsule turbinate, seeds 8–12 in each cell straight ascending. *E. tripet'ala,* Sm.

Margins of ponds and lakes, rare, from Perth to Surrey and Cornwall (not in E. counties), ascends to 1,617 ft. in the Highlands; N. and W. Ireland; fl. July– Sept. *Stems* 1–3 in., matted, flaccid, rooting at the nodes. *Leaves* ⅛–½ in., spathulate. *Flowers* 1/16 in. diam., alternate, axillary. *Sepals* unequal. *Petals* pink, longer than the sepals.– DISTRIB. Europe, from Norway, southd. to Spain, Lombardy, and Hungary, Azores.

2. **E. Hydropi'per,** *L.* ; flowers sessile 4-merous, capsules subglobose, seeds 4 in each cell hooked pendulous.

Muddy ponds, very rare; Surrey, Worcester, and Anglesea; Lough Neagh, and Lagan Canal, Ireland; fl. July–Aug.—Very similar to *E. hexan'dra.* – DISTRIB. Europe to S. Russia (excl. Spain, Greece, Turkey, and Denmark).

ORDER XV. **HYPERICI'NEÆ.**

Herbs, shrubs, or trees. *Leaves* opposite, often covered with pellucid glands, entire or glandular-toothed, exstipulate. *Flowers* terminal, cymose, rarely axillary. *Sepals* 5, rarely 4, imbricate in bud. *Petals* as many, hpyogynous, usually twisted in bud. *Stamens* many, rarely few, more or less connate in bundles; anthers versatile. *Disk* obscure or of interstaminal glands. *Ovary* of 3–5 carpels, 1- or 3-5-celled; styles as many, filiform, stigmas terminal; ovules few or many, on parietal or axile placentas, anatropous, raphe lateral or superior. *Fruit* a septicidal capsule, rarely a berry. *Seeds* exalbuminous ; embryo straight or curved.—DISTRIB. Temp. and mountains of warm regions ; genera 8 ; species 210.—AFFINI-TIES. Close with *Guttiferæ* and *Ternstrœmiaceæ*, less close with *Elatineæ*. PROPERTIES. Drastic purgatives, astringents, and tonics.

1. HYPER'ICUM, *L.* St. John's Wort.

Herbs, shrubs, or small trees. *Leaves* sessile, often gland-dotted. *Flowers* cymose, yellow, not honeyed, homogamous. *Sepals* 5. *Petals* 5, generally very oblique. *Ovary* 1-celled with 3 or 5 parietal, or 3-5-celled with axile placentas ; styles distinct or connate ; ovules many in the cells, rarely few. *Capsule* (rarely a berry) septicidal, placentas adhering to the edges of the valves or to the axis. *Seeds* oblong ; embryo straight or incurved.—DISTRIB. All temp. regions ; species 160.—ETYM. obscure.

SECTION 1. *Sepals* . 5, unequal. *Petals* deciduous. *Stamens* connate in 5 bundles at the very base only, without intervening glands. *Ovary* incompletely 3- or 5-celled.

1. **H. Androsæ'mum**, *L.* ; shrubby, leaves sessile ovate or oblong, cymes corymbose few-fld., petals very oblique, styles 3 recurved. *Tutsan.*
Thickets, from Ross southd. ; Ireland ; Channel Islands ; fl. June-Aug.— Glabrous. *Stem* 1-2 ft., compressed, 4-angled. *Leaves* 1-3 in., obtuse or acute, glands very minute close. *Flowers* ½-¾ in. diam. *Sepals* obtuse, glandular, but not on the margins, about as long as the petals and stamens. *Berry* globose, black, incompletely 3-celled.—DISTRIB. Mid. and S. Europe, N. Africa, W. Asia.

H. CALYCI'NUM, *L.* ; shrubby, leaves sessile oblong obtuse, flowers sub-solitary shortly pedicelled, styles 5 straight.
Hedges and thickets, in various places, naturalized ; fl. July-Sept.—Glabrous, extensively creeping. *Stem* 10-16 in., subsimple, compressed, 4-angled. *Leaves* 2-4 in., coriaceous, glands rather large scattered. *Flowers* 3-4 in. diam. *Outer sepals* orbicular, half as long as the petals. *Capsule* ovoid, 5-celled towards the base.—DISTRIB. S.E. Europe.

SECTION 2. *Sepals* 5, connate at the base. *Petals* persistent. *Stamens* connate in 3 bundles at the very base only, without intervening glands. *Ovary* completely 3-celled. *Capsule* septicidal.

* *Margins of sepals entire or toothed, eglandular.*

2. **H. perfora'tum**, *L.* ; stem erect 2-ridged, leaves linear or oblong obtuse, glands and veins pellucid, reticulations opaque, sepals glandular, styles as long as the capsule.
Copses and hedgebanks, N. to Sutherland ; ascends to 1,000 ft. in Yorks. ; Ireland ; Channel Islands ; fl. July-Sept.—Glabrous. *Stems* 1-3 ft., branched above, slender, strict, light brown. *Leaves* ¾-1 in. *Cymes* corymbose, many-flowered. *Flowers* 1 in. diam. *Sepals* acute, entire, or slightly serrate in a narrow-leaved var. (*angustifolium*, Bab.). *Petals* much

longer. *Capsule* transversely wrinkled; carpels 2-vittate.— DISTRIB. Europe (Arctic), N. Africa, N. and W. Asia to the Himalaya; introd. in U. States.

3. **H. quadran'gulum,** *L.* (in part), *Fries ;* stem erect 4-ridged, leaves ovate oblong or orbicular glands few or 0, upper ½-amplexicaul, veins and reticulations pellucid, sepals glandular, styles as long as the capsules.

Copses and moist places, from Perth and Argyll southd.; Ireland ; Channel Islands ; fl. July–Sept.—Habit of *H. perfora'tum. Flowers* 1 in. diam., homogamous. *Sepals* erect or recurved, oblong or lanceolate, obtuse or acute, sometimes minutely toothed. *Petals* broad, glandular. *Carpels* multivittate.— DISTRIB. Europe (Arctic), N. Africa, N. and W. Asia.

H. quadran'gulum proper (*H. du'bium*, Leers); leaves broad, sepals oblong obtuse entire.—Var. *H. macula'tum*, Bab.; leaves and denticulate sepals narrower.

4. **H. undula'tum,** *Schousb.* ; stem erect narrowly 4-winged, leaves oblong glands copious and reticulations pellucid, sepals lanceolate glandular, styles half as long as the capsule. *H. bœ'ticum,* Boiss.

Bogs, Devon and Cornwall; fl. Aug.–Sept.—Glabrous. *Stem* 1–2 ft., slender· *Leaves* with strong nerves beneath. *Cymes* lax-fld. *Flowers* ¾ in. diam. *Sepals* usually finely acuminate. *Petals* narrow, one half tinged red. *Carpels* multivittate.—DISTRIB. Spain, Portugal, Azores.

5. **H. tetrap'terum,** *Fries ;* stem erect narrowly 4-winged, leaves broadly ovate or oblong glands and reticulations pellucid, sepals lanceolate acuminate eglandular, styles shorter than the capsule.· *H. quadran'gulum,* L. in part and Sm.

Moist places, from Ross southd.; Ireland; Channel Islands ; fl. July–Aug.— Glabrous. *Stem* 1–2 ft. *Leaves* sometimes cordate. *Cymes* dense-fld. *Flowers* ½–⅔ in. diam. *Sepals* usually finely acuminate. *Carpels* multivittate. —DISTRIB. Europe, from Sweden southd., N. Africa, Syria.

6. **H. humifu'sum,** *L.* ; stems many procumbent with 2 raised lines, leaves oblong with pellucid glands, margins often revolute with black glands, cymes forked, sepals unequal.

Roadsides, commons, &c.; ascends to 1,000 ft. in Yorks. ; Ireland ; Channel Islands; fl. July–Aug.—Perennial, glabrous. *Branches* 4–10 in., very many, compressed, curving upwards, leafy. *Leaves* ¼–½ in. *Flowers* ⅓–½ in. diam., homogamous.—DISTRIB. Europe, from Denmark southd. (excl. Turkey), Canaries, Azores.

** *Margins of sepals with glandular teeth.*

7. **H. linarifo'lium,** *Vahl ;* stems ascending subterete, leaves linear obtuse, margins revolute, sepals lanceolate acute.

Rocky banks, Cornwall and Devon; Channel Islands ; fl. June–July.—*Stems* 6–15 in., many from the roots, leafy. *Leaves* ½–1 in. *Cymes* few-flowered. *Flowers* ½ in. diam. *Petals* twice as long as the sepals, with black marginal glands. *Stamens* few. *Styles* short.—DISTRIB. S.W. Europe, Canaries.— A hybrid growing with this, and intermediate between it and *H. humifu'sum* (*H. decum'bens*, Peterm.), occurs in Jersey.

8. H. pul'chrum, *L.* ; glabrous, stems erect slender terete branched above, leaves with pellucid glands cordate very obtuse, upper shorter, sepals small oblong.
Dry copses, heaths, and commons, N. to Shetland; ascending to 2,200 ft. in the Highlands ; Ireland; Channel Islands ; fl. June–July.—Very elegant. *Stems* 1–2 ft., flexuous.—*Cymes* panicled, many-fld. *Leaves* ½ in. *Flowers* ½–¾ in. diam. *Petals* twice as long as the sepals, yellow tinged with red, margins with black glands. *Anthers* red. *Styles* short.—DISTRIB. W. Europe from Norway southd. and E. to Russia.

9. H. hirsu'tum, *L.* ; finely pubescent, stem erect terete subsimple, leaves with pellucid glands very shortly petioled ovate or oblong obtuse, sepals linear-oblong subacute.
Copses, &c., from Ross southd.; ascends to 1,300 ft. in Yorks.; very rare in Ireland; fl. July–Aug.—Rather stout, 1–3 ft., leafy; pubescence curly. *Leaves* 1–2 in., without marginal glands. *Cymes* panicled. *Flowers* ¾ in. diam., pale yellow, homogamous. *Petals* twice as long as the sepals.— DISTRIB. Europe, Siberia, W. Asia.

10. H. monta'num, *L.* ; almost glabrous, stems ascending terete, leaves with marginal black glands sessile oblong obtuse, upper cordate-ovate or linear oblong puberulous beneath, sepals lanceolate acute.
Copses in gravelly or chalky soil, England, Ayrshire, and N.E. Ireland ; Channel Islands ; ascends to 700 ft. in Surrey ; fl. July–Aug.—*Stem* 1–2 ft., rigid, very slender, often leafless above. *Leaves* 1–2 in., membranous. *Cymes* few and dense-fld.; bracts glandular, toothed. *Flowers* ½–¾ in. diam., pale yellow, fragrant. *Petals* eglandular or nearly so, twice as long as the sepals.—DISTRIB. Europe to the Caucasus, N. Africa.

SECTION 3.·*Sepals* nearly equal. *Petals* not oblique, persistent. *Stamens* 15, connate ⅛ way up in 3 stalked bundles which alternate with 2-fid hypogynous scales. *Ovary* incompletely 3-celled.

11. H. elo'des, *Huds.* ; villous, leaves orbicular or oblong-cordate.
Bogs, ditches, and wet moors, S. of England, W. of Scotland, from Argyll southd.; all Ireland ; Channel Islands ; fl. July–Aug.—*Stems* 3–18 in., many, creeping, terete. *Leaves* ½–1 in., ½-amplexicaul, pellucid glands small. *Cymes* irregular, often spuriously axillary, 3-chotomus ; bracts small, deltoid, gland-serrated. *Flowers* ½ in. diam., pale-yellow, homogamous. *Sepals* glabrous, oblong, obtuse, with red glandular serratures.—DISTRIB. W. Europe, from Holland to Spain and Italy ; Azores.

ORDER XVI. **MALVA'CEÆ.**

Herbs, shrubs, or trees, hairs often stellate. *Leaves* alternate, 3- or more-nerved at the base ; stipules deciduous. *Inflorescence* various ; bracteoles when present often connate and with their stipules forming an epicalyx. *Calyx* 5-lobed, valvate in bud. *Petals* 5, adnate at the base to the staminal column, twisted in bud, often oblique. *Stamens* many, filaments combined into a tube ; anthers reniform annular or twisted, 1-celled, bursting outwards. *Disk* small. *Carpels* many, whorled, distinct or connate ; styles distinct or connate, stigmatose on the inner face or top ; ovules 1 or more on the inner angles of the carpels, usually horizontal or ascending. *Fruit* usually of many dry indehiscent or 2-valved loculicidal crustaceous or coriaceous 1- or more-seeded carpels. *Seeds* often woolly, albumen little or 0 ; embryo curved, cotyledons usually thin folded or plaited.—DISTRIB. All regions but very cold ones ; genera 60 ; species 700.—AFFINITIES. With *Sterculia'ceæ, Tilia'ceæ,* and *Euphorbia'ceæ.*—PROPERTIES. Mucilaginous ; the bark yields textiles, and cotton is the covering of the seeds of *Gossypium.*

Bracteoles 6–9, connate at the base..1. Althæa.
Bracteoles 3, distinct, inserted on the calyx.................................2. Malva.
Bracteoles 3, connate at the base...3. Lavatera.

1. ALTHÆ'A, *L.* MARSH-MALLOW.

Herbs, hairy or tomentose. *Leaves* lobed or divided. *Flowers* axillary or racemose. *Calyx* 5-fid ; epicalyx 6–9-fid. *Staminal column* long, filaments distinct at its top only. *Ovary* many-celled ; styles filiform, inner surface stigmatose ; ovules 1 in each cell. *Fruit* a whorl of indehiscent 1-seeded carpels. *Seed* ascending.—DISTRIB. Temp. and warm regions ; species 12.—ETYM. ἄλθω, from its *healing* properties.

1. **A. officina'lis,** *L.* ; softly pubescent, cymes axillary shorter than the leaves. *Marsh-mallow, Guimauve.*

Marshes near the sea, local ; from the Clyde southd.; Ireland; fl. Aug.–Sept. —Perennial. *Stem* 2–3 ft., subsimple. *Leaves* 2–3 in. broad, short'y petioled, ovate-cordate or suborbicular, thick, entire or 3–5-lobed, toothed. *Flowers* 1–2 in. diam., rosy. *Sepals* ovate.—DISTRIB. Europe from Denmark southd., N. Africa, Siberia, W. Asia ; introd. in N. America.

2. **A. hirsu'ta,** *L.* ; hispid, peduncles 1-fld. longer than the leaves.

Woods and fields in N. Somerset (wild) ; naturalized in W. Kent and Herts; fl. July–Aug.—Annual or biennial. *Stems* 6–18 in., many, ascending, slender. *Leaves* long-petioled, reniform, acutely 5-lobed, crenate, upper 3-partite. *Flowers* 1 in. diam., rose-purple. *Sepals* lanceolate.—DISTRIB. Europe from Belgium southd., W. Asia.

2. MAL'VA, *L.* MALLOW.

Hirsute or glabrous herbs. *Leaves* angled, lobed or cut. *Flowers* axillary, honeyed. *Calyx* 5-fid, 3-bracteolate. *Staminal column* long, filaments distinct at its top only. *Ovary* many-celled ; styles stigmatose on the inner surface. *Fruit* a whorl of indehiscent 1-seeded carpels separating from a short conical axis. *Seed* ascending, albumen scantly mucilaginous. —DISTRIB. Europe, temp. Asia and N. Africa, and as weeds of cultivation in other regions ; species 16.—ETYM. μαλάχη, in allusion to its *emollient* properties.

1. M. sylves'tris, *L.* ; erect, hairy, stems many ascending, leaves 3–7-lobed crenate-serrate, peduncles spreading, carpels glabrous reticulate.

Waste places from Ross southd., rare (if native, *Watson*) in Scotland ; Ireland ; Channel Islands ; fl. June–Sept.—Perennial or biennial, 2–3 ft. *Leaves* 2–3 in. diam., lobes shallow acute. *Flowers* 1–1½ in. diam., irregularly fascicled, pale purple or blue, dichogamous, proterandrous ; pedicels slender.—Distrib. Europe, N. Africa, Siberia, W. Asia ; introd. in U. States.

2. M. rotundifo'lia, *L.* ; pubescent, stems many decumbent, leaves reniform obscurely lobed crenate, peduncles decurved, carpels pubescent smooth margins rounded.

Waste places from Aberdeen southd. ; rarer in Scotland and Ireland ; Channel Islands ; fl. June–Sept.—Perennial. *Stems* 6 in.–2 ft. *Leaves* ½–1½ in. diam., often serrate. *Flowers* ¾–1 in. diam., fascicled, pale lilac or whitish, homogamous.—Distrib. Europe, N. Africa, N. and W. Asia to India ; introd. in U. States.

3. M. moscha'ta, *L.* ; hairy, erect, leaves 5–7-partite, segments pinnatifid, peduncles erect in fruit, carpels smooth, back rounded hispid.

Meadows, &c., in dry soil ; from the Clyde southd. ; rare in Ireland ; Channel Islands ; fl. July–Sept.—Perennial. *Stem* 2–3 ft., often purple-spotted. *Leaves* 1–3 in. diam., long-petioled. *Flowers* 1–2 in. diam., rosy rarely white.—Distrib. Europe, eastward to Lithuania ; introd. in the U. States.

3. LAVATE'RA, *L.* Tree-mallow.

Tall, hirsute or tomentose herbs or shrubs. *Leaves* angled or lobed. *Flowers* axillary. *Calyx* 5-fid ; epicalyx 3-fid. *Staminal column* long, filaments distinct at its top only. *Ovary* many-celled ; styles filiform, as many as the cells, inner surface stigmatose ; ovules 1 in each cell. *Fruit* a depressed whorl of indehiscent 1-seeded carpels, separating from the axis, *Seed* ascending.—Distrib. Europe, W. Asia, N. Africa, and 1 Australian ; species 18.—Etym. The brothers Lavater, Swiss physicians.

L. arbo'rea, *L.* ; leaves suborbicular 5–9-lobed plaited crenate.

Maritime rocks, from the Clyde southd. ; rare in Ireland ; Channel Islands ; fl. July–Sept.—Biennial, softly pubescent. *Stem* 3–6 ft., very stout, erect. *Leaves* long-petioled ; lobes broad, short, upper more entire. *Peduncles* crowded, axillary, 1-fid., shorter than the petioles. *Flowers* 1½ in. diam., purple, glossy. *Epicalyx* with 3 very large ovate lobes. *Sepals* deltoid. *Carpels* wrinkled.—Distrib. Coasts of Europe eastwards to Greece.

Order XVII. TILIA'CEÆ.

Trees or shrubs, rarely herbs. *Leaves* alternate, entire or toothed, stipulate. *Flowers* cymose, honeyed, proterandrous ; cymes usually corymbose or panicled. *Sepals* 5, distinct or connate below, valvate in bud. *Petal*.

5 or fewer or 0, æstivation various. *Stamens* many, inserted on the disk, filaments filiform distinct or connate in bundles ; anthers 2-celled, opening by pores or valves. *Disk* tumid. *Ovary* free, 2–10-celled ; styles entire or divided, or stigma sessile ; ovules 1 or more, in the inner angle of the cells, anatropous, usually pendulous, raphe ventral. *Fruit* 1–12-celled, dry or baccate, indehiscent or loculicidal. *Seeds* various, albumen fleshy ; embryo straight, cotyledons foliaceous.—DISTRIB. Chiefly tropical ; genera 40 ; species 330.—AFFINITIES with *Sterculia'ceæ* and *Malva'ceæ.*—PROPERTIES. Usually mucilaginous, liber of many species yields fibre.

1. TIL'IA, *L.* LIME-TREE or LINDEN.

Trees with simple or stellate hairs. *Leaves* oblique, cordate, serrate. *Cymes* axillary or terminal ; peduncle with a leafy decurrent bract. *Sepals* 5. *Petals* 5, with often a scale at the base. *Stamens* many, filaments distinct or connate in bundles at the base. *Ovary* 5-celled ; style simple, stigma 5-toothed. *Fruit* globose, indehiscent, 1–2-seeded. *Seeds* ascending ; cotyledons broad, crumpled with involute margins.—DISTRIB. Europe, N. Asia, N. America ; species 8.—ETYM. The old Latin name.

1. **T. parvifo'lia,** *Ehrh.*; glabrous, leaves glaucous and pubescent in the axils of the nerves beneath, fruit crustaceous pubescent.

Woods, from Cumberland southd.; (a doubtful native, *Wats.*); indigenous, *Borrer ;* fl. July–Aug.—A small tree. *Leaves* 1½–2½ in. diam., ovate-cordate, acuminate, finely serrate, glabrous beneath, upper obscurely lobed. *Flowers* ⅓ in. diam. *Fruit* about ¼ in. diam., globose or ellipsoid, faintly ribbed. —DISTRIB. Europe (excl. Greece and Turkey) and Siberia.

2. **T. platyphyl'los,** *Scop.* ; twigs pilose, leaves downy beneath, fruit obovate globose with 3–5 prominent ribs when ripe. *T. grandifolia,* Ehrh. *Large-leaved Lime.*

Woods, Hereford, Radnor, and W. York ; indigenous (*Wats.*) ; fl. June–July. —A tree 70–90 ft. high, differing very little in foliage and floral characters from *T. parvifo'lia.*—DISTRIB. Europe from Denmark southd.

T. VULGA'RIS, *Hayne ;* glabrous, leaves pubescent in the axils of the nerves beneath, fruit woody pubescent not ribbed when ripe. *T. inter-me'dia,* DC. *Common Lime.*

Plantations, not indigenous ; fl. June–July.—DISTRIB. Europe, Caucasus.

ORDER XVIII. **LIN'EÆ.**

Herbs, shrubs, or trees. *Leaves* alternate, rarely opposite, simple, entire, sometimes stipulate. *Inflorescence* cymose. *Sepals* 4–5, distinct or connate, imbricate in bud. *Petals* 4–5, hypogynous, imbricate or twisted in bud. *Stamens* 4–5 with alternating staminodes, or 10 with the filaments inserted on a hypogynous ring ; anthers versatile. *Disk* 0 or of 5 honeyed glands. *Ovary* free, 3–5-celled ; styles 3–5, stigmas terminal ; ovules

1-2 in each cell, pendulous, raphe ventral. *Capsule* septicidally splitting into 2-valved cocci. *Seeds* compressed, albumen fleshy ; embryo long, cotyledons plano-convex.—DISTRIB. All regions ; genera 14 ; species 135.—AFFINITIES, slight with *Malva'ceæ* and *Gerania'ceæ ;* more close with *Malpighia'ceæ* and *Ternstræmia'ceæ.*—PROPERTIES. Mucilaginous, oily, diuretic ; seeds occasionally purgative ; bark fibrous (as in flax).

Sepals 5, quite entire...1. Linum.
Sepals 4, 2–4-toothed...2. Radiola.

1. LI'NUM, *L.* FLAX.

Herbs or small shrubs. *Leaves* alternate, rarely opposite, narrow, quite entire ; stipules 0 or glandular. *Flowers* in dichotomous panicled racemose or fascicled cymes. *Sepals* 5, entire. *Petals* 5, distinct or connate below, fugacious. *Stamens* 5, hypogynous, connate at the base, alternating with 5 minute staminodes. *Disk* of 5 glands opposite the petals. *Ovary* 5-celled, cells sometimes divided into 2 ; styles 5 ; ovules 2 in each principal cell. *Cocci* 5, 1- or partially 2-celled, 2-seeded. *Albumen* scanty ; embryo straight.—DISTRIB. Temp. and warm regions ; species 80.— ETYM. The classical name.

1. **L. cathar'ticum,** *L.* ; annual, small, leaves opposite, upper alternate, buds nodding, petals distinct. *Purging Flax.*

Heaths and pastures, N. to Shetland ; ascends to 2,400 ft. in the Highlands ; Ireland ; Channel Islands ; fl. June–Sept.—Glabrous, glaucous, 2–10 in., very slender. *Leaves* linear-oblong. *Flowers* ⅛–¼ in. diam., white, homogamous. *Petals* oblong, acute or obtuse.—DISTRIB. Europe (Arctic), Canaries, W. Asia to Persia.

2. **L. peren'ne,** *L.* ; perennial, leaves alternate narrow linear-lanceolate acute, sepals obovate obtuse glabrous 3–5-nerved, petals distinct.

Chalky soils, very rare, from Durham to Essex ; fl. June–July.—Glabrous, *Stems* many, 1–2 ft., wiry, very slender. *Leaves* ½–¾ in. *Cymes* few-flowered, racemose. *Flowers* 1 in. diam., bright blue, dimorphous (long-styled and short-styled). *Sepals* obovate, 3 inner broader, quite entire.— DISTRIB. Mid. and S. Europe, to India.

3. **L. angustifo'lium,** *Huds.* ; annual or perennial, leaves alternate narrow linear-lanceolate, outer sepals ovate acuminate, inner ciliate 3-nerved.

From Lancashire southd. ; Ireland rare ; Channel Islands ; fl. May–Sept.— Glabrous, glaucous. *Stems* 1–2 ft. ; branches few, divaricate. *Leaves* as in *L. peren'ne,* but fewer and smaller. *Cymes* few-flowered. *Flowers* ½–¾ in., pale lilac-blue.—DISTRIB. W. and S. Europe, to W. Asia, N. Africa.

L. USITATIS'SIMUM, *L.* ; annual, leaves alternate linear-lanceolate, sepals ovate acuminate ciliate 3-nerved, petals crenulate. *Common Flax.*

An escape of flax-fields ; fl. June–July.—Larger than any of the preceding. *Stem* corymbosely branched above. *Cymes* broad, many-flowered.—DISTRIB. Wherever flax is cultivated for oil or fibre.

2. RADI'OLA, *Gmelin.* ALL-SEED. FLAX-SEED.

A minute annual, with filiform repeatedly-forked branches. *Leaves* opposite, exstipulate. *Flowers* in corymbose cymes. *Sepals* 4, 2-4-toothed. *Petals* 4, fugacious. *Stamens* 4, scarcely connate; staminodes minute or 0. *Disk* glands inconspicuous. *Ovary* 4-celled; cells divided into 2; styles 4; ovules 2 in each cell. *Capsule* of 4 nearly 2-celled 2-seeded divisible cocci. *Seeds* exalbuminous.—DISTRIB. Europe (excl. Central), N. Africa.—ETYM. *radius,* from the *rayed* ramification.

R. **linoi'des,** *Gmel.*; leaves ovate acute 3-5-nerved. *R. Millegra'na,* Sm.

Gravelly and sandy damp places, uncommon, from the Orkneys to Cornwall; Ireland; Channel Islands; fl. July–Aug.—*Stem* 1-4 in. *Leaves* $\frac{1}{10}$–$\frac{1}{8}$ in., sessile, rather succulent. *Flowers* axillary and in the forks, most minute; peduncles short, erect. *Sepals* connate below, as long as the oblong petals.

ORDER XIX. GERANIA'CEÆ.

Herbs, rarely shrubby. *Leaves* opposite or alternate, usually stipulate. *Inflorescence* various; flowers regular or irregular. *Sepals* 5 or fewer, imbricate or valvate in bud. *Petals* 3-5, imbricate in bud. *Stamens* usually 5 in irregular flowers, and 10 in the regular, some often deformed. *Disk* inconspicuous or glandular. *Ovary* 3-5-lobed, 3-5-celled, produced upwards into a styliferous beak, or with one or more terminal styles; ovules 1-2 or many in each cell, anatropous, pendulous, raphe ventral. *Fruit* septicidal or loculicidal, or separating into cocci. *Seeds* small, albumen scanty or 0; embryo various.—DISTRIB. Temp. and Trop. rarely Arctic regions; genera 10; species 750.—AFFINITIES. With *Rutaceæ* and *Lineæ.* —PROPERTIES. Tribe *Geranieæ* are often astringent, aromatic, and abound in volatile oil. *Oxalideæ* abound in oxalic acid, and some have eatable tubers.

TRIBE I. **GERANIE'Æ.** *Flowers* regular. *Sepals* imbricate. *Stamens* alternating with glands. *Capsule* beaked, of several 1-seeded awned cocci, that separate elastically from the beak.
Stamens 10, all antheriferous..1. Geranium.
Stamens 5, staminodes 5..2. Erodium.
TRIBE II. **OXALIDE'Æ.** *Flowers* regular. *Sepals* imbricate. *Glands* 0. *Capsule* loculicidal; cells 2- or more-seeded........................3. Oxalis.
TRIBE III. **BALSAMI'NEÆ.** *Flowers* irregular. *Sepals* coloured, posticous spurred. *Stamens* 5. *Glands* 0. *Capsule* loculicidal; cells 2- or more-seeded...4. Impatiens.

1. GERA'NIUM, *L.* CRANE'S-BILL.

Herbs, rarely shrubs, nodes swollen. *Leaves* opposite or alternate, usually cut or lobed, stipulate. *Flowers* regular, on 1-2-fld. axillary peduncles. *Sepals* and *petals* 5, imbricate in bud. *Stamens* 10, ripening in 2 sets, rarely 5, hypogynous, honeyed. *Disk* of 5 glands opposite the sepals. *Ovary* 5-lobed, 5-celled, with a long beak terminated by 5 stigmas ; ovules 2 in each cell, superposed. *Fruit* of 5 dehiscent 1-seeded carpels, which terminate upwards in slender tails, and usually separate elastically from the styliferous and placentiferous axis. *Seeds* oblong, albumen scanty or 0 ; cotyledons plicate or convolute ; radicle incumbent. DISTRIB. All temp. regions ; species 100.—ETYM. γέρανος, *a crane,* from the form of the fruit.—The garden Geraniums are *Pelargonia,* having irregular flowers, a spurred sepal, perigynous petals, no glands, and few declinate stamens.

* *Perennial. Peduncles 1-flowered. Sepals spreading.*

1. **G. sanguin'eum,** *L.* ; hairy, leaves orbicular 5–7-partite, segments narrow 3–5-fid to the middle, carpels hairy, seeds wrinkled and dotted.

Dry rocks and sandy shores, from Ross southd.; ascends above 1,000 ft. in the Highlands; local in Ireland ; fl. July-August.—*Rootstock* stout, truncate. *Stems* 1–2 ft., geniculate ; hairs spreading. *Leaves* 1–2 in. diam., segments linear-oblong or lanceolate, obtuse or subacute ; stipules ovate, acute. *Flowers* 1–1½ in. diam., crimson or pink, proterandrous ; peduncles very long, 2-bracteate in the middle. *Sepals* oblong, obtuse, awned. *Claw of petals* bearded.—DISTRIB. Europe, W. Asia.

G. sanguin'eum proper ; suberect, hairs scattered.—VAR. *G. prostra'tum,* Cav.; stems shorter decumbent, hairs more copious, flowers pinkish. *G. lancas'triense,* With. Sands, Walney Island, Lancashire.

** *Perennial. Stem erect. Peduncles 2-flowered. Sepals spreading.*

2. **G. sylvat'icum,** *L.* ; erect, pilose and glandular above, leaves orbicular deeply 7-lobed, lobes cut and serrate, sepals awned, petals notched, carpels smooth hairy, pedicels erect, seeds minutely reticulate.

Copses and moist meadows from Stafford and Carnarvon to Caithness ; ascends to 2,700 ft. in the Highlands ; very rare in Ireland ; fl. June–July.—*Rootstock* truncate, creeping. *Stem* 1–3 ft., branched above. *Leaves* 3–5 in. diam., radical long-petioled, cauline sessile ; stipules ovate. *Flowers* ½–¾ in. diam., blue-purple or rose-coloured, cymose at the ends of the branches, gynodiœcious, proterandrous. *Claw of petals* bearded. *Filaments* filiform. ciliate.—DISTRIB. Europe (Arctic), Siberia, W. Asia.

3. **G. praten'se,** *L.* ; erect, pubescent, hairs reflexed, leaves orbicular 7–9 partite, lobes laciniate coarsely serrate, sepals awned, petals notched, carpels smoth glandular-hairy, pedicels deflexed, seeds minutely reticulate.

Moist meadows, &c., from Aberdeen and Isla southd.; ascends to 1,800 ft. in the Highlands ; N.E. Ireland, very rare ; fl. June–Sept.—*Rootstock* truncate. *Stem* 3–4 ft., branched above. *Leaves* 3–6 in. diam., all petioled, radical very long-petioled ; stipules subulate-lanceolate. *Flowers* many, 1¼ in. diam., proterandrous. *Sepals* with very long awns. *Claw of petals* bearded. *Filaments* glabrous, cuneate at the base.—DISTRIB. Europe (Arctic), Siberia.

4. **G. perenne,** *Huds.* ; erect or ascending, hairy, leaves reniform 7–9-lobed, lobes cuneate 3-fid crenate, sepals mucronate, petals 2-lobed, carpels keeled pubescent, pedicels deflexed, seeds smooth. *G. pyrena'icum,* L.

Meadows, from Perth southd., possibly native in S.W. England and Wales ; Ireland ; Channel Islands ; (a denizen, *Wats.*) ; fl. June–Aug.—*Rootstock* fusiform. *Stems* 1–2 ft., many, decumbent below. *Leaves* 3 in. broad, radical very long-petioled ; lobes contiguous ; stipules ovate-lanceolate. *Flowers* ½ in. diam., bright red-purple, proterandrous. *Sepals* small, oblong. *Claw of petals* densely bearded.—DISTRIB. Europe, N. Africa, W. Asia to India.

G. PHÆ'UM, *L.* ; erect, laxly hairy and glandular above, leaves orbicular or reniform 5–7-lobed. lobes cut and serrate, sepals mucronate, petals waved, carpels hairy wrinkled above, pedicels deflexed, seeds smooth.

Woods near parks and gardens, naturalized : fl. May–June.—*Rootstock* truncate. *Stems* 1–2 ft., many. *Leaves* 3–5 in. broad, radical very long-petioled ; stipules lanceolate. *Flowers* ¾ in. diam., dusky purple, proterandrous. *Sepals* oblong, shortly-awned.—DISTRIB. Central and W. Europe.

*** *Annual or biennial. Stems ascending or decumbent. Peduncles 2-flowered. Sepals spreading.*

5. **G. mol'le,** *L.* ; softly hairy, leaves orbicular 7–9-lobed, lobes contiguous, sepals mucronate generally shorter than the notched petals, claw of pe*t*als bearded, carpels persistent wrinkled keeled glabrous, seeds smooth.

Pastures and waste places, N. to the Shetlands ; ascends to 1,500 ft. in Yorkshire ; Ireland ; Channel Islands ; fl. May–Sept.—*Branches* 8–12 in. *Leaves* 1–2 in. diam., lobes broadly irregularly lobed or crenate at the tip, radical long-petioled ; s*t*ipules ovate. *Peduncles* axillary. *Flowers* ¼–½ in. diam., rose-purple or pink, homogamous.—DISTRIB. Europe, N. Africa, W. Asia.

6. **G. rotundifo'lium,** *L.* ; laxly hairy, leaves as in *G. molle,* sepals mucronate generally shorter than the entire petals, claw of petals naked, carpels keeled not wrinkled hairy, seeds pitted.

Hedges and waste places, rare, from S. Wales and Norfolk southd. ; local in Ireland ; Channel Islands ; fl. June to July.—*Branches* 6–12 in., slender, geniculate. *Leaves* ¾–1 in. broad ; stipules ovate-lanceolate. *Flowers* ¼–½ in. diam., pale pink. *Petals* narrow.—DISTRIB. Europe, N. Africa, Siberia, W. Asia to India.

7. **G. pusil'lum,** *L.* ; softly pubescent, leaves as in *G. molle* but deeper lobed, sepals acute equalling the notched petals, claw of petals subciliate, carpels persistent keeled not wrinkled pubescent, seeds smooth.

Hedgebanks and waste places, from Aberdeen and Isla southd. ; rare in Scotland ; Ireland ; Channel Islands ; fl. June–Sept.—*Branches* 6–18 in. ; stipules ovate-lanceolate. *Peduncles* axillary. *Flowers* ¼–⅓ in. diam., many. pale rose-colour, homogamous. *Perfect stamens* often only 5.—DISTRIB. Europe, N. Africa, W. Asia ; introd. in America.

8. **G. columbi'num,** *L.* ; nearly glabrous, leaves 5–7-partite, lobes distant pinnatifid, segments narrow, sepals large acuminate long-awned equalling the entire petals, claw of petals ciliate, carpels not wrinkled keeled glabrous, seeds pitted.

G

Dry copses and pastures, from Forfar and Ayr southd.; rare in Scotland and
Ireland; Channel Islands; fl. June–July.—*Branches* 8–24 in., slender, hairs
reflexed. *Leaves* ¾–1¼ in. broad, long-petioled; stipules ovate-lanceolate.
Flowers ½–¾ in. diam., few, rose-purple, homogamous; peduncles and
pedicels very long and slender. *Calyx* angular.—DISTRIB. Europe, N.
Africa, Siberia.

9. **G. dissec'tum,** *L.* ; hairy and subglandular, leaves as in *G. columbinum*, peduncles very short, sepals long-awned, petals short obovate
notched, carpels not wrinkled or keeled hairy, seeds pitted. *Dove's-foot.*

Hedges and waste places, from the Orkneys southd.; ascends to near 1,000 ft.
in Yorks.; Ireland; Channel Islands; fl. May–Aug.—Similar to *G. columbinum*, but petioles and peduncles very much shorter. and calyx and capsules
quite different. *Stipules* ovate, long-acuminate. *Flowers* ¼–½ in. diam.,
axillary, bright red.—DISTRIB. Europe, N. Africa, W. Asia, N. America.

**** *Annual or biennial. Stems ascending or decumbent. Peduncles 2-flowered.
Sepals erect in flower, conniving in fruit.*

10. **G. Robertia'num,** *L.* ; glabrous or slightly hairy, leaves 5-foliolate,
leaflets 1–2-pinnatifid. sepals long-awned, petals entire narrow, carpels
wrinkled keeled, seeds smooth. *Herb-Robert.*

Waste places and hedgebanks, from Orkney southd.; ascends to near 2,000 ft.
in Yorks.; Ireland; Channel Islands; fl. May–Sept.—Plant fœtid, reddish.
Branches 6–18 in., brittle, leafy. *Leaves* 1–3 in. broad; petiole ½–1 ft.;
stipules ovate. *Flowers* ½ in. diam.. streaked with dark and light red, sometimes white, homogamous. *Calyx* angular. *Claw of petals* glabrous.
Carpels attached by silky hairs to the axis.—DISTRIB. Europe (Arctic),
N. Africa, Siberia, W. Asia to N.W. India.
G. Robertianum proper; glandular. hairy, blade of petal about as long as the
claw, carpels with deciduous hairs.—VAR. *G. purpu'reum*, Vill. (*G. Lebe'lii*,
Bor., *G. modes'tum.* Jord.), is a small flowered maritime state with more
fleshy leaves more divided. S. of England.

11. **G. lu'cidum,** *L.* ; glabrous, shining, bright red, branches above
with 2 lines of hairs, leaves orbicular 5-lobed, sepals long-awned shorter
than the petals, carpels wrinkled keeled glabrous or nearly so, seeds
smooth.

Hedgerows, old walls. &c., from Orkney southd., local; Ireland; fl. May–Aug.
—*Branches* 6–18 in., succulent, brittle. *Leaves* ¾–1½ in. broad, lobes short
obtusely lobulate at the top; petiole 1–2½ in.; stipules ovate, acute.
Peduncles longer than the petioles. *Flowers* ¼–⅓ in. diam., rose-coloured.
Calyx pyramidal, wrinkled. *Claw of petals* glabrous. *Carpels* separating
wholly from the axis.—DISTRIB. Europe, N. Africa, Siberia, W. Asia to
N.W. Himalaya
VAR. *G. Rai'i*, Lindl., is a maritime form with shaggy stem and calyx, more
succulent leaves and wrinkled fruit.

2. ERO'DIUM, *L'Héritier.* STORK'S-BILL

Herbs with swollen nodes. *Leaves* alternate, or, if opposite unequal,
stipulate. *Flowers* regular, solitary or umbellate (contracted cymes) on

axillary peduncles, proterandrous. *Sepals* 5, imbricate. *Petals* 5, hypogynous, imbricate, 2 upper sometimes deficient. *Stamens* 5, alternating with scale-like staminodes. *Disk* of 5 glands opposite the sepals. *Ovary*, fruit and seed as in *Geranium*, but tails of carpels spirally twisted and usually silky on the inner surface.—DISTRIB. Europe, N. Africa, temp. Asia ; rare in S. Africa and Australia ; species 50.—ETYM. ἐρώδιος, a *heron*, from the form of the fruit.

1. **E. cicuta′rium,** *L'Hérit.* ; leaves 1-2-pinnate, leaflets pinnatifid, segments narrow cut, peduncles few- or many-fld., filaments entire.

Waste places, most frequent by the sea ; ascends to 1,200 ft. in N. Wales ; Ireland ; Channel Islands ; fl. June–Sept.—Annual or biennial ; laxly hairy and glandular. *Stems* at first short, then elongating to 6-24 in., prostrate or decumbent. *Leaves* 6-18 in., oblong ; stipules lanceolate. *Peduncles* longer than the leaves, strict. *Flowers* ⅓-½ in. diam., umbelled, rosy or white. *Sepals* hairy. *Petals* rather unequal, two often with a red spot, entire. *Carpels* hairy with an eglandular subapical pit and usually a shallow curved furrow below the pit ; pedicel reflexed.—DISTRIB. Europe, N. Africa, Siberia, W. Asia to N.W. India.

E. cicuta′rium proper (var. *vulgata*, Syme, *E. pimpinellæfolium*, Cav., *E. commix′tum*, Jord.) ; pinnules short, lobes obtuse or subacute, petals hardly longer than the sepals which have spreading often glandular hairs.–-VAR. *E. chærophyl′lum*, Cav., pinnules with longer more acute lobes, petals twice as long as the eglandular sepals which have appressed hairs.

2. **E. moscha′tum,** *L'Hérit.* ; leaves pinnate, leaflets deeply sharply irregularly serrate, antheriferous filaments toothed at the base.

Waste places, &c., from Worcester and Pembroke to Cornwall and Dorset ; local in Ireland ; Channel Islands ; fl. June–July.—Larger than *E. cicuta-rium*, covered with spreading hairs, smelling strongly of musk. *Stem* 2 ft., stout ; stipules broadly ovate, obtuse. *Flowers* pale rose-purple. *Carpels* hairy with a glandular subapical pit subtended by a deep curved furrow.— DISTRIB. Europe, N. Africa, W. Asia.

3. **E. marit′imum,** *L'Hérit.* ; leaves simple oblong or ovate-cordate, margin lobulate, lobes crenate, peduncles 1-2-fld., filaments entire.

Sandy and gravelly places chiefly near the sea. from Wigton southd. ; all round Ireland ; Channel Islands ; fl. May–Sept.—Small, hairy, annual or biennial. *Stems* 6-16 in... decumben *Leaves* ⅓-½ in., petiole longer ; stipules ovate. *Flowers* ⅓ in. diam., pale pink. *Petals* sometimes 0. *Carpels* hairy with a deep subapical eglandular pit subtended by a straight deep furrow.— DISTRIB. W. Europe, from France to Italy.

3. OX′ALIS, *L.* WOOD-SORREL.

Acid herbs. *Leaves* radical or alternate, stipulate or exstipulate, compound, usually 3-foliolate. *Flowers* on axillary 1- or more-flowered peduncles, regular. *Sepals* 5, imbricate in bud. *Petals* 5, twisted in bud. *Stamens* 10, distinct or connate at the base. *Disk* 0. *Ovary* 5-lobed, 5-celled ; styles 5, stigmas terminal ; ovules 1 or more in each cell.

Capsule loculicidal, valves adhering by the septa to the axis. *Seeds* with an elastic dehiscent fleshy coat, testa crustaceous, albumen fleshy ; embryo straight.—DISTRIB. 3 or 4 species widely dispersed, the rest S. African and S. American ; species 220.—ETYM. ὄξυς, *acid.*—Leaflets pendulous at night, sensitive to light.

O. Acetosel'la, *L.* ; stemless, leaves all radical 3-foliolate, stipules broad membranous, scape 1-flowered. *Wood-sorrel.*

Moist shady places, N. to Orkneys ; ascends to near 4,000 ft. in the Highlands ; Ireland ; Channel Islands ; fl. April–Aug.—Glabrous or hairy. *Rootstock* creeping, scaly. *Petioles* 3–6 in.; leaflets obcordate, ½–¾ in., often purple beneath. *Scape* axillary, slender, 2-bracteate about the middle. *Flower* ½–¾ in. diam., dimorphic, larger ½–¾ in. diam.; smaller cleistogamous. *Sepals* oblong. *Petals* obovate, white veined with purple, rarely rose-purple, erose, cohering above the claw. *Capsule* erect, 5-gonal ; cells 2-3-seeded. *Seeds* ribbed.—DISTRIB. Europe (Arctic), N. Africa, N. and W. Asia to the Himalaya, N. America.

O. CORNICULA'TA, *L.* ; pubescent, stems branched procumbent without runners, leaves all cauline 3-foliolate, stipules adnate, peduncles axillary 2-3-flowered, fruiting pedicels deflexed, capsules downy.

Waste shady places, local, possibly indigenous in S.W. England, not north of it ; Channel Islands ; fl. June–Sept.—Very variable in size and habit; annual or biennial. *Stems* 6-16 in. *Leaves* as in *O. Acetosella.* *Flowers* ¾ in. diam., subumbellate, very long-peduncled, yellow. *Seeds* transversely ribbed.—DISTRIB. Ubiquitous, except in very cold regions.

O. STRIC'TA, *L.* ; subglabrous, stem erect with copious runners at the base, leaves as in *O. corniculata* but often whorled and stipules minute, peduncles 2-8-flowered, capsules glabrous, fruiting pedicels spreading.

A weed in Cheshire and south of it, local, not indigenous ; casual in Ireland ; fl. June–Sept.—Similar to and distribution of *O. corniculata,* of which it is perhaps a sub-species.

4. IMPA'TIENS, *L.* BALSAM.

Herbs, rarely shrubby. *Leaves* opposite or alternate, stipules 0 or glandular. *Flowers* irregular, resupinate, on 1- or more-flowered axillary peduncles. *Sepals* 3, rarely 5, petaloid, imbricate ; 2 anterior (if present) minute ; 2 lateral small, flat ; posterior large, produced into a hollow spur. *Petals* 3 ; anterior external in bud, large ; lateral 2-lobed, each formed by a connate lateral and posterior petal. *Stamens* 5, filaments short broad ; anthers cohering. *Disk* 0. *Ovary* oblong, 5-celled ; stigma sessile, 5-toothed ; ovules many in each cell, 1-seriate. *Capsule* loculicidal, valves 5 elastic separating from the placentas and then twisting. *Seeds* smooth or villous, albumen 0 ; embryo straight.—DISTRIB. Mountains of trop. Asia and Africa ; rare in temp. Europe, N. America, N. Asia, and S. Africa ; species 135.—ETYM. The Latin name, from the ripe capsules bursting when touched.—The anterior lateral sepals occur in a few Indian species. Cleistogamous flowers occur in *I. noli-me-tangere* and *I. fulva.*

I. noli-me-tan'gere, *L.* ; glabrous, leaves oblong obtuse crenate-serrate, peduncles 1–3-flowered, posterior sepal funnel-shaped gradually contracted into a slender spur with an entire tip. *Yellow Balsam.*
Moist mountainous situations, probably wild in N. Wales and Westmoreland ; an escape elsewhere; and in Ireland; fl. July–Sept.—Annual, succulent, 1–2 ft., nodes thickened. *Leaves* 2–4 in., alternate, membranous ; petiole half as long, slender. *Flowers* 1¼ in., drooping, pale-yellow dotted with red, proterandrous. —DISTRIB. Europe, Siberia, W. Asia.

I. FUL'VA, *Nuttall;* habit and characters of *I. noli-me-tangere,* but leaves acute, serratures more shallow, flowers orange, posterior sepal saccate suddenly contracted into an upcurved spur with a notched tip.
Naturalized on river-banks in Surrey, the Clyde, and other places; fl. June–Aug.—A North American plant, naturalized within the last 50 years, and spreading rapidly.

I. PARVIFLO'RA, *DC.* ; leaves elliptic-ovate acuminate serrate, peduncles erect 3–10-flowered, posterior sepal contracted into a short straight spur.
A garden escape, naturalized in several places; fl. July–Nov.—An annual weed, with very small yellow flowers (none cleistogamic).—DISTRIB. Siberia.

ORDER XX. **ILICI'NEÆ** OR **AQUIFOLIA'CEÆ.**

Shrubs or trees. *Leaves* alternate, simple, often evergreen ; stipules minute or 0. *Flowers* small, in axillary cymes, often polygamous. *Calyx* 3–6 parted, imbricate in bud, persistent. *Petals* 4–5, distinct or connate at the base, deciduous, imbricate in bud. *Stamens* 4 or 5, hypogynous, free or adnate to the petals, filaments subulate ; anthers oblong. *Disk* 0. *Ovary* free, 3- or more-celled ; style 0 or short, stigmas terminal ; ovules 1, or 2 and collateral in each cell, pendulous, raphe dorsal ; funicle often cupular. *Drupe* with 3 or more 1-seeded distinct or connate stones. *Seed* with a membranous testa, fleshy albumen, and minute straight embryo.—DISTRIB. Temp. and trop. regions, absent from N.W. America ; genera 3 ; species 150.—AFFINITIES. Differing from *Olacineæ* only in the several-celled ovary. — PROPERTIES. Antiseptic and astringent. Holly berries are purgative and emetic ; bark yields bird-lime. "Maté" or "Paraguay tea" is the leaf of *Ilex paraguayensis.*

1. I'LEX, *L.* HOLLY.

Calyx 4–5-parted, persistent. *Corolla* rotate ; petals connate at the base or distinct. *Stamens* 4, adhering to the base of the corolla. *Ovary* 4–6-celled ; stigmas free or confluent. *Drupe* globose, with 4 stones or a 4–5-celled stone.—DISTRIB. Trop. and temp. regions, abundant in S. America, rare in Africa and Australia ; species 145.—ETYM. doubtful.

I. **Aquifo'lium,** *L.* ; glabrous, shining, leaves ovate spinescent.

Copses and woods, from Caithness southd., often planted; ascends to 1,000 ft. in the Highlands; Ireland; Channel Islands; fl. May–Aug.—A shrub or small tree, 10–40 ft., young shoots puberulous; bark ashy, smooth. *Leaves* glossy, 2–3 in., acute or acuminate, with waved spinous cartilaginous margins, those on the upper branches often entire. *Cymes* umbellate, shortly peduncled, many-fld. *Flowers* ¼ in. diam., white, often subdioecious. *Sepals* ovate, puberulous. *Petals* obovate, concave. *Stigmas* 4, sessile. *Drupe* scarlet, rarely yellow; stones 4, bony, furrowed.—DISTRIB. Europe from S. Norway to Turkey and the Caucasus; W. Asia.

ORDER XXI. **EMPETRA'CEÆ.**

Heath-like shrubs. *Leaves* alternate, exstipulate. *Flowers* small, solitary or clustered, axillary or terminal, regular, polygamous, bracteolate or not. *Sepals* (or bracts) 2–3, distinct, coriaceous, or thin, imbricate in bud. *Petals* (or sepals) 2–3, hypogynous, distinct, persistent. *Stamens* 3–4, alternate with the petals, hypogynous; filaments long, filiform, persistent; anthers deciduous, 2-celled; pollen compound. *Ovary* globose, 3–9-celled; styles short, stigmas subulate or dilated; ovule 1, ascending from the inner angle of each cell, anatropous. *Drupe* depressed-globose, with 2–9 bony 1-seeded connate or distinct stones. *Seed* erect, 3-gonous, testa very thin, albumen fleshy; embryo straight, slender, axile, cotyledons short, radicle inferior.—DISTRIB. N. temp. and Arctic zones, Chili and Fuegia; genera 3; species 4.—AFFINITIES. Very close to *Ilicineæ* (Decaisne); reduced *Ericaceæ* (A. Gray); with *Euphorbiaceæ* (A. DC., &c.).—PROPERTIES unimportant.

1. **EMPE'TRUM,** *L.* CROWBERRY.

Flowers bracteolate. *Sepals* and *petals* 3 each,. quite entire. *Ovary* 6–9-celled; stigmas 6–9, dilated. *Drupe* fleshy; stones free.—DISTRIB. of the Order; species 1.—ETYM. ἐν πέτρον, from growing in *stony* places.

E. ni'grum, *L.*; leaves linear-oblong margins so recurved as to meet over the midrib.

Moors, &c., Shetland to Devon and Somerset (Sussex, extinct); ascends to 4,000 ft.; Ireland; fl. April–June.—Glabrous, tufted; branches 6–18 in., slender, wiry, spreading and trailing. *Leaves* ¼–⅓ in., crowded, obtuse, reddish in age, sides minutely scabrid, the recurved portion concealing the pubescent under-surface, and forming a tube closed at both ends. *Flowers* minute, sessile. *Sepals* rounded, concave. *Petals* scarious, subspathulate, pink, reflexed. *Filaments* very long; anthers red. *Drupe* ¼–⅓ in., black (often purple in N. America, red in S. America), eatable.—The structure of the leaf is very curious.

ORDER XXII.   CELASTRI'NEÆ.

Trees or shrubs, sometimes spinous or climbing. *Leaves* opposite and alternate, simple, stipulate or not. *Flowers* small, cymose. *Calyx* small, 4-5-lobed, imbricate in bud, persistent. *Petals* 4-6, short, imbricate in bud. *Stamens* 4-6, inserted on the flat tumid or lobed disk, filaments subulate honeyed. *Ovary* sessile, 3-5-celled ; style entire or 3-5-fid, stigmas terminal ; ovules 2 in each cell, basal, erect, anatropous, raphe ventral. *Fruit* various, dehiscent or not. *Seeds* erect, usually arillate (the aril produced from the exostome, an *arillode*), albumen fleshy or 0 ; cotyledons large, foliaceous.—DISTRIB. Temp. and trop. ; genera 40 ; species 400.—AFFINITIES. With *Ampelideæ, Sapindaceæ, Ilicineæ,* and *Rhamneæ.*—PROPERTIES. Purgative and emetic.

1. EUON'YMUS, *L.*   SPINDLE-TREE.

Shrubs or trees. *Leaves* opposite, persistent ; stipules caducous. *Flowers* small, in axillary cymes, proterandrous. *Calyx* 4-6-fid. *Petals* and *stamens* 4-6, inserted on a broad, fleshy, 4-6-lobed disk. *Ovary* confluent with the disk, 3-5-celled ; style short, stigma 3-5-lobed. *Capsule* 3-5-lobed and celled, angled or winged, loculicidal, cells 1-2-seeded. *Seeds* with a complete arillode, albumen fleshy ; embryo straight. DISTRIB. Temp. Europe, Asia, and N. America ; species 40.—ETYM. *Euonyme,* mother of the Furies, the fruit being reputed poisonous.

E. europæ'us, *L.* ; leaves ovate- or oblong-lanceolate serrulate.

Copses and hedges, &c., from Roxburgh southd.; rare in Scotland ; local in Ireland ; fl. May–June.—A glabrous fœtid shrub or tree, 5-20 ft.; bark grey, smooth, twigs 4-angled green. *Leaves* 1-4 in., acute or acuminate ; petiole short. *Cymes* dichotomous, 5-10-fid.; peduncles 1-2 in. *Flowers* ½ in. diam., greenish white, polygamous. *Capsule* ½ in. deeply 4-lobed, pale crimson ; arillode orange.—DISTRIB. Europe to the Caucasus, N. Africa, W. Siberia. —Wood hard and tough, used for fine gunpowder, spindles, &c.

ORDER XXIII.   RHAM'NEÆ.

Trees or shrubs, often spiny, erect or climbing. *Leaves* simple, alternate or opposite, often 3-5-nerved ; stipules small. *Flowers* small, sometimes unisexual, green or yellow. *Calyx-tube* coriaceous ; lobes 4-5, triangular, valvate in bud. *Petals* 4, 5, or 0, inserted on the throat of the calyx, minute, usually clawed, hooded. *Stamens* 4-5, inserted with and opposite the petals which often enclose them, filaments subulate ; anthers small, versatile. *Disk* large, annular, cupular or coating the calyx-tube. *Ovary* 3-celled, sessile or sunk in the disk, free or adnate to the calyx-tube ; styles short, simple or 3-lobed, stigmas terminal ; ovules 1 in each cell, basal, erect, anatropous, raphe usually ventral. *Fruit* various, free or girt with the calyx-tube. *Seeds* compressed, sometimes arillate, albumen fleshy or 0 ; embryo large, often green, cotyledons plano-convex.—DISTRIB. Chiefly in warm and temp. regions ; genera 37 ; species

430.—AFFINITIES. With *Celastrineæ*, and certain *Euphorbiaceæ.*—PRO-PERTIES. Purgative. *Rhamnus* yields a green dye.

1. RHAM′NUS, *L.* BUCKTHORN.

Leaves alternate, deciduous. **Flowers** in small **axillary cymes,** often unisexual. *Calyx-tube* urceolate. *Petals* 4–5 or 0. **Stamens** 4 or 5, very short. *Disk* coating the calyx-tube, honeyed. *Ovary* adnate below with the calyx-tube, 3–4-celled ; style 3–4-fid, stigmas obtuse. *Drupe* girt with the calyx-tube ; stones 2–4. *Seeds* obovoid, albumen fleshy ; cotyledons flat or with recurved margins ; radicle short.—Temp. and trop. regions, none Australian ; species 60.—ETYM. The Greek name, from the branched habit. (Position of raphe variable, dorsal lateral or ventral.)

1. **R. cathar′ticus,** *L.* ; branchlets spinous, leaves ovate acutely serrate, nerves divergent, flowers 4-merous diœcious.

Woods and thickets from Westmoreland southd., chiefly on chalk, perhaps not wild N. of Durham ; rare in Ireland ; fl. May–July.—A rigid shrub, 5–10 ft., much branched ; bark blackish. *Leaves* 1–2 in., fascicled at the ends of the shoots, subopposite lower down, shortly petioled, young downy beneath ; stipules subulate, deciduous. *Flowers* ⅙ in. diam., solitary or fascicled in the axils of the fascicles of leaves on the previous year's wood, yellow-green, proterandrous ; pedice's very short. *Calyx* of the male campanulate, female cupular, lobes acute. *Style* 4- rarely 2- or 5-cleft. *Drupe* ¼ in. diam., globose, black ; stones 4, obovoid, grooved at the back. *Seed* curved like a horse-shoe ; embryo obcordate, similarly curved.—DISTRIB. Europe, N. Africa, Siberia ; cult. for hedges in the U. States.

2. **R. Fran′gula,** *L.* ; unarmed, leaves obovate quite entire, nerves parallel, flowers 5-merous bisexual. *Berry-bearing Alder.*

Woods and thickets, common in England ; Ayr and Moray in Scotland ; very rare in Ireland ; fl. May–June.—Shrub 5–10 ft. ; branches slender. *Leaves* alternate, stipules subulate. *Flowers* ⅙ in. diam., few, axillary, greenish-white, pedicels ½ in. *Calyx* campanulate. *Style* entire. *Drupe* ½ in. diam., globose, black when ripe ; stones compressed, broadly obovoid. *Seed* of the same form.—DISTRIB. Europe, N. Africa, Siberia.—Drupes cathartic, when unripe used to dye green. The Black Dogwood of gunpowder-makers.

ORDER XXIV. SAPINDA′CEÆ.

Tribe ACERI′NEÆ.

Trees ; juice often sugary, sometimes milky. *Leaves* opposite, simple or pinnate, deciduous. *Flowers* racemed or corymbose, often polygamous, regular ; the lower or earlier in the raceme generally male, the terminal 2-sexual. *Calyx* 5- rarely 4–12-parted, deciduous, imbricate in bud. *Petals* as many or 0, imbricate in bud. *Stamens* 8, rarely indefinite, inserted on the annular, thick, lobed disk. *Ovary* laterally compressed,

2- rarely 3–4-lobed and -celled ; styles 2, filiform, stigmatose on the inner surface ; ovules 2 in each cell, attached by a broad base, superposed or collateral. *Fruit* of 2, rarely 3 or 4 spreading samaras. *Seed* ascending, compressed, testa membranous, inner coat fleshy ; cotyledons plaited, radicle long.—Distrib. Europe, N. Asia, N. America, Java, the Himalaya, and Japan ; genera 3 ; species 50.—Affinities. Allied to *Celastrineæ.*—The Tribe *Acerineæ* differs from the others of this great tropical Order chiefly in the opposite leaves, and stamens inserted on (not within) the disk.—Properties. Several American maples yield sugar.

1. A'CER, *L.* Maple.

Leaves simple, entire or lobed. *Disk* annular.—Distrib. of the Tribe. —Etym. The Latin name.

A. campes'tre, *L.* ; leaves reniform obtusely 5-lobed entire lobulate or crenate, corymbs erect, wings of fruit horizontal. *Common* or *Small-leaved Maple.*

Thickets and hedgerows, from Durham southd.; naturalized in Scotland; Ireland; Channel Islands; fl. May–June.—Small tree 10–20 ft., with spreading branches; bark rough, fissured; wood beautiful, fine-grained. *Leaves* 2–4 in. diam., pubescent when young; petiole 1–1½ in., slender. *Corymbs* 1–2 in. *Flowers* ¼ in. diam., green, shortly pedicelled. *Sepals* linear-oblong. *Petals* similar, but narrower; wings of fruit linear-oblong, slightly curved, each ⅚ in. long.—Distrib. Europe, from Denmark southd., N. and W. Asia.

A. Pseudoplat'anus, *L.* ; leaves 5-angled, 5-lobed, lobes crenate-serrate, racemes elongate pendulous, wings of fruit divergent. *Great Maple, Sycamore, Plane* of Scotland.

Plantations, &c ; fl. May–June.—An umbrageous tree, 40–60 ft. *Bark* smooth, outer layer deciduous. *Leaves* 4–8 in. diam., glaucous beneath, lobes acute or acuminate. *Flowers* as in *A. campestre,* but pedicels shorter, and stamens longer. *Ovary* villous. *Samaras* 1½ in , scimitar shaped.—Distrib. Mid. Europe and W. Asia.—Sap sugary. Wood much used for turnery, &c.

Order XXV. LEGUMINO'SÆ.

Sub-order Papiliona'ceæ.

Herbs, rarely shrubs. *Leaves* alternate, 3- or more-foliolate, rarely simple ; stipules usually present ; leaflets often stipellate. *Inflorescence* various. *Flowers* irregular, proterandrous. *Calyx* of 5 connate sepals, often 2-lipped, *Petals* 5, very unequal, distinct or 2 or more adherent by their claws to the staminal tube, imbricate in bud ; upper (*standard*) broad, often reflexed, exterior in bud ; 2 lateral (*wings*) parallel, enclosing and sometimes adhering to the 2 lower (*keel*), which are interior in bud, and distinct, or connate by their lower edges. *Stamens* 10, perigynous, filaments united into a sheath, or the upper one distinct, equal or the alternate longer ; anthers usually versatile. *Disk* lining the base of the calyx-tube. *Ovary* of one 1-celled carpel, included in the staminal sheath ; style incurved, stigma simple, oblique or terminal ; ovules 2 or more, 1–2-seriate on the ventral suture, campylotropous or anatropous. *Fruit* a dry pod.*(legume)* dehiscent along one or both sutures or not at all, continuous or septate internally. *Seeds* with a coriaceous testa, and simple or dilated funicle, albumen 0 ; cotyledons plano-convex, radicle incurved, incumbent.—DISTRIB. Chiefly N. temp. but found all over the globe ; few in New Zealand ; genera 295 ; species 4,700.—AFFINITIES. Principally with *Rosaceæ.*—PROPERTIES too numerous to mention here.

The Tribal characters here given apply to British Genera to the exclusion of many exotic ones.

SERIES 1. *Leaves 1- or 3-foliolate, without tendrils.*

TRIBE I. **GENIS'TEÆ.** Shrubs. *Leaves* 0 or 1-foliolate or digitately 3-foliolate ; leaflets quite entire. *Filaments* all united. *Pod* 2-valved.
Calyx shortly 2-lipped, lips deeply toothed ....1. Genista.
Calyx deeply 2-lipped, coloured ....2. Ulex.
Calyx shortly 2-lipped, lips minutely toothed ....3. Cytisus.

TRIBE II. **TRIFO'LIEÆ.** Herbs, rarely shrubs. *Leaves* pinnately rarely digitately 3-foliolate ; veins generally ending in teeth. *Upper filament* usually separate.
Filaments all united ....4. Ononis.
Upper filament separate.
Racemes short. Pod longer than calyx, curved, dehiscent...5. Trigonella.
Racemes short. Pod usually spiral ....6. Medicago.
Racemes long. Pods short, indehiscent. Keel petals free...7. Melilotus.
Flowers capitate. Pod short, 1–4-seeded. Keel petals adnate.
8. Trifolium.

SERIES 2. *Leaves 5- or multi-foliolate, with a terminal leaflet.*

TRIBE III. **LO'TEÆ.** Herbs or shrubs. *Upper filament* separate or not ; alternate filaments often dilated. *Pod* 2-valved, without a longitudinal septum.
Calyx inflated, including the pod ....9. Anthyllis.
Calyx not inflated, pod exserted dehiscent ....10. Lotus.

TRIBE IV. **GALEGEÆ** (sub-tribe ASTRAGALE'Æ). Herbs or shrubs. *Upper filament* separate. *Pod* 2-valved, turgid or flat, with a longitudinal septum.
Keel obtuse...11. Astragalus.
Keel beaked or with an incurved tip ....12. Oxytropis.

TRIBE V. **HEDYSAR'EÆ.** Herbs or shrubs. *Upper filament* separate. *Pod* indehiscent, of 1 or many 1-seeded joints.
Pod cylindric, many-jointed ....13. Ornithopus.
Pod flat, of many curved 1-seeded joints ....14. Hippocrepis.
Pod flat, hard, 1-seeded ....15. Onobrychis.

SERIES 3. *Leaves abruptly pinnate (or leaflets* 0) ; *petiole ending in a tendril or point.*

TRIBE VI. **VIC′IEÆ.** Herbs. *Leaves* pinnate ; petiole ending in a tendril or point. *Leaflets* often toothed. *Upper filament* separate. *Pod* 2-valved. Style filiform, hairy below or all round......................................16. Vicia. Style flattened, hairy on the upper margin only ....17. Lathyrus.

1. GENIS′TA, *L.*

Shrubs, sometimes spinous. *Leaves* 1-foliolate in British species ; stipules minute or 0. *Flowers* racemed, yellow, rarely white, bracteate. *Calyx* shortly 2-lipped ; upper lip deeply 2-fid, lower shorter 3-toothed. *Wings* oblong, gibbous at the base, adnate to the staminal tube, deflexed after flowering. *Keel petals* separating and not resilient after deflection. *Filaments* all united, tube entire ; anthers alternately short and versatile, and long and basifixed. *Style* incurved, stigma oblique ; ovules 2 or more. *Pod* 2-valved or indehiscent, 1-celled.—DISTRIB. Europe, W. Asia, N. Africa ; species 70.—ETYM. obscure.

1. **G. tincto′ria,** *L.* ; unarmed, leaflet oblong or lanceolate nearly glabrous, racemes slender, keel and pod glabrous. *Dyers' Greenweed.*

Meadows and fields, from Wigton and Berwick southd.; Ireland; fl. July–Sept.—*Stem* 1–2 ft., branched, rigid, striate. *Leaflets* ½–1 in., appressed, ciliate ; stipules minute, subulate. *Flowers* ½ in., yellow, not honeyed. *Stamens,* 4 outer ripen first, then 5th, followed by 5 inner. *Calyx* deciduous above the base, teeth acuminate. *Pod* 1–1¼ in., compressed, 5–10-seeded, —DISTRIB. From Gothland southd., N. and W. Asia; introd. in N. America.—Yields a yellow dye.

G. tinctoria proper (VAR. *glabra,* Syme) ; glabrous, branches erect or ascending.— VAR. *humifu′sa,* Syme ; branches decumbent hairy, as are the pedicels calyx and pod. Kynance Cove, Lizard district.

2. **G. pilo′sa,** *L.* ; unarmed, leaflet obovate-lanceolate obtuse complicate recurved silky beneath, racemes short leafy, keel and pod pubescent.

Gravelly heaths, rare and local, Suffolk and S. Wales to Cornwall and Kent; fl. May–Sept.— *Stem* much branched, curved, prostrate, tortuous, woody. *Leaflets* ¼ in., very shortly petioled ; stipules ovate, obtuse. *Flowers* ½ in., yellow. *Calyx* with 2 upper lobes lanceolate, and 3 lower subulate. *Pod* ¾ in., deciduous, flat, valves bulging over the seeds.—DISTRIB. From Gothland southd. to Greece and Tauria.

3. **G. an′glica,** *L.* ; glabrous, spinous, leaflet ovate oblong or lanceolate, racemes short leafy, keel and pod glabrous. *Needle Furze.*

Heaths and moist moors, from Ross southd.; ascends to 2,200 ft. in the Highlands ; fl. May–June.—*Branches* 1–2 ft., slender, spreading, curved ; spines ½–1 in., slender, recurved, simple, rarely branched. *Leaflets* ⅒–⅓ in.; stipules obsolete. *Flowers* ½ in., yellow, shortly pedicelled. *Calyx* persistent ; teeth short, triangular. *Pod* ¾ in., deciduous, inflated, acuminate at both ends.—DISTRIB. W. Europe, from Denmark to France, Germany, and Italy.

2. U'LEX, *L.* FURZE, WHIN, GORSE.

Densely spinous shrubs. *Leaves* 3-foliolate in seedling plants, in mature spinescent or reduced to small scales, exstipulate. *Flowers* yellow, axillary ; bracts small. *Calyx* membranous, coloured, 2-partite, upper lobe 2- lower minutely 3-toothed. *Petals* shortly clawed ; keel-petals and wings obtuse. *Stamens* as in *Genista. Style* smooth, stigma capitate ; ovules many. *Pod* 2-valved, 1-celled.—DISTRIB. W. Europe to Italy, N.W. Africa ; species 12.—ETYM. doubtful.

1. **U. europæ'us,** *L.* ; bracts large ovate lax, calyx hairs spreading, teeth minute, wings longer than the keel.

Heaths, &c., N. to Shetland ; rarer in the North ; ascends to 2,100 ft. in Wales ; Channel Islands ; fl. Feb.–March, and Aug.–Sept.—Bush 2–5 ft., rounded ; spines 1–2 in., straight. *Leaves* small ; leaflets hairy ; spines sometimes furnished with minute 1-foliolate leaves. *Flowers* ¾ in., borne on the spines, bright yellow, odorous. *Calyx* yellow, hairs black. *Pod* ¾ in., black, covered with brown hairs, dotted.—DISTRIB. Denmark to Italy, Canaries and Azores.

U. europæ'us proper ; branches spreading, spines furrowed rigid.—VAR. *U. stric'tus,* Mackay ; branches erect compact, spines soft 4-gonous. Lord Londonderry's park, Down. *Irish Furze.*

2. **U. na'nus,** *Forster ;* bracts minute, calyx with appressed pubescence, teeth lanceolate, wings longer or shorter than the keel.

Heaths and commons,. from Ayr and Northumbd. southd.; Ireland ; fl. July–Nov.—Much smaller than *U. europæus. Stems* 1–3 ft. *Spines* ½–1½ in. *Flowers* ½ in., more racemose. *Pods* persist till next season.—DISTRIB. Belgium and France.

U. NA'NUS proper ; branches procumbent drooping, primary spines weak short. —Chiefly in S. England ; not in Ireland ; Channel Islands. (S.E. France). Sub-sp. U. GAL'LII, *Planch.* ; branches ascending, primary spines rigid.— Heaths and downs in the West, common ; ascends to 2,000 ft. in Ireland. (S.W. France).

3. CYT'ISUS, *L.* BROOM.

Shrubs, rarely spinous. *Leaves* 1–3-foliolate or 0. *Stipules* minute. *Flowers* yellow, purple, or white, not honeyed. *Calyx* 2-lipped ; upper lip minutely 2- lower 3-toothed. *Wings* oblong, and keel obtuse, deflexed (as in *Genista*) after flowering, their claws free. *Stamens* as in *Genista. Style* incurved or coiled, smooth, stigma terminal ; ovules many. *Pod* flat, elongate, 2-valved, many-seeded, somewhat septate. *Seeds* with a tumid funicle.—DISTRIB. Europe, N. Africa, W. Asia.—ETYM. obscure.

1. **C. scopa'rius,** *Link ;* branchlets and obovate leaflets silky. *Spartium,* L. *Sarothamnus,* Koch.

Heaths, commons, &c., from Caithness southd.; ascends to 2,000 ft. in the Highlands ; Ireland ; Channel Islands ; fl. May–June.—Shrubby, 2–6 ft., hairy ; branches green, angular, furrowed. *Leaves* shortly petioled ; leaflets

1-3, $\frac{1}{4}-\frac{1}{2}$ in. *Flowers* 1 in., bright yellow, rarely white, pedicels short. *Style* spiral. *Pod* 1-2 in., black; valves twisted after dehiscence.—A prostrate variety is found at Kynance Cove.—Distrib. From Gothland S., excl. Greece and Turkey, N. Asia, Canaries, Azores.—Twigs diuretic, and used for tanning. Seeds a substitute for coffee.

4. ONO'NIS, *L.* Rest-harrow.

Herbs or small shrubs, with often viscid hairs. *Leaves* pinnately foliolate, nerves ending in teeth ; stipules adnate to the petiole. *Flowers* pink, white or yellow, not honeyed ; bracts minute or 0, upper leaves bracteæform ; peduncles sometimes spinescent. *Standard* broad ; wings oblong ; keel incurved, pointed, not adnate to the staminal tube, returning to position after deflection. *Filaments* all connate, 5 or all dilated above ; anthers uniform, or the alternate smaller. *Style* incurved, smooth, stigma terminal ; ovules 2 or many. *Pod* turgid or terete, 1-celled, 2-valved.—Distrib. Europe, W. Asia, N. Africa ; species 60.—Etym. The Greek name for the genus.

1. **O. spino'sa,** *L.* ; suffruticose, hirsute, usually spiny, pod obliquely ovate or oblong 1-4-seeded.

Dry pastures, fields, and sandy shores, N. to Sutherland ; Ireland ; Channel Islands ; fl. June–Aug.—A very variable undershrub, 1-2 ft., much branched, villous or thinly hairy and glandular, hairs on the branches in 2 lines or all round. *Leaves* often 1-foliolate, leaflets $\frac{1}{4}-\frac{3}{4}$ in. *Flowers* sessile or shortly pedicelled, solitary or in leafy racemes, $\frac{3}{4}$ in. long, or less, pink, proterandrous. *Standard* streaked with red. *Pod* $\frac{1}{3}$ in. long. *Seeds* granulate.— Distrib. Europe, W. Asia, N. Africa.—The two sub-species present no constant characters, and authors are greatly at variance with respect to the names they should bear. Wilkomm and Lange appear to have devoted most time and study to them. They adopt the Linnean names for the two principal forms (as does Boissier for the erect form), and they describe a third intermediate one for *O. arven'sis,* L., to which they refer *O. spino'sa a,* L., *O. iner'mis β,* Huds., and *O. procur'rens,* Wallr., as synonyms. It differs from *repens* in the shorter standard and pods, and is fœtid and viscidly hirsute.

O. spinosa proper; erect, spinous, not fœtid, without stolons, hairs on branches usually fascicled, leaflets linear-oblong, pod usually equalling the calyx. *O. spinosa a,* L., *O. campestris,* Koch and Ziz.—From Forfar and Dumbarton southd. ; not in Ireland.

Sub-sp. O. re'pens, L. ; viscidly villous, prostrate or ascending, stoloniferous, rarely spinous, leaflets ovate or obovate, flowers larger, pod usually shorter than the calyx. *O. iner'mis β,* Huds. *O. procum'bens β, marit'ima,* Gren. and Godr.—Dry pastures and sandy shores, common.—Var. *horrida,* Lange, is a maritime spinous form.

2. **O. reclina'ta,** *L.* ; annual, spreading, viscid and hairy, pod cylindric oblong reflexed.

Sea-cliffs, Devon, Wigton; Alderney ; fl. June–July.—*Stem* 2-3 in. *Leaflets* $\frac{1}{4}-\frac{1}{2}$ in., acutely toothed ; stipules large, $\frac{1}{2}$-ovate. *Pedicels* slender, jointed beneath the flower. *Flower* $\frac{1}{3}$ in., rosy. *Pod* $\frac{1}{2}$ in., glandular and hairy as long as the calyx or longer.—Distrib. W. France, Spain, Italy, Greece.

5. TRIGONEL'LA, *L.* FENUGREEK.

Herbs, often strongly scented. *Leaves* pinnately 3-foliolate ; nerves terminating in teeth ; stipules adnate to the petiole. *Flowers* solitary, capitate, or in dense racemes, white, yellow, or blue. *Calyx* tubular, teeth subequal. *Petals* very persistent ; wings longer than the keel, the claws of both free from the staminal tube. *Filaments* not dilated, upper distinct or nearly -so ; anthers uniform. *Style* glabrous, stigma terminal ; ovules many. *Pod* various in form, dehiscent in the British species, longer than the calyx.—DISTRIB. Europe, temp. Asia, N. and S. Africa, Australia ; species 50.—ETYM. The old Greek name.

T. ornithopodioi'des, *DC.* ; glabrous, prostrate. very slender, leaflets obcordate, keel nearlv as long as the wings. *Trifolium,* L. *Falcatula,* Brot. *Aporan'thus Trifolias'trum,* Bromfield.

Sandy heaths and gravelly places, local, from Fife and Renfrew southd.; F. Ireland, very rare ; Channel Islands; fl. June–Aug.—Annual or biennial, branching from the base; branches 2–8 in., slender. *Leaflets* ½–⅔ in, toothed, strongly nerved ; petiole ½–1 in.; stipules large, ovate, long-acuminate. *Peduncles* solitary, sborter than the petioles, axillary, 1–3-flowered. *Flowers* ¼ in., very shortly pedicelled, white and pink. *Pod* ⅓–½ in., linear-oblong, slightly curved, obtuse, partially dehiscent, 6–8-seeded.—DISTRIB. W. Europe from Denmark to Portugal and Italy.—An anomalous species, on account of the length of the keel; habit of *Trifolium subterraneum.*

6. MEDICA GO, *L.* MEDICK.

Herbs. *Leaves* pinnately 3-foliolate, nerves ending in teeth ; stipules adnate to the petiole. *Flowers* small, yellow or violet, honeyed : bracts small or 0. *Calyx-teeth* 5, nearly equal. *Keel* obtuse, shorter than the oblong wings, not adnate to the stamens, spreading and exposing the anthers. *Filaments* not dilated, upper distinct ; anthers uniform. *Style* subulate, glabrous, stigma subcapitate ; ovules few or manv. *Pod* spirally curved or coiled, verv rarely falcate, often spiny, rarely dehiscent, 1-more-seeded.—DISTRIB. Europe, W. Asia, N. Africa ; species 40.—ETYM. The Greek name.

* *Perennial. Pod dehiscent, falcate. annular, or coiled into an open helix, smooth, many-seeded ; marginal nerve* 0.

1. **M. falca'ta,** *L.* ; erect or decumbent, pod linear compressed falcate or annular downy.

Gravel banks and waste sandy places in the E. counties; fl. June–July.—*Stems* 6–24 in., diffuse. *Leaflets* ¼–⅓ in., narrowly linear, or obovate oblong, toothed, notched and mucronate at the tip; stipules large, subulate. *Flowers* ⅓ in., shortly pedicelled, in contracted racemes. *Pod* ½ in.—DISTRIB. Europe, N. Asia, India.

M. sylves'tris, Fries ; with flowers vellow then dark green, and pod semicircular or annular, is probably a hybrid, *Nyman.*—E. counties.

M. SATI'VA, *L.* ; erect, pod with 2–3 coils downy. *Purple Medick, Lucerne.*

Hedges and fields, not indigenous; fl. May–July.—*Stem* 1–2 ft., fistular, branched. *Leaflets* narrowly obovate-oblong, toothed, tip notched and apiculate. *Flowers* $\frac{1}{3}$–$\frac{1}{2}$ in., in a short dense raceme, yellow blue or purple; peduncles longer than the leaves, pedicels very short. *Pod* $\frac{1}{4}$ in. diam.—DISTRIB. E. Mediterranean region; naturalized elsewhere.

** *Annual or biennial. Pod indehiscent, reniform, tip coiled, smooth,* 1-*seeded; marginal nerve* 0.

2. **M. lupuli'na,** *L.* ; procumbent, flowers in ovoid heads, pod reticulate. *Black Medick.*

Waste places and fields, N. to Shetlands; ascends in Derby to 1,200 ft.; Ireland; Channel Islands; fl. May–Aug.—Pubescent or glabrous. *Stems* 6–24 in., much branched. *Leaflets* $\frac{1}{4}$–$\frac{2}{3}$ in., obovate, toothed, apiculate; petiole very short; stipules $\frac{1}{2}$-cordate. *Flowers* $\frac{1}{8}$ in., peduncles longer than the petioles; pedicels very short. *Pods* $\frac{1}{8}$–$\frac{1}{6}$ in., black.—DISTRIB. Europe, N. Africa, temp. Asia, India; naturalized elsewhere.

*** *Annual. Pod indehiscent, coiled into a closed helix, many-seeded, bordered with spines or tubercles, marginal nerve strong.* (*Stems prostrate. Leaflets toothed. Peduncles few-flowered. Flowers yellow; calyx-teeth subulate, incurved.*)

3. **M. denticula'ta,** *Willd.* ; nearly glabrous, stipules laciniate, pod flat deeply reticulate with 2 or 3 coils and a double row of spines.

E. and S. counties from York and Norfolk to Cornwall, casual elsewhere; (a denizen? *Wats.*); Ireland; Channel Islands; fl. May–Aug.—*Stem* 6–24 in., furrowed. *Leaflets* $\frac{1}{4}$–1 in. *Flowers* $\frac{1}{6}$ in., umbelled, yellow; pedicels very short. *Pod* $\frac{1}{4}$ in. diam.—DISTRIB. Mid. and S. Europe, N. Africa, N. Asia, India; introd. in N. America.

M. denticula'ta proper; spines subulate half the diameter of the pod.—VAR. *M. apicula'ta,* Willd.; spines of pod very short.—VAR. *M. lappa'cea,* Lamk.; spines longer than the semi-diameter of the subglobose pod. Bedford.

4. **M. macula'ta,** *Sibth.* ; nearly glabrous, stipules $\frac{1}{2}$-cordate toothed, pod subglobose faintly reticulate with 3–5 coils and a double row of long curved spines.

Pastures and hedgebanks, from Northumberland southd.; S. Ireland; Channel Islands; fl. May–Aug.—*Stems* 1–2 ft., numerous. *Leaflets* $\frac{1}{4}$–1 in., obovate obcordate or cuneate, with often a black central spot; stipules herbaceous; petiole hairy. *Flowers* $\frac{1}{6}$ in., yellow. *Pod* $\frac{1}{4}$ in. broad; margin with 4 ridges. —DISTRIB. Europe, N. Africa, W. Asia; introd. in N. America.

5. **M. min'ima,** *Desr.* ; downy, stipules $\frac{1}{2}$-cordate faintly toothed, pod subglobose faintly reticulate with 4 or 5 coils and a double-row of close-set hooked spines.

Sandy fields, rare; E. counties, Norfolk to Kent; Channel Islands; fl. May–July.—*Stems* 6–10 in., very many, rigid, prostrate. *Leaflets* $\frac{1}{4}$–$\frac{1}{3}$ in., from narrowly to very broadly obovate. *Flowers* $\frac{1}{8}$ in. *Pod* $\frac{1}{4}$ in. diam., margin keeled.—DISTRIB. From Denmark southd., N. Africa, W. Asia, Kashmir.

96   *LEGUMINOSÆ.*   [MELILOTUS.

7. **MELILO'TUS,** *Tournefort.* MELILOT.

Annual or biennial, fragrant herbs. *Leaves* pinnately 3-foliolate, nerves
ending in teeth ; stipules adnate to the petiole. *Flowers* in axillary
racemes, small, drooping, yellow or white, honeyed ; bracts minute or 0.
Calyx-teeth 5, nearly equal. *Petals* very deciduous ; standard oblong ; keel
shorter than the wings, obtuse, not adnate to the stamens, resilient after
depression. *Filaments* not dilated upwards, the upper distinct, or only so
above the middle ; anthers uniform. *Style* filiform, stigma terminal ;
ovules few. *Pod* short, straight, thick, hardly dehiscent. *Seeds* 1 or few.
—Warm and temp. regions of the Old World ; species 10.—ETYM. *mel*
and *lotus,* from the *honeyed* smell.

1. **M. altis'sima,** *Thuill.* ; erect, petals nearly equal, pod ovoid compressed acuminate reticulate hairy. *M. officina'lis,* Willd.

Fields,&c., from Perth southd.; Ireland very rare; (a denizen,*Wats.*); fl. June–
Aug.—Annual or perennial. *Stem* 2–3 ft., much branched. *Leaflets* ½–1¼
in., obovate- or linear-oblong, toothed; stipules subulate, very slender.
Racemes 3–4 in. *Flowers* ¾ in., secund, deep yellow, pedicels short. *Corolla*
more than twice as long as the calyx. *Pod* black when ripe, 1–2-seeded.—
DISTRIB. Europe, E. and W. Asia, Tibet; introd. in N. America.

2. **M. al'ba,** *Desr.* ; standard longer than the wings or keel, pod ovoid
acute reticulate, glabrous. *M. vulgaris,* Willd. *M. leucantha,* Koch. *White
Melilot.*

Waste places, not rare, from Elgin southd.; Ireland; (an alien or colonist?
Wats.) ; fl. July–Aug.—Very similar to *M. altissima,* but more slender ;
flowers smaller, white, in long racemes; pod more ovoid and glabrous, also
black when ripe.--DISTRIB. Europe, E. and W. Asia, India; introd. in
N. America.

M. OFFICINA'LIS, *Desr.* ; standard longer than the keel as long as the
wings, pod ovoid obtuse mucronate transversely ribbed rugose glabrous.
M. arven'sis, Wallr.

Waste places, not indigenous, chiefly in the E. counties ; Ireland ; fl June–
Aug.– Habit, &c., of the preceding. *Flowers* pale-yellow or white. *Pod*
olive-brown when ripe. –DISTRIB. Europe, N. Africa, N. and W. Asia,
India.

8. **TRIFO'LIUM,** *L.* TREFOIL, CLOVER.

Herbs, usually low. *Leaves* digitately, rarely pinnately 3-foliolate ;
stipules adnate to the petiole. *Flowers* capitate or spiked, rarely solitary,
red, purple, or white, rarely yellow, honeyed ; bracts small or 0, sometimes
forming a toothed involucre. *Calyx-teeth* 5, subequal. *Petals* persistent ;
wing longer than the keel, the claws of both adnate to the staminal tube ;
keel petals resilient after depression, but exposing the stamens. *Upper
stamen* distinct ; filaments all or 5 with dilated tips ; anthers uniform.
Style filiform, stigma oblique or dorsal ; ovules few. *Pod* small, indehiscent, or with the top falling off, rarely 2-valved, 1-4-seeded, nearly enclosed

in the calyx.—DISTRIB. N. temp. regions, rare in S. ; species 150.—ETYM. in allusion to the 3 leaflets.

SECTION 1. *Heads* axillary. *Fertile flowers* few. *Calyx* enclosing the 1-seeded pod, which at length splits ; its throat naked. *Petals* caducous. *Pods* burrowing in the earth when ripening, then covered by the reflexed deformed calyces of the other flowers.

1. **T. subterra'neum,** *L.* ; very hairy, stipules broadly ovate acute, calyx-teeth setaceous as long as the tube, deformed calyces slender with 5 rigid palmate lobes.

Gravelly and sandy pastures, from Chester southd. ; Wicklow ; Channel Islands ; fl. May–June.—Annual ; covered with spreading soft hairs. *Stems* ½–2 ft., very many, prostrate. *Leaflets* ½ in., broadly obcordate. *Heads of flowers* ¼ in. diam., lengthening after flowering. *Flowers* cream-coloured (cleistogamous occur). *Pod* orbicular, compressed. *Seeds* shining.— DISTRIB. From Holland southd., N. Africa, W. Asia, N.W. India.

SECTION 2. *Heads* many-fld., rarely axillary, globose or oblong ; pedicels ebracteate. *Calyx* not inflated ; throat with a ring of hairs or callous constriction (obscure in *T. Bocconi*) ; teeth ciliate, equal or the lower longest. *Petals* usually persistent. *Pod* sessile, 1-seeded.

* *Heads cylindric or oblong during or after flowering.*

2. **T. arven'se,** *L.* ; softly hairy, leaflets narrow obovate-oblong longer than the petiole, stipules with very long setaceous points, heads terminal peduncled cylindric soft, calyx-teeth persistent longer than the corolla plumose. *Hare's-foot Trefoil.*

Dry pastures and fields, from Isla and Ross southd. ; local in Ireland ; Channel Islands ; fl. July–Sept.—Annual. *Stems* many, ascending or suberect. *Leaflets* ½–¾ in. *Heads* ½–1 in., dense. *Flowers* minute, white or pale pink. —DISTRIB. Europe, N. Africa, N. and W. Asia ; introd. in America.

3. **T. Bocco'ni,** *Savi ;* pubescent, leaflets obovate. stipules ovate with setaceous points, heads axillary and terminal cylindric sessile, calyx glabrous teeth straight erect spinescent rather shorter than the petals.

Dry places, W. Cornwall, very rare ; (a native? *Wats.*) ; fl. July.—Annual *Stems* 2–4 in. (10–12 in. in cultivation). *Leaves* shortly petioled ; leaflets ¼–¾ in., glabrous above, variable in breadth. *Heads* ¼–½ in. *Flowers* white. *Calyx* with a very obscure ring in the throat. *Pod* enclosed in the (not ventricose) calyx.—DISTRIB. Mediterranean region.

T. INCARNA'TUM, *L.* ; pubescent or villous, leaflets broadly obovate or obcordate, stipules obtuse, heads peduncled terminal ovoid or cylindric, calyx hairy teeth shorter than the corolla spreading in fruit. *Crimson Clover.*

Cultivated in England ; Channel Islands ; fl. June–July.—Annual. Very variable in size and pubescence. *Stems* rather slender. *Leaves* shortly petioled ; leaflets ¾–1½ in. *Heads* 1–2 in. *Flowers* ½ in. *Calyx ribs* strong. —DISTRIB. S. and W. Europe.

T. incarnatum proper; stem villous with spreading hairs, flowers scarlet. Naturalized only.—VAR. *T. Molinerii*, Balbis; hairs of stem appressed, heads shorter, calyx-teeth glabrous at the tips, flowers pale white or rose.—Lizard Point and Kynance Cove, amongst short grass. Probably the original form, of which *incarnatum* is the cultivated state.

** *Heads ovoid or globose.*

4. **T. ochroleu'cum,** *L.* ; softly pubescent, leaflets obovate or oblong, heads terminal globose at length ovoid, peduncles short subtended by opposite leaves, calyx-teeth spinescent spreading and recurved in fruit, lowest much longest.

Dry pastures, local; E. counties, Norfolk and Bedford to Essex; Channel Islands; fl. June–Aug.—Perennial. *Stems* ascending, 6–18 in. *Leaflets* ½–1 in., tip entire or notched; stipules lanceolate, adnate to the middle. *Heads* ¾–1 in. broad. *Flowers* pale yellow, brown when old. *Calyx* ribbed, half as long as the corolla. *Pod* striate, opening by the conical top falling away.— DISTRIB. From Belgium southd., W. Asia.

5. **T. praten'se,** *L.* ; more or less pubescent, leaflets oblong, stipules membranous free portion appressed to the petiole, heads terminal sessile globose at length ovoid subtended by opposite leaves with much-dilated stipules, calyx-teeth slender setaceous erect or spreading in fruit, the lowest longest. *Red* or *Purple Clover.*

Pastures, roadsides, &c., N. to Shetland; ascends to 1,900 ft. in the Highlands; Ireland; Channel Islands; fl. May–Sept.—Annual or perennial. *Stems* 6–24 in., solid or fistular, robust or slender. *Leaflets* ½–2 in., often with a white spot or lunate band, finely toothed; stipules often 1–1½ in., with long setaceous points. *Heads* ½–1½ in. diam., pink purple or dirty white. *Flowers* proterandrous. *Calyx* strongly nerved, throat with a 2-lipped contraction; teeth not exceeding the corolla, very slender, unequal. *Pod* opening by the top falling off.—DISTRIB. Europe (Arctic), N. Africa, N. and W. Asia, India; introd. in N. America.—Cultivated for fodder; also wild in a small form with a shorter corolla (var. *parviflora*, Bab.).

6. **T. me'dium,** *Huds.* ; slightly hairy, leaflets oblong obtuse or acute, stipules herbaceous free portion spreading, heads terminal subglobose shortly peduncled subtended by opposite leaves, calyx-teeth setaceous spreading in fruit, lowest a little longest. *Meadow Clover.*

Pastures, meadows, &c., from Ross southd.; ascending to 1,300 ft. in the Highlands; Ireland; fl. June–Sept.—Perennial. *Stems* straggling, flexuous. *Leaflets* 1–2 in., rather rigid almost quite entire, ciliate. *Heads* 1–1½ in. diam. *Flowers* ¾ in., rose-purple, proterandrous. *Calyx-throat* with a ring of hairs, tube 10-nerved, glabrous; teeth reaching half-way up the corolla. *Pod* dehiscing longitudinally.—DISTRIB. Europe (Arctic), N. and W. Asia; introd. in N. America.

7. **T. marit'imum,** *Huds.* ; pubescent, leaflets narrowly obovate-oblong obtuse or acute, stipules herbaceous free portion linear-subulate spreading, heads terminal ovoid very shortly peduncled subtended by opposite leaves, calyx-teeth short triangular subulate spreading and herbaceous in fruit, 4 upper shorter than the tube.

Salt marshes and meadows, from Gloster and Lincoln to Somerset ; Ireland ; fl.
June–Aug.—Annual. *Stems* 6–18 in., rigid, decumbent or ascending. *Leaflets*
½–⅔ in., almost quite entire. *Heads* ½ in., terminal, elongating in fruit.
Calyx with a 2-lipped contraction in the throat; tube ribbed ; teeth reaching
half-way up the corolla. *Pod* 2-valved.—DISTRIB. W. and S. Europe,
Holland to Spain, Greece and N. Africa.

8. **T. stria'tum,** *L.*; softly hairy, leaflets obovate, stipules membranous
free portion broadly triangular tip recurved, heads terminal and axillary
ovoid sessile, calyx ventricose, teeth short spinescent triangular-subulate
1-nerved spreading in fruit.

Dry pastures, from Forfar southd.; rarer in Scotland; very rare in Ireland ;
Channel Islands; fl. June–July.—Annual. *Stems* 4–12 in., spreading, prostrate or ascending. *Leaflets* ¼–¾ in., almost quite entire, petiole 2 in. ; stipules
of the leaves under the heads very much dilated. *Heads* ¼–½ in., dense,
broadest at the base. *Flowers* ¾ in. rosy. *Calyx-tube* ovoid, ribbed, mouth
contracted; teeth shorter than the corolla.—DISTRIB. From Gothland southd.,
N. Africa, Caucasus.—VAR. *erecta,* Leight., is an erect luxuriant form.

9. **T. sca'brum,** *L.* ; pubescent, leaflets narrowly or broadly obovate,
stipules rather rigid, free portion short triangular-subulate, heads terminal
and axillary ovoid sessile, calyx-tube ribbed, teeth triangular spinescent
1-nerved erect in flower lengthening spreading rigid and recurved in fruit.

Sandy and stony pastures, &c., in England; E. Scotland, from Forfar southd. ;
E. Ireland ; Channel Islands ; fl. May–July.—Annual. *Stems* 4–10 in., rigid,
stout, prostrate, zigzag. *Leaves* very shortly petioled; leaflets ¼–½ in., rigid,
strongly nerved, toothed. *Heads* ½–⅔ in. long, broadest in the middle.
Flowers minute, white. *Calyx-tube* purplish; teeth equalling the corolla.
Pod minute.—DISTRIB. From Belgium southd., W. Asia, N. Africa.

SECTION 3. *Heads* many-fld., usually axillary, rarely both axillary and
terminal ; pedicels distinct, short or long, bracteate. *Calyx* not inflated ;
throat naked ; teeth equal or the upper longest. *Pod* 2–4-seeded (or 1-seeded
through imperfection).

10. **T. glomera'tum,** *L.* ; glabrous, leaflets obovate acute or obcordate,
stipules ovate with long points, heads axillary and terminal globose sessile,
calyx-teeth short ovate 1-nerved spinescent reticulated spreading in fruit.

Gravelly and sandy pastures, &c., rare ; Norfolk and Kent to Cornwall; Wicklow ; Channel Islands; fl. June.—Annual. *Stems* 6–12 in., prostrate, spreading, slender. *Leaflets* ⅓–⅔ in., nerves fine but ending in strong teeth. *Heads*
¼ in. diam., distant. *Flowers* blue-purple, subsessile. *Calyx-teeth* nearly
equal, shorter than the corolla, contracted at the base. *Standard* persistent,
scarious, striate.—DISTRIB. W. France, Spain, and Mediterranean region.

11. **T. suffoca'tum,** *L.* ; glabrous, petioles very long, leaflets obcordate,
stipules ovate acuminate, heads sessile ovoid, calyx campanulate, teeth not
spinescent recurved in fruit exceeding the corolla.

Sandy and gravelly pastures, especially near the sea, rare ; Anglesea and
Norfolk to Cornwall and Kent; Channel Islands; fl. June–July.—Annual.
Stems 2–6 in., prostrate, spreading, slender. *Leaflets* ¼–¾ in., toothed, nerves

H 2

faint petiole 1–3 in. *Heads* ¼ in. diam., often confluent, axillary and terminal. *Flowers* minute, whitish. *Standard* persistent, scarious. *Calyx* membranous; teeth herbaceous, lanceolate, as long as the tube.—DISTRIB. Mid. Europe and Mediterranean region.

12. **T. stric′tum,** *L.* ; glabrous, petioles very short, leaflets linear-lanceolate, stipules broadly ovate acute toothed, heads peduncled globose, calyx-tube campanulate, teeth subulate spinescent sub-equal spreading in fruit.

Lizard Rocks, very rare ; Jersey ; fl. June–July.—Annual. *Stems* 2–6 in., few, ascending. *Leaflets* ½–1 in., toothed ; nerves slender. *Heads* ¼–½ in. broad, terminal and axillary ; peduncle ½–1 in., strict. *Flowers* ⅛ in., rose-purple. *Corolla* longer than the calyx. *Pod* obliquely orbicular, compressed, beaked, dorsal suture much thickened, 1–2-seeded.—DISTRIB. W. France and Spain to Greece, N. Africa.

T. HY′BRIDUM, *L.* ; almost glabrous, leaflets obovate or oblong, stipules oblong tips triangular, heads axillary peduncled globose, pedicels elongate at length reflexed, flowers drooping, calyx-tube campanulate gibbous, teeth subulate nearly equal unaltered in fruit. *Alsike Clover.*

Fields, &c., introduced with clover (the var. *elegans* only); fl. June–Aug.—Perennial. *Stems* 2–10 in., flexuous. *Petioles* long ; leaflets ½–1½ in., toothed. *Stipules* herbaceous, nerves green. *Heads* ¾–1 in. diam., depressed (from the dropping flowers); peduncles 2–4 in. *Flowers* ⅓ in., white or rosy. *Calyx* white, teeth green. *Standard* twice as long as the calyx, striate, folded over the 2-seeded pod.—DISTRIB. Europe, N. Africa, W. Asia. *T. hy′bridum* proper ; stem stout fistular, stipules few-nerved.—VAR. *T. el′egans,* Savi ; stem weak decumbent solid, leaves more sharply toothed, stipules with several nerves, heads small.

13. **T. re′pens,** *L.* ; glabrous, leaflets obovate or obcordate, stipules lanceolate acuminate, heads all axillary very long-peduncled globose, pedicels at length reflexed, calyx-tube campanulate gibbous, teeth triangular unaltered in fruit. *White or Dutch Clover.*

Meadows and pastures, N. to Shetlands: ascends to 2,700 ft. in the Highlands; Ireland ; Channel Islands ; fl. May–October.—Perennial. *Stems* 1–18 in., creeping. *Leaflets* ¼–1 in., toothed, with often a white semilunar band towards the base; nerves slender; petiole 2–4 in. *Heads* ⅜–1¼ in. diam.; peduncle 3–6 in. *Flowers* nearly ½ in., white or rosy. *Standard* persistent, brown, covering the pod, much longer than the calyx. *Pod* elongate, 4–6-seeded.—A Scilly Is. variety (*Townsen′dii*) has dark rose-purple flowers.—DISTRIB. Europe (Arctic), N. Africa, N. and W. Asia, India, N. America.

SECTION 4. *Heads* many-fld., axillary ; pedicels short, bracteate. *Calyx* 2-lipped, becoming inflated above, membranous, reticulated ; throat naked : teeth equal, the upper pair lengthening. *Corolla* withering, standard deciduous. *Pod* sessile, 1–2-seeded.

14. **T. fragif′erum,** *L.* ; glabrous, leaflets obovate or obcordate, stipules oblong-triangular with a long point, heads dense-flowered long-peduncled

globose, outer bracts lanceolate as long as the calyx, calyx-tube downy above striate vesicular, 2 upper teeth enlarged and deflexed in fruit.

Meadows and ditches, from Fife southd.; local in Scotland and Ireland; Channel Islands; fl. July–Aug.—Perennial. *Stems* 6–12 in., creeping *Heads* ¼–⅔ in. diam. *Flowers* ¼ in., rose-purple. *Pod* ovoid, compressed, reticulate, 1–2-seeded.—DISTRIB. Europe, N. Africa, W. Asia, N.W. India. —Habit when flowering of *T. repens*, but the large bracts and fruit are widely different.

SECTION 5. *Heads* many-fld., axillary, globose, long-peduncled; pedicels short or 0, bracteate. *Flowers* at length pendulous. *Calyx* not inflated; throat naked; teeth equal or upper shorter. *Corolla* yellow; standard persistent, enlarged and bent down over the fruit. *Pod* stalked, 1- rarely 2-seeded.—*Leaves* pinnately or palmately 3- rarely 5-foliolate.

15. **T. procum'bens,** *L.*; stems pubescent, leaves more or less pinnately 3-foliolate, leaflets obovate or obcordate, heads many and dense-fld., standard broad slightly convex arching but not folded over the pods. *T. agra'rium*, Huds. *Hop Trefoil.*

Pastures, roadsides, &c., from Ross southd.; ascends to 1,200 ft. in Derby; Ireland; Channel Islands; fl. June–Aug.—Annual. *Stems* 6–18 in., central erect, lateral decumbent. *Leaflets* ¼–¾ in., toothed; petiole ¼–¾ in., slender; stipules ½-ovate, tip triangular, acute. *Heads* ½–⅔ in. diam. *Flowers* pale yellow, upper pedicelled. *Calyx* campanulate, upper teeth triangular, very short; lower longer, lanceolate. *Standard* ¼ in., brown, shining. *Pod* obovoid; style hooked.—DISTRIB. Europe, N. Africa, N. and W. Asia; introd. in N. America.—This is the *T. procum'bens* of Linn. Herb. (*Syme*).

16. **T. du'bium,** *Sibth.*; nearly glabrous, leaves pinnately 3-foliolate, leaflets narrow-obovate or obcordate, heads dense-flowered, pedicels very short, standard narrow keeled folded over the pod. *T. procum'bens*, L. (in part), Huds. *T. minus*, Sm.

Pastures, roadsides, &c., from Caithness southd.; ascends to 1,350 ft. in Derby; Ireland; Channel Islands; fl. June–Aug.—Annual. *Stems* 10–20 in., straggling, slender. *Leaflets* truncate or notched, finely toothed; petiole very short, slender; upper part of stipules ovate, acuminate. *Heads* 4–20-fld., much smaller than in *T. procumbens;* peduncle long, very slender. *Flowers* yellow, small, turning dark brown. *Calyx* and *pod* much as in *T. procum'bens.*—DISTRIB. Europe, N. Africa; introd. in N. America.—This is the *T. filiforme* of various authors.

17. **T. filifor'me,** *L.*; sparsely hairy, leaflets obcordate or obovate, heads axillary, peduncles slender, flowers few subracemose, pedicels as long as the calyx-tube spreading or reflexed, standard keeled folded over the pod.

Dry pastures often near the sea, rare, from Roxburgh southd.; doubtfully N. of it; Ireland; Channel Islands; fl. June–July.—Annual. *Stems* 4–8 in., very slender, prostrate. *Leaves* pinnately or digitately 3-foliolate; upper part of stipules ovate, acute. *Leaflets* ⅛–¼ in., toothed at the tip; *Heads* very small, 2–6-fld.; peduncle capillary. *Flowers* yellow. *Calyx* and *pod* much as in the two last species.—DISTRIB. Europe to the Caucasus.

9. ANTHYL'LIS, *L.* KIDNEY-VETCH.

Herbs or shrubs. *Leaves* pinnate, with a terminal leaflet ; stipules small or 0. *Flowers* in capitate cymes, sometimes involucrate, yellow white or red-purple ; bracts various or 0. *Calyx* inflated, mouth oblique 5-toothed. *Petals* with long claws, those of the 4 lower adnate to the staminal tube ; standard auricled at the base ; keel incurved, gibbous on each side, resilient after deflection with the anthers retracted. *Filaments* all united, or the upper distinct ; anthers uniform. *Style* smooth, stigma terminal ; ovules 2 or more. *Pod* enclosed in the calyx, obliquely ovoid, dehiscent or not, 1-3-seeded.—DISTRIB. Europe, Asia, N. Africa ; species 20.—ETYM. The old Greek name.

A. Vulnera'ria, *L.* ; silky, stems herbaceous, leaflets 2-6 pairs.

Dry rocky banks, N. to Shetlands, rather local; ascends to 2,400 ft. in the Highlands ; Ireland ; Channel Islands ; fl. June–Aug.—*Rootstock* woody, short, branched. *Stems* 6-16 in., many, leafy, herbaceous, suberect. *Radical leaves* 2-4 in., pinnate, unequal ; leaflets ½-1½ in., narrow-oblong. *Heads* ½-1½ in. diam., in pairs, rarely solitary, or with small accessory ones, the pairs peduncled, but each sessile ; involucre of subsessile linear appressed or spreading leaflets. *Flowers* ½-⅔ in., usually yellow, but variable in colour, proterandrous. *Calyx* membranous, longer than the petals ; teeth minute, ovate, acute, mouth contracted. *Pod* very small, acute, glabrous, reticulated, partially dehiscent, 1-seeded.—DISTRIB. Europe (Arctic), N. Africa, W. Asia.

A. Vulnera'ria proper ; involucral leaves short, flowers yellow. — VAR. *A. Dille'nii*, Schult.; involucral leaves nearly as long as the cream coloured flowers with red tips. Sussex to Cornwall and Wales.

10. LO'TUS, *L.* BIRD'S-FOOT TREFOIL.

Herbs or under-shrubs. *Leaves* pinnately or palmately 4-5-foliolate ; stipules minute or 0. *Flowers* in capitate or umbellate axillary peduncled cymes, yellow red or white, proterandrous ; bracts 3-foliolate. *Calyx* 2-lipped, or with 5-subequal teeth. *Petals* free from the staminal tube ; keel incurved or inflexed, beaked, gibbous on each side, resilient after deflection. *Alternate filaments* dilated upwards, upper distinct ; anthers uniform. *Ovary* sessile ; style inflexed, glabrous, stigma terminal or lateral ; ovules many. *Pod* elongate, cylindric, 2-valved, septate between the seeds.—DISTRIB. Europe, N. and S. Africa, temp. Asia, America, Australia ; species 50.—ETYM. unknown.

1. **L. cornicula'tus,** *L.* ; perennial, decumbent, heads 5-10-fld., peduncles very long, calyx-teeth erect in bud, 2 upper triangular converging with an obtuse sinus.

Pastures and waste places, N. to Shetlands ; ascends to 2,800 ft. in the Highlands; Ireland ; Channel Islands ; fl. June–Sept.—Variable in habit and stature. *Rootstock* short, woody, branched, not or scarcely stoloniferous. *Stems* 4-16 in., tufted at the base. *Leaves* very shortly petioled ; leaflets ¼-½ in.; stipules ovate or lanceolate. *Heads* ½-1½ in. diam., depressed.

Flcwers ½-¾ in., very shortly pedicelled. *Petals* twice as long as the calyx, bright yellow, often streaked with crimson, often greenish or purple-brown when dry. *Pod* ¾-1¼ in.—DISTRIB. Europe (Arctic), N. Africa, N. and W. Asia, India.

L. CORNICULA'TUS proper; almost glabrous, stem short, leaflets membranous obovate obtuse or subacute.—VAR. *L. crassifo'lius*, Pers.; almost glabrous, leaflets fleshy. Chiefly maritime.—VAR. *villo'sa;* covered with spreading hairs. Kent, Devon, Sandgate.

Sub-sp. L. TEN'UIS, *Waldst.* and *Kit.*; stem filiform, leaflets generally linear-lanceolate acuminate, stipules narrower, flowers fewer and smaller. *L. tenuifo'lius*, Reichb., *L. decum'bens*, Forst.—Damp soils, not common, from Forfar southd.; very rare in Ireland.

2. **L. uligino'sus,** *Schk.*; perennial, erect or ascending, heads 5-12-fld., peduncles very long, calyx-teeth spreading in bud, 2 upper triangular or subulate diverging with an acute sinus. *L. ma'jor*, Sm., not of Scop.

Moist meadows, &c., from Isla and Banff southd.; ascends to 1,200 ft. in Yorkshire; Ireland; Channel Islands; fl. July–Aug.—Glabrous or hairy. *Rootstock* elongate, stoloniferous, branched at intervals. *Stems* 6-24 in. *Leaflets* ½-1 in., obliquely obovate. *Flcwers* proterandrous, and *pods* much as in *L. cornicula'tus*.—DISTRIB. Europe, W. Asia, N. Africa.—The *L. ma'jor* of Scopoli, with lanceolate leaflets and bracts, and which grows in dry places, is probably a var. of *L. cornicula'tus*.

3. **L. his'pidus,** *Desf.*; annual, procumbent, laxly villous, peduncles longer than the leaves 3-4-fld., pods ½-⅔ by ₁₀ in. more than twice as long as the calyx subtorulose.

Dry banks by the sea, from Hants to Cornwall, rare; Channel Islands; fl. July–Aug.—*Stems* slender, 6-36 in. *Leaflets* ¼-½ in., elliptic or oblanceolate acute; stipules ½-cordate. *Flowers* ⅓ in. *Calyx-teeth* subulate, straight in bud. *Standard* obovate. *Seeds* subreniform.—DISTRIB. S.W. France, Spain, Portugal, Italy.

4. **L. angustis'simus,** *L.*; annual, procumbent, laxly villous, ped-uncles shorter than the leaves 1-2-fld., pods ¾-1 by ₁₂ in. four times as long as the calyx, torulose. *L. deflex'us*, Sol.; *L. diffusus*, Sm.

Dry banks by the sea, from Kent to Cornwall, very rare; Channel Islands; fl. July–Aug.—*Stems* very slender, 6-12 in. *Leaflets* ¼-⅓ in., elliptic, obovate or oblanceolate, acute or obtuse; stipules ovate-lanceolate. *Flowers* ⅓ in. *Calyx-teeth* subulate, straight in bud. *Standard* elliptic. *Seeds* globose.— DISTRIB. of *hispidus*, and E. to Hungary and W. Asia.

11. ASTRAG'ALUS, *L.* MILK-VETCH.

Herbs or shrubs. *Leaves* pinnate with a terminal leaflet, rarely 3-folio-late; leaflets entire; stipules distinct or connate, free or adnate to the petiole. *Flowers* in axillary racemes or spikes; bracts small. *Calyx* tubular; teeth 5, subequal. *Petals* usually narrow with long claws; keel obtuse. *Upper filament* distinct; anthers uniform. *Ovary* sessile or stalked; style filiform, beardless, stigma terminal; ovules many. *Pod* 2-valved, often longitudinally 2-celled by the inflexion of the suture next to the keel —DISTRIB. Temp. and cold regions; most abundant in Asia; absent from S. Africa and Australia; species 500.—ETYM. doubtful.

1. **A. glycyphyl'los,** *L.* ; glabrous, stipules free, flowers racemed, peduncles much shorter than the leaves, pods suberect exserted elongate many-seeded.

Fields and copses, from Ross southd.; fl. June–Sept.—*Rootstock* short, stout. *Stems* 2–3 ft., prostrate, stout, zigzag. *Leaves* 4–6 in.; leaflets ¾–1¼ in., broadly oblong, obtuse, glabrous above, slightly hairy beneath; stipules 1 in., ovate-lanceolate, acute, lower auricled. *Racemes* 1–2 in., compact, ovoid; bracts subulate, longer than the short pedicels. *Flowers* ½ in., creamy white. *Calyx* campanulate, half as long as the corolla. *Pod* 1–1½ in., linear, terete, curved, acuminate, 2-celled, pale. *Seeds* many, pale, compressed.—DISTRIB. Europe, N. Asia.

2. **A. hypoglot'tis,** *L.* ; hairy, stipules connate, flowers spiked, peduncles usually much longer than the leaves, pods very short ovoid suberect included 2-seeded. *A. danicus,* Retz.

Gravelly and chalky soil, from Sutherland to Essex and Wilts; Isle of Aran, Ireland; fl. June–July.—Clothed with soft white hairs mixed with black above. *Rootstock* very slender, straggling, branching. *Stems* 2–6 in., slender, ascending. *Leaves* 2–4 in., leaflets ⅙–½ in., many, oblong or linear-oblong, obtuse. *Spikes* 1 in., ovoid; peduncle stout. *Flowers* ½–¾ in., blue-purple. *Calyx* with black hairs, longer than the bracts. *Pod* shortly stipitate.— DISTRIB. Arctic and Alpine Europe, N. Asia, N. America.

3. **A. alpi'nus,** *L.* ; decumbent, hairy, stipules free, flowers shortly racemose or subcapitate, peduncles rather shorter than the leaves, pods oblong pendulous exserted few-seeded. *Phaca astragali'na,* DC.

Aberdeen and Forfar, very rare; alt. 2,400–2,600 ft.; fl. July.—Perennial, more slender than *A. hypoglottis;* leaves very similar; peduncles shorter; flowers horizontal or drooping, pale blue tipped with purple; calyx shorter. *Pod* ⅓ in., 1-celled, stalk longer than the calyx, hence wholly exserted, covered with black hairs.—DISTRIB. Arctic and Alpine Europe, N. Asia, W. Tibet, N. America.

12. OXY'TROPIS, *DC.*

Herbs or shrubs. *Leaves* pinnate with a terminal leaflet; leaflets quite entire; stipules free or adnate to the petiole. *Flowers* in axillary spikes or racemes, purple white or pale yellow; bracts membranous. *Calyx* tubular, teeth subequal. *Petals* with long claws; keel erect, with a straight or recurved tooth at the tip. *Upper filament* free; anthers uniform. *Ovary* sessile or stalked; style beardless, stigma minute; ovules many. *Pod* longitudinally more or less 2-celled by the inflexion of the suture next the standard.—DISTRIB. Europe, temp. and cold Asia, and N. America; species 100.—ETYM. ὀξύς and τρόπις, from the *sharp keel.*

Habit, &c., of *Astraga'lus;* distinguished by the mucronate keel, and septum of the pod never produced from the dorsal suture.

1. **O. uralen'sis,** *DC.* ; silky, leaflets ovate-lanceolate, peduncles longer than the leaves, flowers pale purple. *O. Halleri,* Bunge.
Dry and rocky pastures, from Wigton and Fife to Caithness ; ascends to 2,000 ft.; fl. June–July.—*Rootstock* stout, woody ; branches very short. *Leaves* 2–4 in. ; leaflets ¼–⅔ in., many, membranous, close-set ; stipules lanceolate. *Heads* 6–10-fld. ; peduncle stout, erect ; bracts leafy, shorter than the calyx. *Flowers* ¾ in., pale ; keel tipped with dark purple. *Calyx-tube* oblong, hairy and slightly glandular, cylindric ; teeth short, subulate. *Pod* about 1 in., sessile, erect, ovoid, tumid, bursting the calyx, hairy, many-seeded, beak curved.—DISTRIB. Arctic and sub-Alpine Europe, N. Asia, N. America. — I cannot distinguish the Uralian plant from the Scotch, and that found all round the N. temp. and Arctic zones, except by its rather larger size. Boissier has but the one name for the Russian and Alpine plants.

2. **O. campes'tris,** *DC.* ; softly hairy, leaflets linear-oblong or oblong-lanceolate, flowering peduncles shorter than the leaves, flowers pale yellow.
Alpine rocks, Clova Mts. ; alt. 2,000 ft. ; fl. June–July.—Habit of *O. uralen'sis,* but larger ; leaves 4–6 in. ; leaflets usually longer, sometimes 1 in., narrower, more obtuse ; peduncles lengthening after flowering to 8 in. *Flowers* ¾ in., yellow tinged with purple. *Pod* ½–⅔ in,, sessile, ovoid-lanceolate, hairy, half 2-celled, beak curved.—DISTRIB. Arctic and Alpine Europe, Siberia, N. America.

13. **ORNITH'OPUS,** *L.* BIRD'S-FOOT.

Slender, hairy herbs. *Leaves* pinnate with a terminal leaflet ; leaflets small ; stipules membranous. *Flowers* minute, in long peduncled heads or umbels, pink white or yellow. *Calyx-lobes* equal or 2 upper connate. *Keel* obtuse, sometimes very short. *Alternate filaments* dilated upwards, the upper free ; anthers uniform. *Style* inflexed, stigma capitate ; ovules many. *Pod* curved, slender, indehiscent, breaking up into many short 1-seeded joints.—DISTRIB. Europe, N. Africa, W. Asia.—ETYM. ὄρνις and πούς, from the fruits resembling *birds' claws.*

1. **O. perpusil'lus,** *L.* ; bracts pinnate, flowers white, pod much constricted between the seeds.
Sandy and gravelly places from Moray and Dumbarton southd.; E. Ireland, very rare; Channel Islands; fl. May–July.—Annual, grey-green, hairy above. *Stems* 6–18 in., many, prostrate, filiform, leafy, sparingly branched. *Leaves* 1–2 in., upper sessile ; leaflets ⅛–¼ in., 6–14 pair, close-set, oblong or linear-oblong, the lowest pair recurved when at the base of the petiole ; stipules minute. *Peduncles* axillary, slender, strict, longer or shorter than the leaves. *Heads* 3–6-fld. *Flowers* ⅙ in., veined with red ; pedicels very short. *Calyx-tube* subcampanulate ; teeth short. *Pod* ½–1 in., 7–9-jointed, beaked, glabrous or pubescent, reticulated.—DISTRIB. Europe, N. Africa.

2. **O. ebractea'tus,** *Brot.* ; bracts 0, flowers yellow, pod slightly constricted between the seeds. *Arthrolo'bium ebracteatum,* DC.

Sandy places, Scilly and Channel Islands ; fl. June–Aug.—Annual, glaucous, nearly glabrous. *Stems* 6–18 in., filiform, ascending. *Leaves* 1–4 in., all petioled ; leaflets ¼–½ in., oblong, distant. *Heads* 2–5-flowered ; peduncles as long as the leaves. *Flowers* ¼ in., bright yellow with red veins. *Pod* ¾–1 in., very slender, beaked, 10–14-jointed, granulate.—DISTRIB. W. Mediterranean region, Canaries, Azores.

14. HIPPOCRE'PIS, *L.* HORSESHOE VETCH.

Diffuse, glabrous herbs. *Leaves* pinnate with a terminal leaflet ; leaflets many, quite entire ; stipules small or 0. *Flowers* yellow, nodding, honeyed. *Calyx* with 2 upper teeth connate. *Petals* long-clawed ; keel incurved, beaked, resilient after deflection. *Style* inflexed, subulate, stigma minute ; ovules many. *Alternate filaments* slightly dilated, upper free ; anthers uniform. *Pod* flat, curved, upper margin deeply notched opposite each seed, breaking up into 3–6 horseshoe-like joints. *Seeds* curved.—DISTRIB. Europe, N. Africa, W. Asia ; species 12.—ETYM. Ἵππος and κρηπίς, from the shape of the joints of the pod.

H. como'sa, *L.* ; leaflets 4–8 pair oblong-obovate.

Rocky and stony pastures in England, Ayr, and Kincárdine; ascends to 1,800 ft. in Yorkshire; fl. May–Aug.—Nearly glabrous. *Rootstock* branched. *Stems* 6–18 in., very many, branched, ascending. *Leaves* 2–6 in.; leaflets ¼–½ in. *Heads* 6–10-flowered ; peduncle curved, slender, longer than the leaves. *Flowers* ½–½ in., shortly pedicelled, yellow. *Calyx-tube* short, broad. *Petals* at length reflexed. *Pod* 1–1½ in., falcate, granulate.— DISTRIB. W. and S. Europe, N. Africa.

15. ONO'BRYCHIS, *Tournefort.* SAIN-FOIN.

Herbs or shrubs. *Leaves* pinnate with a terminal leaflet ; leaflets quite entire ; stipules scarious. *Flowers* purple red or white, in axillary spikes or racemes, honeyed. *Calyx-lobes* subulate. *Wings* short ; keel obliquely truncate, as long as or longer than the standard, resilient after deflection. *Upper filament* distinct at the base ; anthers uniform. *Style* inflexed, stigma minute. *Pod* compressed, indehiscent, not jointed, often spiny winged or crested, 1–2-seeded.—DISTRIB. Europe, temp. Asia and Africa ; species 50.—ETYM. obscure.

O. sati'va, *Lamk.* ; pod tubercled on the lower margin.

Dry fields and pastures, indigenous? in mid and S.E. England, usually a relic of cultivation ; fl. June–Aug.—A perennial herb, 1–2 ft., pubescent with appressed hairs. *Rootstock* woody, branched. *Stems* ascending, stout, tough, leafy. *Leaves* 3–6 in. ; leaflets ½–¾ in., obovate, or linear-oblong, apiculate, very shortly petioled ; stipules ovate-lanceolate, membranous. *Racemes* ovoid, compact; peduncle slender, erect; bracts subulate. *Flowers* ½ in., bright rosy-red, veins darker; wings very short. *Calyx* short, woolly, teeth subulate. *Pods* ¼–½ in., obliquely semicircular in outline, strongly reticulate, pubescent 1-seeded.—DISTRIB. W. and S. Europe, N. Asia,

16. VIC′IA, *L.* Vetch, Tare.

Climbing or diffuse herbs. *Stems* terete, angled, or ridged. *Leaves* abruptly pinnate ; petiole usually ending in a simple or branched tendril ; leaflets many, entire or toothed at the tip ; stipules ½-sagittate. *Flowers* blue purple or yellow, axillary, racemed, honeyed. *Calyx-teeth* subequal or the lower longer. *Wings* adnate to the keel, which is resilient after deflection. *Staminal tube* abruptly truncate ; filaments filiform, upper more or less free ; anthers uniform. *Style* inflexed, cylindric or flattened, glabrous or downy all round, or bearded below the terminal stigma ; ovules usually many. *Pod* compressed, 2-valved. *Seeds* globose, with a small aril.—Distrib. Temp. N. hemisphere, and S. America ; species 100.—Etym. The old Latin name.

Section 1. **Er′vum,** *L.* (gen.). Annuals. *Leaflets* few. *Flowers* few. *Calyx* equal at the base. *Style* equally pubescent all round.

1. **V. tetrasper′ma,** *Mœnch ;* leaflets 3–6 pair, peduncles 1–2-fld., pod shortly stipitate glabrous 3–8-seeded.

Hedges, cornfields, &c., from Lanark and Forfar southd. ; hardly indigenous in Scotland ; Ireland, very rare ; Channel Islands ; fl. May–Aug.—Almost glabrous. *Stems* 1–2 ft., filiform *Leaflets* variable, ¼–1 in , usually truncate, narrow ; tendrils once or twice forked ; lower stipules 2-fid, upper entire toothed on the base at one side. *Peduncles* ¾–1½ in., produced beyond the flowers ; pedicels slender, curved. *Flowers* ¼ in., pale blue. *Calyx-tube* short, upper teeth shortest. *Pod* ½–¾ in., linear oblong.—Distrib. Europe, N. Africa, W. Asia, India ; introd. in N. America.
V. tetrasper′ma proper ; leaflets 4–6 obtuse mucronate, peduncles as long as the leaves 1–2-flowered, pod 3–4-seeded.
Sub-sp. V. grac′ilis, *Loisel.* ; leaflets 3–4 pair acuminate, peduncles longer than the leaves 1–4 flowered, pod longer 5–8-seeded.—From Warwick and Cambridge to Kent and Devon ; (a native or colonist, *Wats.*).

2. **V. hirsu′ta,** *Koch ;* leaflets 6–8 pair obtuse mucronate, peduncles 1–6-fld. pod sessile hairy 2-seeded. *Common Tare.*

Hedges and waste places, from Caithness southd. ; Ireland ; Channel Islands ; fl. May–Aug.—Habit of *V. tetrasper′ma,* but hairy ; leaflets smaller, more numerous ; stipules often 4-lobed ; pedicels straighter ; flowers smaller, and pods much shorter, sessile, hairy and 2-seeded.—Distrib. Europe (Arctic), N. Africa, N. and W. Asia, N.W. India ; introd. in N. America.

Section 2. **Crac′ca.** Perennials. *Leaflets* many. *Flowers* very many. *Calyx-tube* gibbous at the base, teeth very unequal. *Style* equally pubescent all round. *Pod* rather short, stalked, 3–many-seeded.

3. **V. Crac′ca,** *L.* ; tendrils branched, stipules nearly entire, flowers bright blue.

Hedges, fields, and waste places, N, to Shetlands ; ascends to 2,400 ft. in the Highlands ; Ireland ; Channel Islands ; fl. June–Aug.—Pubescent or slightly silky. *Rootstock* creeping. *Stems* 2–6 ft, angled, scandent or diffuse. *Leaves* 1–4 in., sessile ; leaflets ⅓–1 in., linear-oblong, acute or mucronate. *Racemes* dense, 10–30-flowered, unilateral; peduncle longer than the leaves, pedicels short. *Flowers* ½ in., drooping, proterandrous. *Calyx-tube* short. *Pod* ¾–1 in., obliquely truncate, beaked, many-seeded.— DISTRIB. Europe (Arctic), N. Africa, N. and W. Asia, India, Greenland, N. America.

4. **V. Or'obus**, *DC.*; tendrils 0, stipules slightly toothed, flowers white tinged with purple. *Or'obus sylvat'icus*, L. *Bitter Vetch.*

Western wooded and rocky districts, from Skye and Forfar to Hants and Cornwall; Ireland, very rare; fl. June–Sept.—Sparingly pubescent. *Stem* 1–2 ft., erect, stout, leafless or with reduced leaves below. *Leaves* 2–3 in., sessile; leaflets ¾–1½ in., linear-oblong or oblong-lanceolate, acute, or obtuse and mucronate; petiole produced beyond the leaflets. *Racemes* loose, 6–20-flowered, unilateral; peduncle as long as or longer than the leaves; pedicels much shorter than the calyx-tube. *Flowers* ⅓ in. *Pods* oblong-lanceolate, 1 in., acute at both ends, glabrous, 3–5-seeded.—DISTRIB. Norway, Denmark, S. France, Bavaria.

5. **V. sylvat'ica**, *L.*; tendrils much branched, lower stipules lunate toothed, flowers white with blue veins. *Wood Vetch.*

Rocky woods, local, from Caithness southd.; ascends to near 1,700 ft. in the Highlands; Ireland; fl. June–July.—Glabrous. *Rootstock* creeping. *Stems* 2–4 ft., usually trailing. *Leaves* 2–4 in., sessile; leaflets ½–1 in., oblong, obtuse, mucronate, membranous; stipules with spreading teeth. *Racemes* laxly 6–18-flowered, unilateral; peduncle as long as or longer than the leaves; pedicels nearly as long as the calyx-tube. *Flowers* ¾ in. *Pods* oblong-lanceolate, acuminate at both ends, slightly curved, 3–4-seeded.— DISTRIB. Europe (Arctic), N. Asia.

SECTION 3. **Vic'ia** proper. Annual, rarely perennial. *Leaflets* many. *Flowers* few, sessile, or on very short pedicels. *Style* villous below the stigma on the outer side.

* *Leaflets many. Calyx gibbous at the base.*

6. **V. se'pium**, *L.*; perennial, flowers racemose, calyx-teeth unequal shorter than the tube, pod stalked linear glabrous.

Hedges, copses, &c., N. to Orkney; ascends to 2,000 ft. in Yorkshire; Ireland; Channel Islands; fl. May–July.—Slightly hairy. *Rootstock* creeping, stoloniferous. *Stem* 2–3 ft., climbing or trailing. *Leaves* 2–5 in.; leaflets in 6–8 pairs, smaller upwards, lowest ⅜–1 in., ovate, acute obtuse or truncate, mucronate, membranous; stipules ½-sagittate. *Racemes* 1¼–1½ in., 2–6 fld., subsessile; pedicels shorter than the calyx-tube. *Flowers* ½–⅔ in., dull pale purple. *Pod* linear, 1 in., beaked, black, 6–10-seeded. *Seed* with a linear hilum.—DISTRIB. Europe (Arctic), N. and W. Asia, Kashmir.

7. **V. lu'tea**, *L.*; annual, flowers solitary sessile, upper calyx-teeth shorter lower longer than the tube, pod shortly stalked hairy.

Rocky and pebbly places, rare and local, from Ayr and Forfar southd.; Channel Islands; fl. June–Aug.—Sparingly hairy. *Rootstock* short. *Stems* 6–18 in., tufted, prostrate. *Leaves* 1–2 in.; leaflets ¼–½ in., 5–7 pairs, linear-oblong, obtuse, mucronate; stipules small, ovate, acute, lower ½-hastate. *Flowers* rarely in pairs, suberect, pedicels shorter than the calyx-tube. *Flowers* ¾-1 in., narrow, pale yellow. *Pod* 1-1½ in., hairy, narrow, oblong, acuminate at both ends, beaked 4-8-seeded. *Seeds* with a short hilum.—DISTRIB. From Holland southd., N. Africa, W. Asia.

V. HY'BRIDA, *L.*, formerly found on Glastonbury Tor, which differs in the larger truncate and retuse leaflets and hairy standard, has long been extinct It is a native of W. France, Spain, and the Mediterranean.—V. LÆVIGA'TA, *Sm.*, is another extinct closely allied plant, formerly found on the Weymouth Beach, and differs in being glabrous and having pale blue or whitish flowers A solitary specimen in Smith's Herbarium is all that is known of it.

V. SATI'VA, *L.* ; annual, leaflets obovate or oblong truncate or retuse, flowers 1-2 axillary subsessile, calyx-teeth nearly equal as long as the tube, pod 2-3 in. sessile. *Common Vetch.*

Hedges and roadsides, a casual ; ascends to upwards of 1,600 ft. in Derby ; Ireland ; Channel Islands ; fl. April–June.—Annual, sparsely hairy. *Stems* many, trailing or climbing, stout or slender. *Leaflets* variable, in 5-6 pairs, ⅛-¾ in., ₁⁰₀-½ in. broad. *Stipules* ½-hastate, toothed or entire, often with a dark blotch. *Flowers* ¾ in., pale purple. *Pods* linear, 4-10-seeded, slightly hairy. *Seeds* subglobose, ¼ in. diam., variable in colour, smooth ; hilum linear.—DISTRIB. Mediterranean region ; cultivated in Europe, Asia, and America.

8. **V. angustifo'lia,** *Roth ;* annual, leaflets of upper leaves linear or oblong acute or obtuse, calyx-teeth nearly equal as long as the tube, pods 1-2 in.

Dry places, from Aberdeen and the Clyde southd. ; Ireland ; Channel Islands ; fl. May–July.—An excessively variable plant, of which *V. sati'va* is probably the cultivated form, differing in the much smaller flowers, pods, and seeds ; stem robust or slender, 6-18 in. long ; leaflets ¼-1½ in. ; seeds only ⅛ in. diam.—DISTRIB. Europe, N. Africa, and W. Asia.—The following varieties run into one another and into *V. sati'va.*

V. angustifo'lia proper (*V. angustifo'lia,* Forst., *V. segeta'lis,* Thuill.) ; stout, upper leaves with oblong leaflets, flowers usually 2-nate, pods 1½-2 in. bursting the calyx.—VAR. *V. Bobar'tii,* Forst. (*V. angustifo'lia,* Sm.); slender, upper leaves with linear leaflets, flowers subsolitary, pod 1-1½ in., not bursting the calyx.

** *Leaflets few. Calyx not gibbous at the base.*

9. **V. lathyroi'des,** *L.* ; annual, flowers solitary sessile, calyx-teeth equal nearly as long as the tube, pod sessile glabrous.

Dry pastures and roadsides, local from Ross southd. ; Ireland, very rare ; Channel Islands ; fl. May–June.—Hairy. *Stem* 6-8 in., spreading, slender. *Leaves* ½-1 in. ; leaflets ⅕-¼ in., 2-3 pair, linear-oblong or obovate, acute obtuse or notched ; tendrils simple or 0. *Flowers* ¼-⅓ in., lilac. *Calyx* funnel-shaped ; teeth subulate. *Pods* ½-1 in., linear, tapering at both ends, beaked, 8-12 seeded. *Seeds* obtusely angled, granulate ; hilum very short.—DISTRIB. Europe, N. Africa, W. Asia.

10. **V. bithyn'ica,** *L.* ; perennial, peduncles 1- rarely 2-3-fld., pedicels as long as the calyx, calyx-teeth unequal subulate, pod shortly stalked reticulate hairy.

Bushy places, local, from Flint and York to Kent and Devon; fl. May–June.
—*Stems* 1–2 ft., many from the root, glabrous, except at the tips, trailing
or climbing, flexuous. *Leaves* 1½–2 in., exclusive of the tendril; leaflets
variable, ⅜–2½ in., ⅛–⅓ in. broad, obtuse and mucronate or acute or
acuminate; stipules large; tendrils elongate, branched. *Peduncles* ½–3 in.,
and pedicels hairy. *Flowers* ¾ in., pale purple, wings paler. *Calyx*
hairy, teeth longer than the tube, very slender. *Pod* the largest of the
British species, 1½ in. by ⅓ in. broad, abruptly beaked, 4–6-seeded. *Seeds*
large, globose, speckled, ⅛ in. diam., dark-brown.—DISTRIB. W. and S.
Europe, Asia Minor, N. Africa.
Syme distinguishes 2 vars.; *latifo'lia*, with leaflets ovate or oblong, stipules
all toothed; and *angustifo'lia*, with leaflets linear acuminate, and upper
stipules sparingly toothed.

17. LATH'YRUS, *L.* EVERLASTING PEA.

Herbs with the habit of *Vic'ia*, but fewer leaflets, petals broader, staminal tube obliquely truncate, and the style flattened and longitudinally
bearded on its inner face.—DISTRIB. of *Vic'ia ;* species 100.—ETYM. An
old Greek name.

SECTION 1. **Aph'aca.** Annual. *Stipules* leaf-like. *Leaves* reduced to
tendrils. *Calyx-tube* equal at the base, shorter than the teeth.

1. **L. Aph'aca,** *L.* ; stipules ovate-hastate acute or obtuse, peduncles
elongate 1-fld., flowers erect.

Cornfields, &c., from Warwick and Norfolk to Kent and Devon southd.; (a
colonist, *Wats.*); fl. June–July.—Glabrous. *Stems* 1–3 ft. trailing. *Leaflets*
chiefly seen on seedlings, then oblong, when developed on older plants
linear ½–¾ in.; stipules ½–1 in. broad, quite entire, striated with nerves.
Peduncles slender, stout and 2–3 in. in fruit; pedicels ¼ in., with a
minute bract at the base. *Flowers* ½ in., pale yellow. *Calyx-lobes* green,
linear, nearly as long as the corolla. *Pod* 1–1½ in., ¼ in. broad, ascending,
slightly falcate, beaked, reticulate, 6–8-seeded.—DISTRIB. From Denmark
southd., N. Africa, W. Asia, India.

SECTION 2. **Nisso'lia.** Annual. *Stipules* minute, setaceous ; tendrils 0.
Petioles leaf-like (phyllodes). *Calyx* rather gibbous at the base.

2. **L. Nisso'lia,** *L.* ; phyllodes grass-like, peduncles very slender
1-fld., flowers erect.

Grassy bushy places from Cheshire southd., and as an escape north of it ; fl.
May–June.—Glabrous. *Stems* 1–3 ft., ascending, very slender. *Phyllodes*
3–6 in., quite entire; nerves many, parallel. *Peduncles* 1–4 in., pedicels as
long as the calyx, bracts obsolete. *Flowers* ½ in., crimson. *Calyx-teeth*
lanceolate, lower teeth longest. *Pod* 1–2 in., very slender, slightly compressed, glabrous. *Seeds* granulate.—DISTRIB. From Holland southd., W.
Asia, N. Africa.

SECTION 3. **Lath'yrus** proper. *Petioles* all with leaflets and tendrils.
Calyx gibbous at the base.

* *Leaflets one pair.   Flowering peduncles longer than the leaves.*

3. **L. hirsu'tus,** *L.* ; stem winged, leaflets linear-lanceolate, stipules small ½-sagittate, peduncles 1–3 fld., calyx-teeth ovate-lanceolate longer than the tube, pods tubercled densely silky, seeds papillose.

Fields, York, Kent, Surrey, and Essex, very rare; (a colonist, *Wats.*); fl. June–July.—Annual, almost glabrous except the pod.   *Stems* 2–4 ft., angled and 2-winged, wings herbaceous ⅒ in. broad.   *Leaves* scattered; petiole ¼–½ in.; tendrils stout, branched; leaflets 1–2½ in., nerves parallel; stipules with long subulate auricles.   *Peduncles* 2–3 in.; flowers distant, pedicel as long as the calyx, hairy; bracts minute, subulate.   *Flowers* ½ in.; standard crimson, keel and wings paler.   *Calyx-tube* short, obconic.   *Pod* 1½–2 in., ⅓ in. broad, stipitate, dilated upwards, 8–10-seeded.—DISTRIB. From Belgium southd., W. Asia, N. Africa.

4. **L. praten'sis,** *L.* ; stem acutely angled, leaflets lanceolate, stipules very large lanceolate sagittate, peduncles 3–12-fld., calyx-teeth subulate as long as the tube, pod glabrous or hairy, seeds smooth.

Hedges, copses, and meadows, N. to Shetland; ascends to near 1,600 ft. in the Highlands; Ireland; Channel Islands; fl. June–Sept.—Glabrous or slightly hairy.   *Rootstock* creeping.   *Stems* 1–2 ft.; trailing or climbing. *Leaflets* ½–1 in., nerves indistinct; stipules with spreading subulate auricles ; petiole ½–¾ in.; tendrils short.   *Peduncles* longer than the leaves; pedicels as long as the calyx-tube, hairy; bracts setaceous.   *Flowers* ½–¾ in., racemose, bright yellow.   *Pod* 1–1½ in., sessile, linear, acuminate, many-seeded. —DISTRIB.   Europe (Arctic), N. and W. Asia to the Himalaya; introd. in N. America.

L. TUBERO'SUS, *L.* ; stem angled, leaflets obovate, stipules large ½-sagittate, peduncles 2–5-fld., calyx-teeth triangular as long as the tube, pod glabrous.

Cornfields, Essex; (a denizen, *Wats.*); fl. June–Aug.—Glabrous.   *Rootstock* creeping, rootlets with small tubers.   *Stem* 2–4 ft., climbing and trailing. *Leaflets* 1–1½ in., nerves diverging; petiole and tendrils stout.   *Peduncles* very long; pedicels longer than the calyx; bracts subulate.   *Flowers* ¾ in., racemose, crimson.   *Calyx-tube* broad.—DISTRIB.   Europe, W. Asia, N. Africa.—Root tubers edible.

5. **L. sylves'tris,** *L.* ; stem winged, leaflets large ensiform, stipules falcate ½-sagittate, peduncles 3–10-fld., calyx-teeth triangular, upper shorter than the tube, pod glabrous, seeds faintly remotely tubercled.

Rocky thickets, local, often an escape, from Mull and Forfar southd.; fl. June–Aug.—Glaucous, glabrous.   *Rootstock* creeping.   *Stem* 3–6 ft., wings herbaceous.   *Leaflets* 4–6 in., ¼–½ in. broad; nerves parallel; stipules large, falcate, with long lanceolate auricles; petiole winged or not; tendrils slender, branched.   *Peduncles* 4–6 in.; pedicels longer than the calyx; bracts filiform.   *Flowers* ⅝–¾ in., racemed; standard rosy; wings purplish.   *Calyx-tube* broad, short.   *Pod* 2–3 in., sessile, narrowly winged above, 10–14-seeded. —DISTRIB.   Europe (Arctic), Caucasus, N. Africa.—Very near the Everlasting Pea, which occurs here and there as an escape.

** *Leaflets 2 or more pairs. Flowering peduncles longer or shorter than the leaves.*

6. **L. palus'tris,** *L.* ; stem winged, leaflets 2–3 pair sword-shaped, stipules lanceolate ½-sagittate, peduncles usually longer than the leaves 2–6-fld., pod compressed glabrous.

Boggy meadows and copses, from York and Carnarvon to Somerset and Suffolk, local; Ireland, very rare; fl. June–Aug.—Glaucous, glabrous. *Rootstock* creeping. *Stems* 2–4 ft., climbing or trailing, wings herbaceous. *Leaflets* 2–3 pair, 2–3 in., nerves parallel; tendrils rather short, branched; stipules ¾ in. *Peduncles* 1–4 in.; pedicels shorter than the calyx-tube; bracts minute. *Flowers* ⅝–¾ in., pale blue-purple. *Calyx-tube* short. *Pod* 1½–2 in., stipitate, reticulate, 6–8-seeded. *Seeds* smooth.—DISTRIB. Europe (Arctic), N. Asia, N. America.

7. **L. marit'imus,** *Bigelow ;* stem angled, leaflets 3–5 pair oblong, stipules ovate ½-hastate, peduncles usually shorter than the leaves 5–10-fld., pod turgid glabrous. *Pisum marit'imum,* L.

Pebbly beaches on the E. coast, local; from Shetland to Kent and Dorset; Kerry; fl. June–Aug.—Glaucous, glabrous. *Rootstock* long, stout, black. *Stems* 1–3 ft., creeping at the base, prostrate. *Leaflets* 1–2 in., alternate or subopposite, nerved, upper gradually smaller; petiole 2–4 in.; tendrils short; stipules nearly 1 in. *Peduncles* stout; pedicels shorter than the calyx; bracts minute. *Flowers* ¾–¾ in., purple, fading to blue. *Pod* 1½–2 in., reflexed, straight, 6–8-seeded. *Seeds* smooth.—DISTRIB. Arctic and N. Europe, Asia, and America.

L. marit'imus proper; leaflets broadly oblong obtuse.—VAR. *acutifo'lia,* Bab.; slender, straggling, leaves elliptic-lanceolate acute. Shetlands and Orkneys.

SECTION 4. **Or'obus,** *L.* ; (gen.). Perennial. *Petiole* ending in a short point without tendrils. *Calyx* gibbous at the base.

8. **L. macrorrhi'zus,** *Wimm.* ; stem winged, leaflets 2–4 pairs, stipules ½-sagittate usually toothed below.

Copses, &c., N. to Shetlands, ascends to 2,100 ft. in the Highlands; Ireland; fl. June–Aug.—Glabrous or very slightly hairy. *Rootstock* creeping and forming tubers. *Leaflets* 1–2 in., ¾–1½ in. broad, nerves parallel in the narrow forms, diverging in the broad; petiole ½–1½ in.: stipules ¼ in., rarely quite entire, very variable. *Peduncles* slender, equalling or exceeding the leaves, 2–6-fld.; pedicels shorter than the calyx; bracts minute. *Flowers* ½–¾ in., lurid crimson, fading to green or blue. *Calyx-teeth* triangular, shorter than the tube, upper very short. *Pod* subcylindric. *Seeds* globose. —DISTRIB. Europe.

L. macrorrhi'zus proper; leaflets elliptic-oblong acute or obtuse. *Or'obus tubero'sus,* L.—VAR. *O. tenuifo'lius,* Roth ; leaflets sword-shaped or narrow-linear.

9. **L. ni'ger,** *Wimm.* ; stem angled, leaflets 3–6 pairs oblong-lanceolate acute or mucronate, stipules linear-lanceolate acute.

Subalpine Scotch valleys; Den of Airly, Killiecrankie, Moy House; ascends to 1,200 ft.; (native? *Wats.*); fl. June–Aug.—Glabrous, black when dry.

Rootstock short. *Stems* 1–2 ft., erect, branched, stout or slender. *Leaflets* ¾–1¾ in., nerves diverging; petiole 1–2 in.; stipules ⅓–¼ in. *Peduncles* shorter or longer than the leaves, 2–8-fld.; pedicels about as long as the calyx; bracts filiform or 0. *Flowers* ½ in., livid-purple fading to blue. *Calyx-teeth* very short, triangular. *Pods* 2 in., narrow, turgid, rugose, acuminate at both ends, 6–8-seeded, subseptate. *Seeds* compressed, obtusely angled. —DISTRIB. Europe, Caucasus.

ORDER XXVI. ROSA'CEÆ.

Herbs, shrubs, or trees. *Leaves* alternate, rarely opposite, simple or compound, stipulate. *Inflorescence* various. *Flowers* regular. *Calyx* superior or inferior ; lobes 5, the 5th next the axis, imbricate in bud. *Petals* 5, rarely 0, perigynous, often orbicular and concave, claws very short or 0, deciduous, imbricate in bud. *Stamens* many, rarely 1 or few, inserted with the petals or on the disk, 1-many-seriate, incurved in bud ; anthers small, usually didymous. *Disk* lining the calyx-tube. *Carpels* 1 or more, distinct or connate, free or adnate to the calyx-tube ; styles as many, distinct or connate, terminal ventral or basal, stigma simple rarely feathery or decurrent ; ovules 1, or 2 collateral in each carpel, rarely more, anatropous. *Fruit* various, a pome, or of one or many drupes achenes or follicles, rarely a berry or capsule. *Seeds* ascending or pendulous, albumen scanty or 0 ; cotyledons plano-convex, radicle short.—DISTRIB. Ubiquitous ; genera 71 ; species 1,000.—AFFINITIES with *Leguminosæ* and *Saxifragceæ.*—PROPERTIES astringent in *Potentilleæ* and *Roseæ. Pruneæ* and *Pomeæ* yield hydrocyanic and malic acids.

SERIES 1.   *Ripe carpels not enclosed within the calyx-tube.*

TRIBE I.   **PRU'NEÆ.**   *Calyx* deciduous. *Carpel* 1; ovules 2, pendulous. *Fruit* a drupe..1. Prunus.

TRIBE II.   **SPIRÆ'Æ.**   *Calyx* persistent, ebracteolate. *Carpels* 5 or more ; ovules 2 or more in each carpel, pendulous. *Fruit* a follicle...2. Spiræa.

TRIBE III.   **RU'BEÆ.**   *Calyx* persistent, ebracteolate. *Carpels* many ; ovules 2 in each carpel, pendulous. *Fruit* of many small drupes ....3. Rubus.

TRIBE IV.   **POTENTIL'LEÆ.**   *Calyx* persistent, bracteolate. *Carpels* 4 or more ; ovule 1 in each carpel, ascending. *Fruit* of 4 or more achenes.

* *Style elongating after flowering.*

Leaves simple. Scape 1-fld. Styles of achenes feathery............4. Dryas.
Leaves pinnate. Stem several-fld..5. Geum.

** *Style not elongating after flowering.*

Leaves 3-foliolate. Achenes on a large fleshy receptacle ....6. Fragaria.
Leaves 3–many-foliolate. Achenes on a small dry receptacle...7. Potentilla.

I

Series 2. *Ripe carpels enclosed within the calyx-tube.*

Tribe V. **POTERIE′Æ.** *Petals* 4, 5, or 0. *Carpels* 1–3; ovules 1 in each carpel, erect or pendulous. *Fruit* of 1–3 achenes enclosed in the small dry calyx-tube.
Calyx 4–5-lobed, with 4–5 adnate bracts. *Petals* 0..............8. Alchemilla.
Calyx 5-lobed. *Petals* 5. *Stamens* 12–20...........................9. Agrimonia.
Calyx of 4 petaloid lobes. *Petals* 0. *Stamens* 4–30............10. Poterium.

Tribe VI. **RO′SEÆ.** *Petals* 4–5. *Carpels* many; ovules 1 in each carpel, pendulous. *Fruit* of many achenes enclosed in the fleshy calyx-tube.
                                                                     11. Rosa.

Tribe VII. **PO′MEÆ.** *Petals* 5. *Carpels* 1–5; ovules 2 collateral in each carpel, erect or ascending. *Fruit* fleshy, 1- 2- or 5-celled.
Fruit 2–5-celled, cells with cartilaginous walls........................12. Pyrus.
Fruit a drupe with 1–5 included stones ....13. Cratægus.
Fruit a drupe with 3–5 ½-exserted stones ....14. Cotoneaster.

1. PRUNUS, *L.* Plum and Cherry.

Shrubs or trees. *Leaves* alternate, simple, glandular-serrulate ; petiole 2-glandular. *Flowers* white or red, solitary corymbose or racemed, honeyed. *Calyx* inferior, deciduous in fruit ; lobes 5, imbricate. *Petals* 5. *Stamens* 15–20, perigynous, filaments free. *Carpels* 1 ; style terminal ; ovules 2, collateral, pendulous. *Drupe* with an indehiscent or 2-valved, 1-seeded, smooth, or rugged stone. *Seed* pendulous, testa membranous, albumen scanty or 0.—Distrib. N. temp. regions, rare in the tropics ; species 80. —Etym. The old Latin name.

Section 1. **Pru′nus** proper. *Leaves* convolute in bud. *Flowers* solitary or fascicled, appearing with the leaves or before them. *Drupe* glaucous.

1. **P. commu′nis,** *Huds.* ; leaves ovate or oblong-lanceolate pubescent beneath when young, petals obovate-oblong, flesh of drupe adhering to the stone.

Copses, hedges, &c., ascending to 1,300 ft. in Yorkshire ; Ireland ; Channel Islands; fl. March–April.—A small, rigid, much-branched shrub, 3–8 ft. ; branches usually spinescent ; wood very hard and tough. *Leaves* petioled, ¾–2 in., variable in breadth, acuteness, and length of petiole. *Flowers* white, shortly pedicelled. proterogynous. *Petals* variable in breadth. *Drupe* globose. —Distrib. The Sloe is confined to Europe, the Bullace extends to N. Africa and the Himalaya.
P. communis proper; bark black, branches divaricate all spinescent, leaves finely serrulate at length glabrous beneath, flowers ½–¾ in. diam. preceding the leaves, pedicels solitary or in pairs glabrous, petals obovate, drupe ½ in. diam. black erect very austere. *P. spinosa, L.*—From Sutherland southd. *Sloe, Blackthorn.*
Sub-sp. P. insiti′tia, *L.*; bark brown, branches straight a few spinescent, leaves larger broader more obtusely serrate pubescent beneath, peduncles downy, petals broader, drupe ¾–1 in. diam. globose drooping black or yellow.—From Lanark southd., but doubtfully indigenous in many habitats *Bullace.*

Sub-sp. P. DOMES'TICA, L. ; bark brown, branches straight unarmed, leaves pubescent on the ribs beneath, peduncles glabrous, drupe 1–1½ in. diam., black.—Not indigenous, except in W. Asia. *Wild Plum.*

SECTION 2. **Cer'asus.** *Leaves* conduplicate in bud. *Flowers* solitary or fascicled, appearing with the leaves or after them.

2. **P. Cer'asus,** *L.* ; leaves spreading oblong-obovate or elliptic crenate-serrate glabrous, petiole short, corolla cup-shaped, petals firm suberect obovate, fruit acid. *Wild Cherry ; Dwarf Cherry.*

Copses, &c., from York southd., wild or well established; rare in Ireland ; Channel Islands; fl. May.—A bush or small tree with copious suckers; bark red ; branches slender, pendulous. *Leaves* dark blue-green. *Flowers* homogamous ; buds with scarious outer scales and leafy inner ones. *Calyx-tube* not constricted, lobes crenate. *Petals* notched. *Fruit* red, juice not staining. —Origin of the Morello, Duke, and Kentish cherries.—DISTRIB. Europe to W. Himalaya, Azores, Canaries.

3. **P. Avi'um,** *L.* ; leaves drooping oblong-obovate acutely serrate pubescent beneath, petiole long, corolla open, petals flaccid almost obcordate, fruit sweet or bitter. *Gean.*

Copses and woods from Caithness southd., probably wild only in the S. ; Ireland ; fl. May.—A tree without suckers, branches short, stout, rigid ascending. *Leaves* large, pendulous, pale green. *Flowers* homogamous ; buds with none of the scales leafy. *Calyx-tube* constricted at the top, lobes quite entire. *Fruit* with staining juice.—Origin of the Geans, Hearts, and Bigaroon cherries.—DISTRIB. Europe to W. Himalaya.

SECTION 3. **Laurocer'asus.** *Leaves* conduplicate in bud. *Flowers* in axillary or terminal racemes, appearing after the leaves.

4. **P. Padus,** *L.* ; leaves elliptic or obovate acutely doubly serrate. *Bird Cherry.*

Copses and woods, from Caithness to S. Wales and Leicester ; ascends to 1,500 ft. in Yorkshire ; Ireland ; fl. May.—A tree, 10–20 ft. *Leaves* 2–4 in., unequally cordate at the base, axils of the nerves pubescent; stipules linear-subulate, glandular-serrate. *Racemes* 3–5 in., from short lateral buds, lax-fld. *Flowers* ½–¾ in. diam., white, erect, then pendulous, proterogynous ; pedicels ¼ in., erect in fruit; bracts deciduous, linear. *Calyx-lobes* obtuse, glandular-serrate. *Petals* erose. *Drupe* ⅓ in., ovoid, black, bitter ; stone globose, rugose.—DISTRIB. Europe (Arctic), N. Africa, N. and W. Asia, Himalaya.

2. SPIRÆ'A, *L.*

Perennial herbs or shrubs. *Leaves* alternate, simple or compound ; stipules free or adnate to the petiole, rarely 0. *Flowers* in axillary or terminal cymes, white or red. *Calyx* inferior, persistent ; lobes 4–5, imbricate or valvate in bud. *Petals* 4–5. *Stamens* 20–60, filaments free or connate below. *Disk* fleshy, often hairy. *Carpels* 5 or more, free or connate below ; styles subterminal ; ovules 2 or more, pendulous. *Follicles*

I 2

116   *ROSACEÆ.*   [SPIRÆA.

5 or more, few-seeded.—DISTRIB. Temp. and cold regions of the N. hemisphere; species 50.—ETYM. doubtful.

1. **S. Ulma′ria,** *L.* ; herbaceous, leafy, leaves interruptedly pinnate serrate white and downy beneath, terminal segments large acutely lobed, cymes corymbose very compound, carpels glabrous twisted 2-ovuled. *Meadow-sweet, Queen of the Meadows.*

Meadows and water-sides, N. to Shetlands; ascends to 2,700 ft. in the Highlands; Ireland; Channel Islands; fl. June–Aug.—*Rootstock* short. *Stems* 2–4 ft., erect, furrowed. *Radical leaves* 1–2 in.; terminal leaflets 1–3 in.; lateral entire, alternate very small; stipules leafy, ½-ovate, toothed. *Cymes* 2–6 in. diam., pubescent. *Flowers* ¼–½ in. diam., white, proterandrous, not honeyed. *Calyx-lobes* reflexed. *Carpels* 5–9, twisted together into an almost horizontal plane.—DISTRIB. Europe (Arctic), Asia Minor, N. Asia.

2. **S. Filipen′dula,** *L.* ; herbaceous, leaves interruptedly pinnate glabrous, leaflets sessile deeply cut serrate, cymes panicled, carpels pubescent straight 2-ovuled. *Dropwort.*

Dry pastures, from Caithness southd.; ascends to 1,200 ft. in Yorkshire; W. Ireland; fl. June–July —*Rootstock* short; root-fibres interruptedly tuberous. *Stem* 2–3 ft., erect, grooved, with few small leaves. *Leaves* 4–10 in., chiefly radical; leaflets ⅓–½ in., very many, almost pinnatifid, sessile by a broad base, alternate very small, terminal 3-lobed; stipules of cauline leaves toothed. *Cymes* loose; peduncles slender. *Flowers* ¼–½ in. diam., white or rosy outside, homogamous, not honeyed. *Calyx-lobes* obtuse. *Carpels* 6–12, erect.—DISTRIB. Europe, N. Africa, N. Asia.

S. SALICIFO′LIA, *L.* ; shrubby, leaves oblong-lanceolate serrate glabrous, stipules 0, cymes terminal racemose, carpels glabrous many-ovuled.

Plantations, not indigenous; fl. July–Aug.—*Stems* 3–5 ft., stoloniferous. *Leaves* 2–3 in., equally or unequally serrate. *Cymes* dense, subcylindric. *Flowers* rosy or pink. *Carpels* 5.—DISTRIB. Europe (Arctic), N. Asia, N. America.

3. RU′BUS, *L.* BRAMBLE, RASPBERRY, &c.

Creeping herbs or sarmentose shrubs, almost always prickly. *Leaves* alternate, simple or compound ; stipules adnate to the petiole. *Flowers* in terminal and axillary corymbose panicles, rarely solitary, white or red. *Calyx* inferior, tube broad ; lobes 5, persistent. *Petals* 5. *Stamens* many. *Disk* coating the calyx-tube. *Carpels* many, distinct, on a convex receptacle ; style subterminal ; ovules 2, collateral, pendulous. *Drupes* many, 1-seeded, crowded upon a dry or spongy conical receptacle. *Seed* pendulous.—DISTRIB. Abundant in the N. hemisphere, few in the Southern ; species 100.—ETYM. The old Latin name.

* Stem herbaceous or nearly so.

1. **R. Chamæmo′rus,** *L.* ; stem erect unarmed 1-flowered, leaves few suborbicular-cordate obtusely 5–7-lobed, flowers diœcious. *Cloudberry.*

Peaty alpine and subalpine moors, from Derby and Wales northd.; ascends to 3,200 ft. in the Highlands; N. Ireland, very rare; fl. June–July.—Pubescent. *Rootstock* creeping, branched. *Stem* 4–8 in., simple, sheathed below by obtuse leafless stipules. *Leaves* 1–3 in. diam., petioled, crenate, plaited, rugose; stipules ovate, obtuse. *Flowers* 1 in. diam., white. *Sepals* oblong, obtuse, unequal, villous. *Petals* oblong. *Fruit* ½ in., orange-yellow; drupes few, large, persistent, stone smooth.—Distrib. N. Europe (Arctic), Siberia, N. America.—Berry very grateful, fresh or preserved.

2. **R. saxat'ilis,** *L.* ; barren stems procumbent unarmed or with scattered bristles, flowering shorter erect, leaves 3-foliolate.

Stony banks of subalpine rivulets, copses, &c., from Cornwall, Devon, S. Wales, and Gloster to Sutherland; ascends to 2,700 ft. in the Highlands; Ireland; fl. June–July.—Softly pubescent. *Rootstock* creeping, stoloniferous. *Stems* simple, with leafless obtuse stipules below; leafing 2–3 ft., flowering 6–18 in. *Leaves* few; leaflets 1–3 in., green, membranous, lateral shortly petioled, rhomboid-ovate, obscurely lobed, sharply doubly toothed; stipules linear. *Flowers* ½ in. diam., few, white; peduncles terminal, very short. *Calyx-lobes* ovate. *Petals* very small, linear-obovate. *Drupes* 2–3, globose, scarlet, persistent, stone reticulate.—Distrib. Europe, N. and W. Asia, Himalaya.

** *Stem shrubby, with many suckers. Leaves pinnately 3–5-foliolate.*

3. **R. Idæ'us,** *L.* ; prickles of the stem straight slender, of the flowering shoots curved, leaflets ovate or elliptic acuminate white and hoary beneath, flowers drooping, petals short, drupes deciduous. *Raspberry.*

Woods, from the Orkneys southd.; ascends to near 2,000 ft. in the Highlands; fl. June–Aug.—*Rootstock* short. *Stems* 3–5 ft., erect, biennial, terete, pruinose. *Leaves* variable; leaflets 3–5 in., acutely irregularly serrate; stipules adnate half-way, subulate. *Cymes* ⅓ in. diam., few-fld., axillary and terminal, white. *Flowers* honeyed, homogamous. *Calyx-lobes* ovate-lanceolate, tips long. *Petals* linear-obovate. *Drupes* many, red or yellow, hoary, stone pitted.—Distrib. Europe (Arctic), N. Africa, N. and W. Asia.—*R. Lees'ii,* Bab., is a reduced state with crowded shorter leaflets. Westmoreland, Warwick, Oxford, Devon, and Somerset.

*** *Stem shrubby, without suckers. Leaves 3–7-foliolate, rarely pinnate.*

4. **R. frutico'sus,** *L.* ; stem prickly, flowers in panicled or racemed corymbs or fascicles. *Blackberry, Bramble.*

Copses, hedges, &c.; ascends to near 1,000 ft. in Yorkshire; Ireland; Channel Islands; fl. July–Sept.—*Stem* glabrous or with prickles bristles and gland-tipped hairs in various proportions, best marked on the flowerless shoots, which are suberect, or arched and rooting from a callus at the tip, thus giving rise to new individuals. *Leaves* usually pinnately 3–5-foliolate, subpersistent, glabrous or pubescent; leaflets petioled, overlapping or not, obovate or rhomboid-ovate, coarsely irregularly serrate or toothed, convex, dark green above, paler, often glaucous beneath. *Flowers* white or pink, homogamous, in terminal racemes, the lateral branches corymbose or elongate. *Drupes* black or red-purple.—Distrib. Europe (Arctic), N. Africa, N. and W. Asia, Himalaya.

For the following arrangement of the British forms of *R. frutico'sus* I am indebted to Mr. J. G. Baker.

a Stems with scattered uniform prickles, quite glabrous, i.e. *without bristles or gland-tipped hairs.*

ub-sp. R. SUBEREC'TUS, *Anders.*; barren shoots suberect, tips not rooting, leaflets large membranous bright green glabrous or slightly hairy beneath, sepals green with distinct white edges, drupes numerous claret-coloured or black. *R. umbro'sus,* Lees. *R. fastigia'tus,* W. and N.—Copses, England, Scotland, and Ireland, especially in the north. One of the best-marked forms.—*R. plica'tus,* W. and N., has larger and more hooked prickles. *R. fis'sus,* Lindl., has copious small prickles, leaflets more hairy beneath, sepals sometimes appressed to the fruit. *R. affi'nis,* W. and N. (*lentigino'sus,* Lees, a form), is a connecting link with sub-sp. *rhamnifo'lius.* VAR. *R. hemiste'mon,* Müll.; has subracemose inflorescence, very hairy rachis and calyx appressed to the fruit.

Sub-sp. R. RHAMNIFO'LIUS, *W.* and *N.*; barren stem arched angular tips rooting not glaucous, leaflets often large cordate and reflexed more finely toothed than in the two following, sometimes white and tomentose beneath, sepals not distinctly white edged, fruit large black juicy.—From Mid. Scotland, southd.; Ireland.—*R. cordifolius,* W. and N., has leaflets larger, rounder, more coarsely toothed, terminal more cordate.—*R. incurva'tus,* Bab., differs in the broad basal prickles, leaflets more sharply toothed with incurved waved margins, more hairy beneath, and more densely hairy white calyx.— *R. imbrica'tus,* Hort, is intermediate between *corylifo'lius* and *cordifo'lius;* *R. Grabow'skii,* Weihe, and *R. Coleman'ni,* Bloxam, are intermediate between this sub-sp. and *Kœh'leri.* *R. macrophyl'lus* var. *glabra'ta,* Bab., belongs here. *R. ramo'sus,* Blox., has fewer prickles, broadly ovate convex shining leaflets that do not overlap, a larger laxer panicle and pink petals.

Sub-sp. R. LINDLEIA'NUS, *Lees*; barren stems arching glabrous, leaflets not imbricate terminal obovate or oblong cuspidate, panicle very open compound with patent hairy corymbose branches and many deflexed unequal prickles. *R. nitidus,* Bell Salter, not W. and N.—Common in Britain, unknown on the Continent.

Sub-sp. R. CORYLIFO'LIUS, *Sm.*; barren stem between arched and prostrate not glaucous tip rooting, leaflets 5 membranous hazel-like imbricate, sepals densely tomentose all over the back, drupes few large. *R. sublus'tris,* Lees, *R. purpu'reus,* Bab.—From the Clyde southd.—The type has terete stems and many small slender prickles.—*R.'Balfouria'nus,* Blox., and *althæifo'lius,* Bab., not Host, have sepals appressed to the fruit; the former approaches *cæ'sius* by its corymbose inflorescence. *R. latifo'lius,* Bab., is a rare form intermediate between this sub-sp. and the last. *R. Wahlber'gii,* Arrh. (*conjun'gens,* Bab.), is a form with more angular barren stems and larger prickles.

Sub-sp. R. CÆ'SIUS, *L.*; stem prostrate glaucous, prickles more unequal setaceous, leaflets usually 3 green on both surfaces, sepals appressed, densely tomentose all over the back, drupes few large glaucous.—Hedges and thickets from Perth southd.; ascends to near 1,000 ft. in Yorkshire; Ireland. —*Dewberry.*—Often a well-marked form from its glaucous character. *R. ten'uis,* Bell Salter, *ulmifo'lius, interme'dius,* and *his'pidus* are indistinguishable forms; *R. pseudo-idæ'us,* Lej., is probably a hybrid with *R. Idæ'us.*

β *Stems with scattered uniform prickles, pubescent or hairy, but with few or no gland-tipped hairs or bristles, barren ones arching and rooting at the tip.*

Sub-sp. R. DIS'COLOR, *W.* and *N.*; stem with appressed stellate hairs, prickles strong, leaflets small most persistent of any, bright green above, densely tomentose and white beneath, flowers pink, sepals always reflexed, drupes small with little flesh. *R. abrup'tus,* Lindl.—The most common form except in Scotland (chiefly S. and Mid. Europe.)—*R. thyrsoi'deus,* Wimm., has stem stronger and more arching, flowers white, leaflets larger less white beneath, pubescence of stem and leaves looser.—VAR. *pubi'gera,* Bab., connects this with *leucosta'chys.*

Sub-sp. R. LEUCOSTA'CHYS, *Sm.*; stem angular between arching and prostrate with copious spreading hairs and sometimes a few glandular hairs rachis of panicle densely villous, leaflets finely toothed densely pubescent often white beneath, terminal one roundish with a point, panicle thyrsoid compound, sepals reflexed, petals often deep red. *R. vesti'tus,* Weihe ; *R. Leightonia'nus,* Bab.—Common from Berwick southd. ; Ireland.

Sub-sp. R. VILLICAU'LIS, *Weihe ;* stem angular with copious spreading hairs but no glandular ones, rachis of panicle densely villous, leaflets finely toothed densely pubescent often white beneath, terminal roundish with a point, panicle thyrsoid compound, sepals reflexed, petals pale. *R. carpini-fo'lius,* Bab. not Blox. ; *R. pampino'sus,* Lees. *R. vulga'ris,* W. and N., connects *villicau'lis* with *Rad'ula* by *R. adsci'tus,* Genev., and *R. dera'sus,* Müll.—Common, N. to Sutherland ; Ireland.

Sub-sp.R. SALT'ERI, *Bab.* ; stem angular arching with no glandular hairs, spreading hairs few and deciduous, leaflets coarsely and irregularly toothed shortly grey-pubescent or finally subglabrous beneath, terminal roundish pointed often cordate, panicle thyrsoid compound, rachis finely pubescent with copious strong red prickles, sepals reflexed, petals pale. *R. calva'tus,* Blox.—From York southd.; Ireland.—Connects this group with *rhæmnifo'lius.*

Sub-sp. R. UMBRO'SUS, *Arrh.*; stem angular, hairs few and deciduous none glandular, leaflets sometimes 7 finely toothed finely grey-pubescent beneath, terminal roundish or obovate pointed, panicle thyrsoid compound, rachis finely pubescent, sepals reflexed, petals pale. *R. macrophyl'lus,* var. *umbro'sa,* Bab.; *R. carpinifo'lius,* of many English writers, not Weihe.—*R. hirtifo'lius,* Müll., is an allied form unknown to me.—Common, from the Clyde southd.

Sub-sp. R. MACROPHYL'LUS, *Weihe ;* stem angular with a few spreading hairs and no glandular ones, leaflets coarsely and irregularly toothed finely grey-pubescent beneath, terminal obovate or obovate-oblong pointed, panicle thyrsoid compound, rachis finely pubescent, sepals reflexed, petals pale. *R. Schlechtendahl'ii,* Weihe ; *R. amplifica'tus,* Lees.—From Aberdeen southd.; Ireland.

Sub-sp. R. MUCRONULA'TUS, *Boreau ;* stems between arching and prostrate with a few spreading and often a few glandular hairs, leaflets shortly grey-pubescent beneath, terminal roundish abruptly pointed, panicle sparse often subsimple, rachis densely villous, its prickles few and weak, sepals reflexed, petals pale. *R. mucrona'tus,* Blox. not Seringe.—Local, from the Clyde southd.; Ireland.

Sub-sp. R. SPRENGE'LII, *Weihe ;* stem weak wide-trailing terete with a few spreading and often a few glandular hairs, leaflets coarsely toothed finely and shortly grey-pubescent or finally subglabrous beneath, often only three

on the fully developed leaves, terminal obovate pointed, panicle broad
sparse often subsimple, rachis finely pubescent with few prickles, sepals
ascending often leaf-pointed, petals pink. *R. Borre'ri,* Bell Salter; *R.
rubicolor,* Blox.—A well-marked but local form from York to Hants.

γ *Stems with copious bristles and glandular hairs, prickles unequal, often
very numerous.*

† *Leaves of the barren shoot 5-foliolate.*

Sub-sp. R. DUMETO'RUM, *Weihe;* stem between arching and trailing subcylindric sometimes pruinose, bristles few or many, hairs 0 or few on the
barren shoots, generally fewer on the rachis of the panicle than in all the
following, leaflets broad finely grey-pubescent or subglabrous beneath,
lateral pairs much imbricated, toothing open but not long, terminal suborbicular, sepals reflexed or ascending not leaf-pointed, petals broad, drupes
few large. *R. nemoro'sus* of many, scarcely of Hayne.—The glandular
representative of *corylifo'lius* in fruit, leaves, and prolonged flowering. *R.
tubercula'tus,* Bab., is a variety with subequal prickles, appressed sepals, few
bristles and glandular hairs. *R. diversifo'lius,* Lindl., has more copious and
irregular prickles and sepals mostly reflexed. *R. emersisty'lus,* Müll. (*R.
Briggs'ii,* Blox.), is near *diversifo'lius,* but stems more hairy, leaflets round,
and calyx appressed to the fruit. *R. concin'nus,* Baker, has subequal
prickles, smaller and less coarsely toothed leaflets and reflexed sepals.—
Common in hedges, England and Ireland ; very variable.
Sub-sp. R. RAD'ULA, *Weihe;* stem arching angular, prickles strong subequal,
bristles hairs and glands of barren stem copious, leaflets not imbricated
grey or often white-pubescent beneath, toothing moderately fine in the
typical form, terminal leaflet obovate, sepals reflexed not leaf-pointed,
petals broad.—*R. ru'dis,* Weihe, is a variety with stronger prickles and
leaflets deeply and very irregularly toothed. *R. Leighto'ni,* Lees, *denticula'tus,*
Bab., *muta'bilis,* Genev., and *obli'quus,* Wirtg., are closely allied.—Common
in hedges, from Fife southd. ; Ireland.
Sub-sp. R. BLOXA'MII, *Lees;* stem arching angular, prickles smaller than in
the preceding, subequal, hairs bristles and glands moderately numerous,
leaflets not imbricated moderately coarsely toothed green finely-pubescent
beneath, terminal suborbicular often cordate, sepals reflexed not leaf-pointed,
petals broad.—*R. sca'ber,* Weihe (*Babingto'nii,* Bell Salter), is an allied form
with stronger prickles and the panicle often very large and lax with patent
branches. *R. fusco-a'ter,* Weihe, is a rare form connecting this with
villicau'lis.—Local, from Durham southd.
Sub-sp. R. KŒHL'ERI, *Weihe;* stem trailing nearly terete, prickles very
numerous irregular strongly hooked, bristles numerous, glandular and simple
hairs few, leaflets not imbricate pubescence thin grey, toothing moderately
coarse, terminal orbicular, sepals reflexed not leaf-pointed, petals broad.—
R. cavatifo'lius, Müll., is a subglabrous form with few hairs and bristles,
leaflets not tomentose hairy on the veins beneath terminal cordate, panicle
abrupt with short thick terminal peduncles. *R. infes'tus,* Weihe, is a variety
with an arching stem, prickles and bristles much less dense.—From the
Clyde southd. ; Ireland.

Sub-sp. R. Hys'trix, *Weihe;* stem arching angular, prickles weak subequal, bristles glandular and simple hairs moderately numerous, leaflets not imbricate green thinly pubescent or finely glabrous beneath, terminal obovate, toothing fine, sepals ascending remarkably leaf-pointed, petals broad.—*R. Lejeu'nii*, Weihe, is a variety with a large lax panicle with patent branches. *R. rosa'ceus*, Weihe, is a trailing variety with more irregular prickles and narrow petals.—Local, from Northumberland southd.; Ireland.

Sub-sp. R. pal'lidus, *Weihe;* stem angular wide-trailing, prickles small copious very unequal, bristles simple and glandular hairs copious, leaflets not imbricate finely toothed densely grey or sometimes white-pubescent beneath, terminal obovate, sepals reflexed not leaf-pointed, petals narrow. —*R. humifu'sus*, Weihe, and *R. folio'sus*, Weihe, are varieties with the leaves prolonged into the panicle. *R. hirtus*, Weihe (*fuscus*, Lees), and *Reuteri*, Merc., are closely allied forms. *R. præ'ruptorum*, Boul. (*R. pygmæ'us*, Bab., not Weihe, and var. *Men'kii*, Bab.), is a form between *pal'lidus* and *Blox'ami*. —Common in woods, from Perth and the Clyde southd., local ; Ireland.

†† *Leaves of barren shoot 3-foliolate.*

Sub-sp. R. glandulo'sus, *Bell.* ; stem subterete trailing, prickles small straight weak, bristles simple and glandular hairs copious, leaves green thinly pubescent beneath, panicle thyrsoid compound, bristles of rachis very fine and copious, sepals ascending leaf-pointed densely bristly on the back, petals narrow.—*R. Bellar'di*, Weihe (*R. denta'tus*, Blox.), is a variety with less prickly rachis and sepals and membranous leaves green and nearly glabrous beneath. *R. rotundifo'lius*, Blox., has prickles stronger, leaves more hairy beneath, sepals less distinctly leaf-pointed. *R. Purchasii*, Blox., is an obscure plant of this affinity.—Local, from Aberdeen southd.; Ireland.

Sub-sp. R. saltu'um, *Focke;* stem trailing terete, prickles small very unequal, bristles glandular and simple hairs much fewer than in the last, leaves densely shortly pubescent often white beneath, panicle compound, rachis very wavy, sepals reflexed not leaf-pointed, petals narrow. *R. Gunthe'ri*, Weihe.—Local.

Sub-sp. R. pyramida'lis, *Bab.* ; stem trailing clothed as in the last, leaves green thinly hairy beneath, panicle subracemose, rachis straight, sepals ascending leaf-pointed, petals narrow.—Wales, Worcester, and Devonshire.

4. DRY'AS, *L.*

Prostrate, tufted, scapigerous shrubs. *Leaves* simple, white beneath ; stipules adnate to the petiole. *Flowers* solitary, large white, or yellow, andro-diœceous, proterandrous or subproterogynous. *Calyx* inferior, persistent ; lobes 8–9, valvate in bud. *Petals* 8–9. *Stamens* many, crowded. *Disk* concave, hairy. *Carpels* many, sunk in the calyx-tube ; style terminal ; ovule 1, ascending. *Achenes* many ; styles slender, feathery.— Distrib. Arctic and Alpine regions of N. temp. zone ; species 2 or 3.— Etym. δρυάς, from the *oak*-like foliage.

D. octopet'ala, *L.* ; leaves oblong-ovate coarsely crenate-serrate. *D. depressa*, Bab.

Stony, chiefly limestone and mountain districts, local, from Carnarvon and Stafford to Orkney, ascends to 2,700 ft.; descending to sea-level in N. and W. Ireland; fl. June–July.—*Stem* tortuous, much branched. *Leaves* ½–1 in., crowded, obtuse, hoary beneath, shining above, margins reflexed, midrib hairy and scurfy; scape 1–3 in., glandular and hairy, longer in fruit. *Flowers* 1–1½ in. diam., white. *Sepals* about 8, woolly and with black glandular hairs, obtuse or subacute. *Petals* oblong. *Achenes* hispid; awn 1–2 in.—DISTRIB. of the genus.

5. GE′UM, *L.* AVENS.

Erect perennial herbs. *Radical leaves* crowded, pinnate; terminal leaflet very large; stipules adnate to the petiole. *Flowers* solitary or corymbose, white yellow or red, honeyed. *Calyx* inferior, persistent, with 5 bracteoles above its base; lobes 5, imbricate or valvate in bud. *Petals* 5. *Stamens* crowded. *Disk* smooth or grooved. *Carpels* many, receptacle short or long; styles filiform, straight or bent; ovule 1, ascending. *Achenes* many, on a dry receptacle, ending in filiform straight or bent styles which are often hooked at the tip.—DISTRIB. N. and S. temp. and cold regions; species 30.—ETYM. γεύω, from the aromatic roots.

1. **G. urba′num,** *L.*; flowers erect, head of achenes sessile, awn with a short glabrous hook at the tip, calyx-lobes reflexed in fruit.

Borders of copses, hedgebanks, &c., from Caithness southd., ascends to near 1,700 ft. in the Lake district; Ireland; Channel Islands; fl. June–Aug.— Softly hairy. *Stem* 1–3 ft. *Radical leaves* long-petioled, interruptedly pinnate; terminal leaflet 2–3 in. broad, suborbicular, obscurely lobed, crenate; lateral ¼–½ in., oblong, sessile; cauline leaves variable; stipules foliaceous, lobed and toothed. *Flowers* ½–¾ in. diam., yellow, proterogynous; peduncle slender. *Petals* obovate, spreading, as long as the acute calyx-lobes. *Achenes* hispid, spreading; awn ½ in.; receptacle hispid.—DISTRIB. Europe (Arctic), N. Africa, N. and W. Asia, Himalaya, N. America.

2. **G. riva′le,** *L.*; flowers drooping, head of achenes stalked, awn jointed and hairy beyond the middle, calyx-lobes appressed in fruit.

By streams, in copses, &c., from Devon and Sussex to Orkney; ascends to 2,800 ft. in the Highlands; Ireland; fl. May–July.—*Stem* 1–1½ ft., lower part with soft reflexed hairs, very pubescent above. *Leaves* very variable, much as in *G. urba′num*, but the segments are often numerous, the lateral larger, and all more toothed; stipules small. *Flowers* 1–1½ in. diam., proterogynous, sometimes submonœcious. *Calyx-segments* red-brown, acuminate, pubescent. *Petals* yellow, obcordate. *Achenes* more or less hispid.—DISTRIB. Europe (Arctic), N. and W. Asia, N. and S. America, Australasia.

G. interme′dium, Ehrh., is a hybrid; flowers sometimes erect, petals of *G. urba′num* but deeper coloured, calyx intermediate, not reflexed in fruit, fruit usually sessile.—Damp woods, not uncommon (often with *riva′le*, seldom with *urba′num*, Syme). Bell Salter produced this hybrid artificially, and it proved fertile.

6. FRAGA'RIA, *L.* STRAWBERRY.

Perennial scapigerous herbs, with runners. *Leaves* 3-foliolate (in Britain), pinnate or 1-foliolate; stipules adnate to the petiole. *Flowers* white or yellow, honeyed, proterogynous, often polygamous. *Calyx* inferior, persistent, 5-bracteolate; lobes 5, valvate in bud. *Petals* 5. *Stamens* many, persistent. *Carpels* many, distinct, receptacle convex; styles ventral, persistent; ovule 1, ascending. *Achenes* many, minute, on the surface of the enlarged fleshy receptacle.—DISTRIB. N. temp. regions, Andes, Sandwich Islands, Bourbon; species 3 or 4.—ETYM. The Latin name for the *fragrant* fruit.

F. ves'ca, *L.* ; leaflets usually sessile, pedicels with silky appressed hairs, flowers hermaphrodite. *Wild Strawberry.*

Shady places, N. to Shetland; ascends to near 2,000 ft. in the Highlands; Ireland; Channel Islands; fl. April–May.—Silky and hairy. *Rootstock* short or long, woody, with a terminal tuft of leaves. *Radical leaves* petioled; leaflets 1–2 in., obliquely ovate or oblong, coarsely toothed or serrate, plaited, lateral sometimes cleft; stipules scarious. *Scapes* 1–6 in., axillary. *Flowers* ½–¾ in. diam., in irregular cymes, inclined, white; bract at the base of the cyme leafy, at the pedicel smaller stipuliform; bracteoles ovate, smaller than the acute calyx-lobes. *Petals* obovate. *Receptacle of fruit* obovoid or globose, red or white, covered to the base with achenes, calyx-lobes spreading.—DISTRIB. Europe (Arctic), N. and W. Asia, Himalaya, N. America.

F. ELA'TIOR, *Ehrh.* ; much larger than *F. vesca*, leaflets often shortly stalked, pedicels with spreading hairs, flowers sub-1-sexual.

A garden escape; fl. April–May.—The Haut-bois strawberry, whose origin is probably *F. ves'ca ;* the base of receptacle is without achenes.

7. POTENTIL'LA, *L.* CINQUEFOIL.

Perennial herbs, rarely shrubs. *Leaves* compound; stipules adnate to the petiole. *Flowers* white or yellow, rarely red, solitary or in corymbose cymes, honeyed. *Calyx* inferior, persistent, 5–7- rarely 4-bracteolate; lobes as 'many, valvate in bud. *Petals* as many. *Stamens* many, rarely few and definite. *Disk* annular or coating the calyx-tube. *Carpels* many, rarely 1 or few, on a dry convex or concave receptacle; style persistent or deciduous, ventral or basal; ovule 1, pendulous. *Achenes* many, receptacle dry.—DISTRIB. N. temp. and Arctic regions, 2 are southern; species 120.—ETYM. *Potens*, from the *powerful* medicinal effects attributed to some.

SECTION 1. **Trichothal'amus,** *Lehm.* (gen.). Shrubby. *Petals* 5, orbicular, yellow. *Achenes* many, hairy, on a very hispid receptacle.

1. **P. frutico'sa,** *L.* ; silky, leaves subdigitately-pinnate.

Rocky banks by rivers, local: York, Durham, Cumberland, Westmoreland N. Clare, Galway; fl. June–July.—A much-branched, leafy shrub, 2–4 ft. ;

bark flaking. *Leaves* oblong or lanceolate; leaflets 3–5, $\frac{1}{2}$–$\frac{3}{4}$ in.; margins entire, revolute; stipules entire. *Flowers* 1–1$\frac{1}{2}$ in., few, in terminal subcorymbose cymes, golden yellow, sub-1-sexual, homogamous; bracteoles lanceolate, longer than the ovate calyx-segments.—DISTRIB. N. and Mid. Europe, Alps (Arctic), N. and W. Asia, Himalaya, N. America.—In Teesdale the flowers appear to be functionally 1-sexual; the sexes differ in appearance.

SECTION 2. **Co'marum,** *L.* (gen.). *Petals* 5, small, purple-brown. *Stamens* many. *Achenes* many, glabrous, on a conical dry spongy downy receptacle.

2. **P. Co'marum,** *Nestl.* ; leaves pinnately 5–7-foliolate. *Co'marum palustre,* L.

Bogs and marshes, ascends to 2,800 ft. in the Highlands ; Ireland ; Channel Islands ; fl. June–July.—Sparingly hairy. *Rootstock* long, woody; roots fibrous. *Stems* $\frac{1}{2}$–1$\frac{1}{2}$ ft., ascending, purple-brown. *Leaves* 2–4 in.; leaflets 1$\frac{1}{2}$–2 in., narrow-oblong, obtuse, coarsely serrate, pale beneath; stipules large, membranous, free portion often cut. *Flowers* 1–1$\frac{1}{2}$ in. diam., few. *Bracteoles* smaller than the ovate-lanceolate acuminate sepals, purplish. *Petals* much smaller, dark purplish-brown.—DISTRIB. Europe (Arctic), N. Asia, N. America.—Rootstock powerfully astringent, and yields a yellow dye.

SECTION 3. **Sibbald'ia,** *L.* (gen.). *Petals* 5–7. *Stamens* 4–10. *Achenes* 4–10, glabrous, on a concave pubescent receptacle.

3. **P. Sib'baldi,** *Hall. f.*; leaves 3-foliolate, leaflets obovate truncate tip 3–6-toothed. *P. procum'bens,* Clairv. ; *Sibbald'ia procum'bens,* L.

Stony places on the Scotch alps, from Peebles north to Shetland ; alt. 1,500–4,000 ft. ; fl. July.—More or less hairy and glaucous. *Rootstock* woody, depressed, branches leafy at the tip. *Leaves* 1–3 in.; leaflets $\frac{1}{3}$–1 in. *Flowering stems* 3–5 in., axillary, ascending, leafy. *Flowers* $\frac{1}{4}$ in. diam., few, in terminal close cymes. *Petals* small, narrow, orange-yellow, or 0. *Calyx-segments* lanceolate, acute ; bracteoles linear.—DISTRIB. Alpine and Arctic Europe, N. Asia, Himalaya, N. America.

SECTION 4. **Potentil'la** proper. Petals 5, rarely 4, orbicular or obcordate, usually yellow or white. *Stamens* many. *Achenes* many, on a concave glabrous or hairy receptacle.

* *Flowering stems annual, from below the crown of the perennial rootstock.*

4. **P. Tormentil'la,** *Scop.* ; stem slender rarely rooting, leaves 3-rarely 5-foliolate, flowers usually cymose, petals usually 4 yellow, achenes reticulate.

Heaths, copses, dry pastures, N. to Shetland ; ascends to 3,300 ft. in the Highlands ; Ireland; Channel Islands; fl. June–Sept.—Slightly hairy. *Rootstock* stout, almost tuberous. *Stems* 6–10 in., slender, leafy, clothed with curly hairs. *Leaves* subsessile ; radical petioled, leaflets $\frac{1}{4}$–$\frac{1}{2}$ in., obovate-cuneate, tip 3–4-toothed or -lobed; cauline subsessile, leaflets much narrower; stipules foliaceous, cut. *Flowers* $\frac{1}{2}$–$\frac{3}{4}$ in. diam., yellow, homogamous. *Calyx-*

lobes ovate; bracteoles linear, as long.—DISTRIB. Europe (Arctic), W. Siberia, Azores.—Rootstock strongly astringent, used for tanning.

P. TORMENTIL'LA proper ; erect or suberect, cauline leaves sessile 3-foliolate, flowers cymose. *Tormentilla erecta*, L. ; *T. officina'lis*, Curt. Sub-sp. P. PROCUM'BENS, *Sibth.* ; procumbent, cauline leaves 3–5-foliolate often petioled, flowers few or solitary. *P. nemora'lis*, Nestl.; *Tormentil'la rep'tans*, L.

5. P. rep'tans, *L.* ; stem slender creeping and rooting, leaves digitately 5- rarely 3-foliolate long-petioled, flowers solitary on slender axillary peduncles, petals 5 yellow, achenes granulate.

Meadows, waysides, and pastures, from Banff and Cantire southd.; Ireland ; Channel Islands; fl. June–Aug.—Very variable in size and pubescence ; usually larger than *P. Tormentil'la*, and having besides the above characters bracteoles as broad as the sepals. *Flowers* ¾–1 in. diam., homogamous.— DISTRIB. Europe, from Gothland southd., N. and W. Asia, Himalaya, Canaries, Azores.

P. mixta, Nolte, is a supposed hybrid with *P. Tormentil'la*.

6. P. ver'na, *L.* ; stem prostrate, leaves digitately 5-7-foliolate, leaflets obovate or cuneate truncate deeply crenate or lobulate towards the tip, flowers several yellow, achenes smooth glabrous.

Hilly rocky places, local ; from Forfar to Cambridge and Somerset ; fl. April– June.—More or less hairy. *Rootstock* woody, branched, tufted. *Radical leaves* 2–3 in.; stipules with narrow subulate tips ; leaflets ¼–¾ in., green on both surfaces; terminal tooth short; cauline 1–3-foliolate, their stipules ovate-lanceolate. *Flowers* few, ½ in. diam., homogamous.—DISTRIB. Europe (Arctic), Siberia, W. Asia, Himalaya.

7. P. salisburgen'sis, *Haenke ;* stem ascending, leaves digitately 5-7-foliolate, leaflets obovate or cuneate deeply crenate or serrate usually above the middle, flowers several yellow, achenes smooth glabrous. *P. alpes'tris*, Hall. f. ; *P. au'rea*, Sm. not L. ; *P. macula'ta*, Pourr.

Rocky alpine ledges, local ; from Aberdeen and Argyll to York and Wales ; ascends to 2,700 feet in the Highlands; fl. June–July.—Probably a large form of *P. ver'na*, with ascending stems, 4–10 in., larger less truncate leaflets (but not constantly so), and flowers 1 in. diam.—DISTRIB. Europe (Arctic), N. and W. Asia, Greenland, Labrador.—The name *P. macula'ta* is coeval with *salisburgen'sis.* I have taken the latter because the spotted-petalled form (which occurs on Ben Lawers) is a scarce one.

8. P. anseri'na, *L.* ; stoloniferous, silky, leaves interruptedly pinnate, leaflets many deeply serrate or pinnatifid, the alternate minute, flowers solitary yellow, achenes glabrous smooth. *Silver Weed.*

Roadsides and damp pastures, N. to Shetland ; ascends to 1,200 ft. in Derby ; Ireland ; Channel Islands; fl. July–Aug.—*Rootstock* slender, branched. *Stems* 0. *Leaves* 2–5 in., silvery beneath, stoloniferous from their axils ; leaflets ½–2 in., alternate, close-set, sessile, obovate oblong, obtuse, serratures tipped with silky hairs ; stipules calyptriform, enclosing the buds. *Flowers* ½–¾ in. diam., on solitary axillary slender peduncles, homogamous ; bracteoles

often serrate.—DISTRIB. Arctic and N. and S. temp. regions, Himalaya.—
Rootstock eaten in times of scarcity in the Hebrides.

9. **P. Fragarias'trum,** *Ehrh.* ; leaves 3-foliolate, leaflets obovate,
flowers white, achenes hairy below reticulate. *Fraga'ria ster'ilis,* L.

Waysides, woods, and banks, N. to Caithness; ascends to 2,100 ft. in Wales;
Ireland; Channel Islands; fl. March–May.—Similar to *Fragaria ves'ca,* but has
no runners, nerves of leaflets not sunk above, and fruit very different.—More
or less hairy or silky. *Rootstock* stout, woody, branched, depressed. *Leaves*
2–6 in., tufted; petiole with spreading hairs; leaflets ½–1 in., coarsely
crenate towards the tip, very hairy beneath. *Flowering stems* 1–6 in.,
axillary, slender, naked or 1–2-leaved, 1–3-fld. *Flowers* ½ in. diam.
Receptacle with very long hairs.—DISTRIB. Europe, N. Africa.

** *Flowering stems annual, terminating the branches of the perennial rootstock.*
Receptacle hairy.

10. **P. rupes'tris,** *L.* ; leaves pinnate, radical 5- cauline 3-foliolate,
flowers white.

Rocks, Craig Breidden, Montgomery; fl. May–June.—Hairy, especially below.
Rootstock woody, branched. *Stems* 1–2 ft., erect, branched above. *Radical
leaves* 3–6 in., petiole very slender; leaflets ¾–1 in., unequal at the base,
oblong or obliquely obovate, irregularly crenate; cauline few, subsessile.
Flowers few, ¾–1 in. diam. *Achenes* smooth, glabrous.—DISTRIB. From
Gothland southd., N. and W. Asia, Kashmir.

11. **P. argen'tea,** L. ; leaves digitately 5-foliolate, leaflets cuneate
much cut white beneath, flowers yellow.

Dry pastures and roadsides; from Elgin and E. Scotland southd., local;
Channel Islands; fl. June–July.—More or less covered, especially the leaves
beneath, with white appressed wool. *Rootstock* short, woody. *Stems* 6–18
in., slender, suberect or decumbent, branched, leafy; branches divaricating.
Leaves petioled, upper most sessile; leaflets ½–1½ in., narrowly cuneate,
½-pinnatifid upwards, margins recurved. *Flowers* ¼–½ in. diam., subcorymbose. *Achenes* smooth, glabrous.—DISTRIB. Europe (Arctic), N. and W.
Asia, Himalaya, N. America.

*** *Root annual.*

P. NORVE'GICA, *L.* ; hirsute, leaves palmately 3-foliolate, leaflets oblanceolate coarsely-toothed, flowers yellow.

Middlesex, Hertford, York; naturalized and apparently rapidly spreading;
fl. July.—*Stem* stout, simple below, erect; 8–10 in.; rarely branched above,
rarely at the base and decumbent. *Leaves* 1–2 in. diam., petiole slender,
leaflets 1–2 in., serrate nearly to the base. *Flowers* in crowded terminal
cymes, ½ in. diam. *Petals* stouter than the calyx, obovate. *Achenes* rugose.
—DISTRIB. Europe, temp. and Arctic Asia and America.—Though usually
annual, specimens from the Kolyma river in Siberia have a perennial
rootstock.

8. ALCHEMIL'LA, *L.* LADY'S MANTLE.

Annual or perennial herbs. *Leaves* orbicular, lobed or deeply divided;
stipules sheathing and adnate to the petiole. *Flowers* minute, in scorpioid

cymes, honeyed. *Calyx* urceolate, persistent, 4–5-bracteolate ; lobes 4–5, valvate in bud. *Petals* 0. *Stamens* 1–4, inserted on the mouth of the calyx. *Disk* coating the calyx-tube, its thickened margin all but closing the mouth. *Carpels* 1–5, basal in the calyx-tube ; styles basal or ventral ; ovule 1, basal. *Achenes* 1–4, enclosed in the membranous calyx-tube.—DISTRIB. Europe, India, and America, but chiefly Andean ; species 30.—ETYM. Arabic.

SECTION 1. **Aph′anes,** *L.* (gen.). Annual. *Cymes* leaf-opposed, dense. *Bracteoles* minute or 0. *Antheriferous stamens* 1–2.

1. **A. arven′sis,** *Lamk.* ; leaves cuneate or fan-shaped 3-lobed, lobes cut. *Aphanes,* L.

Fields and waste places in dry soil, N. to Shetland ; ascends to 1,6C0 ft. in the Highlands ; Ireland, Channel Islands ; fl. May–Aug.—Hairy, much branched from the base ; branches 2–8 in., erect or prostrate. *Leaves* $\frac{1}{4}$–$\frac{1}{2}$ in., narrowed into the short petiole, lobes cuneate ; stipules palmately cut. *Flowers* hidden by the stipules. *Calyx* usually 4-cleft. *Achenes* 1–3.— —DISTRIB. Europe, N. Africa, W. Asia ; introd. in N. America.

SECTION 2. **Alchemil′la** proper. Perennial. *Cymes* corymbose or panicled. *Bracteoles* conspicuous. *Antheriferous stamens* usually 4.

2. **A. vulga′ris,** *L.* ; leaves reniform plaited 6–9-lobed green beneath.

Moist pastures and streams, N. to Shetland (absent in Kent) ; ascends to 3,600 ft. in the Highlands ; Ireland ; Channel Islands ; fl. June–Aug.—More or less hairy. *Rootstock* black, stout, short. *Stem* 6–18 in., ascending. *Radical leaves* 2–6 in. diam., lobes serrate, petiole 6–18 in., cauline smaller ; stipules connate, toothed. *Cymes* irregularly racemed or panicled. *Flowers* $\frac{1}{8}$ in. diam., yellow-green, proterandrous, rarely perfect ; pedicel short. *Achenes* 1 or 2, glandular.—DISTRIB. Europe (Arctic), N. and W. Asia, Kashmir, Greenland, Labrador.—Rootstock astringent and edible.—*A. monta′na,* Willd. (*A. hybrida,* Pers.), is a dwarf mountain form with leaves and petioles very pubescent or silky.

3. **A. alpi′na,** *L.* ; leaves 5–7-partite or -foliolate silvery beneath.

Mountain streams and rocks, York to Shetland and Ireland, alt. 400–4,000 ft. ; fl. June–Aug.—More or less clothed with silky hairs, except the upper surface of the leaves. *Rootstock* slender, branched. *Stems* 3–9 in., ascending, slender. *Leaves* on slender petioles, upper 1–2 in. diam., sessile, orbicular-reniform ; leaflets $\frac{3}{4}$–1$\frac{1}{4}$ in., narrow oblong, sharply toothed at the tip ; stipules connate, cleft. *Cymes* interruptedly spiked and panicled. *Flowers* $\frac{1}{6}$ in. diam., yellow-green ; pedicels short, hairy. *Achenes* minutely glandular. —DISTRIB. Europe (Arctic), N. and W. Asia, Greenland.

A. conjunc′ta, Bab. (*A. argentea,* Don), is a sport with subpeltate leaves, the leaflets connate below the middle, found in Forfar and Arran, Faroe Isles, France and Switzerland.

9. **AGRIMO′NIA,** *L.* AGRIMONY.

Slender perennial herbs. *Leaves* pinnate ; leaflets serrate ; stipules partially adnate to the petiole. *Flowers* small, yellow, in terminal spike-

like racemes, not honeyed ; pedicels bracteate at the base, 2-bracteolate.
Calyx inferior, persistent ; tube turbinate, spinous, mouth contracted ;
lobes 5, imbricate. *Petals* 5. *Stamens* 5–10 or more, inserted at the
mouth of the calyx. *Disk* lining the calyx-tube, its margin thickened.
Carpels 2, included in the calyx-tube ; styles exserted, stigma 2-lobed ;
ovule pendulous. *Fruit* pendulous, of 1 or 2 achenes enclosed in the
hardened spinous calyx.—DISTRIB. N. temp. regions, and S. America ;
species 8.—ETYM. obscure.

A. Eupato'ria, *L.* ; leaves interruptedly pinnate, spines of calyx
hooked.

Hedgebanks, copses, and borders of fields ; ascends to 1,200 ft. in Yorkshire ;
Ireland ; fl. June–Aug.—Hairy or villous, erect, leafy, 1½–3 ft., rarely
branched. *Rootstock* woody, short. *Leaves* 3–7 in. ; leaflets 3–10 pairs,
larger upwards, largest 1–3 in., sessile, oblong or lanceolate, deeply coarsely
serrate ; smaller ¼–½ in., obovate or cuneate, 3–5-lobed ; stipules foliaceous,
½-lunate. *Racemes* lengthening and pedicels recurved in fruit ; bracts 3-fid ;
bracteoles close to the calyx. *Flowers* ⅓ in. diam., homogamous. *Calyx-tube*
¼ in., woody in fruit, spines many around the thickened mouth ; lobes
conniving, triangular, acute.—DISTRIB. N. temp. regions, Himalaya, N. and
S. Africa, N. America.—Rootstock astringent, and yields a yellow dye.
A. EUPATO'RIA proper ; calyx-tube obconic deeply furrowed fruiting, lower
spines spreading.—From Sutherland southd.
Sub-sp. **A.** ODORA'TA, *Mill.* ; more branched, resinous-scented, racemes denser,
flowers larger, calyx-tube campanulate scarcely furrowed, lower spines
spreading or reflexed.—From the Clyde and Perth southd., local ; very rare
in Ireland.

10. POTE'RIUM, *L.*

Erect perennial herbs. *Leaves* pinnate ; leaflets petioled ; stipules
adnate to the sheathing petiole. *Flowers* small, in dense long-peduncled
centrifugal heads or spikes, bracteate, 2-bracteolate, often polygamous,
anemophilous. *Calyx tube* turbinate, mouth contracted ; lobes 4, petioled,
deciduous, imbricate in bud. *Petals* 0. *Stamens* 4 or more, inserted at
the mouth of the calyx, filaments slender ; anthers pendulous. *Disk* lining
the calyx-tube and closing its mouth. *Carpels* 1–3, enclosed in the calyx-
tube ; styles filiform, stigmas penicillate ; ovule 1, pendulous. *Achenes*
solitary, enclosed in the hardened 4-angled often winged or muricate
calyx-tube.—DISTRIB. N. temp. regions ; species 20.—ETYM. obscure.

1. **P. Sanguisor'ba,** *L.* ; upper flowers female, lower male or 2-sexual,
stamens 20–30 much exserted, fruiting calyx 4-winged reticulate but not
pitted between the wings, edges of reticulation smooth. *Salad Burnet.*

Dry pastures ; ascending to 1,600 ft. in Yorkshire ; E. Scotland only from
Perth to Berwick, local ; rare in Ireland ; Channel Islands ; fl. June–Aug.
—Glabrous or nearly so. *Rootstock* stout. *Stem* 6–18 in., ascending,
slender, much branched. *Radical leaves* 4–10 in. ; leaflets 5–10 pair, ¼–¾ in.,
broadly oblong, coarsely serrate ; stipules leafy. *Flower-heads* ⅓–⅔ in., on
long peduncles, shortly oblong, purplish. *Calyx-lobes* 1/16 in., oblong ; wings

thin ; bracteoles ciliate. *Stigmas* exserted. *Achene* dark, striate.—Distrib. Europe, N. Africa, N. and W. Asia, Himalaya.

P. murica'tum, *Spach ; flower-heads and flowers as in *P. Sanguisor'ba,* but fruiting calyx with thick entire or toothed wings pitted and reticulated between the wings, ridges muricate and toothed. Cultivated ground in Mid. and S. England; (an alien or colonist, *Wats.*); fl. July.—Chiefly distinguished from *P. Sanguisor'ba* by the larger fruit and calyx.—Distrib. Mid. Europe and Mediterranean region.

2. **P. officina'le,** *Hook. f.* ; flowers 2-sexual, stamens 4 not longer than the calyx-lobes, fruiting calyx 4-winged smooth between the wings. *Sanguisor'ba officina'lis,* L. *Great Burnet.*

Damp meadows, from Ayr and Selkirk southd.; ascends to 1,500 ft. in Yorkshire; W. and N. Ireland; Channel Islands; fl. June–Aug.—Very similar to *P. Sanguisorba,* but rootstock horizontal, stem erect, leaflets fewer, longer, less deeply serrate, usually cordate at the base, and flower-heads often cylindric and 1–1½ in. long. *Flowers* honeyed, homogamous.— Distrib. Europe (Arctic), N. and W. Asia.

11. RO'SA, *L.* Rose.

Erect, sarmentose or climbing prickly shrubs. *Leaves* pinnate ; leaflets serrate ; stipules adnate to the petiole. *Flowers* terminal, solitary or corymbose, white yellow or red, rarely bracteate, not honeyed, homogamous. *Calyx-tube* persistent, globose ovoid or pitcher-shaped, mouth contracted ; lobes imbricate in bud. *Petals* 5. *Stamens* many, inserted on the disk. *Disk* coating the calyx-tube, its thickened margin all but closing the mouth, silky. *Carpels* many, rarely few, sunk in the calyx-tube ; styles subterminal, distinct or connate above, stigma thickened ; ovule 1, pendulous. *Achenes* coriaceous or bony, enclosed in the fleshy or coriaceous calyx-tube.—Distrib. N. temp. regions, rare in America ; Abyssinia, India, Mexico ; species about 30.—Etym. The old Latin name.

The following account of the British roses is condensed from Mr. Baker's monograph (*Linn. Journ.* xi. 197), and revised by himself, most of the species being regarded as sub-species. As with the *fruticose Rubi,* all the so-called species are connected by intermediates ; but whereas, in the *Rubi,* the 4 or 5 most distinct British forms are connected by so many links that various botanists regard them as forms of one species ; in *Rosa,* the five most distinct British forms are connected by so few (comparatively) intermediates, that no botanical authority has reduced them to one species.

1. **R. spinosis'sima,** *L.* ; small, erect, bushy, prickles crowded very unequal nearly straight passing into stiff bristles and glandular hairs, leaves not or slightly glandular, sepals more or less persistent, fruit short, disk small or 0. *Scotch Rose, Burnet Rose.*

Open places especially sandy sea shores, from Caithness southd. ; Ireland ; Channel Islands ; ascends to 1,700 ft. in Scotland ; fl. May–June.—*Shrub* 1–4 ft., much branched. *Leaves* small, eglandular; leaflets 7–9, singly or doubly serrate, usually broad. *Flowers* 1-1½ in. diam., rarely 3 or more, together, white or pink. *Calyx-tube* usually glabrous; limb simple, eglandular. *Styles* free.—Distrib. Europe (Arctic), N. Africa, N. and W. Asia, Himalaya.

2. R. villo'sa, L. ; bush large, branches erect or elongate and arching, prickles uniform scattered slender nearly straight, leaflets very hairy eglandular or nearly so beneath, sepals more or less persistent densely glandular, fruit globose or turbinate densely prickly rarely naked.

Hedges and thickets, N. to Shetland; Ireland; ascends to 1,500 ft. in Yorkshire; fl. June–July.—Chiefly distinguished from *R. spinosissima* by its larger size, equal prickles, fewer very downy leaflets which are more constantly doubly serrate, and the more glandular fruit; and from *R. canina* by the straight prickles, and globose glandular fruit.—Distrib. Europe (Arctic), W. Asia.

R. villo'sa proper; branches arching, sepals copiously pinnate quite persistent, corolla often ciliate and glandular, fruit ripening early, disk obscure. *R. pomif'era*, Herrm.—Stafford and Gloster; not indigenous, common in gardens.

Sub-sp. R. mol'lis, *Sm.*; branches erect, leaflets softly pubescent, sepals sparingly pinnate quite persistent, fruit ripening early, disk obscure. *R. mollis'sima*, Willd. *R. heterophyl'la*, Woods.—From Orkney southd.; Ireland; extends into Arctic Europe.—Var. *cæru'lea*, Baker, has glands and bristles few on the petiole, calyx-tube glandular, fruit broad glabrous, peduncle naked or with few bristles.—*R. pseu'do-rubigino'sa*, Lej., has leaflets nearly glabrous above, glandular beneath, petiole and calyx-tube densely bristly.

Sub-sp. R. tomento'sa, *Sm.*; branches long (6–10) ft., arching, prickles sometimes curved, leaflets pubescent, sepals copiously pinnate not quite persistent, fruit not ripening early, disk distinct.—Common, N. to Shetland; Ireland.—*R. subglobo'sa*, Sm. (*R. Sherrar'di*, Davies); *R. farino'sa*, Rau.; *R. scabrius'cula*, Sm.; *R. sylves'tris*, Woods (*R. Jundzillia'na*, Baker; *R. britan'nica*, Deseg.); and *R. obova'ta*, Baker, are forms differing in pubescence and amount of glands on the leaves petioles peduncles calyx-tube and sepals, amount of double serration of the leaflets, number of flowers in a cluster, and of prickles, their length, strength, and curvature.—Var. *Woodsiana*, H. and J. Groves, is a form allied to *scabriuscula*, but smaller, more compact, with erect persistent sepals.—Wimbledon Common.

3. R. involu'ta, Sm. ; small, erect, branches short, prickles crowded gradually passing into bristles, leaflets doubly serrate glabrous or pubescent and glandular beneath, flowers 1-3, peduncles bristly, sepals persistent densely glandular on the back, fruit erect subglobose red, disk inconspicuous.

Banks and hedges, from Orkney southd.; Ireland; fl. June–July.—Intermediate between *R. spinosissima* and *villosa*, but nearest the latter; excessively variable. *Branches* sometimes arching; prickles scarcely curved. *Leaflets* with very open often compound teeth; petiole and stipules densely glandular and ciliate. *Peduncle* densely bristly. *Sepals* leafy. *Petals* white or pink. *Fruit* sparingly produced, colouring late.— —Distrib. Belgium, Switzerland (very rare).

R. involu'ta proper (var. *Smith'ii*, Baker); dwarf, mature leaflets glabrous above hairy and eglandular beneath, serratures close sharp; flowers solitary, calyx-tube densely acicular, sepals simple.—Var. *R. Sabi'ni*, Woods; prickles ⅓ in. straight, leaflets with copious compound serratures thinly pubescent above, petioles and peduncles densely hairy glandular and bristly, calyx-tube subglobose more or less setose, sepals pinnate, fruit subglobose. *R. gra'cilis*, Woods; *R. niva'lis*, Don; *R. corona'ta*, Crep. The most common form.—Var. *R. Donia'na*, Woods; small, leaflets more densely hairy, flowers solitary, sepals hardly pinnate, calyx-tube and fruit densely prickly. Dry places. Approaches *R. mollis'sima.*—Var. *graciles'cens*, Baker; robust, leaflets thinly hairy on both surfaces eglandular beneath much toothed, terminal 1–1½ in., flowers 3–6, calyx-tube glabrous ellipsoid. Antrim.— Var. *Robertso'ni*, Baker; sepals of *Sabi'ni*, but teeth of leaflets sharper and less compound, upper surface glabrous when mature, calyx-tube sometimes naked. Newcastle, Yorkshire, Antrim.—Var. *Nicholso'ni*, Crep.; densely setose, leaflets broadly ovate glabrous above, glandular beneath with hairy nerves, teeth very compound glandular.—Var. *læviga'ta*, Baker; leaflets of *Sabi'ni*, petiole villous and glandular rarely bristly, peduncle and calyx-tube glabrous, sepals simple eglandular, fruit depressed globose. Yorkshire, Antrim, and Derry.—Var. *Moor'ei*, Baker; prickles very stout ½ in., leaflets densely glandular beneath, petiole peduncle and calyx-tube densely setose and glandular, the larger prickles curved, sepals slightly pinnate. Derry. Approaches *R. rubigino'sa.*—Var. *occidenta'lis*, Baker; near *Wilso'ni*, but leaflets smaller and petioles and peduncles glandular and bristly, calyx-tube globose. Ireland, locality unknown.—Var. *R. Wilso'ni*, Borrer; tinged with purple, prickles as in *Sabi'ni*, leaflets often cordate terminal large glabrous above, ribs thinly hairy and subeglandular beneath, serratures simple, calyxtube almost glabrous, sepals nearly simple, fruit subovoid. Menai Straits, Derry. Approaches *R. rubel'la.*

4. **R. rubigino'sa**, *L.*; bush small, branches erect or arching, prickles stout at the base scattered hooked with often glandular hairs and bristles intermixed, leaflets densely glandular aromatic glabrous or thinly hairy, flowers 1–3, sepals subpersistent, fruit globose ovoid or oblong.

Best distinguished by its suberect habit and copious glandular pubescence, which gives out the strong sweetbriar odour; this, however, becomes fainter in the forms that pass into *R. canina* and *R. villosa.*—Europe, N. and W. Asia to N.W. India; introd. in N. America.

R. RUBIGINO'SA proper; very sweet-scented, erect, branches compact, prickles with a few bristles and glandular hairs intermixed, peduncles densely bristly, leaflets glabrous above, pubescent beneath, sepals densely glandular pinnate, fruit globose. *R. Eglante'ria*, Woods. Chalk hills, S. of England, native?; probably not indigenous in Scotland and Ireland; Channel Islands. *Sweetbriar.*—Var. *R. permix'ta*, Deseg.; leaves and styles glabrous, sepals deciduous, fruit ovoid. Box-hill. Approaches *micran'tha.*—*R. sylvic'ola*, Deseg. and Rip., is less scented, prickles more slender, leaflets hairy and less glandular beneath, styles hairy, fruit ovoid. N. Yorkshire.

Sub-sp. R. MICRAN'THA, *Sm.*; branches long arched, prickles equal, scent faint, leaflets small more pointed glabrous above densely glandular beneath, flower 1 in. diam., sepals deciduous densely glandular with a leafy point and 1–2 leaflets, styles glabrous, fruit urceolate scarlet, disk evident.

Midway between *rubigino'sa* and *cani'na.* From Roxburgh southd.; Ireland; Channel Islands.—Var. *Briggs'ii,* Baker; large and luxuriant, leaflets larger less glandular beneath, peduncles and fruit naked, sepals more pinnate eglandular on the back. Plymouth.—*R. Hys'trix,* Leman, is small, leaflets narrow glabrous, calyx-tube glabrous, peduncle densely aciculate. Box-hill, Oxfordshire, Bristol.

Sub-sp. R. Agrestis, Savi; laxer in habit than *rubiginosa,* prickles with a few bristles and glandular hairs intermixed, leaflets small narrowed to both ends glabrous but densely glandular beneath, peduncle and ovoid fruit naked, sepals subpersistent, disk moderate, styles pubescent. *R. se'pium,* Thuill. Surrey, Sussex; Ireland.—Var. *R. Billiet'ii,* Puget (*R. se'pium,* Borrer); differs only by its leaves rather hairy beneath, and rounded at the base. Warwickshire.—*R. inodo'ra,* Fries. (*R. pulverulen'ta,* Lindl. not M. Bieb.), is much taller, flowers 1½ in. diam., leaflets larger rounded at the base, sepals more copiously pinnate, fruit ovoid. England, local.—Var. *crypto-po'da,* Baker; differs from the last principally by its very short peduncles and round fruit. West Yorkshire.

5. **R. hiber'nica,** *Sm.* ; small, erect, branches short, prickles rather crowded gradually passing into bristles, leaflets simply-serrate glabrous or pubescent beneath wholly eglandular, peduncles naked, sepals persistent naked on the back, fruit erect globose naked, disk moderate.

From Sutherland southd.; Ireland; ascends to 1,000 ft.; fl. June–July.— Intermediate between *spinosis'sima* and *cani'na,* most like the latter; (a hybrid, *Christ.*). *Branches* sometimes arching; prickles stout, curved. *Leaflets* with rarely cut serratures, glaucous green above, nerves beneath thinly hairy; petiole pubescent; stipules nearly naked on the back, auricles gland-ciliated. *Flowers* sometimes 12. *Peduncles* and broad calyx-tube always naked. *Sepals* leafy. *Petals* pale pink. *Fruits* in October. —Distrib. France; very rare on the Continent. *R. hibernica* proper (*gla'bra,* Baker); leaflets glabrous, serratures sharper, peduncle naked. Resembles *R. cani'na.*—Var. *cordifo'lia,* Baker; prickles more slender and denser, leaflets almost glabrous beneath, terminal 1½ in., serratures more open and obtuse, peduncle bristly and glandular. Northumberland.

6. **R. cani'na,** *L.* ; bush large, branches long arching, prickles scattered uniform stout broad hooked base thickened, leaflets eglandular (except rarely the midrib and veins beneath) glabrous or thinly hairy acute very sharply toothed, peduncle usually naked, sepals usually naked reflexed pinnate, styles free or nearly. so hirsute, fruit ovoid urceolate or subglobose, mouth of disk conspicuous. *Dog Rose.*

Thickets, hedges, &c., N. to Orkney; ascends to 1,350 ft. in Yorkshire; Ireland; Channel Islands; fl. June–Aug.—Of the above characters, most disappear in one or other of the following 29 varieties, which Mr. Baker has system-atized with great care. In its common form, this is the largest and freest growing of British roses, and may be distinguished from *spinosis'sima* by the hooked prickles and habit, from *villosa* by being more glabrous, from *rubigi-nosa* by being eglandular, and from *arven'sis* by the free styles.—Distrib. Europe, N. Africa, Siberia.

SERIES 1. *ECRISTA'TÆ. Leaves* eglandular beneath. *Sepals* reflexed after
the petals fall, deciduous before the fruit (which ripens late) colours.

* *Leaves glabrous on both surfaces. Peduncles not bristly.*

R. lutetia'na, Leman ; 10–12 ft., leaflets about 7 green or glaucous, terminal
obovate, serratures simple, flowers 1–4 pink 2 in. diam., sepals naked.
Abundant.—VAR. *R. surculo'sa,* Woods; robust, flowers 10–30, leaflets flat
rounded at the base, teeth open.—VAR. *R. sphæ'rica,* Gren.; like *lutetia'na,*
but leaflets broader, petioles pubescent, fruit globose ⅔ in. diam., styles
villous.—VAR. *R. sentico'sa,* Ach.; slender, flexuous, leaflets 1 in., teeth
acute, fruit small globose.—VAR. *R. duma'lis,* Bechst. (*R. sarmenta'cea,* Sm.;
glaucophyl'la, Winch); petioles glandular, stipules and sepals more densely
gland-ciliated, leaflets doubly-serrate. Very common.—VAR. *R. biserra'ta,*
Merat ; quite like *duma'lis,* but serratures more open and very compound,
petioles very glandular.—*R. vina'cea,* Baker, has leaflets and bracts narrow
acute, and fruit oblong.

** *Leaves glabrous above, hairy on the nerves beneath. Peduncles not bristly.*

R. urb'ica, Leman (*R. colli'na,* Woods; *Forsteri,* Sm.; *platyphyl'la,* Rau.);
like *lutetia'na,* but leaves hairy beneath and petioles pubescent, scarcely
glandular.—VAR. *R. frondo'sa,* Stev. (*R. dumeto'rum,* Woods); leaflets smaller
flatter ovate-oblong more rounded at the base, fruit smaller globose.—
VAR. *arvat'ica,* Baker; like *urb'ica,* but leaves doubly-serrate, fruit ovoid.
Common in the N. of England.

*** *Leaves more or less hairy on both surfaces. Peduncles not bristly.*

R. dumeto'rum, Thuill. (*R. uncinel'la,* Bess.); leaflets green terminal often
large simply-serrate thinly hairy above, softly beneath, fruit large ovoid,
styles villous.—VAR. *pruino'sa,* Baker (*R. cæs'ia,* Borr.); leaflets glaucous
doubly-serrate, petioles glandular. VAR. *R. inca'na,* Woods (*R. canes'cens,*
Baker); leaflets very glaucous above densely pubescent beneath with few
inconspicuous glands doubly-serrate, fruit large oblong.—VAR. *R. tomentel'la,*
Leman; flexuous, leaflets short green above very hairy beneath doubly
serrate, petioles prickly, peduncles very short, flowers small white, fruit
small short.—*R. inodo'ra,* Hook. *Fl. Lond.* (*R. obtusifo'lia,* Desv.), is like
tomentel'la, but the leaflets are simply serrate.

**** *Peduncles more or less bristly and glandular.*

R. andevagen'sis, Bast., is *lutetia'na* with bristly peduncles. S. of England,
rarer in the N. and Scotland.—VAR. *verticillacan'tha,* Merat, is *dumalis* with
ditto. Not uncommon.—VAR. *R. colli'na,* Jacq. (*R. Kosincia'na,* Bess., *R.
asperna'ta,* Deseg.), is *urb'ica* with ditto. Surrey and Devonshire.—VAR.
R. cœsia, Sm., is near *colli'na,* but leaflets grey-green, stipules and bracts
pubescent on the back. Argyll, Northumberland, Leicester. — VAR.
concin'na, Baker ; prickles much hooked, leaflets very small simply serrate,
petioles not setose pubescent, ovary small, styles short thinly hairy.
Devonshire.—VAR. *decip'iens,* Dumort.; like *tomentel'la,* but for the bristly
peduncles, more glandular midrib, pubescent and glandular petiole, and
sepals densely glandular not fully reflexed. Northumberland, Chester,
Leicester.

Series 2. *SUBCRISTA'TÆ. Leaves* eglandular beneath. *Sepals* ascending after the petals fall, not deciduous till after the fruit (which ripens early) colours.—Vars. *R. sclerophylla,* Scheutz, and *R. monticola,* Rap., include various forms of this.

R. Reute'ri, Godet (*R. nu'da,* Woods; *Crepinia'na,* Deseg.); near *lutetia'na* but prickles more slender, peduncles short almost concealed by the bracts, leaflets glaucous, bracts stipules and branches turning red. N. England, common.—Var. *subcrista'ta,* Baker; like *duma'lis,* with characters of *Subcrista'tæ.* N. of England and Scotland.—Var. *Hailstone'i,* Baker; leaves as in *subcrista'ta,* fruit later, styles less villous, sepals sooner deciduous, prickles passing into unequal bristles. Yorkshire.—Var. *R. implex'a,* Gren.; like *urb'ica,* but fruit of this series. Yorkshire.—Var. *R. coriifo'lia,* Fries (*R. bractes'cens,* Woods); like *dumeto'rum,* but fruit of this series, peduncles very short bracts large. From Yorkshire northward.— Var. *Watso'ni,* Baker; differs from *coriifo'lia* in the doubly-serrate leaves, smaller densely gland-ciliated bracts and glabrous back of stipules, petioles pubescent and glandular. N. of England and Scotland.—Var. *celera'ta,* Baker; habit and foliage of *tomentel'la,* but fruit of this series. Northumberland.—Var. *Grovesii,* Baker, connects *Subcrista'tæ* with *R. hiber'nica.*

Series 3. *SUBRUBIGINO'SÆ. Leaflets* glandular beneath on the midrib and principal nerves only (not on the surface, as in *R. rubigino'sa*).

R. Borre'ri, Woods (*R. dumeto'rum,* Engl. Bot.); prickles stout much hooked, leaflets flat doubly-serrate glabrous above, glands obscure, stipules and bracts densely gland-ciliate, backs of these and sepals naked, flowers many, peduncles bristly, calyx-tube ovoid naked, sepals at length spreading or reflexed deciduous on the fruit changing colour. Yorkshire to Sussex.— Var. *R. Bake'ri,* Deseg.; prickles less hooked, leaflets obovate double-serrate, flowers 3–4, peduncles very short, back of sepals bracts and stipules thinly glandular, sepals at length ascending not deciduous till the naked oblong fruit has changed colour. N. Yorkshire.—Var. *R. margina'ta,* Wallr. (*R. Blondæa'na,* Ripart); branches purple, prickles more slender less hooked, leaflets altogether glabrous, glaucous above, pale beneath, veins prominent glandular, flowers 3–4, sepals and fruit much as in *R. Bake'ri.* Arran on the Clyde, N. Yorkshire, N. Wales, Derry.

7. **R. arven'sis,** *Huds.* ; bush large, branches long arching or trailing, prickles uniform stout strongly hooked, leaflets eglandular glabrous or slightly pubescent, flowers 1–6, sepals deciduous, styles glabrous connate into an exserted column, fruit ripening late, disk much thickened.

Hedges and thickets; England and Ireland; fl. June–July.—A low trailing plant in its usual form, with the foliage and hooked prickles of *R. cani'na. Leaflets* glabrous and shining above, rarely downy. *Flowers* rarely solitary, generally white, scent faint or 0. *Fruit* naked; achenes sessile or stalked. —Distrib. W. and Mid. Europe, from Belgium southd.

R. arven'sis proper; bush 2–3 ft., branches trailing purple glaucous, prickles often very large, leaflets quite glabrous glaucous beneath, flowers white throat yellow, calyx purple, sepals naked on the back short broad reflexed after flowering deciduous not much pinnate, fruit subglobose small, styles glabrous as long as the stamens. *R. re'pens,* Scop.—Common in S. of

England, rare to the N. and in Scotland.—Var. *R. bibractea'ta*, Bast., shoots stronger more arching, leaflets large more acute, fruit obovoid, peduncles thinly glandular. (Easily mistaken for *stylo'sa*.)
Sub-sp. R. stylo'sa, *Bast.*; bush tall, rarely low, leaflets pubescent beneath, peduncles elongate more or less bristly and glandular, sepals reflexed much pinnate, styles as long as or shorter than the stamens.—Connects *arven'sis* with *cani'na.—R. stylo'sa* proper; leaflets oblong acute rounded at the base hairy all over beneath, petioles and peduncles with a few glands and bristles, flowers 3–6 white, styles protruded, disk very prominent. Sussex.—Var. *R. sys'tyla*, Bast. (*R. collina*, Engl. Bot., not Jacq. *R. leucochroa*, Desv.), like *stylo'sa* but flowers usually pink, leaflets hairy only on the nerves beneath, and petioles less hairy. Mid. and S. England.—Var. *opa'ca*, Baker; leaflets still more hairy beneath rounded at the base, peduncle shorter naked, flower white 1 in. diam., styles scarcely protruded. Kent.—Var. *gallicoi'des*, Baker; habit and leaflets of *sys'tyla*, but prickles of stem mixed with copious glands and bristles, leaflets almost doubly-serrate, flowers white, fruit narrow, styles as long as stamens. Warwick.—Var. *Monso'niæ*, Lindl.; low, erect, flowers red very large, styles scarcely protruded, fruit subglobose orange-red. Hereford.—Var. *R. fastigia'ta*, Bast.; flowers pink, styles not protruded.

12. PY'RUS, *L.* Pear, Apple, Service, &c.

Trees or shrubs. *Leaves* deciduous, simple or pinnate; stipules deciduous. *Flowers* white or pink, in terminal cymes or corymbs, honeyed; bracts subulate. *Calyx-tube* urceolate; lobes 5, superior, reflexed, persistent or deciduous. *Petals* 5. *Stamens* many, filaments sometimes connate at the base. *Disk* annular, or coating the calyx-tube. *Carpels* 2–5, connate and adnate to the calyx-tube; styles distinct or connate below, stigmas truncate; ovules 2 in each cell, ascending. *Fruit* (a pome) fleshy, 2–5-celled; endocarp cartilaginous or bony often 2-valved, cells 1–2-seeded.—Distrib. N. temp. and cold regions; species 40.—Etym. The old Latin name.

Section 1. **Py'rus** proper. *Fruit* large 5-celled; cells 1–2-seeded; endocarp cartilaginous. *Flowers* umbellate or in simple cymes. *Styles* 5.

1. **P. commu'nis**, *L.*; cymes simple, styles distinct to the base, fruit pyriform. *Wild Pear.*

Woods and thickets, from Yorkshire southd., a relict of gardens?; (a denizen? *Wats.*); fl. April–May. A shrub or small tree, 20–40 ft.; branchlets more or less spinescent and pendulous. *Leaves* 1–1½ in., fascicled on the last year's wood, alternate on the shoots, oblong-ovate, acute, obtusely serrate, more or less pubescent or flocculent below when young, those of the young tree often lobed; petiole slender. *Flowers* 1–1½ in. diam., white, proterogynous. *Fruit* 1–2 in. long.—Distrib. E. Europe to W. Asia, Himalaya.
P. commu'nis proper (*P. Pyras'ter*, L.); leaves shortly acuminate pubescent below when young, base of fruit obconic.—Var. *P. Ach'ras*, Gærtn.; leaves broader acute or cuspidate flocculent on both surfaces when young, fruit rounded at the base. Rarer.—Var. *P. corda'ta*, Desv. (*Brigg'sii*, Syme); leaves ovate base rounded, fruit very small globose or pyriform. Cornwall.

2. **P. Ma'lus,** *L.* ; peduncles umbellate, style connate below, fruit subglobose indented at the base. *Wild* or *Crab-apple.*

Copses and hedges,. from Perth and the Clyde southd.; often a relict of gardens; wild in Ireland; Channel Islands; fl. May.—A shrub or small tree; branches spreading. *Leaves* 1–2 in., oblong rounded acuminate or cuspidate at the tip, glabrous or downy beneath when young. *Flowers* few, 1–1½ in. diam., pink and white, proterogynous. *Calyx-segments* woolly. *Fruit* 1 in. diam., yellow.—DISTRIB. Europe, W. Asia, Himalaya.

P. Malus proper (*P. acer'ba,* DC.); young leaves and tube of calyx glabrous, pedicels slender glabrous or nearly so, fruit drooping.—VAR. *mi'tis;* young leaves tube of calyx and stout pedicels pubescent, fruit erect.

SECTION 2. **Sor'bus,** *L.* (gen.). *Fruit* small, 2–5-celled ; cells 1–2-seeded ; endocarp brittle. *Flowers* in compound corymbose cymes. *Styles* 2–5.

3. **P. tormina'lis,** *Ehrh.* ; leaves 6–10-lobed serrate glabrous when mature on both surfaces. *Wild Service.*

Woods and hedges, local, from Lancashire southd.; fl. April–May.—A small tree, branchlets and young leaves beneath pubescent. *Leaves* 2–4 in., oblong-ovate or cordate ; lobes triangular, serrate, acuminate. *Flowers* ½ in. diam., many, white. *Carpels* usually 2. *Fruit* ½ in., pyriform or sub-globose, greenish-brown, dotted, 2-celled.—DISTRIB. Europe, W. Asia, N. Africa.—Fruit sold in country markets.

4. **P. A'ria,** *Sm.* ; leaves simple or pinnatifid rarely pinnate at the base, deeply lobed white and flocculent beneath. *White Beam.*

Copses and borders of forests, from Sutherland to Kent and Devon, local ; ascends to 1,500 ft. in Yorkshire; fl. May–June.—A bush or small tree, 4–40 ft. *Leaves* 2–6 in., very variable, glabrous above, plaited, coarsely irregularly serrate lobed or pinnatifid. *Flowers* ½ in. diam., in lax corymbs, white. *Fruit* ½ in. diam., subglobose, dotted red, usually 3-celled.—DISTRIB. Europe, N. Africa, N. and W. Asia.

P. A'RIA proper; leaves broad ovate or oblong crenate-serrate lobulate or hardly lobed beyond the middle, permanently snow-white beneath, nerves 8–13 very prominent on each side, fruit ½ in. diam., scarlet.—Mid. England southd., Ireland. *P. rupicola,* Syme, has nerves fewer, fruit smaller.

Sub-sp. P. LATIFOLIA, *Syme;* leaves from ovate-oblong to suborbicular more or less lobed grey-tomentose beneath, lobes deltoid serrate acuminate, nerves 5–9 on each side less prominent beneath. *Sorbus latifo'lia,* Pers. *P. scan'dica,* Bab.—Considered by foreign authors to be a hybrid between *A'ria* and *tormina'lis;* but it is found in Cornwall (*Briggs*) where *A'ria* is not known.

Sub-sp. P. SCAN'DICA, *Syme;* leaves less coriaceous oblong deeply lobed or pinnatifid glabrous above loosely grey-tomentose beneath, lobes oblong or rounded. *Sorbus scan'dica,* Fries. Arran.

P. hyb'rida, L. (*P. pinnatifi'da,* Sm. in part, *Sorbus fen'nica,* Fries), of Arran, which resembles *P. scan'dica,* but with the leaves pinnatifid towards the base, is a supposed hybrid between *A'ria* and *Aucupa'ria,* of which latter it has the sweet-scented flower and other characters.

5. **P. Aucupa'ria,** *Gærtn.* ; leaves pinnate, leaflets serrate glabrous beneath when old or nearly so, fruit globose. *Mountain Ash, Rowan-tree.*

Woods and hillsides; ascends to 2,600 ft. in the Highlands; Ireland; fl. May–
June.—Tree 10–40 ft.   *Leaves* 5–8 in.; leaflets 6–8 pair, 1–1½ in., linear-
oblong, subacute, pale beneath and hairy along the midrib and nerves.
Cymes 4–6 in. diam., compound, corymbose, dense-flowered. *Flowers* ½ in.
diam., cream-white, proterogynous; pedicel and calyx villous. *Fruit* ¼ in.
diam., globose, scarlet, flesh yellow; endocarp usually 3-celled, almost woody.
—DISTRIB. Europe, Madeira, N. and W. Asia, Himalaya, N. America (a
form).

SECTION 3. **Mes'pilus,** *L.* (gen.). *Fruit* large; endocarp bony, 5-
celled; cells 1-seeded. *Flowers* solitary. *Styles* 5.

P. GERMAN'ICA, *L.* (*Mespilus*); leaves obovate or oblong-lanceolate entire
or serrulate. *Medlar.*

Hedges and thickets, Mid. and S. England, Channel Islands, naturalized;
fl. May–June.—A small much-branched spinous tree. *Leaves* subacute,
pubescent beneath. *Flowers* 1½ in. diam., white; peduncle ½ in. *Calyx*
woolly, lobes with dilated foliaceous tips. *Fruit* ½–1 in. diam., globose,
with a large depressed area at the top, and persistent calyx-lobes.—DISTRIB.
Greece, Asia Minor, Persia.

13. CRATÆ'GUS, *L.* HAWTHORN, WHITETHORN.

Shrubs or small trees, often spiny. *Leaves* simple lobed or pinnatifid;
stipules deciduous. *Flowers* in terminal corymbose cymes, white or red,
honeyed, proterogynous; bracts caducous. *Calyx-tube* urceolate or campanulate; mouth contracted; lobes 5, superior. *Petals* 5, inserted at the
mouth of the calyx. *Stamens* many. *Carpels* 1–5, adnate to the calyx-
tube; styles short, stigma truncate; ovules 2 in each cell, ascending. *Fruit*
ovoid or globose, with a bony 1–5-celled stone, or with 5 bony 1- rarely
2-seeded stones.—DISTRIB. N. temp. regions, chiefly American, extending into New Granada; species about 50.—ETYM. κρατός, from the *strong*
wood.

C. **Oxyacan'tha,** *L.*; spinescent, leaves deeply pinnatifid.

Forests and hedges, N. to Shetland, often only where planted; ascending to
1,800 ft. in Yorkshire; Ireland; Channel Islands; fl. May–June.—A small
round-headed tree, 10–20 ft., much branched. *Leaves* 1–2 in., very variable,
cuneate, shortly petioled, lobes cut or crenate; stipules leafy, ½-sagittate,
toothed. *Cymes* corymbose, many-flowered. *Flowers* ¾ in. diam., white;
pedicel and calyx glabrous or pubescent. *Anthers* pinkish-brown. *Carpels*
1–2 very rarely 3. *Fruit* ovoid or subglobose, usually scarlet, rarely yellow
or black.—DISTRIB. Europe, N. Africa, N. and W. Asia, Himalaya; introd.
in N. America.

C. OXYACAN'THA proper; peduncle and calyx-tube glabrous, carpels 2–3. *C.*
oxyacanthoides, Thuill.

Sub-sp. C. MONOG'YNA, *Jacq.*; leaves more deeply lobed or pinnatifid, peduncle
and calyx-lobes pubescent, flowers and fruit smaller (appearing later), carpel
solitary.

14. COTONEAS'TER, *Lindl.*

Shrubs or small trees. *Leaves* coriaceous, often downy ; stipules deciduous. *Flowers* solitary or in few-fld. cymes, small white or pink, sometimes polygamous. *Calyx-tube* turbinate or campanulate ; lobes 5, superior, short, persistent. *Petals* 5. *Stamens* many, inserted at the mouth of the calyx. *Carpels* 2–5, adnate wholly or by their backs only to the calyx-tube ; styles 2–5, distinct, stigma truncate ; ovules 2 in each cell, erect. *Fruit* small, with 2–5 bony 1-seeded stones.—DISTRIB. Europe, temp. Asia, N. Africa, Mexico ; species 15.—ETYM. The Latin name.

C. vulga'ris, *Lindl.* ; leaves broadly elliptic-oblong densely pubescent beneath, cymes lateral few-fld.

Great Orme's Head, on limestone cliffs, very rare; fl. May–June.—A small erect shrub; branchlets pubescent. *Leaves* ¾–1½ in., rounded or acute at the tip; petiole very short; stipules scarious. *Flowers* ¼ in. diam., pink ; pedicels short, decurved, pubescent ; bracts minute. *Calyx* turbinate, lobes obtuse, margins woolly. *Petals* small, persistent. *Styles* about 3. *Fruit* ¼ in. diam., globose, shining, red.—DISTRIB. Europe, N. and W. Asia, Himalaya.

ORDER XXVII. SAXIFRAG'EÆ.

Tribes SAXIFRAG'EÆ proper and RIBESI'EÆ.

Shrubs or herbs. *Leaves* alternate or opposite ; stipules 0, or adnate to the often dilated petiole. *Calyx* free or more or less adnate to the ovary, 5- rarely 4-lobed, valvate or imbricate in bud. *Petals* 5, rarely 4 or 0, imbricate in bud. *Stamens* 5 or 10, rarely 4 or 8, perigynous ; anthers dorsally inserted, connective frequently glandular at the back. *Disk* various. *Carpels* 2 or more, usually connate into a 2- rarely a 1-celled ovary ; styles distinct or combined, stigmas capitellate ; ovules many, 2-seriate, placentas attached to the inner angles of the carpels. *Fruit* a 1–3-celled berry or capsule, or of 2 or more follicles, many-seeded. *Seeds* small, albumen copious fleshy or horny ; embryo minute, terete or clavate. —DISTRIB. (of the British tribes). N. temp. and Arctic regions ; a few occur on the Andes and in the S. temp. zone ; genera 20 ; species 300.—AFFINITIES. So close to *Rosaceæ* and *Crassulaceæ* as to be scarcely separable ; also allied to *Lythraceæ, Rhizophoreæ* and *Droseraceæ.*—PROPERTIES unimportant.

TRIBE I. **SAXIFRAG'EÆ** proper. Herbs. *Flowers* 4–5-merous. *Ovary* 1–3-celled. *Petals* 5. *Stamens* 10. *Fruit* capsular.
Ovary 2-celled ; styles 2..1. Saxifraga.
Petals 0. Stamens 8 or 10. Ovary 1-celled ; styles 2...2. Chrysosplenium.
Petals 5. Stamens 5. Ovary 1-celled ; stigmas 3–4..............3. Parnassia.
TRIBE II. **RIBESI'EÆ.** Shrubs. *Ovary* 1-celled. *Fruit* a berry.
4. Ribes.

1. SAXIF'RAGA, *L.* Saxifrage.

Perennial, rarely annual herbs. *Leaves* various, radical and cauline ; petiole sheathing. *Flowers* cymose, white or yellow, rarely red or purple, honeyed, proterandrous. *Calyx-tube* free or partially adnate to the ovary ; lobes 5, imbricate. *Petals* 5. *Stamens* 10, rarely 5, incurving and anthers opening in succession. *Ovary* superior or partially inferior, 2-lobed, 2-celled ; styles 2. *Capsule* 2-beaked, 2-valved between the beaks, many-seeded. *Seeds* small, smooth or rough.—Distrib. N. temp. and Arctic zones, Andes ; species 160.—Etym. *Saxum* and *frango,* from some species rooting into *rocks* and *breaking* them up.

Section 1. **Porphyr'ion,** *Tausch.* Perennial. *Stems* trailing, leafy. *Leaves* opposite, small, with a pore at the tip, ciliate. *Flowers* solitary, subsessile. *Sepals* erect in fruit.

1. **S. oppositifo'lia,** *L.* ; tufted, leaves small 4-farious ovate-oblong.

Alpine rocks, from N. Wales and Yorkshire to Shetland ; ascends to near 4,000 ft. in the Highlands ; N.W. and N. Ireland ; fl. April–May.—Glabrous, dark green, depressed. *Stems* 6–8 in., creeping. *Leaves* ¼ in., 4-fariously imbricate, thickened and obtuse at the tip, ciliate with stout bristles. *Flowers* ½ in. diam., sessile on short annual shoots, campanulate. *Sepals* obtuse, connate to the middle. *Petals* obovate, bright purple. *Capsule* free, ½ in. ; beaks diverging, subulate.—Distrib. Arctic and Alpine Europe, Asia (to W. Tibet) and America.

Section 2. **Micran'thes,** *Haw.* (gen.). Perennial, stemless, scapigerous. *Radical leaves* petioled. *Cymes* dense-fld. *Sepals* adnate to the base of the carpels. *Petals* white.

2. **S. niva'lis,** *L.* ; leaves broadly spathulate crenate-dentate.

High alps, alts. 2,000–4,300 ft , of Snowdon, the Lake district, and the Highlands ; Sligo ; fl. July–Aug.—Glandular-hairy on the leaf-margins, scape, bracts and pedicels, elsewhere glabrous. *Rootstock* small. *Leaves* ½–1 in. diam., subcoriaceous, red beneath ; petiole 1–2 in. *Scape* 3–6 in., erect, simple. *Flowers* ¼ in. diam., white, in capitate 4–12-flowered cymes. *Bracts* linear. *Calyx-lobes* connate, purplish. *Capsule* with short divergent beaks.—Distrib. Arctic Europe, Silesia, N. Asia, N. America.

Section 3. **Hydat'ica** and **Arabidi'a,** *Tausch.* Perennial, scapigerous. *Stem* short or 0. *Cymes* lax-fld. *Sepals* almost free, reflexed. *Petals* white with 2 purple dots above the base.

3. **S. stella'ris,** *L.* ; stemless, leaves rosulate subsessile cuneate-lanceolate usually coarsely toothed, scape leafless, filaments subulate.

Alpine and subalpine rills, from N. Wales and York t6 Caithness ; ascends to 4,300 ft. ; Ireland ; fl. June–July.—Glabrous or sparsely hairy. *Rootstock* small, branched. *Leaves* ½–1 in., subsucculent, ciliate, casually quite entire (var. *integrifolia,* Hook.). *Scape* 3–8 in. ; cyme panicled. *Flowers* few, ⅓ in. diam., white ; anthers and pistils red ; bracts linear. *Sepals* lanceolate.

140   *SAXIFRAGEÆ.*   [SAXIFRAGA.

Capsule with suberect slender beaks.—DISTRIB. Arctic and Alpine Europe, N. Asia, N. America.

4. **S. umbro'sa,** *L.* ; leaves orbicular obovate or broadly ovate narrowed into the stout petiole coarsely crenate or toothed. *St. Patrick's Cabbage, London Pride.*

W. and S.W. Ireland; ascends to 3,400 ft.; naturalized elsewhere; fl. June–July.—Glabrous or laxly hairy. *Leaves* 1½–2 in. diam., rosulate, coriaceous; petiole ½–1 in., flattened. *Scape* 6–12 in., leafless. *Cyme* panicled, bracts linear. *Flowers* ¼ in. diam., white, sometimes spotted with red. *Sepals* reddish. *Filaments* slightly dilated upwards; anthers red. *Capsule* with short, divergent beaks.—DISTRIB. N. Spain, Portugal, Corsica.—*S. puncta'ta*, Haw., with loosely rosulate orbicular crenate-serrate leaves, and *S. serratifo'lia*, Mackay, with sharply toothed ascending obovate leaves, are slight varieties.—*S. hirsu'ta*, L. (*S. gra'cilis*, Mackay MSS.) is a more hairy form with sharply toothed leaves rounded or obtuse at the base.—*S. el'egans*, Mackay, with acutely toothed leaves abruptly narrowed into the very short petiole is probably a hybrid with *S. Geum* (Ball.).

5. **S. Ge'um,** *L.* ; leaves orbicular or reniform crenate or toothed, base cordate, petiole slender.

Mountains of Kerry and Cork; ascends to about 2,000 ft.; fl. June.—Very similar in habit and floral character to *S. umbro'sa*, and as variable in hairiness, but distinguished by the leaf base. I doubt its being more than a sub-species.—DISTRIB. N. Spain, Pyrenees.

SECTION 4. **Hir'culus,** *Haw.* (gen.). Perennial. *Stem* leafy. *Leaves* alternate, linear-lanceolate or oblong, entire or toothed. *Sepals* free or ½-adnate to the ovary, spreading or reflexed. *Petals* yellow.

6. **S. Hir'culus,** *L.* ; stem subsimple erect leafy stoloniferous, flowers subsolitary, sepals free reflexed, capsule superior.

Bogs and wet moors, rare and local; Chester (formerly), Perth to York (ascending to 2,100 ft.); Ireland, local; fl. Aug.—Pubescent above, 4–8 in., branched from the base. *Radical leaves* ½–1½ in., rosulate, petioled, lanceolate or spathulate; cauline linear, sometimes faintly serrulate. *Flowers* ½–¾ in. diam. *Sepals* ciliate. *Petals* obovate dotted red at the base, where there are 2 tubercles. *Capsule* with short divergent beaks.—DISTRIB. Alpine N. and Mid. Europe (Arctic), N. Asia, Himalaya, N. America.

7. **S. aizoi'des,** *L.* ; stems tufted decumbent leafy much branched, leaves linear-oblong crowded below, scattered on the flowering stems, sepals erect connate adnate below to the ovary.

Stony mountain rills, from York to Orkney; ascends to 3,000 ft. in the Highlands; Ireland; not in Wales; fl. June–July.—Often forming bright green cushions a foot across; branches 3–8 in. *Leaves* ½–2 in., spreading, lower reflexed, often ciliate (*S. autumna'lis*, L.); on the flowering stem narrower, strongly ciliate. *Flowers* 1–10, ½ in. diam., orange or golden yellow, dotted red. *Calyx-tube* obconic. *Petals* narrowly obovate-spathulate, distant. *Ovary* orange, depressed. *Capsule* with erecto-patent, subulate beaks.—DISTRIB. Alpine N. and Mid. Europe (Arctic), N. Asia, N. America.

SECTION 5. **Nephrophyl'lum**, *Gaud.* Annual or perennial. *Leaves* chiefly radical, broad, palmately lobed. *Sepals* suberect, more or less connate and adnate to the ovary. *Petals* white.—Hairs articulate.

8. **S. tridactyli'tes,** *L.* ; annual, stem erect many-flowered glandular-hairy, leaves cuneate 3-5-fid, uppermost entire, petiole broad, calyx-lobes short erect obtuse.

Wall-tops and dry places, E. Scotland, from Caithness to Cornwall and Kent ; ascends to 1,800 ft. in Yorkshire; Ireland; Channel Islands; fl. April–June.—*Stem* 2-6 in., simple or branched above. *Radical leaves* ½-1 in., rosulate; segments linear-oblong, outer cleft. *Cyme* subracemose ; bracts subopposite, pedicels slender. *Flowers* ⅛ in. diam., erect, white. *Calyx-tube* oblong. *Petals* small. *Capsule* inferior; beaks short divergent.— DISTRIB. Europe, N. Africa, N. and W. Asia.

9. **S. rivular'is,** *L.* ; perennial, stems decumbent rooting 1-3-flowered, leaves reniform palmately 5-lobed, petiole slender, calyx-lobes acute.

Alpine wet rocks and streams, alt. 3,000–3,600 ft., Ben Lawers, Braemar, Ben Nevis; fl. July–Aug.—Tufted, slender, succulent, 1-4 in., slightly glandular-hairy. *Leaves* ½-1 in. diam., lobes entire; petiole as long as the stems. *Flowers* 1-2, ¼ in. diam., erect, white; bracts leafy, opposite, usually undivided. *Calyx-tube* ½ as long as the lobes, hemispherical. *Petals* distant, small, obovate-oblong. *Capsule* with short divergent beaks.—DISTRIB. Scandinavia, Arctic Europe, Asia, and America.

10. **S. granula'ta,** *L.* ; glandular-pubescent, stem erect bulbiferous at the base branched and many-flowered above, leaves petioled reniform palmately lobulate, cauline sessile, calyx-lobes erect obtuse.

Sandy banks and meadows, from Elgin to Somerset and Kent; ascends to 1,500 ft. in Yorkshire ; rare in Ireland ; fl. April–May.—Gregarious. *Bulbs* as large as a pea, brown. *Stem* 6-18 in. *Radical leaves* ¾-1¼ in. diam., petiole slender ; cauline deeper and more acutely cut. *Flowers* 1 in. diam., campanulate, inclined or drooping, white. *Petals* large, obovate. *Calyx-lobes* as long as the tube. *Stigmas* large, reniform. *Capsule* with slender beaks.--DISTRIB. Europe, N. Africa, W. Asia, Himalaya.

11. **S. cer'nua,** *L.* ; perennial, stem erect simple 1-3-fld., leaves petioled reniform palmately deeply crenate or lobulate, cauline sessile with axillary scarlet buds, calyx-lobes erect obtuse.

Schistose rocks, Ben Lawers, alt. 4,000 ft.; fl. July.—Sparingly glandular-hairy, 2-6 in., leafy. *Rootstock* with scaly buds at its top. *Radical leaves* ½-¾ in. diam., often tinged red; cauline more deeply lobed. *Flowers* ½-⅔ in. diam., campanulate, drooping, white, rarely produced in Britain. *Calyx-tube* very short. *Petals* large, obovate.—DISTRIB. Arctic and Alpine Europe, N. Asia, America, Himalaya.—Probably a form of *S. granula'ta.*

SECTION 6. **Dactyloi'des,** *Tausch.* Perennial, tufted, with many flowerless leafy shoots (except *S. cæspito'sa*). *Leaves* alternate, palmately-partite or -lobed. *Sepals* connate at the base and adnate to the ovary, suberect. *Flowers* white.—Hair articulate, glandular.

12. **S. hypnoi'des,** *L.* ; barren shoots elongate, leaves 3–5-cleft, lobes divergent acute, those of the shoots often entire, calyx-tube short obconic or hemispheric, lobes usually spreading.

Hilly subalpine districts, from N. Somerset and Glamorgan to Caithness; Ireland; fl. May–July.—Often forming large cushions, glabrous or glandular-hairy. *Leaves* lax or dense, with the broad compressed petiole ¼–1 in.; lobes entire, or the lateral cleft, flat or channelled. *Flowering-shoots* 3–8 in., stout or slender, leafy or nearly naked, their leaves simple, linear, or broad and more or less lobed like the radical. *Flowers* ½–1 in. diam., few or many, campanulate, white. *Sepals* rarely connate above the middle, oblong linear or ovate, obtuse or acute. *Capsule* not concealed in the calyx-tube, beaks sub-erect.—I have repeatedly studied the forms of *S. hypnoi'des* and its allies, and always with the result that the passage from *hypnoi'des* proper to *cæspito'sa* is undefinable. Mr. Baker's exposition of the sequence of the forms (*Seeman's Journ. Bot.* viii. 280) very well expresses their relationship, and is here followed.

S. hypnoi'des proper (*S. leptophyl'la*) ; tufts large loose, barren shoots long with linear entire acute leaves, axils usually bulbiferous (var. *gemmi'fera*, Syme), sepals lanceolate acute. N. England, chiefly on calcareous soil. (W. Europe, from Belgium to Spain.)—VAR. *S. sponhem'ica*, Gmel. (*S. palma'ta*, Lej., *S. læte-vi'rens*, D. Don, *S. elongel'la* and *platypet'ala*, Sm., *S. condensa'ta*, Gmel. *S. quinque'fida, tri'fida, hir'ta,* and *læ'vis*, Haw., *S. affi'nis*, Mack.) ; tufts closer, large, barren shoots long, their leaves with 3–5 acute lobes, sepals lanceolate acute. Wales, N England, Scotland, Ireland. (Färoe Islands.)—VAR. *S. decip'iens*, Ehrh. ; tufts still closer, barren shoots long, their leaves with 3–5 acute lobes, sepals oblong-lanceolate subacute. Carnarvon and Tralee. (Iceland, Greenland, Germany, Switzerland.)—VAR. *S. Sternber'gii*, Willd. (*S. palma'ta*, Sm., *S. hiber'nica*, Haw.) ; robust, barren shoots rather long, their leaves with 3–5 obtuse lobes, sepals oblong-ovate obtuse. Ireland (Arctic regions, Scandinavia to Austria and Switzerland, N. America.)—VAR. *S. cæspito'sa*, L. (*S. grœnlan'dica*, L., *S. incurvifo'lia*, Sm.) ; densely tufted, barren shoots not longer than the flowering, leaves cuneate 3–5-lobed, lobes subparallel, calyx-lobes short erect obtuse. Highest peaks of Welsh, Scotch, and Irish mountains, very rare. (Färoe Islands, Scandinavia, Arctic regions, N. America.)

2. CHRYSOSPLE'NIUM, *L.* GOLDEN SAXIFRAGE.

Small, succulent, annual or perennial herbs. *Leaves* alternate or opposite, exstipulate. *Flowers* minute, green or yellow, in axillary or terminal cymes. *Calyx-tube* urceolate or obconic, adnate to the ovary ; lobes 4–5, obtuse, imbricate in bud. *Petals* 0. *Stamens* 8 or 10, inserted on the margin of an epigynous disk, filaments short. *Ovary* inferior, 1-celled, 2-lobed at the top ; styles short, recurved ; ovules many, placentas 2 parietal. *Capsule* ½-superior, 2-lobed, membranous, opening at the top by a cruciate mouth. *Seeds* oblong or compressed.—DISTRIB. N. temp. and Arctic regions, temp. S. America ; species 15.—ETYM. χρυσός and σπλήν, *golden spleen*, of doubtful application.

1. **C. alternifo'lium,** *L.* ; stem simple erect, leaves alternate.

Banks of streams and wet places, from Argyll and Elgin to Kent and Devon; ascends to near 3,200 ft. in the Highlands; Ireland; Channel Islands; fl. April–June.—Perennial. *Stems* 2–4 in., tufted, glabrous above, clothed below with soft white hairs. *Radical leaves* ¾–2 in. diam., long-petioled, reniform, crenate; cauline few. *Flowers* ⅛ in. diam., 4-merous, yellow, in compact leafy cymes, homogamous. *Calyx-lobes* obtuse, spreading. *Capsule* almost inferior, beaks short. *Seeds* smooth.—DISTRIB. Europe (Arctic), N. and W. Asia to the Himalaya, N. America.

2. **C. oppositifo′lium,** *L.* ; stem creeping below, leaves opposite.

Marshy and shady places, from Orkney to Cornwall and Kent; ascends to 3,300 ft. in the Highlands; Ireland; Channel Islands; fl. May–July.—Size and general habit of *C. alternifo′lium,* but stem more leafy, branched, creeping and rooting at the base; leaves ½–1 in. diam., nearly orbicular, suddenly contracted into a short broad petiole; cauline many. *Flowers* proterogynous.—DISTRIB. Europe, Siberia.

3. **PARNAS′SIA,** *L.* GRASS OF PARNASSUS.

Slender, simple, glabrous, erect, perennial herbs. *Leaves* chiefly radical, quite entire, exstipulate. *Flower* solitary, large, yellow or white. *Calyx-tube* short, free or adnate to the base of the ovary; lobes 5, imbricate, persistent. *Petals* 5, thickish, persistent. *Stamens* 5, hypogynous or perigynous, alternating with 5 large scales. *Ovary* superior or ½-inferior, 1-celled ; style short or 0, stigmas 3–4 opposite the parietal placentas ; ovules many. *Capsule* membranous, loculicidally 3–4-valved, many-seeded. *Seeds* small, testa lax, albumen scanty.—DISTRIB. Arctic and temp. regions from S. India northwards ; species 12.

P. palus′tris, *L.* ; leaves ovate-cordate, stamens hypogynous.

Wet moors and bogs, from Dorset and Surrey to Shetland; ascends to 2,700 ft. in the Highlands; Ireland; fl. Aug.–Sept.—Glabrous. *Radical leaves* 1–2 in., ovate cordate; petiole slender 2–3 in. *Stem* slender, angular, twisted, with one sessile leaf about the middle. *Flower* ½–1 in. diam., honeyed, proterandrous. *Sepals* nearly free, obtuse. *Petals* coriaceous, with strong veins. *Stamens* incurving and anthers ripening in succession. *Scales* obovate, fringed with a comb of capitellate filaments, and with 2 glands on the surface facing the ovary. *Ovary* ovoid, superior; stigmas 4. —DISTRIB. Europe, N. Africa, N. and W. Asia to W. Tibet, N. America.

4. **RI′BES,** *L.* CURRANT, GOOSEBERRY.

Shrubs, often glandular and spinous, buds scaly. *Leaves* alternate, entire or lobed, plaited or convolute in bud ; stipules 0, or adnate to the petiole. *Flowers* solitary or racemose, white red yellow or green, often unisexual, honeyed ; pedicels bracteate (often minutely), 2–3-bracteolate. *Calyx-limb* tubular or campanulate, 4–5-fid, imbricate or subvalvate in bud. *Petals* small and stamens 4–5, inserted in the throat of the calyx. *Ovary* inferior, 1-celled ; styles 2 ; ovules few or many, on 2 slender parietal placentas. *Berry* ellipsoid or globose, 1-celled, few- or many-seeded. *Seeds* horizontal, testa with a gelatinous coat, raphe free, albumen adhering to the testa ; embryo minute.—Distrib. N. temp. regions and Andes of S. America ; species 56.—Etym. *Ribs*, in Danish.

Section 1. **Grossula'ria.** *Branches* spinous. *Leaves* plaited in bud. *Peduncles* 1–3-fld.

1. **R. Grossula'ria,** *L.* ; leaves orbicular 3–5-lobed. *Gooseberry.*

Copses in various counties ; indigenous only in N. England, where it ascends to 1,000 ft.; fl. April–May.—A small spreading shrub with 1–3 spines under the leaf-buds. *Leaves* 1–2 in. diam., fascicled on short lateral branches, lobes irregularly crenate. *Flowers* ¼ in., greenish, drooping, proteranderous ; peduncle short, pubescent, 1–3-bracteate about the middle. *Calyx-lobes* reflexed, purplish, throat and stamens bearded. *Petals* white, erect, minute. *Fruit* ½–1 in.—Distrib. Europe, N. Africa, W. Asia, N.W. Himalaya.

R. Uva-crispa, L., the small form with glabrous fruit, is that found wild in Europe most commonly.

Section 2. **Ribe'sia.** *Branches* not spinous. *Leaves* plaited in bud. *Racemes* many-fld.

2. **R. alpi'num,** *L.* ; diœcious, leaves deeply 3–5-lobed coarsely serrate almost glabrous and shining beneath, racemes glandular erect in flower and fruit, bracts exceeding the pedicels.

Woods in the N. of England, probably indigenous ; not so in Scotland ; fl. April–May.—A small all but glabrous bush ; branches slender. *Leaves* 1½–2 in. diam., broadly ovate, lobes usually 3 acute cut and serrate, slender petiole and both surfaces sparingly hairy. *Racemes* glandular-pubescent ; male 2–2½ in.; 20–30-flowered ; female shorter, 8–10-flowered. *Flowers* ⅛ in. diam., yellowish, females greener ; bracts linear. *Petals* much shorter than the calyx-lobes. *Styles* very short, cleft. *Fruit* ¼ in. diam., globose, scarlet, insipid.—Distrib. Europe, N. and W. Asia, N. America.

3. **R. ru'brum,** *L.* ; leaves 3–5-angled and -lobed, base cordate, lobes triangular crenate, racemes eglandular drooping in fruit, bracts shorter than the pedicels. *Wild Currant.*

Woods and thickets ; indigenous in N. England and the Highlands ; ascends to 1,000 ft. in Yorkshire ; fl. April–May.—*Leaves* 2–4 in. diam., glabrous or pubescent above, usually tomentose beneath ; petiole pubescent or setose. *Racemes* 1–3 in., many-flowered, pubescent or glabrous, never glandular ; bracts ovate. *Flowers* ¼ in. diam., homogamous. *Calyx* glabrous, limb flat. *Petals* minute. *Fruit* ¼ in. diam., red, acid.—Distrib. Europe, N. and W. Asia, N.W. Himalaya, N. America.

R. ru'brum proper (*R. sylves'tre,* Reichb.) ; leaves hairy above tomentose beneath, raceme pubescent usually suberect in flower and drooping in fruit, flowers purplish, filaments very short, fruit contracted at the top, pedicels equalling or exceeding the fruit. Vars. *Smithia'na* and *Bromfieldia'na,* Syme. *R. petræ'um,* Sm. not Wulfen.—Var. *R. spica'tum,* Robson, leaves hairy above when young and tomentose beneath, racemes erect in flower and fruit. Yorkshire and Skye.—Var. *sati'va,* Reichb., the cultivated form, has leaves glabrous on both surfaces when mature, racemes glabrous always drooping, flowers green, fruit globose.

4. **R. ni'grum,** *L.* ; leaves angled 5–7-lobed glandular-dotted beneath, lobes triangular acute serrate, racemes drooping lax-flowered tomentose eglandular, bracts minute, pedicels long. *Black Currant.*

Woods, &c., from Mid. Scotland southd., often a garden escape, but apparently wild in the Lake district and Yorkshire ; fl. April–May.—A stout erect bush, smelling strongly when bruised. *Leaves* 2–3 in. diam., similar to those of *R. rubrum,* but rather deeply lobed ; petiole slender, pubescent. *Racemes* slender, few-flowered. *Flowers* ¼–½ in. diam. ; pedicel ¼ in. *Calyx* campanulate, glandular. *Petals* minute. *Berry* ⅔ in. diam., globose, black.—Distrib. Europe, N. and W. Asia, W. Himalaya.

ORDER XXVIII. **CRASSULA'CEÆ.**

Herbs or shrubs, usually succulent. *Leaves* opposite or alternate, exstipulate. *Flowers* in terminal or axillary cymes, bracteate or not. *Sepals* 3–5, rarely 10–12 or more, distinct or connate. *Petals* as many, distinct or connate, imbricate in bud. *Stamens* perigynous or subhypogynous, as many as the petals, or twice as many, when those opposite the petals are adnate to their bases ; anthers dorsally fixed. *Hypogynous scales* opposite each carpel, rarely 0. *Carpels* 3–5, rarely more, 1-celled, distinct, rarely connate ; styles short or long, stigma small ; ovules many, rarely few, attached to the ventral suture, ascending or pendulous. *Fruit* of 3 or more 1-celled 2- or more-seeded follicles. *Seeds* oblong, minute, albumen fleshy ; embryo terete.—Distrib. Arctic, temp. and warm regions, chiefly S. African ; genera 14 ; species 400.—Affinities. Separable from *Saxifrageæ* by habit and the hypogynous scales.—Properties. Astringent roots, acrid foliage, emetic and purgative qualities; tartaric and malic acids occur.

Leaves opposite. Petals 3–5 free. Stamens 3–5 ....1. Tillæa.
Leaves alternate. Corolla 5-lobed. Stamens 10..................2. Cotyledon.
Leaves alternate. Petals usually 5, free. Stamens usually 10....3. Sedum.
Leaves alternate. Petals 6–20. Stamens 12–40 ....3*. *Semperviʹvum.*

1. TILLÆ'A, *L.*

Small or minute subsucculent heibs. *Leaves* opposite, quite entire. *Flowers* minute, axillary, solitary or cymose, white or reddish. *Calyx* 3–5-lobed or -parted. *Petals* 3–5, distinct or connate at the very base. *Scales* 3–5, linear, or 0. *Carpels* 3–5 ; styles short ; ovules 1 or more. *Follicles* few or many-seeded.—Distrib. Ubiquitous ; species 20.—Etym. M. A. Tilli, an early Italian botanist.

T. musco'sa, *L.* ; leaves oblong, flower solitary subsessile 3-merous.

L

Sany heaths, rare, Norfolk to Hants and Devon ; Channel Islands ; fl. June–July.—Annual. *Stem* 1–2 in., tufted, decumbent, glabrous, reddish ; branches slender, leafy and flowering throughout. *Leaves* thick, concave, obtuse or apiculate. *Flowers* rarely 4-merous. *Sepals* ovate, acuminate, green. *Petals* smaller, subulate, white. *Scales* 0. *Follicles* constricted, 2-seeded.—DISTRIB. W. Europe from Holland southd., N. Africa.

2. COTYLE'DON, *L.* PENNYWORT, NAVELWORT.

Herbs or small shrubs. *Leaves* alternate rarely opposite, sometimes peltate. *Flowers* in terminal spikes or racemes. *Calyx* 5-parted. *Corolla-tube* urceolate or cylindric, terete or 5-angled ; lobes 5, small, twisted in bud. *Stamens* 10, inserted in the tube of the corolla. *Scales* 5, linear-oblong or 4-angular. *Carpels* 5 ; styles filiform ; ovules many. *Follicles* many-seeded.—DISTRIB. W. and S. Europe, all Africa, temp. Asia, Mexico ; species 60.—ETYM. κοτύλη, from the *cup*-like leaf of some species.

C. Umbili'cus, *L.* ; leaves peltate orbicular crenate, raceme long.

Rocks and walls, especially on the W. coasts from Argyll to Kent and Cornwall ; ascends to 1,000 ft. in Wales ; (absent from the E. counties) ; Ireland ; Channel Islands ; fl. June–July.—Glabrous, succulent. *Rootstock* tuberous. *Stem* 6–18 in., simple, stout, terete. *Radical leaves* 1–3 in. diam., petioled, depressed in the centre ; cauline spathulate, upper cuneate. *Raceme* continued almost throughout the stem, sometimes leafy (var. *folio'sa*) ; bracts minute ; pedicel short, slender. *Flowers* close-set, drooping, green. *Corolla* cylindric, shortly 4–5-lobed. *Stamens* adnate to the corolla-tube, included. —DISTRIB. From France southd., W. Asia, N. and trop. Africa.

3. SE'DUM, *L.* ORPINE, STONECROP.

Succulent herbs, erect or prostrate. *Leaves* alternate rarely opposite or whorled. *Flowers* cymose, rarely axillary and solitary, sometimes diœcious, honeyed. *Calyx* 4–5-lobed. *Petals* 4–5, distinct. *Stamens* 8–10. *Scales* 4–5, entire or notched. *Carpels* 4–5, distinct or connate at the base ; styles short ; ovules many. *Follicles* many- or few-seeded.—DISTRIB. N. temp. and cold regions, rare in America ; species 120.—ETYM. *-sedeo*, from the *squatting* habit of the species.

SECTION 1. **Tele'phium.** *Rootstock* stout, perennial. *Stems* annual. *Leaves* broad, flat or concave.

1. **S. Rhodi'ola,** *DC.* ; leaves alternate sessile, flowers 4-merous diœcious. *Rhodi'ola ro'sea,* L. *Rose-root.*

Moist alpine and subalpine rocks, from S. Wales and York to Shetland ; ascends to near 4,000 ft. in the Highlands ; maritime rocks in Scotland ; Ireland ; fl. May–Aug.—*Rootstock* 2–3 in., as thick as the thumb, branched, woody, cylindric, scent of roses ; buds scaly. *Stems* 6–18 in., fleshy. *Leaves* 1–1½ in., glaucous, larger and more crowded upwards, obovate-oblong or lanceolate, acute, toothed at the tip. *Cymes* compact, corymbose. *Flowers* ¼ in. diam., yellow or purplish. *Sepals* narrow. *Petals* linear, smaller or 0 in the female flower. *Scales* notched.—DISTRIB. N. and Mid. Europe (Arctic), Himalaya, N. America.

2. **S. Tele'phium,** *L.* ; leaves subsessile, flower 5-merous hermaphrodite. *Orpine.*

Stony hedgebanks and copses, &c., from Perth southd., local, often a garden escape; ascends to 1,200 ft. in Yorkshire; Derry only in Ireland; fl. July–Aug.—*Rootstock* short, stout; roots many, elongate, tuberous. *Stems* 6–24 in., stout, green or spotted red. *Leaves* 1–3 in., rarely opposite below, ovate or oblong, obtuse, flat or concave, obtusely toothed or serrate. *Cymes* dense, corymbose. *Flowers* ⅓ in. diam., rosy white or speckled, proterandous. *Sepals* ovate-lanceolate. *Petals* twice as long, lanceolate.—DISTRIB. Europe, N. and W. Asia, Himalaya.

S. Tele'phium proper; upper leaves sessile rounded at the base, carpels furrowed at the back. *S. purpuras'cens,* Koch.—VAR. *S. Faba'ria,* Koch; more slender, upper leaves all cuneate at the base, carpels not furrowed. Very local.

SECTION 2. **Cepæ'a.** Annual or biennial. *Stem* simple. *Leaves* sub-cylindric.

3. **S. villo'sum,** *L.* ; glandular-pubescent, leaves sessile ½-cylindric.

Bogs and marshes in hilly districts from York and Westmoreland to Argyll and Elgin; ascends to 2,000 ft. in Yorkshire; fl. June–July.—Biennial. *Stem* with a tuft of leaves the first year, lengthening in the second, then slender, 3–6 in., and flowering. *Leaves* ¼–½ in., scattered, linear, obtuse. *Cyme* few-fld., subscorpioid. *Flowers* 1 in. diam., white or purplish. *Sepals* ovate, obtuse. *Petals* broad, acute.—DISTRIB. Europe (Arctic), Norway to Italy and Hungary, Greenland.

SECTION 3. **Se'dum** proper. Perennial. *Stems* branched, with many flowerless leafy prostrate or ascending shoots. *Leaves* subcylindric or ½-cylindric.

* *Flowers white*

4. **S. al'bum,** *L.* ; glabrous or slightly glandular, leaves alternate subcylindric oblong contracted at the base, petals oblong-lanceolate.

Malvern Hills and Somerset, indigenous (Syme); a garden escape from Forfar southd.; (an alien, *Wats.*); fl. July–Aug.—*Flowerless stems* prostrate; flowering erect, 6–10 in. *Leaves* ¼–½ in., obtuse, bright green. *Cyme* corymbose, glabrous. *Flowers* ⅛–¼ in. diam., proterandrous. *Petals* twice as large as the green sepals.—DISTRIB. Europe, N. and W. Asia, N. Africa.

S. al'bum proper (*S. teretifo'lium,* Haw.); leaves flattened above, sepals and petals obtuse.—VAR. *S. micran'thum,* Bast. ; leaves flattened on both surfaces, sepals rounder, petals more acute.—Naturalized in Sussex, Ireland, &c.

5. **S. ang'licum,** *Huds.* ; glabrous, leaves alternate ovoid-oblong gibbous at the base below, petals lanceolate acuminate keeled.

Rocks and banks N. to Shetland, chiefly by the sea; ascends to 3,300 ft. in N. Wales; Ireland; fl. June–Aug.—Tufts matted, glaucous green or reddish. *Flowering stems* 1–2 in., ascending, leafy. *Leaves* ⅛–¼ in., crowded, tumid at the base. *Cymes* short, scorpioid. *Flowers* ⅓ in. diam., few, crowded at the

L 2

top of the flowering stem, white or pink. *Sepals* obtuse, short. *Carpels* pink.—DISTRIB. W. Europe.

S. DASYPHYL'LUM, *L.* ; glandular-pubescent, glaucous, leaves on the flowerless shoots mostly opposite subglobose or shortly ovoid, equal at the base below.

Old walls, &c., naturalized, rare ; fl. June–July.—Loosely tufted, very glaucous and pink, much branched. *Flowerless stems* short, with rosulate leaves ; flowering 2 in., flexuous, slender. *Leaves* ⅛–¼ in. *Cyme* forked, few-fld. *Flowers* ⅓ in. diam. *Petals* often streaked with pink.—DISTRIB. W. and S. Europe, N. Africa.

** *Flowers yellow.*

6. **S. a'cre,** *L.* ; glabrous, leaves densely imbricate alternate erect terete ovoid-oblong obtuse, sepals slightly gibbous at the base, petals lanceolate acuminate. *Biting Stonecrop, Wall-pepper.*

Rocks, walls, and sandy places, especially near the sea ; ascends to 1,500 ft. in Yorkshire ; Ireland ; Channel Islands ; fl. June–July.—Tufts or cushions 3–10 in. diam. *Stems* 3–5 in. *Leaves* ⅛–¼ in., obscurely 6-seriate, broadest at the base, the gibbosity in contact with the stem. *Flowers* ⅜ in. diam., subsessile, few, golden-yellow, proterandrous. *Sepals* obtuse, not half as long as the petals. *Anthers* yellow.—DISTRIB. Europe, W. Asia, N. Africa, W. Siberia.—Taste acrid ; is a vesicant, emetic and cathartic.

S. SEXANGULA'RE, *L.* ; glabrous, leaves cylindric spreading obtuse gibbous at the base, sepals not gibbous at the base.

Old walls in the E. of England, not indigenous ; fl. July.—*Stems* loosely tufted, flowerless with crowded leaves in about 6 rows ; flowering 3–6 in., laxer, spreading or recurved with leaves ¼ in. ; basal gibbosity of the leaf acute, in contact with the stem. *Cyme* 1–2 in. diam., corymbose. *Flowers* ¼ in. diam., subsessile, yellow. *Sepals* obtuse. *Petals* lanceolate, acute.—DISTRIB. N. and Mid. Europe.

7. **S. rupes'tre,** *Huds.* ; glabrous, leaves linear-lanceolate acute flattened gibbous at the base, sepals oblong not gibbous at the base.

Rocks, S.W. England and Wales, rare ; Ireland ; Channel Islands (naturalized) ; fl. June–July.—*Stems* stout, loosely tufted, green or tinged with pink ; flowerless with closely rosulate leaves ; flowering 6–10 in., with suberect scattered leaves. *Leaves* ⅓–1 in., acute or acuminate, the gibbosity close to the stem. *Cymes* 3–4 in. diam., branches scorpioid. *Flowers* ⅔ in. diam., pedicelled, golden-yellow. *Sepals* obtuse. *Petals* lanceolate, acute. *Anthers* yellow.—DISTRIB. From Belgium southd.

S. RUPES'TRE proper ; glaucous, cyme rather flat-topped. *S. el'egans,* Lej. ; *S. pruina'tum,* Brot.—VAR. *ma'jor ;* stout, 6–12 in., leaves ¾–1 in., cyme 3–5 in. diam. Cheddar Cliffs.—VAR. *mi'nor ;* smaller and more slender. Bristol, Shropshire, Wales, indigenous ; elsewhere in England an escape.

Sub-sp. S. FORSTERIA'NUM, *Sm.* ; more slender, cymes rather round topped sometimes capitate. Varieties *glauces'cens* and *vires'cens,* represent shades of colour in two forms. Wet rocks ; Somerset, Gloster, Salop, Wales.

8. **S. reflex'um,** *L.* ; glabrous, leaves crowded cylindric-subulate spreading and reflexed, flowers pedicelled, sepals not gibbous at the base. *S. rupes'tre,* L. (Nyman).

On rocks and housetops in England, Wales, and Ireland ; fl. July–Aug.— Very similar to *S. rupes'tre,* but usually much larger, the leaves are in about 6 series, and almost cylindric with subulate tips; the flowers are usually bracteate, often 6-merous.—Distrib.   N. and Mid. Europe. *S. reflex'um* proper ; leaves green, those of the flowering shoots reflexed, flowers bright yellow.   The common garden form, not indigenous.—Var. *S. albes'cens,* Haw. (*S. glau'cum,* Sm.); smaller, leaves glaucous, those of the flowering stems not reflexed, flowers pale yellow.   Indigenous ; Mildenhall, Suffolk ; Babbicombe, Devon.

3*. SEMPERVI'VUM, L. House-leek.

Succulent herbs or undershrubs. *Radical leaves* densely rosulate, axils stoloniferous ; cauline alternate. *Flowers* in corymbose or panicled cymes. *Calyx* 6–multi-fid or -partite. *Petals* as many, distinct or connate and adnate to the alternate filaments below, narrow, acute. *Stamens* usually twice as many, the alternate sometimes deformed or transformed into carpels. *Scales* simple, distinct or connate in pairs, 2-fid or fimbriate, rarely 0. *Carpels* as many as petals, free or connate and adnate with the calyx-tube ; styles filiform ; ovules many. *Follicles* many-seeded.— —Distrib.   Europe, N. Africa, especially Madeira and Canaries, W. Asia, Himalaya ; species 40.—Etym. *semper* and *vivo,* from their retention of vitality.

S. tecto'rum, *L.* ; perennial, glandular-pubescent above, leaves ciliate.

Tops of walls and houses, not indigenous; fl. June–July.—Flowerless shoots 2–4 in. diam., in globose tufts ; flowering-stems 1–2 ft., erect, stout, with scattered leaves. *Leaves* 1–2 in., very fleshy, oblong or obovate-lanceolate, mucronate, edged with purple, tips flat. *Cyme* 2–5 in. diam., branches scorpioid. *Flowers* ¾–1 in. diam., dull red-purple. *Sepals* 12, narrow, acute. *Petals* lanceolate, ciliate. *Stamens* 12, with as many imperfect or transformed into carpels. *Scales* minute.—Distrib. Europe, W. Asia.

Order XXIX. DROSERA'CEÆ.

Perennial glandular herbs, rarely shrubby below. *Leaves* radical and rosulate, or cauline and alternate, circinate in bud, stipulate. *Inflorescence* various, often circinate cymes. *Sepals* 4–8, imbricate in bud, persistent. *Petals* 4–8, hypogynous or perigynous, distinct or connate at the base, imbricate, persistent. *Stamens* 4–20, inserted with the petals, rarely adnate to them ; anthers versatile or basifixed, bursting outwards. *Disk* 0, or obscure. *Ovary* free, 1–5-celled ; styles 1–5, simple or divided, stigmas simple or multifid ; ovules many, anatropous. *Capsule* 1–5-celled, loculicidally 2–5-valved, many-seeded.   *Seeds* small, albumen fleshy, testa often lax ; embryo axile, straight.—DISTRIB. Sandy or marshy places ; most common in temp. Australia ; genera 6 ; species 110.—AFFINITIES. Close to *Saxifrageæ.*—PROPERTIES. Yield a deep red-purple dye. For their carnivorous properties, *see* Darwin *On Insectivorous Plants.*

1. DROS'ERA, *L.* SUNDEW.

Slender glandular herbs. *Leaves* alternate or rosulate ; stipules scarious, adnate to the petiole, or 0. *Flowers* in scorpioid revolute cymes, rarely solitary. *Sepals* and *petals* 4–6 or 8. *Stamens* as many, hypogynous or perigynous. *Ovary* free, ovoid or globose, 1-celled ; styles 2–5, distinct or connate below ; ovules in many series, on 2–5 parietal placentas. *Capsule* oblong, 2–5-valved. *Seeds* minute, testa usually lax ; embryo large or small.—DISTRIB. Of the Order ; species 100.—ETYM. δροσερός, from the *dew*-like glands.

1. **D. rotundifo'lia,** *L.* ; leaves horizontal orbicular or broadly obovate, petiole hairy, testa loose reticulate.

Spongy bogs and heaths, N. to Shetland ; ascends to 2,300 ft. in the Highlands ; Ireland ; Channel Islands ; fl. July–Aug.—*Rootstock* slender. *Stem* very short. *Leaves* ½–½ in. diam., rosulate, margin glandular, nearly glabrous above ; petiole 1–1½ in., gradually dilated at the sheathing base. *Scapes* 3–6 in., in the centre of the rosette ; bracts subulate ; pedicels short. *Flowers* ¼ in. diam., many, in 2 series, white, usually 6-merous, homogamous. *Petals* a little longer than the sepals. *Styles* 2-fid, incurved, segments clavate. *Capsule* acute, exceeding the sepals. *Seeds* elongate.— DISTRIB. Europe (Arctic), N. and W. Asia, N. America.

2. **D. interme'dia,** *Hayne ;* leaves erect obovate or oblong-spathulate, petiole glabrous, testa close granulate. *D. lvngifo'lia,* L. in part.

Bogs and moist heaths, from Caithness to Sussex and Cornwall ; local in Scotland ; Ireland ; fl. July–Aug.—*Stem* short, leafy. *Leaves* gradually contracted into the petiole, together 1–2 in. *Scapes* 2–4 in., from the base of the rosette, curved at the base. *Flowers* much as in *D. rotundifo'lia,* usually 5–8-merous. *Capsule* pyriform, equalling the sepals. *Seeds* ovoid. —DISTRIB. Europe (Arctic), W. Asia, America from Canada to Brazil.

3. **D. ang'lica,** *Huds.* ; leaves suberect linear-spathulate, petiole glabrous, testa loosely reticulate. *D. longifolia,* L. in part.

Wet moors, from Orkney to Devon, Dorset, and Suffolk ; ascends to 1,700 ft. in the Highlands ; rare in S. England ; Ireland ; fl. July–Aug.—Very similar to *D. interme'dia,* but larger. *Leaves* ⅛ in. broad ; petiole 2–4 in. *Scapes* 4–8 in., from the centre of the rosette. *Flowers* ⅓ in. diam., 5–8-merous. *Capsule* obovoid, longer than the sepals.—DISTRIB. Europe, N. Asia, America.—*D. obova'ta,* Mert. and Koch, with broader leaves, styles often notched, capsule half as long as the sepals, seeds imperfect, is probably a hybrid with *D. rotundifo'lia.*—Scotch moors.

ORDER XXX. **HALORA'GEÆ.**

Herbs or shrubs, often marsh or aquatic, with much-reduced or imperfect perianths. *Leaves* opposite alternate or whorled, exstipulate. *Flowers* often minute and 1-sexual. *Calyx* superior; lobes 2, 4, or 0, valvate or slightly imbricate in bud. *Stamens* 1–8, epigynous in the 2-sexual flowers, filaments usually short; anthers (except in *Callitriche*) long, 4-angled, basifixed, slits lateral. *Disk* small or 0. *Ovary* inferior, mostly of 2 or 4 connate (rarely of 1) carpels; styles or stigmas as many as the carpels, 2 in *Callitriche;* ovules 1 in each cell, pendulous, anatropous. *Fruit* 2–4-celled, indehiscent, or of 1–4 small 1-seeded drupes. *Seed* pendulous, testa membranous, albumen fleshy; embryo cylindric, axile, or minute. DISTRIB. Widely dispersed; genera 9; species 80.—AFFINITIES. Obscure, probably near *Saxifrageæ* and *Rhizophoreæ.*—PROPERTIES unimportant.

Leaves all whorled, entire. Petals 0.....................................1. Hippuris.
Leaves all or lower whorled, much cut. Petals 2–4.........2. Myriophyllum.
Leaves all opposite, quite entire. Perianth 0 ....3. Callitriche.

1. **HIPPU'RIS,** *L.* MARE'S-TAIL.

Glabrous, aquatic herbs. *Stems* stout, erect, simple, leafy. *Leaves* whorled, narrow, quite entire. *Flowers* minute, solitary, sessile, axillary, sometimes 1-sexual. *Calyx-tube* subglobose, limb entire. *Petals* 0. *Stamen* 1. *Ovary* 1-celled; style subulate, stigmatose throughout its length. *Drupe* ovoid, stone crustaceous. *Seed* oblong, albumen scanty. —DISTRIB. Arctic and temp. N. hemisphere, Chili, Fuegia; species 1 or 2.—ETYM. ἵππος and οὐρά, *horsetail.*

H. vulga'ris, *L.*; leaves 6–10 in a whorl linear acute.

Margins of lakes, ponds, &c., from Shetland to Kent and Cornwall, local; fl. June–July.—*Rootstock* submerged, stout, creeping. *Stem* 6–24 in., terete, very many-jointed, as thick as a goose-quill or less; rarely floating flaccid and flowerless. *Leaves* ½–1½ in., close-set, tips withered. *Flowers* green, sessile. *Anthers* red. *Drupe* minute, smooth, green.—DISTRIB. Of the genus.

2. **MYRIOPHYL'LUM,** *L.* WATER MILFOIL.

Glabrous marsh or aquatic herbs; branches often floating. *Leaves* opposite alternate or whorled. *Flowers* small, axillary, solitary or spiked, anemophilous; upper male, lower female, intermediate often 2-sexual.— MALE fl. *Calyx* 4- rarely 2-lobed or 0. *Petals* 2 or 4, concave. *Stamens* 2, 4, or 8.—FEM. fl. *Calyx-tube* 4-grooved; lobes 4, minute or 0. *Petals* minute or 0. *Ovary* 4- rarely 2-celled; styles 4, very short, plumose. *Drupe* deeply 2-4-lobed. *Seeds* oblong; albumen copious.—DISTRIB. Ubiquitous; species 15.—ETYM. μυρίος and φύλλον, from the finely-divided *leaves.*

1. **M. verticilla'tum,** *L.* ; spike erect in bud, floral leaves all whorled in about fives pinnatifid or pectinate all longer than the flowers, upper axils usually flowerless.

Ditches and ponds, from Cumberland to Kent and Somerset; Ireland, rare; Channel Islands; fl. July–Aug.—*Rootstock* creeping. *Stems* floating, leafy. *Leaves* 1–2 in., in close-set whorls; segments distant, capillary, collapsing when removed from the water. *Spike* elongate. *Flowers* white. *Anthers* linear. *Fruit* subglobose, green, carpels rounded on the back.—DISTRIB. Europe, N. Africa, N. and W. Asia, India, N. America.
VAR. *M. pectina'tum,* DC., has very short floral leaves.

2. **M. alternifio'rum,** *DC.* ; spike curved at the tip in bud, female floral leaves whorled in threes or fours pectinate longer than the flowers, male opposite or alternate entire or serrate shorter than the flowers.

Ponds and ditches, from Shetland to Cornwall and Sussex ; ascends to 1,200 ft. in the Highlands; Ireland; Channel Islands; fl. June–Aug.—Habit of *M. spicatum,* but more slender and flowers fewer, and inhabits lakes in hilly and upland districts.—DISTRIB. Europe (Arctic), N. Africa, Arctic America.

3. **M. spica'tum,** *L.* ; spike erect in bud, floral leaves all whorled in about fours pectinate shorter than the flowers.

Ponds and ditches, from Orkney to Devon and Kent; ascends to 1,200 ft. in the Highlands; Ireland ; fl. June–Aug.—The small floral leaves, giving the inflorescence a more spicate appearance than *M. verticillatum,* is its best character ; the fruit also is less globose.—DISTRIB. Europe (Arctic), N. Africa, N. and W. Asia, India, N. America.

3. CALLIT'RICHE, *L.*   WATER STAR-WORT.

Slender glabrous marsh or aquatic plants. *Leaves* opposite, quite entire, upper often rosulate. *Flowers* unisexual, minute, solitary, axillary.— MALE fl. *Perianth* 0. *Stamen* 1, subtended by two caducous bracts, filaments slender ; anther-cells confluent above.—FEMALE fl. *Bracts* 2 or 0. *Ovary* sessile or shortly peduncled, 4-lobed longitudinally, 4-celled ; styles 2, slender, stigmatose all over. *Fruit* compressed, 4-lobed, 4-celled, lobes angled margined or winged at the back, at length separating, indehiscent. —DISTRIB. Chiefly temp. waters ; species 3 or 4.—ETYM. καλός and θρίξ, from the elegance of its *capillary* ramification.—I have followed Hegelmaier's limitation of the British forms of this very variable genus. Its affinities are very doubtful.

1. **C. ver'na,** *L.* ; leaves not dilated at the base, flowers bracteate, carpels slightly keeled connate for about half their breadth.

Ponds, ditches and sluggish streams, from Shetland southd. ; ascends to 2.200 ft. in the Highlands; Ireland ; Channel Islands ; fl. April–Oct.— Very variable in size and habit, covered with scattered stellate hairs or scales. *Stems* 3–12 in., submerged, terete, sparingly branched. *Leaves* ½–1 in., submerged linear; floating rosulate, obovate, notched, 3-nerved. *Flowers,* male and female often in opposite axils. *Bracts* white, deciduous, incurved. *Filaments* very slender. *Furrow* on the carpels shallow, not extending to the base of the lobes.—DISTRIB. All temp. and cold climates (Arctic).

C. VERNA proper; fruit sessile, carpels turgid sharply keeled sinus shallow, styles erect or spreading, pollen ellipsoid. *C. verna'lis,* Kuetz.; *C. aqua'tica,* Sm.

Sub-sp. C. PLATYCAR'PA, *Kuetz.*; fruit subsessile, carpels large flattish sharply keeled sinus deep, styles at length reflexed persistent, pollen subglobose. *C. stagnalis,* Scop., is a terrestrial form.—Ascends to 1,500 ft. in Derby.

Sub-sp. C. HAMULA'TA, *Kuetz.*; fruit subsessile, carpels flattish shortly broadly keeled sinus shallow, styles long at length reflexed deciduous, pollen subglobose.

Sub-sp. C. OBTUSAN'GULA, *Leg.*; leaves obovate, fruit subsessile, styles spreading, carpels turgid obtusely trigonous at the back.—*C. Lachii,* Warren MSS., has almost linear upper leaves and longer styles; it is intermediate between *obtusan'gula* and *hamula'ta* (Warren).

Sub-sp. C. PEDUNCULA'TA, *DC.*; leaves always linear, fruit peduncled or sessile, carpels flattish shortly sharply keeled, styles long at length reflexed deciduous, pollen subglobose.—Flowers earlier.

2. **C. autumna'lis,** *L.* ; leaves all submerged dilated at the base, bracts 0, carpels keeled or winged connate towards the axis only.

Lakes, rare and local, Orkneys to Devon; Ireland; fl. June–Oct.—*Stem* brittle and leaves without stellate hairs. *Leaves* ¼–½ in., all linear, truncate, dark green. *Fruit* much larger than in *C. ver'na.*– DISTRIB. N. and Mid. Europe (Arctic), Siberia, N. America.

C. autumna'lis proper; fruit nearly sessile, winged.—VAR. *C. truncata,* Guss. fruit shortly pedicelled, keeled.

ORDER XXXI. **LYTHRA'RIEÆ.**

Herbs, shrubs (or trees), branches usually 4-angled. *Leaves* opposite or whorled, quite entire, exstipulate. *Flowers* regular or irregular. *Calyx* inferior, tubular or campanulate, persistent; lobes 3–6, valvate in bud, alternating with as many teeth. *Petals* 3–6 rarely 0, inserted in the calyx-tube, crumpled in bud. *Stamens* usually definite, inserted in the calyx-tube, equal or unequal, inflexed in bud; anthers versatile, often recurved. *Disk* annular, unilateral, or 0. *Ovary* 2–6-celled; style straight or flexuous, stigma capitate; ovules many, on the inner angles of. the cells, anatropous, horizontal or erect. *Capsule* enclosed in the calyx-tube, 2–6-celled, or 1-celled by the septa vanishing; placentas usually forming a central seed-bearing column. *Seeds* various, albumen scanty or 0 ; cotyledons oblong or orbicular, 2-auricled.—DISTRIB. Chiefly trop. ; genera 30 ; species 250.—AFFINITIES. With *Onagrarieæ, Myrtaceæ,* and *Halorageæ* —PROPERTIES. Astringent, acrid, and vesicatory. Pomegranate bark is astringent.

Calyx tubular. Petals exceeding the calyx-teeth..................1. Lythrum.
Calyx campanulate. Petals minute or 0.............................2. Peplis.

1. LY'THRUM, *L.* LOOSESTRIFE.

Herbs or shrubs, branches 4-angled. *Leaves* opposite, whorled or alternate, quite entire. *Flowers* axillary, red or purple, honeyed. *Calyx-tube*
cylindric, straight ; teeth and ribs 8–12. *Petals* 4–6, sometimes unequal
or 0. *Stamens* 8–12, 1–2-seriate in the calyx-tube, filaments filiform often
declinate. *Ovary* sessile, 2-celled ; style filiform, stigma obtuse ; ovules
very many. *Capsule* 1–2-celled, septicidally 2-valved or bursting irregularly.
Seeds plano-convex or angular.—DISTRIB. Temp. and trop. regions ;
species 12.—ETYM. λύθρον, *gore*, from the blood-red flowers.

1. **L. Salica'ria,** *L.* ; leaves opposite or whorled lanceolate cordate at
the base, flowers whorled 3-morphic, stamens 12.

River-banks and ditches, &c., from Argyll and Perth southd.; Ireland; Channel Islands ; fl. July–Sept.—Glabrous and pubescent. *Rootstock* creeping.
Stem 2–5 ft., branched, 4–6-angled or winged. *Leaves* 2–5 in., often 3–4 in a
whorl, acute. *Cymes* glomerate, in terminal spiked racemes. *Flowers* ⅔–1
in. diam., red-purple, homogamous ; bracts small or 0. *Calyx-tube* ¼ in.,
12-ribbed, outer-teeth lanceolate longer than the inner. *Petals* narrow-
oblong, wrinkled. *Capsule* ovoid.—DISTRIB. Temp. N. regions (Arctic),
Australia.—Flowers trimorphic in respect of length of style and of filaments
and of size of pollen in 3 sets of individuals. Of those growing by the
Thames at Kew, the long-styled is glabrous, slender, with small narrow
leaves, and bright flowers ; that with very short styles is a larger, coarser,
very pubescent plant, with dull purple flowers.—The 3 forms have—1, Long
style, medium stamens, medium yellow pollen.—*2*. Long style, short
stamens, small yellow pollen.—3. Medium style, long stamens, large
green pollen.—4. Medium style, short stamens, small yellow pollen.—
5. Short style, long stamens, large green pollen.—6. Short style, medium
stamens, medium yellow pollen.—These admit of 9 modes of cross-fertili-
zation.

2. **L. hyssopifo'lia,** *L.* ; leaves chiefly alternate linear-lanceolate,
flowers solitary homomorphic, stamens about 6.

Moist places, often inundated, very local, Northampt., Cambridge, Norfolk,
Herts, Cornwall ; Channel Islands ; fl. June–Sept.—Glabrous, annual. *Stem*
½–1½ ft., prostrate or ascending. *Leaves* ½–1 in., sessile, cuneate at the base,
very narrow. *Flowers* small, pink. *Calyx* 2-bracteolate ; teeth subulate,
subequal. *Petals* oblong. *Capsule* cylindric.—DISTRIB. From Hanover,
southd., N. and S. Africa, N. and W. Asia, India, America.

2. PEP'LIS, *L.* WATER-PURSLANE.

Small weak annual herbs. *Leaves* alternate and opposite, quite entire.
Flowers minute, axillary, subsessile, 2-bracteolate. *Calyx* campanulate,
6-lobed, with as many alternate spreading teeth. *Petals* 6, in the throat
of the calyx, fugacious, or 0. *Stamens* 6 or 12, in the middle of the calyx-
tube. *Ovary* subglobose, membranous, 2-celled ; style short, stigma capitate ; ovules very many, placentas on the septum semicylindric. *Capsule*
2-celled, 2-valved, or bursting irregularly, many-seeded. *Seeds* minute,

plano-convex.—Distrib. Europe, N. Africa, temp. Asia ; species 3.—
Etym. πέπλιον, the old name for *Portulaca* transferred.

P. Por'tula, *L.* ; leaves obovate obtuse, flowers solitary.
Moist places, from Caithness southd. ; ascends to 1,200 ft. in Yorkshire ;
Ireland ; Channel Islands ; fl. July–Aug.—Glabrous, branched, tufted.
Stems 3–8 in., 4-angled, creeping, fragile. *Leaves* ½–1 in. opposite, short-
petioled. *Flowers* very minute, in almost all the leaf-axils, purplish.
Calyx 12-ribbed, hemispheric ; teeth triangular. *Petals* minute or 0.
Stamens 6 or 12. *Capsule* globose.—Distrib. Europe, N. Africa.

ORDER XXXII. **ONAGRA'RIEÆ.**

Herbs (rarely shrubs or trees). *Leaves* opposite or alternate, exstipulate.
Flowers regular. *Calyx* superior ; lobes 2–4, valvate in bud. *Petals* 2–4,
rarely 0, perigynous, fugacious, twisted in bud. *Stamens* 1–8, 1–2-seriate,
sometimes declinate ; anthers oblong. *Disk* epigynous and coating the
calyx-tube. *Ovary* 4- (rarely 1–6-) celled ; style filiform, stigma entire or
4-lobed ; ovules 1 or more in the inner angle of each cell, pendulous or
ascending, anatropous. *Fruit* a drupe, berry, or capsule. *Seeds* 1 or
more, smooth papillose or hairy, albumen 0 or very scanty ; embryo ovoid,
cotyledons plano-convex.—Distrib. Temp. regions, rarer in tropical ;
genera 22 ; species 300.—Affinities. With *Lythraceæ* and *Melastomaceæ.*
Properties unimportant.

Petals 4, pink or purple. Stamens 8..................................1. Epilobium.
Petals short or 0. Stamens 4...2. Ludwigia.
Petals 4, yellow. Stamens 8 ....2*. *Œnothera.*
Petals 2, white. Stamens 2...3. Circæa.

1. EPILO'BIUM, *L.* Willow-herb.

Herbs or under-shrubs, stolons creeping. *Leaves* alternate or opposite.
Flowers solitary, axillary, or in terminal leafy spikes, pink or purple, rarely
yellow. *Calyx-tube* long, slender ; limb 4-partite, deciduous. *Petals* 4,
usually 2-lobed. *Stamens* 8, the alternate shorter. *Ovary* 4-celled ; style
filiform, stigma obliquely clavate or 4-lobed ; ovules many, 2-seriate,
ascending. *Capsule* elongate, 4-celled, loculicidally 4-valved ; valves
separating from a 4-winged seed-bearing axis. *Seeds* broadest above, tipped
with a long pencil of hairs, minutely tubercled.—Distrib. Arctic, temp.
and cold regions, abundant in New Zealand ; species 50.—Etym. ἐπί and
λόβιον, from the position of the corolla, &c., on the pod.

Hybrids abound in this genus ; the following are proved or suspected :
lanceolatum with *obscurum ; parviflorum* with *montanum ; obscurum* with
parviflorum and *palustre ; palustre* with *alsinefolium ;* and *parviflorum* with
tetragonum. Many others occur on the Continent.

Section 1. **Chamæne'rion.** *Corolla* irregular, rotate. *Calyx-lobes*
free to the base, spreading. *Stamens* declinate.

1. **E. angustifo'lium,** *L.* ; stem tall simple, leaves lanceolate. *Rose-bay* or *French Willow.*

Banks and copses from Shetland to Devon and Kent; ascends to 2,700 ft. in the Highlands ; Ireland ; Channel Islands ; fl. July–Aug.—Glabrous, inflorescence pubescent. *Stem* 2–4 ft., erect, terete. *Leaves* 3–6 in., petioled, alternate, obscurely toothed, glaucous beneath. *Racemes* elongate. *Flowers* 1 in. diam., bracteate, dark rose-purple, honeyed, proterandrous. *Petals* obovate-spathulate, 2 lower smaller. *Style* bent down, stigmas 4, erect then revolute. *Capsule* 2–4 in. *Seeds* obovoid.—DISTRIB. Temp. and Arctic Europe, N. and W. Asia to the Himalaya, and America.
E. brachycar'pum, Leight., with stem 4–6 ft., buds very oblique, capsules 1–1½ in., is the cultivated form, sometimes found as an escape.

SECTION 2. **Lysima'chion.** *Flowers* regular, corolla campanulate or funnel-shaped. *Calyx-lobes* connate at the base. *Stamens* erect.

* *Stem terete. Stigma 4-cleft, lobes erect or revolute.*

2. **E. hirsu'tum,** *L.* ; glandular-pubescent and hirsute, leaves opposite oblong-lanceolate ½-amplexicaul serrulate, buds erect. *Codlins-and-cream.*

Sides of ditches and rivers, from Sutherland southd.; ascends to 1,200 ft. in Derby ; Ireland ; Channel Islands ; fl. July–Aug.—Odorous. *Stolons* subterranean, thick, fleshy, scaly, leafless. *Stems* 3–5 ft., terete with raised lines from the leaf-bases. *Leaves* 3–5 in., teeth incurved. *Flowers* very many, ¼–¾ in. diam., rose-purple, homogamous. *Petals* broad, notched. *Filaments* bearded at the base. *Stigma-lobes* revolute. *Capsule* 2–3 in.—DISTRIB. Europe, N. Africa, N. and W. Asia, Himalaya ; introd. in N. America.

3. **E. parviflo'rum,** *Schreb.* ; villous pubescent or glabrate, leaves mostly alternate sessile lanceolate obscurely toothed, buds erect.

Ditches and river-banks, from Ross and the Hebrides southd.; Ireland ; Channel Islands ; fl. July–Aug.—*Stolons* autumnal, with subsessile rosulate leaves. *Stem* 1–3 ft., terete, branched above. *Leaves* 1–2 in., linear or oblong-lanceolate, rounded at the base. *Flowers* ⅓ in. diam., many, rose-purple, honeyed, homogamous. *Stigma-lobes* short, not revolute. *Capsule* 1½–2 in., nearly glabrous or pubescent.—DISTRIB. Europe, N. Africa, Himalaya.
E. rivula're, Wahl., is an almost glabrous variety ; and *E. interme'dium,* Merat, has most or all the leaves alternate.

4. **E. monta'num,** *L.* ; stem glabrous or pubescent, leaves mostly opposite glabrous oblong-ovate acute toothed, buds drooping.

Shady banks, walls and cottage roofs, from Shetland southd.; ascends to near 1,700 ft. in the Lake district; Ireland ; Channel Islands ; fl. June–July.— *Stolons* autumnal, subterranean and fleshy, or subaërial with suberect rosulate leaves. *Stem* 6–24 in., erect, slender. *Leaves* 1–3 in., sometimes petioled, or whorled in threes. *Flowers* ¼–⅓ in. diam., pale purple, homogamous. *Stigma-lobes* short, not revolute. *Capsule* 2–3½ in., finely pubescent. —DISTRIB. Europe (Arctic), N. and W. Asia, Himalaya.

5. **E. lanceola'tum,** *Sebast.* and *Maur.* ; finely pubescent, leaves mostly alternate petioled oblong-lanceolate toothed, buds inclined.

Roadsides and stony places by streams, from Surrey to Cornwall, Monmouth, Gloster ; Channel Islands ; fl. July–Oct.—*Stolons* autumnal, with spreading rosulate leaves. *Stem* 1–3 ft., erect, branched, terete, pubescent, hairs short recurved. *Leaves* 1½–3 in., lowest opposite, petiole terminating in obscure decurrent lines on the stem. *Flowers* ¼ in. diam., many, pale rose. *Stigma-lobes* short, spreading. *Capsule* 2–3 in., finely pubescent.—Distrib. From Belgium southd., and eastd. to Asia Minor.

** *Stem often more or less 2–4-angled, or with 2–4 raised_lines. Stigma oblique clavate.*

6. **E. ro'seum,** *Schreb.* ; pubescent above, stem with 2 or 4 raised lines, leaves petioled mostly alternate ovate-oblong narrowed above and below toothed glabrous, buds inclined acuminate.

Copses and moist places, Edinburgh to Kent and Cornwall, local ; rare in the north ; ? Ireland ; fl. July–Aug.—*Stolons* autumnal, with loosely rosulate leaves. *Stem* 1–2½ ft., erect, brittle, much branched. *Leaves* 1½–2½ in. *Flowers* many, ¼ in. diam., rose-red. *Capsule* 2–3 in., pubescent.—Distrib. Europe, N. and W. Asia, Himalaya, N.W. America.

7. **E. tetrago'num,** *L.* ; pubescent above, stem usually with 2 or 4 raised lines, leaves sessile oblong- or ovate- or linear-lanceolate toothed, buds erect acute.

Wet places, from the Orkneys southd. ; ascends to 2,100 ft. in the Highlands ; Ireland ; Channel Islands ; fl. July–Aug.—*Stem* 1–2 ft., erect, branched, rather tough, obtusely angled or with 2 or 4 raised lines or almost terete. *Leaves* 1–3 in., narrow, lower or all below the branches opposite, rarely slightly petioled, base decurrent. *Flowers* ¼–⅓ in., erect, rose-lilac. *Pods* 2–4 in.—Distrib. Europe (Arctic), N. and S. temp. zones, Himalaya.

E. tetrago'num proper ; stolons autumnal with rosulate leaves, leaves linear-oblong ‸or -lanceolate shining above, capsule 2–4 in. slightly incurved.

Sub.-sp. E. obscu'rum, *Schreb.* ; stolons æstival with few distant pairs of opposite leaves, leaves ovate-lanceolate not shining above, capsule 1–2 in. suberect or rather spreading. *E. virgatum,* Gren. and Godr.—The most common form ; ascends to 1,500 ft. in Derby.

8. **E. palus'tre,** *L.* ; finely pubescent above, stem terete without raised lines, leaves subsessile mostly opposite lanceolate from a cuneate base, buds nodding obtuse, seeds fusiform, testa produced at the tip.

Bogs and ditches, from Shetland southd. ; ascends to near 2,000 ft. in York-shire ; Ireland ; Channel Islands ; fl. July–Aug.—*Stolons* æstival, sub-terranean, filiform, scaly, bearing in autumn scaly buds. *Stem* 6–24 in., with often two lines of pubescence. *Leaves* 1½–2½ in., almost all opposite, flaccid, spreading, scarcely toothed, tip narrowed but obtuse. *Flowers* horizontal, ¼ in. diam., rose-lilac. *Capsule* 2–2½ in., pubescent. *Seeds* much narrower than in all preceding species, with a distinctly produced testa.—Distrib. N. temp. and Arctic zones, Himalaya.—Var. *ligula'ta,* Baker, with leaves lanceolate faintly toothed, and seeds shorter, is a hybrid with *obscurum.*

9. **E. alsinefo'lium,** *Vill.* ; almost glabrous, stems tufted ascending with 2–4 obscure pubescent lines, leaves usually opposite subsessile ovate or ovate-lanceolate acuminate toothed glabrous shining, buds drooping obtuse, seeds narrow clavate, testa produced at the tip.

Spongy banks of rills, &c., in alpine and subalpine districts, Wales, and from Westmoreland and Durham to Shetland ; ascends to near 2,900 ft.; fl. July.— *Stolons* æstival, subterranean, filiform, scaly, bearing in autumn a scaly bud. *Stem* 4–12 in., flexuous, subsucculent. *Leaves* 1–2 in., as of *E. monta'num,* flaccid, bright green. *Flowers* ½ in. diam., few, bright rose-purple. *Capsule* 1½–2 in., almost glabrous.—Distrib. Europe (Arctic), N. and W. Asia to Himalaya, America.

10. **E. alpi'num,** *L.* ; small, slightly pubescent, stem with 2 pubescent lines, leaves opposite elliptic-oblong obtuse entire or toothed, buds obtuse and flowers pendulous, seeds narrow-obovoid, testa not produced.

Alpine rills, from Durham and Cumberland to Sutherland ; ascends to near 4,000 ft.; fl. July.—*Stolons* æstival, rosulate, or elongate (*E. anagalli-difo'lium,* Lamk.). *Stem* usually ascending, 3–9 in., slender, simple, often curved. *Leaves* ½–⅔ in., few, shortly petioled. *Flowers* 1–3, ⅓–½ in. diam., bright or pale rose-purple. *Capsule* 1–1½ in., almost glabrous.—Distrib. Arctic and Alpine Europe, Asia, Himalaya, N. America.

2. LUDWIG'IA, *L.*

Annual or perennial herbs, sometimes aquatic. *Leaves* opposite or alternate, quite entire. *Flowers* usually axillary solitary and sessile ; peduncles 2-bracteate. *Calyx-tube* cylindric or angled or turbinate ; lobes 3–5, persistent. *Petals* 3–5 or 0, and stamens (3–5) inserted under the margin of an epigynous disk. *Ovary* 4–5-celled ; style short, stigma 3–5-lobed ; ovules many, in many series, on prominent axile placentas. *Capsule* septicidal or dehiscing by terminal pores, or irregularly rupturing longitudinally. *Seeds* minute.—Distrib. Temp. and warm regions, chiefly of N. America ; species 20.—Etym. C. G. Ludwig, a Leipsic botanist.

L. palus'tris, *Elliot;* leaves all opposite ovate or elliptic. *Isnar'dia,* L.

Boggy pools, very rare, Sussex and Hants ; Jersey ; fl. June–July.—Glabrous, perennial. *Stem* 6–10 in., rooting at the nodes, procumbent or floating, 4-angled, branched. *Leaves* ½–1 in., petioled, acute, shining. *Flowers* 4-merous, minute, axillary, sessile, green ; bracts subulate. *Calyx-tube* ⅛ in., oblong, truncate, with 4 green ribs ; lobes triangular, acute. *Petals* 0 (or small and red in American specimens). *Style* short, stigma large capitate. *Seeds* angular.—Distrib. From Hamburg southd., S. Africa, W. Asia, N. America.

2*. ŒNOTHE'RA, *L.* Evening Primrose.

Herbs, rarely shrubby. *Leaves* alternate. *Flowers* axillary, solitary or in leafy spikes or racemes, large, yellow red or purple, honeyed. *Calyx-tube* elongate, 4-angled ; limb cylindric, 4-lobed, deciduous. *Petals* 4.

Stamens 8 ; anthers usually long. *Ovary* 4-celled ; style filiform, stigma capitate entire or 4-lobed ; ovules many, 1–2-seriate, horizontal or ascending. *Capsule* 4- rarely 1-celled, splitting from the top downwards into 4 septiferous valves, usually leaving the seeds on the axis, sometimes indehiscent. *Seeds* many or few, sometimes appendaged.—DISTRIB. Temp. N. and S. America, rarely tropical ; one Tasmanian ; species 100. —ETYM. obscure.

Œ. BIEN′NIS, *L.* ; erect, leaves ovate-lanceolate, capsule oblong subcylindric.

A garden escape in several places; fl. July–Sept.—Annual or biennial, pubescent or hairy, 2–3 ft. *Leaves* 3–6 in., remotely toothed; petiole short, midrib stout white. *Flowers* 3–3½ in. diam., subspicate, sessile, golden yellow. *Calyx-lobes* much longer than the ovary. *Petals* obcordate. *Capsule* 1–2 in., narrowed upwards, obtusely 4-ribbed. — DISTRIB. N. America.

Œ. ODORA′TA, *Jacq.* ; erect, leaves linear-lanceolate waved, capsule elongate cylindric.

Coasts of Somerset and Cornwall ; a garden escape; fl. July–Sept.—Perennial. *Stem* 1–2 ft., usually purplish, branched, clothed with spreading hairs. *Leaves* 3–6 in., lower nearly flat, nerves green or purple. *Flowers* yellow, 3–4 in. diam., fragrant. *Capsule* 2 in., pubescent.—DISTRIB. Patagonia.

3. CIRCÆ′A, *Tourn.* ENCHANTER'S NIGHTSHADE.

Slender erect herbs, with creeping rootstocks. *Stem* simple. *Leaves* opposite, petioled, toothed. *Flowers* small, white, in terminal and lateral peduncled racemes, honeyed. *Calyx-tube* ovoid ; limb 2-parted, reflexed, deciduous. *Petals* obcordate and stamens 2, inserted under the margin of an epigynous disk. *Ovary* 1–2-celled ; style filiform, stigma capitate 2-lobed ; ovules 1 in each cell, ascending, placentas axile. *Fruit* ovoid or pyriform, 1–2-celled, indehiscent, covered with hooked bristles, cells 1-seeded. *Seeds* oblong, attached by the middle.—DISTRIB. Europe, temp. Asia, and N. America ; species 2 or 3.—ETYM. *Circe*, the enchantress.

1. **C. lutetia′na,** *L.* ; glandular-pubescent, leaves ovate faintly toothed not shining, fruit broadly obovoid 2-seeded.

Damp woods, from Argyll and Aberdeen southd. ; ascends to 1,200 ft. in York-shire; Ireland; Channel Islands ; fl. June–Aug.—*Stem* 1–2 ft., erect or ascending, terete, subsimple ; nodes swollen. *Leaves* 1–3 in., petiole almost as long, covered with translucent dots, rounded truncate or cordate at the base. *Flowers* ½ in. diam. in lax erect terminal racemes, white or pink, proterandrous ; pedicels ½ in., slender, jointed at the base, patent, reflexed in fruit; bracts usually 0. *Disk* tumid. *Fruit* ⅙ in.—DISTRIB. Europe, N. Africa, N. and W. Asia to the Himalaya, N. America.

2. **C. alpi′na,** *L.* ; smaller, less hairy, leaves shining more deeply toothed, ovary less hispid, fruit 1-seeded.

Hilly districts from Gloster, N. Wales and Stafford to Sutherland; ascends to 1,300 ft. in the Lake district; Ireland; fl. July–Aug.—Usually a well-marked plant; comparatively stouter, 6–8 in.; leaves with longer and winged petiole; pedicels with minute subulate bracts, but supposed hybrids or inter-mediates are designated as *C. lutetia'na*, var. *interme'dia*, and *C. interme'dia*, Ehrh., according to their affinities with one or the other parent.—DISTRIB. as *C. lutetia'na*, omitting N. Africa, and extending to within the Arctic circle and to mountains of South India.

ORDER XXXIII. **CUCURBITA'CEÆ.**

Tribe CUCUMERI'NEÆ.

Prostrate or climbing, annual or perennial herbs. *Leaves* alternate, exstipulate. *Tendrils* lateral, simple or divided. *Flowers* usually cymose, unisexual. *Calyx* superior, lobes 5, valvate in bud. *Petals* 5, inserted on the calyx-limb, distinct or connate below, valvate or induplicate in bud. *Stamens* 3; filaments and anthers distinct or connate, the latter adnate to the filaments, bursting outwards, one 1-celled, two 2-celled, cells straight curved or flexuous. *Ovary* inferior, 3-celled; placentas 3, fleshy, project-ing to and confluent in the axis of the ovary and thence reflexed to its walls; style simple or divided, stigmas various; ovules 2-seriate, parietal, horizontal, anatropous. *Berry* 1-celled, many-seeded. *Seeds* usually flattened, testa coriaceous or crustaceous, albumen 0, embryo flattened, cotyledons plano-convex or foliaceous, radicle short.—DISTRIB. Of the Tribe, chiefly Indian and African; genera 50; species 360.—AFFINITIES. With *Passifloreæ*.—The above Tribe comprises the mass of the Order.— PROPERTIES. Purgative and bitter, but many yield by cultivation esculent fruits.

1. BRYO'NIA, *L.* BRYONY.

Slender climbing perennial herbs. *Leaves* 3–5-angled or -lobed.—MALE fl. in racemed corymbose or fascicled cymes. *Calyx-tube* campanulate, 5-toothed. *Corolla* rotate or campanulate, 5-partite. *Filaments* 3, rarely 5; anthers distinct or slightly cohering, cells flexuous.—FEMALE fl. solitary or crowded, *calyx* and *corolla* of the male. *Ovary* ovoid or globose; style slender, 3-fid, stigmas simple or 2-lobed. *Berry* spherical, many or few-seeded. *Seeds* tumid or compressed.—DISTRIB. Temp. and trop.; species 12.—ETYM. βρύω, to shoot, from the rapid growth of the shoots.

B. dioi'ca, *L.* ; hispid, dicecious, leaves palmately 5-lobed.

Hedges and thickets in England, rare in the North; Channel Islands; fl. May–Sept —*Rootstock* of very large fleshy tubers; juice nauseous, milky. *Stems* many, annular, slender, angled; tendrils simple. *Leaves* 3–5 in. diam., petioled, suborbicular, cordate, lobes sinuate. *Cymes* of male corym-bose, 3–8-fid.; of fem. umbelled. *Corolla* ½–¾ in. diam., hairy, greenish. *Ovary* smooth, stigmas 2-cleft. *Berry* ¼ in. diam., red, 3–6-seeded.—DIS-TRIB. From Denmark southd., N. Africa, W. Asia.—Root acrid and cathartic.

Order XXXIV. **UMBELLIF'ERÆ.**

Herbs ; internodes usually fistular. *Leaves* alternate, pinnately or 3-nately compound, rarely simple ; petiole dilated at the base. *Inflorescence* of usually simple or compound umbels, with an involucre of whorled bracts at the base of the primary rays, and of bracteoles at the secondary ; rarely capitate. *Flowers* small, usually honeyed and proterandrous, all 2-sexual and similar, or outer in each umbel male with large unequal petals and long stamens, inner female or 2-sexual. *Calyx* superior ; limb 0 or 5-toothed. *Petals* 5, epigynous, usually obovate or obcordate, tip often inflexed, imbricate induplicate or valvate in bud, white, rarely pink yellow or blue. *Stamens* 5, at the base of the disk, filaments incurved ; anthers versatile. *Disk* epigynous, usually of 2 lobes confluent with the bases of the styles. *Ovary* 2-celled ; styles 2, erect or recurved, stigmas obtuse ; ovules 1 in each cell, pendulous, anatropous, raphe ventral. *Fruit* of 2 indehiscent, dorsally or laterally compressed carpels, separated by a commissure ; carpels each 5- or 9-ridged, adnate to or pendulous from an entire or split slender axis (*carpophore*) ; pericarp often traversed by oil-canals (*vittæ*). *Seed* pendulous, usually adherent to the pericarp, testa membranous, albumen copious dense ; embryo minute, next the hilum, cotyledons ovate-oblong or linear, often very unequal.—Distrib. Chiefly N. Europe, N. and W. Asia, and N. Africa ; genera 152 ; species 1,300.—Affinities. Intimate with *Araliaceæ* and *Corneæ.*—Properties. 1. Poisonous, acrid, watery sap in *Conium, Cicuta, Œnanthe.* 2. Esculent in *Angelica, Samphire, Parsley, Celery,* &c. 3. Sugar and starch abound in *Carrot, Parsnip, Pig-nut.* 4. Milky fœtid gum-resins in stems of *Asafœtida, Galbanum,* &c. 5. Essential oils in the fruit of *Anise, Dill, Caraway, Coriander,* and *Cummin.*—The ridges are normally 9 on each carpel ; viz. 5 primary, of which 2 are lateral next the commissure, 1 dorsal, 2 intermediate ; and 4 secondary, alternating with these. The vittæ occur between the ridges, rarely in them ; normally there are 6 in each carpel, 4 between the primary ridges and 2 on the commissural face.

Series 1. HETEROSCIA'DIEÆ. *Umbels simple, or very irregularly compound, or flowers capitate. Vittæ 0 or obscure. See 35. Caucalis.*

Tribe I. **HYDROCOT'YLEÆ.** *Fruit* laterally much compressed, commissure narrow...1. Hydrocotyle.

Tribe II. **SANIC'ULEÆ.** *Fruit* subterete, or dorsally compressed ; commissure broad.
Leaves spinous. Umbels densely capitate.........................2. Eryngium.
Leaves palmate. Bracts very large.........................2*. *Astrantia.*
Leaves palmate. Fruit with hooked spines.........................3. Sanicula.

Series 2. HAPLOZYG'IEÆ. *Umbels compound. Ridges subequal or primary the most conspicuous (except in 22*. Coriandrum). Vittæ usually obvious.*

Tribe III. **AMMI'NEÆ.** *Fruit* laterally compressed ; commissure narrow.
Section 1. **Smyrn'ieæ.** *Fruit short, ovoid or didymous ; ridges not winged. Seed grooved ventrally.*
Vittæ solitary in the furrows ; ridges slender.................4. Physospermum.

M

Vittæ several. Disk-lobes depressed ; ridges elevated..............5. Conium.
Vittæ several. Disk-lobes conical....................................6. Smyrnium.

Section 2. **Ammi'neæ** proper. *Fruit as in 1, but seed flat ventrally.*
* Petals entire, tip acute or shortly inflexed. Vittæ 1-2.
Leaves simple. Flowers yellow..................................7. Bupleurum.
Leaves compound. Flowers white, diœcious................8. Trinia.
Leaves compound. Flowers white, 2-sexual.............9. Apium.

** Petals 2-lobed, tip long inflexed. Vittæ solitary in the furrows.
Calyx-teeth obsolete. Vittæ as long as the fruit.............10. Carum.
Calyx-teeth obsolete. Vittæ very short.......................11. Sison.
Calyx-teeth ovate, acute. Vittæ long..........................12. Cicuta.

*** Petals as in * ; but vittæ several in each furrow (except in *Ægopodium*).
Calyx-teeth acute. Leaves pinnate.............................13. Sium.
Calyx-teeth obsolete. Leaves 2-ternate. Vittæ 0.............14. Ægopodium.
Calyx-teeth obsolete. Leaves various. Vittæ many...........15. Pimpinella.

Section 3. **Scandici'neæ.** *Fruit elongate. Seed grooved ventrally.*
* Vittæ many in each furrow, often faint..................16. Conopodium.
** Vittæ 0, or 1 in each furrow.
Fruit ¾-1 in.; ridges almost winged........................17. Myrrhis.
Fruit 1-3 in.; ridges prominent...............................18. Scandix.
Fruit ¼ in.; ridges vanishing upwards.....................19. Chærophyllum.
Fruit ⅛-½ in.; ridges 0 or obscure.........................20. Anthriscus.

Tribe IV. **SESELI'NEÆ.** *Fruit globose or ovoid, not laterally compressed;* commissure broad; lateral ridges distinct (except in 22*. *Coriandrum*), rarely winged, if so wings of opposite carpels not in contact.

Sub-tribe 1. **Seseli'neæ** proper. *Fruit subterete ; ridges not thickened or corky.*
Calyx-teeth small. Petals white, notched................21. Seseli.
Calyx-teeth obsolete. Petals yellow, entire................22. Fœniculum.

Sub-tribe 2. **Corian'dreæ.** *Fruit globose ; ridges low, secondary broadest.*
                                                22*. *Coriandrum.*

Sub-tribe 3. **Cachry'deæ.** *Fruit subterete ; primary ridges acute ; outer coat of pericarp lax*..23. Crithmum.

Sub-tribe 4. **Œnan'theæ.** *Fruit subterete ; primary ridges thick, lateral forming a corky rim round the carpel.*
Bracteoles whorled....................................24. Œnanthe.
Bracteoles unilateral.................................25. Æthusa.

Sub-tribe 5. **Schultz'ieæ.** *Fruit subterete ; lateral ridges thickened or winged*..26. Silaus.

Sub-tribe 6. **Seli'neæ.** *Fruit dorsally compressed ; primary ridges broad, thick.*
Seed concave ventrally ; vittæ several ....27. Meum.
Seed almost flat ventrally; vittæ many or obscure..........28. Ligusticum.
Seed biconvex; vittæ solitary in the dorsal furrows.............29. Selinum.

Sub-tribe 7. **Angel'iceæ.** *Fruit much dorsally compressed; lateral ridges broadly winged; wings of opposite carpels not appressed*..........30. Angelica.

TRIBE V. **PEUCEDA'NEÆ.** *Fruit* much dorsally compressed; lateral ridges broadly winged, wings of opposite carpels appressed (face to face); other ridges filiform. *Styles* short, stout.

Wings with thin margins ; vittæ as long as the fruit........31. Peucedanum.
Wings with thin margins ; vittæ club-shaped..................32. Heracleum.
Wings with thick margins.....................................33. Tordylium.

SERIES 3. DIPLOZYG'IEÆ. *Umbels compound (sometimes simple in Caucalis); secondary ridges more distinct than the primary (see also 22*. Coriandrum), spinous in the British genera.*

Bracts pinnatifid or laciniate. Seed flat in front ....34. Daucus.
Bracts entire or 0. Seed grooved in front....................35. Caucalis.

ARTIFICIAL KEY TO THE GENERA.

I. Leaves undivided.
Creeping. Leaves peltate ....1. Hydrocotyle.
Erect. Leaves linear...7. Bupleurum.

II. Leaves palmate, or simply 3-nately divided.
Umbels subglobose. Fruit prickly. Leaves palmate...............3. Sanicula.
Umbels in dense heads. Fruit scaly. Leaves spiny.............2. Eryngium.
Umbels many-rayed. Leaves 3-foliolate ....31. Peucedanum (Imperatoria).
Bracts large, coloured. Leaves palmate ....2*. *Astrantia.*

III. Leaves simply pinnate, rarely compound at the base.
α. Fruit dorsally much compressed, winged.
Border of wings very thick. Petals pink ....33. Tordylium.
Border of wings not thickened. Petals white ....32. Heracleum.
β Fruit not much compressed, terete or didymous.
* Fruit spiny. Seed grooved ventrally......................35. Caucalis.
** Fruit glabrous. Seed flat or nearly so ventrally.
† Petals entire.
Carpophore 2-partite ....10. Carum.
Carpophore entire or 2-fid..9. Apium.
†† Petals notched or 2-lobed.
Ridges of fruit slender ; vittæ clavate, solitary, short.................11. Sison.
Ridges of fruit prominent ; vittæ several, long ....15. Pimpinella.
Ridges of fruit slender ; vittæ several, long...............13. Sium.

IV. Leaves 2-3-pinnate or 2-3-ternate.
A. Fruit terete or angled, not much dorsally compressed or flattened.
* Seed grooved ventrally or with involute margins.

M 2

† Fruit muricate or prickly

Ridges of fruit elevated, with strong spines ....35. Caucalis.
Ridges obscure ....20. Anthriscus.

†† Fruit smooth, glabrous, rarely pubescent.

‡ *Fruit short, ovoid, subglobose, or didymous.*

Flowers white. Ridges filiform; vittæ solitary ....4. Physospermum.
Flowers white. Ridges thick, waved; vittæ several ....5. Conium.
Flowers yellow. Ridges stout, even; vittæ several ....6. Smyrnium.

‡‡ *Fruit narrow, but not beaked. Flowers white.*

Ridges slender; vittæ numerous ....16. Conopodium.
Ridges obtuse; vittæ solitary ....19. Chærophyllum.
Primary ridges winged; vittæ 2-3 ....27. Meum.

‡‡‡ *Fruit narrow, beaked. Flowers white.*

Beak short; ridges sharp..17. Myrrhis.
Beak long; ridges obtuse..18. Scandix.
Beak moderate; ridges obscure ....20. Anthriscus.

** Seed flat ventrally or nearly so. (Fruit short in all.)

† Petals white, entire, with an incurved point.

Ridges slender; vittæ 1-3. Carpophore entire or 2-fid.............9. Apium.
Ridges stout, sharp; vittæ numerous ....23. Crithmum.

†† Petals yellow, entire or with an incurved point.

Ridges slender; carpophore 2-partite...................................10. Carum.
Diœcious. Vittæ within the stout ridges ....8. Trinia.
Bracteoles 0. Vittæ solitary...22. Fœniculum.
Bracteoles many. Vittæ many...26. Silaus.

††† Petals white, obcordate, notched or 2-lobed.

¶ *Calyx-teeth distinct.*

Fruit didymous; ridges depressed; vittæ solitary ....12. Cicuta.
Fruit subterete; ridges depressed; vittæ 1-2.........................21. Seseli.
Fruit globose; ridges obscure; vittæ 0 ....22*. Coriandrum.
Fruit subterete; ridges stout; vittæ solitary ....24. Œnanthe.

¶¶ *Calyx-teeth obscure or 0.*

Fruit covered with hooked bristles....................................34. Daucus.
Bracts and bracteoles linear ....10. Carum.
Bracteoles 0. Ridges slender; vittæ 0; styles slender......14. Ægopodium.
Bracts and bracteoles unilateral, deflexed ....25. Æthusa.
Bracts few or 0, bracteoles many. Ridges almost winged...28. Ligusticum.

B. Fruit much dorsally compressed, broadly winged.

* *Lateral wings of the opposite carpels closely contiguous.*

Petals white, notched. Wings thin, vittæ club-shaped ......32. Heracleum.
Petals various. Wings thin, vittæ long ....31. Peucedanum (proper).
Petals white. Wings with a thickened border..................33. Tordylium.

** *Lateral wings of opposite carpels with a space between them.*
Bracts deciduous. Fruit large...30. Angelica.
Bracts 0.   Fruit small ....29. Selinum.

1. HYDROCOT'YLE, *L.*   WHITE-ROT, PENNY-WORT.

Small perennial herbs, often creeping. *Leaves* entire, lobed, or 3–5-foliolate. *Umbels* usually simple ; bracts few or 0. *Calyx-teeth* 5 or obsolete. *Petals* not inflexed, valvate or imbricate in bud. *Fruit* much laterally compressed ; commissure narrow, carpophore undivided 2-fid or 0 ; carpels nearly orbicular, with 1–5 ridges on each side ; vittæ 0 or slender ; styles filiform, on the flattened disk.—DISTRIB. Temp. and trop. ; species 70.—ETYM. ὕδωρ aud κοτύλη, from the *cupped* peltate leaf.

H. vulga'ris, *L.* ; leaves orbicular peltate crenate, petiole hairy.
Marshes, bogs, &c., from Shetland southd. ; ascends to 1,000 ft. in the Lake district ; Ireland ; Channel Islands ; fl. May–Aug.—*Stem* filiform, white, creeping, rarely floating. *Leaves* ½–2 in. diam., 1 or more at the nodes ; petiole 2–6 in. *Umbels* axillary, shortly peduncled. *Flowers* small, capitate, pinkish-green ; bracts minute, triangular, concave. *Fruit* $\frac{1}{12}$ in. diam. ; carpels covered with resinous points, with 2 ridges on each face.—DISTRIB. Europe, W. Asia, N. Africa.

2. ERYN'GIUM, *L.*   ERYNGO.

Rigid, branched, often glaucous, perennial herbs. *Leaves* spinous-toothed, lobed or cut. *Flowers* sessile, in very dense bracteolate heads, surrounded at the base by a whorl of rigid bracts. *Calyx-tube* scaly ; teeth rigid, acute, longer than the petals. *Petals* narrow, deeply notched, point long inflexed. *Disk* concave, crenulate. *Fruit* ovoid ; commissure broad, carpophore 0 ; carpels ½-terete, primary ridges obscure 1-vittate ; styles filiform, slender, erect. *Seeds* flat or subconcave ventrally.—DISTRIB. Temp. and sub-trop. regions, chiefly S. American ; species 100.—ETYM. uncertain.—Very slender vittæ often occur in the endocarp.

E. marit'imum, *L.* ; very glaucous, radical leaves suborbicular 3-lobed spinous, cauline palmate. *Sea Holly.*
Sandy shores, from Aberdeen and Argyll southd. ; Ireland ; Channel Islands ; fl. July–Aug.—*Rootstock* creeping, stoloniferous. *Stems* 1–2 ft., stout, 3-chotomously branched. *Radical leaves* 2–5 in. diam., margins cartilaginous. *Heads* about 3 together, ½–1 in. diam., at length ovoid. *Primary involucre* of 3 bracts ; partial of 5–7 ovate spinous-serrate bracts ; bracteoles 3-fid, equalling the flowers. *Flowers* ½ in. diam., bluish-white.—DISTRIB. Shores of Atlantic, Mediterranean and Black Seas.—Roots formerly candied as a sweetmeat.

E. CAMPES'TRE, *L.* ; pale green, radical leaves pinnately 3–5-foliolate, cauline 2-pinnatifid.
Reported wild in Kent, and formerly in Suffolk ; supposed to be introduced at Plymouth, Weston-super-Mare, the Tyne, Waterford, &c. ; Channel

Islands; fl. July–Aug.—Erect, 1–2 ft., less glaucous and more branched than *E. maritimum.*—DISTRIB. From Denmark southd.; N. Africa, Caucasus, W. Siberia.

2*. *ASTRANTIA, L.*

Erect herbs. *Rootstock* short, creeping. *Leaves* palmately lobed or cut. *Umbels* simple or irregularly compound; bracts many, radiating, often coloured; flowers polygamous, males on shorter pedicels. *Calyx-limb* campanulate; teeth exceeding the petals, with long points. *Petals* notched; point long, inflexed. *Disk* cup-shaped. *Fruit* ovoid or oblong, nearly terete; commissure broad, carpophore 0; carpels dorsally compressed, primary ridges equal, with plaited wrinkled or toothed inflated ribs, furrows 1-vittate; styles filiform.—DISTRIB. Europe, W. Asia; species 4 or 5.—ETYM. ἄστρον, from the *star*-like umbels.

A. MA'JOR, *L.*; leaves with 3–7 ovate-lanceolate serrate lobes.

Naturalized in woods, Ludlow and Malvern; fl. June–July.—*Stem* 1–2 ft. *Radical leaves* 3–4 in. diam., acute, serratures bristle-pointed; petiole 4–10 in. *Bracts* ½–¾ in., ovate-lanceolate, reticulate, white beneath, above dark green tinged with pink, serrulate. *Flowers* white or pink; pedicels filiform. *Fruit* ⅛ in.; styles spreading.—DISTRIB. Mid. and S. Europe, W. Asia.

3. SANIC'ULA, *L.* SANICLE.

Slender, erect, perennial herbs. *Rootstock* stout, short, creeping. *Leaves* palmately cut. *Umbels* small, subglobose, irregularly compound; bracts leafy; bracteoles few; flowers polygamous. *Calyx-teeth* as long as the petals, subherbaceous, pungent. *Petals* minute, deeply notched, point long inflexed. *Disk* dilated. *Fruit* ovoid; covered with hooked prickles; commissure rather broad, carpophore 0; carpels ½-terete, ridges inconspicuous, furrows 1-vittate; styles filiform. *Seed* flat ventrally.—DISTRIB. N. temp. regions; species 10.—ETYM. *sano*, to heal.

S. europæ'a, *L.*; fertile flowers subsessile, males pedicelled.

Copses, &c., from Caithness southd.; ascends to 1,000 ft. in N. England; Ireland; fl. June–July.—Glabrous. *Stem* 1–2 ft., simple, almost leafless. *Radical leaves* 3 in. diam., long petioled, suborbicular, 3–5-lobed or -partite; lobes cuneate, cut, acutely serrate. *Umbel* ½–¾ in. diam., irregular, rays few; bracts 2–5, unequal, simple or pinnatifid, serrate. *Flowers* pink or white, outer male, central few proterandrous. *Fruit* ¼ in.; styles spreading.—DISTRIB. Europe, Himalaya and S. India, N. and trop. Africa.

4. PHYSOSPER'MUM, *Cusson.* BLADDER-SEED.

Erect, perennial herbs; root fusiform. *Leaves* 3-nately compound, segments cuneate. *Umbels* compound; bracts and bracteoles few, linear; flowers white. *Calyx-teeth* small or 0. *Petals* with a long inflexed point. *Disk-lobes* conical. *Fruit* didymous, bladdery, broader than long; commissure narrow, carpophore simple; carpels terete, smooth, primary ridges slender, furrows 1-vittate. *Seed* loose, concave ventrally.—DISTRIB.

Europe, W. Asia; species 2 or 3.—Etym. φύσα and σπέρμα, from the *bladdery fruits.*

P. cornubien'se, *DC.* ; branches panicled, umbels long-peduncled.

Thickets, S. Devon and Cornwall; fl. July–Aug.—Glabrous except the puberulous margins and ribs of the leaf. *Stem* 1–2 ft., erect, striate. *Radical leaves* long-petioled, flat, 2–3-ternate; segments ½–¾ in., deeply laciniate, long petioled. *Umbel-rays* 10–20, 1–3 in., suberect, furrowed.—Regarded by Nyman and others as a sub-species of the S. European and Oriental *P. aquilegifolium,* Koch.—Distrib. S. of France and Spain eastward.

5. CONI'UM, *L.* Hemlock.

Tall, glabrous, biennial herbs. *Leaves* pinnately compound. *Umbels* compound, many-rayed ; bracts and bracteoles many, small ; flowers white, polygamous. *Calyx-teeth* 0. *Petals* obtuse, or the tip shortly inflexed. *Disk* depressed. *Fruit* broadly ovoid, laterally compressed ; commissure constricted, carpophore undivided ; carpels 5-angled, primary ridges prominent obtuse, lateral distinct ; vittæ many, slender, irregular ; styles short, reflexed. *Seed* deeply grooved ventrally.—Distrib. Europe, Asia, N. Africa ; species 2.—Etym. The old Greek name.

C. macula'tum, *L.* ; stem spotted, leaf-segments pinnatifid.

Banks, roadsides, &c., from Orkney southd.; ascends to near 1,000 ft. in Yorkshire ; Ireland ; Channel Islands ; fl. June–July.—Fœtid. *Stem* 2–5 ft., stout, leafy, furrowed, purple-spotted, paniculately branched above. *Leaves* large, deltoid, finely 2-pinnate; segments ½ in., ovate oblong or deltoid, flaccid, lower petioled, ultimate serrate. *Umbels* terminal and axillary, shortly peduncled; bracts reflexed, short, unilateral ; rays 10–20, ½–1 in.; first open flowers small male, later larger female. *Fruit* ⅛ in., greenishbrown.—Distrib. Europe, N. Africa, N. and W. Asia; introd. in N. America.

6. SMYRN'IUM, *L.* Alexanders.

Stout, erect, glabrous, biennial or perennial herbs. *Radical leaves* 3-nately compound, segments broad. *Umbels* compound ; bracts and bracteoles few or 0 ; flowers yellow, polygamous. *Calyx-teeth* minute or 0. *Petals* with a short inflexed point. *Disk-lobes* conical or depressed. *Fruit* ovoid, laterally compressed or didymous ; commissure much constricted, carpophore 2-partite ; carpels subterete or angular, with 3 prominent ribs ; vittæ many ; styles short, recurved. *Seed* deeply grooved ventrally. —Distrib. Europe, W. Asia, N. Africa ; species 6 or 7.—Etym. The old Greek name.

S. Olusa'trum, *L.* ; cauline leaves petioled 3-foliolate serrate.

Waste places, especially near the sea and amongst ruins, from Aberdeen and the Clyde southd.; Ireland ; Channel Islands; a doubtful native; fl. April–June.—Shining. *Root* stout, biennial. *Stem* 1–3 ft., solid, furrowed, panicled, branches often opposite. *Petioles* large, sheathing, margins hairy. *Leaflets* 1½–2 in., broadly obovate or ovate, obtusely serrate or lobed. *Umbels* lateral and terminal, subglobose; rays few or many, long or short. *Fruit* $\frac{1}{3}$ in., dark brown ; ridges variable in prominence and number ; outer coat of the pericarp often loose ; vittæ adhering to the inner.—DISTRIB. From Holland southd., native only in Mediterranean region, *Ball.*—Formerly cultivated as a pot-herb.

7. BUPLEU'RUM, *L.* HARE'S-EAR.

Annual or perennial, glabrous herbs or shrubs. *Leaves* simple, quite entire. *Umbels* compound, many-rayed, or irregular and few-rayed ; bracts and bracteoles many and leafy, or few and small, or 0 ; flowers yellow, sessile or pedicelled. *Calyx-teeth* 0. *Petals* hooded, with an inflexed point. *Disk-lobes* tumid or dilated. *Fruit* laterally compressed ; commissure broad, carpophore 2-fid ; carpels 5-angled, primary ridges prominent or winged or 0 ; vittæ 0 or 1 or more in the furrows, continuous or interrupted ; styles short, reflexed. *Seed* subterete, flat concave or deeply grooved ventrally.—DISTRIB. Europe, temp. Asia, N. and S. Africa, N.W. America ; species 60.—ETYM. obscure.

1. **B. rotundifo'lium,** *L.* ; annual, stem fistular, leaves perfoliate.
Chalky fields, rare, E. and S. counties, from York to Kent and Somerset: fl. June–July.—Glaucous. *Stem* 8–18 in., simple or branched above, terete. *Leaves* 1–2½ in., lower oblong, upper suborbicular perfoliate. *Bracts* 0 ; bracteoles 3–5, ovate, leafy, longer than the many short rays, connate at the base, suberect in fruit. *Fruit* broad, $\frac{1}{6}$ in. ; vittæ 0 ; ridges slender.—DISTRIB. Europe, W. Asia ; introd. in N. America.

2. **B. falca'tum,** *L.* ; perennial, stem erect slender fistular, leaves oblong-lanceolate, nerves many parallel, upper broader ½-amplexicaul.
Hedgerows and fields, Surrey and Essex ; a doubtful native ; fl. Aug.—Rootstock branched. *Stems* 1½–4 ft., simple or branched above. *Radical leaves* 1–3 in., acute ; petiole ½-amplexicaul ; cauline recurved. *Umbels* very small ; bracts 2–5, short, unequal ; bracteoles 4–5, oblong, awned. *Flowers* minute. *Fruit* $\frac{1}{8}$ in., narrow ; ridges prominent ; vittæ in threes.—DISTRIB. From Belgium southd., W. Asia to India and Japan.

3. **B. tenuis'simum,** *L.* ; annual, stem solid, leaves linear-lanceolate acuminate 3-nerved, bracts subulate.
Waste places and salt marshes, local, from Durham to Devon and Kent : fl. Aug.-Sept.—*Stem* 6–18 in., erect or procumbent, flexuous, ribbed. *Leaves* ½–1 in., rigid, lowest slightly dilated upwards. *Umbels* axillary, very small, racemed or subspicate along the branches ; bracts 3–5, unequal ; bracteoles similar. *Fruit* broad, minute, granulate ; ridges prominent ; vittæ 0.— DISTRIB. Europe, N. Africa, W. Asia.

4. **B. arista'tum,** *Bartl.* ; annual, stem short solid, leaves ensiform pungent 3-5-nerved, bracts oblong awned. *B. Odontites,* Sm. not of L.
Sandy and rocky banks, Devon, very rare, E. Sussex ; Channel Islands ; fl. July.—*Stem* 2–8 in., rigid, simple, or forked ; branches stout, divaricate, ribbed. *Leaves* ½–1 in. lowest sometimes petioled. *Bracts* 3–5, concealing

the umbels, rigid; nerves strongly reticulate; margins scarious. *Fruit* minute, oblong; ridges slender, smooth; vittæ solitary.—Distrib. From France southd. and eastd.

8. TRIN'IA, *Hoffmann.* HONEWORT.

Glabrous, branched herbs. *Leaves* ·pinnately compound. *Umbels* compound, few-rayed; bracts and bracteoles 1, 2, or 0; flowers white, usually diœcious, males with narrower petals. *Calyx-teeth* 0 or small. *Petals* acute or with an inflexed point. *Disk-lobes* conical or depressed; margins undulate. *Fruit* broadly ovoid, laterally compressed or didymous; commissure narrow, carpophore 2-partite; carpels subterete or 5-angled, primary ridges subequal, thick, smooth rugose or plaited, with a large vitta in each. *Seed* terete.—Distrib. S. Europe and temp. Asia; species 8.—Etym. *Dr. Trinius,* a Russian botanist.

T. vulga'ris, *DC.*; glabrous, bracts 0 or solitary, ridges smooth. *Pimpinella dioica,* Sm. *P. glauca,* L., in part.

Limestone rocks, rare; S. Devon, N. Somerset; fl. May–June.—Glaucous. *Root* fusiform, biennial, fibrous at the top. *Stem* 3–6 in., branched from the base, solid, stout, deeply grooved; branches divaricate. *Leaves* spreading, petiole and linear segments very slender. *Male umbels* depressed; female irregular, rays longer; bract 3-cleft or 0; bracteoles 2–3, linear. *Flowers* minute. *Fruit* $\frac{1}{10}$ in., ovoid; styles slender; segments of carpophore flattened.—Distrib. From Belgium southd. to Greece.

9. A'PIUM, *L.* (and *Heloscia'dium,* Koch). CELERY.

Annual or perennial, glabrous herbs. *Leaves* pinnate or 3-nately compound. *Umbels* compound, often leaf-opposed, or in the forks; bracts few or 0; bracteoles many or 0; flowers.white. *Calyx-teeth* 0. *Petals* entire, acute, or with a short incurved point. *Disk-lobes* depressed or conical, margins entire. *Fruit* broadly ovoid, laterally compressed; commissure constricted, carpophore simple; carpels 5-angled, primary ridges equal prominent obtuse; vittæ solitary in the furrows. *Seed* subterete.—Distrib. Temp. and subtrop; species 14.—Etym. obscure.

SECTION 1. **A'pium** proper. *Bracteoles* 0. *Petals* much incurved.

1. **A. grave'olens,** *L.*; leaves pinnate or 3-foliolate. *Wild Celery.*

Marshy places, chiefly by the sea, from Perth and Argyll southd.; Ireland; Channel Islands; fl. June–Aug.—Rank-scented. *Root* fusiform, biennial. *Stem* 2 ft., erect, stout, grooved. *Leaves* 6–18 in.; leaflets $\frac{1}{2}$–$1\frac{1}{2}$ in., cuneate obovate or rhomboid, lower petioled, cut or lobed. *Umbels* shortly peduncled or sessile. *Flowers* greenish-white. *Fruit* $\frac{1}{16}$ in., roundish; styles short, recurved, divergent.—Distrib. Europe, N. Africa, W. Asia, N.W. India.

SECTION 2. **Heloscia'dium,** *Koch* (gen.). *Bracteoles* many. *Petals* nearly straight.—Aquatic or subaquatic.

2. **A. nodiflo'rum,** *Reichb.* ; prostrate or creeping, leaves pinnate or 3-foliolate, leaflets slightly lobed serrate.

Marshy places, from Isla, the Clyde, and Fife southd.; Ireland; Channel Islands; fl. July–Aug.—Perennial. *Stems* 1–3 ft., slender. *Leaflets* ½–1½ in., very variable, sessile, oblong, crenate serrate or lobulate. *Umbels* leaf-opposed, sessile or shortly peduncled; rays unequal; bracts usually 0 ; bracteoles many, oblong, scarious. *Flowers* small. *Fruit* ¹⁄₁₂ in.; styles short, divergent.—DISTRIB. From Belgium southd., W. and N. Asia, N. Africa. *A. nodiflo'rum* proper; stem decumbent, flowering branches rooting at the base only, peduncles short, bracts 0 or 1–2.—VAR. *H. re'pens,* Koch (*Sium re'pens,* Sm.) ; smaller, creeping, leaflets sharply toothed, peduncles long, bracts 2–3 unequal unilateral. Rather rare.—VAR. *ochrea'tum,* DC. ; dwarf, creeping, leaflets small obtuse, peduncles ¼–½ in., bracts 1–3 lanceolate. Barnes, Surrey.

3. **A. inunda'tum,** *Reichb.* ; decumbent or floating, submerged leaves 2–3-pinnate, leaflets capillary rarely linear, floating leaves pinnate, lower leaflets deeply 3-cleft.

Wet places, local, from Orkney southd.; ascends to 1,600 ft. in Yorkshire; Ireland; Channel Islands; fl. June–July.—Perennial, flaccid, small, straggling. *Stem* 4–10 in., stout, flexuous. *Leaflets* of upper leaves ⅛ in., cuneate, cut or lobed. *Umbels* very small, leaf-opposed, peduncles short, rays 2–4 unequal; bracts 0 ; bracteoles 4–6, lanceolate, 3-nerved. *Flowers* minute. *Petals* incurved. *Fruit* ¹⁄₁₀ in., subsessile, elliptic-oblong ; styles recurved. —DISTRIB. From Gothland southd. (excl. Greece).

10. CA'RUM, *L.* CARAWAY.

Annual or perennial, glabrous herbs. *Leaves* pinnate or decompound. *Umbels* compound, few- or many-rayed ; bracts few or 0 ; bracteoles more numerous or 0. *Flowers* white or yellow, 2-sexual or polygamous. *Calyx-teeth* minute or 0, sometimes unequal. *Petals* with an inflexed point and usually very deep notch ; of the male flowers often irregular. *Disk-lobes* conical. *Fruit* ovoid or oblong, often hispid, laterally compressed, hardly constricted at the commissure, carpophore 2-fid ; carpels 5-angled, primary ridges obtuse equal, lateral close to the commissure ; vittæ 1 (rarely 2) in the furrows. *Seeds* ½-terete.—DISTRIB. Temp. and subtrop. ; species 50. —ETYM. The old Latin name.

SECTION 1. **Ca'rum** proper. *Root* fusiform or fibrous. *Leaves* 1–2-pinnate. *Calyx-teeth* minute. *Petals* white, deeply notched.

1. **C. verticilla'tum,** *Koch ;* root of fascicled fibres, leaves linear pinnate, leaflets sessile short whorled palmately multifid, segments capillary.

Meadows in the W. counties, from Argyll to Devon and Cornwall, local ; Ireland; Channel Islands; fl. July–Aug.—*Root-fibres* 1–2 in., thickened downwards. *Stem* 1–2 ft., erect, striate. *Radical leaves* 6–12 in., sub-cylindric; leaflets curved upwards, capillary-multifid. *Umbels* regular, flat-topped ; rays 1–2 in., peduncles slender; bracts and bracteoles many, slender, short, reflexed. *Flowers* white or pink. *Fruit* ovoid; ridges strong; vittæ large ; styles recurved.—DISTRIB. W. Europe from Holland southd.

C. CAR'UI, *L.* ; root fusiform, leaves narrow triangular- or linear-oblong 2-pinnate, leaflets cut to the base into linear lobes, bracts 1 or 0, bracteoles 0. *Caraway.*

Waste places, naturalized; fl. June–July.—*Stem* 10–24 in., slender, branched, striate, fistular. *Leaves* 6–10 in.; pinnules opposite, segments acuminate. *Umbels* rather irregular, peduncles slender. *Flowers* white, outer larger irregular. *Fruit* oblong, ridges short; vittæ conspicuous; styles spreading. —DISTRIB. Europe (Arctic), N. and W. Asia, Himalaya.

SECTION 2. **Petroseli'num,** *Hoffm.* (gen.). *Root* fusiform. *Leaves* pinnate or 2–3-pinnate. *Calyx-teeth* obsolete. *Petals* white or yellowish, scarcely notched. (Intermediate between *Apium* and *Carum.*)

2. **C. sege'tum,** *Benth.* ; leaves pinnate, flowers white. *Corn Parsley.*

Hedgebanks and waste places, local, from York southd.; fl. Aug.-Sept.— Glabrous, annual.—*Stem* 2–3 ft., erect, branched, terete, striate, solid. *Leaves* 4–6 in., oblong ; leaflets $\frac{1}{4}$–$1\frac{1}{2}$ in., subsessile, lobed or pinnatifid, seg-ments crenate. *Umbels* small, irregular, rays very unequal, outer $\frac{1}{2}$–1 in. ; bracts and bracteoles 3–5, linear or subulate. *Flowers* minute. *Fruit* $\frac{1}{3}$ in., ovoid ; styles very short, erect.—DISTRIB. W. Europe, from Holland to Portugal.

C. PETROSELI'NUM, *Benth.* ; leaves 3-pinnate, flowers yellow. *Petroseli-num sativum,* Hoffm. *Common Parsley.*

Waste places ; a garden escape; fl. June–Aug.—Glabrous, shining, biennial. *Stem* 1–2 ft., erect, much-branched, terete, striate, solid. *Leaves* deltoid; leaflets many, $\frac{1}{2}$–1 in., close-set, broadly ovate, 3-cleft ; segments cuneate, crenate, of upper leaves few narrow. *Umbels* regular, flat-topped; rays many, 1–2 in. ; bracts 2–3, often divided ; bracteoles many. *Flowers* minute. *Fruit* $\frac{1}{10}$ in., ovoid, green ; styles slender, reflexed.—DISTRIB. Only known as a cultivated plant or an escape.

SECTION 3. **Bu'nium,** *L.* (gen.). *Root* a solitary tuber. *Leaves* 2–3-pinnate. *Calyx-teeth* minute. *Petals* white, deeply notched.

3. **C. Bulbocas'tanum,** *Koch;* leaves broadly triangular 3-pinnate, primary segments petioled, leaflets cut into few slender lobes. *Bulbocasta-num Linnæi,* Schur.

Chalky fields, rare, Herts, Bucks, Bedford, Cambridge ; fl. June–July.—*Root* globose, as large as a chestnut, black. *Stem* erect, striate, much-branched, and petioles flexuous at the base. *Leaves* 4–6 in., pinnules $\frac{1}{2}$–$\frac{2}{3}$ in., tips callous. *Umbels,* many-rayed ; bracts and bracteoles small, narrow ; pedun-cles stout, grooved, angular. *Flowers* white, outer rather larger. *Fruit* nearly $\frac{1}{4}$ in.; ridges stout; vittæ compressed; styles short, recurved.— DISTRIB. From Belgium southd., N. Africa, Himalaya, Siberia.—Pigs feed on the tubers in Hertfordshire, &c.

11. SI SON, *L.*

Characters of *Ca'rum,* but vittæ very short, often obscure, occupying only the upper half of the fruit.—DISTRIB. West Europe, Italy, and the East ; 2 species.—ETYM. unknown.

S. Amo'mum, *L.* ; leaves pinnate or 2-pinnate below, upper smaller 3-lobed toothed or entire.

Moist places, hedgebanks, &c., from York and Chester southd., rare in N. England; (a native? *Wats.*); Channel Islands; fl. Aug.–Sept.—Biennial, glabrous, nauseous-smelling. *Root* fusiform. *Stem* 2–3 ft., erect, branched, slender, leafy, solid. *Leaves* 6–12 in., deltoid-oblong, leaflets 1–3 in., shortly petioled, linear-oblong or ovate, base cuneate. *Umbels* terminal and axillary, compound ; rays few, slender, unequal; bracts and bracteoles 2–4, short, subulate, rarely 0. *Flowers* minute, white. *Petals* broadly obcordate, notch deep, point long inflexed. *Fruit* ovoid or subglobose; ridges strong; vittæ very short, narrowed upwards; styles short, recurved; disk-lobes thick depressed. –DISTRIB. of the genus.—Closely resembles *Ca'rum sege'tum.*

12. CICU'TA, *L.* WATER-HEMLOCK, COWBANE.

Tall, perennial, glabrous herbs. *Leaves* pinnate or decompound. *Umbels* compound, many-rayed ; bracts few or 0 ; bracteoles many, small ; flowers white. *Calyx-teeth* acute. *Petals* with an inflexed point. *Disk-lobes* depressed, entire. *Fruit* orbicular or broadly ovoid, constricted at the commissure, didymous, carpophore 2-partite ; carpels slightly compressed, primary ridges thick broad flat ; vittæ solitary in the furrows. *Seeds* slightly convex ventrally.—DISTRIB. Marshes of the N. hemisphere ; species 3.—ETYM. A Latin name of the Hemlock.

C. viro'sa, *L.* ; root fibrous, leaflets lanceolate doubly serrate.

Watery places, from Dumbarton and Forfar to Suffolk and Somerset; Ireland ; Channel Islands ; fl. July–Aug.—*Rootstock* short, stout, hollow, septate. *Stem* 2–4 ft., stout, leafy, furrowed. *Leaves* large, deltoid, 2–3-pinnate; petiole stout ; leaflets 2–4 in., oblique. *Umbels* terminal and leaf-opposed, 3–5 in. 'diam., long-peduncled, flat-topped, rays long slender ; bracts 0; bracteoles many, short, slender. *Flowers* minute. *Calyx-teeth* ovate. *Fruit* $\frac{1}{12}$ in. broad, broader than long; styles slender, recurved.—DISTRIB. N. and Mid. Europe (Arctic), N. Asia, Himalaya.

13. SI'UM, *L.* WATER-PARSNIP.

Glabrous herbs. *Leaves* pinnate ; leaflets toothed. *Umbels* compound, terminal or lateral ; bracts and bracteoles many ; flowers white. *Calyx-teeth* acute. *Petals* with an inflexed point. *Disk-lobes* thick, conical or depressed. *Fruit* ovoid or oblong, laterally compressed or constricted at the commissure, carpophore undivided ; carpels 5-angled, primary ridges equal prominent obtuse or thickened, lateral next the commissure ; vittæ many, in the furrows. *Seed* subterete.—DISTRIB. N. temp. regions, S. Africa ; species 4.—ETYM. unknown.

1. **S. latifo'lium,** *L.* ; leaflets regularly serrate, umbels terminal.

Watery places, from Stirling and Ayr to Kent and Devon ; Ireland; Channel Islands ; fl. July–Aug.—*Rootstock* short, stoloniferous. *Stem* 5–6 ft., erect, stout, fistular, grooved, branched above. *Leaves* large ; leaflets 4–6, 2–6 in., sessile, linear- or oblong-lanceolate ; submerged sometimes pinnatifid. *Umbels* large, flat-topped, rays many ; bracts and bracteoles often foliaceous, large. *Flowers* small, outer rather larger. *Fruit* ⅙-in., broadly ovoid, ridges prominent ; styles rather slender.—DISTRIB. Europe, N.W. Asia, N.W. America.

2. **S. angustifo'lium,** *L.* ; leaflets of radical leaves regularly of stem-leaves very irregularly serrate, umbels leaf-opposed. *S. erectum,* Huds.

Wet places, from Elgin southd.; Wigton only in W. Scotland; Ireland; Channel Islands ; fl. summer.—*Rootstock* creeping, stoloniferous, leafing at the nodes. *Stem* 1–3 ft., leafy. *Leaves* 4–8 in.; leaflets of lower leaves 5–10, 1–2 in., sessile, ovate-oblong ; of cauline leaves fewer, smaller. *Umbels* with few and unequal rays ; bracts irregularly cut. *Fruit* shorter than in *S. latifolium,* with more immersed vittæ and conical disk-lobes.—DISTRIB. Europe.—*S. erectum,* Huds., is a rather earlier name, but less appropriate.

14. ÆGOPO'DIUM, *L.* GOAT-, GOUT-, or BISHOP'S-WEED.

Stem stout, glabrous. *Rootstock* creeping. *Leaves* 2–3-ternate ; leaflets broad. *Umbels* compound, many-rayed ; bracts and bracteoles few or 0 ; flowers white. *Calyx-teeth* 0. *Petals* broad, unequal, point inflexed. *Disk-lobes* tumid ; styles slender, reflexed. *Fruit* ovoid, laterally compressed, carpophore 2-fid ; carpels 5-angled, primary ridges slender equal distant ; vittæ 0. *Seed* subterete.—DISTRIB. N. and Mid. Europe, W. Asia.—ETYM. αἴξ and πούς, from the likeness of the leaf to a *goat's foot.*

Æ. Podagra'ria, *L.* ; leaves deltoid. *Herb Gerard.*

Waste places near buildings or gardens, from Elgin southd.; Ireland; Channel Islands; a doubtful native; fl. June–Aug.—Glabrous. *Rootstock* white, pungent, aromatic. *Stem* 1–2 ft., fistular, grooved, branched above. *Leaves* 4–5 in., uppermost opposite; leaflets sessile, obliquely lanceolate or ovate-acuminate, irregularly serrate. *Umbels* terminal. *Flowers* small. *Fruit* ⅙ in., narrow-ovoid.

15. PIMPINEL'LA, *L.* BURNET-SAXIFRAGE.

Perennial, rarely annual herbs. *Leaves* pinnate or 3-nately compound. *Umbels* compound ; bracts 0 ; bracteoles few or 0 ; flowers white or yellow. *Calyx-teeth* small or 0. *Petals* deeply notched, point long inflexed. *Disk-lobes* thick, conical. *Fruit* ovoid or oblong, laterally compressed, constricted at the broad commissure, carpophore 2-fid ; carpels 5-angled, primary ridges equal slender ; vittæ many in the furrows ; styles short or long. *Seed* subterete, nearly flat ventrally, usually free from the pericarp.—DISTRIB. N. temp. regions, S. Africa, S. America ; species 70.—ETYM. *bipennula,* from the 2-pinnate leaves.

1. **P. Saxifraga,** *L.* ; stem terete, radical leaves pinnate, leaflets sub-orbicular, cauline 2-pinnate.

Dry pastures from Sutherland southd.; ascends to 1,800 ft. in Yorkshire; Ireland; Channel Islands; fl. July.—Perennial, glabrous or pubescent. *Rootstock* slender, hot, acrid. *Stem* 1–3 ft., slender, furrowed, branched. *Leaflets* 4–8 pair, very variable, serrate lobed or almost pinnatifid or finely cut (*P. dissecta,* Retz.); lobes of cauline much narrower. *Umbels* flat-topped. *Flowers* small, white. *Fruit* ⅛in., glabrous, broadly ovoid ; styles short, reflexed.—DISTRIB. Europe (Arctic), N. and W. Asia, Himalaya.

2. **P. ma'jor,** *Huds.* ; stem angular, leaves all pinnate, leaflets of radical ovate subcordate, of cauline narrower. *P. mag'na,* L.

Bushy waste places, local, from Perth southd. (E. Scotland only); rare in Ireland; fl. July–Aug.—Much larger than *P. Saxifraga,* but similar, 3–4 ft.; leaflets often 1–2 in., membranous and broad ; styles longer and more slender ; outer flowers 2-sexual, inner male.—DISTRIB. Chiefly W. and Mid. Europe, Caucasus.

16. CONOPO'DIUM, *Koch.* EARTH-NUT, PIG-NUT.[1]

Glabrous or hairy herbs. *Rootstock* tuberous. *Leaves* 3-nately divided. *Umbels* compound, many-rayed ; bracts and bracteoles 0 or membranous ; flowers white, polygamous, outer sometimes radiating. *Calyx-teeth* obsolete. *Petals* of outer flowers often irregular, 2-fid, with an inflexed point. *Disk-lobes* conical or depressed. *Fruit* ovoid or oblong, often shortly beaked ; commissure constricted, carpophore 2-fid ; carpels subterete, primary ridges slender ; vittæ several in the furrows, often obscure or inter-rupted. *Seed* deeply groved ventrally.—DISTRIB. Europe, N. Africa, temp. Asia ; species 8.—ETYM. κώνος and πούς, from the *conical* disk-lobes.

C. denuda'tum, *Koch ;* leaf-lobes linear, ·bracts and bracteoles 0. *Bunium flexuosum,* With. *B. denudatum,* DC. *Carum flexuosum,* Fries.

Woods and fields, N. to Shetland ; Ireland ; Channel Islands; fl. summer.— Glabrous. *Rootstock* size of a chestnut, brown. *Stem* 2–3 ft., slender, terete, flexuous. *Leaves* 3-ternate, broadly deltoid ; petiole slender ; seg-ments pinnatifid, the central lobes largest. *Umbels* terminal, drooping when young, 6–10-rayed. *Flowers* small. *Fruit* ⅓ in., narrow-ovoid, ridges obscure ; styles short, erect.—DISTRIB. W. Europe.—Very similar to *Carum Bulbocastanum.*

17. MYR'RHIS, *Scop.* CICELY.

Perennial, tomentose herbs. *Leaves* decompound. *Umbels* compound, many-rayed ; bracts few or 0 ; bracteoles many, membranous ; flowers white, polygamous. *Calyx-teeth* minute or 0. *Petals* with a very short inflexed point. *Disk-lobes* tumid. *Fruit* much elongate, beaked, com-missure broad, carpophore 2-fid ; back of carpels very convex, primary ridges equal, hollow, often rough, prominent ; vittæ in the furrows

solitary, slender, or 0. *Seed* concave or deeply grooved ventrally.—
DISTRIB. Mts. of Europe and temp. S. America; species 2.—ETYM.
The old Greek name.

M. odora'ta, *Scop.* ; leaves whitish beneath, bracteoles lanceolate.
Pastures, usually near houses, from S. Wales and Lincoln to Caithness;
ascends to 1,200 ft. in Derby; not indigenous in Ireland; (a denizen or
alien, *Wats.*); fl. May–June.—Sparingly and finely hairy. *Root* fleshy,
fusiform. *Stem* 2-3 ft., leafy, terete, fistular, grooved, branched above.
Leaves deltoid, 3-pinnate; leaflets pinnatifid, lobes serrate; sheaths large.
Umbels terminal; bracteoles membranous, awned. *Flowers* small, outer
only fertile, latest male only. *Fruit* ¾-1 in., linear, dark brown, ridges
often scabrid; styles very slender, diverging.—DISTRIB. From France
southd. and eastd. to Caucasus.—Aromatic and stimulant; once cultivated
as a pot-herb, still used in salads in Italy.

18. SCAN'DIX, *L.* SHEPHERD'S NEEDLE.

Annuals. *Leaves* pinnately decompound; segments small. *Umbels*
simple or compound; bracts 1 or 0; bracteoles entire or cut; flowers
white, polygamous, outer often radiating. *Calyx-teeth* minute or 0.
Petals often unequal, point short inflexed or 0. *Disk* dilated, undulated.
Fruit slender, subcylindric, produced into a long beak, carpophore undivided or 2-fid; carpels subterete, primary ridges broad or filiform,
secondary 0; vittæ solitary in the furrows, often obscure. *Seed* deeply
furrowed ventrally.—DISTRIB. Europe, N. Africa, temp. Asia; species 8
or 10.—ETYM. The Greek name for a Chervil.

S. Pecten-Ven'eris, *L.* ; fruit ciliate rough dorsally compressed.
A cornfield weed from Caithness southd.; ascends to 1,000 ft. in Yorkshire;
Ireland; Channel Islands; (a colonist, *Wats.*); fl. June–Sept.—Branched
from the base, pubescent with spreading hairs, branches 6–18 in., rarely more.
Leaves oblong, 2-3-pinnate, segments very slender. *Umbels* terminal and
lateral; rays 1-2; bracteoles many, green, sometimes leafy at the point.
Flowers very irregular. *Fruit* 1-3 in., very slender, scabrid; styles very
short.—DISTRIB. Europe, N. Africa, W. Asia to N.W. India.

19. CHÆROPHYL'LUM, *L.* CHERVIL.

Herbs, often hairy. *Leaves* pinnately, rarely 3-nately decompound.
Umbels compound, many-rayed; bracts 1-2 or 0; bracteoles many;
flowers white, rarely yellow, often polygamous. *Calyx-teeth* subulate
or 0. *Petals* with a long or short inflexed point. *Disk-lobes* small. *Fruit*
oblong or linear, not beaked, laterally compressed, commissure constricted, carpophore undivided or 2-fid; carpels subterete, primary ridges
equal obtuse; vittæ solitary in the furrows. *Seed* deeply grooved ventrally.
—DISTRIB. N. temp. regions, species 30.—ETYM. χαίρω and φύλλον, from
the agreeable odour of the *leaf.*

C. tem'ulum, *L.*; stem swollen below the nodes purple-spotted,
fruit glabrous, bracteoles reflexed. *C. temulentum,* Sm.

Fields and waste places from Caithness southd.; ascends to 1,200 ft. in Yorkshire; rare in Ireland; Channel Islands; fl. June–July.—Perennial, laxly
hairy. *Stem* 1–3 ft., slender, solid, grooved, leafy, branched. *Leaves* deltoid,
2-pinnate, petioles and peduncles very slender; leaflets ovate, membranous,
pinnatifid, crenate. *Umbels* lateral, drooping when young; rays unequal,
slender; bracteoles small, oblong-lanceolate. *Flowers* small, white. *Fruit*
¼ in., ovoid, narrowed upward, not beaked; styles very short, spreading.—
DISTRIB. Europe, Caucasus, N. Africa.

20. ANTHRIS'CUS, *Hoffm.* BEAKED-PARSLEY.

Annual or biennial, hairy herbs. *Leaves* deltoid, pinnately or 3-nately
decompound. *Umbels* compound; nodding when young; bracts 1, 2, or 0;
bracteoles many, entire; flowers white, often polygamous. *Calyx-teeth*
minute or 0. *Petals* with an inflexed point. *Disk-lobes* conical or depressed. *Fruit* ovoid or oblong, shortly beaked, commissure constricted,
carpophore undivided or 2-fid; carpels sub- or ½-terete, primary ridges
confined to the smooth or rough upper part; vittæ very slender, solitary
in the furrows, or 0. *Seed* deeply grooved ventrally.—DISTRIB. Temp.
Europe, Asia, N. Africa, N.W. America; species 10. ETYM. diminutive
of ἀνθηρός, *small-flowering.*

1. **A. vulga'ris,** *Pers.*; stem glabrous, umbels peduncled leaf-opposed,
fruit muricate. *Scandix Anthriscus,* L.; *Chærophyllum Anthriscus,* Lamk.

Hedgebanks and roadsides, N. to Shetland; rather rare in Ireland; Channel
Islands; fl. May–June.—Sparingly hairy. *Stem* 2–3 ft., branched, leafy,
fistular, swollen below the nodes. *Leaves* 3-pinnate; leaflets ovate, pinnatifid; segments short, obtuse. *Umbels* of unequal rays; bracts 0;
bracteoles short, oblong. *Flowers* minute. *Fruit* ⅙ in., ovoid, muricate,
beak short glabrous, pedicel with a ring of hairs at the tip; styles very
short.—DISTRIB. Europe, N. Africa, Siberia, W. Asia.—Formerly cultivated
as a pot-herb.

2. **A. sylves'tris,** *Hoffm.*; stem hairy below, umbels peduncled terminal, fruit glabrous. *Chærophyllum sylvestre,* L.

Hedgebanks and woods, N. to Shetland; Ireland; Channel Islands; fl. April–
June.—Hairy. *Stem* 2–3 ft., stout, erect, leafy, fistular, furrowed. *Leaves*
2–3-pinnate; leaflets pinnatifid, ovate, coarsely serrate. *Bracts* 0; bracteoles
oblong-lanceolate, ciliate, green, spreading or reflexed, often pink. *Flowers*
white. *Fruit* ¼–⅓ in.—DISTRIB. Europe (Arctic), Caucasus, N. Asia,
N. Africa.

A. CEREFO'LIUM, *Hoffm.*; stem hairy above the nodes, umbels sessile
lateral and leaf-opposed, fruit glabrous. *Scandix,* L.; *Chærophyllum
sativum,* Gærtn. *Chervil.*

Waste places, rare; always an escape from cultivation; fl. May–July.—Habit
of *A. vulgaris,* but stouter, leaflets broader and flowers larger. *Fruit* ⅓ in.,
very narrow.—DISTRIB. E. Europe, W. Siberia, W. Asia.—*Root* reputed
poisonous.

21. SES'ELI, *L.*

Biennial or perennial, erect, branched herbs. *Leaves* 2–3-pinnate or decompound. *Umbels* compound ; bracts many, few, or 0 ; bracteoles many, entire ; flowers white. *Calyx-teeth* prominent or minute. *Petals* notched, point long inflexed. *Disk* depressed or conic, undulate or crenate. *Styles* very short. *Fruit* ovoid or oblong, subterete, commissure broad, carpophore 2-partite ; carpels dorsally compressed, primary ridges prominent ; vittæ 1 rarely 2 in the furrows. *Seed* flat ventrally.— DISTRIB. Europe, N. Asia, N. Africa, Australia ; species 40.—ETYM. A Greek name.

S. Libano'tis, *Koch ;* glabrous or slightly pubescent, leaves 2-pinnate, leaflets pinnatifid. *Athamanta,* L. ; *Libanotis montana,* All.

Chalk hills, Sussex, Herts, and Cambridge ; fl. July–Aug.—*Rootstock* perennial, crowned with fibres. *Stem* 1–2 ft., stout, erect, furrowed, solid, sparingly branched. *Leaflets* sessile, variable, ovate ; petiole short. *Umbels* rounded in flower ; rays many, pubescent ; bracts and bracteoles many, subequal, subulate, ciliate, reflexed. *Flowers* small ; calyx-teeth subulate, deciduous. *Fruit* $\frac{1}{10}$ in., broadly ovoid, pubescent ; styles slender, recurved.—DISTRIB. Europe (Arctic), excl. Spain, Greece, and Turkey ; W. Asia.

22. FŒNIC'ULUM, *Adanson.* FENNEL.

Tall, glabrous, biennial or perennial herbs. *Leaves* pinnately decompound, segments slender. *Umbels* compound ; bracts and bracteoles 0 ; flowers yellow. *Calyx-teeth* 0. *Petals* with a short obtuse point. *Disk-lobes* large, conical, entire. *Styles* short. *Fruit* ovoid or oblong, subterete, commissure broad, carpophore 2-partite ; carpels $\frac{1}{2}$-terete, primary ridges stout ; vittæ solitary in the furrows. *Seed* furrowed, flat or subconcave ventrally.—DISTRIB. S. Europe, E. Asia, N. Africa ; species 4.—ETYM. The old Latin name.

F. officina'le, *All.* ; leaves shortly petioled, segments slender. *F. vulgare,* Gærtn.

Sea-cliffs, perhaps native from N. Wales and Norfolk to Cornwall and Kent, not N. of it, nor in Ireland ; Channel Islands ; fl. July–Aug.—Perennial. *Stem* 2–3 ft., terete, striate, polished, almost solid. *Leaves* much divided ; segments very many, linear. *Umbels* large, glaucous ; rays very many. *Flowers* small. *Fruit* $\frac{1}{6}$–$\frac{1}{4}$ in. long, ovoid.—DISTRIB. From Belgium southd. N. Africa, W. Asia to India.

22*. CORIAN'DRUM, *L.* CORIANDER.

An annual, slender, branched, glabrous herb. *Leaves* pinnately decompound. *Umbels* compound ; rays few ; bracts 0 ; bracteoles few, filiform ; flowers white or pink, outer often irregular. *Calyx-teeth* acute. *Petals* 2-lobed, point inflexed. *Disk-lobes* conical. *Fruit* subglobose or ovoid, carpophore 2-fid ; carpels $\frac{1}{2}$-terete, ridges depressed slender, secondary broadest ; vittæ obscure, solitary under each secondary ridge. *Seed* globose, dorsally compressed, top and base incurved.—DISTRIB. S. Europe, N. Africa, W. Asia ; species 2.—ETYM. κόρις, from the *bug*-like smell.

N

C. SATI'VUM, *L.* ; leaflets of lower leaves ovate lobed and crenate.

Waste places in S. and E. of England ; an escape from cultivation ; fl. June.—
Stem 1-2 ft., slender, erect, fistular. *Leaves* membranous, lowest 1-2- upper
2-3-pinnate with narrow leaflets. *Umbel* peduncled, rays 5-10 ; bracteoles
short, linear, acute. *Flowers* small, very irregular. *Fruit* ½ in. diam., subglobose ; carpels cohering ; styles slender, flexuous.—DISTRIB. S.E. Europe,
W. Asia.—Three carpels and styles occur. Fœtid of bugs.

23. CRITH'MUM, *L.* SAMPHIRE.

A fleshy, glabrous, much-branched herb, woody at the base. *Leaves*
3-nately compound ; segments quite entire. *Umbels* compound, many-
rayed ; bracts and bracteoles many, short. *Calyx-teeth* 0. *Petals* minute,
broad, fugacious, point long inflexed. *Disk-lobes* thick, depressed or
subconic. *Fruit* ovoid-oblong, terete, commissure broad, carpophore
2-partite, outer layer corky loose ; carpels ½-terete, primary ridges thick,
acute ; vittæ many ; styles short. *Seed* flat ventrally.—DISTRIB. Coasts
of N. Atlantic, Mediterranean and Black Seas.—ETYM. obscure.

C. marit'imum, *L.* ; leaflets linear lax fleshy.

Maritime rocks from Ayr southd.;'Ireland ; Channel Islands ; fl. June-Aug.
Stem 6-10 in., ascending, flexuous, solid, striate. *Leaves* deltoid ; leaflets
few, 1-2 in., terete, subulate or subfusiform ; petiole short, sheaths long
adnate membranous. *Umbels* flat-topped ; peduncle stout fleshy ; bracts
and bracteoles acute, spreading ; flowers small, white. *Fruit* ¼ in., oblong,
dark green or purplish.—Yields the well-known pickled condiment.

24. ŒNAN'THE, *L.* WATER DROPWORT.

Glabrous herbs, often aquatic. *Roots* fibrous or tuberous. *Leaves* 1-2-3-
pinnate, rarely reduced to a fistular petiole. *Umbels* compound ; bracts
or bracteoles many few or 0 ; flowers white, often polygamous and outer
rayed. *Calyx-teeth* acute. *Petals* notched or 2-lobed, point long inflexed.
Disk-lobes conical. *Fruit* ovoid cylindric or globose, commissure
broad, carpophore 0 ; carpels ½-terete, 2 lateral primary ridges grooved
or much thickened, sometimes obscure ; vittæ solitary in each furrow.
Seed flat or convex ventrally.—DISTRIB. N. temp. regions, S. Africa,
Australia ; species 20.—ETYM. οἶνος and ἄνθος, from the *vinous* scent of
the *flowers.*

* *Root-fibres many, fleshy. Umbels terminal or terminal and lateral, peduncled ;
outer flowers of each partial umbel often irregular and male.*

1. Œ. fistulo'sa, *L.* ; leaves pinnate, stem and petioles terete swollen
fistular, fruit narrow obconic angular.

itches and marshes from Ayr and Berwick southd.; Ireland ; Channel
Islands ; fl. July-Sept.—*Roots* burrowing deep. *Stem* 2-3 ft., stoloniferous,
and with whorls of slender root-fibres below, thin-walled, nodes constricted.
Leaves long-petioled ; segments few, narrow, distant. *Peduncles* stout
fistular ; rays short, few ; bracts 0 ; partial umbels ½ in. diam., spherical in

fruit. *Fruits* ¼ in., crowded, angular ; styles long, erect, spinescent ; carpels cohering ; pedicel not thickened at the top.—DISTRIB. Europe, N. Africa, W. Asia.

2. **Œ. pimpinel'loides,** *L.* ; root-fibres usually tuberous beyond the middle, leaves 2-pinnate, segments broad short entire or acutely cut, fruit cylindric grooved and ribbed.

Meadows and banks, rare ; Worcester and Essex to Sussex and Cornwall ; Channel Islands ; fl. June–Aug.—*Root-fibres* slender, their tuber ½ in. or less. *Stem* 1–3 ft., erect, furrowed. *Lower leaves* with broad small segments, upper with few long ones, or reduced to petioles. *Umbels* 6–12-rayed, flat-topped ; bracts 1–8 ; partial umbels crowded ; bracteoles subulate. *Fruit* ₁₀-⅛ in. ; pedicel short, stout, top much thickened ; styles erect, rigid.— DISTRIB. Europe from Belgium southd., N. Africa, Asia Minor.

3. **Œ. Lachena'lii,** *Gmel.* ; root-fibres usually cylindric, leaves 2-pinnate, segments obtusely-lobed, fruit oblong. *Œ. pimpinelloides,* Sm.

Marshes fresh and salt, from Argyll and Haddington southd. ; Ireland ; fl. July–Sept.—Very similar to *Œ. pimpinelloides,* but root-fibres never tuberous ; root-leaves soon withering ; partial umbels not crowded ; fruit ₁₀ in., much broader, rounded at the top ; styles shorter and slender ; pedicel very short, not thickened at the top.—DISTRIB. From Denmark southd., E. to the Caspian.

4. **Œ. peucedanifo'lia,** *Poll.* ; root-fibres usually fusiform, leaves 2-pinnate, segments cut into narrow acute lobes, fruit subcylindric thickened at the base. *Œ. silaifo'lia,* Syme, not Bieb. ; *Œ. Smithii,* Watson.

Moist meadows and ditches, from Notts, Worcester, and Norfolk to Dorset and Kent ; fl. June–July.—Very near *Œ. pimpinelloides,* but larger, stouter ; root-fibres rarely tuberous in the middle ; rays fewer, longer, stouter in fruit ; partial umbels not crowded ; styles short, erect, rigid.—DISTRIB. S. Europe to the Caspian.

5. **Œ. croca'ta,** *L.* ; root-fibres large fusiform, leaves large deltoid 3–4-pinnate, segments cuneate 2–3-lobed, fruit narrow oblong subcylindric.

Marshes and ditches, from Ross southd. ; Ireland ; Channel Islands ; fl. July. —*Root-fibres* as thick as the thumb, juice yellow or colourless. *Stem* 2–5 ft., stout, branched, grooved, fistular. *Petioles* large, sheathing throughout. *Umbels* many, rays long ; bracts and bracteoles 0 or many. *Fruit* ⅛–¼ in. ; styles erect, rigid ; top of pedicel not thickened.—DISTRIB. From France to Spain and Italy.—Poisonous, often mistaken for celery.

** *Aquatics. Root simple, fusiform, with many slender fibres. Umbels lateral or leaf-opposed, subsessile. Flowers all 2-sexual.*

6. **Œ. Phellan'drium,** Lamk. ; erect floating or ascending, leaves 2–3-pinnate finely cut, segments pinnatifid, fruit terete narrow-oblong or ovoid twice or thrice as long as the styles. *Phellandrium aquaticum,* L.

Ponds and ditches, from Haddington southd. ; rare in Scotland ; Ireland ; Channel Islands ; fl. July–Sept.—*Stem* 1–4 ft., very stout. *Leaves* sometimes submerged with capillary segments ; emersed with broad small obtuse seg-ments. *Umbels* 7–10-rayed ; bracts 0 ; bracteoles many ; outer flowers

slightly irregular. *Fruit* variable, ⅛–⅙ in.; styles slender, flexuous; pedicel not thickened at the top.—DISTRIB. Europe, to the Caspian, Siberia.
Œ. PHELLAN'DRIUM proper; erect, leaves 3-pinnate, segments of submerged leaves capillary, fruit twice as long as its styles.
Sub-sp. Œ. FLUVIAT'ILIS, *Colem.* ; ascending, leaves 2-pinnate, segments of submerged leaves obcuneate, fruit three times as long as its styles.—S. half of England; Kildare Canal, Ireland.

25. ÆTHU'SA, *L.* FOOL'S PARSLEY.

An annual, leafy, glabrous herb. *Leaves* 3-nately pinnate. *Umbels* compound, terminal and leaf-opposed ; bracts 1 or 0 ; bracteoles 1–5, deflexed, on the outer side of the umbel ; flowers white, outer often rayed. *Calyx-teeth* small or 0. *Petals* notched, point inflexed. *Disk-lobes* broad, depressed. *Fruit* broadly ovoid, subterete, carpophore slender 2-partite ; carpels dorsally compressed, primary ridges very thick keeled, or the lateral narrowly winged ; vittæ in the furrows solitary ; styles very short. *Seed* flattish ventrally.—DISTRIB. Europe, Siberia; introd. in N. America. —ETYM. αἴθω, because of its supposed *burning* qualities.

Æ. Cyna'pium, *L.* ; leaves deltoid, leaflets pinnatifid.

A weed in cultivated grounds, from Elgin and the Clyde southd.; Ireland : Channel Islands ; fl. July–Aug.—*Root* fusiform. *Stem* 1–2 ft., corymbosely branched, terete, striate, fistular. *Leaves* 6 in.; segments ½–1 in., membranous, cuneate at the base, lobes acute; petiole slender. *Umbels* small ; rays spreading, irregular; bracteoles 3–5, slender. *Flowers* irregular, small. *Fruit* 1/10 in., green.—Odour nauseous.

26. SILA'US, *Besser.* PEPPER SAXIFRAGE.

Perennial, glabrous herbs. *Leaves* pinnately decompound ; segments slender. *Umbels* compound ; bracts 1, 2, or 0 ; bracteoles many, small ; flowers yellowish. *Calyx-teeth* 0. *Petals* with an incurved tip, base broad truncate. *Disk-lobes* depressed, crenate. *Fruit* ovoid or oblong, subterete, commissure broad, carpophore 2-partite ; carpels ½-terete, ridges obtusely winged, vittæ obscure ; styles short, recurved. *Seed* flattish ventrally.—DISTRIB. Europe, Siberia ; species 2.—ETYM. unknown.

S. praten'sis, *Besser ;* leaflets linear-lanceolate entire or 3-lobed.

Meadows and commons, from Fife to Kent and Devon; E. Scotland only; Ireland, rare ; fl. July–Sept.—*Rootstock* elongate. *Stem* 1–3 ft., angular, grooved, solid, leafless above. *Leaves* 1–3-pinnate ; leaflets few, ¼–¾ in. *Umbel-rays* 1–2 in., few or many, incurved ; bracteoles short, margins scarious. *Flowers* small. *Fruit* ⅛ in., dark brown.—DISTRIB. Finland to Hungary.

27. ME'UM, *Jacquin.* MEU, BALD-MONEY, SPIGNEL.

A perennial, glabrous, very aromatic, tufted herb. *Leaves* mostly radical, pinnately decompound ; segments setaceous, densely crowded.

Umbels compound ; bracts linear, 1–3, or 0 ; bracteoles 4–8, small ; flowers white or purplish. *Calyx-teeth* obsolete. *Petals* acute, narrowed to the base, sometimes with a short inflexed point. *Disk-lobes* depressed, margins entire. *Fruit* ovoid-oblong, subterete, commissure broad, carpophore 2-partite ; carpels ½-terete, primary ridges acute ; vittæ many ; styles very short. *Seeds* concave ventrally.—DISTRIB. Mts. of W. Europe. —ETYM. Perhaps the Greek μέον.

M. athamant'icum, *Jacq.* ; stem subsimple, leaves oblong.

Alpine pastures, from Wales and York to Aberdeen and Argyll ; ascends to near 1,400 ft. in the Highlands ; fl. June–July.—*Rootstock* elongate, crowned with fibres. *Stem* 6–18 in. *Leaf-segments* multifid, spreading in all directions ; petiole as long as the blade. *Umbels* many-rayed ; bracts few ; bracteoles membranous, subunilateral ; some flowers often male only. *Fruit* brown, ⅓ in.—Rootstock eaten in Scotland.

28. LIGUS'TICUM, *L.* LOVAGE.

Perennial, glabrous herbs. *Leaves* 1–3-ternately pinnate. *Umbels* compound, many-rayed ; bracts many, few, or 0 ; bracteoles many ; flowers white pink or yellow. *Calyx-lobes* small or 0. *Petals* notched, point long inflexed. *Disk-lobes* conical, thick. *Fruit* ovoid or oblong, subterete or dorsally compressed, commissure broad, carpophore 2-partite ; primary ridges prominent, acute or winged, lateral often broadest ; vittæ many, slender, or obscure. *Seed* flat, or sub-concave ventrally.—DISTRIB. N. temp. regions ; species 20.—ETYM. *Liguria,* where a species abounds.

L. scot'icum, *L.* ; leaves 2-ternately pinnate. *Haloscias,* Fries.

Rocky coasts, local, Northumberland and all Scotland to Shetland ; N. Ireland ; fl. July.—Dark green, shining. *Rootstock* stout, branched. *Stem* 1–3 ft., erect, sparingly branched, grooved, terete, fistular. *Leaflets* 1–3 in., ovate- or orbicular-cordate, 3-lobed or -partite, crenate. *Umbel-rays* 8–12, 1–2 in. ; bracts few, and bracteoles linear-subulate. *Flowers* white or pink, nearly regular. *Fruit* ½ in., brown ; ridges winged ; styles short, recurved.— DISTRIB. Europe (Arctic), from Denmark northd., N. Asia, N. America.— Leaves eaten as a pot-herb, root aromatic and pungent.

29. SELI'NUM, *L.*

Perennial herbs. *Leaves* pinnately decompound. *Umbels* compound, rays many ; bracts few or 0 ; bracteoles many, small ; flowers white or yellowish. *Calyx-teeth* obsolete. *Petals* 2-lobed, point inflexed. *Disk-lobes* entire, conical or depressed. *Fruit* ovoid or oblong or rounded, commissure broad, carpophore 2-partite ; carpels ½-terete, primary ridges winged, lateral broadly ; vittæ 1 to each dorsal furrow ; styles short or long. *Seed* biconvex.—DISTRIB. Temp. N. hemisphere and S. Africa ; species about 25.—ETYM. σελήνη, from the *moon*-shaped carpels.

S. carvifo'lium, *L.* ; nearly glabrous, stem angled furrowed, leaves 3-pinnate, leaflets ovate lower pinnatifid, segments lanceolate. *Milk Parsley.*

Moist copses, &c., N. Lincoln and Cambridge, very rare; fl. July–Aug.—
Rootstock short. *Stem* 2–4 ft., ridges almost winged. *Leaves* 6–12 in.,
leaflets ½–¾ in., with thickened margins; petiole long, very slender. *Umbels*
puberulous, flat-topped; rays 10–20; bracts 0, or very few subulate;
bracteoles subulate. *Petals* white. *Styles* slender, recurved. *Fruit* ⅛ in.
long, lateral winged ridges spreading.—DISTRIB. From Norway southd.,
and eastd. to Russia.

30. ANGEL'ICA, *L.* ANGELICA.

Tall perennial herbs. _Leaves_ ternately 2-pinnate, segments large.
Umbels compound, many-rayed ; bracts few or 0 ; bracteoles usually many,
small ; flowers white or purplish. *Calyx-teeth* small or 0. *Petals* with a
short inflexed point. *Disk-lobes* depressed. *Fruit* ovoid, dorsally compressed,
commissure broad, carpophore 2-partite ; carpels broad, flat, lateral primary
ridges with broad membranous wings; dorsal and intermediate elevated ;
vittæ 1–2 in the furrows. *Seed* dorsally compressed, flat or subconcave
ventrally.—DISTRIB. N. temp. and sub-Arctic regions ; species 18.—
ETYM. *Angelicus,* from its properties.

A. sylves'tris, *L.* ; leaflets petioled obliquely oblong-ovate serrate.
Damp copses and banks of streams, N. to Shetland ; ascends to 2,700 ft. in the
Highlands ; Ireland ; Channel Islands ; fl. July–Aug.—Glabrous, except
the pubescent umbels. *Stem* 1–5 ft., stout, fistular, striate, green or purple.
Leaves 1–2 ft., deltoid ; leaflets 1–2 in. ; sheaths large. *Umbels* large ; rays
very many, 1–3 in. ; bracts 0, or 1–2, deciduous ; bracteoles few, subulate,
persistent ; flowers white or purple, nearly regular. *Fruit* ⅙–⅓ in. ; styles
slender, reflexed.—DISTRIB. Europe (Arctic), N. and W. Asia.—Aromatic
and bitter.

31. PEUCED'ANUM, *L.* HOG'S-FENNEL.

Perennial, rarely annual. *Leaves* pinnately or 3-nately compound.
Umbels compound, many-rayed ; bracts many, few, or 0 ; bracteoles many
or 0 ; flowers white, yellow, or pink, often polygamous. *Calyx-teeth* 0
or small. *Petals* with an inflexed, often 2-fid point. *Disk-lobes* small ;
often expanded, undulate. *Fruit* ovoid, oblong or suborbicular, much
dorsally compressed, commissure very broad ; carpels flattish, lateral
primary ridges of each forming flat contiguous wings, dorsal and intermediate filiform ; vittæ 1–3 in each furrow. *Seed* nearly flat.—DISTRIB.
Trop. and temp. regions ; species 100.—ETYM. obscure.

SECTION 1. **Peuced'anum** proper. Perennial. *Bracts* few or many,
bracteoles many. *Calyx* 5-toothed. *Fruit* with narrow wings.

1. P. officina'le, *L.* ; leaves 3-ternately pinnate, segments long and
narrow, bracts few deciduous, flowers yellow. *Sulphur-wort.*
Salt marshes, very rare, Kent, Essex ; Channel Islands ; fl. July–Sept.—
Glabrous. *Stem* 2–3 ft., terete, solid, furrowed. *Leaves* oblong ; segments
1–4 in., flaccid. *Umbels* on spreading subopposite branches ; rays many,
2–4 in., spreading ; bracteoles short, filiform. *Flowers* minute, central

imperfect, pedicels slender. *Fruit* ¼ in.; wings narrow; styles stout, recurved.—DISTRIB. From Belgium southd., Siberia.—Root yields a stimulant resin; odour of sulphur.

2. **P. palus'tre,** *Mœnch ;* leaves 3-pinnate, leaflets pinnatifid, segments narrow, bracts many persistent, flowers white. *Milk Parsley.*

Marshes, local, from York, Lincoln, E. counties, Somerset; fl. July–Aug.— Glabrous; juice milky. *Stem* 3–5 ft., terete, fistular, grooved. *Leaves* ½–1 in., deltoid; leaflets petioled, lanceolate. *Umbels* 1–2 in.; rays many, stout, scabrid; bract deflexed. *Flowers* minute. *Fruit* ⅛ in., broadly oblong; wings narrow, thick; styles very short.—DISTRIB. Europe, Siberia. —Root yields a yellow fœtid gum-resin.

SECTION 2. **Imperato'ria,** *L.* (gen.). Perennial. *Bracts* 0; bracteoles many. *Calyx-teeth* 0. *Fruit* with broad wings.

P. OSTRU'THIUM, *Koch ;* leaves 1–2-ternate, leaflets ovate or suborbicular inciso-serrate, base unequal, flowers white. *Master-wort.*

Moist meadows, rare, N. England and Scotland, naturalized; fl. July–Aug.— Glabrous. *Stem* 2–3 ft., stout, terete, fistular, furrowed. *Leaves* deltoid; leaflets few, 1–4 in., large, often confluent; petiole very long. *Umbels* large, many-rayed. *Fruit* ⅛ in.; wings very broad; styles short.—DISTRIB. Mid. Europe.—Formerly cultivated as a pot-herb and medicine.

SECTION 3. **Pastina'ca,** *L.* (gen.). *Bracts* and *bracteoles* 0. *Calyx-teeth* 0. *Fruit* with rather narrow wings.

3. **P. sati'vum,** *Benth.* ; leaves pinnate, leaflets sessile ovate incisoserrate, flowers bright yellow. *Wild Parsnip.*

Roadsides and waste places, from Durham and Lancaster southd.; an escape in Scotland; native? Ireland; Channel Islands; fl. July–Aug.—Annual or biennial, pubescent. *Stem* 2–3 ft., stout, angled, furrowed, fistular. *Leaves* shining; leaflets 2–5 pair, 1–3 in. *Umbel-rays* many, stout, long. *Flowers* small. *Fruit* ⅛ in., broadly oblong; styles very short.—DISTRIB. Europe, Siberia; introd. in N. America.—Cultivated since the time of the Romans.

32. HERAC'LEUM, *L.* COW-PARSNIP, HOGWEED.

Biennial or perennial herbs, sometimes gigantic. *Leaves* 1–3-pinnate ; segments broad, lobed and toothed. *Umbels* compound, many-rayed; bracts few many or 0 ; flowers often polygamous and outer rayed, white pink or yellowish. *Calyx-teeth* small or 0. *Petals* often unequal, the larger or all notched or 2-lobed, point inflexed. *Disk-lobes* depressed or conical. *Fruit* orbicular obovate or oblong, much dorsally compressed, commissure very broad, carpophore 2-partite ; carpels flat, lateral primary ridges expanded into flat contiguous membranous wings, dorsal or intermediate slender ; vittæ 1 in each furrow, short, thickened downwards. *Seed* flattened.—DISTRIB. Europe, N. and trop. Africa, temp. Asia, N. America ; species 50.—ETYM. The god *Hercules.*

H. Sphondyl'ium, *L.* ; leaves pinnate, leaflets few large lobed.

Moist woods and meadows, N. to Shetland; ascends to 2,700 ft. in the Highlands; Ireland; Channel Islands; fl. June–Aug.—Rough, hairs spreading, close or scattered. *Stem* very stout, 3–6 ft., fistular, grooved, branched above. *Leaves* 1–3 ft.; segments 2–6 in., variable in size lobing and toothing, sometimes narrow and pinnatifid (*H. angustifolium,* Sm.), terminal confluent; sheath of petiole broad. *Umbel-rays* ½–1½ in., many, stout; flowers large, outer very irregular; petals very broad deeply obcordate, white or pink. *Fruit* ¼–½ in., orbicular or obovoid, retuse; styles short.—DISTRIB. Europe, N. Africa, N. Asia.—Stem eatable.

33. TORDYL'IUM, L.

Annual, hairy or woolly herbs. *Leaves* simple or pinnate. *Umbels* compound; rays many, or few and unequal; bracts and bracteoles linear, small or 0; flowers white or purplish, outer often rayed. *Calyx-teeth* subulate and unequal or 0. *Petals* with an incurved point, the larger or all 2-lobed. *Disk* flat and undulate, or conical. *Fruit* orbicular or oblong, much dorsally compressed; lateral primary ridges appressed, broad, thickened, dorsal and intermediate slender; vittæ 1–3 in each furrow. *Seed* flattened.—DISTRIB. Europe, N. Africa, temp. Asia; species 12.— ETYM. The old Greek name.

T. max'imum, *L.* ; leaves pinnate, leaflets 1–3 pair pinnatifid.

Hedge-banks, Essex, Middlesex, Oxford, and Bucks; (an alien or denizen, *Wats.*); fl. June to July.—Hispid with short hairs, reflexed on the stem. *Stem* 1–2 ft., slender, erect, deeply grooved, fistular. *Leaflets* ¼–1 in., oblong or lanceolate, more or less cut and toothed; petiole with a small sheath. *Umbels* small; rays 6–8, stout, short, hispid; bracts and bracteoles as many, stiff, short. *Flowers* small, white or pink, subsessile. *Fruit* broadly-oblong, hispid, thickened margin glabrous; styles short, stiff, erect.—DISTRIB. From Belgium southd.

34. DAU'CUS, L. CARROT.

Annual or biennial, hispid herbs. *Leaves* pinnately decompound, segments small. *Umbels* compound; rays many, outer arching over the inner, or few and irregular; bracts and bracteoles many or 0, entire or cut; flowers white, outer often rayed. *Calyx-teeth* slender or 0. *Petals* notched, point inflexed, often unequal. *Disk-lobes* depressed or conical. *Fruit* ovoid or oblong, carpophore undivided or 2-fid; carpels convex, secondary ridges more prominent than the primary, all, or the secondary only, with rows of spines; vittæ solitary under each secondary ridge. *Seed* flattish ventrally.—DISTRIB. Europe, N. Africa, W. Asia; species 20.—ETYM. The old Greek name.

D. Caro'ta, *L.* ; leaves 3-pinnate, leaflets ovate cut.

Fields, road-sides, and sea-shores; fl. June–Aug.—Hispid. *Stem* 1–2 ft., branched, solid, furrowed. *Leaflets* very many, small. *Umbels* peduncled, rays 1–2 in.; bracts usually pinnatifid; bracteoles lanceolate. *Flowers*

white, central purplish. *Fruit* ½ in., broadly oblong; styles short, stout,
straight.—DISTRIB. Europe, N. Africa, N. and W. Asia to India; introd. in
N. America.

D. CARO'TA proper; erect, branches above spreading, leaf-segments narrow
subdistant, umbels concave, spines of the fruit distinct usually hooked at
the tip.—Common N. to Shetland.

Sub-sp. D. GUM'MIFER, Lamk.; branches spreading from the base, leaf-segments
broader closer, umbels convex, spines of fruit dilated and connate at the
base. *D. marit'imus,* With.—Shores from Wigton southd.; Ireland
Channel Islands.

35. CAU'CALIS, *L.*

Annual, hispid herbs. *Leaves* 1–3-pinnate. *Umbels* simple or compound,
terminal or leaf-opposed, usually of few rays, sometimes capitate; bracts
few or 0; bracteoles more numerous; flowers white or purplish, polygamous, outer often rayed. *Calyx-teeth* acute or 0. *Petals* often unequal, the
larger notched, point inflexed. *Disk-lobes* thick, conical. *Fruit* ovoid or
oblong, commissure constricted, carpophore undivided or 2-fid; carpels
subterete, ridges with 1 or 2 series of spines; vittæ solitary in each
secondary ridge. *Seed* deeply grooved ventrally.—DISTRIB. Europe, N.
Africa, temp. Asia; species 18.—ETYM. The old Greek name.

SECTION 1. **Cau'calis** proper. *Secondary ridges* very prominent, with
1 row of spreading spines. *Bur-Parsley.*

1. **C. daucoi'des,** *L.*; leaves 2–3-pinnate, segments oblong pinnatifid.
Chalky fields on the E. and S. coasts, from Durham to Kent and Somerset;
Channel Islands; (a colonist, *Wats.*); fl. July.—*Stem* 6–18 in., erect,
nodes hispid, angular, grooved, solid; branches spreading. *Leaves* 3–4 in.,
segments small. *Umbel-rays* 2–5; bracts few or 0; bracteoles linear. *Male
flowers* (outer) white or pink, pedicelled, female subsessile. *Fruit* ¼–½ in.,
oblong; spines hooked, of the secondary ridges longest; styles short,
stout, erect.—DISTRIB. From Denmark southd, N. Africa, W. Asia,
Himalaya.

SECTION 2. **Turge'nia,** *Hoffm.* (gen.). *Secondary ridges* with 2–3 rows
of spreading spines.

C. LATIFO'LIA, *L.*; leaves pinnate, leaflets few subpinnatifid.
Cornfields, very rare, Cambridge to Gloster, Herts, S. Wales, Somerset; fl.
July.—Hispid. *Stem* 6–18 in., simple, terete, striate, fistular. *Leaflets*
narrow-oblong, lobes ¼–¾ in. *Umbel-rays* 2–4, stout; bracts broadly lanceolate, membranous. *Flowers* much as in *C. daucoi'des,* pink. *Fruit* ¼–½ in.,
broad; spines long, nearly equal, rough; styles short, stout, erect.—
DISTRIB. From Belgium southd., N. and W. Asia, Himalaya.

SECTION 3. **Tori'lis,** *L.* (gen.). *Fruit* covered between the primary
ridges with spreading or appressed bristles. *Hedge Parsley.*

2. **C. Anthris'cus,** *Huds.*; leaves 1–3-pinnate, leaflets broad, umbels
terminal compound, bracts 4–6, spines of fruit incurved not hooked.

Hedges and waste places, N. to Caithness; ascends to 1,350 ft. in Yorks.; Ireland; Channel Islands; fl. July–Sept.—Hispid more or less. *Stem* erect, branched, solid, striate, hairs reflexed. *Leaflets* many, close set, ¼–½ in., pinnatifid or lobed. *Umbels* 5–12-rayed; bracts small, subulate. *Flowers* minute, white or pink, outer pedicelled fertile. *Fruit* ⅛ in., ovoid; styles short, straight.—DISTRIB. Europe, N. Africa, W. Asia to N.W. India.

3. **C. arven′sis,** *Huds.* ; leaves 1–2-pinnate, leaflets lax narrow, umbels terminal compound, bracts 0 or 1, spines of fruit spreading hooked. *C. helvetica,* Jacq. ; *C. infesta,* Curt.

Fields and waste places, from York and N. Wales southd.; (a colonist, *Wats.*); fl. July–Sept.—Hispid. *Stem* 6–10 in., much branched, often from the base, angled, solid, leafy. *Leaflets* pinnatifid, or cut and serrate, oblong. *Umbel-rays* 2–8, short; bracteoles linear or setaceous. *Flowers* white or pink, irregular, outer fertile. *Fruit* oblong, covered with spines; styles rather slender.—DISTRIB. From Belgium southd., N. Africa.

4. **C. nodo′sa,** *Scop.* ; leaves 1–2-pinnate, leaflets very small, umbels leaf-opposed simple, spines of fruit spreading hooked and barbed.

Dry banks, from Banff southd.; Ireland ; Channel Islands; fl. May–July.— Hispid. *Stem* 6–18 in., often prostrate, slender, flexuous, angled, solid. *Leaflets* pinnatifid. *Umbels* shortly peduncled, subglobose; pedicels very short, stout; bracts 0. *Flowers* small, regular, pink; female subsessile. *Fruit* ⅙–⅓ in., ovoid, inner of each umbel tubercled, outer with one or both carpels furnished with hooked spines; styles very short.—DISTRIB. From Denmark southd., W. Africa, W. Asia to India.

ORDER XXXV. **ARALIA′CEÆ.**

Erect or climbing shrubs or trees ; pubescence often stellate. *Leaves* alternate, simple or compound ; stipules adnate to the petiole or 0. *Flowers* regular, umbellate or capitate. *Calyx-limb* superior, very short, entire toothed or lobed. *Petals* 5, often coriaceous, very deciduous, valvate or slightly imbricate in bud. *Stamens* 5, filaments inflexed ; anthers didymous, versatile. *Disk* epigynous. *Ovary* 2- or more-celled ; styles or stigmas as many as the cells ; ovule 1 in each cell, pendulous, anatropous, raphe ventral, integuments confluent with the nucleus. *Drupe* or *berry* with 1 or more 1-seeded cells. *Seed* pendulous, testa menbranous, albumen dense fleshy; embryo minute.—DISTRIB. Chiefly trop. ; genera 31 : species 340.—AFFINITIES. Close with *Corneæ* and *Umbelliferæ.*— PROPERTIES unimportant.

1. **HED′ERA,** *L.* IVY.

Climbing shrubs. *Leaves* undivided or lobed, exstipulate. *Umbels* panicled ; bracts minute or 0 ; pedicels not jointed ; flowers polygamous. *Calyx-limb* entire or 5-toothed. *Petals* and *stamens* 5. *Disk* tumid. *Ovary* 5-celled ; styles short, connate, stigmas terminal. *Berry* subglobose, cells with a parchment-like endocarp closely investing the ovoid seed. *Albumen* lobulate.—DISTRIB. Temp. regions of the Old World ; species 2.—ETYM. unknown.

H. He'lix, *L.* ; shrubby, climbing by adhesive rootlets. Rocks, woods, and walls; ascends to 1,500 ft. in Yorkshire; Ireland; Channel Islands; fl. Oct.-Nov.—*Trunk* 4–10 in. diam., trailing and flowerless, or ascending and flowering at the terminal free branches. *Leaves* very variable, 1–3 in. broad, cordate ; lobes 5, deep or shallow, acute or obtuse ; those of flowering branches ovate or lanceolate. *Umbels* subracemose, subglobose, clothed with stellate hairs; bracts small, concave; peduncles ⅓–1 in. *Flowers* yellow-green, ¼ in. diam., proterandrous; calyx-teeth deltoid; petals triangular ovate. *Berry* black, rarely yellow, globose, ¼ in. diam.— DISTRIB. Europe, N. Africa, W. Asia to the Himalaya and Japan.—The small sylvestral form, with longer leaf-lobes and often pale nerves, never flowers. The so-called *Irish Ivy* (*H. canarien'sis*, Willd.), with broad rather fleshy leaves and 8-rayed stellate hairs, is a doubtful native of Ireland.—VAR. *Hodgen'sii,* another doubtful native Irish form, has deeply 5–7-lobed leaves and 12–15 rayed scale-like hairs.

ORDER XXXVI. **CORNA'CEÆ.**

Shrubs or trees, rarely herbs. *Leaves* opposite or alternate, exstipulate. *Flowers* small, regular, in terminal or axillary cymes umbels or heads, sometimes involucrate. *Calyx-limb* superior, small or 0, open or valvate in bud. *Petals* 4–5, at the base of the disk, valvate or imbricate in bud. *Stamens* 4–5, inserted with the petals, free ; anthers adnate or versatile. *Disk* epigynous, annular. *Ovary* 1–4-celled ; style 1, stigma simple or lobed ; ovules solitary in each cell, pendulous, anatropous ; integuments confluent with the nucleus. *Drupe* with a 1–4-celled stone, or 1–4 stones. *Seed* oblong, testa membranous, albumen copious fleshy ; embryo minute or elongate.—DISTRIB. Chiefly N. temp. regions ; genera 12 ; species 76.—AFFINITIES. Close to *Caprifoliaceæ* and *Araliaceæ.*—PROPERTIES unimportant.

1. COR'NUS, *L.* CORNEL, DOGWOOD.

Herbs, trees, or shrubs. *Leaves* opposite, rarely alternate. *Flowers* small, in dichotomous cymes or involucrate umbels or heads, white or yellow, honeyed. *Calyx-teeth* 4, minute. *Petals* 4, valvate in bud. *Stamens* 4. *Disk* tumid or obsolete. *Ovary* 2-celled ; stigma capitate or truncate. *Drupe* ovoid or oblong, areolate at the top, stone 2-celled. *Cotyledons* foliaceous.—DISTRIB. N. temp. and subtrop. regions ; species 25.—ETYM. *cornu,* from the *horny* hardness of the wood.

1. **C. sanguin'ea,** *L.* ; shrubby, cymes corymbose ebracteate. *Dogwoood, Dogberry, Prickwood.*

Copses and hedges from Westmoreland southd.; ascends to 1,050 ft. in
Derby; Ireland, rare; Channel Islands; fl. June–July.—Pubescent, 6–8 ft.,
branchlets and leaves red in autumn. *Leaves* 2–3 in., petioled, ovate, or
ovate-oblong, acute; lateral nerves sub-basal. *Cymes* terminal, peduncled,
subglobose, dense-flowered. *Flowers* ⅓ in. diam., cream-white, homogamous.
Berry small, black.—DISTRIB. Europe, N. and W. Asia, Himalaya.—
Wood used for skewers, formerly for arrows; and by gunpowder makers.
Berries yield an oil used in France for soapmaking.

2. **C. sue'cica,** *L.*; herbaceous, umbels involucrate.

Alpine moors, Yorkshire to Sutherland; ascends to 3,000 ft.; fl. July–Aug.—
Puberulous with appressed hairs. *Rootstock* slender, creeping. *Stem* 6–8
in., erect, forked at the top, 4-angled, scaly beneath. *Leaves* ½–1 in., sessile.
in few pairs, oblong or ovate, acute, 5–7-nerved, glaucous beneath. *Umbel*
in the fork, peduncled; bracts 4, ¼ in., white, ovate, acute. *Flowers* minute,
purplish. *Drupe* ⅙ in. diam., red.—DISTRIB. N. and Arctic Europe, Asia,
N. America.

ORDER XXXVII. **CAPRIFOLIA'CEÆ.**

Shrubs or small trees, rarely herbs. *Leaves* opposite, simple, ternately
cut or pinnate, usually exstipulate. *Flowers* cymose. *Calyx-limb* superior,
3–5-toothed or -lobed. *Corolla* regular or irregular, sometimes 2-lipped;
lobes 5, imbricate in bud. *Stamens* 4, 5, 8, or 10, inserted on the corolla-
tube, equal or unequal; anthers versatile. *Disk* epigynous, glandular or
0. *Ovary* 1–6-celled; style simple or 3–6-lobed or 0, stigmas capitate;
ovules solitary, pendulous from the top of the cell, or many from its inner
angle, or solitary in one cell and several in others, anatropous, integuments
confluent with the nucleus. *Fruit* a berry or drupe, rarely capsular, 1- or
many-seeded. *Seeds* small, testa usually membranous, albumen copious
fleshy; embryo minute, ovoid, rarely large and terete.—DISTRIB. Temp.
and sub-trop. regions of the N. hemisphere; rare in the south, absent from
trop. and S. Africa; genera 14; species 200.—AFFINITIES. With *Corncœ*
and *Rubiaceœ.*—PROPERTIES unimportant.

TRIBE I. **SAMBU'CEÆ.** *Corolla* usually rotate, regular. *Ovary*-cells 1-
ovuled; style short, 2–3-partite, or stigma sessile.
Shrubs. Leaves simple...1. Viburnum.
Herbs, shrubs, or trees. Leaves pinnate............................2. Sambucus.
Herbs. Leaves 3-nately compound.....................................3. Adoxa.

TRIBE II. **LONICE'REÆ.** *Corolla* tubular or campanulate. *Ovary*-cells
1- or many-ovuled; style slender.
Ovary 2–3-celled, cells with several ovules............................4. Lonicera.
Ovary 3-celled, 1 cell one-ovuled, 2 cells many-ovuled.............5. Liunæa.

1. VIBUR'NUM, *L.*

Shrubs or trees; branches opposite. *Leaves* simple; stipules 0 or small.
Flowers in terminal or axillary corymbs or panicles, white or pink, jointed

on the pedicel, 1–2-bracteolate ; outer sometimes male or neuter, with larger petals. *Calyx-tube* turbinate or ovoid ; limb 5-toothed. *Corolla* rotate, tubular or campanulate, 5-lobed. *Stamens* 5. *Disk* 0. *Ovary* 1–3-celled ; style conical, 3-fid, or stigmas 3 sessile ; ovules 1 in each cell, pendulous. *Drupe* dry or fleshy, terete or compressed, 1- or 3-celled, 1-seeded. *Seeds* compressed ; embryo minute.—DISTRIB. Temp. and subtrop. regions of the N. hemisphere, Andes.—ETYM. unknown.

1. **V. Lanta'na,** *L.* ; scurfily pubescent, leaves broadly oblong-cordate serrulate exstipulate, flowers all perfect. *Wayfaring tree.*

Dry copses and hedges, from York southd.; naturalized elsewhere; Channel Islands; fl. May–June.—Shrubby, 6–20 ft.; pubescence stellate. *Leaves* 2–4 in., rugose, obtuse. *Cymes* flat-topped, rays stout. *Flowers* ¼ in. diam., white, 2-bracteolate. *Corolla* shortly funnel-shaped. *Stamens* shortly exserted. *Drupe* flattened, ⅓ in., black. *Seeds* grooved ventrally.—DISTRIB. From Belgium southd., N. and W. Asia, N. Africa.—Bark acrid.

2. **V. Op'ulus,** *L.* ; subglabrous, leaves 3-lobed stipulate, outer flowers larger neuter. *Guelder-rose.*

Copses and hedges, from Caithness southd.; rare in Scotland; Ireland; Channel Islands; fl. June–July.—Shrubby, 6–8 ft., buds scaly; branches slender, lenticellate. *Leaves* 2–3 in., young downy; lobes unequal, serrate; stipules linear, glandular, adnate to the petiole. *Cymes* 2–4 in. diam., subglobose. *Flowers* honeyed, homogamous, outer corollas ¾ in. diam., white, rotate; inner ¼ in. diam., cream-white, campanulate. *Drupe* ½ in., subglobose, red, translucent. *Seeds* compressed, keeled on the faces.—DISTRIB. Europe, N. and W. Asia, N. America.

2. SAMBU'CUS, *L.* ELDER.

Large herbs, shrubs, or trees ; branches stout, pith thick. *Leaves* pinnate. *Flowers* small, in umbellate corymbs or panicles, jointed on the pedicel, bracteolate. *Calyx-limb* 3–5-toothed. *Corolla* rotate or campanulate, 3–5-partite. *Stamens* 5. *Disk* convex. *Ovary* 3–5-celled ; style short, 3–5-partite, or stigmas 3–5 sessile ; ovules 1 in each cell, pendulous. *Drupe* with 3–5 cartilaginous cells. *Seeds* compressed ; embryo long.—DISTRIB. All temp. regions (S. Africa excepted) and trop. mountains ; species 10–12.—ETYM. σαμβύκη, being formerly used for *musical instruments.*

1. **S. Eb'ulus,** *L.* ; herbaceous, stipules leafy serrate, cymes 3-rayed corymbose compact. *Dwarf Elder, Dane-wort.*

Waste places, local, from Caithness southd.; Ireland ; Channel Islands; (a denizen, *Wats.*); fl. July–Aug.—Glabrous. *Rootstock* creeping. *Stems* 2–4 ft., many, stout, ribbed and grooved. *Leaflets* 4–6 in., 4–6 pair, oblong-lanceolate, serrate. *Cyme* 3–4 in. diam. *Corolla* broadly campanulate, white tipped with pink. *Filaments* crumpled. *Berry* small, globose, black. —DISTRIB. Europe, W. Asia, Himalaya, N. Africa.—Plant fœtid, emetic, and purgative. Supposed to have been introduced by the Danes.

2. **S. ni'gra,** *L.* ; a tree, stipules small or 0, cymes 5-rayed. *Elder.*

Hedges and thickets from Ross southd.; ascends to 1,350 ft. in Yorkshire; (a denizen in Scotland, *Wats.*); Ireland; Channel Islands; fl. June.—*Trunk* often as thick as the thigh; bark corky; buds scaly; branchlets angular, lenticellate. *Leaflets* 2–4 pair, 1–3 in., ovate oblong or lanceolate, rarely orbicular (var. *rotundifolia*, Bromf.), serrate. *Cymes* 4–6 in. diam., flat-topped. *Corolla* ¼ in. diam., white, rotate, lobes rounded. *Filaments* slender. *Berry* small, globose, black, rarely green.—DISTRIB. Europe, W. Asia, N. Africa.—Berries used for wine; flowers for perfumes. The "Cut-leaved Elder," a laciniate-leaved variety, occurs as a garden escape.

3. ADOX'A, *L.* MOSCHATEL.

A small glabrous succulent herb. *Rootstock* creeping ; buds scaly ; stem simple, 2-leaved. *Leaves* ternately cut. *Flowers* small, honeyed, green, in a 5-fld. peduncled head, terminal 4- lateral 5-merous. *Calyx-tube* hemispheric ; limb ½-superior, 2–3-lobed. *Corolla* rotate, 4–5-lobed. *Stamens* 8–10, on the corolla-tube, in pairs alternating with its lobes ; anthers peltate, 1-celled. *Disk* 0. *Ovary* 3–5-celled ; style short, 3–5-partite, stigmas terminal ; ovules solitary in each cell, pendulous. *Drupe* girt by the calyx-teeth, with 4–5 compressed cartilaginous cells. *Seeds* obovate ; embryo minute.—DISTRIB. Europe, N. Asia, Himalaya, N. America.— ETYM. à and δόξα, in allusion to its insignificance.—Each pair of stamens is perhaps one, with separate anther-cells.

A. Moschatelli'na, *L.* ; leaflets broadly triangular-ovate.

Damp hedgebanks and tree-roots, rather local, from Ross southd.; ascends 3,300 ft. in the Highlands; Ireland; fl. April–May.—*Stems* 6–8 in., 4-angled. *Radical leaves* 1–2-ternate, leaflets ½–¾ in., irregularly 3-lobed; petiole slender, dilated at the base; cauline 3-foliolate. *Head* ½–¾ in. diam., sub-4-angular, yellow-green. *Corolla* ¼ in. diam. *Fruit* succulent, green.— Odour musky. A Kashmir variety has 5–6-merous flowers.

4. LONICE'RA, *L.* HONEYSUCKLE.

Erect, prostrate, or climbing shrubs, with scaly buds. *Leaves* opposite, entire, exstipulate, of the young shoots sometimes lobed. *Flowers* in peduncled cymes or heads, often connate in pairs by the ovaries, and subtended by connate bracteoles. *Calyx-tube* ovoid or subglobose ; teeth 5, often unequal. *Corolla* tubular, funnel- or bell-shaped ; tube equal or gibbous at the base, honeyed ; limb oblique or 2-lipped, 5-lobed. *Stamens* 5. *Disk* tumid. *Ovary* 2–3-celled ; style filiform, stigma capitate ; ovules many in the inner angle of each cell. *Berry* fleshy, 2–3-celled ; cells few-seeded, septa sometimes wanting. *Seeds* ovoid or oblong, testa crustaceous.—DISTRIB. Temp. and warm regions of the N. hemisphere ; species 80.—ETYM. *A. Lonicer*, a German botanist.

L. Pericly'menum, *L.* ; twining, leaves ovate or oblong upper sessile, flower-heads terminal peduncled. *Woodbine* or *Honeysuckle.*

Hedges and copses, N. to Shetland ; ascends to 1,500 ft. in Durham ; Ireland ; Channel Islands; fl. June-Sept.—Glabrous or slightly pubescent. *Stem*

10–20 ft. *Leaves* 1-3 in., lower shortly petioled, upper sessile, glaucous beneath. *Bracts* small. *Calyx-teeth* persistent. *Corolla* 1-1½ in., glandular-pubescent, dirty red outside, yellow within. *Berries* globose, crimson.— DISTRIB. W. Europe.

L. CAPRIFO′LIUM, *L.* ; twining, upper leaves connate by very broad bases, flower-heads terminal sessile.

Copses in Cambridge and Oxford; naturalized; fl. May–June.—Glabrous. *Leaves* glaucous beneath; lower petioled, broadly ovate or oblong; upper oblong or triangular. *Bracts* leafy, very large, connate. *Calyx-limb* short, persistent. *Corolla* as in *L. Pericly′menum*. *Berries* globose, scarlet.— DISTRIB. Mid. and S. Europe, W. Asia.

L. XYLOS′TEUM, *L.* ; leaves petioled, flowers axillary in pairs.

Copses, Sussex, Hertfordshire, &c.; naturalized; fl. May–June.—Pubescent. *Stem* suberect. *Leaves* 2–3 in., shortly petioled, ovate or obovate. *Peduncles* shorter than the leaves, 2-fld.; bracts 2, linear; bracteoles minute. *Flowers* sessile, ovaries connate. *Calyx-limb* deciduous. *Corolla* ½ in , pubescent, yellow. *Berries* small, crimson.—DISTRIB. Europe, N. Asia.

5. LINNÆ′A, *Gronov.*

A very slender, creeping, evergreen shrub; branches ascending, ending in a slender, erect, 2-fld. peduncle. *Leaves* opposite, exstipulate. *Flowers* nodding, on slender 2-bracteolate pedicels, honeyed. *Calyx-tube* ovoid; lobes 5, narrow, deciduous. *Corolla* subcampanulate; lobes 5, subequal. *Stamens* 4, inserted near the base of the corolla, 2 longer than the others. *Disk* obsolete. *Ovary* 3-celled; style filiform, stigma capitate; ovules many in one cell, solitary in the others, pendulous. *Fruit* subglobose, 3-celled, one cell 1-seeded, the others seedless. *Seed* oblong; embryo cylindric.—DISTRIB. Lapland to N. Italy, cold and Arctic Asia and N. America.—ETYM. Linnæus.

L. borea′lis, *Gronov.* ; leaves broadly ovate obtuse crenate.

Fir forests and plantations, York to Ross; chiefly in Mid. and E. Scotland, ascends to 2,400 ft.; fl. July.—Almost glabrous, except the glandular inflorescence. *Stems* 3–16 in., filiform. *Leaves* in distant pairs, ¼–¾ in., petioled, rarely obcvate or orbicular, coriaceous. *Peduncles* erect, filiform, 2-bracteate at the top. *Corolla* ½ in., pink, sweet-scented. *Fruit* very small.— Fruit very rare; I have taken Wahlenberg's description of it.

ORDER XXXVIII. RUBIA′CEÆ.

Tribe STELLA′TÆ.

Slender herbs, sometimes woody below; stems 4-angled. *Leaves* and foliaceous stipules together forming whorls, entire. *Flowers* very small, in axillary or terminal subsessile or peduncled cymes; pedicels jointed with the flower. *Calyx-limb* superior, annular or 4-6-toothed or 0. *Corolla* rotate, bell- or funnel-shaped ; lobes 3–5, valvate in bud. *Stamens* 3–5, inserted on the corolla-tube, filaments usually short ; anthers didymous. *Ovary* 2-celled ; styles 2, stigmas terminal ; ovules solitary in each cell, attached to the septum, pendulous, amphitropous, integuments confluent with the nucleus. *Fruit* didymous, of 2 plano-convex or globose indehiscent 1-seeded lobes. *Seed* ascending, plano-convex, testa membranous adnate to the pericarp, albumen horny ; embryo axile, cotyledons foliaceous, radicle terete.—This tribe is the N. temp. representative of the enormous Order *Rubiaceæ.*—DISTRIB. All cold and temp. regions ; genera 7 ; species about 300.—AFFINITIES. With *Caprifoliaceæ* and *Valerianeæ.*—PROPERTIES unimportant. *Rubia* yields Madder.

Calyx-limb entire or obsolete.
    Corolla rotate or bell-shaped, 5-lobed. Fruit fleshy.................1. Rubia.
    Corolla rotate, 4-lobed. Fruit dry.....................................2. Galium.
    Corolla bell-shaped. Fruit dry..3. Asperula
Calyx-limb 4–6-toothed. Corolla funnel-shaped. Fruit dry......4. Sherardia.

1. RU'BIA, *L.* MADDER.

Perennial herbs, often woody below. *Cymes* axillary and terminal. *Calyxlimb* 0, or annular. *Corolla* bell-shaped or rotate, 5-lobed. *Styles* 2, short, connate at the base, stigmas capitate. *Fruit* didymous or globose, succulent.—DISTRIB. Chiefly temp. regions ; species about 50.—ETYM. *ruber,* from the *red* dye.

R. peregri'na, *L.* ; leaves 4–6 in a whorl elliptic or oblong.
Rocks and copses, chiefly near the sea, Wales, Hereford, and from Cornwall to Kent; E. and S. Ireland ; Channel Islands ; fl. June–Aug.—Evergreen, shining, 1–2 ft., glabrous except for the recurved prickles on the angles of stem, midrib, and margins of the leaves. *Leaves* 1–3 in. *Cymes* panicled, longer than the leaves. *Corolla* yellowish, ⅓ in. diam., lobes spreading. *Stamens* short. *Fruit* small, black, globose and 1-celled, or didymous.—DISTRIB. W. Europe, N.W. Africa.

2. GA'LIUM, *L.* BEDSTRAW.

Annual or perennial herbs. *Flowers* minute, in axillary or terminal cymes, honeyed. *Calyx-limb* annular. *Corolla* rotate, 4- rarely 5-lobed. *Stamens* 4. *Styles* 2, short, connate at the base ; stigmas capitate. *Fruit* didymous, dry, often hispid or tubercled.—DISTRIB. Chiefly temp. climates ; species about 150.—ETYM. γάλα, from some species being used to curdle *milk.*

    * *Perennial. Flowers yellow. Fruit glabrous, smooth.*

1. **G. ve'rum,** *L.* ; leaves in whorls of 8–12. *Lady's Bedstraw.*
S↵ndy banks, shores, &c., N. to Shetland ; ascends to 2,000 ft. in the Highlands ; Ireland ; Channel Islands ; fl. June–Sept.—Pubescent, black when dry ; rootstock stoloniferous. *Stems* many, 1–3 ft., erect or ascending, 4-angled. *Leaves* ¼–1 in., linear, deflexed, rough above, mucronate, margin

recurved. *Cymes* very compound, axillary and terminal, leafy; flowers 2-sexual, proterandrous. *Fruit* small, black.—DISTRIB. Europe, N. Asia, Himalaya, N. Africa; introd. in N. America.—Flowers used to curdle milk. —A hybrid (var. *ochroleucum*, Syme) with *G. Mollugo* occurs at Deal. It is greenish when dry, and has pale yellow flowers.

2. **G. Crucia′ta,** *Scopoli;* leaves 4 in a whorl elliptic hairy above and beneath. *G. cruciatum*, With. *Valantia Cruciata*, L. *Crosswort.*

Copses and hedges, from Elgin and the mid-Hebrides southd.; Ireland, very rare; fl. April–June.—*Rootstock* creeping. *Stems* 6 in.–2 ft., slender, decumbent, branched at the base. *Leaves* ½–1 in., 3-nerved. *Cymes* axillary, few-fld., peduncles short; flowers ⅛ in. diam., outer male. *Fruit* globose, pedicel recurved. –DISTRIB. From Holland southd., Siberia, W. Asia.

** *Perennial. Cymes both axillary and terminal. Flowers white. Fruit glabrous, smooth granulate or rough, very minute.*

3. **G. palus′tre,** *L.* ; stems lax rough prickly, leaves 4–6 in a whorl obtuse, cymes diffuse, fruit smooth, pedicels divaricate.

Marshes and ditches, N. to Sutherland; ascends to 2,000 ft. in Northumbd.; Ireland; Channel Islands; fl. July–Aug.—Glabrous except the reflexed prickles of the leaf-margins, black when dry, very variable. *Rootstock* ⸱ceeping. *Stems* 6 in.–3 ft., flaccid, decumbent or ascending, branched throughout. *Leaves* ½–1 in., shining, oblong linear or lanceolate-oblong. *Cymes* corymbose, longer than the leaves. *Corolla-lobes* acute.—DISTRIB. Europe (Arctic), N. Africa, Siberia, Persia, Greenland.

G. palus′tre proper; stem smooth, leaves short and narrow, branches of cyme patent or reflexed after flowering, corolla and fruit smaller.—VAR. *G. elonga′tum*, Presl; stem thick, branches of cyme not reflexed after flowering, corolla ⅛ in. diam., fruit ₁⁄₁₀ in. diam.—VAR. *G. Witherin′gii*, Sm.; stem rough with recurved bristles, leaves linear recurved, cyme narrow, its branches short.

4. **G. uligino′sum,** *L.* ; stem rough prickly, leaves 6–8 in a whorl aristate, cymes small few-fld., fruit granular, pedicels erect.

Marshes and ditches, from Caithness southd.; ascends to 1,600 ft in Northumbd. Ireland, very rare; Channel Islands; fl. July–Aug.—Habit and appearance of *G. palus′tre*, L., but green when dry; leaves 6–8 in a whorl (rarely 4), always narrower, more rigid, mucronate; angles of stem more bristly; panicle narrower and fruit smaller.—DISTRIB. N. and Mid. Europe (Arctic), N. Asia.

5. **G. saxat′ile,** *L.* ; tufted, stem prostrate smooth, leaves usually 6 in a whorl mucronate, cymes small, fruit rough, pedicels erecto-patent.

Rocks, heaths, &c., N. to Sutherland; ascends to 3,700 ft. in the Highlands; Ireland; Channel Islands; fl. July–Aug.—Glabrous, black when dry. *Stems* 4–6 in., with many barren shoots, flowering shoots ascending, internodes very short. *Leaves* ⅛–¼ in., obovate or linear-obovate. *Cymes* compact, panicled. *Corolla-lobes* subacute.—DISTRIB. W. Europe, Iceland to N. Italy, W. Siberia.

6. **G. sylves′tre,** *Poll.* ; diffuse, stem smooth glabrous or pubescent below, leaves 6–8 in a whorl narrow ciliate aristate, cymes much longer than the leaves, fruit rough, pedicels spreading.

o

Dry rocky hills and pastures, local from Orkney and Forfar to Somerset and Dorset; absent from W. Scotland, Wales, and E. England; ascends to 2,400 ft. in Yorkshire; Ireland; fl. July–Aug.—Very similar to *G. saxatile*, but less tufted, more erect and rigid; leaves narrower, stiffer, awned.— DISTRIB. Mid. and W. Europe, Iceland to Spain and Servia.

G. sylvestre proper (*G. montanum*, Vill.; *G. pusillum*, Sm. not L.); stem strict glabrous acutely 4-gonous, leaves suddenly acuminate, margins slightly revolute, cymes lax-fld.—VAR. *G. nitid'ulum*, Thuill. (*G. commutatum*, Bab. not Jord.); stem glabrous or hairy below obtusely 4-gonous, leaves narrower, margin strongly revolute, cymes rather compact.

7. G. Mollu'go, *L.* ; stem flaccid glabrous or hairy, leaves 6–8 in a whorl, cymes horizontal, corolla-lobes with slender tips.

Hedges and copses, from Perth and Lanark southd.; ascends to 1,000 ft. in Yorkshire; Ireland, very rare; Channel Islands; fl. July–Aug.—Pale when dry. *Stem* 1–4 ft., erect or decumbent, angles hairy or scabrid. *Leaves* variable, margins with erect or reversed bristles. *Cymes* large, panicled, many-fld. *Fruit* ⅟₇ in. diam., black, shagreened.—DISTRIB. Europe (Arctic), N. Asia, Himalaya, N. Africa.

G. MOLLU'G oproper (*G. ela'tum*, Thuill., *G. sca'brum*, With., *G. insu'bricum*, Gaud.); decumbent, branches divaricate, leaves obovate-lanceolate, cym.⸃ with spreading branches.—VAR. *Bake'ri*, Syme, has linear leaves and few-fld. cymes.

Sub-sp. *G.* EREC'TUM, *Huds.* (*G. aristatum*, Sm. not L.); suberect, leaves linear or lanceolate, cyme with slender ascending branches.—York to Kent and Dorset; Ireland; Channel Islands.

*** *Perennial. Flowers white. Fruit hispid with hooked hairs.*

8. G. borea'le, *L.* ; stem erect, leaves 4 in a whorl lanceolate, cymes axillary and terminal many-fld.

Moist rocks in mountain districts, from York and Brecon N. to Shetland; ascends to 2,800 ft. in the Highlands; Ireland; fl. June–Aug.—Glabrous or pubescent; blackish when dry. *Rootstock* creeping. *Stems* 1–2 ft., tufted, rigid, sparingly branched above. *Leaves* ½–1½ in., 3-nerved. *Cymes* panicled; branches suberect; bracts ovate. *Flowers* ⅙ in. diam. *Fruit* ⅟₇ in. diam. —DISTRIB. N. and Mid. Europe (Arctic), N. Asia, Himalaya, N. America.

**** *Annual. Flowers white or greenish. Fruit often large, usually hispid or tubercled. Angles of stem and margins of leaves prickly.*

9. G. Apari'ne, *L.* ; leaves 6–8 in a whorl, cymes axillary 3–9-flowered, pedicels divaricate, fruit large, usually hispid. *Goose-grass, Cleavers.*

Hedges and waste places, N. to Shetland; ascends to 1,200 ft. in Yorkshire; Ireland; Channel Islands; fl. June–July.—*Stem* 1–5 ft., weak, straggling, often forming matted masses, very rough. *Leaves* ½–2 in., narrow-lanceolate, usually hispid all over. *Fruit* purplish.—DISTRIB. Europe (Arctic), N. Africa, N. and W. Asia to India, temp. N. and S. America.

G. APARI'NE proper; cymes usually 3-flowered, flowers white, fruit tubercled, tubercles crowned by hooked bristles.

Sub-sp. *G.* VAILLAN'TII, *DC.*; cymes 3–9-flowered, flowers very minute greenish, fruit hispid with hooked bristles.—Saffron Walden, in fields.

10. **G. tricor'ne,** *With.* ; leaves 6-8 in a whorl, cymes axillary 3-9-flowered, fruit large granulate pedicels recurved.

Cultivated fields on chalky soil, from Cumberland southd.; (a colonist, *Wats.*); fl. June–Oct.—Habit and appearance of *G. Apari'ne,* but leaves narrower at the tip and the fruiting pedicels very peculiar.—DISTRIB. From Holland southd., N. Africa, India.

11. **G. ang'licum,** *Huds.* ; leaves about 6 in a whorl, cymes axillary and terminal panicled few-fld., fruit minute tubercled.

Walls and sandy places, S.E. England, from Norfolk to Kent, and Cambridge to Sussex; fl. June–July.—*Stem* diffuse, branched, 6–12 in., without barren shoots. *Leaves* ⅛–½ in., narrow, mucronate, finally reflexed, margins rough with prickles that point forwards. *Flowers* greenish-white.—DISTRIB. From Holland southd., Canaries to Persia.

3. ASPER'ULA, *L.*

Herbs or small shrubs. *Flowers* in terminal or axillary cymes, small, honeyed. *Calyx-limb* 4-toothed, deciduous or 0. *Corolla* funnel- or bell-shaped, 4-fid. *Stamens* 4. *Styles* 2, more or less connate, stigmas capitate. *Fruit* dry or rather fleshy.—DISTRIB. Temp. regions ; species 50 ?—ETYM. *asper,* from the *rough* hairs of many.

1. **A. odora'ta,** *L.* ; upper whorls 7–9- lower 2–6-leaved.   *Wood-ruff.*

Shaded hedgebanks, copses, &c., N. to Shetland; ascends to 1,200 ft. in Scotland; Ireland; Channel Islands; fl. May–June.—Almost glabrous, odoriferous in drying, shining. *Rootstock* perennial, creeping, often stoloniferous. *Stems* 6–18 in., subsimple, hairy beneath the nodes. *Leaves* 1–1½ in., oblong-lanceolate, cuspidate, ciliate. *Cymes* subterminal, subumbellate. *Corolla* ¼ in. diam., tube as long as the limb, white, lobes obtuse. *Fruit* small, hispid with hooked hairs.—DISTRIB. Europe, excl. Spain and Portugal, N. and W. Asia.

2. **A. cynan'chica** *L.* ; leaves 4 in a whorl.   *Squinancy-wort.*

Dry banks, local, from York and Westmoreland southd.; S. and W. Ireland; fl. June–July.—Glabrous or nearly so, tufted ; branches 6–10 in., ascending. *Leaves* ¼–1 in., close set, 2 of each whorl much smaller than the others, narrow-linear, mucronate, rigid, recurved, not ciliate. *Cymes* lax-fld. *Corolla* ⅛ in. diam., pink and papillose outside, white inside. *Fruit* minute, papillose.—DISTRIB. Holland to N. Africa, N. and W. Asia, Himalaya.

4. SHERAR'DIA, *Dillen.*   FIELD-MADDER.

Annual or biennial. *Flowers* small, blue or pink, in terminal involucrate heads, honeyed. *Calyx-limb* 4-6-toothed, persistent. *Corolla* funnel-shaped, tube slender ; limb 4-fid. *Stamens* 4, filaments slender. *Style* 2-fid, stigmas capitate. *Fruit* didymous, dry, crowned with the enlarged calyx-limb, separating into 2 plano-convex lobes.—DISTRIB. Europe, N. Asia, Canaries to Persia.—ETYM. James Sherard, an eminent botanist.

S. arven'sis, *L.* ; leaves 4–6 in a whorl, lower often opposite.

o 2

Fields and waste places, N. to Caithness ; Ireland ; Channel Islands ; fl. April–
Oct.— Hispid. *Stems* 6–18 in., prostrate, spreading from the root. *Leaves*
4–6 in a whorl, ¼–¾ in., oblong-lanceolate, aristate. *Corolla* lilac, ⅛ in. diam.
Fruit minute, hispid, crowned by the erect ciliate calyx-teeth.

ORDER XXXIX. **VALERIA'NEÆ.**

Herbs, rarely shrubs. *Leaves* opposite, exstipulate. *Flowers* small,
usually irregular, in dichotomous cymes. *Calyx-limb* superior, lobed, or
a feathery pappus involute in bud. *Corolla* funnel-shaped, base equal
gibbous or spurred ; lobes 3–5, unequal, obtuse, imbricate in bud. *Disk*
small. *Stamens* 1–3 or 5, inserted at the base of the corolla-tube, filaments
slender, anthers exserted, versatile. *Ovary-cells* 3, 2 empty or suppressed,
1 with a solitary pendulous anatropous ovule ; style filiform, stigma obtuse or 2–3-lobed. *Fruit* small, indehiscent ; 1 cell fertile, 2 small empty
suppressed or confluent. *Seed* pendulous, testa membranous, albumen 0 ;
embryo straight, cotyledons oblong, radicle cylindric.—DISTRIB. Temp.
N. zone, Andes, one S. African ; genera 9 ; species 250.—AFFINITIES.
Close with *Dipsaceæ* and *Compositæ.*—PROPERTIES. Aromatic, antispasmodic, sometimes stimulant.

Calyx pappose. Corolla-tube equal or gibbous. Stamens 3...1. Valeriana.
Calyx pappose. Corolla-tube spurred. Stamen 1............1*. *Centranthus.*
Calyx toothed or lobed. Corolla-tube obconic ....2. Valerianella.

1. VALERIA'NA, *L.* VALERIAN.

Perennial, rarely annual, herbs. *Radical leaves* crowded ; cauline opposite or whorled, entire or pinnatifid. *Flowers* in corymbose capitate or
panicled cymes, 1- or 2-sexual, bracteolate, honeyed. *Calyx-limb* annular,
crenulate, developing a feathery deciduous pappus. *Corolla-tube* equal or
gibbous at the base ; lobes 5, rarely 3–4, unequal. *Stamens* 3. *Stigma*
capitate. *Fruit* compressed, ribbed, membranous, 1-celled, 1-seeded.—
DISTRIB. Temp. Europe, Asia, N. and S. America.—ETYM. *valere,* from
its powerful medicinal properties ; species about 130.

1. **V. dioi'ca,** *L.* ; cauline leaves pinnatifid, flowers tetramorphous.

Wet meadows and bogs, local, from Fife and Ayr southd. ; ascends to 2,000
ft. in Northumberland ; fl. May–June.—Glabrous, nodes and leaf-margins
ciliate. *Rootstock* creeping, stoloniferous. *Stem* 6–18 in., ascending. *Leaves*
½–1 in., radical long-petioled, ovate or spathulate ; cauline with a large
terminal lobe and narrow lateral segments. *Cymes* terminal, corymbose ;
flowers ⅛ in. diam., pinkish ; females denser, darker ; bracts linear. *Fruit*
small.—DISTRIB. N. and Mid. Europe, Himalaya.
The forms of flowers are :—1. ♂, corolla large, pistil 0 ; 2. ♂, corolla smaller,
pistil rudimentary ; 3. ♀, corolla smaller still, anthers rudimentary ; 4. ♀,
corolla smallest, anthers 0.

2. **V. officina'lis,** *L.* ; leaves all pinnate, flowers homomorphous. *Cat's Valerian, All-heal.*

Wet meadows and banks of streams, N. to Orkney; ascends to near 2,500 ft. in the Lake district ; Ireland ; Channel Islands ; fl. June–Aug.—Glabrous or hairy below, fœtid. *Rootstock* short, stoloniferous. *Leaves,* radical, long-petioled, soon withering; cauline 2–5 in., sessile ; leaflets ½–2 in., lanceolate, entire or serrate. *Cymes* terminal and axillary, corymbose. *Flowers* proterandrous. *Corolla* ⅙ in., pale pink. *Fruit* small, narrow, ovoid.— Distrib. Europe (Arctic), N. and W. Asia, Himalaya.—Rootstock an antispasmodic.

V. officina'lis proper (*V. Mika'ni,* Wats.) ; leaflets 6–10 pairs, toothed on one side.—Var. *V. sambucifo'lia,* Mikan ; leaflets 4–6 pairs, toothed all round. Very local.

V. **pyrena'ica,** *L.* ; leaves very large cordate deeply toothed.

Naturalized in plantations; fl. June–July.—A large coarse herb. *Stem* 2–4 ft. *Leaves* often a foot in diam., upper with a few basal leaflets. *Flowers* much as in *V. officinalis.*—Distrib. S. France, Spain.

1*. *CENTRAN'THUS, DC.* Spur-Valerian.

Perennial, glabrous, leafy herbs. *Flowers* in terminal unilateral panicled cymes, bracteolate, red or white, proterandrous. *Calyx-limb* annular, crenulate, developing into a feathery deciduous pappus. *Corolla-tube* compressed, elongate, with a longitudinal septum, spurred at the base ; lobes 5, unequal. *Stamen* 1. *Stigma* capitate. *Fruit* membranous, 1-celled, 1-seeded.—Distrib. Europe, N. Africa, W. Asia ; species about 10.—Etym. κέντρον and ἄνθος, from the *spurred corolla.*

C. **ru'ber,** *DC.* ; lower leaves lanceolate, upper triangular ovate.

Old walls and chalk-pits, naturalized, S. England ; Ireland ; rarer northd.; fl. June–Sept.—*Stem* woody below; branches 2–3 ft., erect, terete, fistular. *Leaves* 2–4 in., thick ; lower 2–3 in., petioled, lanceolate or oblanceolate ; upper sessile, entire or base toothed. *Cymes* long ; flowers dense, secund. *Corolla* ½ in., red or white ; spur slender. *Fruit* ¼ in., rough, narrow ovoid, compressed.—Distrib. Mid. and S. Europe, N. Africa, W. Asia.

2. **VALERIANEL'LA,** *Tournef.*

Small annuals, dichotomously branched. *Flowers* solitary or cymose in the forks of the branches, small, bracteate. *Calyx-limb* toothed lobed or 0. *Corolla* funnel-shaped, regular, not spurred ; lobes 5, obtuse. *Stamens* 3. *Stigma* simple or 3-fid. *Fruit* compressed, unequal-sided, grooved, spuriously 2–3-celled, 1-seeded.—Distrib. Europe, N. Africa, W. Asia, N. America ; species about 50.—Etym. Diminutive of *Valeriana.*

1. **V. olito'ria,** *Mœnch ;* cymes capitate, bracts leafy toothed, fertile cell of fruit corky on the back, empty ones contiguous or confluent, calyx-limb 0. *Valeriana Locusta,* L. in part. *Lamb's Lettuce.*

Cornfields and hedgebanks, N. to Shetland; Ireland ; Channel Islands ; (native? *Wats.*); fl. April–June.—Glabrous, flaccid, brittle, 6–12 in. *Leaves*

1–3 in., linear-oblong or oblong-lanceolate, quite entire or toothed, cauline ½-amplexicaul. *Flowers* minute, pale lilac; bracts linear, ciliate. *Fruit* minute, glabrous or hairy.—Distrib. Europe, N. Africa, W. Asia; introd. into N. America.—An excellent salad.

V. carina′ta, *Loisel.* ; cymes capitate, fruit oblong boat-shaped, fertile cell not corky, empty cells contiguous inflated, calyx-limb indistinct. *Corn Salad.*

Cultivated ground, &c.; naturalized; from Yorkshire southd.; Co. Down; Channel Islands; fl. April–June.—Probably a variety of *F. olitoria*, which it resembles in all characters, save those of the fruit given above.—Distrib. From Holland southd., N. Africa, W. Asia.

2. **V. Auric′ula,** *DC.* ; cymes lax, fruit broadly ovoid turgid narrowly grooved in front, fertile cell not corky, empty cells contiguous inflated, calyx with one large unilateral lobe. *V. rimo′sa*, Bast. (oldest name); *V. denta′ta*, DC. *F. tridenta′ta*, Reichb. not Stev.

Cornfields, rare, from Fife southd.; Ireland; Channel Islands; (a colonist, *Wats.*); fl. June–Aug.—Very similar in size, habit, and foliage to *F. olitoria*, but more slender.—Distrib. From Holland southd., W. Asia.

3. **V. denta′ta,** *Poll.*; cymes lax, fruit narrow-ovoid slightly compressed, fertile cell not corky, empty cells remote slender on one side of the fruit, calyx small oblique 3-4-toothed. *V. Morisonii*, DC.

Cornfields, &c., from Moray and Lanark southd.; Ireland; Channel Islands; (a colonist, *Wats.*); fl. June–Aug.—Habit, &c., of *V. Auric′ula.* *Flowers* flesh-coloured *Fruit* glabrous, or hispid (*V. mix′ta*, Dufr.).—Distrib. From Gothland southd., N. Africa, N.W. India.

4. **V. eriocar′pa,** *Desv.*; cymes crowded, fruit ovoid, hairy or glabrous, fertile cell not corky, empty cells remote slender, calyx large campanulate obliquely truncate reticulate toothed.

Fields, Worcester, Dorset, Cornwall; fl. June–July.—Habit of *V. dentata*, differing chiefly in the calyx.—Distrib. From Belgium southd., Canaries.

Order XL. **DIPSA′CEÆ.**

Perennial or biennial herbs. *Leaves* opposite, rarely whorled, exstipulate. *Flowers* small, capitate; outer bracts involucriform; inner 0 or beneath the flowers; floral bracts forming an involucel embracing the calyx-tube. *Calyx-limb* superior, cup-shaped, entire lobed or ciliate. *Corolla* funnel-shaped or cylindric, often curved; lobes 4–5, obtuse, imbricate in bud, the larger or anterior overlapping. *Stamens* 4, inserted on the corolla-tube; filaments filiform, often unequal, incurved in bud; anthers exserted, versatile. *Ovary* 1-celled; style filiform, stigma oblique or notched, ovule solitary, pendulous from the top of the cell, anatropous. *Fruit* indehiscent, enclosed in the involucel, often crowned by the calyx-limb. *Seed* pendulous, testa membranous, albumen fleshy;

embryo axile, straight, cotyledons broad flat, radicle short.—Distrib.
Chiefly Oriental ; genera 5 ; species 125.—Affinities. With *Compositæ.*
—Properties unimportant.

Floral bracts spinescent, exserted, covering the head...............1. Dipsacus·
Floral bracts concealed, scale-like or 0 ....2. Scabiosa

1. DIP'SACUS, *Tournef.* Teasel.

Erect, biennial, hairy or spinulose herbs, stems angular. *Leaves* usually
connate, toothed or cut. *Heads* oblong or cylindric ; receptacle columnar ;
invol. bracts many, rigid, spreading ; floral bracts exserted, spinescent ;
involucel 4-angled.   *Flowers* proterandrous.   *Calyx-limb* discoid or cupular, lobulate.   *Corolla-tube* slightly dilated upwards ; lobes 4, short,
unequal. *Stamens* 4.   *Stigma* dilated, obliquely decurved.—Distrib.
Europe, W. Asia ; species 12.—Etym. obscure.

1. **D. sylves'tris,** *L.* ; leaves sessile simple obovate-lanceolate, heads
oblong, invol. bracts upcurved. *Wild Teasel.*

Copses and hedges from Perth and Dumbarton southd. ; Ireland? ; Channel
Islands; fl. Aug.–Sept.—Glabrous. *Stem* 3–4 ft., stout, rigid, ribs prickly.
Leaves radical on the first year only, spreading ; cauline 6–8 in., oblong-
lanceolate, entire or crenate, midrib prickly. *Heads* 2–3 in. ; bracts linear,
rigid, longer than the head ; floral bracts very long, rigid, subulate, strict,
ciliate ; involucel pubescent. *Calyx-limb* deciduous. *Corolla* purplish.—
Distrib. From Denmark southd., Canaries to Persia.—*D. Fullo'num,* L.
(Fuller's Teasel), known by its hooked bracts, is probably a form of this
plant only known in cultivation.

2. **D. pilo'sus,** *L.* ; leaves petioled with usually a pair of basal leaflets,
heads subglobose hairy, invol. bracts deflexed.

Moist hedges and banks, local, York to Devon and Kent; fl. Aug.–Sept.—
More or less hairy. *Stem* 2–4 ft., slender ; ribs with soft short hair-pointed
prickles. *Leaves,* radical petioled, hairy, crenate ; cauline 6–12 in., oblong,
crenate-serrate.   *Heads* ¾–1 in. diam., drooping in bud ; bracts many,
shorter than the head, linear, toothed and ciliate ; floral bracts obovate.
with long cuspidate ciliate points ; involucel calyx-tube and corolla white,
very hairy.—Distrib. N. and Mid. Europe, Caucasus.

2. SCABIO'SA, *L.*

Perennial herbs. *Leaves* entire or pinnatifid. *Heads* hemispheric or
depressed ; invol. bracts 1–2-seriate ; receptacle hemispheric or columnar,
hairy or with scaly floral bracts ; outer flowers often larger and rayed.
Involucel tubular, 4–8-angled, truncate, or 4–5-lobed. *Flowers* honeyed,
proterandrous. *Calyx-limb* cup-shaped, with 4–16 rigid bristles or teeth.
Corolla curved, oblique or 2-lipped ; lobes 4–5, obtuse. *Stamens* 4. *Stigma* capitellate, notched.—Distrib. Europe, W. Asia, all Africa ; species
about 80.—Etym. *scabies,* from its use in skin-diseases.

SUB-GEN. I. **Scabio'sa** proper. *Receptacle* elongate, bracts scaly.
Involucel 8-furrowed. *Calyx-bristles* 4–5, persistent.

1. **S. succi'sa,** *L.* ; leaves entire, involucral bracts shorter than the
4-lobed subequal corollas. *Devil's-bit-Scabious.*

Pastures and open places, N. to Shetland ; ascends to 2,500 feet in the Highlands ; Ireland ; Channel Islands ; fl. July–Oct.—*Rootstock* short, abruptly
truncate. *Stem* 1–2 ft., branched above. *Leaves* glabrous or hairy ; radical
oblong or obovate, petioled ; cauline few, toothed. *Heads* ⅜–1½ in. diam.,
some ♂ only ; peduncle with appressed hairs ; invol. bracts lanceolate, floral
linear-spathulate. *Flowers* blue-purple or white ; involucel villous, with
4 ovate teeth. *Calyx-bristles* 4, rigid. *Corolla* hairy, tube curved.
Anthers red-brown, opening in succession. *Fruit* oblong.—DISTRIB. N. and
Mid. Europe (Arctic), Siberia, N. Africa.

2. **S. Columba'ria,** *L.* ; cauline leaves pinnatifid, invol. bracts longer
than the 5-lobed corollas, outer corollas larger very irregular.

Dry pastures and banks, from Perth southd. ; ascends to 1,600 ft. in Yorks. ;
absent in W. Scotland ; Channel Islands ; fl. July–Sept.—*Rootstock* tufted,
often woody. *Stem* 1–2 ft., simple or branched above, hairy. *Leaves* glabrous or pubescent, very variable ; radical narrow, petioled, entire or divided ;
cauline segments often cut. *Heads* 1–1½ in. diam. ; peduncle slender ; invol.
bracts 1-seriate, slender, floral linear-spathulate. *Flowers* lilac or blue-
purple ; involucel subcampanulate, white, membranous, many-nerved.
Calyx-bristles 5, rigid, rough. *Corolla* pubescent, of inner flowers regular,
of outer rayed. *Anthers* yellow. *Fruit* narrow-obovoid, shortly beaked.—
DISTRIB. Europe, Mediterranean region.

SUB-GEN. II. **Knau'tia,** *Coulter* (gen.). *Receptacle* hemispheric, hairy.
Involucel 4-furrowed. *Calyx-bristles* 8–16, deciduous.

3. **S. arven'sis,** *L.* ; invol. bracts shorter than the usually 4-lobed
corollas, of which the outer are much larger and 2-lipped.

Dry banks and fields, N. to Orkney ; Ireland ; Channel Islands ; fl. July–Sept.
—*Rootstock* stout. *Stem* 2–5 ft., stout, very hairy, usually branched above.
Leaves variable, hairy ; radical oblong-lanceolate, entire serrate or crenate ;
cauline toothed lobed or pinnatifid. *Heads* 1–1½ in. diam., depressed ;
peduncle long, stout ; invol. bracts broad, leafy, 2-seriate. *Flowers* pale
lilac or blue, sometimes ♀, with imperfect anthers ; involucels villous,
obscurely 4-toothed. *Corolla* hairy, inner redder, outer larger radiating.
Anthers yellow. *Fruit* ovoid, beaked ; calyx-limb deciduous.--DISTRIB.
Europe (Arctic), Caucasus, Siberia, N. Africa.

ORDER XLI. **COMPOS'ITÆ.**

Herbs (most British species). *Leaves* alternate, rarely opposite or whorled,
stipules 0. *Inflorescence* a centripetal head of many small flowers sessile
on the dilated top of the peduncle (*receptacle*), enclosed in an involucre of

whorled bracts ; floral bracts 0, or reduced to paleæ scales or bristles on the receptacle. *Flowers* usually proterandrous and honeyed ; all tubular (head *discoid*), or the outer, or all, ligulate (head *rayed*), 2-sexual, or the inner 2-sexual or male, the outer female or neuter. *Calyx-limb* superior, of hairs (*pappus*) or scales, or 0. *Corolla* of 2 forms : 1st, tubular, or campanulate, 4-5-lobed, lobes valvate with marginal nerves ; 2ndly, ligulate, lobes elongate and connate into a strap-shaped or elliptic ligule. *Disk* epigynous. *Stamens* 4-5, inserted on the corolla-tube, filaments usually free ; anthers basi-fixed, usually connate ; connective produced upwards ; cells simple or tailed at the base ; pollen subglobose, rough. *Ovary* 1-celled ; style 2-fid, arms (sometimes connate) linear, naked or pubescent or tipped by pubescent cones, margins stigmatic ; ovule solitary, basal, erect, anatropous. *Fruit* dry, indehiscent. *Seed* erect, testa membranous, albumen 0 ; embryo straight, cotyledons plano-convex, radicle short.—AFFINITIES. Close with *Dipsaceæ, Valerianeæ,* and *Lobeliaceæ.*—DISTRIB. All regions ; genera 768 ; species 10,000.—PROPERTIES. Too numerous to mention here.

SERIES 1. TUBULIFLO'RÆ. *Flowers all tubular or the outer only ligulate.*
Juice watery.

TRIBE I. **EUPATO'RIEÆ.** *Leaves* mostly opposite. *Flowers* all tubular, 2-sexual. *Anther-cells* not tailed. *Style-arms* slender, $\frac{1}{2}$-terete, papillose, stigmatic lines not continued to the base........................1. Eupatorium.

TRIBE II. **ASTEROI'DEÆ.** *Leaves* alternate. *Ray-flowers* female or neuter, ligulate, rarely all tubular ; style-arms linear, obtuse, glabrous, stigmatic lines confluent. *Disk-flowers* 2-sexual ; anther-cells not tailed ; style-arms linear, glabrous, tipped with a pubescent cone. *Pappus-hairs* or scales rigid or 0.

Ray-flowers purple, 1-seriate, or 0. Pappus rigid ....2. Aster.
Ray-flowers in 2 or more series. Pappus-hairs rigid..............3. Erigeron.
Ray-flowers white or pink. Pappus 0..... ....4. Bellis.
Ray-flowers yellow. Pappus hairs rigid or 0...........................5. Solidago.

TRIBE III. **INULOI'DEÆ.** *Leaves* alternate. *Ray-flowers* ligulate, yellow, or 0; disk-flowers tubular ; anther-cells with slender tails. *Style-arms* and *pappus* as in *Asteroideæ.*

* *Ray-flowers ligulate. Pappus scabrid.*

Pappus simple..6. Inula.
Pappus with an outer row of short scales............................7. Pulicaria.

** *Ray-flowers slender, tubular. Pappus silky.*

Heads 2-sexual. Receptacle flat, naked...........................8. Gnaphalium.
Heads almost dioecious. Pappus of male clavate...............9. Antennaria.
Heads 2-sexual. Receptacle conical; scales few......................10. Filago.

TRIBE IV. **HELIANTHOI'DEÆ.** *Leaves* opposite. *Ray-flowers* 0 or ligulate, yellow, female or neuter. *Disk-flowers* 2-sexual.

Pappus of 2-5 barbed bristles...11. Bidens.
Pappus of broad ciliate scales......................................11*. Galinsoga.

202   *COMPOSITÆ.*

Tribe V. **ANTHE'MIDEÆ.** *Leaves* alternate. *Ray-flowers* ligulate, or tubular and very slender. *Anther-cells* not tailed. *Style-arms* linear with truncate papillose or penicillate tips ; stigmatic margins confluent below. *Pappus* 0 or minute.

'*Outer flowers ligulate, white, or 0. Receptacle with scales or bristles.*
Ligule oblong. Fruit terete or angled................................12. Anthemis.
Ligule broad, short. Fruit compressed, winged....................13. Achillea.
Flowers all tubular and compressed.......................................14. Diotis.

    Outer flowers ligulate, white or yellow. Receptacle naked.
Receptacle conic, often elongating...................................15. Matricaria.
Receptacle flat or convex...16. Chrysanthemum.

    Flowers all tubular. Receptacle usually naked.
Invol. bracts many-seriate. Recept. broad.......................17. Tanacetum.
Invol. bracts few-seriate. Recept. narrow......................18. Artemisia.

Tribe VI. **SENECION'IDEÆ.** *Leaves* alternate. *Flowers* all yellow, tubular and 2-sexual, or outer ligulate. *Receptacle* naked. *Anther-cells* without tails. *Style-arms* connate, or free and obtuse, or tipped with short cones (as in *Asteroideæ*). *Pappus-hairs* usually very soft (rigid in *Doronicum*).

    * *Style-arms of disk-flowers connate.*
Heads racemose ; outer flowers tubular..............................19. Petasites.
Heads solitary ; outer flowers ligulate............................20. Tussilago.

    ** *Style-arms of disk-flowers free.*
Invol. bracts in many series...20*. *Doronicum.*
Invol. bracts in one series...21. Senecio.

Tribe VII. **CYNA'REÆ.** *Leaves* alternate, usually spinous-toothed. *Involucre* often globose, bracts spinous. *Flowers* all tubular, 2-sexual, or the outer female or neuter (diœcious in *Serratula* and some *Cardui*), tube slender, ventricose above, lobes very narrow. *Anthers* much exserted ; connective elongate, stiff. *Style-arms* usually combined into a pubescent 2-fid cylinder, with a ring of hairs or swelling at their base.
Anther-cells tailed.
  Outer bracts hooked...22. Arctium.
  Outer bracts spinous, inner spreading..............................23. Carlina.
  Bracts all unarmed..24. Saussurea.
Anther-cells not tailed or scarcely so.
  Pappus short unequal or 0 ....25. Centaurea
  Pappus-hairs long.
    Bracts unarmed..26. Serratula.
    Bracts spinescent.
      Filaments free. Fruit not angled. Pappus rough........27. Carduus.
      Filaments free. Fruit not angled. Pappus feathery......28. Cnicus.
      Filaments free. Fruit 4-angled, rugose..................29. Onopordon.
      Filaments connate. Fruit terete, rugose....................30. Silybum.

SERIES 2. LIGULIFLO'RÆ. *Flowers all ligulate. Juice milky.*

TRIBE VIII. **CICHORIA'CEÆ.** *Leaves* alternate. *Style* cylindric, pubescent above; arms linear, ½-terete, obtuse, pubescent at the back; stigmatic lines not confluent.

1. *Pappus of small scales or* 0.

Pappus of scales. Flowers blue...31. Cichorium.
Fruit obpyramidal crowned with a ring; pappus 0 ....32. Arnoseris,
Fruit obtuse. Pappus 0 ....33. Lapsana.

2. *Fruit contracted at both ends. Leafy herbs without wool or stellate hairs. Pappus simple or feathery.*

Pappus-hairs feathery ....34. Picris.
Pappus-hairs simple..35. Crepis.

3. *Fruit truncate; pappus-hairs rough, brown. Scapigerous or leafy herbs with stellate hairs*..36. Hieracium.

4. *Fruit contracted below, beaked above; pappus-hairs simple or feathery. Scapigerous herbs.*

Receptacle paleaceous. Pappus-hairs feathery..................37. Hypochæris.
Receptacle naked. Pappus-hairs feathery........................38. Leontodon.
Receptacle naked. Pappus-hairs simple............................39. Taraxacum.

5. *Fruit usually narrowed below and beaked above; pappus-hairs copious, simple. Leafy glabrous or hispid herbs.*

Fruit more or less beaked, ribs smooth ....40. Lactuca.
Fruit not beaked, ribs rough or smooth ....41. Sonchus.

6. *Fruit slender, curved, beak long; pappus-hairs feathery below, tips naked. Invol.-bracts connate below.......................................*42. Tragopogon.

1. EUPATOR'IUM, *L.* HEMP AGRIMONY.

Herbs or undershrubs. *Leaves* usually opposite. *Heads* few-fld., white or purplish, in terminal corymbs; invol. bracts imbricate, 2–3-seriate; receptacle flat, naked. *Flowers* all tubular, 2-sexual, 5-fid. *Anther-cells* without tails. *Style-arms* long, exserted, cylindric, obtuse, grooved in front, pubescent all over. *Fruit* angular or striate; pappus-hairs 1-seriate, hairy or scabrous.—DISTRIB. Chiefly American, rarer in the Old World; species about 400.—ETYM. The classical name.

E. cannabi'num, *L.*; leaves 3–5-foliolate, leaflets lanceolate serrate.

River banks and moist places from Sutherland southd., uncommon in Scotland; Ireland; Channel Islands; fl. July–Sept.—Pubescent, perennial, woody below. *Stem* 2–4 ft., subsimple, terete; branches short. *Leaves* opposite, radical petioled, oblanceolate; cauline subsessile; leaflets 2–4 in. *Heads* in dense terminal 5–6-fld. corymbs, whitish or pale purple; invol. bracts about 10, ¼ in., scarious, linear-oblong, obtuse, outer shorter. *Flowers* 5–6, longer than the involucre. *Pappus* white, scabrous. *Fruit* angled, and corollas covered with resinous points.—DISTRIB. Europe, N. and W. Asia, Himalaya, N. Africa.—Leaves reputed tonic.

2. AS'TER, *L.*

Perennial herbs. *Leaves* alternate or radical, quite entire or toothed. *Heads* solitary or many, usually radiate ; disk yellow, ray white blue or purple ; invol. bracts many-seriate, herbaceous or leafy ; receptacle flat, pitted, edges of the pits toothed. *Ray-fl.* 1-seriate, ligulate, female ; *disk-fl.* tubular, 5-toothed, 2-sexual. *Anther-cells* simple. *Style-arms* of the ray-fl. linear with thickened margins, of the disk-fl. short tipped with papillose cones. *Fruit* compressed ; pappus-hairs many-seriate, persistent, scabrid, unequal.—DISTRIB. Most temp. and cold regions, chiefly American ; species about 150.—ETYM. from the *star*-like flowers.

1. **A. Tripo'lium,** *L.* ; leaves lanceolate or obovate-lanceolate.

Salt marshes, N. to Sutherland; Ireland; Channel Islands; fl. July–Sept.— *Root* fusiform. *Stem* 2–3 ft., erect, sparingly branched, stout. *Leaves* 3–5 in., scattered, fleshy, slightly toothed or not, faintly 3-nerved, upper linear. *Heads* ⅓–⅔ in., corymbose, campanulate ; peduncle slender, bracts small; invol. bracts few, oblong, obtuse, appressed. *Ray-fl.* whitish or purple, many few or 0 (var. *discoideus*). *Fruit* hairy; pappus dirty white.— DISTRIB. Sea coasts and salt regions of Europe (Arctic), N. and W. Asia. A bad substitute for Samphire.

2. **A. Linosy'ris,** *Bernh.* ; leaves linear quite entire. *Chrysoco'ma Linosy'ris,* L. ; *Linosy'ris vulga'ris,* Cass. *Goldielocks.*

Limestone rocks, N. Somerset, S. Devon, Carnarvon ; fl. Aug.–Sept.—Glabrous. *Stems* ½–1½ in., base woody, ribbed, simple, wiry, leafy. *Leaves* 2–3 in., very narrow, acute, rather thick, gradually narrowed from beyond the middle to the base, dotted, 1-nerved. *Heads* ½–¾ in. diam., in terminal, dense, hemi- spheric corymbs; peduncle slender, bracteate ; involucre gummy, puberulous : bracts subulate, much shorter than the flowers. *Pappus* 2-seriate, reddish. —DISTRIB. From the Baltic southd., N. Africa, Caucasus, Asia Minor.

3. ERIG'ERON, *L.* FLEABANE.

Characters of *Aster,* but *ray-flowers* many-seriate ; *fruit* narrower.— DISTRIB. Temp. and cold regions ; species about 80.—ETYM. ἠριγέρων, the name given to groundsel, from its hairy down.

1. **E. alpi'num,** *L.* ; leaves radical, scape with 1 or few broad heads, ligules much longer than the reddish pappus. *E. uniflorus,* Sm. not L.

Alpine rocks, Breadalbane and Clova Mts.; ascends to 3,000 ft.; fl. July–Aug. —Perennial, hispid, hairy. *Rootstock* short. *Radical leaves* spreading, oblong-lanceolate ; cauline few, linear-oblong. *Scapes* solitary or few, 6–8 in. *Heads* 1–3, ¾ in. diam.; peduncle stout, eglandular ; invol. bracts almost villous, subulate-lanceolate. *Ray-fl.* very many, purple, ligule very slender ; *disk-fl.* yellow. *Fruit* hispid.—DISTRIB. Alps and Arctic regions of Europe, Asia, N. America, S. Chili, Fuegia.

2. **E. a'cre,** *L.* ; stem leafy branched above, heads ½–¾ in. diam. panicled, ligules scarcely longer than the reddish pappus.

Dry banks, &c., in England; sandy coasts of Forfar; E. and S. Ireland; Channel Islands; fl. July–Aug.—Annual or biennial, hispid. *Stem* 1–2 ft., panicled above. *Leaves* quite entire; radical 2–3 in., obovate-lanceolate; cauline linear-oblong, obtuse, ½-amplexicaul. *Heads* axillary and terminal; peduncle slender, naked or bracteate; invol. bracts narrow-linear, hispid. *Ray-fl.* narrow, pale purple; disk pale yellow. *Fruit* hispid.—Distrib. Europe (Arctic), temp. and N. Asia, N. America.

E. canaden'se, *L.* ; stem leafy branched above, heads ¼ in. diam. panicled, ligules white scarcely longer than the white pappus.

Waste places in England; introd. from America; common near London; fl. Aug.–Sept.—Annual, sparingly hairy or glabrous. *Stem* 1–2 ft., corymbosely branched. *Leaves* all linear- or oblong-lanceolate, quite entire or sparingly toothed. *Heads* very many, small, peduncle slender. *Invol. bracts* slender, green with scarious margins, glabrous. *Ray-fl.* sometimes faintly purple; disk pale yellow.—Distrib. Most temp. and warm countries.

4. BEL'LIS, *L.* Daisy.

Small herbs. *Leaves* usually all radical, petioled, toothed. *Heads* solitary, disk yellow, ray white or pink ; involucre campanulate, bracts 1–2-seriate, herbaceous ; receptacle conical, papillose. *Ray-fl.* many, 1-seriate, female, ligulate ; style-arms linear, obtuse, margins thickened. *Disk-fl.* tubular, 2-sexual, 4–5-toothed ; anther-cells simple ; style-arms short, thick, tipped by papillose cones. *Fruit* compressed, obovate, subhispid ; pappus 0.—Distrib. Europe, N. Africa, N. America ; species 7 or 8.—Etym. *bellus,* from its pretty appearance.

B. peren'nis, *L.* ; perennial, stemless, leaves obovate-spathulate.

Pastures and meadows, N. to Shetland ; ascends to near 3,000 ft. in the Highlands ; Ireland ; Channel Islands ; fl. all the year.— Glabrous or hairy. *Rootstock* short, fibres stout. *Leaves* 1–3 in., fleshy, obtuse or rounded at the crenate tip, midrib broad. *Scape* 2–5 in. *Head* ¾–1 in. diam.; invol. bracts green, obtuse, often tipped with black. *Ray-fl.* white or pink, disk bright yellow.—Distrib. Europe, except N. Russia and Greece, Asia Minor.

5. SOLIDA'GO, *L.* Golden-rod.

Herbs, often shrubby below. *Leaves* alternate, entire or serrate. *Heads* usually in branched scorpioid cymes, yellow, rayed ; involucre oblong, bracts many-seriate, appressed ; receptacle naked, smooth or pitted. *Ray-fl.* 1-seriate, ligulate, female or 0 ; style-arms slender, linear, obtuse. *Disk-fl.* tubular, 2-sexual, 5-fid ; anther-cells simple ; style-arms ½-terete, tipped with papillose cones. *Fruit* many-ribbed ; pappus-hairs 1–2-seriate, scabrid.—Distrib. Arctic and N. temp. regions, chiefly N. American ; species about 80.—Etym. obscure.

S. Virgau'rea, *L.* ; leaves linear- or lanceolate-oblong.

Thickets, rocky banks, &c.; ascends to 2,800 feet in the Highlands; Ireland ; Channel Islands; fl. July–Sept.—*Rootstock* stout. *Stem* erect, sparingly branched, 4–24 in., glabrous or pubescent with curled hairs. *Leaves* 1–4 in., obscurely toothed, obtuse or acute.   *Heads* crowded, ½ in., shortly peduncled, golden yellow; invol. bracts linear, acute, glabrous, green, margins scarious. *Ray-fl.* 10–12, spreading; *disk-fl.* 10–20. *Fruit* pubescent; pappus white.—DISTRIB. Europe (Arctic), N. Asia, Himalaya, N. America. *S. Virgau'rea* proper; tall, leaves all oblong-obovate quite entire.—VAR. *angustifo'lia*, Gaud.; tall, leaves oblong-lanceolate upper narrower often serrate. —VAR. *S. cam'brica*, Huds.; short, leaves broader ciliate, cyme simple, heads larger.   Usually in mountainous situations.

6. I'NULA, *L.*

Rather rigid herbs. *Leaves* alternate, erect, entire or toothed. *Heads* panicled corymbose or solitary, rayed, yellow; involucre campanulate, bracts in many series, herbaceous, outer often leafy; receptacle flat, naked. *Ray-fl.* female or neuter, 1-seriate, ligulate; style-arms slender, obtuse. *Disk-fl.* tubular, 2-sexual; anther-cells tailed; style-arms short. *Fruit* terete or angled; pappus-hairs 1-seriate, scabrid, or outer row short. —DISTRIB. Europe, temp. and subtrop. Asia; species about 56.—ETYM. The old Latin name.

* *Pappus without an outer series of short bristles or scales.*

1. **I. Cony'za,** *DC.* ; pubescent, leaves ovate-lanceolate, corymbs branched, fruit terete subglabrous. *Conyza squarrosa*, L. *Ploughman's Spikenard.*

Copses and dry banks, from York and Westmoreland southd.; Channel Islands; fl. July–Sept.—Biennial. *Stem* 2–5 ft., erect. *Leaves* 3–5 in., downy beneath, lower petioled, upper subsessile. *Heads* ⅔ in.; invol. bracts very unequal, linear-oblong, outer slightly recurved obtuse, inner accuminate; ligule inconspicuous. *Fruit* sparsely hairy; pappus shining, reddish. —DISTRIB. From Denmark southd., W. Asia.

2. **I. crithmoi'des,** *L.* ; glabrous, leaves linear fleshy, heads few in simple corymbs, fruit terete silky. *Golden Samphire.*

Maritime marshes and rocks, from Essex, Gloster, and Wales to Kent and Cornwall; Wigton and Kirkcudbright in Scotland; S. and E. Ireland; Channel Islands; fl. July–Aug.—Perennial, glabrous, yellow-green. *Stem* 6–18 in., stout, rarely branched, very leafy. *Leaves* 1–2 in., sessile, gradually narrowed to the base, entire or shortly 2–4-lobed. *Heads* 1 in. diam., on long bracteate peduncles; invol. bracts linear-lanceolate; ligule short. *Pappus-hairs* rigid, unequal, dirty white.—DISTRIB. W. coasts from Belgium southd., N. Africa.

3. **I. salici'na,** *L.* ; leaves linear-oblong toothed and ciliate, upper sessile auricled, heads solitary, fruit terete glabrous.

Shores of Lough Derg, Galway; fl. July–Aug.—Perennial, glabrous or slightly hairy. *Stem* 12–18 in., leafy. *Leaves* 2–3 in., rigid, lower obovate-lanceolate, glabrous above, slightly hairy beneath with reticulate nerves. *Heads*

1½·in. diam.; invol. bracts linear-oblong, ciliate, outer leafy reflexed; ligules slender. *Pappus* dirty white.—DISTRIB. Europe, N. and W. Asia.

4. I. Hele′nium, *L.* ; tall, stout, downy, leaves large toothed, heads very large, invol. bracts leafy, fruit 4-angled glabrous. *Elecampane.*

Copses and meadows, from Ross southd. and in Ireland, local, and usually naturalized; wild in Yorkshire (*Baker*); Channel Islands; fl. July–Aug.—*Rootstock* large, succulent. *Stem* 2–5 ft., very stout, branched. *Leaves* velvety beneath; radical 1–1½ ft., oblong-lanceolate, long petioled; cauline sessile, auricled, ovate-cordate, acute. *Heads* 3 in. diam.; peduncle long, stout, naked; invol. bracts broadly ovate, velvety; ligules long, slender. *Pappus* pale reddish.—DISTRIB. From Gothland southd., Siberia; (doubtful if native W. of Russia); introd. in N. America.—Formerly cultivated as an aromatic and tonic; rootstock still used candied.

7. PULICA′RIA, *Gærtn.*

Characters of *Inula,* but pappus with an outer row of short scales.— DISTRIB. Species 24, European and Asiatic.—ETYM. *Pulex,* from being obnoxious to *fleas.*

1. P. dysenter′ica, *Gærtn.* ; woolly or cottony, leaves oblong-cordate ½-amplexicaul, ligules long, scales of pappus connate toothed.

Moist places, from Isla and Haddington southd., rare in Scotland; Ireland; Channel Islands; fl. July–Sept.—*Rootstock* creeping, stoloniferous. *Stem* branched above, very leafy. *Leaves* 1½–2½ in., irregularly waved and toothed. *Heads* 1 in. diam., few, terminal; peduncle naked, and involucre densely woolly; invol. bracts setaceous; ray very slender. *Fruit* silky; pappus-hairs few, unequal, dirty white.—DISTRIB. From Denmark southd., N. Africa, Himalaya.—Bitter, formerly used in dysentery.

2. P. vulga′ris, *Gærtn.* ; pubescent, leaves sessile oblong-lanceolate, ligules very short, scales of pappus free. *Inula Pulicaria,* L.

Moist sandy places, rare, from Montgomery and Norfolk to Cornwall and Kent; Channel Islands; fl. Aug.–Sept.—Annual, slightly glandular. *Stem* 6–8 in., irregularly much branched, leafy. *Leaves* 1–1½ in., obscurely toothed, auricles small. *Heads* ⅓–½ in. diam., subsolitary, terminal; peduncle short, stout; invol. bracts subulate, glandular-pubescent; ligules erect. *Fruit* terete, silky; pappus-hairs few, unequal, dirty white.—DISTRIB. From Gothland southd., N. Africa, N. and W. Asia, Himalaya.

8. GNAPHA′LIUM, *L.* CUD-WEED.

Herbs, sometimes woody below, usually tomentose or woolly. *Leaves* radical, or radical and cauline. *Heads* small, usually in terminal or axillary fascicled cymes or corymbs ; invol. bracts appressed, scarious, as long as the flowers ; receptacle flat, naked. *Outer fl.* female, in 1 or more series, very slender ; style-arms slender, tips truncate, papillose. *Disk-fl.* 2-sexual, limb dilated 5-lobed ; anther-cells tailed ; style-arms short. *Fruit* terete or compressed ; pappus-hairs 1 seriate, very slender.—DISTRIB.

Temp. and subtrop. regions ; species about 100.—Etym. γνάφαλιον, from their *woolly* habit.

* *Stem leafy. Female flowers in many series. Fruit terete.*

1. **G. luteo-al'bum,** *L.* ; annual, stems simple, heads very glistening in dense ebracteate corymbs.

Light soils ; sporadic in Norfolk, Suffolk, and Sussex ; Channel Islands ; fl., July–Aug.—Densely cottony. *Stems* many, 6–12 in., decumbent below, leafy. *Leaves* 1–2 in., linear-oblong, obtuse or acute, ½-amplexicaul, margin sinuate. *Heads* ¼ in. diam., pale yellow ; invol. scales hyaline. *Fruit* papillose.—Distrib. All warm countries.

2. **G. sylvat'icum,** *L.* ; perennial, stems simple, heads in leafy racemes or spikes, or in alternate fascicles along the spikes.

Woods, pastures, and copses ; N. to Shetland ; fl. July–Sept.—White, cottony. *Rootstock* woody. *Stems* 2–12 in. *Leaves* 1–3 in., narrowly linear or obovate-lanceolate, acute 1-nerved ; petiole not ½-amplexicaul. *Heads* subcylindric, ¼ in. ; invol. bracts unequal, yellow or red-brown above, obtuse, outer cottony. *Fruit* puberulous.—Distrib. Europe (Arctic), N. and W. Asia, N. America.

G. sylvat'icum proper ; leaves narrow usually woolly beneath only, spike elongate, pappus white or brown. *G. rectum,* Sm.—N. to Shetland ; Ireland ; Channel Islands.

Sub-sp. norveg'icum, Gunn. ; leaves broader, woolly on both surfaces, floral suddenly smaller, spike short, invol. bracts very dark, fruit longer, pappus white. *G. sylvat'icum,* Sm.—Perth, Forfar and Aberdeen ; ascends to 1,600 ft.

3. **G. uligino'sum,** *L.* ; annual, stems diffuse, heads terminal.

Damp places, especially in light soils, N. to Shetland ; ascends to 2,000 ft. in Ireland ; Channel Islands ; fl. July–Sept.—Cottony above, rarely glabrate. *Stems* 2–6 in., usually many, erect from the decumbent base. *Leaves* narrow, gradually dilated upwards, acute or obtuse ; petiole not amplexicaul. *Heads* ⅛ in. long, sessile, clustered, subtended by long linear leaves ; invol. bracts narrow, subacute, pale brown. *Fruit* very minute.—Distrib. Europe (Arctic), Siberia, N. America.—*G. pilula're,* Wahl., is a var. with papillose fruit, found at Toft in Cambridgeshire.

** *Leaves chiefly radical. Female flowers in one series. Fruit compressed.*

4. **G. supi'num,** *L.* ; perennial, tufted, scapes with 1 or few heads.

Alpine and subalpine rocks, from Stirling and Argyll to Orkney : ascends to near 4,300 ft. ; fl. July–Aug.—Dwarf ; tufts 1–6 in. diam., covered with cottony appressed wool ; roots fibrous, dark brown. *Leaves* ½–1 in., linear-lanceolate, subacute. *Scapes* ½–3 in., slender, with few linear leaves. *Heads* 1–3, ⅛ in. diam., sessile ; invol. bracts scarious, brown above, woolly, outer obtuse, inner acute. *Fruit* pubescent ; pappus white.—Distrib. Alpine and Arctic Europe, N. America, Asia Minor.

9. ANTENN'ARIA, *Brown.*

Characters of *Gnaphalium,* but heads diœcious or nearly so. *Flowers* all tubular ; female filiform, 5-toothed ; style slender, funnel-shaped ;

male tubular, limb dilated above ; anthers partly exserted, cells tailed ;
style undivided. *Fruit* nearly terete ; pappus-hairs 1-seriate, of female
flowers filiform, of males thickened upwards and serrate.—DISTRIB.
Temp. and Arctic Europe, N. Asia, N. America ; species about 10.—
ETYM. From the likeness of the male pappus to the *antennæ* of a
butterfly.

A. dioi'ca, *Br.* ; herbaceous, densely tufted, leaves spathulate, heads in
simple corymbs. *Cat's-foot.*

Heaths and sandy pastures, from Cornwall and Devon in the west and
Suffolk on the east to Shetland; ascends to 2,000 ft. in Scotland; Ireland;
fl. June–Aug.—Perennial. *Barren shoots* many. *Leaves* chiefly radical.
½–1½ in., apiculate, silky beneath. *Flowering stems* 2–8 in., slender, cottony,
with many linear bracts. *Heads* 2–8, crowded; male subglobose, ¼ in.
diam., outer invol. bracts scarious cottony, inner longer with a white or
pink radiating obtuse ligule, stamens exserted ; female twice as large, invol.
scales more numerous, shorter than the flowers. *Fruit* papillose ; pappus-
hairs silky.—DISTRIB. Of the genus.
A. hyperbo'rea, Don, is a var. with leaves broader cottony above.—Skye.

A. MARGARITA'CEA, *Br.* ; half shrubby, corymb compound.
Naturalized in S. Wales and Scotland; Channel Islands; fl. July–Aug.—
Perennial, stoloniferous. *Stem* 2–3 ft., leafy, stout, and leaves beneath and
corymb densely clothed with white or buff cottony tomentum. *Leaves*
3–5 in., narrow-lanceolate, acuminate, sessile, ½-amplexicaul, glabrous above.
Heads ⅓ in. diam., of male fl. globose; female more campanulate; invol.
bracts oblong, brown below, white and radiating above.—DISTRIB. N.
America.

10. FILA'GO, *L.*

Slender, annual, cottony herbs. *Leaves* alternate, quite entire. *Heads*
in axillary and terminal clusters ; invol. bracts imbricate, often superposed
in series, scarious, acuminate, woolly ; receptacle elongate, slender, with
scales under the outer flowers. *Outer fl.* in 1 or more series, female,
outermost usually concealed in the concave inner invol. bracts ; corolla
very slender ; style-arms slender. *Disk-fl.* numerous, 2-sexual or male ;
corolla-limb dilated, 4–5-toothed ; anther-cells tailed ; style-arms short.
Fruit terete, papillose ; pappus-hairs of the central florets very slender, in
1 or more series, of the marginal often 0.—DISTRIB. Europe, N. Africa,
W. Asia ; species 8.—ETYM. *filum,* from the cottony *hairs.*

1. **F. german'ica,** *L.* ; leaves linear-oblong acute waved, heads 20 or
more in terminal spherical clusters, invol. bracts in many opposite series
not spreading acuminate and mucronate.

Dry pastures and banks, from Ross and Dumbarton southd. ; rare in Scotland ;
common in Ireland; Channel Islands; fl. July–Aug.—*Stem* 6–18 in., stiff,
erect or ascending, dichotomously branched, flower-heads in the forks :
branches ascending, leafy. *Leaves* ½–1 in. *Heads* ½ in. broad; invol. bracts
glistening, subulate-lanceolate, longer than the flowers, pale reddish-brown :

P

female flowers in several series. *Fruit* compressed, papillose.—Distrib. From Gothland southd., N. and W. Asia, India; introd. in N. America.
F. germanica proper (*F. canes'cens*, Jord.); heads obscurely 5-angled leafless, invol. bracts folded longitudinally, tips yellowish.—Var. *F. apicula'ta*, G. E. Sm. (*F. lutes'cens*, Jord.); taller, leaves broader, heads acutely 5-angled, invol. bracts purplish boat-shaped, tips reddish. Chiefly E. counties, from S.W. York to Hants, Worcester, and Surrey.—Var. *F. spathula'ta*, Presl (*F. Jussæi*, Coss.); stem short, leaves spathulate, heads 5-angled subtended by acute leaves, invol. bracts boat-shaped, tips pale. S.E. England, from Lincoln to Dorset and Kent.

2. **F. min'ima,** *Fries;* leaves small lanceolate, heads 3–6 in terminal and axillary clusters longer than the subtending leaves, invol. bracts in 1–2 series at length spreading. *F. monta'na,* DC. not L.
Dry places, rather local, from Ross southd.; Ireland; Channel Islands; fl. June–Sept.—Slender, erect, 6–12 in., dichotomous. *Leaves* ¼ in., erect. *Heads* ⅛ in., woolly, sessile; invol. bracts very gibbous and concave at the base, lanceolate, obtuse, glabrous, tips discoloured; female fl. in 1–2 series. *Fruit* terete, papillose.—Distrib. Europe, N. Africa, N. Asia.

3. **F. gal'lica,** *L.*; leaves slender subulate, heads 2–6 in axillary clusters shorter than the subtending leaves, invol. bracts in 1–2 series subacute at length spreading.
Sandy fields, local, Essex, Herts, Bucks, Channel Islands; (a colonist, *Wats.*); fl. July–Sept.—Very slender, much dichotomously branched. *Leaves* ½–⅔ in., ½-amplexicaul. *Heads* ⅛ in., sessile, woolly; invol. scales very gibbous, concave at the base; tips scarious and discoloured. *Fruit* slightly compressed, papillose.—Distrib. From Denmark southd. to N. Africa.

11. BI'DENS, *L.* Bur-Marigold.

Annual herbs, usually glabrous. *Leaves* opposite, entire or divided, or upper alternate. *Heads* solitary or corymbose, rarely rayed, yellow; invol. bracts 2–3-seriate, outer often leafy; receptacle flattish, with a scale under each flower. *Ray-fl.*, if present, in 1 series, ligulate, neuter. *Disk-fl.* campanulate above, 5-toothed; anther-cells simple; style-arms linear, tipped by papillose cones. *Fruit* compressed, ribbed, ribs often aculeate; pappus of 2–5 rigid barbed bristles.—Distrib. Temp. and trop. regions; species about 50.—Etym. From the 2 stiff pappus-bristles of some species.

1. **B. cer'nua,** *L.*; leaves sessile undivided, heads drooping.
Watery places, from Elgin and Dumbarton southd.; Ireland; fl. July–Oct.—Glabrous, or slightly hispid above. *Stem* 1–2 ft., stout, succulent, terete, branched above, leafy. *Leaves* 2–3 in., in connate pairs, oblong-lanceolate, coarsely serrate. *Heads* 1–1½ in. diam.; peduncle ebracteate; outer invol. bracts leafy, spreading or reflexed; inner shorter, broadly-oblong, tips streaked with black. *Ray-fl.* few, short, broad or 0. *Fruit* narrow-obovoid, ribs and pappus-bristles usually 4, barbed.—Distrib. Europe, N. and W. Asia to India, N. America.—The rayed form is rare, and found in England only.

2. **B. triparti'ta,** *L.* ; leaves petioled simple or 3-cleft, heads suberect.
Watery places from Isla and Elgin southd.; Ireland; Channel Islands; fl.
July–Sept.—Habit of *B. cernua,* but more slender; leaves often 3-partite ;
heads smaller; outer invol. bracts narrower, inner acute, and pappus-
bristles 2 shorter.—A Thames form (Putney) has broader heads, more
numerous florets and 3–4 pappus-bristles.— Distrib.   Europe (Arctic), N.
Africa, W. Asia to India, N. America.

11*. *GALINSO'GA,* Ruiz and Pavon.

Annual trichotomously branched herbs. *Leaves* opposite, triple-nerved,
serrate. *Flower-heads* small, yellow, rayed ; invol. bracts 1-seriate, margins
scarious ; receptacle conical, with lanceolate scales under each flower.
Ray-fl. female ; tube short, pubescent ; ligule very broad, short, 3-fid ;
style-arms linear, obtuse.   *Disk-fl.* tubular, 2-sexual, 5-toothed ; anthers
short, with a short terminal appendage, cells shortly tailed. *Fruit* hispid ;
pappus of 1 series of ciliate scales.—Distrib. S. America ; species 5.—
Etym. M. de Galinsoga, a Spanish botanist.

G. parviflo'ra, *Cav.* ; leaves ovate obtusely-serrate ciliate.
Cultivated fields and roadsides; chiefly in Surrey and Middlesex; introd. from
Peru; fl. July–Oct.—More or less pubescent. *Stem* 1–2 ft., slender. *Leaves*
1–2 in., shortly petioled. *Heads* in dichotomous cymes, $\frac{1}{6}-\frac{1}{4}$ in. diam.;
peduncle ebracteate, glandular ; involucre hemispherical ; bracts unequal
oblong, ciliate. *Ray-fl.* 4–6. *Fruit* obovoid, compressed, black; pappus
scales 10–15, narrow-oblong.—Distrib. Trop. S. America.

12. AN'THEMIS, *L.*

Herbs, sometimes shrubby below, strong-scented. *Leaves* alternate,
2-pinnatifid. *Heads* solitary, yellow, with (rarely without) a broad white
ray ; invol. bracts imbricate, in few series ; receptacle flat or conic ; scales
membranous, slender. *Ray-fl.* 1-seriate, ligulate, female or neuter ; style-
arms short, edges thickened, tips obtuse papillose. *Disk-fl.* tubular,
2-sexual, 5-toothed ; anther-cells not tailed ; style-arms as in the female.
Fruit subterete, grooved or striate, crowned with a tumid disk ; pappus 0,
or a short membrane.—Distrib. Europe, N. Africa, W. Asia ; species
about 60. —Etym. The old Greek name.

1. **A. arven'sis,** *L.* ; annual, scales of receptacle mucronate longer than
the flattened disk-fl., ray-fl. female white. *Corn Chamomile.*
Fields and waste places, local, from Caithness southd.; S. and E. Ireland ;
Channel Islands ; (a colonist, *Wats.*) ; fl. June–Aug.—Pubescent or hoary.
Stem 1–2 ft., usually erect from a decumbent base, branched. *Leaf-segments*
$\frac{1}{8}-\frac{1}{4}$ in., linear, acute, not dotted. *Heads* 1–1½ in. broad; peduncle rather
long, slightly swollen upwards ; invol. bracts scarious, pale, obtuse, inner
lacerate. *Fruit* glabrous, subequally ribbed all round, truncate; disk
broad, crenulate.—Distrib. Europe, N. Africa, W. Asia ; introd. in N.
America.

P 2

A. an'glica, Spr. (*A. maritima.* Sm. not L.), is a maritime form, formerly found in Durham, with leaves fleshy pinnatifid, pinnules deeply serrate, and receptacle flat.

2. **A. Cot'ula,** *L.* ; annual, erect, scales of receptacle setaceous shorter than the flattened disk-fl., ray-fl. usually neuter white. *Stinking May-weed.*

Cultivated fields, from Dumbarton and Fife southd.; rare in the N.; Ireland ; Channel Islands; (a colonist, *Wats.*); fl. June–Sept.—Glabrous or hairy, fœtid. *Stem* 8–18 in., corymbosely branched. *Leaves* glandular-dotted, segments very narrow. *Heads* as in *A. arven'sis,* but peduncles more slender and invol. bracts narrower at the tip. *Fruit* faintly ribbed, more strongly on the back.—DISTRIB. Europe, N. and W. Asia, W. India; introd. in N. America.—Acrid, emetic; a troublesome weed; foliage blisters the hand.

3. **A. no'bilis,** *L.* ; perennial, scales of receptacle lanceolate obtuse, ray-fl. female white, disk-fl. cylindric. *Chamomile.*

Pastures and dry soils, in England, not indigenous in Scotland; Ireland; Channel Islands; fl. July–Sept.—Woolly or pubescent, aromatic. *Branches* spreading from the root, leafy. *Leaf-segments* linear. *Heads* 1–1½ in. diam., few; peduncle long, slender ; invol. bracts pubescent, scarious. *Ray-fl.* sometimes 0. *Fruit* obovoid, terete; disk very small, concealed by the inflated base of the corolla.—DISTRIB. W. Europe, N. Africa.—Tonic and febrifuge.

13. ACHILLE'A, L.

Perennial herbs. *Leaves* alternate, entire or divided. *Heads* corymbose, ray white, yellow, or purple ; inner or all invol. bracts oblong, margins sometimes discoloured and scarious. Receptacle narrow, covered with chaffy scales. *Ray-fl.* female ; ligule broad, short. *Disk-fl.* tubular, 2-sexual, compressed, 5-toothed; anther-cells not tailed. *Fruit* oblong, compressed, margined ; pappus 0.—DISTRIB. Europe, N. Asia, N. America ; species about 80.—ETYM. Mythical.

1. **A. Ptar'mica,** *L.* ; leaves linear serrulate, heads few. *Sneeze-wort.*

Meadows and waste places, N. to Shetland ; ascends to 2,200 ft. in the Highlands; Ireland; fl. July–Sept.—Glabrous or pubescent. *Rootstock* creeping extensively. *Stem* 1–2 ft., erect, rigid, ribbed, sparingly branched. *Leaves* 2–3 in., sessile, scattered, teeth cartilaginous. *Heads* corymbose, ⅓ in. diam., hemispheric ; peduncle ebracteate ; receptacle convex ; invol. bracts pubescent, rigid, outer lanceolate acute margins purple, inner oblong obtuse. *Ray-fl.* 8–12; ligule reflexed, broad, as long as the involucre ; *disk-fl.* greenish-white. *Fruit* glabrous, shining.—DISTRIB. Europe, Asia Minor, Siberia ; introd. in N. America.—Rootstock pungent, a sialogogue.

2. **A. Millefo'lium,** *L.* ; leaves 3-pinnatifid. *Yarrow, Milfoil.*

Pastures, N. to Shetland, ascends to 4,000 ft. in the Highlands; Ireland; Channel Islands ; fl. May–Sept.—Glabrous, pubescent or woolly. *Rootstock* extensively creeping, stoloniferous. *Stem* ½–1½ in., erect, furrowed, usually simple, leafy. *Leaves* 2–6 in., linear-oblong, radical petioled ; leaflets and linear-acute segments very close-placed. *Heads* many, $\frac{1}{4}$ in. diam., corymbose, ovoid; peduncle short, stout, ebracteate; invol. bracts oblong, obtuse, rigid, brown-edged. *Ray-fl.* many, white pink or purple; ligule orbicular, shorter than the involucre, reflexed; *disk-fl.* white or yellowish. *Fruit* glabrous, shining.—DISTRIB. Europe (Arctic), N and W. Asia, Himalaya, N. America.—A reputed astringent.

14. DIO'TIS, *Desf.* COTTON-WEED.

An erect, perennial herb, densely clothed with felted white grey or buff wool. *Leaves* alternate, oblong. *Heads* subglobose, discoid, yellow; involucre campanulate, bracts oblong; receptacle flattish, scaly. *Flowers* all tubular and 2-sexual, much compressed, 5-toothed, corky, with 2 wings produced downwards over the ovary as persistent spurs; anther-cells not tailed. *Fruit* angular, crowned by the corolla-base; pappus 0.—ETYM. δίς and οὖς, from the *ear*-like corolla-lobes.

D. marit'ima, *Cass.* ; stems very many, leaves sessile obtuse.

Sandy shores, E. Suffolk, Essex, Kent to Cornwall, Anglesea (extinct in most); Kerry, Wexford, Waterford; Channel Islands; fl. Aug.–Sept.—*Rootstock* creeping, woody. *Stems* 6–12 in., stout, ascending, branched above. *Leaves* $\frac{1}{2}$ in., entire or toothed. *Heads* $\frac{1}{4}$ in. diam., in small dense terminal corymbs; scales of receptacle oblong, acuminate, tips woolly. *Fruit* curved, smooth, glabrous, 5-ribbed.—DISTRIB. Shores of the Mediterranean, Canaries.

15. MATRICA'RIA, *L.*

Annual, rarely perennial, branched herbs. *Leaves* alternate, much divided; lobes narrow. *Heads* yellow, ray white or 0; invol. bracts in few series, nearly equal; receptacle broad, flat or conical, elongate after flowering, naked. *Ray-fl.* 1-seriate, ligulate, female, or 0. *Disk-fl.* tubular, 2-sexual, 4–5-toothed; anther-cells not tailed. *Fruit* ribbed or angled on the ventral face, not winged; disk large; pappus 0.—DISTRIB. Europe, N. and S. Africa, W. Asia; species about 70.—ETYM. Formerly used in *uterine* affections.

1. **M. Chamomil'la,** *L.* ; aromatic, leaves 2-pinnatifid, segments very narrow, invol. bracts without dark edges, fruit 5-ribbed ventrally only. *Wild Chamomile.*

Cultivated ground, from Cumberland southd.; sporadic in Scotland and Ireland; Channel Islands; fl. June–Aug.—Habit and appearance of glabrous forms of *An'themis arven'sis* and of the following, with the scent, &c., of *A. nobilis,* but fainter. *Stem* much branched. *Heads* $\frac{1}{2}$–$\frac{3}{4}$ in. diam., corymbose, ligules reflexed after flowering or 0. *Fruit* small, grey; ribs slender, white; disk oblique.—DISTRIB. Europe, N. and W. Asia to N.W. India.—A tonic.

2. **M. inodo'ra,** *L.* ; inodorous, leaves 2-pinnatifid, segments very narrow, invol. bracts edged with brown, fruit with 3 thick ribs on the ventral face and 2 pits on the dorsal above. *Pyrethrum,* Gærtn.

214   COMPOSITÆ.   [MATRICARIA.

Fields, &c., N. to Shetland; ascends to 1,200 ft. in the Highlands; Ireland;
Channel Islands; fl. June–Oct.—Glabrous, very similar to *M. Chamomil'la*,
but not aromatic, leaf-segments usually large and more slender; heads
larger, 2 in. diam.; ligules narrower, not reflexed till long after flowering;
receptacle much less conical. *Disk* of fruit entire or 4-toothed.—DISTRIB.
Europe (Arctic), N. and W. Asia.
M. inodo'ra proper; annual or biennial, stem erect, barren shoots 0, leaf-lobes
slender.—VAR. *salina*, Bab.; maritime, leaf-segments short fleshy, close-
set, obtuse convex.—VAR. *M. marit'ima*, L.; maritime, more succulent,
perennial, stem ascending, barren shoots 0, heads fewer, receptacle narrower,
spaces between the ribs of the fruit narrower.

16. CHRYSAN'THEMUM, *L.*

Herbs or shrubs. *Leaves* alternate or radical, toothed or cut. *Heads*
solitary or corymbose, ray yellow or white; involucre campanulate;
bracts imbricate, margins scarious; receptacle flat or convex, naked.
Ray-fl. 1-seriate, ligulate, female. *Disk-fl.* tubular, 2-sexual, terete or
compressed, 4-5-toothed; anther-cells simple. *Fruit* of the ray-fl. ribbed
or winged, of the disk-fl. compressed; pappus 0 or a membranous ring.—
DISTRIB. Europe, W. Asia, N. Africa; species about 80.—ETYM. χρυσός
and ἄνθεμον, from the *golden flowers*.

1. **C. seg'etum**, *L.*; annual, ray-fl. golden yellow, invol. bracts very
broad with broad scarious margins. *Corn Marigold.*
Fields and waste places, N. to Shetland; Ireland; Channel Islands; (a
colonist, *Wats.*); fl. June–Sept.—Glabrous, glaucous. *Stem* 1-1½ ft., erect,
sparingly branched. *Leaves* petioled, obovate, toothed and lobed, lower
pinnatifid, upper oblong ½-amplexicaul. *Heads* 2 in. diam.; peduncle stout,
thickened upwards *Ligules* with retuse or lobed tips. *Fruit!* of the
rays ribbed and narrowly 2-winged, of the disk not winged.—DISTRIB.
Europe, N. Africa, W. Asia.

2. **C. Leucan'themum**, *L.*; perennial, ray-fl. white, invol. bracts
narrow with dark purple margins. *Ox-eye Daisy.*
Meadows and waste places, N. to Shetland; ascends to 2,100 ft. in Wales;
Ireland; Channel Islands; fl. June–Aug.—Glabrous or slightly hairy. *Stem*
1-2 ft., erect, simple or branched. *Leaves* obtusely cut or subpinnatifid;
lower spathulate, petioled; upper oblong or lyrate with pinnatifid ½-amplexicaul bases. *Heads* 2 in. diam.; peduncle slender. *Ligules* notched at the
tip. *Fruits* all terete, equally ribbed, of the ray with a small crown.—
DISTRIB. Europe (Arctic), Siberia, N. and W. Asia; introd. in N. America.
—A sport with bilabiate ray-fl. occurs (*Dickson*).

3. **C. Parthe'nium**, *Pers.*; perennial, ray-fl. short white, invol. bracts
broad ribbed downy. *Matrica'ria*, L.; *Pyre'thrum*, Sm. *Fever-few.*
Hedgebanks, &c., N. to Caithness; Channel Islands; probably naturalized only;
(a denizen, *Wats.*); fl. July–Sept.—Perennial, pubescent and branched above.
1-2 ft. *Heads* many, ½-¾ in. diam.; invol. bracts with a scarious border.
Receptacle hemispheric. *Ligules* short, broad. *Disk* of fruit cup-shaped,
membranous.—DISTRIB. Mid. and S. Europe, introduced elsewhere.—Tonic
and bitter.

17. TANACE'TUM, *L.* TANSY.

Herbs, often shrubby below, strong-scented. *Leaves* alternate, usually much divided. *Heads* solitary or corymbose, subglobose, discoid, yellow ; invol. bracts many-seriate, edges scarious ; receptacle convex, naked. *Outer fl.* 1-seriate, female, tubular, 3–4-toothed. *Disk-fl.* tubular, male, 4–5-toothed ; anther cells not tailed. *Fruit* 3–5-angled, disk large ; pappus 0, or an irregular membrane.—DISTRIB. Europe, N. and S. Africa, temp. and cold Asia, N. America ; species about 30.—ETYM. doubtful.

T. vulga're, *L.* ; leaves 1–2-pinnatifid, segments inciso-serrate.

Waste places, from Shetland southd., and in Ireland and Channel Islands ; probably naturalized only ; fl. Aug.–Sept.—Perennial, glabrous or pubescent. *Stem* 2–3 ft., grooved and angled, leafy. *Leaves* 2–5 in., oblong, gland-dotted, upper ½-amplexicaul, lower petioled. *Heads* many, ½ in. diam., corymbose, dull yellow ; peduncle stout, ebracteate ; invol. bracts coriaceous, appressed, outer acute shorter, inner obtuse, edges narrow scarious. *Outer fl.* exceeding the involucre, rarely 0, obliquely truncate. *Fruit* obovoid, 5-ribbed ; disk membranous, lobed.—DISTRIB. Europe (Arctic), Siberia, N.W. America ; introd. in the U. States.—Bitter, tonic, vermifuge and febrifuge.

18. ARTEMIS'IA, *L.*

Herbs, often shrubby below, bitter or aromatic. *Leaves* alternate, often much cut. *Heads* small, racemed or panicled, discoid, yellow or purplish ; invol. bracts few-seriate, margins scarious ; receptacle very narrow, flat or convex, naked hairy or fimbriate. *Flowers* few, all tubular, anemophilous ; outer female with 3-toothed corollas ; the rest male or 2-sexual with 5-toothed minute corollas ; anther-cells not tailed. *Fruit* obovoid or oblong, disk minute ; pappus 0.—DISTRIB. N. temp. zone ; species about 150.— ETYM. Ἄρτεμις, the Greek Diana.

1. **A. campes'tris,** *L.* ; leaves nearly glabrous, segments very slender, heads drooping glabrous, outer flowers only fertile.

Sandy heaths, Norfolk and Suffolk ; fl. Aug.–Sept.—Perennial, not aromatic. *Stem* and branches ascending, very slender, grooved. *Leaves* 1–2-pinnatifid, young silky ; segments very few, ¼–½ in., acute, margins recurved ; floral linear, entire. *Heads* very many, ⅛ in., yellow, in long slender racemes, subsessile, ovoid ; receptacle glabrous. *Ray-corollas* dilated below.—DISTRIB. Europe, temp. Asia.

2. **A. vulga'ris,** *L.* ; leaves broad white woolly beneath, segments broad acuminate, heads erect woolly, flowers all fertile. *Mugwort.*

Hedgebanks, &c., N. to Shetland ; ascends to 1,200 ft. in Northumbd. ; Ireland ; Channel Islands ; fl. July–Sept.—Perennial, aromatic. *Stem* 2–4 ft., erect, reddish, angled, grooved, branched. *Leaves* 2–3 by 1–2 in., glabrous above, margins recurved ; petiole with pinnatifid auricles. *Heads* in crowded, panicled, short, erect, woolly spikes, ovoid, reddish-yellow ; receptacle glabrous. *Ray-corollas* slender, cylindric.—DISTRIB. Europe (Arctic), N. Africa, N. and W. Asia to India and China.—Formerly used to flavour drinks.

3. **A. Absin'thium,** *L.* ; leaves silky on both surfaces, segments oblong obtuse, heads drooping silky, outer flowers only fertile. *Wormwood.*

Waste places, local, from Shetland southd.; ascends (cultivated) to 2,200 ft. in Northumbd.; rare in N. and W. Scotland ; Ireland, native ?; Channel Islands ; fl. Aug.–Sept.—Perennial, very aromatic, silkily pubescent. *Stems* 1–3 ft., ascending, grooved and angled. *Leaves* 1–2 in., dotted, 2–3-pinnatifid ; segments many, spreading. *Heads* hemispheric, subsessile, in panicled leafy racemes, yellow ; receptacle hairy. *Ray-corollas* dilated below.—Distrib. Europe, N. Africa, N. and W. Asia, Himalaya, N. America. —Aromatic, vermifuge, and used to flavour drinks.

4. **A. marit'ima,** *L.* ; leaves white and woolly beneath, segments linear obtuse, heads erect or drooping cottony, flowers all fertile.

Salt marshes and ditches, rare in Scotland, from Wigton and Aberdeen southd.; N.E. Ireland ; Channel Islands ; fl. Aug.–Sept.—Woolly or hoary, scarcely aromatic. *Rootstock* woody, branched. *Stem* 10–18 in., ascending. *Leaves* 1–2 in., 2-pinnatifid; segments many, very narrow, spreading. *Heads* crowded in short erect panicled spikes, reddish, narrow oblong; receptacle glabrous.—Distrib. Coasts of Europe and salt tracts of Asia, India.— *A. gallica,* Willd., is not distinguishable as a well-marked variety, either by its more compact habit or erect heads.

19. PETASI'TES, *Tournef.* BUTTER-BUR.

Perennial herbs. *Leaves* produced after the flowers, large, broad. *Heads* purplish or white, subdiœcious, in a spiciform panicle terminating an erect bracteate scape ; male heads with a few fem. ; ray-fl., female with a few males in the disk ; invol. bracts sub-2-seriate, outer few, small ; receptacle flat, naked. MALE fl. *Corolla* bell-shaped, 5-cleft ; anther-cells simple ; style stout (arms connate), ovoid or clavate, papillose, terminated by 2 short small cones. FEMALE fl. *Corolla* filiform, mouth oblique, minutely toothed ; style much exserted, arms short. *Fruit* cylindric, glabrous ; pappus of female copious, hairs soft slender (of male scanty). —Distrib. Europe, N. Asia, Arctic America ; species about 12.—Etym. πέτασος, an *umbrella*, from the size of the foliage.

P. vulga'ris, *Desf.* ; leaves reniform or orbicular-cordate irregularly toothed. *Tussilago Petasites,* L., and *T. hybrida,* L.

Wet meadows and roadsides, N. to Shetland, but local ; ascends to 1,000 ft. in Northumbd.; Ireland; fl. March–May.—*Rootstock* extensively creeping, fleshy, stout. *Leaves* 3 in.–3 ft. diam., white or cobwebby beneath, young above also ; petiole long, stout. *Stem* 4–18 in., stout, purplish below ; sheaths ending in small leaves. *Panicle* cylindric, 3–10 in., female longest, elongating after flowering ; pedicels slender, shortest in the male ; bracts on pedicels subulate. *Male heads* ⅓, *female* ½ in. *Fruit* striate ; pappus white.—Distrib. Europe, N. Africa, N. and W. Asia.

20. TUSSILA'GO, *Tournef.* COLTSFOOT.

A scapigerous herb, rootstock creeping. *Leaves* large, produced after the flowers. *Heads* yellow, solitary, many-fld. ; invol. bracts 1-seriate, with a few outer shorter ones ; receptacle flat, naked. *Ray-fl.* female, multi-seriate, ligulate, narrow ; *disk-fl.* male, campanulate, 5-toothed. *Anthers* without tails.   *Style* clavate (arms connate), papillose, with 2 very small cones.   *Fruit* of the ray subcylindric ; pappus-hairs very slender, multi-seriate, rough ; of the disk imperfect, pappus 1-seriate.— DISTRIB. Europe (Arctic), N. Africa, N. and W. Asia, Himalaya ; introd. in N. America.—ETYM. *tussis,* from its use as a *cough* medicine.

T. Far′fara, *L.* ; leaves broadly cordate angled or lobed toothed.

Damp heavy soils, N. to Shetland ; ascends to 2,700 ft. in the Highlands ; Ireland ; Channel Islands ; fl. March–April.—*Rootstock* stout; stolons many, burrowing. *Leaves* 3–10 in. broad, cobwebby above, densely tomentose and white beneath. *Scapes* 1 or more, 4–10 in., tomentose, with many oblong appressed scales. *Head* 1–1½ in. diam., bright yellow, drooping in bud. *Pappus* soft, snow-white.—Leaves used for cigar-making and smoked in cases of asthma. Wool made into tinder.

20*. *DORONI′CUM, L.* LEOPARD'S-BANE.

Herbs, rootstocks creeping or tuberous. *Radical leaves* petioled ; cauline alternate, amplexicaul. *Heads* solitary or corymbose, rayed, yellow ; invol. bracts in few series, linear, acuminate, nearly equal ; receptacle conical, naked or pubescent. *Ray-fl.* ligulate, usually female only.; style-arms truncate, tip penicillate. *Disk-fl.* dilated above, 5-toothed ; anther-cells not tailed ; style-arms obtuse. *Fruit* oblong-turbinate, furrowed ; pappus-hairs of the ray 0 or 1–3, of the disk in many series.—DISTRIB. Europe, N .Asia, Mts. of India ; species 10.—ETYM. doubtful.

D. PARDALIAN′CHES, *L.* ; radical leaves ovate-cordate, heads usually 3–5.

Naturalized in plantations; fl. May–July.—Pubescent and hairy. *Rootstock* creeping, stoloniferous. *Stem* 2–3 ft. *Radical leaves* 2–5 in., long-petioled, rounded at the tip; lower cauline ovate with dilated amplexicaul petioles, upper sessile. *Heads* 1½–2 in. diam., long-peduncled ; invol. bracts long, subulate-lanceolate, glandular ; receptacle pubescent. *Fruit* black, ribbed, of the ray glabrous without pappus, of the disk hairy with white pappus.— DISTRIB. Mid. and S. Europe.—Reputed poisonous.

D. PLANTAGIN′EUM, *L.* ; radical leaves ovate, heads usually solitary.

Naturalized in plantations; fl. June–July.—Habit of the preceding, but more slender and glabrous ; leaves narrower, not cordate, usually narrowed into the petiole, repand-toothed, 3–5-ribbed, uppermost oblong.—DISTRIB. W. Europe, from Belgium southd.

21. SENE′CIO, *L.*

Herbs (the British species). *Leaves* alternate. *Heads* solitary or co-rymbose, usually yellow ; invol. bracts 1-seriate with sometimes a few smaller at the base, narrow, appressed, herbaceous, tip usually discoloured ; receptacle naked. *Ray-fl.* 1-seriate, female, or 0 ; style-arms truncate, tips penicillate. *Disk-fl.* tubular, 2-sexual, 5-toothed ; anther-cells not tailed; style-arms obtuse. *Fruit* terete or angled, furrowed; pappus-hairs in many series, soft, slender, equal in length, caducous.—All temp. and cold climates; species about 500.—Etym. *senex*, from the hoary pappus.

Section 1. **Sene'cio** proper. *Involucre* with a few (or 0) small bracts at the base.

　　* *Leaves pinnatifid or 2-pinnatifid; except* S. aquaticus.

1. **S. vulga'ris,** *L.* ; annual, eglandular, heads few drooping, outer invol. bracts many, ligules usually 0. *Groundsel.*

Waste places, N. to Shetland, ascending to 1,600 ft. in Northumbd.; Ireland; Channel Islands; fl. all the year.—Glabrous or cottony. *Stem* 6-15 in., often branched from the base, succulent. *Leaves* pinnatifid, irregularly coarsely toothed. *Heads* ½ in., cylindric, conical after flowering; outer invol. bracts dark, ovate-subulate. *Fruit* ribbed, silky.—Distrib. Europe (Arctic), N. Africa; introd. in all cool climates.—Var. *radia'ta*, Koch; has ray-flowers with short ligules. Channel Islands.

2. **S. sylvat'icus,** *L.* ; annual, glandular-pubescent, heads many spreading narrow, outer invol bracts few or 0, ligules short, fruit silky.

Dry banks and pastures, N. to Orkney; ascends to 1,000 ft. in the Highlands; Ireland; Channel Islands; fl. July-Sept.—Fœtid. *Stem* ½-3 ft., erect, leafy. *Leaves* as in *S. vulgaris*, but more deeply cut. *Heads* ⅓ in., cylindric; peduncle slender. *Fruit* faintly ribbed.—Distrib. Europe, Siberia.—*S. lividus*, Sm. not L., is a form with larger auricles to the upper leaves.

3. **S. visco'sus,** *L.* ; annual, viscid, heads few erect broad, outer invol. bracts few green ½ as long as the inner, fruit glabrous.

Waste dry ground, local, from Banff and Dumbarton to Kent and Sussex; Wales (not in W. or Midland counties); very rare in Ireland; fl. July-Aug.—Fœtid. *Stem* 1-2 ft., stout, rigid, grooved and angled, flexuous, branched. *Leaves* broad, sub-2-pinnatifid. *Heads* campanulate, nearly ½ in. long and ⅓ in. diam. *Fruit* slender, ribbed.—Distrib. Europe, Asia Minor.

4. **S. Jacobæ'a,** *L.* ; perennial, tall, erect, almost glabrous, heads in a dense corymb, fruit of ray ribbed glabrous of disk hairy. *Ragwort.*

Roadsides and pastures, N. to Shetland; ascends to 2,100 ft. in the Highlands; Ireland; Channel Islands; fl. June-Sept.— Glabrous or slightly cottony. *Stem* 1-4 ft., stout, leafy. *Leaves* pinnatifid or sub-2-pinnatifid, lobed and toothed, terminal lobe large or small, upper leaves auricled, sessile; lower petioled. *Heads* ⅔-1 in. diam., bright, yellow, campanulate; outer invol. bracts few, small, subulate; peduncle slender, bracteate.—Distrib. Europe, N. and W. Asia to India.—*S. flosculo'sus*, Jord., a var. without ray, rarely occurs.

5. **S. erucifo'lius,** *L.* ; perennial, tall, erect, cottony or pubescent, heads corymbose, fruits all ribbed hairy. *S. tenuifo'lius,* Jacq.

Roadsides and banks from Berwick and Lanark southd.; E. Ireland; Channel Islands; fl. July-Aug.—Habit of *S. Jacobæ'a*, but more pubescent with curled hairs, especially above; rootstock shortly creeping; leaves simply pinnatifid, lobes narrower; heads larger, and pappus dirty-white.—DISTRIB. From Gothland southd., N. and·W. Asia.

6. **S. aquat'icus,** *Huds.* ; biennial, tall, erect, rarely glabrous, heads in a very lax corymb, fruits all ribbed glabrous.

Sides of rivers, ditches, &c., N. to Shetland; ascends to 1,500 ft. in the Lake district; Ireland; Channel Islands; fl. July–Aug.—Like *S. Jacobæ'a,* but usually of laxer growth, with longer petioles, and larger heads. *Radical leaves* very variable, ovate or oblong, irregularly toothed, undivided or lobed, base auricled or pinnatifid, often purple beneath; upper irregularly lyrate-pinnatifid. *Heads* 1–1¼ in. diam.; peduncle slender.—DISTRIB. Europe, N. Africa, Siberia.—*S. barbareæfo'lius,* Krock. (*S. errat'icus,* Bert.; Bab. Prim. fl. Sarn.), is a form with pinnatifid leaves.

S. SQUAL'IDUS, *L.* ; annual or biennial, glabrous, stem short flexuous leafy, heads in a very lax corymb, fruits all ribbed silky.

Naturalized on old walls, &c., Oxford, Bideford, Warwick, Cork; fl. June–Oct.—*Stem* 8–12 in., rather stout. *Leaves* irregularly lyrate-pinnatifid, lobes long or short, toothed lobulate or subentire, upper auricled and ½-amplexicaul. *Heads* ¾ in. broad; involucre broadly campanulate, bracts narrower than in the other species of this section; outer numerous, small, all usually dark-tipped.—DISTRIB. S. Europe.

*** Leaves undivided, toothed.*

S. SARACEN'ICUS, *L.* ; leaves glabrous or nearly so, ray-fl. few.

Naturalized by river-sides and in moist meadows, from Aberdeen southd. and in Ireland; fl. July–Aug.—*Rootstock* creeping, stoloniferous. *Stem* 3–5 ft., erect, stout, leafy. *Leaves* 5–8 in., linear-oblong, acute, sessile, lower shortly petioled, lowest auricled and ½-amplexicaul. *Heads* ½ in. diam., many, in lax puberulous corymbs; peduncle short, bracteate; involucre broadly campanulate, outer bracts subulate, inner narrowly linear oblong, tipped with brown. *Fruit* glabrous.—DISTRIB. From Holland southd., Siberia.—Used as a styptic by Irish peasants.

7. **S. paludo'sus,** *L.* ; leaves cottony beneath, ray-fl. very many.

Fens of Lincoln, Norfolk, Suffolk, Cambridge (very rare); Channel Islands; fl. June–July.—*Rootstock* short. *Stems* 3–6 ft., stout, erect, branched at the top, glabrous or slightly cottony, leafy. *Leaves* 3–6 in., sessile, narrowly oblong-lanceolate, coarsely serrate. *Heads* 1 in. diam., in lax spreading simple or compound corymbs; peduncle long, bracteate; involucre broadly campanulate, outer bracts long subulate, inner obtuse. *Fruit* glabrous.—DISTRIB. From Gothland southd., Siberia.

SECTION 2. **Cinera'ria,** *L.* (gen.).   *Outer invol. bracts* 0. *Ray-fl.* spreading.

8. **S. palus'tris,** *DC.* ; tall, erect, leafly, pubescent or villous, stem hollow, leaves sessile, heads many, fruit ribbed glabrous.

Fens of the Eastern counties, very rare; fl. June–July.—Biennial. *Stem* 2–3 ft., stout, ribbed, unbranched. *Leaves* 3–5 in., ½-amplexicaul, oblong-lanceolate, sinuate-toothed, acute or obtuse. *Heads* ¾–1 in. diam., pale yellow, crowded in compound corymbs, broadly campanulate, short, as is the peduncle, villous with crisped hairs; invol. bracts many, slender. *Ray-fl.* short.— DISTRIB. From Gothland to France and Austria, N. Asia, N. America (Arctic).

9. **S. campes'tris,** *DC.* ; scapigerous, pubescent and cottony, leaves petioled, heads few, fruit ribbed silky. *Cineraria integrifolia,* With.

Dry banks and chalk downs, York, Lincoln, and from Cambridge to Gloster, and Sussex to Dorset; fl. May–June. *Rootstock* short, fibres thick. *Radical leaves* 1–2 in., spreading, coriaceous, shortly petioled, ovate, obtuse, entire or sinuate-toothed, pubescent under the cottony hairs. *Scape* 4–12 in., stout or slender; bracts narrow, long, appressed. *Heads* 1 in. diam., pale yellow; peduncle stout, erect, bracteate at the base; involucre broadly campanulate; bracts narrow obtuse. *Ray-fl.* as long as the bracts.—DISTRIB. Europe (Arctic) to France and Italy, N. Asia, N. America. VAR. *marit'ima,* Syme (*S. spathulæfo'lius,* Bab. not DC.), is a tall form with broadly-toothed leaves. Maritime rocks, Anglesea; Mickle fell, Yorkshire.

22. ARC'TIUM, *L.* BURDOCK.

Stout, erect, branching, biennial herbs. *Leaves* alternate, the lower very large. *Heads* solitary racemed or corymbose, not rayed, purple or white ; involucre globose ; bracts very many, imbricate, coriaceous, appressed below, with long, stiff, spreading, hooked tips ; receptacle flat ; scales rigid, subulate. *Corolla-tube* narrow, limb campanulate ; lobes 5, slender. *Filaments* papillose ; anthers with a long terminal appendage, cells with subulate tails. *Style-arms* connate, pubescent below, obtuse. *Fruit* large, oblong or obovoid, laterally compressed, transversely wrinkled, base areolate ; pappus-hairs multi-seriate, short, free, filiform, scabrid.—DISTRIB. Europe, N. and W. Asia ; introd. into N. America ; species 6 or 7.—ETYM. ἄρκτος, *a bear,* from its coarse appearance.

A. Lap'pa, *L.* ; leaves ovate-cordate entire or sinuate-toothed.

Waste places; fl. July–Aug.—Glabrous or cottony, 2–4 ft., very variable. *Leaves* often 1 ft., glabrous above, usually densely cottony beneath. *Heads* ¾–1½ in. diam., webbed or not; peduncle very stout; invol. bracts slender, angled, rigid, spreading. *Corolla* and stamens purple, styles white. *Fruit* compressed, angled, ribbed, grey mottled with black.—Young stalks formerly eaten boiled, and as salad.—The following forms present no constant characters.

A. LAP'PA proper; petioles hollow, heads sub-corymbose hemispherical glabrous all green, corolla-tube longer than the limb. *A. ma'jus,* Schkuhr. From York and Lancaster southd., Channel Islands.—VAR. *subtcmentosa,* Lange (*A. tomento'sum,* Bab.), has more spherical and webbed heads.

Sub-sp. A. MI'NUS, *Schkuhr ;* heads subracemose more ovoid glabrous or cottony, inner invol. bracts purplish, corolla-tube as long as the limb. Advances North to Skye. — VAR. *mi'nus* proper; root-leaves coarsely

toothed, petioles hollow, heads ½–¾ in. diam., subsessile cottony. Common.
—Var. *A. interme'dium,* Lange (*A. pubens,* Bab.); root-leaves crenate,
petioles with a slender tube, heads arachnoid, lower ones ¾–1 in. diam.,
of the raceme peduncled purple.—Var. *A. nemoro'sum,* Lej.; root-leaves
narrower coarsely crenate, crenatures apiculate, heads subsessile globose.

23. CARLI'NA, *L.* Carline-thistle.

Rigid, spinous herbs. *Leaves* pinnatifid. *Outer invol. bracts* leafy,
spinous-toothed, spreading; inner longer, narrower, scarious, coloured,
shining; receptacle flat, deeply pitted, edges of the pits bristly. *Corollas*
all tubular, glabrous, erect; limb campanulate, 5-toothed. *Filaments*
glabrous; anthers with a terminal appendage, cells with short plumose
tails. *Style-arms* connate into a pubescent cone. *Fruit* oblong, terete,
silky with 2-fid hairs; pappus-hairs 1-seriate, feathery, connate in threes or
fours at the base.—Distrib. Europe, N. and S. Africa, W. Asia; species
about 14.—Etym. After Charlemagne, who used it medicinally.

C. vulga'ris, *L.*; heads 2 or more, invol. bracts ciliate.

Dry fields and pastures, from Elgin and Arran southd.; ascends to 1,200 ft.
in Northumbld.; local in Ireland; Channel Islands; fl. June–Oct.—Biennial.
hoary, root tapering. *Stem* 6–18 in., stout, simple or branched above, purple,
Radical leaves 3–5 in., spreading, lanceolate, spinous, cottony beneath;
cauline many, shorter, ½-amplexicaul. *Heads* ¾–1½ in. diam.; outer bracts
cottony, spreading; middle purplish; inner ½ in., narrow, rigid, acute, yellow,
spreading, erect when moist. *Bristles* of receptacle rigid, yellow, longer
than the soft pappus. *Flowers* purple. *Fruit* brown.—Distrib. Europe,
N. Africa, N. and W. Asia.

24. SAUSSU'REA, *DC.*

Herbs. *Leaves* entire or divided. *Heads* corymbose, purple or violet;
invol. bracts multi-seriate, imbricate, obtuse or acute; receptacle flat,
covered with chaffy scales. *Corollas* all tubular, ventricose above, 5-fid.
Filaments glabrous; anthers terminated by a long acute appendage, cells
with ciliate tails. *Style-arms* connate below, pubescent, with a ring of
hairs at the base. *Fruit* glabrous; pappus-hairs 2-seriate, outer filiform,
rough, usually persistent; inner feathery, connate at the base, deciduous.
—Distrib. N. temp. regions; species about 60.—Etym. *De Saussure,*
the Swiss philosopher.

S. alpi'na, *DC.*; leaves oblong-lanceolate toothed cottony beneath.

Alpine rocks, N. Wales, Lake district, Dumfries; Highlands, N. to Shetland;
ascends to 4,000 ft.; W. Donegal; fl. August.—*Rootstock* short, stoloniferous.
Stem 6–8 in., stout, erect, leafy, simple, cottony. *Leaves,* lower petioled, 4–
7 in., acuminate; upper smaller, sessile. *Heads* 2–3 in. in dense corymbs;
involucre ovoid; bracts oblong, obtuse, woolly, inner longer. *Flowers* exserted, purple; anthers bluish. *Fruit* brown, ribbed; pappus dirty white.
—Distrib. Scandinavia (Arctic), N. Russia, Alps of Mid. Europe, N. Asia,
N. America (a form).

222   *COMPOSITÆ.*   [Centaurea

25. CENTAURE'A, *L.*

Herbs of various habit. *Leaves* entire or cut, often spinous-toothed.
Involucre ovoid or globose ; bracts appressed, imbricate, entire and scarious,
or spinous, or dilated fringed or toothed ; receptacle flattish, bristly.
Corollas all tubular, oblique or 2-lipped, ventricose above ; outer usually
larger, neuter, inner 2-sexual ; lobes 5, slender. *Filaments* glandular ;
anthers with a long terminal coriaceous appendage, cells tailed or not.
Fruit compressed, basal areole oblique, top broad ; pappus-hairs short,
slender, scabrid, usually in many series, inner smaller often scaly, rarely
0.—Distrib. Europe, W. Asia, N. Africa, America ; species 320.—Etym.
Mythical.

Section 1. **Ja'cea.** *Invol.* bracts with a dilated broad appendage.

1. **C. ni'gra,** *L.* ; leaves hispidulous lanceolate entire or distantly lobed,
peduncles leafy, pappus-scales short unequal or 0. *Knapweed.*

Meadows and pastures, N. to Shetland ; ascends to 1,600 ft. in Northumbld. ;
Ireland ; Channel Islands ; fl. June–Sept.—Perennial. *Stem* ½–3 ft., slender,
grooved, simple or branched. *Leaves* scattered, variable, lower petioled,
uppermost quite entire. *Heads* 1–1½ in. diam. ; invol. appendage very
variable, pale or dark brown, orbicular, pectinate on the margin or to the
axis. *Flowers* purple, outer often larger. *Fruit* ·grey.—Distrib. W.
Europe ; introd. in N. America.
C. ni'gra proper ; peduncles thickened, appendages of bracts dark brown
deeply pectinate.—Var. *C. decip'iens,* Thuill. (*C. nigres'cens,* Bab.) ; peduncles slender, appendages paler, less deeply pectinate, spines much shorter,
pappus 0.—S. counties.

Section 2. **Cy'anus.** *Invol.* bracts not appendaged, their upper part
and margins scarious and pectinate or ciliate.

2. **C. Scabio'sa,** *L.* ; erect, leaves deeply pinnatifid segments entire or
lobed, peduncles glabrous, involucre globose, bracts with brown pectinate
tip and margins, pappus as long as the fruit. *Hard-heads.*

Dry pastures and waste places, from Sutherland southd. ; Ireland ; Channel
Islands ; fl. July–Sept.—*Rootstock* woody. *Stem* 2–3 ft., grooved, sparingly
branched, clothed with soft hairs. *Leaves* 4–10 in., almost pinnate, segments obovate. *Heads* 1½–2 in. diam., rayed ; invol. bracts with a broad
brown tip which is decurrent on the sides of the bracts ; flowers bright
purple. *Fruit* grey, pubescent.—Distrib. Europe, Siberia, W. Asia to
Persia.

3. **C. Cy'anus,** *L.* ; erect, leaves narrow entire or lobes few spreading,
peduncles cottony, involucre ovoid, margins of bracts deeply toothed
scarious, pappus shorter than the fruit. *Bluebottle, Cornflower.*

Cornfields, from Caithness southd. ; ascending to 1,000 ft. in the Highlands ;
Ireland ; Channel Islands ; (a colonist, *Wats.*) ; fl. June–Sept.—Annual or
biennial. *Stem* 1–2 ft., slender, grooved, sparingly branched. *Leaves* 2–4
in., variable, sessile, acute, lower ½–1 in., upper ⅒ in. broad. *Heads* ½–1 in.
diam., cobwebby ; teeth of bracts triangular, spreading, of outer white, of

inner brown and white; bristles of receptacle silvery   *Flowers* of ray few, large, bright blue; of disk smaller, purplish.   *Fruit* grey, silky; pappus dirty white.—DISTRIB.   Europe, N. Africa, W. Siberia, N.W. India; introd. in N. America.

4. C. paniculata, *L.* ; erect, paniculately branched, woolly, leaves bipinnatifid, lobes linear, upper narrow, entire, involucre ovoid, bracts spiny-toothed or ciliate apiculate, pappus of very short scale-like bristles.

Jersey; fl. July.—Biennial, dwarf or tall. *Stem* and branches slender, acutely angled. *Leaves* 1–3 in., lobes acute. *Heads* ⅔–1 in. long; involucre ovoid; teeth of deeply striate bracts often flexuous. *Flowers* purple. *Fruit* white, silvery, glabrous.—DISTRIB.   S. Europe.

SECTION 3. **Serid'ia.**   *Invol.* bracts tipped by spreading reflexed spines.

5. C. asp'era, *L.* ; ascending, leaves linear-oblong entire toothed or lyrate, peduncles leafy, involucre globose, bracts tipped with 5 palmately spreading reflexed spines, pappus shorter than the fruit. *C. Isnardi,* L.

Vazon Bay, Guernsey, very rare; fl. July–Sept.—Perennial. *Stem* 1–2 ft., slender; branches spreading, sparsely pubescent, tips cottony. *Leaves* very variable, lower sessile or petioled, lanceolate or oblong. *Heads* 1 in. diam.; bracts coriaceous, yellow-brown, spines $\frac{1}{10}$–$\frac{1}{8}$ in.; bristles of receptacle white. *Fruit* grey, pubescent; pappus white.—DISTRIB.   W. and S. Europe to Italy.

SECTION 4. **Calci'trapa.**   *Invol.* bracts ending in a long strong spine.

6. C. Calci'trapa, *L.* ; rigid, diffusely branched from beneath the heads, leaves pinnatifid, lobes recurved aristate, spines of bracts long spreading, with a few smaller basal, pappus 0. *Star-thistle.*

Dry waste places, rare, from Norfolk and S. Wales to Cornwall and Kent; Channel Islands; probably a denizen; fl. July–Sept.—Biennial, 1–2 ft., cottony or glabrous, branches leafy, spreading, stout. *Leaves* 1–3 in., often interruptedly pinnatifid, lobes distant, slender. *Heads* lateral and sessile, or terminating leafy branches, ½ in. diam.; spines as long, yellow and channelled above. *Flowers* rose-purple. *Fruit* white or mottled brown.— DISTRIB.   From Holland southd., N. Africa, N.W. India; introd. in N. America.

C. SOLSTITIA'LIS, *L.* ; erect, branched, cottony, stem winged, lower leaves lyrate, upper linear entire decurrent, spines of upper bracts long spreading with a few smaller ones at the base, pappus soft.

Fields in E. and S. England, rare; introd. with lucern, &c.; fl. July–Sept. —Annual, much branched, 1–2 ft. *Stem* rigid, branches twiggy, terminated by peduncled heads. *Leaves* very variable. *Heads* globose, ½ in. diam., spines as long, not channelled, those of the outer bracts very small. *Flowers* yellow. *Fruit* white; pappus copious, white, as long as the fruit.—DISTRIB. Mediterranean region (naturalized in all warm climates.)

26. SERRAT'ULA, *L.* SAW-WORT.

Perennial herbs. *Leaves* alternate, radical simple, cauline usually pinnatifid. *Heads* solitary, corymbose, sometimes diœcious, purple or white ; invol. bracts many, imbricate, outer shorter, inner more or less scarious at the tip ; receptacle scaly. *Corollas* regular, tubular, limb ventricose ; lobes 5, narrow, oblique. *Filaments* papillose ; anther-cells simple or shortly tailed. *Style* tumid or papillose or with a ring of hairs at the tip ; arms free or connate. *Fruit* oblong, compressed, glabrous, smooth ; pappus-hairs many-seriate, rigid, scabrid, coloured, outer shorter, deciduous.—DISTRIB. Europe, Asia, N. America ; species about 30.—ETYM. *serrula*, from the *serrate* foliage.

S. tincto'ria, *L.* ; leaves lyrate-pinnatifid, lobes distant serrate.

Copses, &c., from Dumfries and Northumbd. southd.; ascends to 1,000 ft. : Channel Islands; fl. Aug.—Glabrous or nearly so. *Stem* 2–3 ft., slender, erect, grooved, leafy, corymbosely branched. *Leaves* 5–9 in., rarely entire : lobes linear-oblong, acute or acuminate ; cauline sessile. *Heads* ½–¾ in., corymbose, or subsessile (*S. montic'ola*, Boreau), cylindric-ovoid, subdiœcious, female largest ; invol. bracts ciliate, rigid, striate ; outer oblong or ovate, acute ; inner linear-oblong, purplish. *Flowers* red-purple; male with blue anthers and contiguous style-arms ; female with white anthers and spreading style-arms. *Fruit* grey, glabrous ; pappus dirty white.—DISTRIB. Europe, W. Siberia.

27. CAR'DUUS, *L.* THISTLE.

Erect herbs. *Leaves* usually spinous-toothed. *Heads* sometimes diœcious ; involucre ovoid or globose ; bracts many, imbricated, appressed, narrow, rigid, acuminate or spinous-tipped ; receptacle deeply pitted, covered with bristles. *Corollas* all tubular, red or purple, rarely white, tube short, ventricose above, oblique ; lobes 5, narrow, long. *Filaments* free or connate, hairy or glandular ; anthers terminated by a linear appendage, cells usually with toothed tails. *Style-arms* connate into a cylindrical 2-fid pubescent column, with a ring of hairs at the base. *Fruit* oblong, compressed or terete, glabrous ; pappus-hairs many-seriate, filiform, scabrid, connate at the base, deciduous.—DISTRIB. Chiefly Eur e and W. Asia ; species about 30.—ETYM. doubtful.

1. **C. nu'tans,** *L.* ; wings of stem interrupted, head large solitary hemispheric drooping, invol. bracts subulate-lanceolate, outer spreading and reflexed. *Musk-thistle.*

Waste places, from Skye and Elgin southd.; ascends to 1,600 ft. in Yorks.; indigenous in Scotland and Ireland ; Channel Islands; fl. July–Sept.— Biennial, rarely branched, more or less cottony. *Stem* 1–3 ft., grooved ; wings sinuous, very spiny. *Leaves* 6–12 in., variable, waved, entire or 1-2-pinnatifid. *Heads* 1–2 in. diam.; peduncle slender, and involucre cottony ; outer bracts ⅔ in., green, spinous tip long. *Flowers* crimson. *Fruit* pale brown, glabrous, granulate.—DISTRIB. Europe, N. Africa, N. and W. Asia, Himalaya; introd. in N. America.

2. **C. cris'pus,** *L.* ; wings of stem continuous, heads small erect fascicled, peduncles leafy, involucre webbed, bracts erect very slender.

Hedgebanks and waste places, from Ross southd. ; ascends to 1,200 ft. in Derby; indigenous (?) in Scotland ; S. Ireland ; Channel Islands ; fl. June–Aug.—Annual or biennial. *Stem* 1–3 ft., erect, cottony or pubescent above ; branches ascending ; wings narrow, waved. *Leaves* variable, cottony beneath, usually pinnatifid ; lobes broad, lobulate. *Heads* variable, ⅓–⅔ in. diam.; involucre ovoid, bracts subulate webbed. *Flowers* purple or white. *Fruit* pale, shining, furrowed.—DISTRIB. Europe (Arctic), N. Asia, N.W. Himalaya.—A hybrid with *nu'tans* occurs.
C. cris'pus proper; leaves downy beneath, heads small, crowded subglobose, bracts with a slender spine.—VAR. *C. polyan'themos*, Koch ; leaves pubescent on the nerves beneath, heads small crowded ovoid.—VAR. *C. acanthoi'des*, L. ; leaves broader, heads fewer much larger, bracts with a stout spine, fruit with an angled crown.

3. **C. pycnoceph'alus,** *Jacq.* ; wings of stem continuous, heads small fascicled, involucre glabrous narrow, bracts broadly subulate-lanceolate with recurved spines.

Sandy places, especially on the coast, from Forfar southd. ; rare in W. Scotland ; common in Ireland ; Channel Islands.—Annual or biennial, erect, 1–4 ft., branched, hoary. *Leaves* oblong-lanceolate, pinnatifid ; lobes broad, sinuate-toothed. *Heads* ¾–1 in., sessile; invol. bracts few. *Flowers* pale purple. *Fruit* grey, shining, minutely pitted, not furrowed.—DISTRIB. Europe, from Denmark southd.; N. Africa.—*C. tenuiflo'rus*, Curt., is hardly distinguished from *C. pycnoceph'alus* by its broader-winged stem, usually more numerous and smaller heads, and outer invol. bracts with a narrow scarious border.

28. CNICUS, *L.*

Characters of *Carduus*, but often subdiœcious and pappus feathery.— DISTRIB. Chiefly European and Oriental ; species 150.—ETYM. *κνῆκος*, the Greek name for a thistle.

* Upper surface of leaves scabrid.

1. **C. lanceola'tus,** *Hoffm.* ; stem winged, leaves pinnatifid, heads fascicled, involucre ovoid cottony, bracts lanceolate spreading. *Spearthistle.*

Waste places, N. to Shetland ; ascends to 1,500 ft. in Derby ; Ireland ; Channel Islands ; fl. July–Oct.—Stout, erect, annual or biennial, 2–5 ft. *Leaves* ½–1 ft., obovate-lanceolate, setose above, cottony beneath ; lobes few, large, 2-fid, toothed, with long stout spines. *Heads* ¾–1½ in. diam., few, erect ; peduncle short ; invol. bracts very many, subulate, midrib strong, spines long. *Flowers* purple. *Fruit* striped, smooth, shining.—DISTRIB. Europe, N. Africa, Siberia ; introd. in America.

2. **C. erioph'orus,** *Hoffm.* ; stem not winged, leaves pinnate, heads very large woolly, involucre globose, bracts ciliate, spines slender recurved.

Q

Waste dry places, local, from Durham to Somerset and Kent; fl. July–Sept.—
Tall, stout, handsome, woolly, biennial, 3–5 ft., branched above. *Leaves*
1–2 ft., copiously setose above and cottony beneath ; lobes distant, slender,
usually 2-partite, the divisions spreading up and down, margins ciliate and
spinous. *Heads* 2–3 in. diam.; invol. bracts very many. *Flowers* pale
purple ; anthers blue. *Fruit* shining, smooth, mottled.—DISTRIB. From
Holland southd.—Young parts eatable as salad, and cooked.

** Upper surface of leaves hairy or pubescent.

† *Stem branched, very leafy or 0. Leaves harsh, rigidly spinous.*

3. C. acau′lis, *Hoffm.* ; stem usually very short, leaves pinnatifid,
segments 3–4-lobed, heads sessile or on naked peduncles, involucre ovoid
glabrous, bracts appressed ciliate mucronate.

Gravelly and chalky pastures from York to Devon and Kent; Channel
Islands ; fl. July–Sept.—Perennial, glabrous or pilose, stemless with one
subsessile head, or with a leafy branched stem 8–18 in., and several peduncled
heads (*C. du′bius,* Willd., possibly a hybrid with *arven′sis*). *Leaves* sessile or
petioled, oblong-lanceolate, rigid, very spinous. *Heads* 1–2 in.; involucre
ovoid in flower, campanulate in fruit ; bracts ovate-lanceolate, mucronate,
inner very long linear. *Flowers* crimson. *Fruit* smooth, glabrous, brown ;
pappus dirty white.—DISTRIB. From Gothland southd., N. and W. Asia.
—A troublesome weed in pastures.

4. C. arven′sis, *Hoffm.* ; erect, subdiœcious, rootstock creeping, leaves
pinnatifid, heads many, male involucre subglobose, female ovoid, outer
bracts with short spreading spines, inner acuminate.

Fields and waste places, N. to Shetland: ascends to 2,000 ft. in Northumbd.;
Ireland ; Channel Islands; fl. July–Sept.—Perennial, very spinous, 2–4 ft.;
male and female plants in separate large patches. *Stem* angled and
grooved, more or less cottony, rarely glabrous. *Leaves* oblong-lanceolate,
lower petioled, upper slightly decurrent, sinuate lobed or pinnatifid, setose
or spinous. *Heads* ½–1 in. diam., corymbose ; peduncle short; involucre
½–¾ in., bracts appressed, ciliate, tips rigid spinous; inner obtuse, tips
toothed. *Flowers* dingy purple. *Fruit* smooth, shining ; pappus dirty white.
—DISTRIB. Europe, N. and W. Asia, India, N. Africa ; introd. in N. America.
—The commonest pest of agriculture. A hybrid occurs between this and
C. acaulis.

C. ARVEN′SIS proper ; stem flexuous, leaves pinnatifid very spinous, upper
½-amplexicaul. VAR. *horridus,* Koch.
Sub-sp. C. SETO′SUS, *Bess.* ; stem less branched strict, leaves sessile oblong-
lanceolate obtuse sinuate-lobed or subpinnatifid, margins setose.—Orkney,
Fife, &c., casual in Ireland ; very rare, always introduced?

5. C. palus′tris, *Hoffm.* ; stem winged, leaves decurrent pinnatifid, lobes
2–3-fid, segments acuminate spinescent, heads in leafy clusters, involucre
cottony, bracts appressed, outer mucronate, inner acuminate.

Wet meadows, ditches, &c., N. to Shetland ; ascends to 2,400 ft. in the Highlands ; Ireland ; Channel Islands ; fl. July–Sept.—Biennial, soft, stout, erect,
2–4 ft. branched, very spinous. *Leaves* very decurrent, hairy on both
surfaces ; lobes narrow. *Heads* ½ in. diam.; involucre ½ in., ovoid, bracts

purplish green. *Flowers* dark purple. *Fruit* pale, narrow, smooth; pappus
dirty white.—Distrib. Europe (Arctic), Siberia.—Stalks formerly eaten.
A hybrid with *C. pratensis* (*C. Forsteri*) occurs in bogs in Kent, Surrey, and
Sussex.

†† *Stem usually simple, not winged. Leaves soft, spines few. Heads 1 or few.*

6. **C. praten'sis**, *Willd.* ; cottony, stoloniferous, roots fibrous, leaves
lanceolate sinuate-toothed or subpinnatifid, heads usually solitary, involucre
hemispherical.   *Cir'sium ang'licum*, Lamk.

Wet meadows, rare, from York (ascending to 1,200 ft.) and N. Wales southd.;
Ireland; fl. June–Aug.—Perennial. *Stem* 10–18 in., terete, cottony. *Leaves*
few, 6–10 in., lower long-petioled, lobes angled not deep or long; upper
½-amplexicaul, auricled. *Heads* 1–1½ in., peduncled ; involucre cottony ;
bracts appressed, outer mucronate, inner slender purple acuminate. *Flowers*
dark purple. *Fruit* pale, slender, smooth; pappus dirty white.—Distrib.
Holland, Spain, France.— *C. Woodwards'ii*, Wats., is, according to Syme,
probably a hybrid with *acau'lis*. Wilts, Glamorgan.

7. **C. tubero'sus**, *Hoffm.* ; not stoloniferous, root of fusiform tubers,
leaves deeply pinnatifid, lobes remote narrow 2–4-cleft, heads 1–3, involucre
ovoid depressed at the base.

Meadows, Boyton, Wilts, and near Swindon; fl. Aug.-Sept.—Very closely
allied to *C. pratensis*, and regarded by Naegeli as a var. of it.—Distrib.
France, Germany, and southd.

8. **C. heterophyl'lus**, *Willd.* ; stoloniferous, roots fibrous, leaves
lanceolate serrulate ciliate white beneath, upper entire or pinnatifid, heads
few large intruded at the base. *Melancholy Thistle.*

Subalpine pastures and rivulets, from Caithness to S. Wales, Stafford, and
Derby; ascends to 2,700 ft. in the Highlands; fl. July-Sept.—*Rootstock*
creeping. *Stem* 2–3 ft., white, cottony, furrowed. *Leaves* soft, radical long-
petioled, 8–18 in.; upper often ovate, ½-amplexicaul, base cordate. *Heads*
1½–2 in. diam.; involucre ovoid, bracts finely pubescent, appressed, outer
mucronate, inner linear. *Flowers* red-purple. *Fruit* smooth, small, brown ;
pappus-hairs brownish.—Distrib. N. and Mid. Europe (Arctic), N. America.
—*C. Carolo'rum*, Jenner, is a hybrid with *C. palus'tris*.

29. ONOPOR'DON, L. Cotton Thistle.

Tall, erect, branched herbs. *Stems* broadly winged. *Leaves* alternate,
decurrent, spinous-toothed. *Heads* large ; involucre subglobose ; bracts
very many, imbricate, coriaceous, spinescent ; receptacle fleshy, pitted,
edges of the pits membranous toothed. *Corollas* all tubular, ventricose
above, purple, rarely white ; lobes 5, long, slender. *Filaments* nearly
glabrous ; anthers with a terminal appendage, cells shortly tailed. *Style-
arms* connate into a 2-fid cylinder, with a ring of hairs at the base. *Fruit*
obovoid, compressed, 4-ribbed, rugose ; pappus-hairs many-seriate, filiform
or flattened, barbed or toothed, connate at the base.—Distrib. S. Europe,
W. Asia, N. Africa ; species 12.—Etym. The old Greek name.

o. **Acan'thium** *L.* ; leaves sinuate-pinnatifid woolly.

Waste dry places, from Fife southd.; Channel Islands; (an alien or denizen, *Wats.*); fl. July–Sept.—A stout, hoary or cottony biennial, 2–5 ft. *Stem* spinous-winged to the top. *Leaves* decurrent, toothed and strongly spinous. *Head* 1½–2 in. diam.; involucre much contracted at the mouth, very cobwebby; bracts subulate, spinescent, recurved, green. *Flowers* pale purple. *Fruit* ¼ in., mottled grey; pappus-hairs white.—DISTRIB. Europe, Siberia; introd. in N. America.

30. SILYBUM, *Gærtn.* MILK-THISTLE.

A spinous glabrous shining herb. Characters of *Carduus*, but filaments glabrous and connate into a sheath.—ETYM. σίλλυβον, a white spotted thistle.

S. MARIA'NUM, *Gærtn.* ; leaves sinuate or pinnatifid, nerves white.

Waste places near gardens, &c., not indigenous; fl. July–Sept.—Erect, annual or biennial. *Stem* 1–4 ft., grooved, not winged. *Leaves* large, spines stout. *Heads* globose, 1–2 in. diam.; invol. bracts coriaceous, closely appressed, very broad, each with one very stout recurved terminal subulate spine ½–¾ in., and several shorter ones at its base, outermost merely spinous-toothed, mucronate; receptacle fleshy, hairy, not pitted. *Flowers* rose-purple. *Fruit* ¼ in., oblong, transversely wrinkled, black or grey; pappus white.—DISTRIB. From Holland southd., but indigenous only in the Mediterranean region and East.—Stems formerly eaten.

31. CICHOR'IUM, *L.* CHICORY.

Perennial herbs, with spreading branches; juice milky. *Leaves* radical and alternate, toothed or pinnatifid. *Heads* axillary; involucre cylindric; bracts in 2 series, inner erect connate at the base, outer shorter appressed; receptacle flattish, naked, pitted or bristly. *Corollas* all ligulate, blue or yellow; anther-cells not tailed; upper part of style and its slender arms hairy. *Fruits* crowded on the indurated receptacle, firmly embraced by the rigid invol. bracts, smooth, obovoid or turbinate, not beaked; pappus in 1–2 series of short obtuse scales.—DISTRIB. Europe, W. Asia; species 3. —ETYM. An old Greek name.

C. In'tybus, *L.* ; lower leaves runcinate, heads usually in pairs.

Waste places, roadsides, &c., throughout England ; rare, if native, in Scotland and Ireland; Channel Islands; fl. July–Oct.—Rather hispid. *Root* fleshy, tapering. *Stem* 1–3 ft., angled and grooved; branches straight, rigid. *Leaves* glandular-ciliate, oblong-lanceolate, upper ovate-cordate amplexicaul. *Heads* many, 1–1½ in. diam.; peduncle thickened in the middle; invol. bracts herbaceous, outer linear-lanceolate gland-ciliate. *Flowers* bright blue, rarely white; ligule rather broad, truncate, 5-toothed. *Fruit* angled, pale, mottled.—DISTRIB. Europe, N. Africa, Siberia, N.W. India; introd. in N. America.—The origin of the cultivated Chicory; the roots are boiled and eaten, or dried and used as Coffee.

32. **ARNOS'ERIS,** *Gærtn.* LAMB'S or SWINE'S SUCCORY.

A small, annual, scapigerous herb ; juice milky. *Leaves* all radical.
Heads few, small ; peduncles clavate, fistular ; invol. bracts in 1 series,
many, after flowering arching over the fruit ; receptacle flat, naked, pitted.
Corollas all ligulate, yellow ; anther cells not tailed ; upper part of style
and its short obtuse arms hairy. *Fruit* obpyramidal, furrowed and ribbed,
not beaked, crowned by a coriaceous angular ring.—DISTRIB. Europe to
Mid. Russia.—ETYM. ἄρνος, *a lamb,* and σέρις, *succory.*

A. pusil'la, *Gærtn.* ; leaves obovate-spathulate or -lanceolate toothed.
Dry pastures and fields, on the E. side of the Island, from Elgin to Dorset
and Kent, local ; (a colonist, *Wats.*) ; fl. June–July.—Glabrous or slightly
hairy. *Leaves* 2–4 in., narrow. *Scapes* 4–12 in., many, slender, rigid,
sparingly branched above. *Heads* campanulate, ½ in., inclined in bud ;
invol. bracts herbaceous, puberulous, linear-lanceolate, tips contracted
obtuse. *Fruit* pale brown, rugose between the ribs.

33. **LAP'SANA,** *L.* NIPPLEWORT.

Slender, erect, annual, branched herbs ; juice milky. *Leaves* alternate,
petioled, lower lyrate, upper toothed or entire. *Heads* small ; invol. bracts
few, 1-seriate, erect, outermost small ; receptacle flat, naked, dotted.
Corollas all ligulate, yellow ; anther-cells not tailed ; tip of style and its
linear obtuse arms hairy. *Fruit* slightly compressed, curved, striate, not
beaked ; pappus 0. – DISTRIB. Europe, W. Asia, N. Africa ; species 4.—
ETYM. An old Latin name.

L. commu'nis, *L.* ; lower leaves lyrate-pinnatifid, upper-entire.
Waste and cultivated ground, N. to Orkney ; ascends to 1,300 ft. in Northumbd.;
Ireland ; Channel Islands ; fl. July–Sept.—*Stem* 1–2 ft., paniculately branched,
hairs spreading. *Leaves* contracted into the petiole, membranous, terminal
lobe very large, sinuate-toothed, lateral small. *Heads* many, ¼ in.; peduncle
short, slender, naked ; invol. bracts 8–10, linear, rigid, keeled, green, glandular or glabrous, tip contracted obtuse. *Fruit* pale.—DISTRIB. Europe
(Arctic), N. Africa, N. and W. Asia, Himalaya ; introd. in N. America.—
Formerly eaten as a salad.

34. **PI'CRIS,** *L.*

Erect, branched, hispid herbs ; juice milky. *Leaves* alternate, entire
or sinuate-toothed. *Heads* solitary or corymbose, yellow ; invol. bracts
many, unequal, outer spreading ; receptacle flat, naked, pitted. *Fl.* all
ligulate ; anther-cells shortly tailed ; upper part of style and its slender
obtuse arms hairy. *Fruit* curved, grooved, transversely rugose, beaked
or not ; pappus-hairs 2-seriate, deciduous, feathery, outer fewer slender,
inner broader at the base.—DISTRIB. Europe, temp. Asia, Australasia ;
species about 24.—ETYM. πικρός, from its *bitterness.*

* *Outer invol. bracts narrow ; fruit not beaked.*

1. **P. hieracioi'des,** *L.* ; leaves obovate-lanceolate, upper ½-amplexicaul.

Waste places in a stiff soil. from Roxburgh southd.; Channel Islands; fl. July–Sept.—Hispid with stiff straight curled or hooked hairs. *Stem* 2–3 ft., stout, corymbosely branched. *Leaves* 4–12 in., narrow, lower petioled, upper sessile. *Heads* 1 in. diam., corymbose, or subumbellate (*P. arva'lis*, Jord.); peduncles bracteate; involucre subcampanulate; outer bracts short, clothed with black hairs. *Fruit* red-brown, angled; pappus snow-white.—Distrib. Europe, Asia, Australasia.

** *Outer invol. bracts cordate, fruit beaked.* Helminthia, *Juss.*

2. **P. echioi'des,** *L.*; lower leaves sinuate-toothed, upper cordate. Ox tongue.

Waste places in stiff soil, from Durham southd.; Haddington to Berwick in Scotland; E. Ireland, rare; Channel Islands; fl. June–Oct.—Stout, erect, 2–3 ft., branched, hispid and setose, hairs with tumid bases. *Leaves* oblong-lanceolate, radical petioled, upper amplexicaul. *Heads* 1 in. diam.; peduncles stout, stiff, rather swollen, diverging, naked; involucre hemispheric; outer bracts foliaceous, inner acuminate. *Ligules* short. *Fruit* red-brown, long, curved; pappus snow-white.—Distrib. From Holland southd., N. Africa.

35. CRE'PIS, *L.* Hawk's-beard.

Branched herbs; juice milky. *Leaves* chiefly radical. *Heads* panicled or corymbose, small; invol. bracts many, linear, equal, with a few smaller at their base; receptacle flat, naked, pitted, margins of the pits hairy or toothed. *Corollas* all ligulate, yellow or purplish; anther-cells not tailed; style-arms slender and upper part of style hairy. *Fruit* terete, striate, beak long short or 0; pappus-hairs in many series, slender, simple, white, silky, brown in *C. paludo'sa*.—Distrib. N. hemisphere, rare in the tropics; species about 130.—Etym. The classical name.

Sub-gen. 1. **Cre'pis** proper. *Peduncles* slender; buds erect. *Fruit* not beaked. *Pappus* white, silky.

1. **C. vi'rens,** *L.*; glabrous below, lower leaves toothed runcinate or lyrate, upper linear sagittate, inner bracts glabrous within as long as the pappus. *C. tectorum*, Sm. not L.

Waste and cultivated ground, cottage roofs, &c., from Caithness southd. ascends to 1,350 ft. in Derby; Ireland; Channel Islands; fl. June–Sept.—Annual, very variable. *Stems* 1 or more, 1–3 ft., furrowed, much branched; inflorescence usually glandular-hairy. *Leaves* often pinnatisect with narrow acute or obtuse lobes, petioled. *Heads* ½–¾ in. diam., campanulate; outer bracts subulate, inner linear. *Fruit* red-brown, ribs 10 or more, smooth.—Distrib. From Denmark southd., Canaries.

2. **C. bien'nis,** *L.*; hispid, leaves all runcinate-lyrate, upper simple sessile, inner bracts pubescent within shorter than the pappus.

Dry pastures, &c., in E. and Midland counties, from York to Kent and Sussex; Aberdeen; Dublin; Channel Islands; fl. June–July.—Biennial. *Stem* 1–4 ft., stout, channelled, ribbed. *Leaves* 6–13 in., lobes very irregular. *Heads* ¾–1 in. diam.; involucre campanulate, bracts glabrous externally.

outer spreading. *Fruit* ⅓ in., red-brown; ribs close set, prominent, rough.—
DISTRIB. Europe.—Resembles *C. taraxacifolia.*

3. **C. hieracioï'des,** *Waldst.* and *Kit.* ; glabrous or hairy, lower leaves
oblong-spathulate, upper linear-oblong ½-amplexicaul, bracts with gland-
tipped hairs. *C. succisæfo'lia,* Tausch. ; *Hiera'cium mol'le,* Sm.
Mountain woods, from York to Dumbarton and Banff; ascends to 1,200 ft. in
Northumbd.; fl. July–Aug.—Slender, perennial, variable in pubescence.
Leaves 2-4 in., obtuse, entire or sinuate-toothed. *Heads* ¾-1 in. diam.,
few; involucre subcylindric, outer bracts appressed short about as long as
the pappus. *Fruit* contracted at the base and tip; ribs many, smooth.—
DISTRIB. Mid. Europe to the Caucasus.

SUB-GEN. 2. **Barkhau'sia,** *Mœnch* (gen.). *Peduncles* slender or
thickened upwards. *Fruit* beaked, many-ribbed, minutely hispid.

4. **C. fœ'tida,** *L.* ; hispid, stem branched from below, leaves runcinate-
pinnatifid, peduncles bracteate incurved thickened upwards, buds drooping,
invol. bracts tomentose inner hardening and enclosing the outer fruits.

Chalky and gravelly banks, rare, from Cambridge and Norfolk to Sussex
and Kent; fl. June–July.—Annual or biennial. *Stem* 1-2 ft., terete, faintly
furrowed, usually with many ascending corymbose branches. *Heads* ¾ in.
diam., bright yellow; peduncles long; invol. bracts often setose and
glandular, outer erect or spreading. *Fruit* yellow-brown, beak of outer
shorter than the bracts, of inner longer.—DISTRIB. From Belgium southd.,
Himalaya, N. Africa.

5. **C. taraxacifo'lia,** *Thuill.* ; hispid, stem branched above, leaves
runcinate-pinnatifid, peduncles very slender not thickened, buds erect,
inner invol. bracts not enclosing the outer fruits.

Dry banks and chalky pastures, local, from Yorkshire to Cornwall and Kent;
Carnarvon ; fl. June–July.—Habit and foliage of *C. biennis,* biennial. *Stem*
1-2 ft., ribbed and furrowed. *Heads* ¾-1 in. diam.; involucre cylindric-
campanulate, often glandular as well as tomentose, outer bracts spreading,
not hardening. *Flowers* yellow, outer striped with brown on the back.
Fruits yellow-brown, all long-beaked and very slender.—DISTRIB. W.
Europe from Belgium southd., N. Africa.

C. SETO'SA, *Haller fil.* ; hispid, lower leaves sinuate-toothed or runcinate-
pinnatifid, upper amplexicaul, peduncles slender not thickened, buds erect,
invol. bracts prickly, inner not enclosing the outer fruits.

A casual in clover fields; fl. Aug.—Biennial. *Stem* 1-3 ft., erect, branched
from the base, angled and furrowed, leafy. *Leaves* very variable. *Peduncles*
rigid, deeply grooved. *Heads* ½ in. diam., campanulate; involucre contracted
in fruit; bracts very rigid, slender, keeled, outer subulate spreading, inner
linear shorter than the pappus, hardening. *Fruits* all slender with long
beaks.—DISTRIB. Mid. and S. Europe.

SUB-GEN. 3. **Ara'cium,** *Monn.* (gen.). *Peduncles* very slender; buds
erect. *Fruit* slender, not beaked, many-ribbed, quite smooth. *Pappus*
of dirty-white fragile hairs.

6. C. paludo'sa, *Mœnch;* radical leaves obovate-lanceolate, petiole slender, cauline sessile amplexicaul auricled. *Hieracium paludosum,* L.

Moist mountain meadows, copses, &c., from S. Wales and Salop to Caithness; ascends to 2,000 ft. in the Highlands; N. Ireland; fl. July–Sept.—Perennial; glabrous, inflorescence covered with black glandular hairs. *Stem* 1–3 ft., slender, furrowed. *Leaves* membranous, runcinate-toothed, long-acuminate; radical 3–5 in.; cauline contracted in the lower third. *Heads* few, ⅔–1 in. diam., corymbose; bracts of peduncles minute, subulate with toothed bases; ligules yellow; styles livid. *Fruit* cylindric, strongly ribbed, obscurely contracted at the top, pale.—DISTRIB. Europe (Arctic), W. Siberia.—A *Crepis* with the pappus of *Hieracium.*

36. HIERACIUM, *L.* HAWKWEED.

Perennial herbs, often covered with glandular or stellate hairs; juice milky. *Leaves* radical and cauline, alternate. *Heads* solitary corymbose or panicled; invol. bracts many, imbricate, unequal; receptacle flattish, naked, pitted, margins of the pits toothed or hairy or fimbriate. *Corollas* all ligulate, yellow, rarely orange; anther-cells not tailed; style-arms slender, and upper part of style hairy. *Fruit* angled or striate, not beaked; pappus-hairs 1-seriate, simple, rigid, unequal, tawny or brownish, brittle, often girt with a short crenulate ring.—DISTRIB. N. temp. and Arctic regions; species about 150.—ETYM. *lépaξ, a hawk.*

In the following attempt to classify the British *Hieracia,* I have been guided by Mr. Baker. I believe that there are no characters whereby the 9 forms, from *alpinum* to *boreale* inclusive, can be more than approximately defined; of these 9 Bentham makes 4, Nyman 21, and Backhouse, followed by Babington (having regard to a considerable amount of constancy under cultivation), makes 30, of which only 16 bear the same name in Nyman. Variable as the genus is, the sequence of its forms is so natural as to have been recognised by all botanists. This sequence represents to a considerable extent the spread of the forms in altitude and area in the British Isles.

SECTION 1. **Piloselloi'dea.** Rootstock stoloniferous. *Stem* scape-like. *Invol. scales* irregularly imbricate. *Ligules* glabrous. *Fruit* minute, furrowed, crowned with a crenate disk; pappus-hairs slender, equal.

1. H. Pilosel'la, *L.*; stolons long, scape leafless, heads solitary, invol. bracts acute, ligules pale yellow, styles yellow. *Mouse-ear Hawkweed.*

Banks, wall-tops, &c., N. to Orkney; ascends to 2,400 ft. in Yorks.; Ireland; Channel Islands; fl. May–Aug.—Villous with long soft hairs. *Stolons* creeping. *Leaves* 2–4 in., oblong-lanceolate or obovate-spathulate, stellately downy beneath. *Scapes* 2–10 in., with 1–3 leaves or 0. *Heads* ⅔–1 in. diam.; involucre subcampanulate and top of scape stellately downy and with glandular hairs; ligule often striped with brown on the back. *Fruit* ${}_{10}^{1}$ in., dark.—DISTRIB. Europe, N. Africa, N. and W. Asia. *H. Peleteria'num,* Mer. (*H. pilosis'simum,* Fries), is a more densely silky form, with stolons shorter, heads larger.

H. AURANTI'ACUM, *L.* ; rootstock creeping, stolons short or 0, heads corymbose, invol. bracts obtuse, ligules orange, styles brown.
Naturalized in copses in the N. of England and Scotland; fl. June–July.—A larger plant than *H. Pilosella;* leaves not pubescent beneath; flowers orange-red.—DISTRIB. Scandinavia, Pyrenees to the Carpathians.

SECTION 2. **Pulmona'rea.** *Stolons* 0. *Rootstock* forming in autumn buds which in the following year develop rosettes of persistent leaves and a naked or 1–few-leaved *scape. Invol. bracts* irregularly imbricate. *Ligules* hairy at the back and tip, or tip only. *Fruit* short, furrowed, without a crenate disk ; pappus-hairs rigid, unequal.

2. **H. alpi'num,** *L.* ; green, not glaucous, softly hairy and shaggy, scape 4–10 in. with 1–2 small leaves or 0, invol. bracts softly silky, outer lax or spreading, ligules hairy on the back or tip.

Alps, N. Wales, Westmoreland to Sutherland, alt. 2,000–4,000 ft., rare ; fl. July-Aug.—*Heads* 1–1½ in. diam., bright or pale yellow.—DISTRIB. N. and Arctic zones, Alps of Mid. and S. Europe, exclusive of Pyrenees.—The following varieties appear to be very local.
H. alpi'num proper (*H. melanoceph'alum,* Tausch.) ; leaves lanceolate or spathulate nearly glabrous above, head solitary, invol. hemispheric, hairs black, bracts all acute, outer lax, style yellow. Forfar, Aberdeen. —VAR. *H. holoseric'eum,* Backh. (*H. alpi'num,* Engl. Bot.); leaves spathulate or lanceolate obtuse entire hairy on both surfaces, heads solitary, invol. turbinate, hairs long white, outer bracts broad obtuse, inner acute appressed, style yellow. Scotland and Cumberland.—VAR. *H. exim'ium,* Backh.; (*H. villosum,* Engl. Bot.); taller, 6–15 in., leaves lanceolate hairy on both surfaces, heads 1 or few, invol. truncate below, hairs black, bracts many slender acute, style yellow or livid. *H. tenellum,* Lond. Cat., is a slender form of this. Scotland.—VAR. *H. calenduliflo'rum,* Backh.; stem simple or branched, leaves broadly obovate obtuse toothed, primordial orbicular, invol. rounded at the base, hairs silky black, bracts acute, style livid. Scotland. The handsomest form, indicating a passage to *H. nigrescens.*

3. **H. nigres'cens,** *Willd.* ; more or less covered with scattered soft hairs, scape 6–18 in. with 1–2 small leaves or 0, involucre villous, hairs black glandular, outer bracts lax, ligules glabrous or nearly so.

Alps, York to Sutherland; ascends to 4,500 ft.; fl. Aug.-Sept.—Intermediate between *H. alpinum* and *Lawsoni* in size, altitudinal distribution and area. Usually larger and more slender than *alpinum,* with broader, narrower, coarsely toothed leaves, much shorter soft hairs, and blacker bristles and glands on the involucre ; heads as large.
H. nigres'cens proper (*H. pulmona'rium,* Sm.); leaves with large irregular teeth, cauline few, involucre dark green ovoid at the base, bracts rather broad outer obtuse tips woolly, flowers deep yellow, style dark. Scotland.—VAR. *H. gracilen'tum,* Backh.; green, root-leaves lanceolate, cauline few large, involucre ventricose black with soft hairs and glands, bracts broad woolly at the tips, ligules nearly glabrous on the back, styles livid. Scotland. Very near *alpinum* indeed.—VAR. *H. globo'sum,* Backh.; glaucous or green,

radical leaves ovate, cauline few small bract-like or 0, buds globose, heads
large, involucre rounded at the base at last spherical, bracts appressed,
styles yellow. Scotland.—VAR. *H. chrysan'thum,* Backh. (*H. atratum,* Bab.);
green, radical leaves ovate sharply coarsely toothed long-petioled, cauline
minute narrow petioled, involucre rounded at the base, hairs and glandular
hairs short, bracts many, styles yellow or faintly livid. (*B. microcephalum,*
Lond. Cat., is a small-headed form.) Frequent in Scotland, rare in Cumberland.—VAR. *H. senes'cens,* Backh.; green, radical leaves lanceolate evenly
toothed, cauline linear-lanceolate petioled, heads 2 or more, involucre ovoid
at the base, bracts woolly at the tip incurved in bud, styles yellow. Scotland.—VAR. *H. lingula'tum,* Backh. (*H. saxif'ragum,* Bab.; *H. divarica'tum,*
Don); green, 15–24 in., radical leaves few toothed coarsely hairy above,
petiole short, cauline few sessile, heads several, involucre ¯broad, base at
length truncate, dark with hairs, bracts straight in bud, styles livid.
Scotland.

4. **H. ang'licum,** *Fries ;* glaucous green, stem 1-2 ft. more or less
leafy slightly hairy or glabrous below, radical leaves ovate-lanceolate,
petioles shaggy, cauline oblong or ovate broad sessile amplexicaul, heads
1–5, 1–1½ in. diam., ligules many, styles livid. *H. Lawso'ni,* Sm. in part,
not Villars.

Mountain districts, York to Orkney; ascends to 2,700 ft.; Ireland; fl. July-
Aug.—A handsome species, best characterised by its size, ovate-lanceolate
leaves, shaggy petioles, several (rarely 1) large bright-coloured heads, and
livid styles.—DISTRIB. Pyrenees.
H. ANG'LICUM proper (*H. cerinthoi'des,* Backh. not L.; *H. decipiens,* Syme);
petioles long shaggy winged, cauline leaves 1–2 ovate, heads 2–5, involucre ventricose rounded at the base, bracts slender, ligules hairy at the tip
(*b. amplexicau'le* and *c. acutifo'lium* of Lond. Cat. are forms of this).
Sub-sp. I'RICUM, *Fries* (*H. Lapeyrou'sii,* Bab. not Frœl.); more robust and
leafy, radical leaves in a less marked rosette or scattered, petioles shorter,
cauline broadly ovate not contracted above the base, involucre truncate at
the base constricted after flowering, ligules glabrous. This, which is confined to Britain and Ireland, bears much the same relation in foliage to
Lawso'ni that *tridenta'tum* does to *vulga'tum.*

5. **H. muro'rum,** *L.* ; green or glaucous, stem 1-2 ft. glabrous or hairy
below, primordial leaves suborbicular, radical in a distinct rosette ovate
acute cordate or rounded at the base entire or toothed, cauline 0 or very
few, heads 2–6 ¾–1 in. diam., peduncles floccose and covered with scattered
simple and gland-tipped hairs.

Woods, heaths, walls and rocks, N. to Shetland; ascends to 2,000 ft.; Ireland;
fl. July–Sept.—The commonest *Hieracium* of Britain except *Pilosel'la* and
sylvat'icum, and best distinguished from *H. ang'licum* by the smaller heads
and less robust habit; and from *H. nigres'cens* by its large size and less
villous or hairy stem and involucres: but there is every transition between
these and the following.—DISTRIB. Europe (Arctic), N.W. Asia, Himalaya,
N. America.—I can make nothing of the following sub-species and varieties,
which are adopted from Backhouse's monograph.

H. MURO′RUM proper; green, radical leaves toothed slightly hairy, petioles slender, cauline often large and petioled, heads many small, peduncles short, involucre rather ventricose more or less villous with black and gland-tipped hairs, styles livid or yellow. The common form.—VAR. *H. nit′idum*, Backh.; radical leaves dark green lanceolate coarsely toothed, involucre more ventricose, style yellow. Aberdeen and Argyll. The passage to *H. nigres′cens*.

Sub-sp. H. CÆ′SIUM, *Fries* (*H. muro′rum*, Sm.); dull glaucous green, radical leaves coriaceous usually narrowed to a toothed base glabrous above, cauline 0 or very small and sessile, heads few large, involucre subglobose almost glandless and less hairy hoary with stellate down, ligules bright yellow glabrous, styles livid. English, Welsh, and Scotch Mts. The passage to *H. sylvat′icum.*—VAR. *H. flocculo′sum*, Backh. (*H. stellig′erum*, Backh. not Frœl.); ashy-green, stem floccose throughout rather leafy, radical leaves more or less toothed narrowed into long petioles stellately downy on both surfaces, cauline large ½-amplexicaul, involucre ovoid hoary with whitish hairs, bracts acuminate. Clova Mts.—VAR. *H. obtusifo′lium*, Backh.; yellow green, stem 1-leaved, petioles densely villous, peduncles spreading. involucre ventricose truncate at the base, ligules hairy at the tips. Clova Mts.

Sub-sp. H. PAL′LIDUM, *Fries;* very glaucous, radical leaves ovate or lanceolate conspicuously fringed with long hairs, cauline sessile or the lower petioled, heads 2–6, involucre ventricose base ovoid constricted above with few simple or gland-tipped hairs, bracts appressed acute, styles yellow. England, Scotland, and Ireland. Very near *H. ang′licum.*—VAR. *H. cineras′cens*, Jord. (*H. lasiophyl′lum*, Bab.); still more hairy and glaucous, stem hairy fragile, peduncles and involucres more densely setose and hoary.—VAR. *H. argen′-teum*, Fries; very glaucous, almost glabrous, stem fragile fistular and subentire radical leaves almost glabrous, cauline sessile or the lower petioled, peduncles long rigid and small involucres with scattered hairs.—VAR. *H. Gibso′ni*, Backh. (*H. hypochœroi′des*, Gibs.); stem wiry often forked. radical leaves broadly ovate, base obtuse spotted with purple, teeth small, petioles slender short, peduncles rigid floccose and setose, invol. bracts short broad obtuse margins downy, styles yellow. Yorkshire and Ireland on mountain limestone.—VAR. *H. aggrega′tum*, Backh.; radical leaves broader obtuse toothed below glabrous or hairy beneath and ciliate. cauline subsessile, peduncles erect crowded umbellate and narrow turbinate involucres densely floccose, bracts obtuse. Aberdeen and Forfar alps.

6. **H. sylvat′icum**, *Sm.* ; green or glaucous, stem 1–3 ft. nearly glabrous below, primordial leaves lanceolate, radical petioled distinctly alternate lanceolate sharply toothed or subpinnatifid, teeth pointing forwards, cauline 2–8, heads ¾–1 in. diam. panicled or corymbose and peduncles floccose and with simple and gland-tipped hairs rarely naked, ligules glabrous, styles livid.

Banks and copses, from Ross southd.; ascends to 3,500 ft. in the Highlands; rare in Ireland; fl. July–Sept.—Best distinguished from *H. muro′rum* by the narrower leaves less crowded in a rosette, and more leafy stem; but some vars. of *muro′rum*, as *cæ′sium* and *nit′idum*, show this foliage. The most leafy states, as *goth′icum* and *tridenta′tum*, show a passage to *borea′le* and *croca′tum*.
—DISTRIB. Europe (Arctic), N. Asia, N. America.

H. SYLVAT′ICUM proper. (*H. vulga′tum*, Fries; *H. macula′tum*, Sm.); green or glaucous, stem 1–1½ ft. hardly leafy, radical leaves petioled in a persistent loose rosette toothed in the middle or nearly entire often spotted, cauline petioled upper sessile, heads many, peduncles straight, involucre cylindric in bud floccose and with gland-tipped hairs, bracts equal alternate subacute. Very common.—VAR. *cine′reum*, Backh.; ashy green or glaucous, stem branched, radical leaves few, cauline subentire, heads nearly glabrous, bracts broad more obtuse. Orme′s Head.—VAR. *rubes′cens*, Backh.; green, stem robust purplish, leaves ovate, cauline 1–2, heads few large, bracts broad subacuminate. Settle, Yorkshire.—*H. nemoro′sum* is a leafy form passing into *tridenta′tum; monta′num* I do not know; *macroceph′alum* is a large-headed alpine form.

Sub-sp. H. GOTH′ICUM, *Fries;* dark green, stem 2–4 ft. rigid slender leafy simple or branched above, radical leaves withering in summer shortly petioled, cauline ovate or lanceolate acute toothed in the middle, upper sessile, heads small few, involucre subglobose dark green glabrous or nearly so, bracts imbricate broad obtuse, ligules glabrous, styles yellow or with livid hairs.—Subalpine districts, N. Wales, York to Aberdeen; Ireland.— Very distinct at first sight, but intermediates occur with *H. sylvat′icum,* as also with *borea′le.*—VAR. *latifo′lium*, Backh., is a Clova Mt. plant with more numerous and broader leaves.

Sub-sp. H. TRIDENTA′TUM, *Fries;* green, stem 2–5 ft. rigid leafy subcorym-bosely branched, radical leaves 0 or withering in summer obtuse, cauline ovate or lanceolate sparingly toothed in the middle acute rounded at the base, branches and peduncles slender leafless, involucre constricted in the middle after flowering.—Hilly districts, York to Devon and Kent; N. Wales.—Forms the passage to the *Accipitri′na* group.

Sub-sp. H. DEWA′RI, Syme; bright green, stem 1–3 ft. sparingly leafy corym-bosely branched, lower leaves elliptic petiole winged, cauline ovate-lanceolate ½-amplexicaul, heads few loosely panicled and peduncles sparingly hairy and setose, involucre cylindric from a conic base, bracts dark green obtuse, outer few short appressed, inner with pale margins.—Scotland.

SECTION 3. **Accipitri′na.** *Stolons* 0. *Rootstock* forming in autumn buds which develop in the following year early withering radical leaves and an erect very leafy stem. *Invol. bracts* imbricate in 2 or many series. *Ligules* glabrous or tip minutely hairy. *Fruit* short, furrowed, without a crenate disk ; pappus-hairs rigid, unequal.

7. **H. prenanthoi′des,** *Villars ;* stem 2–3 ft., leaves oblong or linear-oblong reticulate and glaucous beneath denticulate, lower petioles am-plexicaul, upper cordate and auricled, heads in branched leafy corymbose panicles usually thickly clothed with black gland-tipped hairs, peduncles short floccose, ligules hairy at the tip, styles dark or yellowish. *H. denti-cula′tum*, Sm.

Subalpine regions, York to Orkney; ascends to 2,400 ft.; Wicklow in Ireland; fl. Aug.–Sept.—*Stem* rigid, hairy or glabrous, leafy, often much branched. *Leaves* ciliate, hairy on both surfaces. *Peduncles* spreading, short, hoary. *Heads* ¾ in. diam.; involucre cylindric in bud; bracts few, outer short.— DISTRIB. Europe (Arctic and Alpine), Siberia, Himalaya.

Of *H. stric′tum*, Fries, I have seen no British specimens; Fries′ specimen differs from *prenanthoi′des* chiefly in the larger heads.

8. H. umbella'tum, *L.* ; stem 1-2 ft. wiry hairy or shaggy below, leaves sessile narrowly linear or oblong-lanceolate base narrowed toothed, heads subumbellate, and peduncles stellately downy but hairless, bracts many tips recurved, ligules glabrous, styles yellow.

Thickets, &c., in dry and rocky places, from Roxburgh and Dumfries to Cornwall and Kent; ascends to 1,200 ft. in Yorks.; local in Ireland; Channel Islands; fl. July–Sept.—Usually very distinct, from its short stem, wiry habit, many narrow leaves, large subumbellate glabrous heads and recurved bracts, but varieties pass into *croca'tum. Leaves* generally uniform, teeth distant, reticulate, stellately downy beneath. *Involucre* ovoid or subcylindric, dark or pale ; peduncles slender, rigid.—DISTRIB. Europe (Arctic), N. and W. Asia to the Himalaya, N. America (*H. canaden'se,* Fries).

VAR. *filifo'lium,* Backh. ; stem leafy throughout, leaves very narrow margins revolute quite entire. Lough Neagh.

9. H. croca'tum, *Fries;* stem 2-4 ft., glabrous or hairy, leaves lanceolate or oblong sessile base rounded narrowed or truncate toothed or entire often glaucous beneath, heads and peduncles downy glabrous or hairy, bracts appressed obtuse, ligules glabrous, style livid or yellow, pappus almost white or discoloured.

Mountain districts, from Wales and York to Orkney ; ascends to 1,200 ft. ; Ireland, rare ; fl. July–Sept.—Intermediate between *umbella'tum* and *borea'le.* —*H. jura'num,* Fries (*H. Borre'ri,* Syme), a plant from Selkirk ? with evident root leaves and few broad stem leaves, seems intermediate between this and *sylvat'icum.* Syme refers it to *prenanthoi'des,* but to me it appears quite different.—DISTRIB. Arctic, Mts. of N. and Mid. Europe, Himalaya.

H. CROCA'TUM proper ; leaves sessile base rounded lower narrowed below, nerves obscure, heads simple subcorymbose, base broad, invol. bracts usually glabrous. *H. inuloi'des,* Tausch.—Mountain districts.

Sub-sp. H. CORYMBO'SUM, *Fries;* more robust, branches spreading or ascending, leaves ovate irregularly toothed hairy above glaucous and loosely reticulate beneath, heads many panicled, involucre ultimately truncate sparsely hairy, pappus tawny. *H. rig'idum,* Backh. not Fries.—York to Orkney ; Antrim.—*Leaves* of *H. borea'le,* but upper amplexicaul more glaucous and nerved.

10. H. borea'le, *Fries;* stem 2-4 ft. very hairy below stellately downy above, leaves broad at the base ovate or ovate-lanceolate toothed lower petioled upper broader sessile, heads many in leafy panicles or corymbs, peduncles floccose, involucre ovoid dark nearly glabrous, bracts broad obtuse, ligules glabrous, style livid. *H. sabau'dum,* Sm. ; *H. heterophyl'lum,* Bladon.

Hedgebanks, copses, &c., Skye, and from Banff and Dumbarton southd. ; rarer in Scotland ; Ireland ; fl. Aug.-Oct.—*Stem* often reddish, leafy throughout. *Leaves* with teeth pointing forwards, lowest with villous petioles. *Heads* ⅔-1 in. diam.—DISTRIB. Mid. and S. Europe, rare in Scandinavia.— Allied to *goth'icum,* but without radical leaves, heads smaller, invol. bracts more numerous.

37. HYPOCHŒ'RIS, L. CAT'S-EAR.

Annual or perennial scapigerous herbs ; juice milky. *Leaves* radical, pinnatifid. *Heads* on simple or dichotomously branched scapes ; invol. bracts in many series, imbricate ; receptacle flat, with narrow membranous scales. *Corollas* all ligulate ; anther-cells shortly tailed ; upper part of style and its short obtuse arms hairy. *Fruit* striate, scabrous, beak of the outer very short or 0, of the inner long, slender ; pappus of one row of feathery hairs with usually an outer row of short, stiff bristles.—DIS-TRIB. Europe, W. Asia ; species 30.—ETYM. doubtful.

1. **H. gla'bra,** *L.* ; annual, leaves narrowly obovate-oblong toothed or sinuate, scapes many branched, involucre as long as the flowers, outer pappus short.

Dry fields, &c., from the Clyde and Elgin to Devon and Kent ; rare in Scotland ; Channel Islands ; fl. June–Aug.—Almost glabrous. *Leaves* spreading, 2–4 in., rarely pinnatifid, obtuse or subacute. *Scapes* several, 6–12 in., slender or stout, naked or with 1–2 scale-like leaves ; peduncles erect. *Heads* ½–⅔ in., yellow, cylindric ; involucre subcylindric ; bracts very unequal, few, green, linear, acute. *Fruit* red-brown ; pappus dirty white, longer than the involucre.—DISTRIB. From Gothland southd, N. Africa, W. Asia.

H. gla'bra proper ; marginal fruits not beaked.—VAR. *H. Balbis'ii,* Lois. ; all the fruits beaked.—Kent, Shropshire, Channel Islands.

2. **H. radica'ta,** *L.* ; perennial, leaves narrowly obovate-oblong sinuate or runcinate-pinnatifid, scapes many branched, involucre shorter than the flowers, outer pappus short.

Meadows, waste places, &c., N. to Orkney ; ascends to 1,600 ft. in the Highlands ; Ireland ; Channel Islands ; fl. June–Sept.—Hispid, rarely glabrous. *Leaves* many, 3–10 in., hispid on both surfaces. *Scapes* 6–18 in., stout; peduncles with small scale-like leaves. *Heads* 1–1½ in. diam., yellow ; involucre subcampanulate ; bracts many, green, attenuated to the tip, often strongly ciliate on the margin and back. *Fruits* red-brown, all beaked, strongly muricate.—DISTRIB. Europe, N. Africa.

3. **H. macula'ta,** *L.* ; perennial, leaves obovate-spathulate sinuate-toothed, scapes 1 or few, involucre shorter than the flowers ciliate with curly hairs, outer pappus 0. *Achyroph'orus macula'tus,* Scop.

Chalk and limestone pastures, rare, Westmoreland, N. Wales, Cornwall, Cambridge, Suffolk, Essex ; Channel Islands ; fl. July–Aug.—Hispid. *Leaves* sessile, 1–4 in., often spotted above. *Scape* stout, glabrous above, rarely forked, with 1–2 scale-like leaves. *Heads* 1 in. diam., yellow ; involucre broadly campanulate ; bracts many, slender. *Fruits* all beaked ; faintly muricate ; pappus white.—DISTRIB. Europe, N. Asia.

38. LEON'TODON, L. HAWKBIT.

Perennial, scapigerous herbs ; juice milky. *Leaves* all radical, obovate, sinuate-toothed or pinnatifid. *Heads* on simple or branched scapes, yellow ; invol. bracts many, in several series, outer smaller ; receptacle flat, naked. *Flowers* all ligulate ; anther-cells not tailed : style-arms linear obtuse, and upper parts of style hairy. *Fruit* terete, grooved, transversely rugose ; beak short ; pappus-hairs rigid, 1–2-seriate, slightly dilated at the base, outer simple and rough, or the outer of toothed scales and the inner of one series of feathery hairs.—DISTRIB. Europe, W. Asia ; species about 25.—ETYM. λέων and ὀδούς, from the *toothed* leaves.

SECTION 1. *Pappus* of outer flowers of toothed scales, of inner of feathery hairs. *Buds* drooping.

1. **L. hir′tus,** *L.* ; leaves hispid oblong or lanceolate. *Hedyp′nois,* Sm. *Thrin′cia,* Roth.

Gravelly pastures, &c., from Durham southd. ; Edinburgh to Roxburgh only in Scotland ; Ireland ; Channel Islands ; fl. July–Aug.—A hispid biennial ; hairs often forked. *Leaves* 3–6 in., long-petioled. *Scapes* very many, 4–8 in. *Heads* ½–¾ in. diam. ; involucre campanulate ; bracts lanceolate, subacute, keeled, glabrous, edges and back hispid or ciliate. *Fruit* ⅓ in., closely grooved and minutely muricate, outer stouter curved almost smooth ; pappus white.—DISTRIB. From Gothland southd.

SECTION 2. *Pappus* of all the flowers 2-seriate, slender ; outer short, scabrid ; inner longer, base dilated, feathery. *Buds* drooping.

2. **L. his′pidus,** *L.* ; leaves hispid oblong-lanceolate. *Hedyp′nois,* Sm. *Apar′gia,* Willd.

Meadows, &c., on dry soil, from Isla and Forfar southd. ; ascends to near 2,000 ft. in Northumbd. ; Ireland ; Channel Islands ; fl. June–Sept.—A hispid biennial ; rootstock truncate ; hairs often forked. *Leaves* 3–5 in. *Scapes* 6–18 in., few, slender, swollen at the top. *Heads* 1½ in. diam. ; involucre obconic ; bracts linear lanceolate, tips woolly, obtuse. *Fruit* slender. ribs very muricate above ; pappus brownish-white.—DISTRIB. Europe.

SECTION 3. *Pappus* of all the flowers 1-seriate, feathery, base dilated. *Buds* erect. *Oporin′ia,* Don.

3. **L. autumna′lis,** *L.* ; leaves glabrous lanceolate. *Apargia,* Willd.

Pastures and waste places, N. to Shetland ; ascends to 3,000 ft. in the Highlands ; Ireland ; Channel Islands ; fl. July–Sept.—Glabrous below ; hispid above ; rootstock truncate. *Leaves* 4–10 in., variable, entire or pinnatifid. *Scape* ascending, usually solitary, branched, rarely simple, bracteate ; peduncles swollen above. *Heads* ½–1½ in. diam., involucre obconic or campanulate ; bracts glabrous, hispid, or in subalpine districts clothed with black hairs (*L. praten′sis,* Koch. ; *Apar′gia Tarax′aci,* Hornem.), outer subulate, inner linear obtuse. *Fruit* very slender, ribbed, slightly muricate ; pappus brownish-white.—DISTRIB. Europe (Arctic), N. and W. Asia, Greenland ; introd. in N. America.

39. TARAX′ACUM, *Juss.* DANDELION.

Perennial, scapigerous herbs ; juice milky. *Leaves* all radical, entire or pinnatifid. *Heads* solitary, scapes fistular leafless ; invol. bracts imbricate ; inner equal, erect ; outer often recurved ; receptacle flat, naked, pitted. *Corollas* all ligulate, yellow ; anther-cells not tailed ; style-arms slender and upper part of style pubescent. *Fruit* compressed, ribbed, muricate above the middle, abruptly beaked ; pappus-hairs in many series, simple, white.—DISTRIB. All temp. and cold climates ; species about 6.—ETYM. ταράσσω, from its *alterative* effects.

T. officina'le, *Web.* ; leaves toothed sinuate or runcinate-pinnatifid. *Leon'todon Tarax'acum,* L.

Meadows and waste places, N. to Shetland; Ireland; Channel Islands; fl. March–Oct.—Glabrous, or cottony at the crown and involucre. *Root* long, stout, black. *Leaves* oblong-obovate or spathulate, lobes usually toothed. *Scapes* 1 or more, ascending or erect. *Head* ½–2 in. broad, bud erect; involucre campanulate,. outer bracts more or less recurved, inner erect. *Corollas* bright yellow, outer often brown on the back. *Fruit* brown, with a beak of equal length.—DISTRIB. Arctic and N. and S. temp. regions.—A well-known medicine.

T. officina'le proper (*T. Dens-leo'nis,* Desf.) ; leaves bright green runcinate-pinnatifid, outer bracts recurved, fruit pale. The common form in cultivated ground ; ascends to 2,700 ft. in the Highlands.—VAR. *T. erythrosper'mum,* Andrz. ; leaves dull green often glaucous runcinate-pinnatifid, outer bracts spreading, inner appendiculate below the tip, fruit dark brown. Dry places. —VAR. *T. læviga'tum,* DC. ; leaves dull green pinnatifid, outer bracts erect broader, inner appendiculate or gibbous below the tip, pale. Sandy places. —VAR. *T. palus'tre,* DC. ; leaves sinuate-toothed or pinnatifid, lobes broad, outer bracts ovate spreading or erect, inner simple at the tip, fruit pale.— Moist moorlands ; ascends to 4,000 ft. in the Highlands.

40. LACTU'CA, *L.* LETTUCE.

Erect, annual or perennial, leafy, branched herbs : juice milky. *Leaves* alternate, upper often sagittate. *Heads* corymbose, small, few-fld. ; involucre narrow, cylindric ; bracts few, in several series ; receptacle flat, naked. *Corollas* all ligulate, yellow, purple or blue ; anther-cells shortly tailed ; style-arms slender and upper part of style hairy. *Fruit* flattened terete or angled, beak short or long slender ; pappus of many soft, slender, silvery, fugacious hairs.—DISTRIB. N. temp. regions, S. Africa ; species about 60.—ETYM. The classical name.

* *Leaves with the keel usually bristly. Beak as long as the bordered fruit.*

1. **L. viro'sa,** *L.* ; sparsely scabrous, leaves spreading, radical obovate-oblong sinuate-toothed, cauline amplexicaul with deflexed auricles, branches of panicle long spreading, fruit black.

Hedgebanks and waste places, from Perth and Ayr southd.; rare in Scotland; Channel Islands; fl. July–Aug.—Erect, 3–6 ft., prickly, glaucous, biennial. *Leaves* 6–18 in., radical petioled, often spotted with black; cauline oblong, hardly narrowed at the base. *Heads* ½–¾ in., pale yellow, in slender panicles, subsecund ; peduncles slender, bracteate ; involucre narrow, conical ; bracts few, green, tips red. *Fruit* with a thick cellular wing and ribbed faces.—

Distrib. From Belgium southd., N. Africa, W. Siberia.—Juice fœtid acrid, narcotic, used as an opiate.

2. **L. Scari'ola,** *L.* ; rather scabrous below, leaves suberect, radical ob-ovate-oblong sinuate-toothed or runcinate, upper sagittate amplexicaul, auricles acute spreading, branches of panicle long spreading, fruit grey.

Waste places, rare, Worcester, Norfolk, Cambridge, Essex, Kent, and Surrey ; (native ? *Wats.*); fl. July–Aug.—Closely allied to *L. viro'sa*, but prickly only towards the base ; branches more erect ; leaves usually more run-cinate ; heads smaller ; fruit narrower. — Distrib. Europe, Siberia, Himalaya ; introd. in N. America.—Apparently the origin of the garden lettuce.

3. **L. salig'na,** *L.* ; almost glabrous, leaves entire or runcinate acute, cauline hastate amplexicaul, auricles spreading acute, uppermost narrow entire, branches of panicle very short erect, fruit grey.

Waste grounds, Suffolk and Hunts to Sussex and Kent, especially near the sea, rare ; fl. July–Aug.—More slender than the preceding, much less bristly, with the flowers often fascicled on short erect branches, subspicate.— Distrib. From Holland southd., N. Africa, W. Asia.

** *·Leaves not bristly. Beak shorter than the terete or 4-gonous fruit, or 0.*

4. **L. mura'lis,** *Fresen.* ; glabrous, leaves broad lyrate-pinnatifid, upper amplexicaul auricled, flowers yellow, fruit terete. *Prenanthes muralis,* L.

Old walls and rocky copses, in England, rare ; ascends to 1,300 ft. in Yorks. ; Perth and Stirling in Scotland ; Wicklow and Louth in Ireland ; fl. June–Aug.—Tall, slender, annual or biennial, 1-3 ft. *Leaves* membranous, glaucous beneath, narrow ; radical with a winged petiole ; lobes few toothed, terminal large 3-angular sinuate-lobed. Branches of *panicle* slender diverg-ing. *Heads* ⅔ in., yellow ; peduncles slender, bracteate ; invol. bracts few, linear, green and purplish. *Fruit* black, ribbed and muricate.—Distrib. Europe, W. Asia.

5. **L. alpi'na,** *Benth.* ; nearly glabrous, leaves sagittate lyrate or run-cinate toothed, terminal lobe very large deltoid, flowers blue, fruit 4-gonous. *Sonchus,* L. ; *S. cœruleus,* Sm. ; *Mulgedium,* Less.

Alpine rocks, Forfar, Aberdeen, alt. 2,000–3,000 ft. ; fl. Aug.—Glabrous except the glandular-pilose inflorescence. *Stem* 3-4 ft., simple, stout, succulent, grooved. *Leaves* 4-8 in. broad, membranous, narrowed into ½-amplexicaul auricled winged petioles ; upper broadly ovate or triangular-cordate, acute, shortly petioled. *Heads* 1 in. diam., pale blue, in erect simple or branched racemes ; peduncles ascending, bracteate ; involucre subcylindric, bracts linear. *Fruit* slightly compressed.—Distrib. Arctic and Alpine Europe, W. Siberia.

41. SON'CHUS, *L.* Sowthistle.

Annual or perennial, often succulent brittle herbs ; juice milky. *Leaves* alternate, toothed or pinnatifid. *Heads* corymbose or subumbellate, in-volucre conical after flowering, scales in many series, imbricate ; receptacle

R

flat, naked, pitted. *Corollas* all ligulate, yellow ; anther-cells shortly tailed ; style-arms slender, and upper part of style hairy. *Fruit* truncate, much compressed, grooved, ribbed, usually transversely rugose ; pappus-hairs in many series, simple, silky.—Distrib. N. and S. temp. regions ; species about 24.—Etym. doubtful.

1. **S. arven'sis**, *L.* ; rootstock creeping stoloniferous, leaves sharply toothed, lower runcinate, upper oblong-lanceolate ½-amplexicaul, auricles obtuse, inflorescence usually glandular-hispid. *Corn Sowthistle.*

Cultivated fields, N. to Shetland ; ascends to 1,000 ft. in Northumbd.; Ireland ; Channel Islands ; fl. Aug.–Sept.—*Stem* 2–4 ft., fistular, angled, simple or branched. *Leaves* very variable, margins waved, almost spinous ; glaucous beneath. *Heads* 1–2 in. diam. ; involucre broadly campanulate. *Fruit* light-brown.—Distrib. Europe (Arctic), N. Africa, temp. Asia, India ; introd. in America.—A robust form (*glabra*, Lond. Cat.) occurs with eglandular inflorescence.

2. **S. palus'tris**, *L.* ; rootstock branched, leaves minutely toothed, lower runcinate with few segments, upper entire sessile sagittate, auricles acute, inflorescence glandular-hispid.

Marshes in England, very rare, Norfolk, Suffolk, Cambridge, Hunts, Kent and Essex, now all but extinct ; fl. July–Sept.—*Stem* 5–9 ft., strict, stout, fistular, angled, leafy. *Leaves* long-acuminate, lower very large and long, often reduced to one sagittate blade and a broad winged petiole, very glaucous beneath. *Heads* ¾–1 in. diam., pale yellow, subumbellate ; peduncle stout, very hispid and glandular. *Fruit* 4-ribbed, pale.—Distrib. From Denmark southd.

3. **S. olera'ceus**, *L.* ; annual, leaves lanceolate ½-amplexicaul sharply toothed entire or pinnatifid, auricles rounded, inflorescence subumbellate glabrous rarely glandular.

Fields and waste places, N. to Shetland ; ascends to 1,200 ft. in Yorkshire ; Ireland ; Channel Islands ; fl. June–Sept.—*Stem* 2–3 ft., erect, usually branched, tubular, grooved. *Leaves* sessile or petioled, glaucous beneath, usually much lobed, often spinous-toothed. *Heads* crowded, ¾–1 in. diam.; peduncle sometimes cottony. *Fruit* pale brown.—Distrib Europe, N. and W. Asia, India, N. Africa, S. Australia, N. Zealand ; introd. in America.

S. olera'ceus proper ; auricles hastate, fruit ribbed and transversely wrinkled.

Sub-sp. S. as'per, *Hoffm.* ; leaves waved spinous, auricles suborbicular deflexed and recurved, fruit ribbed but not transversely wrinkled.

42. TRAGOPO'GON, *L.*

Erect, usually simple glabrous herbs, with biennial or perennial tap-roots, juice milky. *Leaves* alternate, entire, amplexicaul. *Heads* solitary ; invol. bracts 1-seriate, usually exceeding the flowers, narrow, nearly equal, connate at the base ; receptacle convex, naked or fimbriate. *Corollas* all ligulate, yellow or purple ; anther-cells tailed ; upper part of style hairy, arms slender obtuse. *Fruit* slender, muricate, beak long ; pappus-hairs

in many series, rigid, feathery with naked tips, the hairs of the bristles
horizontal and interlacing.—Distrib. Europe, N. Africa, W. Asia ;
species about 30.—Etym. τράγος and πώγων, *goat's beard.*

T. praten'sis, *L.* ; leaf-sheaths much dilated, stem scarcely thickened
upwards, flowers yellow. *Goat's beard.*

Meadows, pastures, and waste places, from Lanark and Caithness southd. ;
rare in Scotland and Ireland ; Channel Islands ; fl. June–July.—Glabrous or
slightly cottony on the involucre, glaucous. *Stem* 1–2 ft., stout, erect,
simple or sparingly branched above. *Leaves* flexuous, slender, gradually
contracted upwards from above the dilated sheath, tip linear ; radical with
shorter sheaths, keeled, channelled above, midrib tubular. *Heads* ½–2 in.
diam., yellow ; involucre obconic, bracts about 8, 2-seriate, flat, often
streaked with brown. *Fruit* variable in roughness ; beak about as long
as the body.—Distrib. Europe, N. and W. Asia, Himalaya.—Root edible.
Flowers close at noon.
T. praten'sis proper ; flowers as long as the invol. bracts.—Var. *T. mi'nor,*
Fries ; flowers ½ as long as the invol. bracts. The most common form.—
Var. *grandiflo'ra,* Syme ; flowers much longer than the invol. bracts.
Kent and Surrey, rare.

T. porrifo'lius, *L.* ; leaf-sheaths slightly dilated, peduncle much
thickened upwards, flowers purple. *Salsify.*

Wet meadows, rare and local, nowhere wild ; fl. May–June.—Habit and chief
characters of *T. pratensis,* but usually larger, with more muricate fruits ;
ligule as long as or shorter (var. *parviflora,* Bosw.) than the invol. bracts.—
Distrib. N. and Mid. Europe, N. and W. Asia.—Occasionally cultivated.

Order XLII. **CAMPANULA'CEÆ** (*including* **LOBELIA'CEÆ**).

Herbs, rarely shrubs ; juice milky. *Leaves* alternate, exstipulate.
Calyx-limb 5-cleft, ½ or wholly superior. *Flowers* honeyed, proterandrous.
Corolla epigynous, usually persistent, tube entire or cleft posteriorly ;
limb regular or oblique or 2-labiate, 5-lobed, valvate or induplicate in bud.
Stamens 5, epigynous or epipetalous : anthers conniving round the style,
basifixed, equal or unequal, naked or tipped with a pencil of hairs. *Ovary*
2–8-celled ; style simple, tipped with a ring of hairs or clothed with
deciduous hairs ; stigmas 2–8, hairy on the back, coherent till protruded
beyond the anthers, then spreading ; ovules many, horizontal, anatropous,
placentas axile, fleshy. *Fruit* a berry or capsule, 2- or more-celled, many-
seeded. *Seeds* minute, testa usually pitted, albumen fleshy ; embryo
subcylindric.—Distrib. All regions, most abundant in temp. climates ;
genera 53 ; species 1,000.—Affinities. With *Goodeniaceæ* and *Cichoraceæ.*
—Properties. Acrid and often poisonous, diuretic.

Sub-order I. **Lobeli'eæ.** *Corolla* irregular. *Anthers* cohering in a tube.
1. Lobelia.

R 2

SUB-ORDER II. **Campanule'æ.** *Corolla* regular. *Anthers* usually free.
Capsule dehiscing within the calyx-lobes.
Corolla 5-partite, segments narrow. Flowers capitate ....2. Jasione.
Corolla campanulate, 5-lobed ; lobes broad ....3. Wahlenbergia.
Capsule dehiscing at the sides, below the calyx-lobes.
Corolla 5-partite, segments narrow. Flowers capitate ....4. Phyteuma.
Corolla 5-toothed or lobed. Ovary short, broad..............5. Campanula.
Corolla rotate or campanulate. Ovary long, narrow ....6. Specularia.

1. LOBE'LIA, *L.*

Perennial herbs. *Flowers* in terminal racemes. *Calyx-tube* ovoid or
obconic, limb 5-fid. *Corolla* very irregular ; tube split at the back ; limb
2-labiate, 2 upper lobes smaller erect or recurved, 3 lower pendulous.
Stamens 5, epipetalous ; anthers connate, all, or the 2 lower only, bearded
at the tip. *Ovary* 2–3-celled ; style filiform, included, girt with a ring of
hairs, stigmas 2 broad spreading. *Capsule* 2–3-celled, top exserted and
loculicidally 2–3-valved. *Testa* pitted.--DISTRIB. All but very cold
regions ; species about 200.--ETYM. *Mathias Lobel,* a Flemish botanist.

1. **L. Dortman'na,** *L.* ; leaves all radical submerged subcylindrical
2-fistular, scape slender cylindric fistular, flowers drooping.

Gravelly mountain lake-bottoms, ascending to 1,650 ft.; Wales, Shropshire,
and from Westmoreland to Shetland; Ireland; fl. July–Aug.—*Rootstock* short,
stoloniferous; root-fibres white, cellular. *Leaves* 2–3 in., linear, obtuse,
recurved. *Scape* 1–2 ft., bracteate near the base; raceme lax, emersed.
Flowers ¾–1 in., pedicelled ; bracts short, oblong, obtuse. *Calyx* obconic,
terete ; lobes short, obtuse. *Corolla* pale lilac ; lobes linear obtuse, upper
erect, lower longer. *Anthers* included. *Capsule* clavate, inclined.—DISTRIB.
N. Europe from France to Mid. Russia, N. America.

2. **L. u'rens,** *L.* ; stem leafy, leaves obovate or oblong-spathulate obtuse
sinuate-toothed, upper linear decurrent, flowers erect or spreading.

Heaths, Dorsetshire and Cornwall; fl. Aug.–Sept.—Glabrous or puberulous. *Stem* 1–2 ft., erect, slender, angular. *Leaves* 2–3 in., subsessile,
teeth callous. *Racemes* 4–8 in., pedicels very short; bracts lanceolate,
exceeding the 5-angled narrow obconic calyx. *Flowers* ⅔ in. *Calyx-lobes*
subequal, subulate, shorter than the tube. *Corolla* blue or purple; lobes
subequal, lanceolate, acute. *Capsule* erect.—DISTRIB. W. France, Spain,
Madeira.—Acrid and pungent.

2. JASI'ONE, *L.* SHEEP'S-BIT.

Annual or perennial herbs. *Radical leaves* usually rosulate ; cauline
narrow, alternate. *Flowers* small, in terminal centripetal heads ; involucre of many bracts. *Calyx-tube* short, limb 5-fid. *Corolla* regular,
5-cleft to the base. *Stamens* 5, epigynous : anthers subconnate, tips free.
Ovary 2-celled ; style clavate, with 10 hairy ridges, stigmas 2 short.
Capsule 2-valved at the top. *Seeds* minute, testa shining—DISTRIB.
Europe, N. Africa, W. Asia ; species about 10.—ETYM. doubtful.

J. monta'na, *L.* ; hispidly pubescent, leaves obovate-oblong.
Heathy pastures and light soils, from Kincardine and the Clyde southd.;
Ireland; Channel Islands; fl. June–Sept.—Habit of *Scabiosa.* Annual, or
biennial (var. *littoralis*). *Stem* 6–18 in., branched from the base. *Leaves*
½–1 in., radical petioled; cauline linear-oblong, obtuse, waved or crenate,
ciliate. *Peduncles* long, slender, naked. *Heads* 1–6, ½–¾ in. diam., hemispheric; bracts ovate, acute, as long as the lilac-blue flowers. *Calyx-tube*
turbinate; lobes subulate. *Corolla-lobes* at first coherent, persistent. *Capsule*
small, subglobose.—DISTRIB. Europe, N. Africa, W. Asia.

3. WAHLENBER'GIA, *Schrad.*

Characters of *Campanula*, but capsule dehiscing within the calyx-lobes.
—DISTRIB. Trop. and temp. regions; species 80.—ETYM. G. Wahlenberg, Professor of Botany, Upsala.

W. hedera'cea, *Reichb.* ; glabrous, very slender, creeping, leaves all
petioled orbicular or cordate angled or obscurely lobed, peduncles 1-fld.
leaf-opposed, corolla cylindric campanulate, lobes short. *Campanula*, L.
Bogs and damp woods, in the W. and S. counties, from York to Kent and
Cornwall; Argyll to Ayr only in Scotland; S. and E. Ireland; Channel
Islands; fl. July–Aug.—*Rootstock* creeping. *Stems* filiform. *Leaves* ¼–½ in.
diam., membranous, subacutely 5-angled or -lobed, upper often opposite;
petiole slender, dilated upwards. *Peduncles* much longer than the petioles.
Calyx-tube shortly turbinate; lobes triangular-subulate, erect. *Corolla*
⅓ in., pale blue; lobes obtuse, recurved. *Capsule* subglobose, membranous.
—DISTRIB. W. Europe, from France to Spain.

4. PHYTEU'MA, *L.*

Perennial herbs. *Radical leaves* petioled, cauline usually narrow, sessile.
Flowers in dense spikes or heads; involucre of several bracts. *Calyx*
ovoid or obconic; limb superior, 5-fid. *Corolla* curved in bud, 5-cleft,
segments linear, tardily opening, sometimes at length free. *Stamens* 5,
epigynous, filaments linear dilated at the base; anthers free. *Ovary* 2-3-
celled; style filiform, hairy above, stigmas 2-3 short filiform. *Capsule*
ovoid, dehiscing below the middle by longitudinal valves. *Seeds* many,
ovoid or compressed; embryo straight, cotyledons divaricating.—DISTRIB.
Europe, W. Asia; species about 30.—ETYM. obscure.

1. **P. orbicula're,** *L.* ; glabrous or slightly hairy, heads globose in
flower oblong in fruit, stigmas usually 3.
Chalk downs, from Kent to Surrey and Wilts, local; fl. July–Aug.—*Rootstock*
tuberous below. *Stems* 6–18 in., several, erect or ascending. *Radical leaves*
1–2 in., long-petioled, oblong- or ovate-lanceolate, rarely cordate, crenate;
cauline few, smaller upwards. *Heads* ¾–1½ in. diam.; bracts short, oblong,
acute. *Flowers* deep blue. *Calyx-tube* short; lobes triangular. *Corolla-lobes*
at length free to the base. *Capsule* short, 2-3-celled.—DISTRIB. From
Belgium southd., S. Russia.

2. **P. spica'tum,** *L.* ; glabrous, heads oblong or cylindric in flower
elongate in fruit, styles very long, stigmas 2.

Woods and thickets, E. Sussex; (native? *Wats.*); fl. May–June.—Taller and more robust than *P. orbicula're;* stem 2–3 ft., ribbed. *Radical leaves* cordate at the base, ovate or oblong; cauline spreading and recurved. *Heads* 1–4 in., yellowish. *Capsule* usually 2-celled.—Distrib. Europe to S. Russia.— Formerly cultivated and the root eaten.

5. CAMPAN'ULA, *L.*

Perennial, rarely annual herbs. *Radical leaves* usually petioled, cauline alternate. *Flowers* spiked or racemed, white blue or lilac. *Calyx-tube* ovoid or subglobose; limb 5-fid, lobes flat or folded at the sinus. *Corolla* campanulate or rotate, 5-lobed. *Stamens* 5, epigynous, filaments short, bases broad dilated; anthers linear, free. *Ovary-cells* 3–5, opposite the sepals; style clavate, with rows of deciduous hairs opposite the anther-cells, stigmas 3–5 filiform. *Capsule* ovoid or turbinate, 3–5-celled, cells dehiscing below the calyx-limb by pores or valves. *Seeds* usually flattened. —Distrib. All temp. and most trop. climates; species about 230.— Etym. *campanula,* from the *bell*-shaped corolla.

* *Stem-leaves linear or linear-lanceolate, usually nearly entire. Terminal flower of the raceme opening first.*

1. **C. rotundifo'lia,** *L.* ; glabrous, stem angled slender, cauline leaves lanceolate, upper narrow linear quite entire acute, flowers solitary or racemed drooping, corolla campanulate, lobes short recurved. *Hare-bell.*

Pastures, heaths, &c., N. to Shetland; ascends to 3,500 ft. in the Highlands; Ireland; fl. July–Sept.—Glabrous or slightly pubescent. *Rootstock* slender. *Stem* ½–2 ft., ascending, simple or branched, sparingly leafy. *Primordial leaves* long-petioled, broadly ovate-cordate, crenate. *Pedicels* slender; bracts minute or 0; buds erect. *Calyx-lobes* erect, subulate. *Corolla* ½–1 in., blue, sometimes white, lobes subacute. *Capsule.* subglobose, valves basal.— Distrib. Europe (Arctic), N. Africa, N. Asia, N. America.—Var. *lancifolia,* Koch (*montana,* Syme), has broader cauline leaves and subsolitary flowers; and Var. *speciosa,* More, from W. Ireland, is a beautiful form with much larger flowers and shorter calyx-lobes.

2. **C. Rapun'culus,** *L.* ; hispid or glabrous, stem angled, cauline leaves oblong- or obovate-lanceolate obscurely toothed, flowers panicled erect, calyx-lobes very long subulate, corolla broadly campanulate 5-lobed almost to the middle, lobes recurved. *Rampion.*

Gravelly roadsides and hedgebanks, from Fife southd., but rare and doubtfully wild; fl. July–Aug.—Biennial; root fleshy. *Stem* 2–3 ft., simple or branched. *Leaves* 1–3 in., variable, sessile; primordial broadly ovate, long-petioled. *Pedicels* slender. *Calyx-lobes* 1 in. after flowering. *Corolla* ¾ in., red purple or blue. *Capsule* short, erect, valves close under the calyx-lobes.—Distrib. From Denmark southd. (excl. Greece), N. Africa, W. Siberia.

3. **C. pat'ula,** *L.* ; scabrid, stem slender angled, cauline leaves linear or lanceolate, flowers subcorymbose erect, calyx-lobes linear, corolla broadly campanulate 5-lobed to the middle, lobes spreading.

Copses and hedges, in the W. and S. counties, rare, from York to Somerset
and Kent; fl. July–Sept.—Variable in duration ; root slender.   *Stem* 1–3 ft.,
branched above.   *Radical leaves* obovate or oblong, petioled, obscurely
crenate, cauline subsinuate.   *Pedicels* long, slender.   *Calyx-tube* obconic,
lobes very variable in width.   *Corolla* 1–1½ in. diam., purple.   *Capsule* erect,
obconic, valves close under the calyx-lobes.—DISTRIB.   Europe (excl. Greece,
Turkey), W. Siberia.

C. PERSICIFO'LIA, *L.* ; glabrous, stem terete, leaves coriaceous, cauline
linear-oblong obtuse finely serrate, flowers few racemed, calyx-lobes
triangular-lanceolate, corolla-tube hemispheric, lobes short erect.
Naturalized in woods, Banff, York; fl. July–Aug.—Perennial.   *Rootstock*
creeping, branched, stoloniferous.   *Leaves* 1–3 in., radical sessile, spathulate-
lanceolate, nerves obscure.   *Pedicels* stout, 2-bracteate at the base.   *Calyx-
tube* small, subglobose ; lobes broader than in *C. patula* and coriaceous.
Corolla 1 in. diam., blue, sometimes white.   *Capsule* erect, ovoid, pores large
under the calyx-lobes.—DISTRIB.   Europe, N. Africa, N. and W. Asia.

** *Stem-leaves ovate, toothed or serrate.*

4. C. latifo'lia, *L.* ; pubescent, stem tall stout furrowed, cauline leaves
oblong-ovate, lower-petioled, upper sessile obtusely serrate, flowers large,
lower bracts large leafy.

Copses and woods, from Banff and Isla to N. Wales, Gloucester, and Surrey ;
ascends to 1,200 ft. in York ; absent on S. coast ; fl. July–Aug.   *Rootstock*
stout, woody.   *Stem* 3–4 ft.   *Radical leaves* long-petioled, triangular-ovate,
hispidly pubescent beneath.   *Raceme* very leafy ; flowers many, erect or inclined.   *Calyx-tube* broad, short, 5-ribbed ; lobes large, ovate-lanceolate.
Corolla blue or white, lobes suberect acuminate.   *Capsule* short, valves basal.
—DISTRIB.   Europe (Arctic), N. and W. Asia, Himalaya.

C. RAPUNCULOI'DES, *L.* ; puberulous, stem tall nearly terete, cauline
leaves ovate, lower petioled, upper sessile obtusely serrate, flowers large,
bracts all small linear.

Fields, &c., from Aberdeen southd. to Dorset and Hants, naturalized ;
rare and local ; casual in Ireland ; fl. July–Aug.—*Rootstock* creeping ;
stolons subterranean.   *Stem* 1–2 ft., rather slender, usually simple.   *Radical
leaves* cordate.   *Racemes* long, simple, secund ; flowers drooping ; pedicels
slender, short.   *Calyx-tube* obconic, 5-ribbed ; lobes short, reflexed, linear or
ovate.   *Corolla* bright blue, lobes recurved.   *Capsule* subglobose, valves
basal.—DISTRIB.   Europe, W. Siberia.—The lowest flower opens first.

5. C. Trache'lium, *L.* ; hispid, stem tall angled, leaves all petioled
ovate-lanceolate from a broad base irregularly coarsely obtusely serrate,
bracts leafy.

Woods and copses in dry soil, from Lanark and Fife to Cornwall and Kent ;
Kilkenny, Ireland ; fl. Sept.–Oct.—*Rootstock* short, stout.   *Stem* 1–3 ft.,
hairs reversed.   *Leaves* doubly serrate ; radical long-petioled, cordate ;
cauline gradually attenuated from the base, acute.   *Racemes* panicled,
peduncles short ; bracteoles lanceolate ; flowers 1 or many, erect or inclined.
Calyx-tube hispid, very short and broad ; lobes ovate-lanceolate.   *Corolla*

½–¾ in., blue-purple, broadly campanulate; lobes suberect, acute. *Capsule* drooping, valves basal.—DISTRIB. Europe, N. Africa, Siberia.—The uppermost flower in each raceme opens first.

6. **C. glomera'ta,** *L.* ; pubescent, stem short terete leafy, leaves crenate, radical very long-petioled oblong- or ovate-cordate, cauline sessile ovate ½-amplexicaul, flowers erect in leafy heads or irregularly spiked.

Chalky and dry pastures, from Forfar to Somerset and Kent; rare in the W. counties, and in Ireland ; fl. Sept.–Oct.—*Rootstock* short, stout. *Stem* 6–18 in., slender. *Cauline leaves* ½–¾ in. *Flowers* bright blue, mostly in a terminal head with a few axillary buds below it. *Calyx-tube* short, obconic, 5-ribbed ; lobes ovate. *Corolla* ½–¾ in., lobes recurved, acute. *Capsule* short, valves basal.—DISTRIB. Europe, N. and W. Asia.—The central flower opens first.

6. SPECULA'RIA, *Heist.*

Characters of *Campanula,* but ovary and capsule very long and slender. —DISTRIB. Temp. Europe and Asia, N. and S. America ; species 8.— ETYM. *Speculum,* a mirror, from the form of the corolla-limb.

S. hy'brida, *DC.* ; annual, hispid or puberulous, leaves oblong waved, flowers axillary subsessile, calyx-lobes exceeding the corolla.

Cornfields and dry soils, from Durham to Cornwall and Kent, chiefly in the E. counties ; introduced in Scotland ; (a colonist, *Wats.*); fl. June–Sept.— *Stem* 6–10 in., erect or decumbent, simple or branched from the base, angled. *Leaves* small ; radical with broad petioles, ovate or spathulate ; cauline sessile, obtuse. *Calyx-tube* longer than the floral leaves, sharply angled ; lobes linear-oblong, subfoliaceous. *Corolla* rotate, blue inside, lilac outside, cleft to near the middle ; lobes acute. *Capsule* 1–1½ in., angled. *Seeds* polished, ovoid.—DISTRIB. From Holland southd., N. Africa.

ORDER XLIII. ERICA'CEÆ.

Shrubs, rarely herbs or trees, mostly evergreen. *Leaves* alternate opposite or whorled, simple, often articulate with the stem, exstipulate. *Flowers* regular or nearly so. *Calyx* superior or inferior, 4–5-fid or -partite. *Corolla* hypogynous or epigynous, campanulate, 4–5-toothed or -lobed, rarely of 5 petals ; lobes imbricate in bud. *Stamens* 4, 5, 8, or 10, hypogynous or epigynous ; anthers 1–2-celled, cells obtuse or with tubular tips, opening by terminal pores or slits, often with basal or dorsal awns. *Disk* annular and lobed or of glands or scales. *Ovary* 4–5-celled ; style terminal, stigma small simple or 4–5-lobed ; ovules many in each cell, rarely few or solitary, pendulous. *Fruit* 3–5-celled, a berry or 3–5-valved capsule ; cells many- rarely 1-seeded. *Seeds* small, testa reticulate, albumen fleshy ; embryo minute, clavate.—DISTRIB. Chiefly temp. and cold climates ; genera about 87 ; species about 1,300.—AFFINITIES. With *Epacrideæ.*—PROPERTIES unimportant.

SUB-ORDER I. **Vaccinie'æ.** *Buds* clothed with scales. *Stamens* epigynous. *Ovary* inferior.

Corolla campanulate, urceolate, or rotate ....1. Vaccinium.

SUB-ORDER II. **Eri'ceæ.** *Buds* naked or scaly. *Stamens* hypogynous. *Ovary* superior.

TRIBE I. **ARBUTE'Æ.** *Buds* naked. *Leaves* usually broad. *Corolla* deciduous. *Fruit* a berry or drupe.

Ovary-cells many-ovuled..2. Arbutus.
Ovary-cells 1-ovuled ....3. Arctostaphylos.

TRIBE II. **ANDROMEDE'Æ.** Shrubs. *Buds* clothed with scales. *Leaves* usually broad. *Corolla* deciduous. *Capsule* loculicidal.
4. Andromeda.

TRIBE III. **ERICI'NEÆ.** *Buds* naked. *Leaves* small. *Flowers* 4-merous. *Corolla* persistent. *Anthers* cohering in bud.

Corolla 4-fid. Capsule loculicidal, cells many-seeded ....5. Erica.
Corolla 4-partite. Capsule septicidal, cells few-seeded...........6. Calluna.

TRIBE IV. **RHODORE'Æ.** *Buds* clothed with scales. *Leaves* usually broad. *Flowers* 4–5-merous. *Corolla* deciduous. *Capsule* septicidal.

Corolla urceolate. Anthers 8, opening by pores..................7. Daboecia.
Corolla urceolate. Anthers 10, opening by pores ....8. Phyllodoce.
Corolla campanulate. Anthers 5, opening by slits............9. Loiseleuria.

TRIBE V. **PYROLE'Æ.** Herbs. *Buds* scaly. *Leaves* chiefly radical. *Petals* 5, free or connate, concave. *Capsule* loculicidal. *Testa* produced at both ends far beyond the nucleus ....10. Pyrola.

1. VACCIN'IUM, *L.*

Shrubs ; buds clothed with usually persistent scales. *Leaves* alternate, often evergreen. *Flowers* solitary or racemose, white or red, drooping, honeyed. *Calyx-tube* short ;' limb 4–5-toothed. *Corolla* epigynous, globose or campanulate, 4–5-fid. *Stamens* 8–10, epigynous ; anther-cells with tubular tips, awned or not. *Ovary* 4–5-celled ; style filiform, stigma obtuse ; placentas prominent, many-ovuled. *Berry* 4-5-celled, areolate. *Seeds* angled, testa reticulate.—DISTRIB. Europe, temp. and subtrop. Asia and America ; species about 100.—ETYM. obscure.

* Corolla globose or campanulate.

1. **V. Myrtil'lus,-***L.*; erect, glabrous, branches angular, leaves deciduous ovate serrate, peduncles 1-fld., anthers with dorsal awns. *Whortleberry, Bilberry, Blaeberry.*

Woods, copses, and heaths, N. to Shetland (except Suffolk and Cambridge), chiefly in hilly districts ; ascends to 4,000 ft.; Ireland ; fl. April–June.— *Rootstock* creeping. *Stems* many, 6–24 in., rigid. *Leaves* ½–1 in., nerves reticulate, young rosy. *Peduncles* ¼ in., naked. *Calyx-tube* turbinate ; lobes 5, short. *Corolla* ¼ in. diam., globose, rosy tinged with green. *Berry*

$\frac{1}{3}$ in. diam., dark blue, glaucous.—DISTRIB. Europe (Arctic), N. and W. Asia, N.W. America.—Berries used for preserves, &c.

2. **V. uligino'sum,** *L.* ; procumbent, glabrous, branches terete ascending, leaves deciduous oblong or obovate quite entire glaucous beneath, peduncles 1-fld., anthers with dorsal awns.

Mountain bogs and copses, from Westmoreland and Durham to Shetland; ascends to 3,500 ft. in the Highlands; fl. May–June.—*Stems* 6–10 in., woody, rigid, naked below. *Leaves* $\frac{1}{2}$–1 in., coriaceous, obtuse or acute, veins reticulate beneath. *Peduncles* 1–3 together, $\frac{1}{3}$ in. *Calyx-tube* hemispheric; lobes 4–5, broad, obtuse. *Corolla* $\frac{1}{6}$ in., pale pink, subglobose. *Berry* smaller than in *V. Myrtillus,* of the same colour.—DISTRIB. Europe (Arctic), N. and W. Asia, N. America.

3. **V. Vitis-Idæ'a,** *L.* ; procumbent, branches pubescent, leaves evergreen obovate dotted beneath, margins revolute, racemes short terminal drooping, anthers awnless. *Cowberry.*

Woods and heaths, chiefly in mountain districts, from Devon, S. Wales, and Notts to Shetland (absent in S.E. half of England); ascends to 3,300 ft. in the Highlands; Ireland; fl. June–July.—*Stems* wiry, tortuous, branched and naked below; branches 6–18 in., trailing or ascending. *Leaves* $\frac{1}{2}$–1$\frac{1}{4}$ in., glossy green above (like box), bifarious, very coriaceous, margins thickened entire or obtusely serrulate, pale beneath; nerves not reticulate. *Flowers* crowded; pedicels 2-bracteolate. *Calyx-tube* hemispheric; lobes 4, broadly ovate, ciliolate. *Corolla* campanulate. *Berry* $\frac{1}{3}$ in. diam., globose, red, acid.—DISTRIB. Europe (Arctic), N. Asia, N. America.

** *Corolla rotate.*

4. **V. Oxycoccos,** *L.* ; creeping, leaves evergreen ovate-oblong, base cordate, margins strongly recurved. *Oxycoccus palustris,* Pers. *Cranberry.*

Peat bogs, usually amongst *Sphagnum,* from Somerset and Sussex to Shetland, local; ascends to 2,700 ft. in the Highlands; Ireland; fl. June–Aug.—*Stems* puberulous. *Leaves* $\frac{1}{6}$–$\frac{1}{3}$ in., scattered, spreading, shortly petioled, deep green above with a median channel, glaucous beneath with reticulate nerves. *Peduncles* 1–3, 1 in., capillary, erect, puberulous. *Flowers* $\frac{1}{3}$ in. diam., red. *Calyx-limb* minute. *Corolla-lobes* linear-oblong. *Stamens* exserted, filaments pubescent purple; anthers yellow. *Berry* $\frac{1}{4}$ in. diam., globose, dark red.—DISTRIB. Europe (Arctic) (excl. Turkey), N. Asia and America.—The berries are an excellent antiscorbutic.

2. AR'BUTUS, *L.*

Shrubs. *Leaves* alternate, evergreen. *Flowers* in terminal panicled racemes, bracteate, white or pale red. *Sepals* 5. *Corolla* hypogynous, globose or subcampanulate; lobes 5, reflexed. *Stamens* 10, on the base of the corolla, filaments short dilated below; anthers deflexed, ovoid, opening by 2 pores, awns 2 reflexed. *Disk* annular. *Ovary* 5-celled; style simple, stigma obtuse; ovules many, placentas pendulous from the upper angles of the cells. *Berry* globose, granulate; cells 5, 4–5-seeded. *Seeds* angled, testa coriaceous.—DISTRIB. N. temp. regions; species about 10. —ETYM. The old Latin name.

A. Une'do, *L.* ; leaves obovate- or oblong-lanceolate acute doubly-serrate, panicles drooping many-fld. *Strawberry-tree.*

Woods at Killarney, Muckross, and Bantry ; fl. Sept.-Oct.—A small rounded much-branched evergreen tree, 8–10 ft.; bark rough ; branchlets and petioles hairy and glandular. *Leaves* 2–3 in., petioled. *Panicles* glabrous, lax ; pedicels short; bracts deciduous. *Sepals* short, rounded. *Corolla* ½ in , creamy. *Berry* ⅔ in. diam., orange-scarlet, subglobose, muricate.—DISTRIB. S. France, Spain, Mediterranean region.—Berries eatable when perfectly ripe ; made into a wine in Corsica.

3. ARCTOSTAPH'YLOS, *Adanson.* BEARBERRY.

Characters of *Arbutus,* but cells of ovary 5–10, 1-ovuled ; disk of 3 fleshy scales ; and fruit a drupe with 5–10 stones.—DISTRIB. of *Arbutus,* but more abundant in America; species about 15.—ETYM. ἄρκτος and σταφυλή, the *fruit* being a food of *bears.*

1. **A. alpi'na,** *Spreng.* ; branches depressed, leaves deciduous obovate or spathulate crenate-serrate above the middle, nerves netted, drupe black.

Dry barren Scotch Mts., rare, from Forfar, Perth, and Inverness to Shetland ; ascends to 2,700 ft.; fl. May–July.—Forms woody patches; branches stout, leafy, interlaced; bark scaly. *Leaves* ½–1½ in.; petiole short, wrinkled above, ciliate. *Flowers* 2–3, appearing with the young leaves, shortly pedicelled ; bracts ciliate. *Calyx* minute. *Corolla* ¼ in. broad, white; lobes 4–5, pubescent within. *Filaments* subulate ; anthers brown, awns very short. *Berry* ¼ in. diam.—DISTRIB. Scandinavia (Arctic), Mts. of Mid. Europe, cold and Arctic Asia and America.

2. **A. Uva-ur'si,** *Spreng.* ; branches trailing, leaves evergreen obovate or spathulate quite entire very coriaceous, drupe red.

Heathy rocky places from York and Derby to Shetland ; ascends to near 3,000 ft. in the Highlands; N.W. Ireland; fl. May–June.—Forms depressed trailing masses; branches 1–2 ft., stout, woody, young puberulous ; bark dark, scaling. *Leaves* ½–1 in., petiole and margin finely woolly, deep green, nerves reticulate on both surfaces. *Racemes* crowded, very short, few-fld.; scales and bracteoles persistent, ciliate ; pedicels very short. *Sepals* short, broad. *Corolla* ⅔ in., pink, urceolate ; teeth 4–5, hairy within. *Anthers* with long awns. *Berry* ¼ in. diam.—DISTRIB. Arctic and Alpine Europe (excl. Greece, Turkey), Siberia, N. America.

4. ANDROM'EDA, *L.*

A small shrub ; buds clothed with scales. *Leaves* alternate, evergreen. *Flowers* in subterminal umbels, white or pink. *Sepals* 4. *Corolla* hypogynous, globose ; limb 4-fid, reflexed. *Stamens* 10, hypogynous, included; filaments bearded ; anthers obtuse, dorsally 2-awned. *Ovary* 5-celled ; style simple, stigma obtuse or dilated ; placentas many-ovuled, pendulous. *Capsule* subglobose, loculicidally 5-valved. *Seeds* attached to a central column, small, testa hard, smooth, raphe thickened.—DISTRIB. Arctic and Alpine Europe (excl. Greece, Turkey), Siberia, N. America.—ETYM. Mythical.

A. polifo'lia, *L.* ; leaves elliptic-lanceolate glabrous shining.

Peat bogs, Wales, N. Somerset, W. Norfolk, and from Hunts and Salop to Perth; local in Ireland; fl. May–Aug.—*Stems* long, ascending, rooting at the base; branches 6–12 in., suberect, twiggy; bark smooth, brown. *Leaves* shortly petioled, ⅜–1½ in., acute, thickly coriaceous, glaucous beneath, margins strongly recurved. *Racemes* short. *Flowers* drooping; pedicels ½–1 in., slender, red. *Sepals* small, obtuse. *Corolla* ¼ in., purplish; lobes 5, revolute. *Filaments* bearded; anthers short. *Capsule* erect, 5-lobed. *Seeds* turned in all directions, ovoid.

5. ERI'CA, *L.* HEATH.

Rigid, much-branched, evergreen shrubs. *Leaves* whorled, rarely alternate or scattered, narrow, rigid. *Flowers* usually nodding; pedicels 2–3-bracteate. *Sepals* 4. *Corolla* hypogynous, ovoid globose campanulate or tubular, 4-lobed, persistent, honeyed. *Stamens* 8, inserted on the disk; anther-cells awned or not, opening by pores or slits. *Ovary* 4-celled; styles filiform, stigma capitate dilated 4-lobed; ovules many. *Capsule* 4-celled, loculicidally 4-valved, many-seeded. *Seeds* attached to a central column, ovoid or compressed, testa smooth or reticulate.—DISTRIB. Europe, N. Asia, N. Africa, abundant in S. Africa; species about 400.— ETYM. Classical.

1. **E. Tet'ralix,** *L.* ; pubescent and sometimes glandular, leaves 4 in a whorl ciliate, flowers subumbellate, corolla regular ovoid, anthers included with subulate awns. *Cross-leaved Heath.*

Heaths, abundant, N. to Shetland; ascends to 2,400 ft. in the Highlands; Ireland; Channel Islands; fl. July–Sept.—Sometimes almost woolly; glandular hairs stiff or slender or 0. *Stems* 12–18 in., wiry; branches slender. *Leaves* ⅛–¼ in., in close or distant whorls, spreading, linear, obtuse, rarely glabrous; margins revolute to the midrib. *Flowers* drooping, homogamous; pedicels short, bracteolate in the middle. *Sepals* oblong-lanceolate. *Corolla* ⅓ in., mouth scarcely oblique, rose-red, upper side darkest. *Ovary* usually woolly and with gland-tipped hairs.—DISTRIB. N. and W. Europe to Russia.

E. TET'RALIX proper; pubescent and usually glandular, irregularly branched, leaves linear-oblong pubescent above, ovary pubescent.

Sub-sp. E. MACKAYI, *Hook.* (*E. Mackaiana*, Bab.) ; nearly glabrous, more corymbosely branched, leaves ovate-oblong glabrate, flowers laxly umbelled, ovary glabrous.—Galway. (Spain.)

2. **E. cine'rea,** *L.* ; glabrous, leaves 3 in a whorl, flowers whorled, corolla ovoid mouth regular, anthers included with toothed appendages at the base.

Heaths and commons, N. to Shetland; ascends to 2,200 ft. in the Highlands; Ireland; Channel Islands; fl. July–Sept.—*Stems* 1–2 ft., much-branched; branches slender, ultimate pubescent. *Leaves* ⅛–¼ in., irregularly whorled, with short leafy branches in the axils, linear, acute, margins revolute. *Flowers* drooping or horizontal, homogamous; pedicels short, puberulous; bracteoles under the calyx. *Sepals* ovate-lanceolate. *Corolla* ¼–⅓ in.,

crimson-purple. *Ovary* glabrous.—Distrib. W. Europe to Germany and N. Italy.

3. **E. cilia'ris,** *L.* ; ciliate and glandular, leaves 3–4 in a whorl, flowers in unilateral racemes, corolla ovoid, anthers included awnless.

Sandy heaths, Cornwall, Dorset; fl. June–Sept.—*Stem* 12–18 in., slender; branches many, erect, pubescent, flowering elongate. *Leaves* ⅛ in., close set, subsessile, ovate, pubescent above, beneath glaucous with minute scales, margins recurved. *Flowers* inclined; pedicels very short, bracteolate in the middle. *Sepals* ovate, pubescent and ciliate. *Corolla* ⅓–½ in., curved, crimson, mouth small oblique. *Style* exserted. *Ovary* glabrous.—Distrib. W. France, Spain, Portugal.—A hybrid with *Tetralix* (*E. ciliaris,* var. *Watsoni,* Benth.) occurs in Cornwall and Dorset.

4. **E. va'gans,** *L.* ; glabrous, leaves 3–4 in a whorl, flowers long-pedi-celled in dense axillary racemes, corolla campanulate, anthers exserted awnless. *Cornish Heath.*

Heaths in W. Cornwall; fl. July–Aug.—*Stem* 1–3 ft , stout, woody ; branches stiff, erect, fascicled, densely leafy. *Leaves* ⅓ in., linear, recurved, margins reflexed over the midrib. *Flowers* erect ; pedicels slender, bracteolate about the middle. *Sepals* ovate, ciliolate. *Corolla* ⅛ in., pink; lobes short. *Ovary* glabrous.—Distrib. W. France, Spain, Portugal.

5. **E. mediterra'nea,** *L.* ; stem erect, leaves 4 rarely 3 in a whorl, flowers in dense racemes, corolla subcampanulate, anthers ½-exserted awnless. *E. mediterra'nea,* var. *hiber'nica,* Hook. ; *E. hiber'nica,* Syme.

Boggy heaths, Mayo and Galway; fl. April–May.—Bushy, 2–5 ft., glabrous. *Leaves* ¼–⅓ in., crowded, shortly petioled, linear, margins revolute to the midrib and connate to the under-surface. *Racemes* terminal and axillary ; pedicels solitary or in pairs, short, 2-bracteolate in the middle. *Sepals* ovate-lanceolate. *Corolla* pink, cylindric-campanulate, lobes broad. *Anthers* 2-fid at the tip.—Distrib. S.W. France, Spain.

6. **CALLU'NA,** *Salisb.* Ling.

A much-branched, gregarious shrub. *Leaves* minute, decussate, imbricate (whence the branches appear tetragonal). *Flowers* on axillary peduncles with 2 pairs of opposite bracts under the calyx, honeyed, homogamous. *Sepals* 4, scarious, coloured. *Corolla* shorter than the sepals, hypogynous, 4-partite, persistent. *Stamens* 8, inserted on the disk ; anthers awned, slits short lateral. *Ovary* 4-celled ; stigma dilated ; ovules 2, pendulous from the top of each cell. *Capsule* 4-celled, septicidally 4-valved, septa attached to the axis. *Seeds* 1–2 in each cell, attached to the central column.—Distrib. Europe (Arctic), (excl. Greece, Turkey,) W. Siberia, Azores, Greenland, N. America (very rare).—Etym. καλλύνω, the twigs being used for *sweeping.*

C. vulga'ris, *Salisb.* ; leaves 3-gonous gibbous at the base.

Heaths and moors, N. to Shetland ; ascends to 3,300 ft. in the Highlands; Ireland; Channel Islands; fl. July–Sept.—*Stem* 1–2 ft., woody, inclined;

branches elongate. *Leaves* $\frac{1}{12}$ in., linear-oblong, glabrous (*E. glabrata,* Seem.) or ciliate or pubescent or hoary (var. *incana*). *Flowers* very many; pedicels very short; outer bracts leaf-like, inner scarious. *Sepals* $\frac{1}{10}$ in., concave, obtuse, rose-purple or white, shining. *Corolla-lobes* triangular. *Anthers* short, dorsally fixed, 2-fid. *Style* exserted. *Ovary* pubescent.

7. DABEO'CIA, *D. Don.* ST. DABEOC'S HEATH.

A small shrub; buds scaly. *Leaves* alternate, evergreen. *Flowers* terminal, racemose. *Sepals* 4. *Corolla* hypogynous, urceolate; lobes 4, reflexed. *Stamens* 8, hypogynous, included; anthers obtuse, cells opening by pores, awnless. *Ovary* 4-celled; style simple, stigma dilated; ovules many. *Capsule* 4-celled, septicidally 4-valved, axis persistent. *Seeds* small, testa smooth or pitted.—DISTRIB. Ireland, W. France, Spain. Portugal, Azores.—ETYM. St. Dabeoc.

D. polifo'lia, *Don ;* leaves elliptic-ovate obtuse ciliate.

Boggy heaths, Connemara and Mayo; fl. Aug.—*Stem* 1-2 ft., decumbent; bark flaking; branches slender, erect, hairy and glandular. *Leaves* $\frac{1}{3}$-$\frac{1}{2}$ in., shortly petioled, margins recurved, bright green, glossy above, white and woolly beneath. *Raceme* terminal, lax, 5-10-fld., glandular; flowers drooping; pedicels stout, curved; bracts leafy. *Sepals* ovate-lanceolate, ciliate. *Corolla* $\frac{2}{3}$ in., crimson purple or white, lobes short. *Filaments* flat; anthers sagittate, 2-fid, purple. *Ovary* villous and glandular.

8. PHYLLODO'CE, *Salisb.*

Small glandular shrubs; buds scaly. *Leaves* evergreen, scattered. *Flowers* solitary or umbelled. *Sepals* 5. *Corolla* urceolate or campanulate, 5-fid. *Stamens* 10, hypogynous, filaments slender; anthers truncate, opening by pores, awnless. *Ovary* 5-celled; style slender, stigma capitate. *Capsule* 5-celled, septicidally 5-valved, many-seeded. *Seeds* shining.—DISTRIB. Arctic regions; species 3.—ETYM. Mythological.

P. cæru'lea, *Bab.* ; leaves shortly petioled crowded spreading linear obtuse denticulate glabrous. *P. taxifolia,* Salisb. ; *Menziesia,* Wahlb.

Heathy moors, Sow of Atholl, Perthshire, alt. 2,400 ft., extremely rare; fl. June.—Short, depressed, much branched, woody; branches 5-10 in., tubercled. *Leaves* $\frac{1}{4}$-$\frac{1}{2}$ in., rigidly coriaceous, glossy and channelled above, margins reflexed to the puberulous midrib. *Flowers* few, drooping, in terminal umbellate corymbs; pedicels $\frac{1}{2}$-$1\frac{1}{2}$ in. *Sepals* ovate-lanceolate, glandular-pubescent. *Corolla* $\frac{1}{4}$ in., lilac, urceolate. *Anthers* purple. *Ovary* glandular-pubescent.—DISTRIB. Scandinavia (Arctic), Mts. of W. France, Pyrenees, N. Asia, N. America.

9. LOISELEU'RIA, *Desvaux.*

A small, glabrous, depressed, branching, rigid, evergreen shrub; buds scaly. *Leaves* small, opposite, quite entire, margins revolute. *Flowers* on terminal 1-fld. peduncles, pink, small. *Sepals* 5. *Corolla* hypogynous, broadly campanulate, 5-fid. *Stamens* 5, hypogynous, slightly adnate to the corolla ; anthers short, included, slits lateral. *Ovary* subglobose, 2-3-celled ; style short, straight, stigma capitate ; ovules many. *Capsule* 2-3-celled, septicidally 2-3-valved, valves 2-fid. *Seeds* several, ovoid, testa thick pitted.—DISTRIB. Arctic and Alpine Europe, Asia, America.—ETYM. *Loiseleur-Deslongchamps*, a French botanist.

L. procum'bens, *Desv.* ; leaves rigidly coriaceous recurved linear-oblong obtuse. *Azalea*, L.

Scotch Alps, alt. 1,500–3,600 ft., from Ben Lomond to Shetland; fl. May–June. —Forms flat patches with interlaced rigid woody branches. *Leaves* $\frac{1}{8}-\frac{1}{4}$ in., crowded, petioled, deep green, glossy, deeply channelled above, beneath densely pubescent, midrib very stout. *Flowers* usually 2–3, inclined, sub-corymbose, proterogynous ; peduncle red, $\frac{1}{8}-\frac{1}{4}$ in., stout, ebracteolate. *Sepals* ovate-lanceolate, red. *Corolla* $\frac{1}{8}$ in. diam., pink ; lobes obtuse. *Capsule* minute.

10. PY'ROLA, *Tournef.* WINTER-GREEN.

Biennial or perennial herbs ; rootstock slender, creeping ; stems short, almost woody ; buds scaly. *Leaves* chiefly radical, broad petioled, ever-green. *Flowers* secund, racemose, rarely solitary, white pink or yellowish, nodding ; scapes bracteate ; pedicels bracteolate at the base. *Sepals* 5. *Corolla* globose, rarely spreading, of five free or subconnate orbicular petals. *Stamens* 10, in pairs opposite the petals, hypogynous, erect or declinate, filaments subulate, tip incurved ; anther-cells opening by ter-minal pores. *Ovary* 5-celled ; style erect or declinate, stigma capitate 5-lobed or -cleft ; ovules many. *Capsule* globose, 5-celled, loculicidally 5-valved ; valves septiferous, usually with tomentose edges ; axis bearing the seeds on fungous placentas. *Seeds* very minute, elongate, testa pro-duced loose.—DISTRIB. N. temp. zone ; species about 15.—ETYM. *Pyrus*, from a supposed similarity in the foliage to that of the *pear*.

SUB-GEN. 1. **Py'rola** proper. *Flowers* racemose. *Petals* 5, free, incurved. *Anther-cells* short, obtuse. *Stigmatic lobes* short. *Valves of capsule* cohering by fibres.

1. **P. mi'nor,** *Sw.* ; leaves orbicular-ovate obscurely crenate, stamens erect as long as the short straight style, stigma-rays large.

Woods and heaths, from Sutherland to Kent and Devon ; ascends to 1,500 ft. in the Highlands; Ireland, rare ; fl. June–Aug.—*Stem* 1–3 in., ascending. *Leaves* usually in a rosette, but sometimes alternate, coriaceous, 1–1½ in., blade contracted into the longer petiole. *Scape* 8–12 in., slender. *Raceme* short, bracteoles subulate-lanceolate. *Flowers* ¼ in. diam., drooping, globose, white tinged with rose. *Stamens* very short. *Style* without a ring below the stigma, not lengthening in fruit. *Capsule* drooping.—DISTRIB. Europe (Arctic) (excl. Turkey), N. America.

2. **P. me'dia,** *Sw.* ; leaves orbicular-ovate crenate, stamens erect shorter than the straight or slightly decurved style, stigma-lobes minute erect.

Woods and heaths, local, from Warwick and Worcester to Shetland; ascends to 1,800 ft. in the Highlands; N. and W. Ireland; fl. July–Aug.—Very like *P. minor*, but larger and stouter; leaves broader, sometimes orbicular and 1¾ in. diam., and bracts larger. *Flowers* ½ in. diam. *Style* with a ring round the base of the stigma.—DISTRIB. Europe (excl. Greece, Turkey).

3. **P. rotundifo'lia,** *L.* ; leaves orbicular-obovate, stamens ascending shorter than the long decurved style, stigma-lobes minute erect.

Moist woods and copses, rare, E. Kent, and from Norfolk and Salop to Aberdeen and Inverness; ascends to 2,500 ft. in Scotland; Westmeath in Ireland; Channel Islands; fl. July–Sept.—Habit and stature of *P. minor*, but petioles all longer and very slender; raceme usually longer; bracts larger, spreading; flowers ½ in. diam., pure white; style with a ring below the stigma.—DISTRIB. Europe (Arctic) (excl. Greece, Turkey), N. Asia, Himalaya, N. America.

P. rotundifo'lia proper; bracts on the scape few, pedicels a little longer than the lanceolate sepals.—VAR. *arena'ria*, Koch (*P. maritima*, Kenyon); leaves smaller, bracts many on the scape, pedicels as long as the ovate sepals. Sand-hills, Lancashire. (Belgium.)

4. **P. secun'da,** *L.* ; leaves ovate acute serrate, raceme secund, stamens incurved shorter than the long ascending style, stigma broad lobed.

Rocky mountain woods, rare, Monmouth, York to Ross; ascends to 2,400 ft.; N.E. Ireland; fl. July.—*Stem* straggling, branches 1–4 in., ascending. *Leaves* 1–1½ in., rosulate or alternate, rather thin, reticulate, petioles shorter. *Scape* slender, 2–5 in., with 1–5 bracts. *Racemes* 1–2 in., bracteoles linear-oblong. *Flowers* horizontal, ¼ in. diam. *Sepals* obtuse, erose. *Petals* concave, greenish-white. *Style* without a ring below the stigma, elongate in fruit. *Capsule* drooping.—DISTRIB. Europe (Arctic), N. and W. Asia, N. America.

SUB-GEN. 2. **Mone'ses,** *Salisb.* (gen.). *Flowers* solitary. *Petals* slightly adherent at the base, spreading. *Anther-cells* with tubular tips. *Stigmatic lobes* long. *Valves* of capsule free.

5. **P. uniflo'ra,** *L.* ; scape 1-bracteate at the top, leaves orbicular spathulate sinuate-serrate. *Moneses grandiflora*, Salisb.

Woods chiefly of pine, in the N. Highlands, from Perth and Aberdeen to Sutherland; fl. June–July.—*Stem* ½–2 in., leafy. *Leaves* ½–1 in., alternate, narrowed into a short petiole, rather membranous. *Flowers* ¾ in. diam., nodding, then erect. *Sepals* broad, obtuse, ciliate. *Petals* almost flat, white, spreading. *Filaments* curved. *Style* without a ring. *Capsule* erect; style and elongate stigmas persistent.—DISTRIB. Europe (Arctic) (excl. Greece, Turkey), Siberia, N. America.

ORDER XLIV. **MONO'TROPEÆ.**

Leafless simple erect brown or red root-parasites or saprophytes. *Stem* scaly, upper often passing into bracts. *Flowers* solitary spicate or racemose. *Sepals* or bracts 2–6, erect, deciduous. *Petals* 3–6, rarely connate, erect or spreading, imbricate in bud. *Stamens* 6–12, hypogynous, filaments free or connate below; anthers various. *Ovary* 4–6-lobed, 1- or 4–6-celled; style simple, stigma capitate peltate lobed or funnel-shaped; ovules numerous, minute, naked; placentas thick, parietal or in the inner angles of the cells. *Capsule* membranous, 4–6-lobed, 4–6-celled. *Seeds* minute, testa lax reticulate rarely coriaceous; embryo minute, in fleshy albumen.—DISTRIB. N. temp. regions.—AFFINITIES. Close with Ericeæ. —PROPERTIES 0.

1. **HYPOPI'THYS,** *Scop.* BIRD'S-NEST.

Flowers racemose, upper 5–6-merous, the rest 4-merous. *Petals* saccate at the base. *Stamens* 8–10; anther-cells confluent, valves very unequal. *Disk* of 8 or 10 recurved glands. *Ovary* 4–5-celled, 8–10-furrowed; style columnar, stigma discoid; placentas fleshy, filling the cells. *Capsule* loculicidally 5-valved. *Testa* loose, reticulate, produced at both ends.— DISTRIB. N. temp. regions; species 3 or 4.—ETYM. ὑπὸ and πίτυς, from growing in pine woods.

H. multiflo'ra, *Scop.*; flowers racemose, sepals and petals erose obtuse. *Monot'ropa Hypopi'thys,* L.

Woods near roots of fir and beech, from Aberdeen and Inverness to Kent and Somerset; Mid. Ireland, very rare; fl. July–Aug.—A glabrous or pubescent saprophyte, feeding on decayed vegetable matter. *Stem* 6–12 in., stout, fleshy, cream-white. *Scales* oblong. *Raceme* drooping till after fertilization, then erect. *Flowers* ½ in., many, drooping; pedicels short, erect in fruit; bracts scale-like. *Sepals* irregularly disposed. *Filaments* incurved, the alternate longer. *Style* short. *Fruit* ovoid or globose.—DISTRIB. Europe (Arctic) (excl. Turkey), N. Asia, N. America.

ORDER XLV. **PLUMBAGIN'EÆ.**

Herbs, often maritime, rarely shrubs. *Leaves* alternate or radical, exstipulate. *Flowers* regular, 2-sexual; bracts and bracteoles usually 3. *Calyx* inferior, tubular; limb 5-cleft, plaited, often scarious and coloured. *Petals* 5, hypogynous, claws long free or cohering, twisted in bud, or corolla monopetalous imbricate. *Stamens* 5, opposite the corolla-lobes, epipetalous or hypogynous, filaments filiform; anther-cells bursting inwards. *Ovary* free, 1-celled; styles 5, rarely 3–4, distinct or cohering, stigmas capillary or linear; ovule solitary, anatropous, suspended from a basal funicle. *Fruit* enclosed in the calyx-tube, membranous, bursting irregularly or 5-valved. *Seed* pendulous, albumen scanty floury; embryo straight, cotyledons flat, radicle cylindric superior.—DISTRIB. All regions ; genera 10 ; species about 200.—AFFINITIES. Nearest to *Primu-laceæ*, but distant.—PROPERTIES. Astringent, unimportant.

Flowers in a bracteate head...1. Armeria.
Flowers in panicled unilateral cymes...2. Statice.

1. ARME'RIA, *Willd.* THRIFT, SEA-PINK.

Perennial herbs. *Leaves* all radical, very narrow. *Flowers* pedicelled, fascicled in small bracteate cymes, which are collected into a dense hemispheric head ; involucre (formed of the connate downward prolongation of the bases of the outer bracts) tubular, scarious, sheathing the top of the scape downwards. *Calyx* funnel-shaped, scarious. *Petals* cohering at the very base, persistent and covering the utricle. *Filaments* inserted on the petals, base dilated. *Ovary* obovoid ; stigma capillary, papillose. *Utricle* with 5 hard bosses at the top, dehiscing transversely or irregularly below. —DISTRIB. Alpine, Arctic and maritime N. temp. regions, Chili ; species 30.—ETYM. The monkish Latin *Flos Armeriæ*, applied to a Pink.

1. **A. vulga'ris,** *Willd.* ; pubescent or ciliate, leaves linear 1–3-nerved. *A. marit'ima*, Willd. ; *A. pubes'cens*, Link ; *A. pubig'era, β scot'ica*, Boissier ; *A. durius'cula*, Bab. ; *Stat'ice Arme'ria*, L.

Rocky and stony sea-shores and on lofty mountains, N. to Shetland ; ascends to 3,800 ft. in the Highlands ; Ireland ; Channel Islands; fl. April–Oct.—*Rootstock* woody, branched. *Leaves* densely fascicled, 1–6 in., $\frac{1}{20}$–$\frac{1}{10}$ in. broad, obtuse or acute, covered with impressed points on both surfaces, broader in the alpine form (var. *planifo'lia*, Syme). *Scape* 3–12 in., hairs spreading or reflexed ; heads $\frac{1}{2}$–1 in. diam.; involucre $\frac{1}{2}$–1 in., lacerate, outer bracts shorter than the head, ovate, acute or produced into long brown or green points ; inner oblong, obtuse, scarious ; pedicel $\frac{1}{4}$–$\frac{1}{3}$ in., equalling the calyx. *Calyx* decurrent on the pedicel, with 5 pubescent ribs, lobes cuspidate. *Petals* $\frac{1}{2}$ in., rose-pink or white, limb obovate. *Utricle* exceeding the calyx-tube.—DISTRIB. Europe (Arctic), Asia, N. America, Chili.

2. **A. plantagin'ea,** *Willd.* ; glabrous, leaves narrow-lanceolate usually 3–5-nerved, calyx-lobes awned.

Sandy banks, &c., Jersey ; fl. June–Aug.—More rigid, stouter and larger than *A. vulga'ris ;* leaves $\frac{1}{10}$–$\frac{1}{4}$ in. broad, narrowed into long points, margins cartilaginous and undulate when dry ; scapes taller ; involucre longer and less deeply cut ; outer bracts with usually a long hervaceous point exceeding the head ; flowers darker, on shorter pedicels.—DISTRIB. Mid. and S. Europe. —Syme describes a probable hybrid with *A. vulga'ris*, as growing with them.

2. STAT'ICE, *L.* SEA LAVENDER.

Perennial herbs. *Leaves* all radical. *Flowers* shortly pedicelled, in 1- or more-fld. 3-bracteate cymes, which are alternately distichously or secundly arranged in branched panicles. *Calyx* obconic ; limb scarious, 5-lobed. *Petals* free or united at the very base. *Filaments* inserted on the petals. *Ovary* obovoid or oblong ; styles free or connate at the base, stigmas capillary papillose. *Fruit* as in *Armeria.*—Distrib. Chiefly saline districts and shores of temp. seas, most common in W. Asia ; species 50–60.—Etym. The Greek name for some *astringent* herb.

    * *Leaves pinnately-nerved. Calyx-lobes with intermediate teeth.*

  1. S. Limo'nium, *L.* ; leaves usually long-petioled 1-ribbed, scape branched above, branches nearly all flowering.

Muddy shores, from Fife and Dumfries to Kent and Devon ; Ireland ; Channel Islands ; fl. July–Nov.—Glabrous. *Rootstock* stout, woody, creeping, branched. *Leaves* 2–4 in., oblong- or obovate-lanceolate, variable in breadth, often mucronate ; petiole usually slender. *Scapes* many, 6–18 in., angular, corymbosely branched. *Spikelets* 2–3-fld., ⅓ in.. alternate, erect, secund ; outer bracts herbaceous margins scarious, intermediate scarious, inner scarious and green at the back. *Calyx* purplish-green ; lobes triangular-ovate, often jagged, much enlarged after flowering. *Corolla* ⅓ in., blue-purple. —Distrib. Europe, N. Africa, N. America.

S. Limo'nium proper ; inner bract about twice as long as the intermediate. Absent from Ireland.—Var. *S. Be'hen,* Drejer ; spikelets on corymbose compact or spreading or recurved branches.—Var. *S. serot'ina,* Reichb.; spikelets in pyramidal panicles with flexuous spreading branches, spikelets rather lax short. S. coast.

Sub-sp. S. rariflora, *Drejer ;* spikelets lax on distant erect or incurved branches, inner bract once to once and a half as long as the intermediate. *S. bahusien'sis,* Fries.

    ** *Leaves 1–3-nerved. Calyx-lobes without intermediate teeth.*

  2. S. auriculæfo'lia, *Vahl ;* leaves shortly petioled 3-nerved at the base, scape branched from about the middle, branches nearly all flowering, calyx-lobes obtuse. *S. spathula'ta,* Hook. not Desf.

Sea-shores, from Lincoln on the E. and Wigton on the W. to Cornwall and Kent ; Ireland ; Channel Islands ; fl. July–Aug.—Glabrous. *Rootstock* short, stout, branched. *Leaves* 1–4 in., often mucronate. *Scapes* 4–10 in., paniculately branched ; spikelets unilateral, often imbricate ; bracts often coloured, inner twice as long as the intermediate ; flowers as in *S. Limo'nium. Calyx-limb* white. *Corolla* blue-purple.—Distrib. Coasts of France, Spain, and the Mediterranean.—Boissier (who has examined Vahl's plant) refers this (in *Herb. Hook.*) to auriculæfo'lia, and in DC. *Prodr.* he refers the latter to his *oxyle'pis.*

S. auriculæfo'lia proper (*S. occidenta'lis,* Lloyd ; *S. binervo'sa,* G. E. Sm.) ; leaves obovate-lanceolate, branches of panicle ascending, lower sometimes flower-less, spikelets slender erect or ascending.—Var. *S. interme'dia,* Syme ; leaves obovate-lanceolate, branches of panicle ascending usually all flowering, spikelets stout spreading or ascending.—Var. *S. Dodar'tii,* Gir. ; leaves spathulate, branches of panicle short spreading often in pairs all flowering, spikelets crowded stout spreading. Portland.

  3. S. bellidifo'lia, *Gouan ;* leaves shortly petioled 1-ribbed, scape branched from near the base, branches flexuous most of them flowerless, calyx-lobes mucronate. *S. cas'pia,* Willd. ; ? *S. reticula'ta,* L.

                                        s 2

Salt marshes, Norfolk, Suffolk, Cambridge; fl. July–Aug.—Glabrous, except
the scaberulous panicle. *Rootstock* much branched. *Leaves* ¾–1½ in., few,
obtuse. *Scape* 4–8 in., rather slender, excessively branched; branches
spreading, recurved; spikelets unilateral, usually fascicled at the ends of
the branchlets; bracts with broad scarious margins, inner ½ longer than the
intermediate; flowers much smaller than in *S. Limo'nium*, ⅓ in., pale lilac.
—DISTRIB. Atlantic, Mediterr., and Black Seas, E. Asia.—The commonest
Mediterranean species, I doubt not the *Limo'nium reticula'tum* of Ray, and if
so it is the *S. reticula'ta* of Linnæus, founded on a Maltese plant.

ORDER XLVI. **PRIMULA'CEÆ.**

Perennial, rarely annual herbs. *Leaves* radical or cauline, exstipulate.
Flowers 2-sexual, regular. *Calyx* inferior, superior in *Samolus*, 5- (rarely
4–9-) cleft. *Corolla* usually hypogynous (0 in *Glaux*), rotate bell- or
funnel-shaped, 5- (rarely 4–9-) cleft. *Stamens* inserted in the corolla-tube
opposite its lobes, with sometimes alternating staminodes (hypogynous in
Glaux), filaments usually short; anthers bursting inwards by slits, rarely
by terminal pores. *Ovary* 1-celled; style simple, stigma undivided;
ovules many, amphitropous rarely anatropous, placenta free-central.
Capsule 1-celled, dehiscing by simple or 2-fid valves or transversely.
Seeds sunk in cavities of the placenta, testa thin, hilum usually ventral,
albumen fleshy or horny; embryo terete parallel to the hilum.—DISTRIB.
N. temp., Arctic and mountain regions, Chili and Fuegia; genera 21;
species about 250.—AFFINITIES. With *Myrsineæ* and *Plumbagineæ.*—
PROPERTIES unimportant.

TRIBE I. **PRIMULE'Æ.** *Ovary* superior. *Capsule* valvular. *Hilum*
ventral.

Leaves radical. Corolla-lobes incurved or spreading ....1. Primula.
Leaves radical. Corolla-lobes reflexed ....1*. Cyclamen.
Leaves cauline. Calyx 5-partite. Corolla yellow...............2. Lysimachia.
Leaves cauline. Calyx 5–9-partite. Corolla white ....3. Trientalis.
Leaves cauline. Calyx campanulate, coloured. Corolla 0 ....4. Glaux.

TRIBE II. **ANAGAL'LIDEÆ.** *Ovary* superior. *Capsule* opening transversely. *Hilum* ventral.

Calyx 4-partite. Filaments glabrous ....5. Centunculus.
Calyx 5-partite. Filaments villous ....6. Anagallis.

TRIBE III. **HOTTO'NIEÆ.** *Ovary* superior; ovules anatropous. *Capsule*
valvular. *Hilum* basal ....7. Hottonia.

TRIBE IV. **SAMOLE'Æ.** *Ovary* inferior; ovules anatropous. *Capsule*
valvular. *Hilum* basal ....8. Samolus.

1. PRI'MULA, *L.*

Scapigerous, perennial herbs. *Flowers* in involucrate umbels, white, yellow rose, or purple, honeyed. *Calyx* 5-toothed or -fid. *Corolla* funnel- or salver-shaped, throat naked or with 5 swellings ; lobes 5, incurved or spreading. *Stamens* 5, included. *Ovary* ovoid or globose ; style filiform, stigma capitate ; ovules many, amphitropous. *Capsule* 5-valved at the top, many-seeded, valves simple or 2-fid. *Seeds* plano-convex, peltate. —Distrib. Of the Order ; species about 80.—Etym. *primus*, from flower- ing early.—Flowers usually dimorphic, having long styles with anthers deep in the tube, or the reverse.

* *Leaves not mealy beneath, wrinkled and toothed. Calyx-tube 5-angled.*

1. **P. vulga'ris,** *Huds.* ; leaves and umbels subsessile, flowers spreading or suberect, calyx-teeth acuminate, corolla pale yellow, limb flat, throat contracted with thickened folds. *P. acau'lis*, L. *Primrose.*

Copses, pastures, and hedgebanks, N. to Shetland ; ascends to 1,600 ft. in Yorkshire ; Ireland ; Channel Islands ; fl. April–May.—*Rootstock* stout. *Leaves* 3–6 in., obovate-spathulate, beneath and inflorescence softly hairy. *Umbels* so sessile that the pedicels resemble scapes, which are about as long as the leaves ; bracts linear. *Calyx* ½–¾ in., a little inflated, 5-angled. *Corolla* 1–1½ in., rarely white lilac or purplish, greenish when dry ; lobes orbicular, notched. *Capsule* as long as the calyx-tube, ovoid, on prostrate pedicels.—Distrib. Europe, except the N.E. ; N. Africa.—Varies in the inflorescence being sessile (*P. acau'lis*, Jacq.) or a peduncled umbel (vars. *caules'cens* and *interme'dia*), if these be not hybrids, as suggested by Baker (see *P. ve'ris*). The origin of the Polyanthus.—Rootstock emetic.

2. **P. ela'tior,** *Jacq.* ; petioles winged, umbels peduncled, flowers hori- zontal or drooping, calyx-teeth acuminate, corolla pale yellow, limb concave, throat open without folds. *Oxlip.*

Copses and meadows in Bedford, Suffolk, Cambridge, Essex ; fl. April–May.— Intermediate between *P. vulgaris* and *veris*, differing from the former in the less inflated calyx, shorter pedicels, inodorous flowers, and capsule longer than the calyx-tube ; it hence resembles hybrids between *P. veris* and *offici- nalis*, but differs from those by the more villous calyx, paler flowers, and absence of folds at the mouth of the corolla.—Distrib. From Gothland southd. (excl. Greece, Turkey), Siberia.

3. **P. ve'ris,** *L.* ; petioles winged, umbels peduncled, flowers drooping, calyx-teeth obtuse, corolla buff-yellow funnel-shaped, limb much cupped, throat opened with obscure folds. *P. officina'lis*, Jacq. *Cowslip, Paigle.*

Meadows, pastures, and hedgebanks, from Caithness southd. ; rare in Scotland ; Ireland ; Channel Islands ; ascends to 1,600 ft. in Northumbd. ; fl. April– May.—Besides the characters given above, *P. ve'ris* differs from *P. ela'tior* in the short and often glandular pubescence of the shorter pedicels and calyx ; odorous flowers ; much smaller corolla-limb ; and capsule much shorter than the calyx-tube ; corolla rarely scarlet or orange-brown.— Distrib. Europe, W. Asia, N. Africa ?—Hybrids with *P. vulga'ris* include *P. ela'tior* of older English botanists, probably *P. ve'ris*, β *ela'tior*, L., and *P. variabilis*, Goupil, often taken for the Oxlip.

** *Leaves very mealy-beneath, not wrinkled. Calyx-tube terete.*

4. P. farino'sa, *L.* ; calyx-lobes longer than its tube a little shorter than the corolla-tube, corolla-lobes lilac distant, capsule cylindric-oblong twice as long as the calyx. *Bird's-eye Primrose.*

Bogs and meadows from York and Lancashire to the border ; Peebles (ascends to 1,800 ft.) ; fl. June–July.—*Leaves* 1½-2 in., obovate-spathulate, variable in breadth, crenulate, obtuse or subacute, glabrous above, meal below white or sulphur-coloured. *Scape* stout, 2-8 in., rarely 0. *Flowers* erect or spreading, crowded, dimorphic ; bracts small, saccate at the base. *Calyx-tube* oblong-obovoid, mealy ; lobes linear-oblong, obtuse. *Corolla* ¼-½ in. diam.; lobes flat, wedge-shaped, 2-fid; mouth yellow, contracted, with rounded folds.—DISTRIB. Europe (Arctic), N. Asia, Tibet, Greenland, N. U. States.

5. P. scot'ica, *Hook.* ; calyx-lobes equalling its tube much shorter than the corolla-tube, corolla-lobes blue-purple contiguous, capsule shortly oblong scarcely exceeding the calyx.

Pastures in Orkney, Caithness, and Sutherland ; fl. June–Sept.—Perhaps only a sub-species of *P. farinosa,* but smaller ; bracts less saccate ; leaves broader in proportion, calyx shorter and flowers homomorphic.—DISTRIB. Lapland (Arctic), Norway and Sweden.

1*. CYC'LAMEN, *Tournef.* SOWBREAD.

Herbs with large tuberous rootstocks. *Leaves* all radical, petioled, broad. *Scapes* 1-fld., erect, naked, coiled spirally when fruiting. *Flowers* nodding, white, pink, or lilac. *Calyx* 5-partite. *Corolla-tube* short, throat thickened ; lobes 5, large, reflexed. *Stamens* 5, inserted at the base of the corolla, included ; anthers cuspidate. *Ovary* globose ; style short, stigma simple ; ovules many, amphitropous. *Capsule* 5-valved, many-seeded, valves reflexed. *Seeds* subglobose, angled, peltate ; embryo with one cotyledon.—DISTRIB. S. Europe, N. Africa, W. Asia ; species 8. —ETYM. κύκλος, from the spiral peduncle.

C. HEDERÆFO'LIUM, *Willd.* in part ; tuber fibrous all over, leaves and flowers autumnal. *C. europæum,* Sm. not L. ; *C. neapolita'num,* Tenore.

Hedgebanks and copses in Kent, Sussex ; (a denizen, *Wats.*); fl. Sept.— *Tuber* 1-3 in. diam., turnip-shaped. *Leaves* appearing after the flowers, ovate-cordate, crenulate, 5-9-angled, dark green with a whitish mottled border, often purple beneath. *Scapes* 4-8 in. *Calyx-lobes* ovate-lanceolate, acute. *Corolla-lobes* 1 in., pink, red at the base, or white (var. *ficariifolium,* Syme). *Fruit* ½ in. diam.—DISTRIB. Central and S. Europe.—Root acrid and purgative.

2. LYSIMA'CHIA, *L.* LOOSE-STRIFE.

Erect or procumbent herbs. *Leaves* alternate opposite or whorled, quite entire, sometimes glandular-dotted. *Flowers* axillary or terminal, solitary racemed or spiked, yellow (British sp.). *Calyx* 5-6-partite. *Corolla* rotate ; segments 5-6, spreading or conniving. *Stamens* 5-6, inserted on the corolla-throat. *Ovary* subglobose ; style filiform, persistent, stigma obtuse ; ovules many, amphitropous. *Capsule* subglobose, 5-valved, many-seeded. *Seeds* plano-convex, peltate.—DISTRIB. N. temp. zone, S. Africa, Australia ; species 60.—ETYM. obscure.

SECTION 1. **Lysimas'trum,** *Duby. Flowers* axillary, solitary or in terminal panicled cymes. *Corolla-lobes* without alternating teeth. *Stamens* included. *Seeds* margined.

1. **L. vulga'ris,** *L.* ; erect, pubescent, leaves opposite and whorled ovate-lanceolate punctate, cymes panicled, filaments connate below.

River-banks, &c., from Mull and Aberdeen southd. ; rare in Scotland ; local in Ireland ; fl. July–Aug.—*Rootstock* creeping, stoloniferous. *Stem* 2–4 ft. *Leaves* 2–4 in., sessile, with black glands, glabrous or pubescent beneath. *Cymes* panicled in the upper axils ; bracts linear. *Flowers* dimorphic. *Calyx-lobes* lanceolate, ciliate, margins red. *Corolla* ½ in., subcampanulate, deep yellow with orange dots inside ; lobes ovate, not ciliate. *Capsule* globose. *Seeds* 3-gonous, rough.—DISTRIB. Europe (Arctic), N. Africa, N. Asia, with a closely allied Australian representative.

2. **L. nem'orum,** *L.* ; procumbent, leaves opposite shortly petioled ovate acute, peduncles slender solitary axillary 1-fld., sepals linear-subulate, filaments free eglandular. *.Yellow Pimpernel.*

Copses, hedgebanks, &c., from Caithness southd. ; ascends to 2,500 ft. in the Highlands ; Ireland ; Channel Islands ; fl. May–July.—Glabrous, shining, eglandular. *Stems* slender, 3–12 in. *Leaves* 1–2 in. *Peduncles* capillary, curved, as long as the leaves or longer. *Corolla* rotate, ½–⅔ in. diam. ; lobes spreading, not ciliate. *Filaments* very slender, glabrous. *Capsule* small, globose.—DISTRIB. Europe (excl. Russia, Greece).

3. **L. Nummula'ria,** *L.* ; prostrate, leaves opposite ovate-cordate or orbicular obtuse, peduncles axillary 1-flowered, sepals broad, filaments connate at the base glandular. *Creeping Jenny, Moneywort.*

Moist shaded places, from York and Durham to Devon and Kent (never seeding, native ?) ; not native in Scotland or Ireland ; fl. June–July.—Glabrous, shining. *Stems* 1–2 ft., rather stout, pendulous. *Leaves* ¾–1½ in., shortly petioled, gland-dotted. *Peduncles* stout, about equalling the leaves, solitary, rarely in pairs. *Flowers* homogamous. *Sepals* ovate-cordate, acute. *Corolla* ¾ in. diam., cup-shaped ; lobes obtuse, ciliate and glandular.— DISTRIB. Europe.

SECTION 2. **Naumbur'gia,** *Mœnch* (gen.). *Flowers* in dense axillary racemes. *Corolla-lobes* with minute alternating teeth. *Stamens* and slender style exserted. *Seeds* scarcely margined.

4. **L. thyrsiflo'ra,** *L.* ; erect, leaves opposite sessile lanceolate.

Marshes and canal-banks, local, from Dumbarton and Forfar to York, Lancaster and Notts ; fl. June–July.—Glabrous or nearly so. *Rootstock* creeping, stoloniferous. *Stem* 1–3 ft., stout, simple. *Leaves* rarely whorled, lower small, upper 2–3 in., covered with black dots. *Racemes* ¾–1½ in., from the lower axils, cylindric, peduncles long, suberect ; pedicels ⅛ in., as long as the calyx; bracts and calyx-lobes linear, dotted. *Flowers* protero-gynous. *Corolla* ¼ in., campanulate, lobes 5–6, narrow, dotted. *Filaments* connate at the very base. *Capsule* ovoid.—DISTRIB. Europe N. of the Alps, N. Asia, N. America.

3. TRIENTA'LIS, *L.* CHICKWEED WINTER-GREEN.

Erect, simple, glabrous herbs; rootstock slender, creeping. *Leaves* in one whorl of 5–6, with a few small alternate scales beneath the whorl. *Flowers* white, ebracteate, solitary on slender peduncles. *Calyx* 5–9-par-tite. *Corolla* rotate, 5–9-partite. *Stamens* 5–9, filaments filiform. *Ovary* globose; style filiform, stigma obtuse; ovules few, amphitropous. *Capsule* globose, 1-celled, few-seeded, valves 5 revolute. *Seeds* flattened, peltate, crowded, cohering by the membranous epidermis of the testa.—DISTRIB. A European and a N. American species.—ETYM. doubtful.

T. europæ'a, *L.* ; leaves obovate or obovate-lanceolate.

Subalpine woods, from York to Shetland, local; ascends to 2,800 ft. in the Highlands; fl. June–July.—*Stem* wiry, slender, 4–8 in., leafy at the top. *Leaves* 1½–2½ in., shining, rigid, obtuse or acute, narrowed into short petioles. *Flowers* erect, few, honeyless, proterogynous. *Peduncles* 1–3 in., filiform. *Sepals* linear-subulate. *Corolla* ½–¾ in. diam.; lobes ovate, acute. *Capsule* size of a pea; valves very deciduous. *Seeds* hexagonal, testa grey-white punctate.—DISTRIB. Europe (Arctic) N. of the Alps, N. Asia.

4. GLAUX, *Tournef.* SEA MILKWORT.

A small succulent glabrous herb; rootstock creeping, stoloniferous. *Leaves* decussate. *Flowers* small, axillary, sessile, white or pink. *Calyx* 5-partite, coloured. *Corolla* 0. *Stamens* 5, hypogynous, alternate with the calyx-lobes. *Ovary* subglobose; style filiform, stigma obtuse; ovules few, amphitropous. *Capsule* globose, 5-valved, few-seeded. *Seeds* peltate. —DISTRIB. Europe (Arctic), N. and W. Asia, Tibet, N. America.—ETYM. γλαυκός, of a *sea-green* colour.

G. marit'ima, *L.* ; glabrous, leaves 4-farious sessile quite entire.

Sea-shores and estuaries, N. to Shetland; salt districts of Worcester and Stafford; Ireland; Channel Islands; fl. June–July.—*Stem* 6–10 in., pro-cumbent, rarely suberect. *Leaves* ¼–¾ in., linear- or obovate- or ovate-oblong. *Flowers* suberect. *Calyx* ⅛–¼ in., campanulate, margins of the obtuse lobes scarious. *Anthers* subexserted. *Capsule* small, globose, cuspidate. *Seeds* plano-convex, testa rough.

5. CENTUN'CULUS, *L.* BASTARD PIMPERNEL.

Very small annual herbs. *Leaves* subopposite or alternate. *Flowers* minute, solitary, axillary, subsessile, ebracteate, white or pink. *Calyx* 4–5-partite. *Corolla* short, urceolate, persistent; limb 4–5-partite. *Stamens* 4–5, on the corolla-throat, exserted, filaments flattened. *Ovary* subglobose; style filiform, stigma obtuse; ovules many, amphitropous.

Capsule globose, bursting transversely, many-seeded. *Seeds* peltate.—
DISTRIB. Europe, Asia, America ; species 3. —ETYM. doubtful.

C. min'imus, *L.* ; glabrous, leaves ovate or ovate-lanceolate.
Wet turfy and sandy places, local, from Inverness and Elgin southd.; rare in
Ireland ; fl. June–Oct.— *Stem* 1–3 in., branched from below. *Leaves* $\frac{1}{10}$–$\frac{1}{2}$ in.,
very shortly petioled, ovate or ovate-lanceolate. *Flowers* numerous, white
or pink, homogamous. *Sepals* lanceolate, longer than the erect usually
4-lobed corolla. *Filaments* naked. *Capsule* mucronate. *Seeds* 3-gonous,
testa areolate.—DISTRIB. Europe (excl. Greece, Turkey), Azores, Siberia,
Andes, Brazil, Australia.

6. ANAGAL'LIS, *Tournef.* PIMPERNEL.

Slender annual or perennial herbs. *Leaves* opposite, quite entire.
Flowers on axillary 1-fld. peduncles, ebracteate, red or blue, rarely white.
Calyx 5-partite. *Corolla* rotate or funnel-shaped, 5-partite. *Stamens* 5,
on the base of the corolla-tube ; filaments filiform, villous. *Ovary* globose ;
style simple, stigma obtuse ; ovules many, amphitropous. *Capsule* globose,
bursting transversely, many-seeded. *Seeds* plano-convex, peltate.—DIS-
TRIB. Europe, temp. Asia, N. Africa, S. America ; species 12.—ETYM.
The Greek name.

1. **A. arven'sis,** *L.* ; annual, erect or procumbent, leaves sessile ovate
or lanceolate dotted beneath, sepals almost equalling the rotate corolla.
Poor Man's Weather-glass, Scarlet Pimpernel.
Fields and waste places, Elgin southd., rare or absent in W. Scotland ;
Ireland ; Channel Islands ; (a colonist, *Wats.*) ; fl. May–Oct.—Glabrous,
glandular-dotted. *Stem* 6–18 in., branched from the base, 4-angled; branches
ascending. *Leaves* $\frac{1}{2}$–1$\frac{1}{2}$ in., rarely in whorls of 3–4, sometimes cordate, acute.
Peduncles 1–2 in., slender, erect in flower, decurved in fruit. *Flowers*
homogamous. *Sepals* narrow, acuminate. *Corolla* $\frac{1}{3}$–$\frac{1}{2}$ in. diam., opening in
clear weather, lobes often ciliate. *Capsule* size of a small pea. *Seeds*
3-gonous.—DISTRIB. Europe, N. Africa, N. and W. Asia, India ; introd. in
many countries.
A. arven'sis proper (*A. phœnic'ea,* Lamk.); corolla scarlet rarely pink, or
white with a purple eye (var. *pal'lida*), lobes usually glandular-ciliate.—VAR.
A. cæru'lea, Schreb.; more erect, corolla bright blue, lobes rarely ciliate.
Rare.

2. **A. tenel'la,** *L.* ; perennial, creeping, leaves shortly petioled broadly
ovate or orbicular not dotted, sepals much shorter than the funnel-shaped
corolla. *Bog Pimpernel.*
Marshes and wet meadows, N. to Shetland; Ireland ; Channel Islands ; fl.
July–Aug.—Glabrous, rather succulent. *Stems* 3–4 in., 4-angled. *Leaves*
$\frac{1}{8}$–$\frac{1}{4}$ in. bifarious, opposite or subopposite. *Peduncles* 1–2 in., rather stout.
Sepals linear-subulate. *Corolla* $\frac{1}{2}$ in. diam., rosy with dark veins. *Filaments*
united at the base. *Capsule* very small. *Seeds* as in *A. arven'sis.*—DISTRIB.
From Belgium southd., E Siberia, N. Africa, temp. S. Africa.

7. HOTTO'NIA, *L.* WATER VIOLET.

Floating herbs. *Leaves* submerged, imperfectly whorled, pectinate and multifid. *Flowers* racemose, dimorphic, white or lilac, honeyed ; pedicels whorled. *Calyx* 5-partite. *Corolla* salver-shaped, throat thickened ; limb 5-partite, fringed at the base. *Stamens* 5, included. *Ovary* globose ; style filiform, persistent, stigma obtuse ; ovules many, anatropous. *Capsule* 5-valved ; valves cohering at the top, many-seeded. *Seeds* angled, hilum basilar.—DISTRIB. Europe, N.W. Asia, N. America ; species 2.— ETYM. Pierre Hotton, an early Leyden professor of botany.

H. palus'tris, *L.* ; sepals subacute equalling the corolla tube.

Ponds and marshes, local, from Durham and Westmoreland to Somerset and Kent ; Co. Down in Ireland ; fl. May–June.—Perennial, pale green, glabrous, except the slightly glandular-hairy inflorescence. *Branches* as thick as a goosequill, succulent, leafy, 6–10 in., radiating from the base of the scape, floating and rooting ; joints not inflated. *Leaves* 1–2 in., segments slender, linear, acute. *Scape* 1–2 ft., stout, erect. *Whorls* 4–8-fld. ; pedicels ¾–1½ in., decurved in fruit ; bracts linear. *Flowers* dimorphic, some cleistoga-mous. *Calyx-lobes* ½ in., linear. *Corolla* ¾ in. diam., lilac with a yellow eye ; lobes obtuse. *Capsule* the size of a small pea.—DISTRIB. Europe (excl. Spain, Greece, Turkey), W. Siberia.

8. SA'MOLUS, *Tournef.* BROOK-WEED.

Annual or perennial herbs. *Leaves* alternate or mostly radical. *Flowers* in terminal racemes or corymbs, white ; pedicels bracteate. *Calyx* ½-superior, limb 5-fid. *Corolla* perigynous, subcampanulate, 5-partite. *Stamens* 5, on the tube or throat of the corolla, alternating with staminodes, filaments very short. *Ovary* subglobose ; style short, stigma obtuse or capitate ; ovules many, anatropous. *Capsule* ½-inferior, ovoid, 5-valved above the calyx-tube, many-seeded. *Seeds* angular, hilum basilar.— DISTRIB. Temp. climates ; species about 8.—ETYM. obscure.

S. Valeran'di, *L.* ; leaves obovate or spathulate obtuse or apiculate, quite entire, radical rosulate.

Wet ground and ditches, often near the sea, from Skye and Elgin southd. ; Ireland ; Channel Islands ; fl. June–Sept.—Glabrous, rather shining. *Root-stock* short. *Stem* 1–2 ft., erect, with prostrate or ascending sometimes rooting branches from the base. *Leaves* 1–4 in., cauline alternate. *Racemes* ¾–⅔ in., erect ; pedicels ascending ; bracts adnate to above their middle, small, lanceolate. *Calyx-tube* hemispherical ; lobes deltoid, acute. *Corolla* ⅙ in. diam., lobes short, obtuse. *Stamens* included. *Capsule* ¼ in. diam., globose. *Seeds* rough.—DISTRIB. Temp. N. hemisphere, Himalaya.

Order XLVII. **OLEA'CEÆ.**

Shrubs or trees; branches opposite; buds scaly. *Leaves* opposite, simple, or impari-pinnate, exstipulate. *Flowers* in 3-chotomous cymes, small, white or greenish, 1–2-sexual; pedicels opposite, 2-bracteolate. *Calyx* inferior, 4-lobed or 0. *Corolla* hypogynous, regular, 4-partite, deciduous, rarely 0, or of 4 free petals, valvate in bud. *Stamens* 2, epipetalous or hypogynous. *Disk* 0. *Ovary* 2-celled; style simple or 0, stigma entire or 2-fid; ovules 2 collateral in each cell, rarely 3 pendulous from the septum, anatropous. *Fruit* 1–2-celled, cells 1- rarely 2-seeded. *Seed* pendulous, testa sometimes winged, albumen fleshy or horny; embryo straight, cotyledons flat thin.—Distrib. Trop. and temp. regions, chiefly northern; genera 18; species 280.—Affinities. With *Jasmineæ.*—Properties. *Olea europœa* yields oil in its drupe, a bitter bark, and hard durable wood. The flowers of *O. fragrans* used to scent China tea. Manna is the produce of several Ashes. Ash bark is cathartic.

Corolla funnel-shaped, 4-lobed. Fruit a berry.....................1. Ligustrum.
Corolla 0 (in the Brit. sp.) Fruit a samara............................2 Fraxinus.

1. **LIGUS'TRUM,** *Tournef.* Privet.

Shrubs, rarely small trees. *Leaves* often evergreen, quite entire. *Flowers* in terminal thyrsoid cymes, honeyed, homogamous. *Calyx* shortly tubular, 4-toothed, deciduous. *Corolla* funnel-shaped, 4-lobed. *Stamens* 2, rarely 3, included in the corolla-tube. *Ovary* ovoid; stigma 2-fid lobes obtuse; ovules 2 in each cell. *Berry* globose, 2-celled, cells 1–2-seeded, flesh oily. *Seeds* ovoid or angled, albumen hard; cotyledons ovate-lanceolate.—Distrib. Europe, N. Asia, especially Japan; species about 25.—Etym. *ligare,* to bind, from a use of the twigs.

L. vulga're, *L.* ; shrubby, leaves oblong-lanceolate quite entire.

Thickets in England, from Forfar and Lanark southd.; (naturalized except in chalk districts and coast cliffs, *Wats.*); wild in S. Ireland; Channel Islands; fl. June–July.—A glabrous bush, 4–10 ft., almost evergreen; branches slender, bark smooth. *Leaves* 1–2 in., shortly petioled, acute. *Panicles* 1–3 in. *Corolla* ⅙–¼ in. *Berry* ⅓ in. diam., globose, purple-black.—Distrib. Europe, N. Africa.—Berries yield a rose-dye and a bland oil used for cooking in Germany.

2. **FRAX'INUS,** *Tournef.* Ash.

Deciduous trees. *Leaves* simple or pinnate, leaflets opposite with an odd one, toothed or serrate. *Flowers* polygamous or diœcious. *Calyx* 4-fid or 0. *Petals* 0, or 4 connate at the base. *Stamens* 2, hypogynous. *Ovary* oblong; stigma subsessile 2-fid; ovules 2–3 in each cell. *Fruit* a compressed 1–2-celled samara, winged at the tip, cells 1-seeded. *Seed* compressed, albumen fleshy; embryo straight, cotyledons broad.—Distrib. Europe, N. Asia, N. America; species about 30.—Etym. doubtful.

F. excel'sior, *L.* ; leaflets oblong-lanceolate serrate, perianth 0.

268 *OLEACEÆ.* [Fraxinus.

Woods and hedges, N. to Shetland, but generally planted; ascends to 1,350 ft. in Yorkshire; Ireland, Channel Islands; fl. April–May.—*Tree* 50–80 ft.; bark pale; branchlets stout; buds large, black. *Leaflets* 1–3 in., 4–7 pair. *Flowers* small, polygamous, in dense small axillary panicles. *Stamens* purple-black. *Samaras* 1½ in., in large drooping panicles, pedicelled, linear oblong, notched at the tip, nucleus oblong ribbed. *Seed* ½ in.—Distrib. Europe, N. Africa.—Wood excellent.—*F. heterophyl'la,* Vahl, is a 1-foliolate state.

Order XLVIII. **APOCYNA'CEÆ.**

Trees or shrubs, rarely herbs, often climbing; juice milky. *Leaves* opposite, rarely whorled, quite entire; stipules 0 or rudimentary. *Flowers* regular, solitary or cymose. *Calyx* 4–5-fid. *Corolla* hypogynous, funnel- or salver-shaped; throat naked or with scales; lobes usually oblique, contorted in bud. *Stamens* 4–5, on the tube or throat of the corolla, filaments very short; anthers basifixed, free, or connate and adhering to the stigma, cells sometimes obliterated below; pollen granular. *Disk* 0 or annular. *Ovary* of 2 free or connate carpels; style short, dilated, stigma entire or 2-fid often constricted in the middle; ovules many, placentas marginal. *Fruit* of 2 many-seeded follicles, or a berry or drupe. *Seeds* compressed, sometimes winged, hilum basilar or lateral, with often a pencil of silky hairs at the hilar (rarely at the other) end, albumen fleshy or hard or 0; embryo straight, cotyledons flat thin.—Distrib. Chiefly trop.; genera 103; species 900.—Affinities. With *Asclepiadeæ,* and *Rubiaceæ.* — Properties. Usually poisonous, drastic purgatives, or febrifuges.

1*. *VIN'CA, L.* Periwinkle.

Perennial herbs or slender decumbent undershrubs. *Leaves* evergreen. *Flowers* solitary, white blue or purple. *Calyx* 5-partite, lobes acuminate, glandular at the base inside. *Corolla* salver-shaped; tube hairy inside; throat thickened, angled; lobes 5, oblique. *Stamens* 5, filaments short; anthers inflexed, tipped with a bearded membrane; pollen glutinous. *Disk* of 2 glands alternating with the carpels. *Ovary* of 2 carpels; style terminated by a cup-shaped reflexed membrane, within which is the short conical entire or 2-lobed stigma. *Follicles* 2, slender, terete, many-seeded. *Seeds* subcylindric, testa black tuberculate, hilum lateral, albumen fleshy; embryo axile.—Distrib. Europe, Asia, Africa; species about 10.— Etym. *vincere,* from a use of the stems in *binding.*

V. mi'nor, *L.*; flowerless-stems prostrate rooting, leaves elliptic-ovate margins glabrous, calyx-lobes glabrous ⅓ the length of the corolla-tube.

Woods, copses, and hedgebanks, from Mull and Elgin southd.; (a denizen, *Wats.*); not indigenous in Scotland, Ireland, or the Channel Islands; fl. April–May.—*Stems* 1–2 ft. trailing, tough, flowering ones short erect. *Leaves* 1–1½ in., very shortly petioled. *Flowers* 1 in. diam.; peduncles not as long.   *Corolla* blue-purple.   *Fruit* rarely found.—Distrib.   From Denmark southd. (excl. Greece), W. Asia.

V. ma'jor, *L.* ; flowerless-stems prostrate not rooting, leaves ovate or cordate at the base ciliate, calyx-lobes ciliate equalling the corolla-tube.

Copses and hedges, naturalized; fl. April–May.—Much larger in all its parts than *V. mi'nor;* flower less-stems not tough, rooting at the tip only ; flowering erect in flower, elongating afterwards.—Distrib.   Mid. and S. Europe, N. Africa.

ORDER XLIX. **GENTIA'NEÆ.**

Annual or perennial herbs.   *Leaves* opposite, quite entire, rarely whorled (alternate and 3-foliolate in *Menyanthes,* alternate and floating in *Limnanthemum*) ; exstipulate.   *Flowers* regular, solitary, or in 2–3-chotomous cymes.   *Calyx* inferior, 4–8-toothed or lobed ; lobes twisted or valvate in bud.   *Corolla* hypogynous, often persistent, rotate campanulate or funnel-shaped ; throat naked fimbriate or scaly ; lobes 4–8, usually contorted in bud.   *Stamens* 4–8, inserted on the corolla-tube, filaments filiform ; anthers introrse, often extrorse during flowering.   *Ovary* 1–2-celled ; style simple or 0, stigma 2-fid or 2-lamellar ; ovules many, 1- or more-seriate on 2 opposite placentas, horizontal, anatropous.   *Capsule* 1- or incompletely 2-celled, septicidally 2-valved, rarely indehiscent, many-seeded.   *Seeds* minute, testa reticulate, albumen copious fleshy ; embryo minute, cotyledons small.—Distrib.   Cold and temp. regions ; genera 50 ; species 520. —Affinities.   Close with *Apocyna'ceæ.*—Properties.   Bitter and tonic.

Tribe I.   **CHIRONIEÆ.**   *Leaves* opposite.   *Corolla-lobes* twisted.   *Style* slender.

Corolla rotate.   Stamens 6–8.   Leaves perfoliate ....1. Chlora.
Corolla funnel-shaped.   Stamens 4.   Stigma peltate ....2. Microcala.
Corolla salver-shaped.   Stamens 4.   Stigma 2-lamellate ....3. Cicendia.
Corolla funnel-shaped.   Stamens 5.   Anthers twisted............4. Erythræa.

Tribe II.   **SWERTIEÆ.**   *Leaves* opposite.   *Corolla-lobes* twisted.   *Style* short.
Corolla-tube subclavate.   Anthers straight.   Stigmas 2 ....5. Gentiana.

Tribe III.   **MENYAN'THEÆ.**   *Leaves alternate.*   *Corolla* induplicate.

Leaves 3-foliolate ....6. Menyanthes.
Leaves orbicular, floating...7. Limnanthemum.

-----

1. **CHLO'RA,** *L.*   Yellow-wort, Yellow Centaury.

Annual or biennial, erect, glaucous herbs.   *Leaves* broadly connate at the base.   *Flowers* yellow, in 3-chotomous cymes.   *Calyx* 6–8-partite.   *Corolla* rotate, persistent, 6–8-partite.   *Stamens* 6–8, inserted on the throat of the corolla.   *Ovary* 1-celled ; style 2 fid, deciduous, stigmas oblong obtuse.   *Capsule* septicidally 2-valved, many-seeded.   *Seeds* minute, sunk in the placentas.—Distrib.   Europe, N. Africa, W. Asia ; species 2.— Etym. χλωρός, *yellow.*

C. perfolia'ta, *L.* ; radical leaves obovate-spathulate, cauline ovate.
Chalk or clay banks and pastures from Westmoreland and Durham southd.;
Ireland ; fl. June–Sept.—Glabrous. *Stems* 1 or more from the root, 6–18 in.,
terete. *Radical leaves* 1–2 in., rosulate, obtuse ; cauline smaller, broadly
ovate, acute. *Flowers* ½–⅛ in. diam., many, proterogynous. *Sepals* slender,
lanceolate-subulate. *Corolla-lobes* bright yellow, oblong, obtuse ; tube
finally ruptured by the capsule. *Stigma* 2-fid.—DISTRIB. From Belgium
southd., N. Africa, W. Asia.

2. MICROCA'LA, *Link* et *Hoffm.*

Small branched annuals. *Leaves* oblong or filiform. *Flowers* small,
yellow or pink, in 3-chotomous cymes or fascicled. *Calyx* tubular, 4-lobed.
Corolla salver-shaped, tube short, throat naked, lobes 4. *Stamens* 4,
inserted on the corolla-tube. *Ovary* 1-celled ; style deciduous, stigma
capitellate ; ovules many, on 2 parietal placentas. *Capsule* 1- or almost
2-celled, 2-valved, many-seeded. *Seeds* minute, immersed in the placenta.
—DISTRIB. S. Europe, N. America ; species 2.—ETYM. μικρός and καλος,
small and *pretty.*

M. filifor'mis, *Link ;* leaves subulate, calyx campanulate, teeth deltoid
acute, corolla-lobes obtuse equalling the tube. *Ex'acum,* Sm. ; *Cicen'dia,*
Reichb.
Sandy bays, rare, Pembroke and Cornwall to Sussex ; Killarney and Cork in
Ireland ; fl. July–Oct.—Glabrous. *Stem* 4–8 in., angled, simple or sparingly
branched above, very slender, branches suberect. *Leaves* ½–¼ in. *Flowers*
yellow ; pedicels stout, very long, ½–2½ in. *Capsule* ovoid, ⅛ in.—DISTRIB.
From Denmark southd. (excl. Russia), Azores.

3. CICEN'DIA, *Adans.*

A small branched annual ; branches divaricate. *Leaves* linear. *Flowers*
small, in the forks, pedicelled. *Calyx* 4-partite, segments narrow. *Corolla-
tube* cylindric, lobes spreading. *Stamens* 4, inserted on the corolla-tube ;
anthers short, straight. *Ovary* 1-celled ; style filiform, stigma 2-lamel-
late. *Capsule* 1-celled, 2-valved, many-seeded ; placentas intruded. *Seeds*
minute.—DISTRIB. S.W. France, Spain, Italy.—ETYM. doubtful.

C. pusil'la, *Griseb.* ; calyx-lobes subulate, corolla-lobes mucronate ½ as
long as the tube.
Sandy commons, Channel Islands ; fl. July–Sept.—Smaller and more slender
than *Microca'la filifor'mis. Stems* several. *Leaves* ¼ in. *Flowers* sometimes
5-merous, pink ; peduncles slender. *Calyx-lobes* erect. *Capsule* fusiform.

4. ERYTHRÆ'A, *Pers.* CENTAURY.

Annuals. *Stems* erect, angular. *Leaves* connate. *Flowers* small, pink
white or yellow, in terminal 3-chotomous cymes, not honeyed. *Calyx*
4–5-partite. *Corolla* funnel-shaped, persistent ; tube cylindric, throat
naked ; lobes 4–5. *Stamens* 4–5, on the corolla-tube ; anthers spirally

twisted, exserted. *Ovary* almost 2-celled ; style deciduous, stigmas 2-
lamellate ; ovules many. *Capsule* linear 1- or almost 2-celled, 2-valved,
many-seeded. *Seeds* minute.—DISTRIB. N. temp. regions ; species 15.
ETYM. *ἐρυθρός*, from the *red* flowers.

E. Centau′rium, *Pers.* ; radical leaves ovate or oblong-spathulate.

Dry pastures and sandy coasts, N. to Shetland ; Ireland ; Channel Islands ;
fl. June–Sept.—Erect, glabrous, 6–18 in., usually branched above. *Radical
leaves* ¼–2 in., upper sometimes linear. *Flowers* ½–½ in. diam., many, red
or pink (heterostyled with dimorphic pollen, *H. Müller*). *Calyx-lobes*
linear-subulate. *Corolla-lobes* oblong, obtuse or subacute. *Capsule* slender.
—DISTRIB. From Gothland southd., N. Africa ; introd. in N. America.
I am quite unable to follow critical authors in their endeavours to limit the
British forms of this variable plant. Of these Babington makes 5 and
Nyman 7 species. Regarding those of the former author as sub-species,
they are the following. (The comparative length of calyx and corolla-tube
must be observed on the expansion of the flower.)
E. CENTAU′RIUM proper ; branched above, leaves 3–7-nerved, leaves oblong
upper acute, cymes lax or compact, flowers subsessile, corolla-tube about
twice as long as the calyx, lobes ovate.—VAR. *capita′ta,* Koch (*E. latifo′lia,*
Engl. Bot.), is a stunted var. with capitate flowers. Pastures.
Sub-sp. E. LATIFO′LIA, *Sm.* ; stem shorter subsimple, leaves broad lowest
rounded 5–7-nerved, flowers in subcapitate cymes, corolla-tube about
equalling the calyx, lobes lanceolate.—Shores near Liverpool.
Sub-sp. E. LITTORA′LIS, *Fries ;* stem or stems simple, leaves ovate-oblong
obtuse, radical crowded spathulate, cymes dense-fld., corolla-tube equalling
the calyx, lobes oblong obtuse. *E. chloo′des,* Gen. and Godr.—Sandy shores,
N. to Shetland.—This is referred to *linarifo′lia,* Pers., by Nyman, who further
puts *chloo′des* with *littora′lis,* Sm., Engl. Fl., into *confer′ta,* Pers.
Sub-sp. E. PULCHEL′LA, *Fries ;* stem simple or branched tetraquetrous, leaves
oblong-lanceolate radical few, cymes lax-fld., flowers all pedicelled, corolla-
tube longer than the calyx, lobes oblong obtuse.—Sandy ground, from
Dumfries and Haddington southd.—VAR. *E. tenuiflo′ra,* Link, from the I. of
Wight, has a long corymbose inflorescence. It is referred to *latifo′lia,* Sm.,
by Nyman.
Sub-sp. E. CAPITA′TA, *Willd.* ; stem short, leaves oblong or subspathulate
obtuse, radical rosulate 3–5-nerved, cymes dense-fld. sessile, corolla-tube
equalling the calyx, top not constricted, not elongating, lobes oblong obtuse,
stamens inserted at the base of the tube, capsule ½ protruded.—Downs,
I. of Wight and Eastbourne.—The insertion of the stamens being anomalous
in the Order, suggests this being a heteroclite form of *E. Centau′rium.*
It has been found in Norway and Prussia. It is omitted in Nyman.

5. GENTIA′NA, *L.* GENTIAN.

Herbs of various habit. *Leaves* opposite. *Flowers* solitary, or in terminal cymes. *Calyx* 4–5-fid or -partite, or spathaceous. *Corolla* 4–5-lobed,
angles sometimes folded and produced, throat naked bearded or with five
scales. *Stamens* 4–5, inserted within the corolla-tube, included. *Disk*
annular. *Ovary* 1-celled ; stigmas 2, persistent, recurved ; ovules many.

Capsule septicidally 2-valved, many-seeded. *Seeds* immersed in the broad membranous placentas.—DISTRIB. Temp. and alpine regions, rare in Arctic ; species about 180.—ETYM. The classical name.

1. **G. campes'tris,** *L.* ; annual, calyx ebracteate 4-partite, lobes very unequal, corolla-tube subcylindric, throat ciliate.

Moist pastures, especially in the N., from Shetland southd.; ascends to 2,400 ft. in the Highlands; Ireland ; fl. July–Oct.—*Stem* 4–12 in., simple below. *Radical leaves* obovate-spathulate; cauline ovate-oblong or lanceolate, acute, 3–7-nerved. *Flowers* 1 in., cymosely panicled, pedicelled, sub-proterogynous. *Calyx-tube* short; outer lobes oblong-ovate, acuminate, inner narrower. *Corolla* pale lilac, rarely white; lobes oblong acute. *Capsule* subsessile.—DISTRIB. Europe (excl. Turkey), W. Siberia.

2. **G. Amarel'la,** *L.* ; annual, calyx ebracteate 5-lobed, corolla-tube subcylindric, throat ciliate. *Felwort.*

Dry pastures, from Shetland southd.; ascends to 2,100 ft. in Yorkshire; Ireland ; fl. July–Sept.—Habit and stature of *G. campes'tris,* but calyx less deeply divided, lobes 5, much smaller, subulate-lanceolate. –DISTRIB. Europe (Arctic), N. Asia.

G. AMAREL'LA proper ; calyx-lobes subequal from half as long to as long as the corolla-tube.—VAR. *præ'cox,* Raf. (*G. uligino'sa,* Willd.), is an early flowering state with 4-merous flowers and unequal calyx-lobes.

Sub-sp. G. GERMAN'ICA, *Willd.* ; larger, stouter, flowers larger, calyx-lobes unequal, 2 broader and more acute, much shorter than the corolla-tube.— York, Pembroke, Herts, Berks, Surrey, Hants.

3. **G. Pneumonan'the,** *L.* ; perennial, calyx 2-bracteate, lobes 5 equal, corolla narrow-campanulate, throat naked.

Moist heaths, from Cumberland to Dorset, and from York and Norfolk to Surrey and Anglesea, local ; fl. Aug.–Sept.—*Rootstock* short. *Stems* 1–2 ft., few, slender, scaly below, simple or nearly so, leafy above. *Leaves* 1–1½ in., linear-oblong, obtuse, 1–3-nerved. *Flowers* 1–2 in., few, axillary and terminal, shortly pedicelled, proterandrous ; bracts 2, long, linear. *Calyx-tube* obconic; lobes linear, obtuse. *Corolla* bright blue within. *Capsule* stipitate. —DISTRIB. Europe (excl. Turkey), N. Asia, N. America (a form).

4. **G. ver'na,** *L.* ; perennial, calyx 2-bracteate, lobes 5 equal, corolla salver-shaped, throat with a 2-fid scale between each lobe.

Wet subalpine limestone rocks of Westmoreland, York, Durham ; ascends to 2,400 ft. ; Mayo, Galway, Clare ; fl. May–June.—Tufted, stoloniferous. *Radical leaves* rosulate, ovate or ovate-oblong, obtuse or subacute, 1-nerved ; cauline few, smaller, oblong. *Stem* 1–2 in., curved or ascending. *Flowers* 1 in. diam., solitary, sessile ; bracts foliaceous. *Calyx-tube* large, 5-winged ; lobes acute. *Corolla* bright blue ; lobes ovate, obtuse. *Capsule* subsessile.— DISTRIB. Europe (excl. Turkey), N. and W. Asia.

5. **G. niva'lis,** *L.* ; annual, calyx 2-bracteate, lobes 5 equal, corolla funnel-shaped, throat with a 2-fid scale between each lobe.

Breadalbane and Clova Mts., from 2,400–3,300 ft., very rare ; fl. Aug.–Sept.— *Stem* slender, 2–8 in., simple or branched. *Radical leaves* ½–¾ in., few, obovate-oblong, obtuse or acute; cauline smaller, in distant pairs, all 3–5-nerved. *Flowers* ½–¾ in., solitary or few, shortly pedicelled; bracts small, oblong. *Calyx* narrow-campanulate, 5-angled and ribbed; lobes subulate. *Corolla* ¼ in. diam., blue; lobes ovate, obtuse. *Capsule* subsessile.—DISTRIB. Arctic, and Mts. of N. and Mid. Europe, Arctic America.

6. MENYAN'THES, *Tournef.* BUCK- OR BOG-BEAN.

Perennial scapigerous marsh herbs. *Rootstock* creeping. *Leaves* alternate, 3-foliolate, petiole sheathing. *Flowers* racemose, dimorphous. *Calyx* 5-partite. *Corolla* fleshy, funnel-shaped, deciduous, limb 5-partite, disk fimbriate, induplicate-valvate in bud. *Stamens* 5, inserted on the corolla-tube. *Disk* of 5 hypogynous glands. *Ovary* 1-celled; style filiform, persistent, stigma 2-lobed; ovules uniseriate on 2 parietal placentas. *Capsule* globose, 1-celled, obscurely loculicidally 2-valved, many-seeded. *Seeds* small, testa polished.—DISTRIB. Europe (Arctic), N. Asia, N.W. India, N. America; species 2.—ETYM. obscure.

M. trifolia'ta, *L.* ; leaflets subsessile oblong or obovate.

Marshy and spongy bogs, N. to Shetland; ascends to 1,800 ft. in the Lake district; Ireland; Channel Islands; fl. May–July.—*Rootstocks* stout, matted. *Leaflets* 1½–3 in., obtuse, quite entire, ultimate nerves with free tips within the larger areoles; petiole 3–7 in., sheath long narrow. *Scape* longer than the petioles, many-fld. *Flowers* ⅔ in. diam., white or pink; pedicels ¼–½ in., stiff, spreading; bracts broad, short, obtuse. *Sepals* oblong, obtuse. *Corolla-lobes* recurved, subacute. *Stamens* reddish. *Capsule* apiculate.—Bitter, reputed tonic and febrifuge; used to add bitterness to beer; rootstock full of starch, hence eaten.

7. LIMNAN'THEMUM, *S. P. Gmel.*

Perennial water-herbs. *Leaves* alternate or opposite, floating, peltate or cordate. *Flowers* yellow in sessile umbels that terminate short axillary branches, dimorphous. *Calyx* 5-partite. *Corolla* rotate, membranous, deciduous; segments 5–8, erose, with 5–8 fimbriate scales at the base, margins broadly inflexed in bud. *Stamens* 5–8, inserted on the corolla-tube. *Disk* of 5–8 hypogynous glands. *Ovary* 1-celled; style persistent, stigmas 2 simple or lobed; ovules many, placentas 2 parietal. *Fruit* bursting irregularly, few- or many-seeded. *Seeds* small, testa smooth or muricate.—DISTRIB. Temp. and trop. regions; species about 12.—ETYM. λίμνη, *a pool*, and ἄνθος, *a flower.*

L. peltatum, *Gmel.* ; leaves opposite on the flowering-stems. *L. nymphæoi'des*, Link. *Villars'ia nymphæoi'des*, Vent.

Still waters, rare, from Norfolk and Oxford to Sussex; naturalized further north, and in Scotland and Ireland; fl. July–Aug.—*Rootstock* creeping, with alternate leaves. *Flowering-stems* floating; their branches short, in the axils of opposite leaves *Leaves* all petioled, orbicular, base deeply cordate, quite entire, green, shining, purple-spotted above, opaque purplish and studded with glands beneath; petioles of radical leaves long, slender, not sheathing; of floating leaves shorter, stout, sheathing. *Peduncles* 1–3 in., crowded

T

Flowers 1 in. diam., subumbellate, opening one at a time. *Sepals* linear-oblong, obtuse. *Corolla* bright yellow; lobes erose or fimbriate. *Fruit* flagon-shaped, green. *Seeds* few, compressed, winged, ciliate.—DISTRIB. From Denmark southd., N. and W. Asia, N.W. India.

ORDER L. **POLEMONIA′CEÆ.**

Annual or perennial herbs (rarely shrubs). *Leaves* alternate, or the lower opposite, entire or divided, exstipulate. *Flowers* in terminal dichotomous cymes, usually blue or white. *Calyx* inferior, 5-lobed, imbricate in bud. *Corolla* subperigynous, regular or nearly so, 5-partite; lobes contorted in bud. *Stamens* 5, inserted on the corolla-tube, usually unequal; anthers 2-celled; pollen subglobose, reticulate. *Disk* fleshy. *Ovary* 3-celled; style simple, stigmas 3 linear revolute; ovules 1 or more, attached to the inner angle of each cell, amphitropous. *Capsule* 3-celled, loculicidally 3-valved; valves separating from the persistent axis; cells 1- or many-seeded. *Seeds* angled or plano-convex, testa spongy sometimes winged, hilum ventral, albumen fleshy; embryo axile, cotyledons subfoliaceous.—DISTRIB. Chiefly Arctic and temp., especially W. American; genera 8; species about 150.—AFFINITIES. With *Convolvulaceæ.*—PROPERTIES unimportant

1. POLEMO′NIUM, *L.* JACOB'S LADDER.

Perennial herbs. *Leaves* alternate, pinnate. *Flowers* corymbose, usually ebracteate. *Calyx* campanulate, 5-lobed. *Corolla* rotate; lobes 5, obovate. *Stamens* declinate, inserted on the corolla-throat, filaments dilated and hairy at the base. *Disk* cup-shaped, crenate. *Ovary* ovoid; style filiform, stigma 3-fid; ovules many in each cell, 2-seriate. *Capsule* ovoid, 3-celled, many-seeded, loculicidally 3-valved. *Seeds* ovoid, angled, with a short wing or 0, testa thick, abounding in spiral vessels. DISTRIB. N. temp. and Arctic regions; species about 8.—ETYM. obscure.

P. cæru′leum, *L.* ; leaflets 6–12 pairs subsessile. *Greek Valerian.*
Copses and streams, apparently indigenous from Stafford and Derby to the Cheviots, doubtfully elsewhere; an escape in Scotland and Ireland; fl. June–July.—*Rootstock* short, creeping. *Stems* 1–3 ft., glabrous or pubescent and glandular above, angular, fistular, leafy. *Leaves* 4–18 in.; petiole very slender, winged; leaflets ¾–1½ in., quite entire, ovate or oblong-lanceolate, acute. *Flowers* many, drooping, ½–1 in. diam., blue or white, proterandrous. *Calyx* campanulate; lobes oblong, acute. *Corolla-lobes* spreading, subacute. *Capsule* erect, included in the calyx. *Seeds* compressed, angular; testa ribbed and rugose, shortly winged.—DISTRIB. Europe (Arctic), N. of Alps to Russia, N. Asia, N.W. Himalaya, N. America.

ORDER LI. **BORAGIN'EÆ.**

Herbs, rarely shrubs, usually hispid scabrid or pilose. *Stems* terete. *Leaves* alternate, quite entire or sinuate; nerves usually strong, very prominent beneath ; exstipulate. *Flowers* regular or irregular, bracteate or not, in simple forked spiked or racemed often scorpioid cymes, rarely axillary. *Calyx* persistent, 5-lobed or partite, valvate in bud. *Corolla* hypogynous, rotate tubular campanulate or salver-shaped ; throat often closed by hairs or hollow folds opposite the 5 lobes, imbricate in bud. *Stamens* 5, inserted on the corolla-tube or throat, filaments usually short ; anthers 2-celled, often subulate. *Disk* 0, or confluent with a tumid receptacle. *Ovary* of 2 2-lobed 2-celled carpels connate at the very base ; style simple, arising from the base of the carpels, stigma simple or 2-fid ; ovules solitary in each cell, suspended, anatropous or ½-anatropous. *Fruit* of 4 indehiscent 1-seeded nutlets, inserted on the receptacle, which is continuous with the base of the style. *Seed* straight or curved ; hilum basal or ventral, often concave with thickened margins, testa membranous, albumen 0 or scanty and fleshy ; embryo straight or curved, cotyledons foliaceous, radicle superior.—DISTRIB. All climates, abundant in S. Europe and E. Asia ; genera 68 ; species 1,200.—AFFINITIES. With *Verbenaceæ, Labiatæ,* and *Convolvulaceæ.*—PROPERTIES. Mucilaginous and emollient ; often abounding in alkalies. Roots yield purple or brown dyes.—The above character does not include the tropical sub-order *Heliotropeæ,* chiefly distinguished by the obscurely lobed ovary.

TRIBE I. **ECHIE'Æ.** *Corolla* irregular; throat usually naked. *Nutlets* inserted by flat bases on the flat receptacle. *Stamens* exserted...1. Echium.

TRIBE II. **ANCHU'SEÆ.** *Corolla* regular; throat closed with scales. *Nutlets* inserted by broad cup-shaped bases on the flat receptacle.

Corolla rotate. Anthers exserted, conniving in a cone..............1*. Borago.
Corolla tubular, 5-toothed. Anthers included...................2. Symphytum.
Corolla salver-shaped. Anthers included....................3. Anchusa.

TRIBE III. **LITHOSPER'MEÆ.** *Corolla* regular; throat naked or closed by scales. *Nutlets* inserted by small flat bases to the flat receptacle.

Calyx-tube 0. Stamens included. Nutlets stony...........4. Lithospermum.
Calyx-tube short. Stamens protruding. Nutlets fleshy.........5. Mertensia.
Calyx-tube long, funnel-shaped. Stamens included. Nutlets smooth.
6. Pulmonaria.
Calyx-tube long. Corolla salver-shaped. Nutlets smooth ......7. Myosotis.

TRIBE IV. **CYNOGLOS'SEÆ.** *Corolla* regular; throat naked or closed with scales. *Nutlets* inserted by broad ventral surfaces on an elevated receptacle.

Calyx-lobes leafy. Nutlets granulate ....7*. Asperugo.
Calyx-lobes not leafy. Nutlets with hooked bristles.........8. Cynoglossum

1. E'CHIUM, *Tournef.* Bugloss.

Herbs, sometimes shrubby, usually large, stout, hispid or scabrous with tuberous-based hairs. *Leaves* entire. *Flowers* white red purple or blue, in spiked or panicled racemes, honeyed, proterandrous. *Calyx* 5-partite. *Corolla-tube* cylindric or funnel-shaped ; throat dilated ; limb unequally 5-lobed. *Filaments* unequal, adnate to the corolla below, exserted. *Style* filiform, stigma 2-lobed. *Nutlets* 4, ovoid or turbinate, wrinkled, scabrid, bases flat, receptacle flat.—Distrib. Chiefly S. Europe and Oriental ; species 20.—Etym. ἔχις, *a viper*, of disputed application.

1. **E. vulgare,** *L.* ; cauline leaves lanceolate or oblong base rounded, cymes short, calyx exceeding the corolla-tube, 4 stamens protruded, 5th included. *E. italicum,* Huds. not L. *Viper's Bugloss.*

Waste ground on light soils from Sutherland southd.; (an alien or colonist in Scotland, *Wats.*) ; S.E. Ireland ; Channel Islands ; fl. June–Aug.—*Root* fusiform, annual or biennial. *Stem* 1–3 ft., erect or ascending below, stout, leafy. *Radical leaves* petioled, 4–8 in.; cauline sessile, acute, rounded at the base. *Cymes* 1 in. or more, recurved, lengthening in fruit, panicled ; bracts and sepals linear. *Corolla* ¾ in., red-purple in bud, then bright blue, rarely white. *Nutlets* angular, rugose.—Distrib. Europe, N. Africa, W. Siberia ; introd. in N. America.

2. **E. plantagin'eum,** *L.* ; cauline leaves linear-oblong cordate at the base, calyx much shorter than the corolla-tube, cymes elongate, stamens slightly protruded. *E. viola'ceum,* Hook. and Arn., not of L.

Cornwall ; Jersey ; fl. June–Aug.—*Root* fusiform, annual or biennial. *Stem* 1–3 ft., erect or ascending, diffusely branched. *Leaves,* radical 4–6 in., lanceolate, petioled ; cauline spreading, obtuse, sometimes dilated at the base. *Cymes* 4–6 in., spreading, curved. *Sepals* subulate-lanceolate. *Corolla* 1 in., dark blue-purple. *Nutlets* as in *E. vulgare.*—Distrib. W. France, Mediterranean region.

1*. BORA'GO, *Tournef.* Borage.

Annual or perennial herbs, hispid with tuberous-based hairs. *Flowers* in lax forked cymes, bracteate, blue, honeyed, proterandrous. *Calyx* 5-partite. *Corolla* rotate, throat closed by notched scales, lobes acute. *Stamens* 5, on the throat of the corolla, filaments stout, concave, with an obtuse tooth ; anthers elongate, mucronate, conniving, exserted. *Style* filiform, stigma capitate. *Nutlets* 4, rugose, base truncate concave receptacle flat fleshy.—Distrib. S. Europe, N. Africa ; species 3.—Etym. doubtful.

B. officina'lis, *L.* ; stem erect stout, sepals linear connivent.

Waste ground, near habitations, England, Channel Islands ; an alien or escape ; fl. June–July.—*Root* annual or biennial. *Stem* 1–3 ft., stout, succulent, leafy, branched. *Leaves* waved or sinuate-toothed, subacute ; radical 4–6 in., ovate-lanceolate, petiole broad, winged ; cauline sessile or contracted towards the auricled base, upper oblong. *Cymes* axillary and terminal, few-fld., branched; pedicel 1–1½ in., decurved; bracts linear or lanceolate. *Sepals* subulate-lanceolate. *Corolla* ¾ in. diam., bright blue; lobes triangular-ovate. *Anthers* purple-black, spurred at the back. *Nutlets* ⅙ in.—Distrib. Mid. and S. Europe, N. Africa; introd. in America.—Used as a cordial, but has no sensible properties.

2. SYM'PHYTUM, *Tournef.* COMFREY.

Perennial coarse hispid herbs ; roots tuberous or fascicled. *Leaves*, radical petioled, cauline sessile or decurrent, upper often opposite. *Flowers* in terminal forked cymes, drooping, bracteate, white blue purple or yellow, honeyed, proterandrous. *Calyx* 5-partite or -toothed. *Corolla* tubular, dilated above the middle, shortly 5-toothed, throat closed by elongate ciliate scales. *Stamens* 5, on the middle of the corolla, filaments slender ; anthers long, included. *Style* slender, stigma capitate. *Nutlets* 4, ovoid, smooth, base broad excavated, receptacle flat.—Distrib. Europe, W. Asia ; species 16.—Etym. doubtful.

1. **S. officina'le**, *L.* ; stem broadly winged above, leaves decurrent.

River-banks and watery places, from Caithness southd.; (a denizen in N. Britain, *Wats.*); Ireland; Channel Islands ; fl. May–June.—Hispid and hairy. *Rootstock* branched; roots fleshy, fibrous. *Stem* 1–3 ft., stout, angular, branched. *Leaves* ovate-lanceolate, radical 4–8 in.; petiole long, winged ; cauline shortly petioled. *Cymes* scorpioid ; pedicels ¼–½ in. *Sepals* small, narrow-lanceolate. *Corolla* ¾ in., yellow, red or purple. *Nutlets* ⅛ in., shining.—Distrib. Europe, W. Siberia; an escape in the U. States.— An old styptic ; young leaves sometimes cooked and eaten.
S. officina'le proper ; corolla ochreous.—Var. *S. pa'tens*, Sibth.; rougher, corolla purple.

2. **S. tubero'sum**, *L.* ; stem hardly winged, leaves scarcely decurrent.

Copses in wet places, N. Wales, Stafford, and Bedford, to Isla and Elgin; fl. June–July.—Hairy, not hispid, often glandular. *Rootstock* short, horizontal ; root-fibres slender. *Stem* 1–2 ft., rather slender, leafy. *Radical leaves* much as in *S. officinale* in form, but longer petioled. *Flowers* rather smaller, ochreous.—Distrib. Mid. Europe from France to Turkey.

3. ANCHU'SA, *L.* ALKANET.

Annual or perennial herbs, usually villous and hispid. *Flowers* in scorpioid cymes, drooping, usually bracteate, blue or purple, rarely white or yellow, honeyed. *Calyx* 5-fid or -partite. *Corolla-tube* straight or curved, throat closed by hairs or scales ; limb oblique or spreading, 5-partite. *Stamens* included. *Nutlets* 4, rugose or granulate, base broad deeply concave, receptacle flat.—Distrib. Europe, W. Asia ; species 30. —Etym. doubtful.

Section 1. **Lycop'sis**, *L.* (gen.). *Corolla-tube* curved, equalling or exceeding the oblique limb. *Nutlets* with the ring equal at the base.

A. **arven'sis**, *Bieb.* ; hispid, with tuberous-based bristles. *Bugloss.*

278 *BORAGINEÆ.* [ANCHUSA.

Corn-fields and waste places in light soils, N. to Shetland ; ascends to 1,000 ft.
in the Highlands; Ireland; Channel Islands ; fl. June–July.—Annual ;
root fusiform. *Stem* simple below, $\frac{1}{2}$–$1\frac{1}{2}$ ft., angular, rather slender. *Leaves*,
radical 1–4 in., petioled, obovate-lanceolate ; cauline linear-oblong, sessile,
acute, margin waved and toothed, upper $\frac{1}{2}$-amplexicaul. *Cymes* 4–5 in.,
terminal, simple or forked, short, at length elongate, drooping, recurved ;
bracts leafy ; flowers subsessile. *Sepals* $\frac{1}{2}$ in., narrow. *Corolla* $\frac{1}{4}$ in. diam.,
bright blue, scales white. *Nutlets* small, reticulate. –DISTRIB. Europe,
N. and W. Asia to N.W. India; introd. in the U. States.—A. de Candolle
regards it as indigenous only in S. Europe.

SECTION 2. **Anchu'sa** proper. *Corolla-tube* straight, equalling or exceeding the limb. *Nutlets* with the basal ring not produced.

A. OFFICINA'LIS, *L.* ; densely softly hispid, leaves narrow-lanceolate.

Ballast hills, rare, Northumberland, Glasgow ; (an alien, *Wats.*) ; fl. June–
July.—*Root* stout, biennial. *Stem* 1–2 ft., angled, simple or branched.
Leaves, radical 3–6 in., gradually narrowed into long winged petioles ;
cauline sessile, oblong-lanceolate or linear-oblong, uppermost $\frac{1}{2}$-amplexicaul.
Cymes forked or in pairs, 1–2 in., lengthening to 4–6 in. ; bracts and sepals
ovate-lanceolate ; flowers subsessile. *Flowers* homogamous. *Corolla* $\frac{1}{4}$ in.
diam., violet-blue, scales white papillose. *Nutlets* small, brown.—DISTRIB.
Europe, W. Asia.

SECTION 3. **Caryolo'pha**, *Fisch. et Traut.* (gen.). *Corolla-tube* straight,
shorter than the limb. *Nutlets* with the ring produced towards the style.

A. SEMPERVI'RENS, *L.* ; hispidly hairy, leaves ovate acute.

Hedges and waste places from Caithness southd., rare ; Ireland ; Channel
Islands ; (an alien, *Wats.*); fl. May–June.—*Root* stout, perennial. *Stem*
1–2 ft., simple. *Leaves*, radical 8–12 in., oblong-ovate, long-petioled ;
cauline ovate, shorter petioled. *Cymes* in axillary pairs, very hispid, subcapitate, long-peduncled, 2-bracteate, not lengthening much ; peduncles 2–3
in., spreading, very slender ; bracts $\frac{1}{2}$–1 in., ovate-lanceolate ; flowers shortly
pedicelled. *Sepals* linear. *Corolla* $\frac{3}{4}$ in. diam., bright blue, scales white.
Nutlets small, reticulate.—DISTRIB. From Belgium and Spain to Lombardy.

4. **LITHOSPER'MUM**, *Tournef.* GROMWELL.

Annual or perennial herbs, or shrubs, hispid or hairy. *Flowers* in
bracteate cymes. *Calyx* 5-partite. *Corolla* funnel- or salver-shaped, throat
naked or with 5 tumid folds ; lobes 5, spreading. *Anthers* oblong, included in the corolla-tube. *Style* simple, stigma capitate obscurely lobed.
Nutlets bony or stony, smooth or rugose, base truncate, receptacle flat.—
DISTRIB. Europe, temp. Asia and America ; species about 40.—ETYM.
λίθος and σπέρμα, from the *stony nutlets*.

1. **L. officina'le**, *L.* ; perennial, stems many all erect branched and
flowering, leaves $\frac{1}{2}$-amplexicaul narrow-lanceolate, nutlets smooth white.

Copses, hedgebanks, &c., from Ross southd.; Ireland ; fl. June–July.—
Rootstock stout, woody. *Stem* 1–3 ft., rough with the tuberous bases of the
stiff hairs, very leafy. *Leaves* 2–4 in., pubescent above, strigose beneath.

Cymes capitate, small, strigose, on short leafy axillary branches; bracts longer than the calyx; flowers subsessile. *Sepals* equalling the corolla. *Corolla* ½ in. diam., yellow-white. *Nutlets* 1–2, narrowed upwards, shining. —Distrib. Europe, N. and W. Asia; introd. in N. America.

2. **L. arven′se,** *L.* ; annual, stems solitary erect branched, leaves sessile ½-amplexicaul linear-oblong, nutlets grey shining wrinkled.

Cornfields and waste places from Ross southd.; Ireland; Channel Islands; fl. May–June.—*Root* tapering. *Stem* 10–16 in., stout, flexuous, shortly hispid. *Leaves*, radical obovate-lanceolate, petioled; cauline 2–3 in., obtuse. *Cymes* short, terminal; bracts large, leafy; pedicels very short. *Flowers* honeyed, homomorphous. *Sepals* narrow-linear, almost equalling the corolla. *Corolla* ½ in., cream-white. *Nutlets* narrowed upwards.—Distrib. Europe, N. Africa, N. and W. Asia, N.W. India; introd. in the U. States.—Yields a red dye.

3. **L. purpu′reo-cæru′leum,** *L.* ; perennial, barren stems creeping, flowering erect, leaves subsessile narrow-lanceolate, nutlets white smooth.

Copses on limestone and chalk, very rare, Wales, Devon to Kent; fl. June– July.—*Rootstock* creeping, woody, slender. *Stems* 1 ft., scabrid, flowering 1–2 ft., leafy, rigid, simple or branched. *Leaves* 1½–3 in., rather softly strigose. *Cymes* terminal, few-fld.; bracts large, leafy; pedicels very short. *Sepals* very slender, much shorter than the corolla. *Corolla* ¾ in., bright blue-purple. *Nutlets* 1–2, nearly globose, shining.—Distrib. From Belgium southd.

5. **MERTEN′SIA,** *Roth.*

Perennial herbs. *Leaves* usually obovate, lower petioled, upper sessile. *Flowers* in terminal cymes, blue-purple, dimorphic. *Calyx* 5-fid or -partite. *Corolla-tube* cylindric, limb campanulate, 5-fid or -partite, throat naked or with 5 transverse folds. *Stamens* 5, towards the top of the corolla-tube. *Style* filiform, lengthened after flowering, stigma obtuse. *Nutlets* 4, rather fleshy, smooth or rough, base contracted, receptacle small 2-4-lobed.—Distrib. N. and Arctic Europe, Asia, and America; species 15.—Etym. Prof. F. C. Mertens, a German botanist.

M. marit′ima, *Don ;* glabrous, glaucous, leaves ovate or obovate.

Sea-shores, Wales and Berwick to Shetland; Ireland; fl. May–June.—Succulent. *Rootstock* fleshy, stoloniferous. *Stems* 1–2 ft., decumbent, leafy, much branched. *Leaves* 2-farious, 1–3 in., lower petioled, upper sessile, with prominent callous points when dry. *Cymes* dichotomous, with 2 opposite leafy bracts at the base; pedicels short, decurved in fruit. *Calyx* angular in fruit; lobes ovate. *Corolla* ¼ in. diam., 5-lobed to the middle, pink then blue, throat with 5 folds. *Nutlets* flattened, large, fleshy, outer coat becoming inflated and papery, back rounded.—Distrib. Lapland to Denmark, N. and Arctic shores of Asia and America.

6. **PULMONA′RIA,** *Tournef.* Lungwort.

Perennial herbs ; rootstock creeping, usually terminating in sterile branches. *Flowering-stems* simple. *Cymes* terminal. *Flowers* often polygamous or dimorphic, purple, white or pink in bud, honeyed. *Calyx* 5-angled at the base, 5-fid, after flowering campanulate, lobes erect. *Corolla* funnel-shaped, 5-cleft, with 5 pencils of hairs between the stamens. *Stamens* 5, included. *Stigma* subglobose, 2-lipped. *Nutlets* 4, turbinate, smooth, base truncate, receptacle flat.—DISTRIB. Europe, N. Asia ; species 4.—ETYM. *pulmo,* from its former use in *lung* complaints.

P. angustifo'lia, *L.* ; leaves narrow-lanceolate cauline sessile.

Copses, &c., on clay soil, very rare; Hants, Dorset; fl. April–June.—*Rootstock* short, stout; root-fibres fleshy. *Stem* 1–1½ ft., hairy, hardly hispid, brittle. *Leaves,* radical 6–10 in., petioled, often spotted with pale green ; cauline sessile, much smaller, more oblong, acute, ½-amplexicaul. *Cymes* short, much incurved, bracts leafy ; pedicels rather slender. *Flowers* dimorphous, the short-styled with larger flowers, a smaller ovary and less honey. *Calyx-lobes* lanceolate, ⅓ in., enlarging in fruit. *Corolla* ¾ in., pink, then bright blue. *Nutlets* smooth, black.—DISTRIB. Europe (excl. Spain, Greece, Turkey).

P. OFFICINA'LIS, *L.* ; leaves ovate or ovate-lanceolate.

Woods and copses, S. Scotland and England, rare, naturalized.—Habit, &c. of *P. angustifo'lia,* but the leaves very different,'and always blotched with pale green, and flowers pale purple.—DISTRIB. Europe.

7. MYOSO'TIS, *L.* SCORPION-GRASS.

Annual or perennial strigose herbs. *Radical leaves* petioled ; cauline sessile, linear-oblong. *Cymes* terminal, scorpioid ; flowers small, bracteate or not. *Calyx* 5-toothed or -cleft. *Corolla* salver- or funnel-shaped, throat closed by 5 short notched scales ; limb 5-fid, lobes contorted in bud. *Anthers* included, connective slightly produced. *Style* short, stigma capitate. *Nutlets* 4, minute, usually highly polished, compressed or 3-gonous, base small, receptacle small.—DISTRIB. N. and S. temp. regions, most common in Europe and Australia ; species about 30.—ETYM. μῦς and οὖς, from the leaves resembling a *mouse's ear.*

SECTION 1. Perennial. *Hairs* of stem appressed or spreading. *Pedicels* slender, longer than the calyx. *Calyx* campanulate, strigose with straight appressed hairs only.

1. **M. palus'tris,** *With.* ; hairs scanty spreading rarely appressed, corolla ⅓–½ in. diam., style nearly equalling the calyx. *Forget-me-not.*

Wet places, N. to Shetland ; Ireland; Channel Islands; fl. May–July. Ligh[t] green, rather shining. *Rootstock* creeping; stolons creeping, with small leaves. *Stem* 1–2 ft., erect or ascending, rather stout, flexuous. *Leaves* 1–3 in., linear-oblong or narrowly spathulate, obtuse, shining ; upper sessile or shortly decurrent. *Cymes* variable, flowering pedicels ⅛–¼ in. *Corolla* sky-blue, disk yellow, lobes retuse. *Nutlets* small, black, bordered, hard keeled in front.—DISTRIB. Europe (Arctic), N. Asia; introd. in N. America.

M. PALUS'TRIS proper ; stolons subterranean, calyx-lobes triangular shorter than the corolla-tube. Lowlands, N. to Orkney, rarer Scotland.—VAR. *M. strigulo'sa,*

Reich. ; more erect with more copious appressed hairs, leaves sessile, flowers smaller.

Sub-sp. M. RE′PENS, *D. Don ;* stolons above ground, calyx-lobes lanceolate exceeding the corolla tube.—More northern and upland, ascends to 2,200 ft. in Yorkshire, Channel Islands to Shetland.

2. **M. cæspitosa,** *Schultz ;* hairs appressed, calyx-lobes triangular nearly as long as the tube, corolla ⅛ in. diam., style much shorter than the calyx.—*M. lingula′ta*, Lehm.

Wet places, N. to Shetland ; ascends to 1,600 ft. in Yorkshire ; Ireland ; Channel Islands ; fl. May–Aug.—Light green, tufted, rather shining. *Rootstock* short, without stolons. *Stem* 6–18 in., much branched from the base, branches slender. *Leaves*, radical spathulate-oblong, polished ; cauline 1–2¼ in., linear-oblong, tip rounded, narrowed to the sessile base. *Cymes* usually long and slender, pedicels ¼–½ in. *Sepals* oblong-ovate, obtuse. *Corolla* sky-blue, disk yellow. *Nutlets* black, short, broad, bordered, not keeled in front.—DISTRIB. Europe (Arctic), N. Africa, N. and W. Asia, Himalaya, N. America.—This and the preceding are the only species with bright green shining foliage.

SECTION 2. *Hairs of stem spreading. Calyx with spreading and hooked hairs.*

* *Lower leaves petioled. Pedicels usually much longer than the calyx.*

3. **M. sylvat′ica,** *Hoffm.* ; perennial, stolons 0, calyx campanulate cleft ¾ of the way with few straight and many incurved or hooked hairs, closed in fruit, corolla ⅓ in. diam. flat.

Dry woods, &c., from Forfar and Dumfries to Kent, Hants, and Wales ; ascends to 1,200 ft. in Yorkshire ; Channel Islands ; fl. May–Sept.—Biennial or perennial. *Rootstock* 0 or short. *Stem* ½–2 ft., branched from the base, erect or the lateral branches ascending. *Leaves* 1½–2½ in., acute or apiculate, hairs spreading. *Cymes* very lax in fruit ; pedicels about twice as long as the calyx. *Flowers* homomorphous, odorous in the evening ; corolla bright blue, tube very short. *Nutlets* bordered, keeled at the tip in front.—DISTRIB. Europe (Arctic), Canaries, N. and W. Asia.

M. SYLVAT′ICA proper ; leaves long-petioled subacute, calyx base rounded, tube with hooked hairs, nutlets brown.

Sub-sp. M. ALPES′TRIS, *Schmidt ;* leaves subsessile, calyx with many straight and few incurved or hooked hairs, open in fruit, base acute, corolla ½ in. diam. flat, fruiting pedicels shorter, nutlets black. *M. rupic′ola*, Sm. —Moist rocks, Ben Lawers, alt. nearly 4,000 ft. ; Teesdale, alt. 2,400 ft. ; Westmoreland.

4. **M. arven′sis,** *Hoffm.* ; annual or biennial, calyx shortly campanulate cleft ½ way or lower with few appressed and many spreading hooked hairs, closed in fruit, corolla ⅛ in. diam. usually concave. *M. interme′dia*, Link.

Fields and waste places, N. to Shetland ; ascends to 1,200 ft. in Yorkshire ; Ireland ; Channel Islands ; fl. June–Aug.—Closely allied to *M. sylvat′ica*, but pedicels usually much longer, slender, and flowers very small and paler blue. *Nutlets* brown, bordered, keeled in front.—DISTRIB. Europe (Arctic), N. Africa, N. and W. Asia to India, native of N. U. States.

M. arven'sis proper; annual, corolla-limb concave.—VAR. *umbro'sa,* Bab.; biennial, corolla larger, limb flatter. Shaded places.

** *Lower leaves subsessile. Pedicels usually shorter than the calyx.*

5. **M. colli'na,** *Hoffm.* ; annual, calyx shortly campanulate cleft ½ way with many spreading hooked and few straight hairs, open in fruit, corolla ⅛ in. diam. usually concave bright blue, tube short.

Field banks and waste grounds, N. to Shetland; ascends to 1,000 ft. in the Highlands; E. Ireland; Channel Islands; fl. May–July.—*Stems* 3–12 in., usually branched from the base. *Leaves* ½–1 in., linear-oblong, obtuse or apiculate, strigose. *Cymes* very long and slender in fruit, often exceeding the leafy part of the stem. *Style* not half as long as the calyx. *Nutlets* turgid, brown, scarcely bordered.—DISTRIB. Europe, N. Africa, W. Asia. *M. Mitte'ni* is a variety with the flowers pale, and the lower in the cyme bracteate.

6. **M. versic'olor,** *Reichb.* ; annual, calyx shortly campanulate cleft ½ way with many spreading hooked and few straight hairs, closed in fruit, corolla ₁/₁₀ in. diam. usually concave yellow then dull blue, tube long.

Waste grounds, N. to Shetland; ascends to 1,500 ft. in Northumbd.; Ireland; Channel Islands; fl. April–June.—*Stems* 3–12 in., usually much branched from the base. *Leaves* ½–1 in., linear-oblong, subacute, strigose. *Cymes* elongate; flowers not secund, lowest sometimes bracteate. *Flowers* homo-gamous; corolla-tube elongating till the anthers reach the stigma. *Style* nearly as long as the calyx. *Nutlets* black, bordered.—DISTRIB. Europe, N. Africa, W. Asia; introd. in the U. States.

7*. *ASPERUGO, Tournef.* MADWORT (MADDERWORT).

An annual hispid procumbent herb. *Radical leaves* petioled ; cauline alternate subopposite or whorled. *Cymes* axillary, 1–3-fld. ; flowers on short recurved pedicels, small, blue. *Calyx* deeply 5-lobed, with alter-nating teeth ; lobes leafy, spreading, veined, enlarged after flowering, and forming a compressed 2-lipped laciniate covering to the fruit. *Corolla* funnel-shaped, throat closed by scales ; lobes 5, rounded. *Stamens* included. *Stigma* subcapitate. *Nutlets* laterally compressed, subacute, tubercled, attached by the edge to an elevated receptacle.—DISTRIB. Europe (Arctic), W. Asia to N.W. India.—ETYM. *asper,* from the *rough* leaves.

A. PROCUM'BENS, *L.* ; stem prickly, leaves linear-oblong.

Waste places, rare and casual, from Sutherland to Kent; fl. May–July.— *Stem* 1–2 ft., stout or slender, soft, simple or branched, sharply ridged, prickles scattered short hooked. *Leaves* 2–5 in., lower petioled, uppermost sessile, variable in form, obtuse or acute, thin, hispid. *Corolla* ⅛ in. diam., blue-purple. *Fruiting-calyx* ¾ in. broad; lips unequal, fan-shaped, pal-mately lobed ; pedicels very short, decurved. *Receptacle* of nutlets with 2 membranous scales formed of the detached cuticle of the calyx.

8. CYNOGLOS'SUM, *Tournef.* HOUND'S-TONGUE.

Coarse hispid villous or silky biennials. *Flowers* small, blue purple or white, in forked cymes, usually ebracteate. *Calyx* 5-partite. *Corolla* funnel-shaped, mouth closed by prominent scales; lobes obtuse. *Stamens* included. *Style* rigid, persistent, stigma entire or notched. *Nutlets* 4, depressed or convex, covered with hooked or barbed bristles, peltately attached to a thickened conical receptacle.—DISTRIB. Temp. and trop. regions, especially Asiatic; species about 60.—ETYM. κύων and γλῶσσα, *dog's tongue*, from the texture of the leaf surface.

1. **C. officina'le**, *L.* ; hoary with soft rather appressed hairs, nutlets with a thickened border.

Fields and waste places, not common, E. Scotland, from Forfar to Kent and Cornwall; S.E. Ireland, rare; Channel Islands; fl. June–July.—*Root* fleshy, tapering. *Stem* 1-2 ft., stout, erect, branched, leafy. *Leaves* radical, 8–10 in., long-petioled, oblong or oblong-lanceolate; cauline sessile, linear-oblong or lanceolate, obtuse, base rounded or cordate. *Cymes* lengthening to 6–10 in.; pedicels recurved, stout, lower often bracteate. *Sepals* oblong, obtuse, enlarged to $\frac{1}{3}$ in. in fruit. *Corolla* $\frac{1}{2}$ in. diam., dull red-purple. *Nutlets* $\frac{1}{4}$ in., face flat ovate with short hooked spines; border thickened.—DISTRIB. Europe, N. Africa, N. and W. Asia; introd. in U. States.—Narcotic and astringent; smells like mice.

2. **C. monta'num**, *Lamk.* ; scabrid with short spreading hairs, nutlets without a thickened border. *C. sylvat'icum*, Hænke.

Copses and waste places in Mid. and E. England, rare, from Salop and Norfolk to Kent and Surrey; Dublin; fl. May–July.—Habit, &c., of *C. officina'le*, but greener, more slender, with linear sepals $\frac{1}{2}$ in. long in fruit, bluer corollas, and the marginal spines of the nuts largest.—DISTRIB. From France and Germany southd. (excl. Greece).

ORDER LII. CONVOLVULA'CEÆ.

Herbs or shrubs, usually twining (rarely trees); juice often milky. *Leaves* alternate, 0 in *Cuscuta*, exstipulate. *Flowers* in axillary or terminal racemes, cymes, or heads, rarely solitary, often large, of all colours. *Sepals* 5, persistent. *Corolla* hypogynous, regular, tubular bell- or funnel-shaped; limb 5-lobed or -angled, plaited induplicate or imbricate in bud. *Stamens* 5, inserted on the corolla-tube, filaments often unequal and dilated at the base; anthers sagittate, basifixed, often twisted after flowering. *Ovary* 2-4- (rarely 1-) celled; style slender, 2–4-fid, stigmas capitate linear or lamellar; ovules 1 or 2 in each cell, erect from its base, 4 in the 1-celled ovaries. *Capsule* 1-4-celled, 2-4-valved, or bursting transversely at the base. *Seeds* basal, erect; testa coriaceous or membranous, often villous, albumen scanty mucilaginous (fleshy in *Cuscuta*); embryo curved, cotyledons broad thin folded, radicle short (embryo spiral and undivided in *Cuscuta*).—Distrib. Chiefly trop. ; genera 32 ; species about 800.—Affinities. With *Boragineæ* and *Hydrophyllaceæ.*—Properties. Often purgative ; some (*Batatas*) yield esculent roots.

Sub-order I. **Convolvu'leæ.** Leafy. *Corolla* plaited in bud. *Albumen* scanty. *Cotyledons* foliaceous ....1. Convolvulus.

Sub-order II. **Cuscu'teæ** (*Presl*, Order). Leafless parasites. *Corolla* imbricate in bud. *Albumen* copious, fleshy. *Embryo* filiform spiral.
2. Cuscuta.

1. CONVOL'VULUS, *L.* BINDWEED.

Slender, often perennial twining herbs ; juice milky. *Leaves* alternate, often cordate or sagittate. *Flowers* axillary, solitary or corymbose, white pink purple or blue, bracteate or ebracteate, honeyed. *Sepals* 5. *Corolla* funnel- or bell-shaped, limb 5-angled, plaited and twisted in bud. *Stamens* 5, inserted at the bottom of the corolla-tube, filaments dilated at the base. *Ovary* 2-celled ; style filiform, stigmas 2 oblong or linear ; ovules 2. *Capsule* 2-celled, the dissepiment sometimes imperfect. *Seeds* 2 in each cell, erect, testa hard.—Distrib. Temp. and trop. regions ; species about 160.—Etym. *convolvo*, to *entwine.*

Sub-gen. 1. **Convol'vulus** proper. *Bracts* small, placed low on the peduncle. *Stigmas* slender.

1. **C. arven'sis,** *L.* ; leaves hastate or sagittate entire or sinuate. *Small Bindweed.*

Fields and waste places, N. to Caithness; ascends to 1,200 ft. in Derby ; local in Scotland ; Ireland ; Channel Islands ; fl. June–Sept.—Glabrous or pubescent. *Rootstock* slender, extensively creeping underground. *Stems* many, 6–24 in., trailing or twining, slender. *Leaves* 1–3 in , very variable, apiculate, lobes acute. *Flowers* 1 in. diam., white or pink, proterandrous, odorous; peduncle 1–4-fld. ; pedicels 4-gonous, recurved in fruit; bracts 2, small, linear. *Sepals* unequal, broadly oblong, obtuse. *Capsule* 2-celled, globose, apiculate. *Seeds* 4, obtusely 3-gonous, muricate.—Distrib. Europe, N. Africa, N. and W. Asia, N.W. India.—A pest to agriculture.

Sub-gen. 2. **Calyste'gia,** *Br.* (gen.). *Bracts* 2, large, enclosing the calyx. *Stigmas* broad.

2. **C. se'pium,** *L.* ; stem twining, leaves hastate or sagittate.

Hedges and thickets, from Berwick and the Clyde southd.; ascends to 1,200 ft. in Derby ; Ireland ; Channel Islands ; fl. June–Aug.—Glabrous, rarely pubescent. *Rootstock* tuberous, creeping. *Stems* twining, 3–5 ft., slender. *Leaves* 3–5 in., membranous, entire, obtuse or acute, deeply cordate, lobes rounded or angled. *Flowers* 2 in. diam., white or pale pink, inodorous, open in rain, closed at night ; peduncle solitary, 1-fld., 4-gonous ; bracts ovate-cordate or triangular. *Sepals* subequal, ovate-lanceolate. *Capsule* ½ in. diam., globose, apiculate, 1-celled above, 2-celled below. *Seeds* smooth, obtusely 3-gonous.—Distrib. Europe, N. Asia, N. Africa, temp. N. and S. America, Australasia.—Rootstock purgative.

3. **C. Soldanel'la,** *L.* ; stem procumbent, leaves orbicular or reniform.
Sandy sea-shores from Isla and Forfar southd.; Ireland; Channel Islands;
fl. June–Aug.—Glabrous. *Rootstock* slender, running extensively. *Stems*
6–12 in., slender, rarely twining. *Leaves* ½–1½ in. diam., fleshy, usually
much broader than long, lobes rounded; petiole 1–3 in., suddenly dilating
at the tip. *Flowers* 1–1½ in. diam., pale purple or pink; peduncle solitary,
1-fld., 4-quetrous; bracts ¼–⅔ in., broadly oblong, obtuse, shorter than
the similar or retuse unequal sepals. *Capsule* large, incompletely 2-celled.
Seeds ¼ in. diam., obtusely 3-gonous, smooth, black.—DISTRIB. From
Belgium southd., N. Africa, W. Asia, S. temp. regions.

2. CUS'CUTA, *Tournef.* DODDER.

Slender, twining, leafless, pink yellow or white, annual parasites.
Flowers in bracteate heads (in British species), rarely spiked. *Calyx* 4–5-
fid. *Corolla* urceolate, persistent ; limb 4–5-fid ; tube naked, or with a
ring of scales below the stamens. *Stamens* 4–5, inserted on the corolla-
tube. *Ovary* 2-celled ; styles 2, free or connate, stigmas acute or capitate ;
ovules 2 in each cell, erect. *Capsule* 2-celled, circumsciss at the base,
cells 2-seeded. *Seeds* angled ; embryo filiform, undivided, spirally coiled
round the fleshy albumen, radicle thickened —DISTRIB. Temp. and trop.
regions ; species about 80.—ETYM. doubtful.

1. **C. europæ'a,** *L.* ; sepals erect obtuse, tips spreading, corolla twice
as long ventricose above, scales short or 0, stamens included.
On nettles, vetches, &c., from York to Sussex and Devon, rare; fl. July–Sept.
—*Stems* as thick as twine, reddish or yellow. *Flower-heads* ½–¾ in. diam.,
tinged with red. *Sepals* fleshy at the base only. *Corolla-lobes* about as
long as the tube, obtuse, spreading. *Scales* appressed to the corolla-tube,
remote, 2-fid. *Styles* included.—DISTRIB. Europe, N. Africa, Siberia.

2. **C. Epi'thymum,** *Murr.* ; sepals suberect acute shorter than the
cylindric corolla-tube, scales converging toothed, stamens exserted.
On Furze, Thyme, Ling, &c., from Ayr southd.; Channel Islands; fl. July–
Oct.—*Stems* filiform, very slender, reddish. *Heads* ¼–¾ in. diam., variable
in colour. *Corolla-lobes* spreading. *Scales* large, contiguous. *Styles* exserted.
—DISTRIB. From Denmark southd., N. Africa, W. Asia.—*C. Trifo'lii*, Bab.,
is a variety with shorter distant scales, found sporadically in clover fields.

C. EPI'LINUM, *Weihe;* sepals acute appressed to the equally long in-
flated corolla-tube, scales small distant toothed, stamens included.
Sporadic on Flax in England, Scotland, and Ireland; fl. July–Aug.—About
as large and stout as *C. europæ'a*, but usually paler and more succulent.
Sepals fleshy, triangular-ovate. *Corolla-lobes* obtuse. *Scales* incurved as in
C. Epi'thymum, but smaller, distant, often 2-fid. *Styles* short, included.—
DISTRIB. Europe.

ORDER LIII. **SOLANA'CEÆ.**

Herbs or shrubs (rarely trees). *Leaves* alternate, or in pairs, or subopposite, simple lobed or pinnatisect, exstipulate. *Flowers* regular, on 1-
or more-fld. supra-axillary or axillary ebracteate cymes. *Calyx* inferior,
5-fid, usually persistent, often enlarged in fruit. *Corolla* hypogynous,
rotate campanulate or salver-shaped ; lobes 5, imbricate plicate or induplicate-valvate in bud. *Stamens* 5, rarely unequal, inserted on the corolla-
tube, short ; anthers connivent or cohering by their tips, opening inwards
by slits or terminal pores. *Disk* annular. *Ovary* 2- or incompletely
4-celled ; style simple, stigma simple or lobed ; ovules many, amphitropous, placentas on the septum. *Fruit* a many-seeded capsule or berry.
Seeds small, usually compressed, reniform, hilum ventral or lateral, testa
thick, albumen fleshy ; embryo terete, straight or curved, in- or out-side
the albumen, radicle next the hilum.—DISTRIB. Chiefly tropical ; geneɪa
66 ; species about 1,250.—AFFINITIES. With *Convolvulaceæ* and *Scrophularineæ*.—PROPERTIES. Narcotic and excitant, or tonic and bitter,
pungent or stimulant.

Corolla subcampanulate. Capsule 2-celled ....1. Hyoscyamus.
Corolla rotate, anthers with pores. Berry 2-celled ....2. Solanum.
Corolla subcampanulate, anthers with slits. Berry 2-celled........3. Atropa.

1. HYOSCY'AMUS, *Tournef.* HENBANE.

Annual or biennial, heavy-scented herbs, often viscid. *Leaves* toothed or
sinuate-pinnatifid. *Flowers* axillary, or in bracteate scorpioid cymes,
honeyed. *Calyx* urceolate, 5-toothed. *Corolla* irregular, bell- or funnel-
shaped ; lobes 5, unequal, obtuse, plaited in bud. *Stamens* 5, inserted at
the base of the corolla-tube, declinate ; anthers with slits. *Ovary* 2-celled ;
style simple, stigma capitate ; ovules many. *Capsule* hidden in the calyx-
tube, constricted in the middle, 2-celled, membranous, circumsciss at the
crown, many-seeded. *Seeds* reniform, punctate.—DISTRIB. Warm and
temp. Europe, Africa, and Asia ; species about 9.—ETYM. obscure.

H. niger, *L.* ; pubescent, leaves angled toothed or subpinnatifid.
Sandy waste places, from Forfar and Dumbarton southd.; not native in
Scotland ; Ireland ; Channel Islands ; fl. June–Aug.—Fœtid and viscid ; hairs
pale, soft, glandular. *Stem* 1-2 ft., stout, branching, terete. *Radical leaves*
6-8 in., petioled, ovate ; cauline oblong, amplexicaul, with few large lobes
or teeth. *Flowers* 2-seriate, subsessile. *Calyx-tube* ovoid; limb subcylindric, 5-toothed. *Corolla* 1-1¼ in. diam., lurid yellow, veined with purple
(or not, *H. pal'lidus*, Kits.) ; lobes broad, subequal. *Anthers* purple. *Calyx-
tube* globose in fruit, ¾ in. diam., veined.—DISTRIB. Europe, N. Africa, N.
and W. Asia, India.—Anodyne and antispasmodic.

2. SOLA'NUM, *Tournef.* NIGHTSHADE.

Herbs or shrubs (rarely trees). *Leaves* scattered or in pairs, entire or
divided. *Flowers* solitary fascicled or cymose, white or blue, honeyless,

homogamous. *Calyx* 5–10-fid. *Corolla* rotate ; lobes 5–10, plaited in bud. *Stamens* 5, inserted on the corolla-throat, exserted ; filaments very short ; anthers conniving, or connate, pores terminal. *Ovary* 2- rarely 3–4-celled ; style simple, stigma obtuse ; ovules many. *Berry* 2- rarely 4-celled, many seeded. *Seeds* reniform.—Distrib. An immense tropical genus ; species probably 700.—Etym. doubtful.

1. **S. Dulcama'ra,** *L.* ; perennial, stem flexuous, leaves ovate-cordate or 3–5-partite, cymes panicled leaf-opposed or lateral. *Bitter-sweet.*

Hedges and copses from Isla and Ross southd. ; Ireland ; Channel Islands ; fl. June–Aug.—Glabrous, pubescent, or tomentose. *Rootstock* extensively creeping. *Stem* 4–6 ft., trailing. *Leaves* 1–3 in., acuminate, cordate, or upper hastate, or with 2 auricles or petioled pinnules at the base. *Flowers* many, drooping, homogamous ; pedicels slender. *Calyx-lobes* broad, obtuse. *Corolla* ½ in. diam., purple or white, lobes revolute. *Anthers* yellow, cohering in a cone. *Berry* ½ in., ovoid, mucronate, red, rarely yellow-green.—Distrib. Europe, N. Africa, W. Asia to India ; introd. in N. America.
Var. *mari'num,* Bab. ; stem prostrate branched, leaves fleshy. S. coast.

2. **S. ni'grum,** *L.* ; annual, stem erect angled usually tubercled, leaves rhomboid-ovate narrowed into the petiole, cymes umbellate lateral.

Waste places from Wigton and Northumbd. southd. ; casual in Scotland and Ireland ; Channel Islands ; fl. July–Oct.—Glabrous or pubescent. *Stem* 6–24 in., rarely more. *Leaves* 1–3 in., sinuate or toothed. *Flowers* few, drooping, homogamous ; pedicels slender. *Calyx-lobes* broad, obtuse. *Corolla* ¼–⅓ in. diam., white ; lobes ciliate, recurved. *Berries* ¼ in. diam., globose, black yellow or red.—Distrib. All temp. and trop. regions.
S. ni'grum proper ; hairs usually upcurved, leaves sinuate, berry black.—Var. *S. minia'tum,* Bernh. ; hairs usually straight, leaves toothed, berry scarlet. Kent, Channel Islands.—Var. *S. luteo-virescens,* Gmel. ; berry bright green. Mortlake.

3. AT'ROPA, *L.* Dwale.

A branched herb. *Leaves* scattered or in pairs, quite entire. *Flowers* solitary or few, peduncled, lurid violet or greenish. *Calyx* 5-partite. *Corolla* campanulate, regular ; lobes 5, plaited in bud. *Stamens* 5, inserted at the bottom of the corolla-tube, filaments filiform ; anthers with slits. *Ovary* 2-celled ; style simple, stigma peltate ; ovules many. *Berry* 2-celled, subtended by the spreading calyx, many-seeded. *Seeds* reniform, minutely pitted.—Distrib. From Denmark southd., N. Africa ; introd. in N. America.—Etym. Ατροπος, one of the Fates.

A. Belladon'na, *L.* ; leaves ovate acuminate. *Deadly Nightshade.*

Waste places, probably indigenous on chalk and limestone ; oftenest naturalized near ruins, from Westmoreland southd. ; near houses in Scotland ; Ireland ; Channel Islands ; fl. June–Aug.—Glabrous or pubescent and glandular. *Rootstock* stout, fleshy, stoloniferous. *Stem* 2–3 ft., stout. *Leaves* usually in unequal pairs, larger 3–8 in., contracted into the petiole. *Flowers* axillary supra-axillary and from the forks, drooping ; peduncles ½–1 in.,

288   *SOLANACEÆ.*   [ATROPA.

slender. *Sepals* ½-¾ in., broadly ovate. *Corolla* 1 in., greenish-purple;
lobes subequal, spreading, obtuse. *Filaments* subequal, tip incurved;
anthers pale, included. *Berry* spheroidal, obscurely 2-lobed.—A sedative,
poisonous in overdoses.

ORDER LIV. **PLANTAGINE'Æ.**

Annual or perennial scapigerous herbs. *Leaves* usually all radical with
parallel ribs, rarely cauline and opposite or alternate. *Scapes* axillary.
Flowers small, green, usually spiked, regular bisexual (1-sexual in *Littorella*). *Sepals* 4, persistent, imbricate in bud. *Corolla* hypogynous,
salver-shaped, scarious; lobes 4, spreading, imbricate in bud. *Stamens*
4, inserted on the corolla-tube (hypogynous in *Littorella*), filaments capillary, inflexed in bud, pendulous in flower, persistent; anthers large,
versatile, deciduous. *Ovary* free, 2-4-celled (1-celled and 1-ovuled in
Littorella); style filiform, with 2 lines of stigmatic hairs; ovules solitary and
basal, or many peltately attached to the septum. *Fruit* a 1-4-celled, 1-
or more-seeded, membranous circumsciss capsule (bony and indehiscent in
Littorella), seed-bearing septum free. *Seeds* usually peltate, albumen
fleshy; embryo transverse, cylindric, cotyledons oblong or linear, radicle
nferior.—DISTRIB. All temp. regions; genera 3; species about 50.—
AFFINITIES. Doubtful; probably reduced form of *Scrophularineæ.*—
PROPERTIES. Mucilaginous.

Terrestrial. Flowers spiked, 2-sexual ....1. Plantago.
Aquatic. Flowers few, 1-sexual ....2. Littorella.

1. PLANTA'GO, *L.* PLANTAIN, RIB-GRASS.

Flowers 2-sexual, in terminal spikes or heads, anemophilous, usually
proterogynous. *Sepals* 4. *Stamens* 4, inserted on the corolla. *Ovary*
2-4-celled. *Capsule* circumsciss. *Seeds* with a mucilaginous testa.—
DISTRIB. Of the Order; species about 48.—ETYM. The old Latin name.

1. **P. ma'jor,** *L.*; leaves petioled oblong or ovate-oblong toothed 3-7-
ribbed, scape short not furrowed, spike very long, sepals free, capsule 2-
celled 8-16-seeded, seeds flat in front.

Pastures and waste places, N. to Shetland; ascends to 2,000 ft. in Northumbd.;
fl. May-Sept.—Glabrous or hairy. *Rootstock* stout, truncate. *Leaves* 2-5 in.,
petiole broad, short, teeth very irregular. *Scape* about as long as the leaves;
spike slender, longer than the scape; bracts equalling the calyx, oblong-ovate,
concave, obtuse, glabrous. *Sepals* ¼ in., obtusely keeled, margins scarious.
Corolla-tube glabrous. *Filaments* short; anthers purple. *Seeds* black, rough.
—DISTRIB. Europe, N. Africa, N. and W. Asia, Himalaya; introd. in N.
America.—Seeds used for feeding cage-birds.

VAR. *P. interme'dia,* Gilib., is a dwarf very downy form with ascending
scapes.

2. **P. me'dia,** *L.* ; leaves subsessile elliptic-oblong toothed 5-9-ribbed, scape not furrowed, spike short, sepals free, capsule 2 celled 2-seeded, seeds flat in front.

Roadsides and waste places, usually on a dry soil, from Aberdeen and Ayr southd.; ascends to 1,600 ft. in Northumbd.; introduced? in Scotland and Ireland; Channel Islands; fl. June–Oct.—Pubescent. *Rootstock* tapering. *Leaves* 6-10 in., very variable. *Scape* 6–12 in.; spike 1–3 in.; bracts often purple, concave, obtuse, edges silvery, shorter than the glabrous unkeeled sepals. *Corolla-tube* glabrous. *Filaments* long; anthers whitish. *Seeds* brown, rough.—DISTRIB. Europe, N. Africa, N. and N.W. Asia, Himalaya. —Dimorphic : (1) scape long, corolla-lobes spreading acute, filaments white; (2) scape shorter, corolla-lobes rounded, filaments red, stigma shorter. *Delpino.*

3. **P. lanceola'ta,** *L.* ; leaves petioled lanceolate entire or toothed 3–6-ribbed, scape deeply furrowed, spike short, 2 anterior sepals often connate, capsule 2-celled 2-seeded, seeds concave in front. *Ribwort.*

Pastures and waste places, N. to Shetland; ascends to 2,200 ft. in the Highlands; Ireland; Channel Islands; fl. May–Oct.—Glabrous or pubescent. *Rootstock* tapering, crown woolly. *Leaves* 1–12 in., very variable. *Scape* as long ; spike ½–3 in., ovoid globose or cylindric; bracts acuminate. *Sepals* hairy at the tip, 2 dorsal keeled. *Corolla* longer than the calyx, glabrous. *Stamens* long, all white. *Seeds* black, shining.—DISTRIB. Europe, N. Africa, N. and W. Asia, Himalaya; introd. in N. America.—Leaves used for dressing sores in Scotland. Trimorphic, gynodiœcious : (1) scape tall, anthers broad white (anemophilous) ; (2) scape shorter; (3) dwarf (entomophilous). *Delpino.*
P. lanceola'ta proper ; bracts and sepals blackish at the tips.—VAR. *P. Timba'li,* Jord.; bracts and sepals with broad silvery margins. Fields of clover, &c., not indigenous.

4. **P. marit'ima,** *L.* ; leaves narrow linear fleshy faintly 3–5-ribbed, scape not furrowed, spike short or long, 2 dorsal sepals connate, capsule 2-celled 2-seeded, seeds flat in front.

Salt marshes, N. to Shetland; mountain streams, in York and Perth : ascends to 1,800 ft.; fl. June–Sept.—Glabrous or hairy (*b. hirsu'ta,* Lond Cat.). *Rootstock* woody, branched, crown rough. *Leaves* 1–12 in., ⅓–½ in. diam., very variable, sometimes narrowly lanceolate, quite entire or remotely toothed. *Scapes* as long as or longer than the leaves; spikes ¼–3 in.; bracts ovate-lanceolate. *Flowers* homomorphous. *Sepals* nearly glabrous, 2 dorsal with a toothed keel, margins narrowly scarious. *Corolla-tube* pubescent. *Stamens* pale yellow. *Seeds* brown, faintly winged at one or both ends.—DISTRIB. Europe (Arctic), N. Africa, N. and W. Asia, Himalaya, N. America.

5. **P. Coro'nopus,** *L.* ; leaves narrow linear 1-ribbed toothed or 1-2-pinnatifid, scape not furrowed, spike short or long, 2 dorsal sepals with a winged scarious keel, capsule 3-4-celled 3-4-seeded, seeds flat in front.

Sandy and gravelly places, N. to Shetland; Ireland; Channel Islands; most common near the sea; fl. June–Aug.—Annual or biennial, usually pubescent with long hairs. *Leaves* 1-12 in., very variable. *Scapes* usually ascending equalling or exceeding the leaves; spikes ¼–6 in.; bracts ovate, long acuminate, much exceeding the obtuse sepals. *Corolla-tube* pubescent. *Stamens* pale yellow. *Seeds* pale brown.—DISTRIB. Europe, N. Africa, W. Asia.

2. LITTOREL'LA, *L.* SHORE-WEED.

A creeping aquatic perennial herb. *Leaves* all radical, ½-cylindric. *Scapes* short, few-fld. ; flowers monœcious. MALE fl. solitary. *Sepals* 4. *Corolla-tube* cylindric, lobes 4. *Stamens* 4, hypogynous. *Ovary* rudimentary. FEMALE fl. usually 2 at the base of the male scape. *Sepals* 3–4, unequal. *Corolla* urceolate, mouth 3–4-toothed. *Stamens* 0. *Ovary* flagon-shaped, 1-celled; style very long, rigid; ovule 1 rarely 2, erect, campylotropous, flanked by a column of placental tissue. *Fruit* bony. *Seed* erect, testa membranous.—DISTRIB. Europe (Arctic) (excl. Turkey, Greece), Azores.—ETYM. *littus,* from growing near *shores.*

L. lacus'tris, *L.* ; leaves fleshy lacunose internally.

Sandy or gravelly edges of lakes and ponds, N. to Shetland; ascends to 1,600 ft. in the Highlands; Ireland; Channel Islands; fl. Aug.—Glabrous or puberulous, often forming a submerged matted turf. *Rootstock* creeping, white, stoloniferous. *Leaves* 1–4 in., linear-subulate, sheathing at the base. MALE fl. *Scape* axillary, much shorter than the leaves, 1–2-bracteate below the middle, papillose. *Sepals* ⅛ in., obtuse, green, edges scarious. *Stamens* long; anthers very large, pale. FEMALE fl. subsessile, enclosed in lanceolate bracts.

ORDER LV. SCROPHULARI'NEÆ.

Herbs or shrubs. *Leaves,* lower opposite or whorled, upper alternate, rarely all opposite or alternate and stipulate. *Inflorescence* various, flowers usually irregular, peduncles 2-bracteate at the forks. *Calyx* inferior, usually persistent, 5-merous. *Corolla* hypogynous; lobes 4–5, imbricate or subvalvate in bud. *Stamens* 4, rarely 2 or 5, inserted on the corolla-tube, with or without a rudimentary fifth; anthers 1- or 2-celled. *Disk* annular glandular or cup-shaped. *Ovary* 2-celled; style simple, stigma capitate or 2-lobed; ovules many, very rarely 2 in each cell, anatropous or amphitropous, placentas axile. *Fruit* a many-seeded capsule, rarely a berry, dehiscence various. *Seeds* small, testa various, hilum lateral or ventral, albumen fleshy; embryo straight, rarely curved, radicle next the hilum or lateral.—DISTRIB. All climates; genera 157; species about 1,900. AFFINITIES. With *Orobancheæ* and *Solaneæ.*—PROPERTIES. A few are purgative, or emetic, or intensely bitter, or very poisonous.

SUB-ORDER I. **Antirrhi'nideæ.** *Corolla* with the posticous (upper) lobes external in bud.

TRIBE I. **VERBAS'CEÆ.** *Leaves* alternate. *Inflorescence* centripetal. *Corolla* rotate. *Stamens* 5, declinate, unequal.............1. Verbascum.

Tribe II. **ANTIRRHINE'Æ.** Lower leaves or all opposite or whorled. *Inflorescence* centripetal. *Corolla* personate, tube saccate or spurred. *Stamens* 4. *Stigma* 2-lobed. *Capsule* opening by pores.
Corolla spurred at the base..2. Linaria.
Corolla saccate at the base...3. Antirrhinum.

Tribe III. **CHELONE'Æ.** *Inflorescence* composite. *Corolla* not spurred or saccate. *Stamens* 4. *Stigma* notched. *Capsule* 2–4-valved.
4. Scrophularia.

Tribe IV. **GRATIOLE'Æ.** *Inflorescence* centripetal. *Corolla* not spurred or saccate. *Stigma* 2-lamellate. *Capsule* 2-valved.........4*. *Mimulus.*

Sub-order II. **Rhinan'thideæ.** *Corolla* with the posticous (upper) lobes never exterior in bud.

Tribe V. **SIBTHORPIE'Æ.** *Flowers* axillary. *Leaves* alternate. *Corolla* short, subregular. *Stamens* 4, nearly equal. *Stigma* entire.
Leaves fascicled, linear. Anthers 1-celled........................5. Limosella.
Leaves alternate, orbicular. Anthers sagittate, 2-celled......6. Sibthorpia.

Tribe VI. **DIGITALE'Æ.** *Inflorescence* centripetal. *Leaves* alternate, lower petioled. *Stamens* 2 or 4. *Stigma* 2-lobed.............7. Digitalis.

Tribe VII. **VERONICE'Æ.** *Inflorescence* centripetal. *Leaves* opposite. *Corolla* almost regular. *Stamens* 2, diverging. *Stigma* capitate.
8. Veronica.

Tribe VIII. **EUPHRASIE'Æ.** *Inflorescence* centripetal. *Leaves* usually opposite. *Corolla* 2-lipped. *Stamens* 4, converging. *Stigma* unequally 2-fid.
Leaves opposite. Seeds many, small, not winged.
  Upper corolla-lip entire or notched...............................9. Bartsia.
  Upper corolla-lip with 2 spreading or reflexed lobes......10. Euphrasia.
Leaves opposite. Seeds few, compressed, winged..........11. Rhinanthus.
Leaves alternate. Seeds many, small, not winged..........12. Pedicularis.
Leaves opposite. Seeds 1–2 in each cell, not winged....13. Melampyrum.
Leafless parasite..14. Lathræa.

1 VERBAS'CUM, *L.* Mullein.

Tall, erect, usually biennial, tomentose or woolly herbs. *Leaves* alternate. *Flowers* in simple or compound racemes, red, yellow, or purple, rarely white, honey scanty. *Calyx* 5-partite. *Corolla* rotate; segments 5, nearly equal. *Stamens* 5, 3 posterior or all the filaments bearded; anther-cells confluent. *Stigma* undivided or 2-lamellate. *Capsule* septicidally 2-valved, many-seeded. *Seeds* pitted.—Distrib. Chiefly Europe and W. Asia, species about 100.—Etym. A corruption of *Barbascum,* the Latin name.

The following hybrids occur; their names indicate their parentage: *Thapso-Lychnitis, Thapso-nigrum, nigro-pulverulentum, nigro-Lychnitis.*

1. **V. Thap'sus,** *L.*; densely woolly, eglandular, stem terete simple, leaves very decurrent, flowers in a dense simple woolly spike.

u 2

Waste dry places, local, from Argyll and Elgin southd.; Ireland; Channel
Islands; often an escape; a denizen in Scotland; fl. June–Aug.—*Stem* 2–3
ft., stout. *Radical leaves* 6–18 in., obovate-lanceolate, entire or crenate;
cauline oblong, acute, upper acuminate. *Spike* 6–10 in.; bracts longer
than the flowers. *Corolla* ¾–1 in. diam., woolly externally. *Filaments* with
white hairs; anthers of long stamens slightly decurrent. *Seeds* ribbed.—
DISTRIB. Europe, N. and W. Asia, Himalaya; introd. in N. America.—
Wool formerly used for lamp-wicks.

2. **V. Lychni'tis,** *L.* ; stem angled, leaves stellately-pubescent, racemes
panicled narrow, flowers small whitish, hairs of filaments white.

Waste places, Denbigh, and from Herts and Stafford to Worcester, and
Kent to Somerset; a doubtful native; fl. July–Aug.—*Stem* 2–3 ft. *Radical
leaves* 4–10 in., petioled, oblong-lanceolate, obtuse, coarsely crenate, green
above, white beneath; cauline sessile, ovate, acuminate. *Racemes* erect,
many-fld. *Flowers* ½ in. diam., several to each bract. *Calyx* small, very
woolly. *Anthers* not decurrent. *Style* slender. *Capsule* small, ovoid.—
DISTRIB. From Denmark southd., W. Asia; introd. in N. America.

3. **V. pulverulen'tum,** *Vill.* ; mealy, stem terete, leaves stellately
pubescent, racemes panicled pyramidal, flowers yellow, hairs of filaments
white.

Waste places Norfolk and Suffolk; (native? *Wats.*) ; fl. July.—Habit of *V.
Lychni'tis,* but leaves much broader, sesssile, with small crenatures and more
matted with woolly hairs; cauline cordate. *Flowers* ½–¾ in. diam., several
to each bract, bright yellow. *Sepals* small, lanceolate. *Anthers* not decurrent. *Capsule* small, ovoid.—DISTRIB. From Belgium southd. (excl.
Greece, Russia).

4. **V. ni'grum,** *L.* ; stem angular, leaves stellately pubescent, racemes
nearly simple, flowers yellow, hairs of filaments purple.

Waste places, fields, &c., from Notts, Derby, and Carnarvon southd., but
often an escape; fl. June–Oct.—*Stem* 2–3 ft., whole plant covered with long
hairs, not so matted as in the former species. *Radical leaves* petioled,
sometimes 1 ft., ovate-oblong or oblong-lanceolate, often cordate, crenate;
cauline, except the upper, petioled, ovate-cordate, hardly white beneath.
Raceme 1–1½ ft., slender, erect. *Flowers* ½–¾ in. diam., many to each bract,
pedicelled. *Sepals* small, lanceolate, tomentose. *Anthers* not decurrent.—
DISTRIB. Europe, Siberia.—An Alderney var. (*tomento'sum,* Bab.), has
more woolly leaves beneath, and smaller flowers.

5. **V. Blatta'ria,** *L.* ; nearly glabrous, stem subangular, branches of
panicle slender glandular, flowers yellow, hairs of filaments purple.

Waste places, rare, from Norfolk and Stafford southd.; S. and W. Ireland;
Channel Islands; a denizen or alien; fl. June–Oct.—*Stem* 8 in. to 4 ft.,
rather slender, simple or branched. *Radical leaves* 4–10 in., oblong-lanceolate, obtuse, crenate lobulate or subpinnatifid; cauline small, sessile, ovate,
or oblong, acute or acuminate, sometimes cordate, irregularly toothed or
subcrenate. *Flowers* ¾–1¼ in. diam., bright yellow, rarely cream-coloured,
lax or dense; lower bracts leafy; peduncles ¼–1 in. *Sepals* oblong, often

large. *Anthers* of long stamens decurrent. *Capsule* nearly globose.—
DISTRIB. From Holland southd., N. Africa, N. and W. Asia, Himalaya;
introd. in N. America.

V. BLATTA'RIA proper; upper leaves not decurrent, racemes lax-fld., pedicels
solitary slender longer (often much) than the calyx.—Native (?) of S.W.
England, probably introd. elsewhere.

Sup-sp. V. VIRGA'TUM, *With.*; more glandular, upper leaves shortly decurrent,
racemes dense-fld., pedicels more fascicled shorter than the calyx.

2. LINA'RIA, *Tournef.* TOAD-FLAX.

Herbs, rarely shrubby. *Lower leaves* opposite whorled or alternate.
Flowers in bracteate racemes or spikes, or axillary and solitary. *Calyx*
5-partite. *Corolla* personate, tube spurred; upper lip erect, mid-lobe of
lower smallest; palate sometimes closing the throat. *Stamens* 4 fertile,
5th 0 or rudimentary; anthers oblong. *Stigma* notched or 2-lobed. *Capsule* ovoid or globose; cells subequal, dehiscing by simple or toothed
pores. *Seeds* angled or rugose, sometimes discoid and winged.—DIS-
TRIB. Europe, W. Asia; species 130.—ETYM. *linum, flax,* which some
species resemble.

SECTION 1. **Cymbala'ria,** *Chav.* Trailing and creeping. *Peduncles*
axillary, 1-fld. *Spur* short; palate not projecting. *Capsule* dehiscing by
small 3-fid valves.

L. CYMBALA'RIA, *Mill.*; glabrous, leaves petioled subsucculent broadly
reniform irregularly 3-7-angled or -lobed. *Ivy-leaved Toad-flax.*

Old walls, from Perth southd.; Ireland; an alien; fl. May–Sept.—Perennial.
Branches 6–24 in., slender. *Leaves* ½–¾ in. diam.; lobes acute. *Flowers*
⅓ in., blue-purple or white, homogamous; peduncles slender. *Sepals* linear-
lanceolate. *Palate* yellow, closing the throat. *Capsule* small, globose.
Seeds wingless, testa wrinkled.—DISTRIB. From Holland southd. (excl.
Turkey).—Eaten as a salad in S. Europe.

SECTION 2. **Elatinoi'des,** *Chav.* Diffuse. *Peduncles* axillary, 1-
fld. *Spur* as long as the corolla; palate projecting. *Capsule* dehiscing
by pores which are furnished with deciduous valves.

1. **L. spu'ria,** *Mill.*; hairy or villous and glandular, leaves shortly
petioled ovate or orbicular entire or obtusely toothed. *Male Fluellen.*

Sandy and chalky cornfields, from Norfolk, Lincoln, and S. Wales southd.;
Channel Islands; (a colonist, *Wats.*); fl. July–Oct.—Annual. *Stem* 4–18 in.,
erect; branches many, prostrate, slender. *Leaves* ½–1 in., opposite or
alternate, rarely cordate, acute or apiculate. *Peduncles* longer than the
leaves. *Sepals* oblong or ovate-cordate. *Corolla* ½ in., yellow, throat
purplish above; spur at right angles to the tube. *Fifth stamen* a small
scale. *Capsule* subglobose. *Seeds* with broad deep pits.—DISTRIB. From
Holland southd., N. Africa, W. Asia; introd. in N. America.

2. **L. Elat'ine,** *Mill.*; hairy or villous and slightly glandular, leaves
shortly petioled, cauline alternate broadly hastate acute.

Sandy and gravelly cornfields, from N. Wales and York southd.; Ireland,
rare; Channel Islands; (a native or colonist, *Wats.*); fl. July–Oct.—Annual.
Branches 6–30 in., many from a very short stem, prostrate, very slender.
Leaves, radical and lower opposite, ovate; cauline ¼–1 in., sometimes toothed
towards the very acute basal lobes. *Peduncles* capillary, longer than the
leaves, curved at the top. *Sepals* oblong-lanceolate. *Corolla* ¼–⅓ in., yellow;
upper lip purple within. *Copsule* globose. *Seeds* much as in *L. spu'ria.*—
DISTRIB. From Denmark southd., N. Africa, W. Asia, Himalaya; introd.
in N. America.

SECTION 3. **Linarias'trum,** *Chav. Flowers* racemed. *Spur* long or
short; palate prominent. *Capsule* 4–10-valved at the top.

3. **L. vulga'ris,** *Mill.* ; perennial, erect, almost glabrous, glaucous,
leaves linear or lanceolate, sepals ovate or lanceolate shorter than the spur
or capsule, corolla yellow.

Waste ground, from Elgin and the Clyde southd., rare in Scotland and
Ireland; Channel Islands; fl. July–Oct.—*Rootstock* creeping. *Stem* 1–2 ft.,
stout or slender, leafy. *Leaves* 1–3 in., often whorled. *Raceme* dense-fld.;
pedicels variable; bracts linear. *Corolla* ¾–1 in.; spur parallel to and as
long as the tube, acute. *Capsule* broadly oblong. *Seeds* scabrous, winged.
—DISTRIB. Europe (Arctic), N. Asia; introd. in N. America.—A reputed
purgative and diuretic.—A Peloria form occurs with regular flowers, 5
spurs and corolla-lobes, and 5 stamens.
L. vulga'ris proper; leaves obscurely 3-nerved, raceme glandular-pubescent,
sepals ovate-lanceolate.—VAR. *latifo'lia,* Bab. (*L. specio'sa,* Ten.); leaves
3-nerved, raceme glabrous, bracts foliaceous, sepals lanceolate, corolla 1½ in.
—Isle of Wight, Kent.

4. **L. Pelisseria'na,** *Mill.* ; annual, erect, glabrous, leaves linear,
sepals subulate shorter than the slender spur longer than the capsule, corolla
purple.

Jersey; fl. June–July.—Small, 6–10 in., with short barren stolons at the base.
Leaves on the stolons oblong, ¼–½ in., opposite and whorled in threes,
cauline ¾ in., scattered, linear, obtuse. *Raceme* short, few-fld.; bracts
longer than the pedicels. *Corolla* ¾ in.; spur parallel to the tube, acute.
Capsule broad, 2-lobed. *Seeds* flat, winged, tubercled on one face, wing
fimbriate.—DISTRIB. From Belgium southd., W. Asia.

5. **L. re'pens,** *Ait.* ; perennial, creeping, glabrous, sepals lanceolate as
long as the spur shorter than the capsule, corolla violet. *L. stria'ta,* DC.

Waste places, from Westmoreland and Mid. Wales southd., rare; naturalized
north of this; Ireland, very rare; Channel Islands; fl. July–Sept.—*Rootstock* slender, creeping. *Stems* 1–3 ft., many, very slender, branched, leafy.
Leaves ⅔–1½ in., whorled or scattered, linear-lanceolate. *Racemes* elongating;
bracts very small. *Corolla* ½ in.; spur almost parallel to the tube, obtuse.
Capsule broad, compressed. *Seeds* angled, wrinkled transversely.—DISTRIB.
W. Europe.—*L. se'pium,* Allman, is a hybrid with *L. vulga'ris.*

SECTION 4. **Chænorrhi'num,** *DC. Flowers* axillary or racemed. *Spur*
short; palate depressed; upper lip horizontal.

6. **L. mi'nor,** *Desf.* ; leaves alternate linear oblong or lanceolate.
Cornfields in chalky and sandy soils, from Lanark and Berwick southd , local ;
Ireland, very rare ; Channel Islands ; (a colonist, *Wats.*) ; fl. May–Oct.—
Annual, glandular-pubescent; *Stem* 6–18 in., slender ; branches ascending.
Leaves ½–1 in. *Peduncles* axillary, longer than the leaves. *Sepals* linear-
oblong. *Corolla* scarcely exceeding the calyx, ¼ in., pale purple ; lower lip
whitish ; palate yellow ; spur short. *Capsule* gibbous at the base, cells
subequal, opening by ragged pores. *Seeds* truncate, furrowed.—DISTRIB.
Europe, N. Africa.

3. ANTIRRHI'NUM, *Tournef.* SNAPDRAGON.

Annual or perennial herbs. *Leaves* entire, rarely lobed, lower opposite,
upper alternate. *Flowers* solitary and axillary, or racemose and bracteate.
Calyx 5-partite. *Corolla* personate, tube saccate, compressed ; upper lip
erect ; lower spreading, mid-lobe smallest ; palate broad, bearded, closing
the throat. *Stamens* 4 fertile, 5th rudimentary or 0 ; anther-cells oblong.
Stigma shortly 2-lobed. *Capsule* 2-celled, upper cell bursting by one pore,
lower by 2 many-toothed pores, rarely globose with 1 pore to each cell.
Seeds minute, oblong, truncate, rugose or pitted.—DISTRIB. Europe, W.
Asia ; species 25.—ETYM. ἀντί and ῥίν, from the *snout*-like flower.

A. Oron'tium, *L.* ; low, annual or biennial, leaves linear-lanceolate,
raceme leafy or flowers axillary, sepals longer than the corolla.
Cornfields, from Cumberland southd. ; Ireland, very rare ; Channel Islands ;
(a colonist, *Wats.*) ; fl. July–Oct.—Glabrous below, usually glandular-pubescent above. *Stem* 6–18 in., much branched from the base. *Leaves* 1–2 in.,
sessile, sometimes ciliate. *Bracts* leafy ; pedicels short, erect. *Sepals* ½–1
in., very narrow, spreading. *Corolla* ½–⅔ in., rose-purple. *Capsule* ½ in.,
pubescent. *Seeds* compressed, one face concave, the other 1-ribbed, margin
thickened.—DISTRIB. From Denmark southd., N. Africa, N. and W. Asia,
N.W. India ; introd. in N. America.

A. MA'JUS, *L.* ; tall, perennial, leaves lanceolate oblong or linear, raceme
bracteate glandular-pubescent, sepals short.
Old walls ; an alien ; fl. July–Sept.—Erect, branched, shrubby and glabrous
below, above glandular-pubescent. *Leaves* 1–3 in., very variable. *Racemes*
dense-fld. ; bracts ovate, acuminate ; pedicels erect. *Sepals* unequal, very
obtuse. *Corolla* 1½ in., purple, white, yellow, or crimson ; palate yellow,
spur hairy within. *Capsule* ½ in., glandular. *Seeds* ribbed, muricate.—
DISTRIB. From Holland southd. ; introd. in N. America.—Seeds yield oil in
Russia. Leaves, &c., bitter and stimulant.

4. SCROPHULA'RIA, *Tournef.* FIG-WORT.

Herbs, often fœtid. *Leaves* opposite, or the upper alternate, entire or
divided, often with pellucid dots. *Flowers* in panicled thyrsoid cymes,
greenish-purple or yellow ; peduncles glandular-pubescent. *Calyx* 5-partite.
Corolla-tube oblong or ventricose ; lobes short, 4 upper erect, lowest spreading. *Stamens* 4, declinate, 5th usually a scale ; anther-cells adnate to

the filaments, confluent, bursting transversely. *Disk* oblique. *Stigma* notched. *Capsule* acute, septicidally 2-valved ; valves entire or 2-fid. *Seeds* ovoid, rugose.—DISTRIB. Europe, temp. Asia, N. Africa, rare in America ; species about 120.—ETYM. In reference to its former use in *Scrofula.*

SECTION 1. **Scrophula'ria** proper. *Sepals* obtuse. *Corolla* purplish, throat not contracted, upper lobes longer than the lateral. *Stamens* 4, included, 5th reduced to a scale on the upper lip.

1. **S. nodo'sa,** *L.* ; glabrous below, stem acutely 4-angled, leaves ovate or triangular-cordate acutely doubly-serrate, border of sepals narrow.

Shady places, from Caithness southd.; ascends to 1,500 ft. in Yorkshire; Ireland ; Channel Islands ; fl. July–Oct.—*Rootstock* tuberous, nodose. *Stem* 1–3 ft., simple. *Leaves* 2–4 in., shortly petioled, acute or acuminate, nerves strong, basal teeth largest. *Cymes* lax ; bracts linear, acute, lower leafy ; pedicels erect, slender, glandular at the base. *Flowers* ⅓ in., green or brownish, proterogynous ; scale notched. *Capsule* broadly ovoid, acuminate. *Seeds* rugose, brown.—DISTRIB. Europe, N. and W. Asia, N. America.

2. **S. aquat'ica,** *L.* ; glabrous below, stem 4-winged, leaves oblong-lanceolate obtuse or acute doubly crenate-toothed, petiole winged, border of sepals broad.

Ditches, edges of ponds, &c., from Berwick southd.; Ireland; Channel Islands; fl. July–Sept.—*Rootstock* stout, creeping. *Stems* 2–4 ft. *Leaves* 3–8 in.. petiole winged and lobed or leaves pinnatisect below, lower cordate at the base, glabrous or pubescent, teeth largest upwards. *Panicles* large, erect ; cymes opposite, dichotomous ; lower bracts small, linear. *Flowers* proterogynous. *Corolla* ¼–½ in., greenish below, brown above, rarely white ; upper lip 2-fid, scale broad. *Capsule* small, ovoid or subglobose pointed.— DISTRIB. From Denmark southd., N. Africa, N. and W. Asia, Himalaya.

Sub-sp. S. AQUATICA proper ; cymes lax, many-fld., leaves crenate, scale reniform. *S. Balbis'ii,* Hornem.—VAR. *S. cinerea,* Dum., has an entire scale.

Sub-sp. S. UMBROSA, Dum.; cymes contracted few-fld., leaves more toothed, scale 2-lobed. *S. Ehrhar'ti,* Stevens.

3. **S. Scorodo'nia,** *L.* ; glandular-pubescent, stem obscurely 4-angled, leaves ovate or triangular-ovate doubly crenate, petiole not winged, border of sepals broad.

Shaded places, Cornwall, S. Devon ; Kerry ; Channel Islands ; fl. July–Aug. —*Rootstock* creeping. *Stem* 2–4 ft., simple. *Leaves* 1½–4 in., coarsely crenate-serrate, usually deeply cordate, much wrinkled. *Cymes* corymbose, in long lax panicles ; bracts leafy ; pedicels slender. *Corolla* ½ in., dull purple ; upper lip 2-partite ; scale broad, entire or notched. *Capsule* sub-globose or ovoid, acuminate.—DISTRIB. Belgium, W. France, S. Europe, N. Africa.

SECTION 2. **Ceraman'the,** *Reichb.* *Sepals* linear-oblong, subacute. *Corolla* yellow, throat contracted ; lobes nearly equal. *Stamens* 4, ex-serted ; 5th absent.

S. VERNA'LIS, *L.* ; glandular-hairy, stem obscurely 4-angled, leaves broadly ovate or deltoid, petiole not winged, sepals not bordered.

Waste places, very local ; a denizen ; fl. April–June.—Pale green, flaccid, *Rootstock* creeping. *Stems* 1½–4 ft. *Leaves* 2–3 in., deeply double-crenate or lobulate. *Cymes* subumbellate, on axillary peduncles arranged in a terminal leafy panicle; pedicels short or slender; bracts and bracteoles leafy. *Corolla* urceolate. *Capsule* broadly ovoid, acuminate.—DISTRIB. Europe (excl. Spain, Greece).

4*. *MIM'ULUS,* L.

Erect or decumbent herbs. *Leaves* opposite. *Flowers* solitary, axillary. *Calyx* tubular, 5-angled, 5-toothed. *Corolla* 2-lipped ; upper lip erect or reflexed, 2-lobed ; lower spreading, 3-lobed ; petals usually with 2 swellings ; lobes flat, rounded. *Stamens* 4 ; anther-cells subconfluent. *Stigma* with 2 equal lamellæ. *Capsule* loculicidally 2-valved, valves separating from a seed-bearing column, many-seeded. *Seeds* minute, oblong.— DISTRIB. American, a few Australian and New Zealand ; species 40.— ETYM. μιμώ, an *ape*, from the form of the corolla.

M. LU'TEUS, *L.* ; suberect, leaves ovate-oblong coarsely toothed.

River-sides, &c., ascending to 1,000 ft., from Skye southd.; Ireland ; naturalized from N. America ; fl. July–Sept.—Glabrous, or glandular-pubescent. *Stems* ½–3 ft., with many prostrate barren shoots, stout, hollow, terete. *Leaves* ½–3½ in., 6–9-nerved ; lower petioled with often a few lobes on the petiole; upper sessile. *Flowers* 1–2 in., yellow ; peduncles slender, exceeding the leaves. *Calyx-teeth* short, upper longest. *Stigmatic* plates sensitive, closing when touched on inner surface. *Capsule* enclosed in the inflated calyx, ovoid, compressed.

5. LIMOSEL'LA, *L.* MUDWORT.

Small, tufted, creeping, glabrous, annual, aquatic herbs. *Leaves* narrow, fascicled, rarely alternate, petiole dilated at the base. *Flowers* minute, ebracteate, axillary, solitary. *Calyx* campanulate, 5-toothed. *Corolla* subcampanulate, tube short, limb 5-fid. *Stamens* 4 ; anther-cells confluent. *Style* short, stigma clavate. *Capsule* septicidally 2-valved, valves entire, septum incomplete. *Seeds* grooved and transversely rugose.—DISTRIB. Temp. and cold regions ; species 5–6.—ETYM. *limus*, mud.

L. aquat'ica, *L.* ; leaves narrow oblong-lanceolate or spathulate.

Edges of ponds from Forfar and Ayr to Somerset and Sussex ; ascends to 1,500 ft. in Yorks.; fl. July–Sept.—*Rootstock* filiform. *Leaves* 1–2 in., tufted; petiole long. *Peduncles* shorter than the leaves, recurved in fruit. *Corolla* ¹⁄₁₀ in., pink or white. *Capsule* globose.

6. SIBTHOR'PIA, *L.*

Slender, creeping, hairy herbs. *Leaves* petioled, alternate or fascicled, reniform, lobulate. *Flowers* yellow or pink, very small, axillary, solitary, ebracteate. *Calyx* 4–8-fid. *Corolla* subrotate, 5–8-fid. *Stamens* as many

298   *SCROPHULARINEÆ.*   [SIBTHORPIA.

as the corolla-lobes or one fewer ; anthers sagittate, cells contiguous at the top, not confluent. *Stigma.* capitate. *Capsule* membranous, compressed, loculicidally 2-valved. *Seeds* dorsally convex, ventral face flat or concave, hilum ventral.—DISTRIB. Europe, N. and W. trop. Africa, Andes ; species 6.—ETYM. *Dr. Sibthorp,* an Oxford Professor of Botany.

S. europæ'a, *L.* ; leaves 7–9-lobed, calyx 5-lobed, stamens 4.

Moist shady banks, rare; S. Wales, Sussex to Cornwall; Channel and Scilly Islands; Kerry; fl. July–Oct.—*Hairs* flaccid, jointed. *Stem* 6–14 in., filiform, creeping. *Leaves* ¼–¾ in. diam., membranous; lobes broad, rounded, or retuse. *Flowers* minute; peduncle short. *Calyx-lobes* lanceolate. *Corolla* pink, 5-lobed, two smaller lobes yellowish. *Capsule* very small. —DISTRIB. W. France, Spain, Portugal.

7. DIGITA'LIS, *Tourn.* FOXGLOVE.

Tall, biennial or perennial herbs. *Lower leaves* crowded, petioled ; upper alternate. *Flowers* in terminal racemes, purple orange yellow or white, spotted inside, honeyed, proterandrous ; throat bearded. *Calyx* 5-partite. *Corolla* declinate, tube campanulate or ventricose, constricted above the base ; upper lobe short, spreading, notched or 2-fid ; lower longer, horizontal. *Stamens* 4, ascending ; anthers in pairs, cells at first parallel contiguous, then divaricate. *Stigma* 2-lobed. *Capsule* septicidally 2-valved, valves entire with inflexed margins, separating from the seed-bearing column. *Seeds* minute, oblong, angled.—DISTRIB. Europe. N. Africa, W. Asia ; species 18.—ETYM. *digitus,* a *finger.*

D. purpu'rea, *L.* ; leaves ovate-oblong or lanceolate crenate.

Copses, banks, &c., ascends to near 2,000 ft. in the Highlands; Ireland ; Channel Islands; fl. July–Sept.—Glandular-pubescent and hoary. *Stems* 2–4 ft., stout, erect, rarely branched. *Leaves* 6–12 in., rugose above, radical petioled ; upper cauline sessile. *Raceme* 1–2 ft., elongate, secund, dense-fld.; bracts leafy ; pedicels short. *Flowers* pendulous. *Calyx-lobes* oblong-lanceolate. *Corolla* 1½–2½ in., speckled with purple ocellated rarely white spots ; lobes ciliate. *Capsule* ovoid, exceeding the calyx. *Seeds* alveolate. —DISTRIB. W. Europe.—A well-known sedative drug.

8. VERON'ICA, *Tourn.* SPEEDWELL.

Herbs or shrubs. *Leaves* lower or all opposite, rarely whorled. *Flowers* in axillary or terminal racemes, rarely solitary, usually blue, never yellow. *Calyx* 4-, rarely 5-partite. *Corolla* rotate or subcampanulate ; limb 4-, rarely 5-fid, spreading, lateral lobes usually narrower. *Stamens* 2, inserted on the corolla-tube at the sides of the upper lobe, exserted ; anther-cells diverging or parallel, tips confluent. *Stigma* subcapitate. *Capsule* compressed or turgid, septi- or loculicidal. *Seeds* ovoid or orbicular, peltate smooth or rugulose.—DISTRIB. North temp. regions, Australia, New Zealand, Chili ; species about 160.—ETYM. obscure.

SECTION 1. **Omphalo'spora.** Annual. *Peduncles* 1-fld., solitary, axillary in alternate leaf-like bracts. *Seeds* cup-shaped, or deeply grooved on one side.

1. **V. agres'tis,** *L.* ; prostrate, slightly hairy and glandular, leaves petioled ovate-cordate coarsely serrate, sepals ovate or oblong, cells of the capsule compressed 2-lobed 4-10 seeded.

Waste places, N. to Shetland; ascends to 1,200 ft. in Northumbd.; Ireland; Channel Islands; fl. April–Sept.—*Branches* 4–8 in., slender. *Leaves* $\frac{1}{3}$–$\frac{2}{3}$ in., obtuse; floral similar, about as long as the decurved pedicels. *Flowers* honeyed, homogamous. *Sepals* $\frac{1}{4}$ in., ciliate, 3-nerved, when in fruit $\frac{1}{2}$ in. and unequal. *Corolla* $\frac{1}{6}$–$\frac{1}{3}$ in. diam., shorter or longer than the sepals. *Capsule* $\frac{1}{4}$ in. diam., hairy or ciliate.—DISTRIB. Europe, N. Africa, N. and W. Asia, Himalaya; introd. in N. America.

V. AGRES'TIS proper; leaves $\frac{1}{3}$–$\frac{2}{3}$ in., serratures regular not deep, sepals linear-oblong, corolla pale-blue or white, seeds 4–5 in each cell.

Sub-sp. V. POLI'TA, *Fries* ; leaves $\frac{1}{4}$–$\frac{1}{2}$ in. broader, serratures deeper and rather irregular, sepals ovate subacute, corolla bright blue, seeds 8–10 in each cell.

2. **V. Buxbau'mii,** *Ten.* ; prostrate, hairy, eglandular, leaves shortly petioled oblong or ovate-cordate coarsely serrate, sepals lanceolate subacute spreading in fruit, cells of the capsule diverging compressed 2-lobed 5–8-seeded. *V. persica,* Poir. (the oldest name).

Fields from Caithness and the Clyde southd.; ascends to 1,000 ft. in Northumbd.; Ireland; (a colonist since 1825, *Wats.*); fl. April–Sept.—*Branches* 6–12 in., tips often ascending. *Leaves* $\frac{1}{3}$–$1\frac{1}{3}$ in., obtuse, petiole variable; floral similar, shorter than the decurved pedicels. *Sepals* usually large, ciliate, 5–7-ribbed and reticulate in fruit. *Corolla* $\frac{1}{2}$ in. diam., bright blue. *Capsule* $\frac{1}{2}$ in. diam., reticulate, glandular-pubescent.—DISTRIB. From Belgium southd., N. Africa, W. Asia, Himalaya; introd. in N. America.

3. **V. hederæfo'lia,** *L.* ; prostrate, pubescent, leaves petioled very broadly-ovate 5–7-lobed, sepals cordate ciliate exceeding the corolla, cells of the biglobose capsule 1–2-seeded.

Cultivated ground, N. to Shetland; Ireland; Channel Islands; fl. March–Aug. — *Branches* 6–18 in. *Leaves* $\frac{1}{3}$–$\frac{2}{3}$ in. broad, lobes rounded, petiole dilated upwards; floral similar, about equalling the decurved or straight peduncles. *Flowers* appearing in succession as the branch lengthens, homogamous. *Sepals* acuminate, membranous. *Corolla* $\frac{1}{6}$ in. diam., pale blue. *Seeds* large, rugose, with a small deep pit on the inner face.—DISTRIB. Europe, N. Africa, W. Asia, Himalaya; introd. in N. America.

SECTION 2. **Veronicas'trum.** *Leaves* all opposite, or floral alternate. *Flowers* in terminal racemes. *Capsules* flat, valves adhering to the axis. *Seeds* plano-convex, biconvex, or flattened.

4. **V. triphyl'los,** *L.* ; annual, suberect, glandular-pubescent, leaves small, upper incised, lower petioled, sepals linear-oblong, cells of the obcordate capsule many-seeded.

Sandy fields, Norfolk and Suffolk, York; fl. May–June.—*Stem* 3–8 in., rigid, branches few. *Leaves* $\frac{1}{4}$–$\frac{1}{2}$ in., long and broad, palmately 3–7-lobed; lobes entire, obtuse, oblong or linear; floral subdigitate. *Flowers* few, subracemose; bracts 3–5-partite; peduncles slender, ascending. *Sepals* subequal, obtuse. *Corolla* $\frac{1}{4}$ in. diam., dark blue. *Capsule* $\frac{1}{4}$ in. diam., as broad as long, deeply-lobed, glandular. *Seeds* rugose.—DISTRIB. From Gothland southd., N. Africa, N. and W. Asia, N.W. India.

5. **V. arven'sis,** *L.* ; annual, erect or ascending, pubescent, leaves ovate-cordate irregularly crenate-serrate, bracts alternate, pedicels short, capsule obcordate, seeds 6–7 in each cell nearly flat.

Dry sandy, &c., places, N. to Shetland; ascends to near 2,000 ft. in Yorkshire; Ireland; Channel Islands; fl. May–Oct.—*Branches* 4–18 in., stiff, 2-fariously pubescent, often very long and simple. *Leaves* $\frac{1}{3}$–$\frac{3}{4}$ in., lower petioled, serratures few obtuse. *Bracts* often in dense leafy cylindric racemes, linear or oblong-lanceolate, obtuse, entire or obscurely lobed, exceeding the flowers. *Sepals* narrow, obtuse, ciliate. *Corolla* minute, pale blue with a white eye. *Style* very short. *Capsule* $\frac{1}{6}$–$\frac{1}{5}$ in. broad, glandular, shorter than the sepals.—DISTRIB. Europe, N. Africa, N. and W. Asia, Himalaya; introd in N. America.

VAR. *exim'ia,* Towns. (? var. *perpusil'la,* Bromf.), from Hants, is prostrate, tufted, with shorter leaves bracts and sepals.

6. **V. ver'na,** *L.* ; annual, erect, glandular-pubescent, leaves lobed or pinnatifid, bracts lanceolate entire, pedicels short, capsule deeply obcordate, seeds 6–7 in each cell flat.

Sandy fields, Norfolk and Suffolk; fl. May–June.—Habit of *V. arven'sis,* but always small, 2–4 in., densely leafy and more glandular-pubescent; flowers always much crowded; capsule and seeds much the same in both.— DISTRIB. Europe, N. and W. Asia, N.W. India.

7. **V. serpyllifo'lia,** *L.* ; perennial, ascending, stem glandular-pubescent, leaves subentire, racemes many-flowered, pedicels short erect, style as long as the didymous flat capsule, seeds plano-convex minute.

Fields, moist waste places, &c., N. to Shetland; ascends to 2,500 ft. in the Highlands; Ireland; Channel Islands; fl. May–June.—Much branched from the base; branches 8–10 in. *Leaves* $\frac{1}{2}$–$\frac{3}{4}$ in., sessile, ovate-rotundate or oblong, obtuse, subcoriaceous. *Raceme* 1–4 in.; bracts alternate, quite entire, lower leaf-like, upper narrow. *Flowers* proterogynous or homogamous. *Sepals* oblong-obovate, obtuse, ciliate, shorter than the corolla. *Corolla* $\frac{1}{4}$ in. diam., white or lilac. *Capsule* shorter than the sepals, glabrous.—DISTRIB. Europe (Arctic), N. Africa, N. and W. Asia, Himalaya, N. America.—*V. humifu'sa,* Dicks., is an alpine variety, decumbent, rooting at the nodes, with leaves entire, flowers few, corolla blue, capsule glandular-pubescent. Ascends to 3,700 ft. in the Highlands.

8. **V. alpi'na,** *L.* ; perennial, erect, glandular-pubescent above, leaves ovate entire or serrulate, raceme corymbose few-fld., bracts alternate, pedicels short erect, capsule obovate notched, seeds plano-convex.

Springs and rills, highest Scotch Alps, rare, from 1,600 to 3,700 ft.; fl. July–
Aug.—Resembles *V. serpyllifo'lia*, but more erect; leaves rather larger;
flowers fewer, in a denser raceme; sepals narrower, subacute, half as long as
the obscurely notched capsule; corolla dark blue; style short, and seeds
larger.—DISTRIB. Arctic and Alpine Europe, Asia, America.

9. **V. saxat'ilis,** *L.* ; perennial, decumbent, subglabrous, lower leaves
obovate, upper oblong, raceme subcorymbose few-fld., bracts subopposite,
pedicel long erect, style long, capsule oblong, seeds nearly flat.

Highest Scotch Alps, alt. 1,600–3,000 ft., rare; fl. July–Sept.—*Stem* woody;
branches many 2–4 in., ascending, with leafy barren shoots. *Leaves* ¼–½ in.,
coriaceous, teeth very few or 0. *Sepals* linear-oblong, obtuse, shorter than
the capsule. *Flowers* ½ in. diam., bright blue, very beautiful, honeyed,
homogamous; pedicels ⅓–½ in., stiff. *Style* short, slender. *Capsule* ¼ in.,
exceeding the sepals.—DISTRIB. Arctic, N. and Alpine Europe, Greenland.

SECTION 3. **Chamæ'drys.** Perennial. *Branches* diffuse, ascending.
Leaves all opposite. *Racemes* axillary. *Capsule* flat, notched or 2-lobed,
valves adhering to the axis. *Seeds* plano-convex or turgid.

10. **V. officina'lis,** *L.* ; glandular-pubescent or subglabrous, stem hairy
all round, leaves shortly petioled obovate-oblong or orbicular serrate,
pedicels very short, capsule triangular or obcordate truncate or retuse.

Banks and pastures, N. to Shetland; ascends to near 3,000 ft. in the Highlands; Ireland; Channel Islands; fl. May–July.—Decumbent; branches
2–18 in., ascending. *Leaves* ¾–1 in., contracted into the petiole, sharply or
obtusely serrate. *Racemes* slender, many-fld.; pedicels erect; bracts lanceolate or subspathulate, obtuse. *Sepals* linear-oblong. *Corolla* ⅛ in. diam.,
pale blue or lilac. *Style* very long. *Capsule* ¼ in., much longer than the
sepals. *Seeds* nearly flat.—DISTRIB. Europe, N. and W. Asia, Himalaya
N. U. States.—Leaves bitter and astringent.

V. hirsu'ta, Hopkirk (*V. salig'na,* D. Don.), is a very hairy small variety with
leaves narrower smaller, capsule seedless entire at the tip. Ayrshire.

11. **V. Chamæ'drys,** *L.* ; hairy, stem pubescent on opposite sides,
leaves subsessile ovate-cordate deeply serrate, pedicels slender, raceme long
lax, capsule obcordate shorter than the calyx.

Copses, pastures, banks, &c., N. to Shetland; ascends to 2,700 ft. in the Highlands; Ireland; Channel Islands; fl. May–June.—*Branches* 8–24 in., slender,
ascending. *Leaves* ½–1½ in. *Raceme* with its slender peduncle 2–5 in.;
bracts linear, much shorter than the pedicels. *Flowers* honeyed, homogamous. *Sepals* linear-lanceolate, acute. *Corolla* ⅓–½ in. diam., bright blue.
Capsule broader than long, pubescent and ciliate. — DISTRIB. Europe
(Arctic), Siberia; introd. in N. America.

12. **V. monta'na,** *L.* ; hairy, stem pubescent all round, leaves petioled
ovate-cordate serrate, pedicels slender, racemes short lax, capsule orbicular
longer than the calyx glabrous, margin subcrenulate ciliate.

Moist woods from Skye and Banff southd.; ascends to 1,000 ft. in Yorkshire;
Ireland; fl. June–Sept.—Habit of *V. Chamæ'drys,* but sepals much broader

corolla not ¼ in. diam., paler blue; capsule glabrous except the margins; and foliage darker (though not black) in drying.—DISTRIB. From Gothland southd. (excl. Greece, Turkey), N. Africa, W. Siberia.

13. **V. scutella'ta,** *L.* ; suberect, leaves sessile linear-lanceolate faintly toothed, recemes subopposite, pedicels slender deflexed in fruit.

Bogs, edges of ditches, &c., N. to Shetland, not very common; ascends to 2,200 ft. in Yorkshire; Ireland; Channel Islands; fl. July–Aug.—Glabrous, rarely hairy (var. *pubes'cens*), stoloniferous. *Stem* 6–24 in., slender, brittle, decumbent and rooting below, sparingly branched. *Leaves* 1–2 in., ½-amplexicaul. *Racemes* many; peduncles filiform, usually as long as the leaves; pedicels secund, spreading; bracts small. *Sepals* ovate-oblong, acute. *Corolla* ¼ in. diam., white or pinkish. *Capsule* ⅛ in. diam., broader than long, deeply 2-lobed, margins ciliate.—DISTRIB. Europe (Arctic), N. Africa, N. and W. Asia, N. America.

SECTION 4. **Beccabun'ga.** Perennial. *Leaves* opposite, serrate. *Racemes* axillary, opposite, many-fld. *Capsule* flat, loculicidal ; valves 2-fid, falling away from the seed-bearing axis.

14. **V. Beccabun'ga,** *L.* ; procumbent, glabrous, succulent, leaves petioled oblong obtuse serrate, capsule orbicular notched. *Brooklime.*

Margins of brooks, ditches, &c., N. to Shetland; ascends to 2,800 ft. in the Highlands; Ireland; Channel Islands; fl. May–Sept.—*Stem* hollow, rooting below; branches 1–2 ft., spreading. *Leaves* 1–2 in., sometimes obovate. *Racemes* 2–4 in.; bracts narrow, usually shorter than the pedicels. *Flowers* honeyed. *Sepals* small, ovate-oblong, subacute, glabrous. *Corolla* ⅛ in. diam., bright blue or pink. *Capsule* turgid, a little exceeding the sepals. *Seeds* minute, biconvex.—DISTRIB. Europe, N. Africa, N. and W. Asia, Himalaya.—A reputed antiscorbutic.

15. **V. Anagal'lis,** *L.* ; erect, glabrous, stout, succulent, stoloniferous, leaves sessile ½-amplexicaul ovate- or oblong-lanceolate serrate, capsule orbicular notched.

Watery places, N. to Shetland; ascends to 1,050 ft. in Derby; Ireland; Channel Islands; fl. July–Aug.—Stoloniferous. *Stem* 1–3 ft., simple or sparingly branched. *Leaves* 2–5 in., subacute, sometimes auricled at the base, teeth small. *Racemes* ascending, 4–10 in., elongating in fruit; bracts lanceolate, about equalling the pedicels. *Sepals* ovate-lanceolate, subacute, glabrous or glandular. *Corolla* ⅛–¼ in. diam., pale lilac or white. *Style* moderate. *Capsule* shorter than sepals, rather turgid. *Seeds* as in *V. Beccabun'ga.*—DISTRIB. Europe, N. Africa, N. and W. Asia, Himalaya, N. America.

SECTION 5. **Pseudo-Lysima'chia.** Perennial. *Leaves* opposite or whorled. *Racemes* terminal. *Corolla-tube* cylindric. *Capsule* slightly compressed, valves adhering to the axis. *Seeds* plano-convex.

16. **V. spica'ta,** *L.* ; pubescent, leaves oblong, spike dense.

Chalky pastures, Cambridge, Suffolk, and Norfolk, and limestone rocks in the W. of England and Wales; fl. July–Aug.—*Rootstock* creeping. *Stem* 6–18 in., stout, erect from a decumbent base, leafy. *Leaves* 1–1½ in., coriaceous, sessile or petioled; lower ovate, obtuse, narrowed below, crenate-serrate. *Spike* 1½–3 in.; bracts ovate-lanceolate, exceeding the obtuse ciliate sepals. *Corolla* ¼ in. diam., bright blue, tube as long as broad, throat bearded; lobes narrow, acute. *Stamens* very long; anthers large, purple. *Style* long. *Capsule* equalling the sepals, ovoid, pubescent. — Distrib. Europe, N. Asia.

V. spica'ta proper; leaves subsessile narrow-oblong or oblong-lanceolate serrated above the middle. E. counties, very rare.—Var. *V. hyb'rida*, L.; larger, stouter, leaves petioled broader serrate throughout. W. counties, from Wales and Westmoreland to Bristol.

9. BART'SIA, *L.*

Erect herbs, parasitic on roots. *Leaves* opposite or upper alternate, crenate or serrate. *Flowers* in bracteate spikes; bracts leafy. *Calyx* tubular or campanulate, 4-fid. *Corolla* tubular, 2-lipped; upper lip arched, entire or notched, its sides not reflexed; lower as long or shorter, tip 3-fid. *Stamens* 4, didynamous, hidden by the upper lip; anther-cells mucronate. *Stigma* obtuse, or 2-lobed, posticous lobe very small. *Capsule* ovoid or oblong, compressed, loculicidally 2-valved, septa and placentas persistent. *Seeds* transversely ovoid.—Distrib. N. temp. regions; species 60.—Etym. *John Bartsch,* a Prussian botanist.

Section 1. **Bart'sia** proper. *Capsule* ovoid, acuminate. *Seeds* many, large, ribbed or winged on the back; hilum lateral.

1. **B. alpi'na,** *L.* ; perennial, glandular-pubescent, leaves sessile ovate obtusely serrate, upper cordate amplexicaul.

Subalpine meadows and wet banks, York, Durham, Westmoreland; Mts. of Perth, Inverness, and Ross; ascends to near 3,000 ft. in Scotland; fl. June–Aug.—Black when dry. *Rootstock* woody. *Stem* 4–8 in., erect, simple, terete, leafy. *Leaves* ¼–⅜ in., rather coriaceous, obtuse. *Spikes* short, few-fld.; bracts purplish. *Calyx-lobes* ovate-lanceolate. *Flowers* proterogynous. *Corolla* ½–¾ in., dull blue-purple, glandular, lips small. *Anthers* exserted, bearded. *Capsule* longer than the sepals.—Distrib. Europe (Arctic), from Gothland northd., Alps, Greenland, Labrador.

Section 2. **Eufra'gia,** *Griseb.* (gen.). *Capsule* oblong, acute. *Seeds* very minute, faintly granulate; hilum basal.

2. **B. visco'sa,** *L.* ; annual, viscid, leaves sessile ovate or oblong-lanceolate coarsely serrate.

Meadows, &c., S. England, Sussex to Cornwall; Wales; Chester to Argyll; W. Ireland; Channel Islands; fl. June–Oct.—*Stem* 6–18 in., terete, usually simple, often flowering throughout much of its length. *Leaves* ½–1½ in., rather scabrid, nerves prominent beneath. *Calyx-tube* curved; lobes triangular-lanceolate, acuminate. *Corolla* ¾ in., yellow, glandular; lower lip

large. *Anthers* yellow, slightly bearded   *Capsule* pubescent, shorter than
the calyx.—DISTRIB.   W. Europe, N.W. Africa, W. Asia.

SECTION 3. **Odonti'tes,** *Persoon* (gen.)   *Capsule* ovate or oblong,
compressed, obtuse.   *Seeds* strongly ribbed, pendulous ; hilum basal.

3. **B. Odonti'tes,** *Huds.* ; annual, pubescent, leaves linear-lanceolate
distantly serrate. *Euphra'sia Odonti'tes,* L.   *Odonti'tes rubra,* Pers.

Fields and waste places, N. to Orkney ; ascends to 1,200 ft. in the Highlands ;
Ireland ; Channel Islands ; fl. June–Aug.—*Stem* 6–18 in., erect or ascending,
wiry, 4-gonous, paniculately branched ; branches opposite, terminating in
slender leafy racemes. *Leaves* ½–2 in., sessile, very variable. *Spikes* subsecund ; lower bracts leafy, upper narrower. *Flowers* proterogynous,
honeyed. *Calyx* campanulate ; lobes ovate, acute, as long as the tube.
Corolla ½ in., pink, pubescent ; upper lip long, entire. *Anthers* usually
exserted, yellow, almost glabrous. *Style* moderate. *Capsule* ¼ in., rather
exceeding the calyx. *Seeds* narrow-oblong.—DISTRIB. Europe, N. Asia, N.
Africa, Himalaya.—Often placed in *Euphrasia,* and with reason.

VAR. *O. ver'na,* Reichb. ; branches ascending, leaves rounded at the base
bracts longer than the flowers, calyx-teeth equalling the tube.—VAR. *O.
seroti'na,* Reichb. (*Euphra'sia rotunda'ta,* Ball) ; branches flexuous upcurved,
leaves narrow at the base, bracts shorter than the flowers, calyx-teeth
equalling the tube.—VAR. *O. diver'gens,* Jord. ; much branched, branches
widely spreading.

10. EUPHRA'SIA, *Tournef.* EYEBRIGHT.

Annual or perennial herbs, parasitic on roots. *Leaves* opposite, toothed
or cut. *Flowers* in dense secund or interrupted bracteate spikes, white
yellow or purple. *Calyx* tubular or campanulate, 4-fid. *Corolla* tubular,
2-lipped ; upper lip concave, lobes 2, broad spreading ; lower spreading
3-fid. *Stamens* hidden by the upper lip ; anthers-cohering by hairs, cells
mucronate. *Stigma* dilated, obtuse. *Capsule* oblong compressed, loculicidally 2-valved. *Seeds* few or many, pendulous, oblong or fusiform,
furrowed.—DISTRIB. Temp. regions ; species 20.—ETYM. εὐφραίνω, to
gladden, in allusion to its former use as an eye-medicine.

E. officina'lis, *L.* ; annual, lower leaves crenate, upper cut.

Meadows, heaths, &c., N. to Shetland ; ascends to 3,600 ft. in the Highlands ;
Ireland ; Channel Islands ; fl. May–Sept.—Glabrous or glandular-pubescent.
Stem 1–10 in., erect, wiry, usually with many opposite branches. *Leaves*
⅓–½ in., sessile, ovate or lanceolate. *Spikes* terminal ; bracts leafy ; flowers
minute. *Flowers* dimorphous, larger proterogynous, smaller proterandrous.
Calyx about equalling the bracts, tube ribbed ; lobes acute. *Corolla* ⅙–½ in.,
white or lilac, purple veined ; mid-lobe of lower lip yellow. *Anthers* brown,
pubescent, one cell spurred. *Capsule* included or exserted. *Seeds* very
variable in form and colour.—DISTRIB. Europe (Arctic), N. and W. Asia,
Himalaya, N. America.—Of this plant there are a multitude of forms.

E. officina'lis proper ; bracts triangular-ovate base broad, lower corolla-lip
equalling or exceeding the tube.—VAR. *E. grac'ilis,* Fries ; more slender,
bracts broader, base narrowed, lower corolla-lip shorter than the tube.—

VAR. *marit'ima;* capsule much longer than the calyx. Shores of Shetland.

11. RHINAN'THUS, *L.* YELLOW-RATTLE.

Annual erect herbs, black when dry, parasitic on roots. *Leaves* opposite, narrow, serrate. *Flowers* in secund spikes, yellow spotted with violet, honeyed ; bracts broad, cuspidate-toothed. *Calyx* ventricose, compressed, 4-toothed. *Corolla* 2-lipped, tube subcylindric ; upper lip obtuse, compressed, entire, with a toothed appendage on each side the tip ; lower shorter, lobes 3 spreading. *Stamens* 4, hidden by the upper lip ; anthers cohering by hairs, not spurred. *Style* filiform, tip inflexed, stigma subcapitate. *Capsule* orbicular, compressed, loculicidally 2-valved ; valves membranous, entire, bearing the placentas in the middle. *Seeds* suborbicular, compressed, hilum lateral ; embryo small.—DISTRIB. Europe (Arctic), N. Asia, N. America ; species 2-3.—ETYM. ῥίν, the *nose,* and ἄνθος, *flower,* in allusion to the form of the corolla.

R. Crista-gal'li, *L.* ; erect, leaves deeply crenate-serrate.

Damp pastures and wet places, N. to Shetland; ascends to 2,500 ft. in the Highlands ; Ireland ; Channel Islands; fl. May–July.—Glabrous or glandular pubescent. *Stem* 6–18 in., simple or with opposite branches, 4-gonous. *Leaves* distant, 1–2 in., oblong-lanceolate or linear-oblong, obtuse or acute. *Bracts* longer than the calyx, ovate, taper-pointed. *Flowers* dimorphic as in *Euphrasia. Calyx* $\frac{1}{4}$–$\frac{1}{2}$ in., mouth small, teeth triangular. *Corolla* $\frac{1}{2}$–1 in., yellow; lobes of upper lip blue. *Anthers* bluish. *Capsule* included in the bladdery calyx-tube. *Seeds* with a broad or narrow wing.—DISTRIB. Of the genus.

R. CRISTA-GAL'LI proper ; stem subsimple, bracts green, spikes few- and laxfld.,corolla-tube hardly exceeding the calyx,lobes of upper lip short roundish. R. *mi'nor,* Ehrh. Meadows and pastures.

Sub-sp. R. MA'JOR, *Ehrh.*; taller, much branched, bracts yellowish, spikes many and dense-fld., corolla larger tube exceeding the calyx, lobes of upper lip oblong, seed-wing broad (var. *platyp'tera,* Fr.) or narrow (*stenop'tera,* Fr.), or 0 and seed ribbed (*ap'tera,* Fr., R. *Reichenbach'ii,* Drej.). Fields, sporadic, not in Ireland or Channel Islands.

12. PEDICULA'RIS, *Tournef.* LOUSEWORT.

Herbs, black when dry ; parasitic on roots. *Leaves* alternate whorled or opposite, toothed or pinnatisect. *Flowers* in bracteate spikes or racemes, white red purple or yellow, honeyed. *Calyx* tubular or campanulate, split anteriorly ; teeth 2–5, unequal, entire lobed or crested. *Corolla* 2-lipped, tube cylindric or throat dilated ; upper lip compressed, entire or notched, or with 2 teeth below the tip ; lower 3-lobed. *Stamens* 4, concealed by the upper lip, filaments or the 2 posterior only hairy ; anthers cohering by hairs, cells obtuse. *Capsule* compressed, loculicidally 2-valved, seeds few, in its lower part. *Seeds* ovoid, testa firm or lax rugose or smooth ; embryo short or long.—DISTRIB. N. temp. regions ; species 120.—ETYM. *pediculus,* being supposed to encourage *lice* in sheep.

X

1. **P. palus'tris**, *L.* ; stem erect branched above, calyx ovoid compressed hairy ribbed, lobes subequal crenate.

Bogs and marshes, N. to Shetland, ascends to 1,800 ft. in the Highlands; Ireland; Channel Islands; fl. May–Sept.—Annual, glabrous or sparingly hairy. *Stem* 6–18 in., stout. *Leaves* 1–3 in., linear-oblong, pinnate; segments oblong, obtuse, crenate or pinnatifid, ultimate rounded. *Bracts* leafy. *Calyx* ¼–⅓ in., reddish green, ventricose in fruit. *Corolla* 1 in., dull pink; upper lip obtuse 3-toothed, lower broad. *Capsule* exceeding the calyx, curved.— Distrib. Europe (Arctic) (excl. Spain, Greece), N. Asia.

2. **P. sylvat'ica**, *L.* ; stem branched at the base, branches spreading, calyx oblong angled glabrous, lobes unequal foliaceous crenate.

Copses, heaths, and damp meadows, N. to Shetland ; ascends to 2,000 ft. in the Highlands; Ireland; Channel Islands; fl. April–July.—Perennial, glabrous. *Stems* many, 3–10 in., decumbent and ascending from a very short rootstock, leafy. *Leaves* and bracts much as in *P. palus'tris*, but segments acute. *Spike* lax-fld. *Calyx* ½ in., 5-lobed, upper lobe entire, the rest foliaceous. *Corolla* 1 in., rose-cold., tube slender, upper lip 2-toothed. *Capsule* longer than the calyx, obliquely truncate.—Distrib. Europe (excl. Greece, Turkey).

13. MELAMPY'RUM, *Tournef.* Cow-wheat.

Erect, branched, annual herbs, black when dry, parasitic on roots. *Leaves* opposite, cauline narrow entire. *Flowers* axillary and solitary or in leafy spikes, honeyed. *Calyx* tubular or campanulate ; teeth 4, acuminate. *Corolla* 2-lipped, tube cylindric, dilated above ; upper lip short, compressed, truncate, margins recurved ; lower shorter, 3-lobed, palate prominent. *Stamens* 4, under the upper lip, anthers oblong, cohering by hairs, cells mucronate. *Disk* hypogynous, anticous. *Stigma* small, obtuse ; ovules 2, collateral near the base of each cell, one subsessile erect, the other stalked fixed laterally. *Capsule* compressed, ovate, oblique or falcate, loculicidally 2-valved, 1–4-seeded. *Seeds* oblong, testa smooth, hilum thickened ; embryo small.—Distrib. Europe, temp. Asia ; species 6.—Etym. μέλας and πυρός, *black wheat.*

1. **M. praten'se**, *L.* ; flowers axillary secund, bracts entire or toothed, corolla much longer than the calyx, lips closed, lower straight.

Copses, heaths, and pastures, N. to Orkney ; ascending to 3,000 ft. in the Highlands ; Ireland; fl. June–Sept.—Glabrous or pubescent, very variable. *Stem* 6–24 in., terete, wiry ; branches spreading. *Leaves* ½–3 in., ciliate, linear- or ovate-lanceolate, sessile or very shortly petioled, quite entire. *Bracts* ciliate, often toothed laciniate or pinnatifid, sometimes hastate. *Calyx-teeth* lanceolate ascending. *Corolla* pale yellow, horizontal, tube straight. *Capsule* ovoid, deflexed.—Distrib. Europe (Arctic) (excl. Spain, Italy, Greece Turkey), Siberia.

M. praten'se proper ; leaves linear-lanceolate acute, bracts toothed or pinnatifid tips slender.—Var. *latifo'lia ;* leaves ovate-lanceolate from a broad base, bracts deeply toothed, the upper with short tips. Chalk, &c. districts, Oxford, Monmouth, I. of Wight.—Var. *M. monta'num,* Johnst. ;

leaves linear-lanceolate, bracts entire tips long slender.—VAR. *ericeto'rum,* D. Oliv. ; hispid, leaves linear-lanceolate, bracts toothed.

2. **M. sylvat'icum,** *L.* ; flowers axillary secund, bracts quite entire, corolla not much exceeding the calyx, lips not closed, lower deflexed.

Subalpine woods, rare, from York to Caithness ; ascends to 1,000 ft. in the Highlands ; N.E. Ireland, rare ; fl. July–Aug.—Similar in habit and foliage to *M. praten'se,* but flowers suberect ; calyx-teeth spreading ; corolla smaller, shorter, deep yellow rarely pale (var. *pallidiflo'ra*), its tube curved, and capsule not deflexed.—DISTRIB. Europe (Arctic) (excl. Greece), Siberia.

3. **M. arven'se,** *L.* ; flowers spiked, bracts straight broad pinnatifid, corolla-tube longer than the long slender calyx-teeth, lips closed.

Cornfields, very local, Norfolk to Essex, Herts, I. of Wight ; (casual or a colonist, *Wats.*); fl. July–Aug.—*Stem* 1–2 ft., obtusely 4-gonous, stout, erect, branched, scaberulous. *Leaves* 1–3 in., lanceolate, quite entire or the upper toothed at the base. *Spike* 3–4 in., stout, obscurely 4-gonous ; bracts leafy, ¼–1 in., rose purple. *Calyx-teeth* nearly equal, subulate. *Corolla* ¾ in., erect, tube curved, puberulous, rosy, throat yellow, lips dark pink. *Capsule* shorter than the calyx, ovoid, cells 1-seeded. *Seeds* oblong, like black wheat-grains.—DISTRIB. Europe, W. Asia.

4. **M. crista'tum,** *L.* ; flowers densely spiked, bracts ovate-cordate acuminate recurved finely pectinate, corolla-tube much longer than the unequal acute calyx-teeth, lips closed.

Copses and fields in the E. counties, rare, from Hants to Norfolk, and Suffolk to Bedford; (a colonist? *Wats.*); fl. Sept.–Oct.—Puberulous. *Stem* 6–18 in., rigid, erect, obtusely 4-angled. *Leaves* 2–4 in., spreading, narrow linear-lanceolate. *Spike* 1½–2 in., oblong, obtuse, acutely 4-angled with hollow faces ; bracts densely imbricate, cordate, purple, teeth very slender, margins folded enclosing the calyx. *Calyx-tube* short, teeth very unequal, upper long subulate. *Corolla* ½ in., tube bent, yellow tipped with purple. *Capsule* exceeding the calyx.—DISTRIB. Europe, Siberia.

14. LATHRÆ'A, *L.* TOOTHWORT.

Fleshy root-parasites. *Flowers* 2-seriate in a secund raceme, bracteate. *Calyx* campanulate, regular, 4-fid. *Corolla* gaping, upper lip arched entire, lower smaller 3-toothed. *Stamens* 4, anther-cells spreading at the base. *Disk* glandular, broad, anticous. *Ovary* 1-celled, placentas broad subconfluent in pairs, stigma large capitate 2-lobed. *Capsule* 2-valved, many-seeded.—DISTRIB. Europe, N. and W. Asia, Himalaya ; species 3.— ETYM. λαθραῖος, *hidden,* from its locality.

L. squama'ria, *L.* ; raceme decurved in bud, flowers subsessile.

On roots chiefly of hazel in shady places, from Perth and Inverness southd.; Ireland ; fl. April–May.—Perennial, white or purplish, glabrous or pilose, fleshy, black when dry. *Rootstock* branched, scaly, rootlets attached by tubercular bases to the rootlets of the plant it preys upon. *Stems* 4–10 in., stout, scaly. *Raceme* 4–6 in.; bracts like the scales, broadly oblong.

Flowers ½ in. *Calyx* 2-lipped, lobes broad subacute, shorter than the dull purple arched corolla, which has small erect lobes. *Anthers* coherent, pubescent. *Style* exserted, decurved, stigma purple. *Capsule* ovoid.— DISTRIB. Europe, N. and W. Asia, Himalaya.

ORDER LVI. OROBAN'CHEÆ.

Leafless brownish root-parasites. *Rootstock* often tuberous, naked or scaly. *Stem* usually stout, solitary, scaly. *Flowers* in lax or dense spikes or racemes. *Sepals* 4–5, inferior, free or connate. *Corolla* hypogynous, irregular, tube curved ; limb 2-lipped ; upper lip arched, lower 3-fid ; throat with 2 villous folds. *Stamens* 4, didynamous, inserted on the corolla-tube ; anthers 2-celled, cells spurred at the base, opening by lateral slits or basal pores. *Disk* unilateral, or 0. *Ovary* 1-celled, of 2 carpels ; style simple, curved at the tip, stigma capitate 2-lobed ; ovules many, rarely few, anatropous, placentas 2 pairs free or confluent parietal. *Capsule* 1-celled, 2-valved, few- or many-seeded. *Seeds* minute, testa thick pitted or tubercled, albumen fleshy ; embryo ovoid, undivided or 2-fid.— DISTRIB. Temp. and trop. ; chiefly S. Europe and E. Asia ; genera 11 ; species 150.—AFFINITIES. With *Scrophularineæ* and *Cyrtandraceæ*, and possibly parasitic forms of these.—PROPERTIES. Astringent and bitter.

See *Lathræa* in Scrophularineæ.

1. OROBAN'CHE, *L.* BROOM-RAPE.

Flowers usually glandular-pubescent, 1-3-bracteate. *Sepals* 4, usually connate in pairs, with sometimes a small 5th. *Corolla* gaping ; upper lip erect, 2-lobed ; lower spreading, 3-lobed. *Stamens* included, filaments flattened below ; anther-cells spreading. *Disk* glandular or 0. *Ovary* ovoid ; style usually glandular, stigma 2-lobed. *Capsule* imperfectly 2-valved, many-seeded ; valves cohering at the base and usually at the top also.—DISTRIB. Of the Order ; species about 100.—ETYM. ὄροβος, *a vetch,* and ἄγχειν, *to strangle,* from its parasitic habit.

SECTION 1. **Ospro'leon** proper. *Stem* usually quite simple. *Flowers* 1-bracteate, ebracteolate. *Calyx-segments* 2, 2-fid, rarely entire. *Capsule* with the valves coherent above.

1. **O. ma'jor,** *L.* ; brown, sepals 1-3-nerved nearly as long as the corolla-tube, upper lip entire, mid-lobe of lower lip longer than the lateral, filaments glabrous below glandular above, stigma yellow. *O. Ra'pum,* Thuill.

On roots of shrubby *Papilionaceæ,* from Dumfries southd. ; Ireland ; Channel Islands ; fl. June–Aug.—*Stem* stout, 1–2 ft. *Spike* rather dense-fld. ; bracts equalling or exceeding the corolla. *Sepals* entire or 2-fid. *Corolla* 1 in., yellow and purplish, subcampanulate, curved, lobes small, waved, scarcely

toothed; upper lip arched, scarcely notched.—DISTRIB. From Holland
southd. (excl. Greece, Turkey), N. Africa.—Reichenbach figures the lobes of
lower lip as small and equal, and flower as red-brown. I take this to be *O.
ma'jor* of Linnæus, who states that it is parasitic especially on *Leguminosæ*,
and quotes for it the *Rapum genistæ* of Lobel. There are no specimens in
Linn. Herb.

2. **O. ela'tior,** *Sutt.* ; yellow, sepals 2-fid 2-3-nerved as long as the
corolla-tube, lobes of lip subequal acute, filaments glandular below glabrous above, stigma yellow.

Parasitic on *Centaurea Scabiosa*, chiefly in the E. counties from York and
Durham to Sussex and Somerset, rare; S. Wales; fl. June–Aug.—Habit of
O. major, of which it is probably a sub-species; but bracts shorter than
the corolla, which is narrower, more compressed above, the lobes more
toothed, and the upper lip deeply notched, with inflexed margins.—DISTRIB.
From Denmark southd., Caucasus, Siberia.

3. **O. caryophylla'cea,** *Sm.* ; brown, sepals 2-fid several-nerved shorter
than the corolla-tube, lobes of lower lip subequal waved and toothed,
filaments hairy below glandular above, stigma purple. *O. Gal'ii,* Duby.

On *Galia, Rubi,* &c., Kent; fl. June.—*Stem* 6–12 in., stout. *Spike* lax-fld.;
bracts shorter than the corolla. *Corolla* 1 in., tube broad, curved; lobes
large, reddish-brown or purplish, spreading; under lip arched, notched.—
DISTRIB. From Denmark southd. (excl. Greece), N. and W. Asia.

4. **O. ru'bra,** *Sm.* ; red, sepals entire 1-nerved subulate longer than
the corolla-tube, lips toothed crisped, of mid-lobe lower longest, filaments subpilose below, glandular above, stigma pale red. *O. Epithy'-
mum,* DC.

On *Thymus*, chiefly on the W. coast; W. Ross to Cornwall; Ireland; fl. June–
Aug.—*Stem* 4–8 in. *Spike* lax-fld. *Flowers* odorous. *Corolla* ½–¾ in.,
equalling or shorter than the bracts, dull red; tube broad, curved; lobes
small; upper lip arched, almost entire.—DISTRIB. From Belgium southd.,
W. Asia.

5. **O. mi'nor,** *Sutt.* ; yellow-brown or purplish, slender, sepals 1- or
more-nerved with long slender points, corolla-tube contracted in the
middle, filaments more or less hairy below, stigma subglobose.

On various plants in many counties, from the Border southd.; Ireland
(one sub-sp.); fl. June–Oct.—*Stem* ½–2 ft., rather slender. *Spike* elongate,
many-fld., lax or dense. *Flowers* ½–¾ in.—DISTRIB. From Denmark
southd., N. Africa, W. Asia.
O. MI'NOR proper; bracts equalling or exceeding the corolla whose tube is
gently curved, limb white or yellowish, lobes of upper lip spreading, of
lower nearly equal toothed and waved, stigmas purple.—On clover.
Sub-sp. O. AMETHYS'TEA, *Thuill.*; corolla much curved in the lower third,
upper two-thirds nearly straight, mid-lobe of lower lip much the largest.
O. Eryn'gii, Duby.—Cornwall to Kent; Channel Islands.
Sub-sp. O. PI'CRIDIS, *F. Schultz ;* very pale, bracts about equalling the flowers,
sepals entire or 2-cleft exceeding the slightly-curved corolla-tube 1-nerved,

upper lip retuse, tip inflexed, lobes of lower lip toothed nearly equal, stigma purple.—On *Picris*, Cambridge, Kent, Surrey, Hants, Tenby.
Sub-sp. O. HED′ERÆ, *Duby*; spike more lax, sepals 1-nerved, stigma yellowish. *O. barbata*, Eng. Bot. Suppl., not Poir.—On Ivy, Wales, Gloster, Kent to Cornwall; Ireland; Channel Islands.

SECTION 2. **Trionychon.** *Stem* simple or branched. *Flowers* bracteate and 2-bracteolate. *Calyx* tubular, 4–5-lobed. *Capsule* with the valves free above.

6. **O. cœru′lea,** *Vill.* ; blue-purple, stem simple, calyx-lobes 4 lanceolate, corolla-tube curved, lobes of both lips acute, filaments almost glabrous, suture of anthers glabrous, stigma white.

Herts and Norfolk, and Hants to Cornwall, local; Channel Islands; fl. June-Oct.—*Stem* 6–12 in., slender, tough, not tumid below. *Spikes* usually dense; bracts lanceolate. *Corolla-tube* ½–1 in., curved, upper lip 2-lobed, lobes pale blue with darker veins.—DISTRIB. From Holland southd., W. Asia.

Sub-sp. O. ARENA′RIA, *Bork.*, found in Alderney, is distinguishable from *O. cœru′lea* by the hairy anthers.

ORDER LVII. **LENTIBULARI′NEÆ.**

Scapigerous herbs, chiefly aquatic or marsh. *Leaves* radical, crowded or whorled, undivided or multifid, exstipulate. *Flowers* solitary racemed or corymbose, irregular, bracteate. *Calyx* free, persistent, 2-labiate or 5-partite. *Corolla* hypogynous, deciduous, personate or 2-lipped, tube short; upper lip short 2-fid, lower entire or 3-fid; palate convex. *Stamens* 2, opposite the lateral sepals, hypogynous or inserted on the corolla-tube, filaments short arching; anthers adnate, 1-celled, transversely 2-valved. *Disk* 0. *Ovary* free, 1-celled; style short, thick, stigma 2-lipped, upper lip short, lower dilated; ovules many, anatropous, placenta free basal globose. *Capsule* 2-valved or bursting irregularly, many-seeded. *Seeds* minute, oblong or peltate, placenta spongy, testa striate or pitted sometimes hairy, albumen 0; embryo straight, sometimes undivided.—DISTRIB. Chiefly temp. and cold regions; genera 4; species about 180.—AFFINITIES. With *Scrophularineæ.*—PROPERTIES unimportant.

Terrestrial. Stamens on the base of the corolla-tube..........1. Pinguicula.
Aquatic. Stamens on the base of the lip of the corolla ......2. Utricularia.

1. **PINGUI′CULA,** *Tournef.* BUTTERWORT.

Perennial, succulent, simple herbs. *Leaves* rosulate, quite entire, margins incurved; surface cellular, insectivorous. *Scapes* axillary, ebracteate, 1-fld. *Flowers* yellow, white, or purple, inclined. *Calyx* unequally 5-partite. *Corolla* ringent. *Stamens* hypogynous, filaments stout ascending;

anthers terminal. *Ovary* subglobose. *Capsule* erect, 2–4-valved. *Seeds* oblong, testa rugose.—DISTRIB. N. temp. regions, Fuegia; species 20. —ETYM. *pinguis*, from the *greasy* texture.

1. **P. vulga'ris,** *L.* ; calyx-lobes ovate-oblong obtuse, corolla violet, lips very unequal, spur slender about equalling the lower lip.

Wet bogs, &c., Shetland to Hants and Devon; rare in S. England; ascends to near 3,000 ft. in the Highlands; Ireland; Channel Islands; fl. May–July.—Glabrous, except the glandular top of the scape and calyx. *Leaves* 1–3 in., appressed to the ground, oblong, obtuse, succulent; petiole broad, very short. *Scapes* several, 4–6 in., purplish. *Calyx-lobes* very variable. *Corolla* ⅔–1 in.; lower lip much longer and broader than the upper; segments broad, obtuse; spur straight or incurved, variable in length. *Stamens* 2 anterior, and sometimes 2 imperfect lateral. *Capsule* ovoid or subglobose. *Cotyledon* solitary.—DISTRIB. Europe (Arctic), N. Asia, N. America.—Leaves insectivorous, used to curdle milk in Lapland.
P. VULGA'RIS proper; calyx-lobes ovate sometimes subacute, corolla ⅔ in., lobes of lower lip not overlapping, spur entire at the tip, capsule ovoid subacute.
Sub-sp. P. GRANDIFLO'RA, *Lamk.* ; larger, leaves broader, calyx-lobes more oblong, tip rounded, corolla 1 in., lobes of lower lip very broad overlapping, spur often 2-fid, capsule subglobose.—Bogs, Cork and Kerry; (Penzance, introduced).— DISTRIB. W. France, Alps, Pyrenees, Spain, Portugal.—The Irish is an extreme form; Alpine and Pyrenean intermediates are numerous.

2. **P. lusitan'ica,** *L.* ; calyx-lobes suborbicular, corolla lilac, lips nearly equal, throat yellow, spur short stout conical incurved.

Bogs, S.W. England, local, from Hants to Cornwall; W. Scotland, Orkney to Wigton; Ireland, ascending to 1,500 ft. in Mayo; fl. June–Oct.—Glabrous. *Leaves* ½–¾ in., oblong, shortly petioled, thin, succulent, obtuse. *Scapes* very slender. *Corolla* ½ in.; lips nearly equal, lower pouched from without, lobes short broad; spur very broad, obtuse. *Capsule* globose. *Cotyledons* 2.—DISTRIB. W. France, Spain, Portugal.

3. **P. alpi'na,** *L.* ; calyx-lobes broadly ovate obtuse, corolla white, lips unequal, throat yellow, spur very short conical.

Bogs, Skye, Ross; fl. May–June.—Similar to *P. lusitanica,* but larger, scapes shorter; corolla ½ in., throat hairy; lower lip longer than the upper; spur broader and more obtuse; capsule ovoid, acute.—DISTRIB. Europe, N. of the Alps (Arctic), N. Asia, Himalaya, Greenland, Fuegia.

2. UTRICULA'RIA, *L.*

Aquatic or terrestrial herbs, often floating and propagated by hybernacula. *Leaves* (of Brit. species) floating, multifid; segments very slender, furnished with minute pitchers, which entrap animalcules. *Flowers* solitary spiked or racemed, naked or bracteate. *Calyx* 2-partite, lobes subequal. *Corolla* personate, palate protruded. *Stamens* inserted on the upper lip of the corolla, filaments incurved; anthers subterminal, free or coherent, simple or constricted in the middle. *Style* short, stigma unequally 2-lobed. *Capsule* globose, bursting irregularly. *Seeds* oblong or peltate, striate,

pitted or with capitate or glochidiate hairs.—DISTRIB. Widely dispersed ;
species 150.—ETYM. *utriculus,* from the *bladder*-like pitchers.

1. **U. vulga′ris,** *L.* ; leaves spreading pinnately multifid, pitchers at
the bases of and upon the leaf-segments, upper corolla-lip exceeding the
palate or not, spur conic.

Pools and ditches, N. to Shetland, not common; ascends to 1,500 ft. in the
Highlands ; Ireland ; Channel Islands ; fl. July-Aug.—*Stems* 6–18 in.,
leafy. *Leaves* ¾–1 in., broadly ovate ; segments very slender, obtuse, remotely toothed with tufts of cilia at the sinus ; pitchers ⅛–¼ in., shortly
stalked. *Scape* 4–8 in., 2–8-fld. ; pedicels 2–3 times as long as the calyx,
reflexed after flowering. *Corolla* ½–¾ in., yellow, upper lip broad short,
palate prominent ; spur appressed to the under lip, honeyed. *Anthers*
cohering. *Stigma* irritable.—DISTRIB. Europe, N. Africa, Siberia, N.
America.

U. VULGA′RIS proper ; stem and scapes stout, upper corolla-lip about equalling
the palate, spur obtuse.

Sub-sp. U. NEGLEC′TA, *Lehm.* ; stem and scapes slender, leaves smaller more
remote, upper corolla-lip exceeding the palate, spur conic ascending acute.—
Pools, rare, Essex, Gloster, Surrey, Kent, Hants ; fl. June–Aug.—*Stems*
capillary, 6–8 in. *Leaves* rather remote, nearly orbicular, segments subulate, quite entire, ciliate here and there ; pitchers very shortly stalked.
Scape very slender ; pedicels many times longer than the corolla, ascending
in fruit. *Corolla* pale yellow. *Anthers* conniving.—DISTRIB. W. Europe.

2. **U. interme′dia,** *Hayne ;* leaves distichous dichotomously multifid,
pitchers on leafless branches, upper corolla-lip far exceeding the palate,
spur conic acute.

Pools and ditches, rare, Dorset, Hants, Norfolk, Westmoreland to Sutherland ; Ireland ; fl. July–Sept.—*Stems* slender, 4–8 in. *Leaves* close-set,
¼–½ in. broad, orbicular ; segments subulate, distantly ciliate ; pitchers ⅛–¼
in. long, on slender stalks. *Scape* rather stout, 3–4-fld. ; pedicels equalling
or much exceeding the calyx. *Corolla* ½ in., pale yellow ; upper lip twice
as long as the prominent palate ; lower broad, flat ; spur appressed to the
under lip. *Anthers* free.—DISTRIB. Europe N. of the Alps, N. Asia,
N. America.

3. **U. mi′nor,** *L.* ; leaves dichotomously multifid, pitchers on the leaf-
axils, spur minute obtuse.

Pools and ditches, from Orkney southd. ; Ireland ; fl. June–Sept.—*Stems*
capillary, 3–10 in. *Leaves* lax, ⅛–¼ in. broad, orbicular ; segments subulate,
quite entire ; pitchers 1⁄12 in., on slender stalks. *Scapes* 2–6 in., 2–6-fld. ;
pedicels 2–3 times as long as the calyx, decurved in fruit. *Corolla* ⅙ in.,
pale yellow ; upper lip as long as the inconspicuous curved palate, lower
lip broadly ovate. *Anthers* free.—DISTRIB. Europe (excl. Spain, Greece,
Turkey), N. Africa, N. and W. Asia, Himalaya, N. America.

Flowerless specimens of *U. Brem′ii,* Heer, have been found in Nairn and
Moray. It differs from *U. mi′nor* in its more robust habit and orbicular lip.

ORDER LVIII. **VERBENA'CEÆ.**

Herbs, shrubs, or trees. *Leaves* opposite or whorled, exstipulate.
Flowers cymose, irregular, bracteate. *Calyx* inferior, tubular, cleft or
toothed, persistent, imbricate in bud. *Corolla* hypogynous, tubular,
usually 2-lipped, imbricate in bud. *Stamens* usually 4, didynamous,
inserted on the corolla-tube ; anthers 2-celled. *Ovary* 2–4-celled ; style
simple, terminal, stigma simple or 2-fid ; ovules solitary or 2 collateral in
each cell, erect and anatropous or ascending and ½-anatropous. *Fruit* a
2- or 4-celled berry, or a drupe with 2–4 1–2-celled stones, or of 4 nutlets.
Seeds ascending, exalbuminous ; cotyledons foliaceous, radicle inferior.—
DISTRIB. Chiefly tropical ; genera 59 ; species 700.—AFFINITIES. With
Boragineæ and *Labiatæ.*—PROPERTIES unimportant.

1. VERBE'NA, *L.* VERVAIN.

Herbs or undershrubs. *Stem* 4-gonous. *Leaves* opposite or 3-nate,
simple pinnatifid or 3-partite. *Flowers* bracteate, in terminal spikes or
racemes. *Calyx* tubular, 5-ribbed, unequally 5-toothed. *Corolla* salver-
shaped ; tube straight or curved, villous within ; limb oblique, 2-lipped,
5-fid. *Stamens* 4, 2 or all perfect, included. *Disk* annular. *Ovary* 4-
celled ; style slender, 2-lobed, one lobe stigmatic ; ovules solitary in each
cell, erect. *Fruit* of 4 ribbed nutlets.—DISTRIB. Almost wholly American ;
species 80.—ETYM. Classical, obscure.

V. officina'lis, *L.* ; leaves opposite, flowers spiked.

Dry waste ground, local, from Northumbd. southd.; Ireland, local ; Channel
Islands; fl. July–Sept.—Perennial, hispidly pubescent. *Rootstock* woody.
Stems 1–2 ft., rigid, branched above. *Leaves* oblong, pinnatifid or 3-partite,
lobes acute or obtuse, upper narrower. *Spikes* dense-fld., afterwards
elongating ; bracts ovate, acute, ½ as long as the calyx, which is ½ as long as
the corolla-tube. *Corolla* lilac, limb ⅛ in. diam. *Nutlets* truncate, granulate.—DISTRIB. From Denmark southd., N. Africa, W. Asia, Himalaya;
introd. in N. America.—An object of much superstition amongst the
ancients.

ORDER LIX. **LABIA'TÆ.**

Glandular herbs or shrubs ; branches 4-angled, opposite or whorled.
Leaves opposite or whorled, entire or divided, exstipulate. *Flowers*
solitary or in axillary opposite centrifugal often crowded (falsely whorled)
cymes, irregular, 2-bracteate and bracteolate, proterandrous. *Calyx* inferior, persistent, 5-cleft, ribbed. *Corolla* hypogynous, deciduous, 5-merous,
2-lipped, imbricate in bud, upper lip outermost. *Stamens* inserted on the
corolla-tube, perfect usually 4, 5th and sometimes 2 lateral imperfect or
0, anthers polymorphous. *Disk* annular. *Ovary* of 2 connate deeply
2-lobed carpels, hence 4-partite, 4-celled ; style from between the lobes,

slender, stigma simple or 2-fid ; ovules solitary in each cell, erect, anatropous. *Fruit* of 1–4 1-seeded nutlets. *Seed* erect or ascending, testa thin, albumen 0 ; embryo straight, rarely curved, cotyledons fleshy, radicle next the hilum.—DISTRIB. All warm and temp. regions, rare in Arctic and alpine ; genera 136 ; species about 2,600.—AFFINITIES. With *Boragineæ* and *Verbenaceæ.*—PROPERTIES. Stimulant, fragrant, aromatic.

TRIBE I. **SATUREINE'Æ.** *Corolla-lobes* flat or margins recurved. *Stamens* 2–4, remote, spreading or conniving under the upper lip, 2 upper shorter or 0 ; anther-cells contiguous or confluent. *Nutlets* free, smooth, or nearly so.

* Corolla subregular. Stamens spreading ; anthers 2-celled.
Perfect stamens 4 ....1. Mentha.
Perfect stamens 2 ....2. Lycopus.

** Corolla 2-lipped. Stamens 4, distant.
Erect, leaves broad. Calyx equally 5-toothed ....3. Origanum.
Procumbent, leaves small. Calyx 2-lipped............................4. Thymus.

*** Corolla 2-lipped. Stamens 4, conniving under the upper lip.
Corolla-tube straight ; upper lip flat ....5. Calamintha.
Corolla-tube curved, ascending ; upper lip concave.................5*. *Melissa.*

TRIBE II. **MONARDE'Æ.** *Stamens* 2, erect or ascending ; anthers 1-celled, or if 2-celled cells remote. *Nutlets* free, smooth, or nearly so.
Calyx 2-lipped..6. Salvia.

TRIBE III. **NEPETE'Æ.** *Stamens* 4, 2 upper longer, ascending or diverging ; anther-cells 2, parallel or nearly so. *Nutlets* smooth or tubercled.
Upper lip of corolla truncate...7. Nepeta.

TRIBE IV. **STACHYDE'Æ.** *Stamens* 4, parallel, 2 upper shorter, ascending under the concave upper lip or included in the tube. *Nutlets* free, smooth or tubercled.

* Calyx 2-lipped, not inflated, lips closing over the fruit.
Filaments 2-fid, anthers all 2-celled.....................................8. Brunella.
Filaments simple, 2 lower anthers 1-celled.........................9. Scutellaria.
**Calyx inflated or 2-lipped. Anthers exserted.........................10. Melittis.
***Calyx tubular. Anthers included.................................11. Marrubium.

**** Calyx 5-toothed, subcampanulate, equal or oblique.

Calyx 5-toothed. Anthers glabrous. Nutlets obtuse.............12. Stachys.
Calyx-teeth 5, spinous. Anthers ciliate. Nutlets compressed ..13. Galeopsis.
Calyx-teeth 5, spinous. Anthers glabrous. Nutlets 3-quetrous, truncate.
13*. *Leonurus.*
Calyx 5-nerved. Anthers hairy. Nutlets 3-quetrous, truncate...14. Lamium.
Calyx-limb spreading ; teeth broad. Anthers glabrous. Nutlets obtuse.
15. Ballota.

Tribe V. **AJUGOIDE'Æ.** *Stamens* 4, parallel, ascending, exserted, 2 upper shorter. *Nutlets* connate, base oblique, reticulate and rugose.
Calyx tubular, 5-toothed. Upper corolla-lip 2-partite..........16. Teucrium.
Calyx ovoid, 5-cleft. Upper corolla-lip entire or notched..........17. Ajuga.

1. MEN'THA, *L.* MINT.

Strong-scented perennial herbs ; rootstock stoloniferous, creeping. *Whorls* many-fld., axillary, or forming terminal spikes ; bracts subulate or foliaceous ; bracteoles small or 0 ; flowers small. *Calyx* campanulate or tubular, 5-toothed ; throat naked or villous. *Corolla-tube* short, limb campanulate 4-lobed ; lobes subequal, upper broader. *Stamens* 4, equal, erect, distant, glabrous ; anther-cells parallel. *Style* shortly 2-fid. *Nutlets* dry, smooth.—Distrib. N. temp. regions ; species about 25.—Etym. The old Greek name.—Species often variable, hybridizing and difficult to discriminate.

* *Whorls* in terminal spikes ; bracts minute. *Throat of calyx glabrous.*

† *Leaves all sessile, or the lower only petioled.*

1. **M. sylves'tris,** *L.* ; leaves broadly or narrowly oblong-obovate or lanceolate subacute serrate smooth above hoary beneath, spike continuous, calyx-teeth lanceolate, corolla hairy glabrous within. *Horse-mint.*
Moist waste places, rare, from Forfar and the Clyde southd. ; Ireland ? native ; fl. Aug.-Sept.—*Stem* 2-3 ft., robust, tomentose with white hairs. *Leaves* 1-3 in., ¾-2 in. broad, rounded or cordate at the base. *Spikes* 1-3 in., ¼-½ in. broad, dense ; bracts lanceolate ; bracteoles subulate ; pedicels hairy. *Corolla* lilac, about ⅛ in. diam.—Distrib. Europe, N. Africa, N. and W. Asia, N.W. India.
M. sylves'tris proper ; leaves oblong-lanceolate acute, base rounded, slightly hairy above hoary beneath, spikes slender.—Var. *M. nemoro'sa,* Willd. ; leaves broadly oblong acute, base rounded, slightly hairy above tomentose beneath, spike stouter.—Var. *M. mollis'sima,* Bock., leaves broadly ovate acute, base subcordate, finely serrate hoary above felted beneath, spikes stout. Rare.—Var. *M. alopecuroi'des,* Hull (*veluti'na,* Bab.) ; leaves broad large coarsely serrate, base subcordate, somewhat wrinkled above very hairy beneath, spikes short stout, bracts broader. Kent, Essex, Norfolk, Perth, rare ; (an escape ? *Wats.*). The transition to *M. rotundifo'lia.*

2. **M. rotundifo'lia,** *L.* ; leaves broadly ovate-oblong very obtuse crenate much wrinkled above shaggy or deeply tomentose beneath, spikes interrupted, calyx hairy teeth subulate, corolla hairy glabrous within.
Wet places, from Forfar and the Clyde southd. ; (indigenous only in S. England, *Wats.*) ; Ireland ; Channel Islands ; fl. Aug.-Sept.—Habit of *M. sylves'tris,* but usually much branched. *Leaves* often densely woolly beneath. *Spikes* dense, cylindric. *Flowers* white or pink.—Distrib. From Belgium southd., N. Africa, N. and W. Asia.—An escape in N. America.

M. vir'idis, *L.* ; glabrous or nearly so, leaves oblong-lanceolate sub-acute serrate smooth above, spikes slender, corolla wholly glabrous. *Spear-mint.*

Wet places, naturalized in England and Scotland, rare; possibly indigenous
in W. York; fl. Aug.-Sept.—Probably a cultivated form of *M. sylves'tris*,
easily distinguished by its pungent smell.—DISTRIB. Cultivated for culi-
nary purposes.—VAR. *cris'pa* is a garden form with crisped foliage.

†† *Leaves petioled.*

3. **M. piperi'ta,** *Huds.* ; glabrous, leaves ovate or oblong-lanceolate
acute serrate upper smaller, spikes cylindric interrupted below, pedicels
and flowers glabrous or very sparingly hispid. *Peppermint.*

Damp places, from the Clyde and Forfar to Sussex and Cornwall; Ireland;
(a doubtful native, *Wats.*); fl. Aug.-Sept.—Usually smaller and more
slender than the preceding. *Leaves* 1-4 in., acute or obtuse at the base,
coarsely serrate, smooth above, rarely sparingly hairy on the nerves beneath
uppermost sometimes bracteiform. *Calyx* often red.—DISTRIB. Europe;
introd. in America. Nowhere indigenous, *Nyman.*—Probably a garden
form of *M. aquatica*, Bentham.

Two forms occur: *M. officina'lis*, Hull; leaves acute or rounded at the base,
spikes elongate, and *M. vulga'ris*, Sole; leaves rounded or subcordate at the
base, spikes shorter.

4. **M. aquat'ica,** *L.* ; usually softly hairy, leaves ovate-oblong or
cordate, upper bracteiform, spikes oblong continuous or interrupted below,
pedicels and flowers usually hairy or villous.

River-sides, marshes, N. to Orkney, &c., ascends to 1,500 ft. in Yorkshire;
Ireland; Channel Islands; fl. Aug.-Sept.—Very common and variable;
some forms are with difficulty distinguished from *M. sati'va*. *Stem* 1-5 ft.
Leaves 1-3 in. *Spikes* ¾-1 in. diam., long or short, usually stout. *Flowers*
lilac or purplish. *Bracts* and *bracteoles* lanceolate-subulate. *Calyx-teeth*
slender.—DISTRIB. Europe, N. Africa, N. and W. Asia; introd. in America.
M. AQUAT'ICA proper; leaves ovate rounded or subcordate at the base serrate
more or less hairy on both surfaces, spikes axillary and terminal ovoid or
subglobose, calyx-teeth ½-⅔ the length of the tube. *M. hirsuta*, L.
Common.—*M. hirsu'ta* proper; leaves tomentose on both surfaces, calyx
pedicels and corolla hairy. VAR. *M. subgla'bra*, Baker; leaves narrower
glabrous except on the nerves beneath, calyx pedicels and corolla hairy.—
VAR. *M. citra'ta*, Ehrh. (*Bergamot Mint*); leaves glabrous on both surfaces,
calyx pedicels and corolla glabrous. Staffordshire, Wales, &c., rare.
Sub-sp. M. PUBES'CENS, *Willd.*; leaves ovate-oblong or lanceolate pubescent
above tomentose or woolly beneath sharply serrate, spikes cylindric stout
dense, calyx-teeth ⅔ the length of the tube. Mid. and S. England only.—
VAR. *M. palus'tris*, Sole; leaves ovate-oblong tomentose above woolly
beneath.—VAR. *M. hirci'na*, Hull; leaves ovate-oblong green and subglabrous above hairy beneath.

** *Whorls* in axillary clusters, shorter than the leaves; bracts foliaceous.
Throat of calyx glabrous.

5. **M. sati'va,** *L.* ; leaves petioled ovate- or oblong-lanceolate acutely
serrate, upper smaller, bracteoles usually shorter than the flowers acuminate calyx-teeth lanceolate acuminate ½-⅔ the length of the tube.

Wet waste places, from Argyll and Elgin southd.; Ireland; fl. July-Sept.—
Probably a form of *M. aquat'ica*, distinguished by the inflorescence alone.—

DISTRIB. Europe, Canaries; introd. in N. America.—The forms here enumerated are not constant; I have taken their diagnosis from Baker's and Syme's works.

M. SATI′VA proper (*b. riva′lis*, Lond. Cat.); green, 2–3 ft., leaves hairy on both surfaces, pedicels calyx and corolla hairy. Common in England, rare in Scotland and Ireland.—VAR. 1, hairy, whorls all separate, bracts all foliaceous, upper sometimes flowerless.—VAR. *M. paludo′sa*, Sole; hairy, upper whorls collected into a spike with smaller bracts. The passage to *M. aquat′ica.* —VAR. *subgla′bra*, Baker; almost glabrous, whorls all separate, bracts all foliaceous.

Sub-sp. M. RU′BRA, *Sm.*; stem 3–5 ft., and nerves of leaves purple, leaves glabrous or sparingly hairy, calyx hairy, pedicels and corolla glabrous. Not uncommon in England, rare in Scotland and Ireland.

Sub-sp. M. GRAC′ILIS, *Sm.*; slender, green, leaves oblong-lanceolate glabrous or sparingly hairy, whorls all separate, bracts all smaller and narrower than the leaves, bracteoles equalling and exceeding the flowers, calyx ciliate, pedicels and corolla glabrous.—VAR. *grac′ilis* proper, stem hairy below, lower bracts shortly petioled 5–6 times as long as the rather remote whorls. Wiltshire (gathered by Sole only).—VAR. *cardi′aca*, Baker; nearly glabrous, bracts sessile 2–4 times as long as the crowded whorls. From Middlesex N. to the Tyne; often cultivated; smells of Basil.

Sub-sp. M. PRATEN′SIS. *Sole;* leaves drooping rounded at both ends finely serrate hairy above, glabrous except on the close-set nerves beneath, whorls in the upper leaves only all separate, calyx-teeth ciliate, pedicels and corolla glabrous. New Forest, gathered by Sole only in 1789.

Sub-sp. M. GENTI′LIS, *L.*; leaves spreading ovate acute serrate slightly hairy above and on the (few) nerves especially beneath, bracts all leaf-like uppermost flowerless, pedicels and corolla glabrous, calyx-teeth ciliate. Common in England, rare in Scotland.—VAR. *genti′lis* proper; stem hairy, leaves rather coriaceous most hairy beneath, calyx-teeth densely hairy.—VAR. *M. Wirtgenia′na*, F. Schultz (a hybrid, *Nyman*); stem subglabrous, leaves slightly hairy above, but on the nerves only beneath, calyx-teeth sparingly hairy.—VAR. *M. Paulia′na*, F. Schultz; stem subglabrous, leaves as in *Wirtgenia′na*, calyx-teeth densely hairy (a var. of *arven′sis*, Nyman).

6. M. arven′sis, L.; leaves petioled ovate- or oblong-lanceolate obtusely serrate, upper smaller, bracteoles shorter than the flowers acute, calyx hairy, teeth triangular, corolla hairy without and within.

Cultivated fields and waste places, N. to Orkney; ascends to 1,000 ft. in the Highlands; Ireland; Channel Islands; fl. Aug.–Sept.—Usually a low branched plant, 1–2 ft., very variable, chiefly distinguished from *M. sati′va* by the short calyx-teeth. *Flowers* honeyed, dimorphous, larger 2-sexual, smaller males.—DISTRIB. Europe (Arctic), N. and W. Asia, Himalaya; introd. in N. America.—The following are inconstant varieties.

M. arven′sis proper; stem short, hairs dense reflexed, leaves smooth hairy all over, calyx very hairy.— *M. nummula′ria*, Schreb.; stem long, hairs few, leaves and calyx sparingly hairy.—*M. agres′tis*, Sole; stem long hairy, leaves coarsely serrate broad often cordate wrinkled and calyx very hairy, bracts smaller upwards.— *M. præ′cox*, Sole; stem stout erect slightly hairy, leaves smooth sparingly hairy, bracts much smaller upwards, calyx-teeth longer. Approaches *M. sati′va.*—*M. Allio′nii*, Bor.; stem tall slightly hairy above nearly glabrous below, leaves smooth thinly hairy, bracts uniform, calyx-teeth very short.—*M. parietariæfo'lia,* Beck. ; subglabrous, stem long, leaves smooth serrated towards the tip, bracts smaller upwards, calyx-teeth short.

*** *Whorls* axillary, distant, none towards the ends of the branches. *Calyx* 2-lipped; throat closed with hairs.

7. M. Pule'gium, *L.* ; leaves small shortly petioled ovate or oblong subserrate, calyx-teeth ciliate, corolla hairy without glabrous within. *Penny-royal.*

Pools, wet heaths, &c., from Ayr and Berwick southd.; Ireland; Channel Islands; fl. Aug.–Sept.—Glabrous or more or less tomentose. *Stems* 4–10 in., prostrate or erect (var. *erec'ta*), much branched, very leafy. *Leaves* $\frac{1}{4}$–$\frac{3}{4}$ in., spreading and recurved, base acute. *Whorls* all separate; bracts foliaceous, upper sessile flowerless; bracteoles 0, or obovate and shorter than the flowers. *Flowers* proterogynous. *Calyx* and pedicels pubescent or hispid.—Distrib. Europe, N. Africa, N. and W. Asia.—Formerly much used medicinally.

2. LYC'OPUS, *Tournef.* GIPSY-WORT.

Marsh herbs. *Leaves* toothed or pinnatifid. *Whorls* axillary, dense, many-fld. ; bracts foliaceous ; bracteoles minute. *Flowers* small, sessile honeyed, proterandrous. *Calyx* campanulate, equal, 4–5-toothed ; throat naked. *Corolla* short, campanulate, equal, 4–5-fid. *Stamens,* 2 upper imperfect, with capitate anthers or 0 ; 2 lower fertile, distant, anther-cells parallel. *Style* 2-fid, lobes flattened. *Nutlets* dry, smooth, truncate, narrowed below, margins thickened.—Distrib. Temp. regions ; species probably 2 with many varieties.—Etym. λύκος and πούς, *wolf's foot,* of doubtful application.

L. europæ'us, *L.* ; stem acutely 4-angled, calyx-teeth 5 subulate.

Ditches and river-banks from Ross southd.; rarer in Scotland; Ireland; Channel Islands; fl. June–Sept.—Glabrous or slightly pubescent. *Rootstock* creeping or stoloniferous. *Stem* 1–3 ft., tough. *Leaves* subsessile, elliptic-oblong, coarsely serrate or pinnatifid. *Whorls* many; bracts smaller upwards; flowers sessile. *Corolla* $\frac{1}{8}$ in., bluish-white, dotted with purple, hairy within.—Distrib. Europe, N. Africa, N. and W. Asia, India, N. America, Australia.

3. ORIG'ANUM, *Tournef.* MARJORAM.

Aromatic herbs or undershrubs. *Leaves* entire or toothed. *Flowers* crowded, in corymbose cymes, honeyed ; bracts large, imbricating. *Calyx* subcampanulate, 10–13-nerved, 5-toothed or 2-lipped, upper lip entire or 3-toothed, lower 2-toothed truncate or 0. *Corolla* obscurely 2-lipped ; upper lip notched or 2-fid ; lower spreading, 3-fid. *Stamens* 4, ascending, distant ; anther-cells distinct, spreading. *Style-lobes* acute. *Nutlets* dry, smooth. —Distrib. Temp. N. regions ; species about 20.—Etym. ὄρος and γάνος, from affecting *hilly* localities.

O. vulga're, *L.* ; erect, corymbosely branched, leaves broadly ovate.
Dry copses and hedgebanks, &c., from Caithness southd.; Scotland, rare;
ascends to 1,300 ft. in Yorkshire; Ireland, local; Channel Islands; fl. July--
Sept.—More or less pubescent, bifariously on the branches. *Rootstock*
short, stoloniferous. *Stems* many, 1–3 ft., stout. *Leaves* ½–1 in., shortly
petioled, rhombic-ovate, lower early withering, entire or obtusely serrate.
Cymes ¼–1 in., ovoid, 4-gonous; bracts ¼ in., green or purple, ovate, acute,
longer than the calyx. *Flowers* proterandrous, dimorphic, larger 2-sexual,
purple, stamens long, smaller female paler. *Calyx* yellow-dotted, teeth
short; throat closed with hairs.—Distrib. Europe (Arctic), N. Africa,
N. and W. Asia, Himalaya; introd. in N. America.—Aromatic, bitter, and
balsamic.—*O. megasta'chyum,* Link, is a large state with 4-gonous spikes.

4. THY'MUS, *L.* Thyme.

Small shrubs, often hairy, much branched, very aromatic. *Leaves*
small, quite entire ; margins often revolute. *Whorls* few-fld., in lax or
dense spikes ; bracts minute ; flowers purple, rarely white, honeyed,
proterandrous. *Calyx* ovoid, 10–13-nerved, 2-lipped, upper lip 3-toothed,
lower 2-fid ; throat villous. *Corolla-tube* naked within, obscurely 2-lipped ;
upper lip straight, flattish, notched ; lower 3-fid. *Stamens* 4, usually
exserted, straight, distant, the lower longer ; anther-cells parallel or
diverging. *Style-lobes* subequal, subulate. *Nutlets* nearly smooth.—
Distrib. Temp. Old World ; species 40.—Etym. The Greek name.

T. Serpyl'lum, *L.* ; prostrate, leaves green flat quite entire.
Hills and dry grassy places, N. to Shetland; ascends to 3,500 ft. in the Highlands; Ireland; Channel Islands; fl. June–Aug.—Glabrous or hairy, hairs
often reflexed and in lines on the stems. *Rootstock* woody. *Stems* decumbent. *Leaves* ⅛–¼ in., shortly petioled, ovate or obovate-lanceolate, obtuse.
Whorls capitate ; bracts foliaceous ; pedicels very short; flowers dimorphic,
males large, bisexual smaller. *Calyx* purplish, teeth ciliate. *Corolla* ¼–⅜
in., rose-purple.—Distrib. Europe (Arctic), N. and W. Asia, Himalaya,
Greenland ; introd. in N. America.
T. Serpyl'lum proper ; flowering-branches ascending from trailing shoots
that are barren at the tip, leaves often obovate, whorls in one head, upper
lip of corolla oblong.—Commonest form in mountain districts.
Sub-sp. T. Chamæ'drys, *Fries;* flowering- and barren-branches ascending
from the crown of the rootstock, leaves usually ovate, whorls in many
axillary heads, upper lip of corolla short and broad.

5. CALAMIN'THA, *Mœnch.*

Herbs or shrubs. *Whorls* dense and axillary or loose and panicled ;
flowers purple white or yellow. *Calyx* tubular, 13-nerved, 2-lipped, upper
lip 3-toothed, lower 2-fid ; throat naked or villous. *Corolla-tube* straight ;
throat naked, often inflated ; upper lip erect, flattish, lower spreading,
3-lobed. *Stamens* 4, ascending under the upper lip, upper sometimes
imperfect ; anther-cells parallel or diverging. *Style-lobes* equal, or the

lower larger. *Nutlets* smooth.—DISTRIB. Temp. N. hemisphere ; species about 40.—ETYM. καλός and μίνθα, *beautiful mint.*

SUB-GEN. 1. **Calamin'tha** proper. *Whorls* many, compound, lax, lower many-fld. ; bracts minute. *Calyx-tube* straight. *Corolla* with mid-lobe of lower lip notched.

1. **C. officina'lis,** *Mœnch;* perennial, leaves ovate, cymes secund. *Calamint.*

Waste places in dry soil from Westmoreland southd. ; Ireland ; Channel Islands ; fl. July–Sept.—Hairy. *Rootstock* more or less creeping, stoloni-ferous. *Stem* 1–3 ft., erect or decumbent below ; branches usually long, straggling and ascending. *Leaves* crenate-toothed, very variable in size and depth of serratures but pretty uniform in shape, almost glabrous or downy or nearly woolly. *Cymes* more or less unilateral, peduncles and pedicels variable.—DISTRIB. From Belgium southd., N. Africa, W. Asia ; introd. in N. America.

C. OFFICINA'LIS proper ; branches long ascending, leaves 1–1½ in. crenate-serrate green beneath, cymes few-fld., calyx bent on the pedicel 2-lipped, teeth with long bristles, upper triangular, lower subulate much longer, hairs of throat included, corolla ½ in., mid-lobe of lower lip longest. *C. menthæfo'lia,* Host.—VAR. *Brigg'sii,* Syme ; larger, more hispid, peduncles of lower whorls equalling or exceeding the pedicel of the central flower.—Devonshire.

Sub-sp. C. NEP'ETA, *Clairv.* ; branches short erect, leaves usually ½–1 in., subentire pale beneath, cymes about 10-fld., calyx erect on the pedicel, teeth nearly equal, upper triangular recurved, lower subulate, hairs of throat prominent, corolla ⅓ in., mid-lobe of lower lip broad truncate.—York to Kent.

Sub-sp. C. SYLVAT'ICA, *Bromf.* ; branches 0 or long ascending, leaves 1–3 in., deeply crenate-serrate, cymes loose, flowers large, calyx bent on the pedicel 2-lipped ciliate, 3 upper teeth abruptly recurved, 2 lower twice as long incurved, hairs of throat concealed in the tube, corolla ¾–1 in., mid-lobe of lower lip about equalling the lateral.—Chalk banks, I. of Wight, Hants, Devon.—This is the true *C. officina'lis,* Mœnch, according to Nyman.

SUB-GEN. 2. **Clinopo'dium,** *L.* (gen.). *Whorls* few, compound, dense-fld., surrounded by many linear bracteoles. *Calyx-tube* slightly curved. *Corolla* with mid-lobe of lower lip notched.

2. **C. Clinopo'dium,** *Benth.* ; perennial, softly hairy, leaves ovate obscurely toothed. *Clinopo'dium vulga're,* L. *Wild Basil.*

Copses and rocky places, Banff and Renfrew, southd. ; rare in Scotland and Ireland ; ascends to 1.050 ft. in Derby ; fl. July–Sept.—*Rootstock* woody, stoloniferous. *Stem* 1–3 ft., slender, flexuous, subsimple. *Leaves* remote, 1–2 in., subacute. *Whorls* terminal and axillary, ¾–1 in. diam., depressed ; bracts equalling the calyx ; pedicels slender ; flowers crowded. *Calyx* ⅓–½ in., striate, bristly. *Corolla* ¾–1 in., purple, hairy ; lips subequal.—DISTRIB. Europe, N. Africa, N. and W. Asia, Himalaya, Canada ; introd. in the U. States.

SUB-GEN. 3. **A'cinos,** *Mœnch* (gen.). *Whorls* many, simple, lower 5-6-fld. ; bracteoles few, minute. *Calyx-tube* curved. *Corolla* with the mid-lobe of the lower lip almost entire.

3. **C. A′cinos,** *Clairv.* ; annual or biennial, branched, leaves petioled ovate acuminate. *Thy′mus A′cinos,* L. *Basil Thyme.*

Banks and fields, on dry soil, from Elgin and Inverness southd.; rare, indigenous? in Scotland; N.E. Ireland, very rare ; Channel Islands; fl. July–Aug.—More or less pubescent. *Stem* 3–6 in., ascending, slender, leafy. *Leaves* ¼–½ in., narrowed into the petiole, variable in breadth, entire or slightly serrate. *Bracts* leafy ; bracteoles shorter than the pedicels. *Calyx* bent on the pedicel ; tube much enlarged below in fruit ; throat closed with hairs ; 3 upper teeth recurved. *Corolla* ½ in., blue-purple, spotted white and darker purple.—Distrib. Europe, N. Africa.

5*. *MELIS′SA,* Tournef. BALM.

Herbs, sometimes shrubby. *Whorls* few-fld., axillary, secund ; bracts few, subfoliaceous ; flowers white or yellowish. *Calyx* coriaceous, curved, subcampanulate, ribs 13, 5 very strong, 2-lipped ; upper lip flattish, 3-toothed, lower 2-fid. *Corolla-tube* ascending, more or less recurved ; throat dilated, naked ; upper lip concave, notched ; lower 3-fid, lobes flat. *Stamens* 4, converging under the upper lip ; anther-cells at length spreading. *Style-lobes* subequal, subulate. *Nutlets* smooth.—Distrib. Europe, W. Asia, Himalaya ; species 4.—Etym. μέλισσα, from bees affecting the plant.—Genus hardly differing from *Calamintha.*

M. officina′lis, *L.* ; leaves ovate crenate-toothed, flowers white.

Naturalized in the S. of England; fl. July–Aug.—More or less hairy. *Rootstock* short. *Stems* 1–2 ft., many, erect. *Leaves* 1–3 in., petioled, wrinkled above. *Whorls* shortly stalked; bracteoles small, oblong. *Calyx-teeth,* upper deltoid recurved with setaceous points ; lower longer, slender, straight. *Corolla* ½ in., white or spotted with rose.—Distrib. Mid. and S. Europe, W. Asia ; a garden escape in N. America.

6. SAL′VIA, *L.* SAGE.

Herbs or shrubs. *Whorls* usually racemed or spiked ; bracts leafy or small. *Calyx* tubular or campanulate, 2-lipped, upper lip entire or 3-toothed, lower 2-fid ; throat naked. *Corolla-tube* naked or with a ring of hairs or smooth processes inside ; upper lip erect, entire or notched ; lower 3-lobed, mid-lobe entire or notched, lateral spreading. *Stamens* 2, filaments short jointed on the long slender arched connective which bears at one end a perfect anther-cell, at the other a rudimentary one. *Disk* forming a large gland anteriorly. *Style* ascending, 2-fid, lobes subulate or dilated. *Nutlets* 3-quetrous, usually shining.—Distrib. All temp. and trop. regions ; species about 450.—Etym. *salvo,* from the healing properties of Sage.

1. **S. Verbena′ca,** *L.* ; stem leafy, corolla ½ in. glabrous inside. *Clary.*

Dry pastures and waste places, E. Scotland from Ross southd.; all England ; Ireland; Channel Islands; fl. June–Sept.—Subglabrous below, glandular-hairy above. *Rootstock* woody. *Stem* 1–2 ft., erect. *Leaves* 2–4 in., wrinkled, radical petioled, oblong, obtuse, irregularly crenate or serrate; upper cauline sessile, oblong or deltoid-ovate. *Whorls* 6-fld., in long bracteate spikes; bracts ovate-cordate. *Calyx* campanulate, upper lip with recurved edges and minute spinescent teeth, lower teeth subulate. *Corolla* blue-purple; upper lip short, compressed. *Connective* dilated.—DISTRIB. From Denmark southd., N. Africa, W. Asia.—The nutlets become mucilaginous in water; formerly used for eye complaints.

S. clandesti'na, L., is a smaller more slender variety, with leaves narrower, upper calyx-teeth less spiny, corolla longer more purple, upper lip longer arched.—Channel Islands.

2. **S. praten'sis**, *L.* ; stem-leaves few, corolla 1 in. lip viscid.

Dry fields, very rare, Cornwall, Kent, Oxford; (a denizen, *Wats.*); fl. June–Aug.—Glandular-hairy, especially above. Habit of *S. Verbena'ca*, but larger. *Leaves* wrinkled, 3–6 in., radical oblong or ovate-cordate, long-petioled, obtuse, sometimes 2-lobed at the base, crenatures large irregular; cauline similar or more oblong, smaller, shorter petioled. *Whorls* about 4-fld., in spikes 1–1½ ft.; bracts small, ovate-cordate, long acuminate and calyx coloured. *Flowers* dimorphous, honeyed, larger 2-sexual, proterandrous, smaller female. *Corolla* bright blue, glabrous inside; upper lip long, compressed, much arched; lower broad. *Connective* dilated.—DISTRIB. From Belgium southd., W. Asia.

7. **NEP'ETA,** *L.*

Erect or prostrate herbs. *Whorls* axillary or terminal; flowers blue yellow or white. *Calyx* tubular, 15-ribbed, 5-toothed, teeth equal or unequal. *Corolla-tube* slender below, throat dilated, naked; upper lip straight, notched or 2-fid; lower 3-fid, mid-lobe large. *Stamens* 4, ascending under the upper lip, upper pair longest; anther-cells diverging. *Style-lobes* subulate. *Nutlets* smooth.—DISTRIB. Temp. Europe, N. Africa and Asia; species about 120.—ETYM. The Latin name.

SUB-GEN. 1. **Nep'eta** proper. *Whorls* subterminal; upper bracts small. *Corolla-tube* short; mid-lobe of lower lip suborbicular, concave. *Anthers* subparallel, both cells opening by one slit.

1. **N. Cata'ria**, *L.*; erect, leaves ovate-cordate inciso-serrate white and pubescent beneath, flowers white. *Cat-Mint.*

Banks and waste places, from Northumbd. southd., rare; introduced only in Scotland; Ireland; Channel Islands; (a doubtful native, *Wats.*); fl. July–Sept.—Hoary pubescent. *Rootstock* stout. *Stem* 2–3 ft., branched, very leafy. *Leaves* 1–3 in., deeply lobed at the base. *Whorls* shortly stalked, upper sessile, many and dense-fld., in broad heads 1 in. long; bracts leafy; bracteoles longer than the short pedicels. *Calyx* pubescent; teeth subulate, upper longest. *Corolla* ½ in., dotted with purple, tube curved. *Nutlets* granulate.—DISTRIB. Europe, N. and W. Asia, Himalaya; introd. in N. America.

Sub-gen. 2. **Glecho'ma,** *L.* (gen.). *Whorls* axillary ; bracts all large and leaf-like. *Corolla-tube* long ; mid-lobe of lower lip obcordate, flat. *Anthers* conniving and forming a cross ; cells each with a slit.

2. **N. Glecho'ma,** *Benth.* ; procumbent, leaves ovate- or orbicular-reniform deeply crenate, flowers blue-purple. *Glecho'ma hedera'cea,* L. *Ground Ivy.*

Hedges and copses, from Caithness southd. ; ascends to 1,350 ft. in Derby ; Ireland; Channel Islands; fl. March–June.—Perennial, more or less pubescent. *Stems* 6–18 in., rooting at the base, slender, branched. *Leaves* ⅔–1½ in. diam., petiole ½–2 in. *Whorls* 3–6-fld.; bracteoles subulate, equalling the short pedicels. *Flowers* dimorphic, larger 2-sexual, smaller female. *Calyx-teeth* short, recurved. *Corolla* ½–1 in., tube very variable in length. *Nutlets* granulate.—Distrib. Europe (Arctic), N. and W. Asia ; introd. in U. States.—Bitter and aromatic, formerly used for beer, occasionally for tea.

8. BRUNEL'LA, *L.* Self-heal.

Small, hairy, perennial herbs. *Whorls* about 6-fld., in dense terminal heads, surrounded by orbicular leaf-like bracts ; flowers purplish, rarely white. *Calyx* subcampanulate, reticulate, 2-lipped, closed in fruit ; upper lip flat, 3-toothed, lower 2-lobed ; throat naked. *Corolla-tube* broad, ascending, with a short hairy basal ring inside ; upper lip erect, concave, lower spreading, lateral lobes deflexed, mid-lobe concave. *Stamens* 4, exserted, filaments glabrous 2-toothed at the tip, lower tooth antheriferous ; anthers conniving in pairs under the upper lip, cells diverging. *Disk* erect, symmetrical. *Style-lobes* subulate. *Nutlets* oblong, smooth.— Distrib. N. and S. temp. regions ; species 3.—Etym. Doubtful.

B. vulga'ris, *L.* ; leaves ovate-oblong, corolla not twice as long as the purplish calyx. *Prunel'la vulgaris,* L.

Pastures and waste places, N. to Shetland ; ascends to 2,400 ft. in Yorkshire ; Ireland; Channel Islands; fl. July–Sept.—More or less hairy. *Rootstock* creeping. *Stems* 4–12 in., erect or ascending, branches often abbreviated. *Leaves* 1–2 in., petioled, uppermost sessile, ovate-oblong or oblong-lanceolate, entire toothed or subpinnatifid. *Whorls* in cylindric 1–3 in. spikes ; bracts broadly ovate-cordate, ciliate, green with purple edges ; bracteoles 0 ; pedicels very short. *Flowers* dimorphic, larger 2-sexual, proterandrous, smaller female. *Calyx-teeth* minute, mucronate. *Corolla* ½–¾ in., purple, rarely rosy or white.—Distrib. Europe (Arctic), N. Africa, temp. Asia, America, Australia.

9. SCUTELLA'RIA, *L.* Skull-cap.

Slender herbs, rarely shrubs. *Flowers* solitary or in pairs, axillary or in terminal spikes or racemes. *Calyx* campanulate, 2-lipped, tube dilated opposite the posterior lip into a broad flattened hollow pouch, lip and pouch deciduous in fruit, mouth closed after flowering, persistent. *Corolla-tube* long, naked inside, throat dilated ; upper lip entire or notched : lower dilated, lateral lobes free spreading. *Stamens* 4 ; anthers

Y 2

conniving in pairs, ciliate, lower 1- upper 2-celled. *Disk* elongate, curved. *Ovary* oblique. *Style* with the upper lobe very short. *Nutlets* smooth or tubercled.—DISTRIB. N. temp. and subtrop. regions, abundant in America; species about 90.—ETYM. *scutella*, from the *dish*-like pouch of the calyx.

1. **S. galericula'ta,** *L.* ; leaves crenate-serrate, flowers ⅔ in.

Marshy places, river-banks, &c., from Shetland and Harris southd.; rare in Ireland; Channel Islands; fl. July–Sept.—Glabrous or puberulous. *Root-stock* creeping. *Stems* 6–18 in., simple or branched. *Leaves* ½–2½ in., shortly petioled, ovate lanceolate, base cordate, obtuse or subacute, crenatures rather remote, upper often entire. *Flowers* secund, pubescent, solitary ; bracts leaf-like; bracteoles minute, setaceous; pedicels very short. *Calyx-lips* short, broad. *Corolla* blue variegated with white inside, 3–4 times as long as the calyx; tube curved ; lips short. *Stamens* and style included. *Nutlets* granulate, enclosed in the 2-valved calyx.—DISTRIB. Europe (Arctic), N. Africa, N. and W. Asia, India, N. America.—A hybrid ? with *S. minor* is common at Virginia Water, Surrey, intermediate in habit and flowers between the two.

2. **S. mi'nor,** *L.* ; leaves with 1–2 crenatures near the base, flowers ⅓ in.

Swampy heaths and sides of ditches, &c., from Inverness and W. Scotland southd.; ascends to 1,000 ft. in Devon ; Ireland ; Channel Islands ; fl. July–Oct.—Habit of *S. galericula'ta*, but only 4–6 in., more slender, often much branched ; leaves shorter-petioled or sessile, ¼–1 in., obtuse, upper quite entire ; flowers much smaller, ⅓ in., pale pink-purple.—DISTRIB. W. Europe, W. Asia, N.W. Himalaya.

10. MELIT'TIS, *L.* BASTARD-BALM.

An erect perennial herb. *Whorls* axillary, 2–6-fld. ; flowers large. *Calyx* broadly campanulate, membranous, nerved ; upper lip broad, orbicular, irregularly 2–3-lobed, lower with 2 rounded lobes. *Corolla-tube* broad, naked inside, orbicular ; lower lip broadly 3-lobed. *Stamens* 4, ascending under the upper lip ; anthers conniving in pairs, cells diverging. *Style-lobes* ovate. *Nutlets* smooth or reticulate.—DISTRIB. From France southd., excl. Greece.—ETYM. Same as *Melissa*.

M. Melissophyl'lum, *L.* ; leaves ovate or oblong crenate-serrate. *M. grandiflo'ra*, Sm.

Copses, Worcester, Wales, Cornwall to Sussex; fl. May–June.—Sparingly hairy. *Rootstock* long, creeping. *Stem* 1–2 ft., erect, simple or branched. *Leaves* subsessile or petioled, nerves hairy beneath. *Bracts* leaf-like ; bracteoles 0 ; pedicels short, stout. *Calyx* very open, lobes all short broad. *Corolla* 1–1½ in., cream-white, spotted pink or purple, tube nearly straight, mouth oblique, lips diverging.—DISTRIB. From France southd.

11. MARRU'BIUM, *L.* WHITE HOREHOUND.

Perennial, tomentose or woolly herbs. *Whorls* axillary ; bracts leaf-like ; flowers small. *Calyx* tubular, 5–10-nerved ; teeth 5–10, subspinous, erect or spreading. *Corolla*-short; tube naked or with a ring of hairs inside; upper lip erect; lower spreading, 3-fid, mid-lobe broadest. *Stamens* 4, included; anthers glandular, cells diverging, subconfluent. *Style-lobes* short, obtuse. *Nutlets* obtuse.—DISTRIB. Temp. and warm regions of the Old World; species 30.—ETYM. The old Latin name.

M. vulga're, *L.* ; leaves broadly ovate crenate, whorls dense-fld.

Waste places, Elgin and E. Scotland, rare; all England; not native except in I. of Wight; Ireland, rare; Channel Islands; fl. July-Nov.—Hoary, almost woolly. *Rootstock* short, stout. *Stem* 1-1½ ft., stout, branched, leafy. *Leaves* ½-1½ in., base cordate or cuneate, nerves stout usually diverging from the broad rather long petiole, much wrinkled, leathery. *Whorls* of innumerable partial ones, depressed, axillary, villous. *Calyx* oblong; teeth 10, short, spinous, hooked at the tip. *Corolla* ½ in., white; tube slender; upper lip long, 2-fid.—DISTRIB. Europe, N. Africa, W. Asia, N.W. India; introd. in N. America.—Aromatic and bitter.—Much used as a cough medicine.

12. STA'CHYS, *L.* WOUNDWORT.

Herbs, rarely shrubs. *Leaves* crenate or serrate. *Whorls* 2- or more-fld., usually in terminal racemes; flowers honeyed, proterandrous. *Calyx* subcampanulate, 5-10-nerved, 5-toothed, teeth usually equal. *Corolla-tube* cylindric, with usually a ring of hairs inside, often incurved above; throat not dilated; upper lip erect or spreading; lower spreading, 3-lobed, mid-lobe largest, lateral often reflexed. *Stamens* 4, ascending, 2 lower longest; anthers conniving, cells parallel or diverging. *Style-lobes* nearly equal, subulate. *Nutlets* obtuse.—DISTRIB. Chiefly N. temp. and Oriental regions; species about 160.—ETYM. στάχυς, from the *spiked* inflorescence.

SECTION 1. **Sta'chys** proper. *Whorls* in elongate interrupted spikes or racemes; lower or all the bracts foliaceous. *Anther-cells* diverging.

1. **S. sylvat'ica,** *L.* ; perennial, hispid, cauline leaves long-petioled ovate broadly cordate coarsely serrate, whorls 6-12-fld.

Shady places, N. to Shetland; ascends to 1,500 ft. in Northumbd.; Ireland; Channel Islands; fl. July-Aug.—Fœtid when bruised; softly hispid. *Rootstock* creeping, stoloniferous. *Stem* 1-3 ft., rather slender, solid, simple or branched. *Leaves* 2-4 in., petiole often longer, radical withering early. *Spikes* 4-8 in., hairy and glandular; lower bracts serrate, upper lanceolate quite entire; bracteoles minute; pedicels short. *Calyx-teeth* triangular-subulate, spinescent. *Corolla* ½-⅔ in., red-purple; tube equalling or exceeding the calyx; lower lip variegated with white.—DISTRIB. Europe (Arctic), Siberia, N.W. Himalaya.

2. **S. palus'tris,** *L.* ; perennial, hairy, cauline leaves shortly petioled or sessile ovate- or oblong-lanceolate, whorls 8-10-fld.

River-banks and moist places, N. to Shetland; ascends to 1,500 ft. in Northumbd.; Ireland; Channel Islands; fl. July-Sept.—Habit of *S. syl-vat'ica*, but leaves much narrower, stem stouter hollow, hairs less coarse, odour less, flowers paler.—DISTRIB. Europe (Arctic), N. and W. Asia, Himalaya, N. America.

S. ambig'ua, SM., is a hybrid nearer *sylvat'ica* than *palus'tris*, with leaves shortly petioled, fruit never maturing. Cultivated ground, not uncommon. Other hybrids nearer to *palus'tris* are more common.

3. **S. german'ica,** *L.* ; biennial, shaggy with white silky hairs, cauline leaves shortly petioled ovate-oblong or lanceolate, spikes stout, whorls very dense-fld. lower remote.

Fields and roadsides on dry soil, very rare, Hants, Oxford, Kent; Channel Islands; (an alien or denizen, *Wats.*); fl. July–Aug.—*Rootstock* stoloniferous. *Stem* 1–3 ft., very stout, branched. *Leaves* coarsely crenate-serrate, often cordate, wrinkled under the matted hairs; radical 2–5 in., tufted, rather long-petioled. *Whorls* interrupted; upper bracts lanceolate; bracteoles subulate; pedicels very short. *Calyx* villous; upper lip longest; teeth triangular spinescent. *Corolla* ¾ in., pale rose-purple, pubescent, lower lip spotted.—DISTRIB. From Belgium southd., W. Asia, N.W. India.

4. **S. arven'sis,** *L.* ; annual, sparingly hairy, leaves ovate or oblong obtuse crenate, whorls 4–6-fld., corolla very small.

Fields and waste places, N. to Sutherland; rare in Scotland; Ireland; Channel Islands; fl. April–Nov.—Branched from the base; branches 6–18 in., weak, ascending, often rooting below. *Leaves* small, ¼–1 in., base cordate or cuneate, upper sessile; petiole very variable. *Whorls* ½–¾ in. diam.; bracts sessile, subacute; bracteoles 0; pedicels short. *Calyx-teeth* longer than the tube, lanceolate-subulate, spinous-tipped. *Corolla* equalling the calyx, ½ in., pale pink variegated with white.—DISTRIB. Europe, N. Africa, N. and W. Asia, N.W. India; introd. in N. America.

SECTION 2. **Beton'ica,** *L.* (gen.). *Whorls* in short dense terminal spikes ; lowest bracts leafy, upper minute. *Anthcr-cells* parallel.

5. **S. Beton'ica,** *Benth.* ; hairs deflexed, leaves petioled oblong-cordate obtuse deeply crenate, whorls all close or the lower only separate, calyx-lobes triangular spinescent. *Beton'ica officina'lis*, L. *Wood Betony.*

Copses, woods, roadsides, &c., from Perth southd.; rare in Scotland and Ireland; ascends to 1,200 ft. in Northumbd.; Channel Islands; fl. June–Aug.—*Rootstock* woody. *Stems* 6–24 in., ascending or erect, simple or sparingly branched from the base. *Leaves* 1–4 in., coriaceous, radical tufted on slender petioles, crenatures large rounded ; cauline few, much narrower, more toothed or serrate. *Whorls* in an oblong, obtuse, long-peduncled spike, 1–3 in.; bracteoles as long as the calyx, oblong-lanceolate, awned ; pedicels short. *Calyx-teeth* as long as the tube. *Corolla* ¾ in., red-purple, hairy, tube exserted; upper lip erect, lower 3-lobed.—DISTRIB. Europe, N. Africa, W. Siberia.—Formerly much used medicinally.

13. GALEOP'SIS, *L.* HEMP-NETTLE.

Annual herbs ; branches diverging. *Leaves* toothed. *Whorls* many-fld., dense, axillary and terminal ; bracts foliaceous ; flowers red orange or variegated, honeyed, proterandrous. *Calyx* subcampanulate, subregular,

5-nerved ; teeth 5, spinescent. *Corolla-tube* straight, naked inside, throat dilated ; upper lip ovate, arched ; lower 3-fid, mid-lobe obcordate or 2-fid ; palate with erect teeth at the union of the lobes. *Stamens* 4, exserted ; anther-cells transversely 2-valved, inner valve rounded ciliate, outer larger naked. *Nutlets* rounded, compressed, obscurely reticulate.— Distrib. Temp. Europe, W. Asia ; species 3.—Etym. γαλέη and ὄψις, from the resemblance of the corolla to a *weasel's* head.

1. **G. Lad'anum,** *L.* ; softly pubescent, nodes not thickened, calyx-teeth not exceeding the tube, corolla red.

Cornfields and waste places, from Elgin southd. ; ascends to 1,050 ft. in Derby ; rare in Scotland ; E. Ireland, local ; (a colonist, *Wats.*) ; fl. July–Oct.—*Stem* 6–18 in. ; branches many, ascending. *Leaves* petioled, 1–2 in., ovate-oblong or linear-lanceolate, acute, serrate. *Whorls* few- or many-fld. ; bracts sessile ; bracteoles linear, spinescent ; pedicels very short. *Calyx-teeth* subulate-lanceolate. *Corolla* ¾–1 in., hairy, rosy, lower lip mottled.—Distrib. Europe, N. and W. Asia ; introd. in N. America.

G. Lad'anum proper (*L. Herb.*) ; leaves ovate or ovate-lanceolate serrate throughout, whorls all separate, tube of corolla equalling the calyx. *G. intermedia,* Vill.—Denbigh, Moray.

Sub-sp. G. angustifo'lia, *Ehrh.* ; leaves narrower connate at the base interruptedly serrate or subentire, upper whorls approximate, tube of corolla much longer than the calyx. *G. canes'cens,* Schultz.

2. **G. du'bia,** *Leers ;* glandular-pubescent, nodes not thickened, calyx-teeth not exceeding the tube, corolla often yellow. *G. villo'sa,* Huds. ; *G. ochroleu'ca,* Lamk.

Sandy cornfields, rare, Durham, York, Lincoln, Carnarvon, Notts, Essex ; (a colonist, *Wats.*) ; fl. July–Aug.—Habit of *G. Lad'anum,* but glandular, leaves broader, more deeply serrate ; whorls dense, almost silky ; bracteoles much smaller ; calyx-teeth less spinescent ; corolla 1–1¼ in., pale yellow, rarely white or purple.—Distrib. From Denmark southd.

3. **G. Tet'rahit,** *L.* ; hispid, nodes thickened, calyx-teeth as long as or longer than the tube, corolla yellow and purple or white.

Cornfields and waste places, N. to Sutherland ; ascends to 1,300 ft. in Northumbd. ; Ireland ; Channel Islands ; (colonist ? *Wats.*) ; fl. July–Sept.—Usually much larger and stouter than the two preceding, with spreading and deflexed hairs. *Stem* ½–3 ft., succulent, nodes very hispid. *Leaves* 1–4 in., rather long-petioled, ovate or ovate-lanceolate, acute or acuminate, very coarsely serrate. *Whorls* dense-fld. ; bracts leafy ; bracteoles shorter than the calyx. *Calyx* ½–¾ in. ; teeth very long, straight, subulate. *Corolla* ¾–1¼ in.—Distrib. Europe (Arctic), N. and W. Asia, N.W. India ; introd. in N. America.

G. Tet'rahit proper ; corolla ¾ in., rosy or white, upper lip flattish longer than broad, lower flat, nutlets slightly convex above ventrally.—*G. bif'ida,* Bœnn., is a slender var. with the mid-lobe of the lower lip notched, margins at length reflexed.

Sub-sp. G. specio'sa, *Miller ;* corolla larger broader 1–1¼ in. yellow and purple, tube much exceeding the calyx, upper lip arched as broad as long, nutlets very convex above ventrally. *G. versic'olor,* Curt.

13*. *LEONU'RUS, L.* MOTHER-WORT.

Erect herbs. *Leaves* lobed. *Whorls* axillary, dense-fld., scattered ; bracteoles subulate : flowers small, pink or white. *Calyx* 5-nerved, turbinate, truncate, with 5 subspinescent spreading teeth. *Corolla-tube* naked, or with an oblique ring within ; upper lip entire, erect ; lower 3-fid, midlobe obcordate. *Stamens* 4; anthers conniving in pairs, cells transverse. *Style-lobes* subulate or obtuse. *Nutlets* smooth, 3-quetrous, truncate.— DISTRIB. Europe, Asia, America ; species 10.—ETYM. λέων and οὐρά, *lion's tail.*

L. CARDI'ACA, *L.* ; pubescent, leaves palmately lobed, bracts subtrifid, corolla-tube with a ring of hairs inside.

Hedges and waste places, England, Scotland, Ireland, Channel Islands, rare, not indigenous ; fl. July–Sept.—*Rootstock* stout. *Stem* 2–4 ft., stout, erect, angles prominent, very leafy. *Leaves* very close, radical with slender long petioles, ovate or orbicular-cordate, lobed and toothed; cauline 2–3 in., petioled, cuneate or obovate-oblong, lower multifid, upper 3-fid palmately 3–many-nerved, lobes acute. *Whorls* very many ; bracts large, leaf-like, petioled ; bracteoles small, subulate, pungent; flowers sessile. *Calyx* ⅓ in.; teeth broadly triangular. *Corolla* ½ in., woolly, pale rose, upper lip nearly straight. *Nutlets* villous-tipped.—DISTRIB. Europe, N. and W. Asia, Himalaya ; introd. in N. America.

14. LA'MIUM, *L.* DEADNETTLE.

Annual or perennial hairy herbs. *Whorls* many-fld., axillary, or in leafy bracteate heads ; bracteoles 0 or subulate ; flowers red purple white or yellow, honeyed, homogamous. *Calyx* tubular or subcampanulate ; teeth 5, equal, or the upper longer. *Corolla-tube* naked, or with a ring of hairs within, throat dilated ; upper lip arched ; lower 3-fid spreading, lateral lobes sometimes toothed at the base ; mid-lobe broad, base contracted. *Stamens* 4 ; anthers conniving in pairs, cells diverging. *Style-lobes* subulate. *Nutlets* 3-quetrous, truncate, smooth scaly or tubercled.— DISTRIB. Temp. Europe, Asia, N. Africa ; species 40.—ETYM. λαιμός, from the *throat*-like corolla.

SECTION 1. **Lamiop'sis.** Annual (the British sp.). *Corolla-tube* nearly straight, naked or with a ring of hairs, not constricted below the ring ; throat very wide. *Anthers* hairy. *Nutlets* with white scales.

1. **L. purpu'reum,** *L.* ; leaves petioled cordate crenate, whorls subterminal crowded, calyx slightly hairy, teeth spreading in fruit about as long as the tube.

Fields and waste places, N. to Shetland ; ascends to near 2,000 ft. in Northumbd. ; Ireland; Channel Islands; fl. April–Oct.—Silkily hairy, or subglabrous. *Stem* 6–18 in., decumbent below, branched from the base, often purplish. *Leaves* ½–2 in., obtuse, petiole as long or longer. *Bracts* crowded, bases not overlapping, upper subsessile, together forming a flat-topped head. *Calyx* about ⅓ in.; teeth triangular, tips spinous. *Corolla* ½–¾ in.,

purple, rarely white; lateral lobes generally 1-2-toothed at the base.
—DISTRIB. Europe, Canaries, N. and W. Asia; introd. in N. America.
L. PURPU'REUM proper; corolla-tube longer than the calyx-teeth, with a ring
of hairs within.—VAR. *decip'iens*, Sonder; has leaves and bracts deeply
crenate.
Sub-sp. L. HY'BRIDUM, *Vill.*; leaves more deeply crenate, floral with cuneate
bases, calyx-teeth erect; corolla-tube shorter, ring of hairs within obscure.
L. dissec'tum, With.; *L. inci'sum*, Willd.

2. **L. interme'dium,** *Fries ;* leaves petioled orbicular-cordate deeply
crenate, whorls crowded, calyx slightly hairy, teeth spreading in fruit
much longer than the tube.
Cultivated ground, Scotland, N. England; Ireland, rare; fl. June–Sept.—
Intermediate between *purpureum* and *L. amplexicaule*, of which it has the
habit, overlapping upper bracts, and foliage; stouter and more succulent
than either; the long calyx-teeth distinguish it from both. The tube of
the corolla slightly exceeds the calyx and has a very obscure ring of hairs;
lateral lobes toothed.—DISTRIB. N.W. Europe.

3. **L. amplexicau'le,** *L.* ; leaves petioled orbicular inciso-crenate,
whorls distant, calyx small densely pubescent, teeth converging in fruit
about equalling the tube, corolla-tube long slender. *Henbit Deadnettle.*
Waste sandy places, N. to Orkney; Ireland; Channel Islands; fl. May–Aug.
—Hairy or almost glabrous. *Stem* 4–10 in., branched from the base. *Leaves*
small, ¼–¾ in., lower long-petioled, almost lobulate, base rounded or cordate.
Bracts sessile, broader than long, many-lobed, bases overlapping. *Corolla*
sometimes imperfect, often ¾ in., very slender, rosy, pubescent.—DISTRIB.
Europe, N. Africa, N. and W. Asia, N.W. India; introd. in N. America.

SECTION 2. **Lamioty'pus.** Perennial. *Corolla-tube* curved, ascending,
with a ring of hairs inside, constricted below the ring. *Anthers* hairy.
Nutlets without scales.

4. **L. al'bum,** *L.* ; leaves all petioled cordate crenate or serrate, calyx-
teeth narrow straight longer than the straight tube, corolla white, tube
gibbous at the base below, ring of hairs oblique.
Fields and waste places, N. to Orkney; rare and local in Scotland and Ireland;
Channel Islands; fl. May–Dec.—More or less hairy. *Rootstock* creeping,
branched, stoloniferous. *Stem* 6–18 in., rooting and branched from the
base, then erect. *Leaves* 1–3 in., sometimes blotched with white, lower
long-petioled, subacute or acuminate, rarely deeply incised. *Whorls* 6–10-
fld., upper crowded, lower remote; bracts shortly petioled. *Calyx* glabrous
or hairy; teeth triangular-subulate, points long slender. *Corolla* ¾–1 in.,
throat gradually dilated; upper lip vaulted, villous; lateral lobes variable.
Stamens, sometimes outer, at others inner pair longest.—DISTRIB. Europe,
N. Africa, N. Asia; introd. in N. America.

L. MACULA'TUM, *L.* ; leaves all petioled cordate crenate or serrate, calyx-
teeth broad recurved as long as the oblique tube, corolla usually purple,
tube equal at the base below, ring of hairs transverse. *L. læviga'tum*,
Engl. Bot.

Waste places England, Scotland, not indigenous; fl. June–Sept.—Very closely
allied to *L. al'bum*, but the calyx and corolla are different; flowers fewer,
rarely white; leaves more wrinkled, and almost always with a median
white stripe, and corolla-throat suddenly dilated.—DISTRIB. Europe, N.
Africa, N. and W. Asia.

SECTION 3. **Galeob'dolon**, *Huds.* (gen.). Perennial. *Corolla-tube*
curved, ascending, with a ring of hairs inside, constricted below the ring;
upper lip stipitate. *Anthers* glabrous. *Nutlets* without scales.

5. **L. Galeob'dolon**, *Crantz ;* leaves petioled ovate acuminate doubly-
crenate or -serrate, calyx-teeth shorter than the oblique tube, corolla yellow,
ring of hairs oblique. *Galeob'dolon luteum*, Huds. *Yellow Archangel.*

Hedges and copses, from Cumberland southd., local; E. Ireland, local; fl.
May–June.—Hispid or subglabrous, hairs often reflexed. *Rootstock* short,
stoloniferous. *Stems* 6–18 in., flowering erect; barren, elongate, prostrate.
Leaves 1–2 in., petioles variable. *Whorls* remote, 6–10-fld.; upper bracts
sometimes lanceolate. *Calyx-teeth* mucronate. *Corolla* ¾–1 in., tube short,
gibbous at the base below; lips long, lower spotted red-brown.—DISTRIB.
Europe, W. Siberia.

15. **BALLO'TA**, *L.* BLACK HOREHOUND.

Perennial, hairy or woolly herbs or undershrubs. *Whorls* axillary,
dense-fld. ; bracts subulate ; flowers small, proterandrous. *Calyx* tubular
or funnel-shaped, 10-nerved ; teeth 5–10, dilated at the base or connate
into a spreading limb. *Corolla-tube* with a ring of hairs inside ; upper
lip erect, concave ; lower as long, 3-lobed, spreading. *Stamens* 4, ascending under the upper lip ; anthers conniving in pairs, cells at
length diverging. *Style-lobes* subulate. *Nutlets* obtuse.—DISTRIB.
Europe, N. and S. Africa, temp. Asia ; species 25.—ETYM. The Greek
name.

B. ni'gra, *L.* ; erect, hairy, calyx-teeth exceeding the corolla-tube.

Hedgebanks, &c., from the Forth and Clyde southd.; rare and seldom indigenous in Scotland and Ireland; Channel Islands; fl. July–Aug.—Dull green,
hoary or woolly, foetid. *Rootstock* stout, short. *Stem* 2–3 ft., stout, erect,
much branched, hairs usually reflexed. *Leaves* 1–2 in., petioled, ovate- or
orbicular-cordate, crenate or almost lobulate. *Whorls* many ; cymes
peduncled, 3–6-fld.; bracts leaf-like, bracteoles small ; flowers sessile.
Calyx ½ in., slightly enlarged in fruit; tube cylindric, strongly ribbed ; limb
short, expanded ; teeth 5, very variable, nerved, spinescent. *Corolla* ½–¾ in.,
pale red-purple; upper lip hairy outside and in; mid-lobe of lower obcordate. *Nutlets* obtusely 3-gonous, brown, smooth, shining.—DISTRIB.
Europe, N. Africa, W. Asia ; introd. in N. America.
B. ni'gra proper (*B. ruderd'lis*, Sw.); calyx-teeth ovate-lanceolate, tips
ascending. Northumbd., Oxford, Hereford.—VAR. *B. al'ba*, L. (*B. fœ'tida*,
Lamk.); stouter, calyx-teeth deltoid, tips spinous spreading or reflexed.

16. TEU'CRIUM, *L.* GERMANDER.

Herbs. *Whorls* with leafy bracts or in leafy unilateral racemes or spikes ; flowers proterandrous, honeyed. *Calyx* tubular or campanulate ; teeth 5, equal or the upper broader and reflexed. *Corolla-tube* short, naked within ; limb obliquely 5-lobed ; 2 upper lobes very small, 2 lateral larger, lowest largest, rounded or oblong, often concave. *Stamens* 4, 2 lower longest, protruded between the upper corolla-lobes ; anther-cells confluent. *Style-lobes* subequal. *Nutlets* subglobose, smooth reticulate or pitted, base obliquely truncate.—DISTRIB. Temp. and warm regions ; species 100.—ETYM. The ancient name.

SECTION 1. **Scorodo'nia,** *Mœnch* (gen.). *Whorls* 2-fld., in terminal branched 1-sided racemes. *Upper lip* of calyx much dilated.

1. **T. Scorodo'nia,** *L.* ; leaves all shortly petioled ovate-cordate crenate, bracts short, calyx gibbous at the base. *Wood Sage.*

Stony copses, heaths, hedges, and soils, N. to Orkney ; ascends to 1,500 ft. in Northumbd.; Ireland; Channel Islands; fl. July–Sept.—Perennial, finely pubescent or hairy. *Rootstock* woody, stoloniferous. *Stems* 8–24 in., tufted, usually ascending, rigid. *Leaves* 1–1½ in., in distant pairs, rarely laciniate. *Racemes* 3–6 in., branched at the base ; bracts petioled, green ; bracteoles 0 ; pedicels short. *Calyx* broadly campanulate, reticulate in fruit; lobes cuspidate, 4 lower small incurved subulate. *Corolla* ½ in., ochreous. *Stamens* purplish, 2 longer deflexed after dehiscence. *Nutlets* subglobose, smooth.— DISTRIB. Europe, except Russia, N. Africa.—Bitter, aromatic, tonic ; a substitute for hops.

SECTION 2. **Scor'dium,** *Benth.* *Whorls* 2-6-fld. ; bracts leaf-like. *Upper lip* of calyx equal to or rather larger than the lower.

2. **T. Scor'dium,** *L.* ; leaves sessile oblong coarsely serrate, calyx nearly equal at the base, teeth nearly equal. *Water Germander.*

Wet meadows, chiefly in the E. counties, very rare ; York, Lincoln to Suffolk, Berks, Northampton, Cambridge, Devon ; Ireland, very rare; Channel Islands; fl. July–Aug.—Perennial, fœtid, hairy woolly and glandular. *Rootstock* creeping, stoloniferous. *Stem* 4–10 in., branched from the base, erect or prostrate, leafy. *Leaves* ½–1½ in., base narrowed rounded or cordate. *Bracts* leaf-like, sometimes auricled at the base (*T. scordioi'des,* Bab., not Schreber); bracteoles 0 ; pedicels short. *Calyx-teeth* straight, short, triangular. *Corolla* ½ in., rose-purple ; lower lip spotted. *Nutlets* wrinkled. —DISTRIB. Europe, N. Africa, N. and W. Asia, N.W. Himalaya.

3. **T. Bo'trys,** *L.* ; leaves all petioled triangular-ovate pinnatifid, calyx saccate at the base, teeth equal.

Chalky fields, Surrey, very rare; (a colonist? *Baker*); fl. Aug.—Annual, pubescent and glandular with long hairs. *Stem* 4–8 in., erect, much branched from the base, leafy. *Leaves* 1–1 in., segments 3–5 pair, linear, obtuse, lower again lobed; nerves prominent beneath. *Bracts* leaf-like. *Calyx* large, glandular, inflated and reticulate in fruit; lobes triangular, mucronate. *Corolla* ⅔ in., rose-purple, lower lip spotted white and red, *Nutlets* deeply pitted.—DISTRIB. From Belgium southd. (excl. Greece), N. Africa.

SECTION 3. **Chamæ'drys,** *Benth. Whorls* 2–6-fld., all or the upper only in lax terminal racemes. *Calyx-teeth* subequal.

T. CHAMÆ'DRYS, *L.* ; leaves petioled ovate incised-crenate, bracts sessile leaf-like, calyx nearly equal at the base, teeth nearly equal.

Old walls in England, Scotland, and sandy fields in Ireland, rare; a garden escape; fl. July–Sept.— Perennial, almost hispidly hairy. *Rootstock* creeping, stoloniferous. *Stem* 6–18 in., ascending, much branched, leafy. *Leaves* ½–1½ in., gradually narrowed into the petiole, nerves prominent beneath. *Whorls* about 6-fld., in the axils of leafy bracts, or subsecund in leafy terminal spikes; bracts exceeding the calyx, quite entire, acuminate, often purple, upper smaller; pedicels very short. *Calyx-teeth* straight, triangular, spinescent. *Corolla* ¾ in., rosy, lower lip spotted white and red. *Nutlets* nearly smooth. –DISTRIB. From Holland southd.; W. Asia.—A reputed tonic and famous old gout medicine.

17. A'JUGA, *L.* BUGLE.

Annual or perennial herbs. *Whorls* few- or many-fld. ; bracts leaf-like ; flowers blue purplish or yellow, proterandrous. *Calyx* subcampanulate, 5-fid or 5-toothed. *Corolla-tube* usually with a ring of hairs within, straight or twisted ; upper lip short, notched ; lower longer, spreading, 3-fid, lateral lobes oblong, middle broader notched or 2-fid. *Stamens* 4, ascending, protruded beyond the upper lip, 2 lower longer ; anther-cells diverging, at length confluent. *Style-lobes* subequal. *Nutlets* reticulate or rugose.—DISTRIB. Temp. regions of the Old World, from Europe to Australia ; species 30.—ETYM. doubtful.

1. **A. rep'tans,** *L.* ; almost glabrous, stoloniferous, leaves repand-crenate, whorls in a loose spike with spreading bracts, flowers blue.

Copses, woods, and pastures, N. to Shetland; ascends to 2,000 ft. in the Highlands; Ireland ; Channel Islands; fl. May–July.—Perennial, subglabrous or pilose, lines of hairs bifarious on the stem. *Rootstock* short, stout; stolons slender, leafy, tips ascending. *Flowering-stem* 6–12 in. *Leaves,* radical 2–3 in., long-petioled, narrowly obovate, obtuse ; those on the stolons small, obovate-spathulate; cauline few, sessile, oblong, obtuse. *Spike* 3–8 in. ; bracts subentire, obtuse, much shorter than the upper flowers, upper often purplish ; whorls 6–10-fld.; pedicels very short. *Calyx* small, teeth triangular acute ciliate. *Corolla* ½–¾ in., rarely white or rosy, mid-lobe of lower lip broadly obcordate.—DISTRIB. Europe.—A form without stolons (var. *pseudo-alpi'na*) has been mistaken for *A. alpina,* an exotic species.

2. **A. pyramida'lis,** *L.* ; pilose with soft-jointed hairs, leaves obscurely crenate, whorls in a compact ·pyramidal spike, upper bracts appressed, flowers blue. *A. geneven'sis,* L. var. Benth.

Mountain woods and streams, very rare, Westmoreland, Argyll to Orkney and Hebrides; W. Ireland; fl. May–July.—Similar to *A. rep'tans,* but stolons produced late only; radical leaves shortly pet:oled; calyx woolly with longer teeth; lower corolla-lip smaller, mid-lobe less cordate.—DISTRIB. Europe (excl. Greece).

3. **A. Chamæ'pitys,** *Schreber ;* annual, villous, cauline leaves 3-partite, flowers solitary in the axils of leaf-like bracts yellow. *Ground Pine.*

Chalky fields, local, Bedford, Herts, Cambridge, Essex, Surrey, Kent, and Hants; (a colonist, *Wats*); fl. May–Sept. *Hairs* long, scattered. *Stem* 3–6 in.; branches ascending from the root, densely leafy. *Leaves,* radical early withering, petioled, ovate-lanceolate, entire or toothed; cauline 1–1½ in., spreading, lobes narrow-linear. *Whorls* many, 2-fld.; bracts many times longer than the flowers. *Calyx* hispid; teeth narrowly triangular. *Corolla* ½ in.; lower lip spotted with red. *Nutlets* large, oblong, deeply pitted.— DISTRIB. From Belgium southd., N. Africa, W. Asia.

ORDER LX. ILLECEBRA'CEÆ,

Annual or perennial generally small, often tufted herbs. *Leaves* oppo-site or alternate, quite entire (serrulate and exstipulate in *Scleran'thus*); stipules scarious. *Flowers* very small, cymose, 2-sexual. *Sepals* 4–5, distinct or connate, persistent, closing over the fruit. *Petals* small or 0. *Stamens* hypogynous or perigynous, filaments short distinct or connate anthers small. *Disk* 0 or annular. *Ovary* free, ovoid, 1-celled; style 2-rarely 3-fid, stigmas decurrent; ovule 1, erect, or pendulous from a basal funicle. *Utricle* enclosed in the perianth, 1-seeded. *Seed* globose reniform or lenticular, ·testa smooth, hilum ventral or lateral, albumen floury; embryo straight curved or annular.—DISTRIB. Chiefly warm and dry regions; genera 17; species 60.—AFFINITIES. Close to *Caryophylleæ* and *Amaranthaceæ.*—PROPERTIES unimportant.

Leaves alternate. Petals 5. Stigmas 3...........................1. Corrigiola.
Leaves when opposite not connate. Sepals green, obtuse........2. Herniaria.
Leaves opposite. Sepals white, concave, with long points......3. Illecebrum.
Leaves opposite, connate at the base. Petals 0..................4. Scleranthus.

1. CORRIGI'OLA, *L.* STRAPWORT.

Annual or perennial prostrate glabrous herbs. *Leaves* alternate, linear or oblong; stipules scarious. *Flowers* minute. *Sepals* 5, connate at the base, obtuse; margins membranous. *Petals* 5, small, white. *Stamens* 5, perigynous. *Style* short, 3-partite; ovule suspended from a basal funicle. *Fruit* crustaceous, 3-gonous, dotted or rugose, testa membranous; embryo annular.—DISTRIB. Europe, Africa, temp. America; species 5.—ETYM. Corrigiola, *a little strap.*

C. littora'lis, *L.* ; leaves linear-lanceolate, stipules ½-sagittate.
Sandy places, very rare, Cornwall, Devon; fl. July–Sept.—Annual. *Stems* many from the root, 4–8 in., slender, prostrate or ascending. *Leaves* ⅓–½

in., narrowed into an obscure petiole; stipules small. *Flowers* in crowded terminal cymes. *Petals* as long as the sepals.—DISTRIB. From Denmark southd.

2. HERNIA'RIA, *L.* RUPTURE-WORT.

Annual or perennial prostrate herbs. *Leaves* opposite and alternate, narrow. *Flowers* minute, green, crowded, axillary, 1–2-sexual. *Sepals* 4–5, connate at the base, obtuse, equal or unequal. *Petals* 4–5, setaceous, minute or 0. *Stamens* 3–5, inserted on an annular disk. *Style* 2-fid or -partite; ovule erect. *Utricle* indehiscent. *Seed* subglobose or reniform, testa crustaceous shining; embryo annular.—DISTRIB. Europe, N. and S. Africa, W. Asia; species 8 or 10.—ETYM. The classical name.

1. **H. gla'bra,** *L.* ; leaves oblong glabrous or ciliate.

Sandy soils, rare, Lincoln, Norfolk, Suffolk, Cambridge; Channel Islands; fl. July–Aug.—*Root* woody, often perennial. *Stems* many, 4–6 in., tufted, glabrous or slightly pubescent. *Leaves* ⅛–¼ in.—DISTRIB. Europe, N. and W. Asia.—Probably a var. of *H. hirsu'ta,* L., a more southern plant. VAR. *H. cilia'ta,* Bab.; perennial, stouter, forming larger tufts, leaves broader, stipules larger whiter.—Lizard Point; Guernsey.—Syme remarks that this keeps its green colour during the winter of Middlesex, which *H. gla'bra* proper does not.

2. **H. hirsu'ta,** *L.* ; leaves elliptic-oblong hirsute.

Sandy soil, Christchurch, Hants; fl. July–Aug.—Very near *H. gla'bra,* but hirsute all over.—DISTRIB. From Belgium southd., and eastd. to W. Asia.

3. ILLECE'BRUM, *L.*

A small diffuse glabrous annual herb. *Leaves* opposite. *Flowers* minute, white, crowded in all the leaf-axils, 1–2-sexual. *Sepals* 5, white, corky, compressed laterally, keeled, tips awned. *Petals* 5, setaceous, very minute. *Stamens* 5, hypogynous. *Style* very short, stigmas 2 capitate; ovule erect. *Utricle* fissured at the base, included in the hardened calyx. *Seed* oblong; embryo curved, lateral.—DISTRIB. W. Europe from Denmark southd., N. Africa.—ETYM. doubtful.

I. **verticilla'tum,** *L.* ; leaves ovate-oblong or spathulate.

Moist sandy places, Devon, Cornwall; Channel Islands; fl. July–Aug.— Branched from the root, very slender, prostrate; branches 4–8 in., ascending, covered throughout with leaves and tufts of white flowers. *Leaves* ⅛–¼ in. *Flowers* subsessile, shorter than the leaves. *Sepals* opaque.

4. SCLERAN'THUS, *L.* KNAWEL.

Low, tufted, annual or perennial herbs. *Leaves* opposite, connate, subulate, pungent, often serrulate; stipules 0. *Flowers* minute, green, in axillary and terminal cymes or fascicles, honeyed. *Calyx-tube* funnel-shaped or urceolate, hardening over the fruit; lobes 4–5, short, erect. *Petals* 0. *Stamens* 1, 2, 5, or 10, inserted on the calyx-mouth. *Styles*

2, filiform, stigmas capitate ; ovule pendulous from a filiform basal funicle. *Utricle* indehiscent. *Seed* lenticular, testa smooth ; embryo annular.— DISTRIB. Europe, E. Asia, Africa, Australia, N. Zealand ; species 10.— ETYM. σκληρός, from the *indurated* perianth.

1. **S. an'nuus,** *L.* ; calyx-lobes suberect in fruit acute with narrow membranous margins.

Fields and waste places, N. to Caithness ; ascends above 1,000 ft. in the High-lands ; Ireland ; Channel Islands ; fl. June–Sept.—Annual, rarely biennial (*S. bien'nis,* Reut.). *Stem* 2–8 in., slender, green, sometimes puberulous. *Leaves* ¼–½ in., recurved, base often ciliate. *Flowers* solitary in the lower axils, and fascicled in terminal dichotomous cymes. *Calyx-tube* 10-grooved in fruit.—DISTRIB. Europe to the Caucasus, N. Africa, N. and W. Asia ; introd. in U. States.

2. **S. peren'nis,** *L.* ; calyx-lobes incurved obtuse with broad scarious margins.

Sandy fields, from Radnor, Warwick, and Norfolk to Cornwall and Dorset; fl. June–Aug —Very similar to *S. an'nuus,* but more glaucous, with shorter bracts and pubescent calyx-tube.—DISTRIB. Europe to the Caucasus Siberia.

ORDER LXI. **CHENOPODIA'CEÆ**

Herbs or shrubs. *Leaves* simple, alternate, exstipulate. *Flowers* 1-2-sexual, small, regular, often dimorphic ; bracts 1–3 or 0. *Calyx* inferior, of 3–5 free or connate sepals, imbricate in bud. *Petals* 0. *Stamens* usually 5, opposite the sepals, perigynous or hypogynous ; anthers 2-celled. *Ovary* ovoid globose or depressed, 1-celled ; stigmas 2–4 ; ovule solitary, basal or lateral, campylotropous. *Fruit* usually a utricle, enclosed in the often enlarged or fleshy calyx. *Seed* horizontal or vertical, testa crustaceous, inner coat membranous, albumen floury fleshy or 0 ; embryo curved annular or spiral.—DISTRIB. All climates ; genera 70 ; species about 450.—AFFINITIES. With *Amaranthaceæ, Caryophyllæ,* and *Illecebreæ.*—PROPERTIES. Chiefly known as pot-herbs.

A. **CYCLOLOBEÆ.** Embryo annular.

TRIBE I. **EUCHENOPODIEÆ.** *Stem* leafy. *Flowers* bisexual, or if uni-sexual perianths of males and females similar

Flowers 2-sexual. Utricle membranous ....1. Chenopodium.
Flowers 2-sexual. Utricle striate and hard above ....2. Beta.

TRIBE II. **ATRIPLICEÆ.** *Stem* leafy. *Flowers* 1-sexual, male 3–5- female 2-sepalous..3. Atriplex.

TRIBE III. **SALICORNIEÆ.** *Stem* leafless, jointed. *Flowers* 2-sexual. *Albumen* scanty ; embryo conduplicate ....4. Salicornia.

B. SPIROLOBEÆ. Embryo spiral.

TRIBE IV. **SUÆDEÆ.** *Stem* leafy. *Sepals* 4–5, not winged at the
    back ....5. Suæda.

TRIBE V. **SALSOLEÆ.** *Stem* leafy. *Sepals* 4–5, transversely winged in
    fruit ....6. Salsola.

1. CHENOPO'DIUM, *Tournef.* GOOSE-FOOT.

Erect or prostrate, very variable herbs, usually littoral or on made soil.
Stem angled, often striped white or red and green. *Leaves* entire lobed or
toothed. *Flowers* minute, 2-sexual, ebracteolate; clusters axillary, or
in simple or panicled cymes. *Sepals* 3–5. *Stamens* 2–5, perigynous,
filaments subulate. *Disk* 0. *Ovary* free, depressed or compressed; styles
2–3. *Utricle* membranous, often enclosed by the calyx. *Seed* horizontal
or vertical, testa crustaceous, albumen floury; embryo annular.—DISTRIB.
All climates; species about 50.—ETYM. χῆν and ποὺς, *goose-foot.*

SECTION 1. Annual. *Flowers* 5-merous. *Styles* short. *Seeds* horizontal.

* Leaves quite entire.

1. **C. Vulva'ria,** *L.* ; mealy, diffuse, leaves deltoid-ovate, sepals not
keeled covering the utricle. *C. ol'idum,* Curtis.

Roadsides and waste places, Edinburgh, and from Northumbd. southd.; S. and
    E. Ireland (? extinct); Channel Islands; fl. Aug.–Oct.—Annual, fœtid.
    Branches 6–18 in., opposite, divaricate. *Leaves* ½–1 in., greasy to the touch,
    acute, grey-green; petiole as long or shorter. *Spikes* small, ¼–½ in., dense,
    axillary and terminal. *Seeds* black, punctulate.—DISTRIB. From Denmark
    southd., N. Africa.—Odour of stale salt fish.

2. **C. polysper'mum,** *L.* ; glabrous, erect or ascending, leaves ovate,
sepals not keeled shorter than the utricle. *C. acutifo'lium,* Sm.

Cultivated ground, manure-heaps, &c., from Berwick southd.; Channel
    Islands; fl. Aug.–Oct.—Annual. *Stem* 6–18 in.; branches many, spreading,
    leafy. *Leaves* ½–1½ in., shortly petioled, membranous, obtuse or acute.
    Cymes ½–1 in., axillary and terminal, simple or panicled; branches very
    slender, spreading; flowers very minute. *Seeds* minute, dark brown, rough.
    —DISTRIB. Europe, N. Asia; introd. in N. America.

** Leaves more or less toothed or lobed.

3. **C. al'bum,** *L.* ; erect, mealy, leaves rhombic or deltoid-ovate subentire or irregularly toothed, upper oblong entire, sepals keeled covering
the utricle narrowly scarious or all green, seed smooth keeled.

Waste, especially cultivated ground, N. to Shetland; ascends to 1,000 ft. in
    Yorkshire; Ireland; Channel Islands; fl. July–Sept.—Very variable. *Stem*
    1–3 ft.; branches erect or ascending. *Leaves* 1–3 in., tip acute obtuse or
    rounded, base cuneate often 3-nerved; petiole usually long and slender.
    Spikes terminal and axillary, simple or panicled, leafy below. *Seed* almost
    black, hardly dotted.—DISTRIB. Europe (Arctic), temp. Asia; introd. in
    N. America.

C. album proper (*C. can'dicans*, Lamk.); leaves usually much toothed mealy, spikes simple shorter than the leaves in a slender terminal panicle, calyx very mealy.—VAR. *C. vir'ide*, L.; leaves subentire sparingly or hardly mealy, spikes lax axillary subcorymbose, branches recurved, calyx almost glabrous.—VAR. *C. paga'num*, Reichb. (*vi'rens*, Lond. Cat.); lower leaves obtusely serrate glabrous or sparingly mealy below only, spikes lax erect axillary simple or in terminal panicles, calyx sparingly mealy.

4. **C. ficifo'lium,** *Sm.* ; erect, mealy, flaccid, leaves oblong-hastate sinuate-toothed upper entire, basal lobes ascending, sepals covering the utricle, seed dotted not keeled.

Waste places, usually in rich soil, chiefly in the E. and S. of England, York to Kent and Sussex; Ireland rare; Channel Islands; fl. Aug–Sept.—Very near *C. al'bum*, but more flaccid, flowers later, with more oblong-hastate leaves whose basal lobes spread more, and above which the leaf is often contracted, inflorescence almost leafless, seeds smaller.—DISTRIB. Europe.

5. **C. ur'bicum,** *L.* ; erect, sparingly mealy, leaves triangular acute deeply toothed or subentire, spikes erect leafless simple axillary or in a terminal erect panicle, sepals not keeled nor covering the utricle broadly scarious, seed large punctulate not keeled.

Waste places, indigenous from York southd. only; Ireland; fl. Sept.–Oct.—*Stem* 6–36 in., stout, sparingly branched. *Leaves* 1–3 in., often as broad as long; petiole usually shorter, winged above; lateral nerves spreading. *Spike* ¾–2 in., rather dense-fld.—DISTRIB. Europe, Canaries, N. and W. Asia introd. in N. America.

C. ur'bicum proper (*C. deltoi'deum*, Lamk.); leaves deltoid shortly toothed or subentire, spikes erect longer than the leaves, panicle leafless above.—VAR. *C. interme'dium*, Mert. and Koch ; leaves rhombic-triangular deeply sinuate-toothed, spikes shorter than the leaves, panicle leafy almost to the top.

6. **C. hy'bridum,** *L.*; erect, almost glabrous, leaves large long-acuminate with 2–4 broad lobes on each side, spikes in lax axillary almost leafless corymbs, sepals obtusely keeled not covering the utricle broadly scarious, seed large opaque coarsely pitted not keeled.

Fields and waste places, from Lancashire and Norfolk to Somerset and Kent, local ; (a native ? *Wats.*); fl. Aug.–Sept.—Odour heavy. *Stem* 1–3 ft., stout, branched. *Leaves* 3–5 in., almost shining, broadly ovate, pale green, membranous, 3–5-nerved near the usually cordate base. *Clusters* of flowers rather large.—DISTRIB. Europe, N. Africa, N. and W. Asia, N.W. India; introd. in N. America.

7. **C. mura'le,** *L.* ; nearly glabrous, leaves bright green rhombic- or deltoid-ovate acute entire at the cuneate base, upper narrower serrate, spikes short densely panicled, sepals slightly keeled almost covering the utricle narrowly scarious, seed sharply keeled.

Waste places, near houses, from Northumbd. southd., rare in England; Ireland, very rare ; Channel Islands; fl. Aug.–Sept.—Rather fœtid. *Stem* 6–18 in. erect or ascending; branches decumbent. *Leaves* ¾–3 in. broad, rather shining, teeth sharp ; petiole shorter than the blade. *Spikes* ½–¾ in. *Seed* rather opaque, dotted.—DISTRIB. Europe, N. Africa, W. Asia, N.W. India ; introd. in N. America.

SECTION 2. Annual. *Lateral flowers* of each cluster usually 2–4-merous, seed vertical ; terminal 5-merous, seed horizontal or vertical. *Styles* short.

8. **C. ru'brum,** *L.* ; glabrous, shining, leaves deltoid or rhombic-ovate, spikes leafy panicled, sepals not keeled covering the utricle narrowly scarious, seed mostly vertical minute brown shining obscurely keeled.

Waste places, ditches, salt marshes, &c., from Aberdeen and Clyde southd. ; Ireland, very rare ; Channel Islands ; fl. Aug.–Sept.—*Stem* 1–3 ft., erect or ascending. *Leaves* excessively variable, entire irregularly toothed or serrate, obtuse or acute, base 3-nerved. *Spikes* very short in terminal and axillary panicles.—DISTRIB. Europe, N. and W. Asia.
C. ru'brum proper ; leaves sinuate-serrate not fleshy, panicle leafy to the top, spikes short compact dense-fld. often almost capitate.—VAR. *pseudo-botryo'des,* Wats., is smaller, often reddish, stem slender prostrate, leaves rhomboid almost entire, panicles much reduced.—VAR. *C. botryo'des,* Sm.; leaves subentire more triangular fleshy, panicle leafless above.

9. **C. glau'cum,** *L.* ; prostrate, leaves mealy beneath oblong or ovate-oblong sinuate-lobed, spikes short dense leafless, sepals keeled nearly covering the utricle narrowly scarious, seed acutely keeled.

Waste ground, sporadic and very scarce, Fife to Hants ; indigenous only in S. England ; (native? *Wats.*) ; fl. Aug.–Sept.—*Stem* 6–18 in., usually spreading, widely branched, shining, glabrous. *Leaves* ½–1 in., obtuse or rounded, base cuneate. *Spikes* ¼–1 in., simple or compound, terminal and axillary. *Seeds* variable, very small, the horizontal largest.—DISTRIB. Europe, N. and W. Asia, Himalaya, N. America, S. Chili, Australasia.

SECTION 3. Perennial, glabrous, or nearly so. *Flowers* all 5-merous, or lateral 2–3-androus. *Seeds* nearly all vertical. *Styles* very long.

10. **C. Bo'nus-Henri'cus,** *L.* ; leaves triangular-hastate subacute, spikes mostly in a compound leafless panicle, sepals not keeled toothed at the tip broadly scarious, seed large tumid black not keeled. *All-good.*

Waste places, often near houses, from Caithness southd.; ascends to 1,200 ft. in N. England ; common in Ireland ; Channel Islands ; (a native? *Wats.*) ; fl. May–Aug.—*Rootstock* stout, fleshy, branched. *Stem* stout, erect or ascending, 1–3 ft., papillose. *Leaves* 2–4 in., succulent, papillose beneath, variable in shape, entire or sinuate-toothed, petiole of lower long, basal lobes often large acute and spreading. *Spikes* 1–2 in., dense-fld. *Sepals* shorter than the utricle. *Stamens* 2–5, rarely 0. *Seed* punctulate.—DISTRIB. Europe, Siberia ; introd. in N. America.—Cultivated as a pot-herb in Lincolnshire, and called " Mercury."

2. BE'TA, *L.* BEET.

Herbs. *Leaves* almost entire. *Flowers* 2-sexual, in axillary spiked or cymose fascicles, cohering in fruit by the enlarged hardened bases of the sepals. *Calyx* urceolate. *Stamens* 5, perigynous, filaments subulate. *Disk* fleshy, annular. *Ovary* sunk in the disk, depressed ; style short, stigmas 2–4 subulate. *Fruit* adnate to the disk and calyx-base. *Seed* horizontal, testa thin, albumen floury ; embryo annular.—DISTRIB. N. temp. Europe and Asia ; species about 9.—ETYM. Uncertain.

B. marit'ima, *L.* ; decumbent, clusters of flowers spiked.

Muddy, &c., sea-shores, from Fife and Argyll southd.; Ireland; Channel Islands ; fl. June–Oct.—Glabrous, perennial. *Rootstock* branched, tapering into a fleshy root. *Stems* many, 1–2 ft., branched, angular, striped, tips ascending. *Leaves* 2–4 in., fleshy, shining, lower rhomboid-ovoid, acute, upper lanceolate ; petiole broad. *Spikes* 3–6 in., slender, panicled, clusters 2–3-fld., sessile; bracts linear-lanceolate, acute, lower ½–1 in. ; flowers ⅛ in diam., green. *Sepals* incurved, obtuse, keel entire, edges scarious.— DISTRIB. From Denmark southd., N. Africa, W. Asia, India.—Probably the origin of the Beet and Mangold Wurzel. An excellent spinach.

3. A'TRIPLEX, *Tournf.* ORACHE.

Herbs or shrubs, mealy or scaly. *Leaves* alternate or opposite, petioled, often hastate, entire or sinuate-toothed. *Flowers* small, 1-sexual, ebracteate ; clusters usually in branched cymes.—MALE. *Sepals* 3–5. *Stamens* 3–5, hypogynous, filaments filiform. *Ovary* rudimentary.— FEMALE. *Sepals* 2, free or connate. *Stamens* 0. *Styles* 2, filiform, connate at the base. *Utricle* compressed, enclosed in the enlarged calyx. *Seed* compressed, vertical or horizontal, albumen floury ; embryo annular. —DISTRIB. Shores and waste places ; species 100.—ETYM. The old Latin name.

SUB-GEN. 1. **A'triplex** proper. Annuals. *Flowers* monœcious ; sepals of female united below. *Pericarp* not adherent to the sepals. *Testa* crustaceous. *Radicle* basal or sublateral.

1. A. pat'ula, *L.* ; mealy, stem erect or ascending striped, branches spreading, leaves deltoid hastate or rhombic, floral usually broad, fem. sepals rhombic or deltoid usually toothed and tubercled or rarely hardened.

Waste places, manure-heaps, &c., N. to Shetland; Ireland ; Channel Islands; fl. June–Oct.—Very variable, 6 in.–3 ft., rarely prostrate, less mealy than *A. lacinia'ta. Female fl.* mostly mixed with males, but axillary ones occur separately. *Fruiting sepals* usually ⅛–⅙ in. diam., except in sub-sp. *Babingto'nii*, excessively variable in form, length of free portion, toothing and sculpture ; base hastate truncate or deltoid. *Seeds* of two forms, largest ¹⁄₁₀–⅙ in. diam., dark brown, much compressed; smallest ¹⁄₁₈–¹⁄₁₂, smooth, shining.—DISTRIB. Most cool (Arctic) parts of the globe, native or naturalized.

A. PAT'ULA proper ; erect or ascending, deep green, sparingly mealy, lower leaves opposite rhombic or rhombic-hastate with ascending cusps acute entire or serrate, spikes dense simple leafy below, sepals united at the base only deltoid entire or serrate (var. *erec'ta,* E. B.), smooth or muricate, seeds all vertical. *A. erec'ta,* Huds.—Common, extending to India ; ascends to

z 2

1,000 ft. in N. England.—*A. angustifo'lia*, Sm. ; is a weak procumbent state with branches divaricate, leaves subentire, spikes long lax panicled, sepals usually smooth.

Sub-sp. A. HASTA'TA, *L.* ; erect or decumbent, dark green, mealy, lower leaves opposite hastate-deltoid with horizontal cusps subacute entire or toothed, upper lanceolate, spikes simple or panicled interrupted leafy at the base, sepals deltoid united at the base only, seeds dimorphic, larger brown rough, smaller black smooth. *A. pat'ula*, Sm.; *A. Smith'ii*, Syme. Common, extending to India; ascends to 1,300 ft. in N. England.—*A. triangularis*, Willd. (*A. prostra'ta* and *A. deltoide'a*, Bab.), is a var. with upper leaves hastate, spikes dense, terminal of the panicle short, sepals truncate but little longer than the utricle, seeds mostly small.

Sub-sp. A. BABINGTO'NII, *Woods;* usually pale and very mealy, branches spreading ascending, leaves mostly opposite deltoid or rhombic-ovate entire or sinuate-toothed, upper usually similar, clusters of flowers remote, spikes simple lax leafy, sepals connate at the often hardened base or united nearly to the middle, seeds all vertical large pale rather rough. *A. ro'sea*, Bab., not L. Sea-shores, abundant.—Very variable; as green as sub-sp. *hasta'ta*, or almost as white as *A. lacinia'ta*, from which the striped stems distinguish it.—N. to Shetland; Ireland ; Channel Islands.

2. **A. littora'lis**, *L.*; mealy, stem erect striped, branches ascending, leaves linear- or elliptic-oblong usually quite entire upper very narrow, fem. sepals rhombic or deltoid toothed tubercled not hardened.

Salt and brackish marshes, banks, &c., from Perth to Dorset and Kent (excl. W. Scotland); Ireland ; Channel Islands; fl. July–Sept.—Best distinguished from *A. pat'ula* by the narrower usually quite entire leaves, which are never hastate and hardly ever rhombic. *Clusters of flowers* in slender terminal spikes. *Seeds* nearly smooth, shining, all vertical.—DISTRIB. Of *A patula.*

A. mari'na, L. (*A. serra'ta*, Huds.), is a var. with leaves serrate or lobed, tips of fruiting sepals appressed.

3. **A. lacinia'ta**, *L.* ; clothed with persistent silvery scales, stem not striped reddish, lower leaves opposite rhombic-ovate, upper similar or hastate, floral sessile, fem. sepals connate at the swollen hardened base. *A. arena'ria*, Woods, not Nuttall ; *A. farino'sa*, Dumort.

Sandy sea-coasts, from Sutherland southd.; Ireland; Channel Islands; fl. July–Oct.—Silvery-white all over. *Stem* angled, branched from the base ; branches 4–10 in., diffuse, stout or slender. *Leaves* 1–1½ in., petiole short, acute or obtuse, base cuneate, subentire or irregularly acutely or obtusely lobed toothed or serrate. *Male flowers* in short dense subpanicled spikes, female axillary with a few males intermixed. *Fruiting sepals* ½–¾ in. diam., united to the middle, often broader than long, rhombic, acute or acuminate, entire lobed or toothed; disk often prominently veined or wrinkled, rarely tubercled. *Seed* large, ⅛ in. diam., much compressed, rough, red-brown.—DISTRIB. W. Europe, Norway to France.

SUB-GEN. 2. **Obi'one**, *Gœrtn.* (gen.). Annual or perennial. *Flowers* monœcious or diœcious. *Pericarp* adherent to the cup-shaped perianth. *Testa* coriaceous or crustaceous ; radicle superior. *Hal'imus*, Wallroth.

4. A. portulacoi'des, *L.* ; shrubby, mealy, leaves obovate- or spathu-late-lanceolate quite entire, fem. perianth sessile compressed 2–4-lobed.

Muddy maritime cliffs and marshes, from Ayr and Northumbd. southd.; Ireland, very rare ; Channel Islands ; fl. Aug.–Oct.—Covered with minute persistent greyish-white scales. *Rootstock* woody, branched. *Stem* 1–3 ft., woody below, flexuous, decumbent ; branches 12–18 in., erect. *Leaves* 1–3 in., mostly opposite, tip rounded, upper linear. *Spikes* in terminal inter-rupted panicles, leafy below. *Fruiting perianth* ⅛–½ in., cupular ; lobes rounded, unequal. *Seed* chestnut, rough.—DISTRIB. From Denmark southd., N. Africa, W. Asia.

5. A. peduncula'ta, *L.* ; herbaceous, mealy, leaves narrowly obovate-oblong quite entire, fem. perianth pedicelled, lobes 2 recurved.

Muddy maritime marshes, rare, Lincoln, Norfolk, Suffolk, Kent ; fl. Aug.–Oct.—Annual, mealy like *A. portulacoi'des.* *Stem* 3–8 in., flexuous, slender, terete, simple or sparingly branched. *Leaves* ½–1½ in., tip rounded, shortly petioled. *Spikes* terminal and reduced to axillary fascicles ; flowers sub-sess'le, one or few fruiting in each fascicle, when the pedicel elongates to ⅓–½ in., and is very spreading. *Fruiting perianth* ¼ in. diam., campanulate with 2 lateral lobes and 2 intervening small teeth. *Seed* as in *A. portula-coi'des.*—DISTRIB. From Gothland southd., N. Africa, Siberia.

4. SALICOR'NIA, *Tournef.* MARSH SAMPHIRE.

Annual or perennial leafless herbs. *Stems* cylindric, very succulent, jointed ; branches opposite. *Flowers* 2-sexual, minute, 2 together sunk in pits at the nodes. *Perianth* turbinate, compressed, fleshy, 3–4-lobed or truncate, mouth contracted. *Stamens* 1–2, perigynous. *Styles* 2. *Utricle* compressed, included in the swollen perianth. *Seed* vertical, testa membranous hairy, albumen scanty fleshy or 0 ; embryo conduplicate green, radicle inferior incumbent, cotyledons ½-terete thick.—DISTRIB. Salt districts ; species 8.—ETYM. *sal* and *cornu,* from the *horn-*like branches.

1. S. herba'cea, *L.* ; annual, root slender, stem ascending, branches more or less fusiform all flowering.

Salt marshes, N. to Shetland ; Ireland ; Channel Islands ; fl. Aug.–Sept.— *Stem* 6–18 in., ¼–⅓ in. diam. at the thickest part, which is above the base ; internodes ½ in., usually contracted above and below, 2-lobed at the top when dry, lower woody slender, upper fleshy slightly compressed ; branches spreading or ascending ; flowering internodes in short spikes. *Flower-bearing* cavities 2 at each node, opposite. *Stamens* inserted at various heights, if 2 successively protruded. *Fruiting perianth* narrowly winged at the top. *Seed* ovoid or oblong, greenish, covered with curled hairs.—DISTRIB. Europe, N. Africa, N. and W. Asia, India, N. America.—Formerly burnt for Barilla, and sometimes pickled.

S. herba'cea proper ; green, glaucous, ascending, branches suberect, spikes many-fld.—VAR. *S. procum'bens,* Sm. ; red, decumbent, branches cruciate, spikes few-fld.—VAR. *ramosis'sima,* Woods ; grass-green, erect, much-bran¬hed, spikes few-fld. Hayling Island.—VAR. *pusil'la,* Woods; very small, spikes about ¼ in. few-fld.

2. **S⸺ radi'cans,** *Sm.* ; rootstock perennial woody creeping sending up herbaceous terete barren and flowering branches.
Salt marshes, from York southd. to Devon; fl. Sept–Oct.—Much more branched and tufted than *S. herba'cea,* colour browner.—*Stems* ½–2 ft., 1–1½ in. diam., spikes thicker and more obtuse.—DISTRIB. W. Europe from Denmark southd.—Erroneously referred to *Arthrocne'mum frutico'sum* by Moquin Tandon (in DC. *Prodr.*).
Imperfectly known species are *S. ligno'sa,* Woods (*Linn. Soc. Proceed.*, 1851, p. 111), with the growth, &c., of *S. radi'cans,* but stem shorter thicker more woody below; and *S. megasta'chya,* Woods, with tubercled hairless seeds, which is possibly an *Arthrocne'mum.*

5. SUÆ'DA, *Forsk.* SEABLITE.

Saline herbs or shrubs. *Leaves* fleshy, alternate, terete or ½-terete. *Flowers* 1–2-sexual, small, green, axillary, minutely 3-bracteolate. *Calyx* 5-partite ; segments obtuse, not keeled or winged. *Stamens* 5, hypogynous. *Styles* 3–5, compressed. *Utricle* enclosed in the fleshy or dry calyx. *Seed* horizontal or vertical, testa crustaceous, inner coat thin, albumen 0 or fleshy and scanty ; embryo in a flat spiral, radicle inferior.— DISTRIB. Salt marshes and shores ; species about 40.—ETYM. unknown.

1. **S. marit'ima,** *Dumort.* ; annual, stem procumbent or ascending branched, leaves subacute tapering at the base, styles 2, seed horizontal. *Schobe'ria,* C. A. Meyer ; *Chenopodi'na,* Moq. Tand.
Salt marshes, from Shetland southd.; Ireland; Channel Islands; fl. July–Oct· —Glabrous, glaucous, reddish in winter, usually branching from the base; branches 3–24 in , straggling, slender. *Leaves* ¼–1 in. or more. *Flowers* 3–5 together, rarely solitary, subsessile. *Seed* shining, striate, brownish-black, beaked.—DISTRIB. Europe (Arctic), N. Africa, N. and W. Asia, India, N. America.

2. **S. frutico'sa,** *Forsk.* ; stem perennial woody, leaves rounded at the base and tip, styles 3, seed vertical. *Schobe'ria,* C. A. Meyer.
Sandy and pebbly beaches, Norfolk, Suffolk, Essex, Dorset, rare and local; fl. July–Oct.—Glabrous, rather glaucous. *Stem* 1–3 ft., ½ in. diam. at the base; branches erect or ascending. *Leaves* ¼–⅜ in., crowded, fleshy, dotted with white. *Flowers* solitary or 2–3 together, subsessile. *Seed* shining, black.—DISTRIB. From Spain eastd., N. Africa, W. Asia, India.

6. SAL'SOLA, *L.* SALTWORT.

Herbs or shrubs. *Leaves* alternate or opposite, sessile, subcylindric or subulate, fleshy rigid or spinescent. *Flowers* small, axillary, sessile, dichogamous, 2-bracteate. *Sepals* 5, rarely 4, with a broad transverse dorsal wing that forms after flowering. *Stamens* 5, rarely 3, hypogynous ; filaments linear free or connate below. *Ovary* subglobose ; style elongate, stigmas 2-3 compressed or subulate. *Utricle* depressed, enclosed in the stellately 5-winged much enlarged calyx. *Seed* horizontal, testa membranous, albumen 0 ; embryo forming a conical helix.—Distrib. Saline districts ; species about 40.—Etym. *sal,* from yielding alkalies.

S. Ka'li, *L.* ; herbaceous, rigid, leaves spinous-pointed.
Sandy sea-shores from Caithness southd. ; Ireland ; Channel Islands ; fl. July–Aug.—Annual, pubescent or scabrid, glaucous. *Stem* 6–18 in., erect or procumbent, striped ; branches many, spreading, flexuous *Leaves* ½–1½ in., spreading and recurved, fleshy, ovate-subulate, ½-amplexicaul. *Flowers* 1–3 ; bracts spinescent. *Wings of fruiting perianth* very variable, broad or narrow, scarious, often rose-coloured. *Seed* brown, adherent to the pericarp. —Distrib. Europe, N. and S. Africa, N. and W. Asia, India, N. and S. America, Australia.—Formerly burnt for Barilla.

Order LXII. **POLYGONA'CEÆ.**

Herbs, rarely shrubby. *Leaves* alternate, simple, quite entire or serrulate ; margins revolute in bud ; petiole dilated ; stipules sheathing, scarious. *Flowers* usually 2-sexual, pedicels jointed. *Sepals* 3–6, petaloid or herbaceous, free or connate, persistent, imbricate in bud. *Stamens* 5–8, rarely more or less, perigynous or hypogynous, opposite the sepals ; anthers 2-celled. *Disk* glandular annular or 0. *Ovary* free, ovoid, 3-gonous or compressed ; styles 1–3, stigmas capitate or penicillate ; ovule 1, basilar, orthotropous. *Fruit* indehiscent, hard, usually enveloped in the perianth. *Seed* erect, testa membranous, albumen floury ; embryo straight and axile, or lateral and curved, cotyledons various, radicle superior.— Distrib. Chiefly temp. regions ; genera 30 ; species about 600.—Affi-nities. With *Amaranthaceæ* and *Chenopodiaceæ.*—Properties. Root often astringent or purgative ; some yield oxalic and malic acids ; the leaves or seeds of others are alimentary.

Sepals 5, subequal. Fruit compressed or 3-gonous, wingless..1. Polygonum.
Sepals 6, 3 inner much larger. Fruit 3-gonous..........................2. Rumex.
Sepals 4, 2 inner larger. Fruit winged....................................3. Oxyria.

1. POLYG'ONUM, *L.*

Herbs. *Leaves* alternate ; stipules tubular. *Flowers* 2-sexual, in panicled racemed or spiked clusters ; bracts ochreate. *Sepals* 5, 3 outer sometimes enlarging in fruit. *Disk* usually glandular. *Stamens* 5–8 ; anthers versatile. *Ovary* compressed or 3-gonous ; styles 2–3, stigmas capitate. *Fruit* 3-quetrous or compressed. *Embryo* axile or lateral. —Distrib. All climates ; species 150.—Etym. πολύς and γόνυ, from the *many nodes.*

SECTION 1. **Bistor'ta,** *Tournef.* *Rootstock* perennial. *Stem* simple, erect. *Stipules* truncate. *Racemes* solitary, spike-like ; pedicels jointed at the top. *Stamens* 8. *Fruit* 3-quetrous ; embryo lateral, cotyledons thin flat.

1. **P. Bistor'ta,** *L.* ; leaves obtuse or cordate at the base, petiole winged, raceme dense cylindric. *Bistort, Snake-root.*

Wet meadows, wild from Renfrew and Edinburgh southd., often introduced N. of it ; ascends to 1,050 ft. in Derby ; Ireland, rare, native? ; fl. June–Sept.— Glabrous, except the leaf-nerves beneath. *Rootstock* woody, twisted, creeping ; roots tuberous. *Stem* 1–2 ft., strict, slender. *Leaves,* radical 3–6 in., oblong-ovate, obtuse, waved, glaucous beneath ; petiole 6–12 in., broadly winged above ; cauline subsessile, broader at the base ; stipules ½–3 in. *Raceme* 1½–2 in. ; bracts cuspidate. *Flowers* ⅓ in. long, white or pink, honeyed, proterandrous. *Stamens* exserted. *Fruit* brown, shining.—DISTRIB. Europe (Arctic), N. and W. Asia, Himalaya.—Rootstock astringent ; used as food in famine-times, and formerly medicinally.

2. **P. vivip'arum,** *L.* ; leaves narrowed at the base, petiole not winged, raceme spike-like bulbiferous below.

Mountain pastures and wet alpine rocks ; from Carnarvon and York to Shetland ; ascends to 4,000 ft. ; W. Ireland ; fl. June–Aug.—Glabrous. *Rootstock* slender. *Stem* 4–16 in., slender. *Leaves* 1–2½ in., ⅓–½ in. broad, radical narrow linear-oblong, petiole as long or shorter, subacute, glaucous beneath, margins revolute ; cauline few, shorter petioled ; stipules ½–1½ in. *Racemes* 1–3 in., slender, obtuse. *Flowers* white or pink, polygamous ; bulbils purple. *Fruit* rarely ripening.—DISTRIB. Alps of N. temp. and Arctic regions.

SECTION 2. **Persica'ria,** *Meissn.* Rarely perennial. *Stipules* truncate, subentire. *Racemes* spike-like ; pedicels jointed at the top. *Stamens* 4–8. *Fruit* compressed or 3-quetrous ; embryo lateral, cotyledons thin, flat.

* *Spikes short, usually dense, not or rarely interrupted, not leafy at the base.*

3. **P. amphib'ium,** *L.* ; perennial, creeping or floating, leaves oblong or lanceolate, racemes subsolitary dense-fld., peduncles hairy, sepals eglandular, stamens 5, styles 2 united half way, fruit ovoid, faces convex.

Damp and watery places, N. to Shetland ; Ireland ; Channel Islands ; fl. July– Aug.—*Rootstock* creeping, slender, woody, branched. *Stem* very variable in length. *Leaves* floating and long-petioled or aërial and subsessile, obtuse or acute, serrulate or ciliate, eglandular ; stipules large, appressed, glabrous or hispid, mouth entire. *Racemes* 1–3, ½–2 in. ; peduncle stout ; pedicels short ; bracts obtuse acute or cuspidate. *Sepals* ⅙ in., not nerved, pale or bright rose-red, much longer than the shining fruit.—DISTRIB. N. temp. and Arctic regions.

4. **P. lapathifo'lium,** *L.* ; annual, leaves ovate or oblong-lancolate, racemes subcylindric, peduncle rough and sepals glandular, stamens 5–6, styles 2 free, fruit orbicular, faces concave.

Fields and waste places, N. to Shetland; ascends to near 1,000 ft. in York; Ireland; Channel Islands; fl. July–Aug.—*Stem* 1–4 ft., decumbent and rooting below, much branched, green red or spotted; nodes stout, swollen. *Leaves* 4–6 in., shortly petioled, acuminate, ciliate, glabrous pubescent scaberulous or cottony above and beneath, sometimes glaucous beneath, or with a black blotch above, punctate and sparingly glandular; stipules loose, often ciliate. *Racemes* 1–3, ¾–1½ in., often panicled, obtuse, stout; bracts broad, obtuse or cuspidate. *Sepals* nerved, equalling the obtuse dark ·fruit.—DISTRIB. Europe, N. Africa, N. and W. Asia, India; introd. in America.

P. LAPATHIFO'LIUM proper; racemes remote, sepals shorter than the fruit. Common.

Sub-sp. P. MACULA'TUM, *Dyer* and *Trimen;* racemes crowded, sepals larger than the smaller fruit. *P. nodo'sum,* Reichb.; *P. lax'um,* Bab.

5. **P. Persica'ria,** *L.* ; annual, racemes usually short dense, leaves ovate or lanceolate, peduncle glabrous, sepals subeglandular, stamens 5–8, styles 2–3 united half way, fruit plano-convex or 3-gonous. *Persicaria.*

Waste moist places, N. to Shetland; ascends to 1,300 ft. in N. England; Ireland; Channel Islands; fl. July–Oct.—*Stem* 6–18 in., branched, erect or ascending, nodes usually swollen. *Leaves* subsessile, subacute, ciliolate, often with a black blotch, pubescent and punctate below, eglandular; stipules ciliate. *Racemes* ½–1½ in., with sometimes a leaf at the base, erect or suberect, peduncled, lateral sessile. *Flowers* homogamous; anthers, outer introrse, inner extrorse. *Sepals* red or white, equalling the fruit.—DISTRIB. Europe (Arctic), N. Africa, N. and W. Asia, India.

P. *Persica'ria* proper; branches divaricate, racemes remote short stout, or slender (*P. bifor'me,* Wahlb.) obtuse cylindric.—VAR. *P. nodo'sum,* Pers.? branches erect, racemes slender rather lax attenuated upwards, young crowded. *P. bifor'me,* Bab.

*** Spikes long, lax, slender, interrupted and leafy below.*

6. **P. mi'te,** *Schrank ;* annual, suberect, racemes slender erect, sepals eglandular, stamens 5–6, styles 2–3 united half way, fruit roughish plano-convex or 3-gonous as long as the sepals.

Wet places, from York southd., local; fl. Aug.–Sept.— *Stems* 1–2 ft., erect, decumbent at the rooting base, branched, slender. *Leaves* 2–4 in., shortly petioled, lanceolate or elliptic-lanceolate, subacute, ciliolate, eglandular; stipules loose, strongly ciliate. *Racemes* ½–3 in., solitary, lax. *Sepals* white or pink, nerves faint. *Fruit* black, rather narrow.—DISTRIB. Europe (excl. Greece).

7. **P. Hydrop'iper,** *L.* ; annual, suberect, racemes very slender tips drooping, sepals with few very large glands, stamens 6 (rarely 8), styles 2–3 free, fruit plano-convex or 3-gonous as long as the sepals. *Water-pepper.*

Watery places from Skye southd.; ascends to 1,300 ft. in the Lake district; Ireland; Channel Islands; fl. Aug.–Sept.—*Stem* 1–3 ft., creeping and rooting at the base, much branched. *Leaves* 2–4 in., shortly petioled, lanceolate or elliptic-lanceolate, subacute, ciliolate, minutely glandular beneath; stipules short, inflated, ciliate or not. *Racemes* 3–8 in., curved, lax. *Sepals* green

and rose. *Fruit* black, punctulate.—DISTRIB. N. temp. hemisphere.—
Very acrid, and a reputed diuretic.

8. **P. mi'nus,** *Huds.* ; annual, racemes very slender straight, sepals
usually 5 with minute glands at the base only, styles 2–3 united half way,
fruit smooth plano-convex or 3-gonous as long as the sepals.

Marshy places, local from Perth and Renfrew southd.; Ireland, rare; fl. Aug.-
Sept.—*Stem* 6–24 in., usually very slender, much branched, erect or
ascending. *Leaves* 1–3 in., narrow-lanceolate, ciliolate, eglandular; stipules
not inflated, short, ciliate. *Racemes* 1–3 in., solitary or panicled, usually on
slender peduncles. *Sepals* $\frac{1}{10}-\frac{1}{8}$ in., very small. *Fruit* pitchy-black, shining,
acute.—DISTRIB. Europe, N. and W. Asia, India.

SECTION 3. **Avicula'ria,** *Meissn.* Annual or biennial. *Leaves* narrow ;
stipules silvery, at length lacerate. *Flowers* axillary, solitary or fascicled ;
pedicels jointed at the top. *Stamens* usually 8. *Styles* usually 3. *Fruit*
3-quetrous ; embryo lateral, cotyledons thin flat.

9. **P. avicula're,** *L.* ; prostrate, nerves of leaves obscure beneath,
stipules small, nerves few simple. *Knotgrass.*

Fields and waste places, N. to Shetland; ascends to 1,800 ft. in Northumbd.;
Ireland; Channel Islands; fl. May–Oct.—Annual, glabrous, eglandular,
branched from the base; branches $\frac{1}{2}$–3 ft., straggling, grooved, angular
above, leafing and flowering throughout. *Leaves* $\frac{1}{2}$–1$\frac{1}{2}$ in., $\frac{1}{10}-\frac{1}{2}$ in. broad,
sessile or shortly petioled, linear-lanceolate or -oblong, narrowed at both
ends, rarely broadly elliptic or almost filiform, acute or obtuse, margins
flat or recurved; stipules $\frac{1}{8}-\frac{1}{4}$ in., white, red at the base, lacerate. *Flowers*
$\frac{1}{10}-\frac{1}{8}$ in., white, pink, crimson, or green, clustered in the axils, homogamous.
Fruit brown, minutely striate and punctate.—DISTRIB. Europe (Arctic),
N. and W. Asia; introd. in N. America.

P. AVICULA'RE proper; leaves rather thin, fruit dull included.—VAR. *P. littora'le,* Link ; leaves rather fleshy, fruit more shining, tip exserted. Littoral.
The passage to *P. marit'imum.*—VAR. *agresti'num,* Jord., is the common
robust field form, *arenas'trum,* Boreau, a sand-loving prostrate one; *micro-
sper'mum,* Jord., a small fruited one; and *ruriva'gum,* Jord., a wayside one
with narrow very acute leaves.

Sub-sp. P. ROBER'TI, Loisel; fruit longer than the sepals. *P. Rai'i,* Bab.
in part.—Sandy shores.

10. **P. marit'imum,** *L.* ; prostrate, nerves of the leaves reticulate
beneath, of the stipules few or many, sepals shorter than the fruit.

Sea-shores, Hants, Devon, Cornwall; Channel Islands; fl. July–Sept.—Similar
to *P. avicula're,* but perennial, much stouter, more rigid and woody, darker
when dry; leaves thicker, often glaucous beneath; stipules larger more
scarious and nerved; flowers and fruit much larger.—DISTRIB. France,
Spain, Mediterranean, W. Asia, N. America.

SECTION 4. **Tinia'ria,** *Meissn.* Annual (the British species), usually
twining. *Leaves* cordate or sagittate ; stipules truncate, mouth entire.

Flowers in racemose clusters. *Sepals* enlarging. *Stamens* 8. *Styles* 3, united. *Fruit* 3-quetrous ; embryo lateral, cotyledons narrow flat.

11. **P. Convol'vulus,** *L.* ; leaves cordate-sagittate, 3 outer sepals obtusely keeled rarely winged, pedicels short jointed above the middle, fruit dull striate granulate. *Black Bindweed.*

Fields and waste places, from Caithness southd. ; ascends to 1,350 ft. in Derby; Ireland; Channel Islands; fl. July–Sept.—*Root* fibrous. *Stem* 1–4 ft., angular, twining climbing or prostrate, slender, angles puberulous. *Leaves* 1½–4 in., petiole shorter, slender, gradually acuminate, lateral angles obtuse or acute, eglandular, puberulous beneath; stipules short. *Racemes* erect, terminal and axillary, slender, pedicels recurved. *Sepals* 5, obtuse, green, margins white, 3 outer rough at the back, at length ¼ in., and covering the fruit.—DISTRIB. N. temp. and Arctic regions ; introd. in America.

P. Convol'vulus proper ; clusters 4–6-fld., outer sepals obtusely keeled in fruit. —VAR. *pseudo-dumeto'rum,* Wats.; clusters 5–10-fld., outer sepals broadly winged in fruit.

12. **P. dumeto'rum,** *L.* ; leaves cordate-sagittate,. 3 outer sepals with broad membranous wings, pedicels very slender jointed below the middle, fruit smooth highly polished.

Hedges and thickets, Monmouth and Essex to Kent and Devon, rare, soon disappearing ; fl. July–Aug.—Habit of *P. Convol'vulus,* but pedicels capillary (often ¾ in.) and seed polished. The stems are described as terete, but I find them as much angled as in *P. Convol'vulus.*—DISTRIB. Europe, N. and W. Asia, N.W. India.

2. RU'MEX, *L.* DOCK.

Biennial or perennial herbs. *Rootstock* stout, tapering into the root. *Stems* usually grooved. *Leaves* alternate ; stipules tubular. *Flowers* 1–2-sexual, in panicled or racemed whorls, anemophilous. *Sepals* 6, 3 inner enlarging. *Stamens* 6 ; anthers basifixed. *Ovary* 3-quetrous ; styles 3, filiform, stigmas penicillate. *Fruit* 3-quetrous. *Embryo* lateral.— DISTRIB. All temp. climates ; species about 100.—ETYM. The old Latin name.

Hybrids are common in this genus; those most known to cross are said to be *pulcher, crispus,* and *conglomeratus.*

SECTION 1. **Lap'athum,** *Meissn.* *Leaves* not hastate. *Flowers* 2-sexual (monœcious in *R. alpinus*). *Inner sepals* coriaceous in fruit.

* Inner fruiting sepals usually strongly toothed.

1. **R. obtusifo'lius,** *L.* ; radical leaves oblong-ovate cordate obtuse, panicle leafy below, inner fruiting sepals elongate triangular obtuse usually strongly toothed at the base, upper or all with an ovoid tubercle.

Fields, waste grounds, N. to Shetland; ascends to 1,600 ft. in N. England ; Ireland ; Channel Islands ; fl. Aug.–Sept.—Perennial, 2–3 ft., stem stout and leaves beneath puberulous. *Leaves* 6–12 in., subacute or obtuse, margin creuulate waved, upper oblong-lanceolate ; petiole rather slender. *Panicle* narrow ; pedicels ½–twice as long as the reticulate fruiting sepals, ⅛–¼ in. ;

tubercle red or brown.—DISTRIB. Europe, N. Africa, N. and W. Asia,
N.W. India; introd. in N. America.

R. obtusifo'lius proper (*R. Fries'ii*, Gren. and Godr.); inner sepals with spread-
ing subulate teeth, oblong one tubercled, apex entire.—VAR. *R. sylves'tris,*
Wallr.; all the inner sepals tubercled nearly entire in fruit. Thames at
Putney.

2. **R. acu'tus**, *L.* ; radical leaves linear or oblong-lanceolate waved,
panicle leafy below, inner fruiting sepals unequal triangular or cordate
with short broad teeth near the top, upper with an ovoid or lanceolate
tubercle. *R. praten'sis*, Mert. and Koch.

Roadsides, &c., from Orkney southd.; ascends to 1,200 feet in N. England;
fl. June–July.—Similar to *R. obtusifo'lius*, but leaves narrower, and fruiting
inner sepals much broader, with more and shorter teeth.—Syme and Koch
regard it as a hybrid between *R. cris'pus* and *obtusifo'lius;* Watson as a
medley of intermediate forms.—DISTRIB. Europe from the Alps northd.,
Spain.

R. consper'sus, Hartm., found in a few Scotch counties from Orkney to
Berwick, differs only in the more crisped leaves and equal inner fruiting
sepals.

3. **R. pul'cher**, *L.* ; leaves oblong-cordate or fiddle-shaped obtuse,
upper acute, panicle leafy to the top, inner fruiting sepals oblong deeply-
toothed to above the middle, tubercle oblong often muricate. *Fiddle
Dock.*

Waste places in dry soil from N. Wales and Notts southd.: Ireland; Channel
Islands; fl. June–Oct.—Biennial or perennial, glabrous or nearly so. *Stem* 6–
24 in.,flexuous; branches slender,spreading,tips often decurved. *Leaves* 3–6 in.,
soon withering, always contracted above the base, crenulate; petiole slender.
Panicle with spreading branches, whorls remote; pedicels stout, shorter
than the fruiting sepals, jointed below the middle. *Fruiting sepals* $\frac{1}{4}$ in.,
pale, obtuse truncate or cuneate at the base, deeply pitted and reticulate;
teeth short, straight. *Fruit* $\frac{1}{10}$ in.—DISTRIB. From Belgium southd.
N. Africa, W. Asia.

4. **R. marit'imus**, *L.* ; leaves linear- or oblong-lanceolate, panicle
leafy to the top, inner fruiting sepals triangular or rhomboid acuminate,
teeth 2–4 very long, tubercle linear-oblong very tumid. *Golden Dock.*

Marshes, &c., rare, from Northumbd. to Kent and Somerset; Ireland; Channel
Islands; fl. July–Aug.—Biennial, puberulous, yellow-green. *Stem* 1–2 ft.;
branches ascending. *Leaves* 3–10 in., base acute obtuse or cuneate, shortly
petioled, margins slightly waved. *Panicle* with spreading densely flowering
branches, whorls often confluent; pedicels jointed at the base, variable in
length, rarely twice as long as the fruiting sepals, which are reticulate, $\frac{1}{10}$–$\frac{1}{8}$
in., orange-yellow, spines as long; tubercle often almost concealing the sepal.
Fruit small, pale chestnut.—DISTRIB. Europe, N. and W. Asia, N.W. India,
N. America.—*R. Knaf'i*, Celak, is a hybrid with *conglomera'tus* of which only
a single specimen has been seen; it was first figured as *R. marit'imus*, L.
(forma hybrida ?) *Warren'ii*, by Trimen in Journ. Bot. iii. 161, t. 146.

R. MARIT'IMUS proper; whorls confluent, inner fruiting sepals triangular, teeth
often longer than the sepal.

Sub-sp. R. palus'tris, *Sm.*; whorls laxer usually distinct fewer-fld., inner fruiting sepals oblong triangular or rhomboid, teeth shorter usually fewer, fruit much larger. *R. Stein'ii*, Becker.

** Inner fruiting sepals quite entire or minutely toothed.

† *One or all the inner sepals with a prominent tubercle on the midrib.*

5. **R. cris'pus,** *L.* ; leaves lanceolate or oblong-lanceolate subacute much waved and crisped, panicle leafy below, inner fruiting sepals oblong-ovate or cordate obtuse subentire, upper with a broad smooth tubercle.

Waste places, N. to Shetland.; ascends to 2,000 ft. in Northumbd.; Ireland; Channel Islands; fl. June–Oct.—Perennial, glabrous or puberulous. *Stem* 1–3 ft., branched. *Leaves* 6–10 in., base obtuse rounded or acute; petiole moderate. *Panicle* with erect branches; whorls crowded; pedicels jointed at the base, twice as long as the fruiting sepal or shorter. *Fruiting sepals* ⅛–¼ in., green or reddish, reticulate; inner entire or crenulate; tubercle small, smooth. *Fruit* brown.—Distrib. Europe, N. Africa, temp. Asia; introd. in N. America.

Var. *trigranula'ta*, Syme, has the panicle very dense, branches short appressed, inner fruiting sepals all tubercled. Orkney, Annan, Fife.—Var. *subcorda'ta*, Warren, has a lax panicle, and inner fruiting sepals larger and more triangular. Lewes.

R. elonga'tus, Guss., is a var. with flat leaves and laxer panicles from wet places by the Thames and Wye; it attains 6 ft.

6. **R. sanguin'eus,** *L.* ; leaves oblong-lanceolate fiddle-shaped sparingly waved, base of panicle leafy, inner fruiting sepals oblong obtuse base rounded entire, upper (or all) with a large smooth tubercle.

Roadsides and hedges from Isla and Elgin southd.; ascends to 1,200 ft. in York; Ireland; Channel Islands; fl. July–Aug.—Perennial, glabrous. *Stem* 1–4 ft., slender, simple or sparingly branched. *Leaves* 6–10 in., base usually cordate, nerves red or green (*R. vir'idis*, Sibth.; *R. nemoro'sus*, Schrad.); petiole shorter. *Panicle* lax, usually leafless; whorls distant, many-fld.; pedicels usually equalling the fruiting sepals, rarely twice as long, jointed at the base. *Fruiting sepals* oblong or oblong-lanceolate tubercle on the outer larger, subglobose, on the others small or 0. *Fruit* brown, shining.—Distrib. Europe, W. Asia; introd. in N. America.

7. **R. conglomera'tus,** *Murray ;* leaves oblong-lanceolate base rounded or cordate, panicle leafy almost to the top, inner fruiting sepals linear-oblong subacute rounded at the base quite entire, all with oblong tubercles. *R. acu'tus*, Sm. and L. Herb.

Wet meadows and waste places, from Skye and Aberdeen southd.; Ireland; Channel Islands; fl. June–Oct.—Closely allied to *R. sanguin'eus*, differing in the leaves never contracted above the base, pedicels jointed below the middle, longer tubercles, and in the characters given above.—Distrib. Europe, N. Africa, W. Asia; introd. in N. America.

R. rupes'tris, Le Gall, is a more upright var. with panicle tapering, root-leaves narrower, bracts few and narrow, fruiting sepals larger more obtuse. —Sea coasts, Sussex to Cornwall; Channel Islands.

8. **R. Hydrolap'athum,** *Huds.* ; leaves broadly oblong-lanceolate,
panicle almost leafless, inner fruiting sepals deltoid-ovate acute or obtuse
quite entire or faintly toothed, all with oblong tubercles.

Ditches and river-sides from Perth and Isla southd.; Ireland; Channel
Islands; fl. July–Aug.—Perennial, glabrous. *Stem* 3–6 ft., erect, branched.
Leaves 1–2 ft., acute, base rounded cordate or acute, margins flat crenulate;
petiole 6–10 in., flat above, not winged. *Panicle* very large; whorls rather
crowded; fruiting pedicels as long or twice as long as the sepals, jointed
near base. *Fruiting sepals* ¼–½ in., reticulate, base truncate or cuneate.
Fruit pale chestnut.—DISTRIB. Europe.—The largest British species. Root
astringent.

R. Hydrolap'athum proper; petioles flat, base of inner fruiting sepals narrowed.—VAR. *latifo'lia*, Borr. (? *R. max'imus*, Schreb.) ; margins of petioles
raised, base of inner fruiting sepals truncate or cordate. Essex, Hants,
Sussex, Cornwall, Scilly Is.

†† *Fruiting sepals without tubercles on the midrib.*

9. **R. aquat'icus,** *L.* ; lower leaves oblong-lanceolate crisped and
waved, panicle leafy at the base only, inner fruiting sepals cordate waved
membranous reticulate. *R. longifo'lius*, DC.

Wet meadows and ditches, from York to Shetland; ascends to 1,600 ft. in
the Highlands; fl. July–Aug.—Perennial, glabrous. *Stem* 1–3 ft., very
stout. *Leaves* 3–4 in. diam. *Panicle* with erect branches; whorls confluent; pedicels usually the length of the sepals, jointed below the middle.
Fruiting sepals ¼–½ in. diam., obtuse, green, strongly reticulate; midrib
slightly thickened. *Fruit* broad, small, pale brown.—DISTRIB. Scandinavia (Arctic), France, Germany, N. and W. Asia, Himalaya, N. America.

R. ALPI'NUS, *L.* ; leaves broadly ovate-cordate obtuse, panicle leafy at
the base only, inner fruiting sepals triangular-ovate obtuse faintly reticulate. *Monk's Rhubarb.*

Roadsides, near cottages, &c., N. England and Scotland, rare, naturalized; fl.
July–Aug.—Perennial, puberulous with cellular hairs. *Rootstock* very stout.
Stem 2–4 ft., stout. *Leaves* 6–24 in., not so broad, margins waved; petiole
long, stout. *Panicle* with very many erect branches; whorls very many,
not confluent; flowers monœcious; fruiting sepals, ⅙–¼ in.; pedicels twice
as long, jointed below the middle. *Fruit* grey.—DISTRIB. N. and Alpine
districts of S. Europe, excluding Russia.—Root formerly used medicinally,
and leaves as a pot-herb.

SECTION 2. **Aceto'sa,** *Tournef.* *Leaves* hastate or sagittate. *Flowers*
monœcious or diœcious.

10. **R. Aceto'sa,** *L.* ; diœcious, lower leaves sagittate, upper sessile,
outer fruiting sepals reflexed, inner enlarged orbicular entire scarious
tubercled at the base. *Sorrel.*

Meadows and pastures, N. to Shetland, ascends to 4,000 ft.; Ireland; Channel
Islands; fl. May–Aug.—Perennial, glabrous, acid, rather succulent. *Rootstock* tufted, slender. *Stem* 1–2 ft., simple, slender. *Leaves*, radical 3–6 in.,
very long-petioled, basal sinus rounded or angled, glaucous beneath; stipules

brown. *Panicle* with erect branches, leafless; male whorls densely 4–8-fld.; pedicels jointed below the middle. *Sepals* of male fl. herbaceous, margins white or pink, scarious; of female, ¼ in. when in fruit, pink or crimson. *Fruit* brown, shining.—DISTRIB. N. temp. and Arctic zones.—A salad and pot-herb; abounds in binoxalate of potash.

11. **R. Acetosel'la,** *L.*; diœcious, lower leaves hastate, uppermost sessile, outer fruiting sepals appressed, inner hardly enlarged oblong-ovate entire herbaceous, midrib thickened at the base. *Sheep's Sorrel.*

Dry pastures, N. to Shetland, ascends to 2,500 ft. in York; Ireland; Channel Islands; fl. May–Aug.— Perennial, acid, glabrous, often bright red in autumn. *Rootstock* creeping, much branched. *Stems* 3–20 in., often many and tufted, decumbent at the base, slender. *Leaves* ½–2 in., long-petioled, variable in breadth, often 3-lobed; stipules silvery, torn. *Panicle* leafless, branches erect; male flowers largest; pedicels as long as the fruiting sepals, jointed at the top, length variable. *Fruiting sepals* ₁₁₂ in., obtuse, closely investing the yellow-brown fruit.—DISTRIB. N. temp. and Arctic zones; introd. into the S.

3. OXYR'IA, *Hill.* MOUNTAIN SORREL.

Characters of *Rumex*, but sepals 4; anthers versatile; ovary compressed; stigmas 2; fruit lenticular, broadly winged, and embryo axile.— DISTRIB. Arctic regions and Alps of the N. temp. zone; species 1.— ETYM. ὀξύς, from the acidity of the leaves.

O. digyna, *Hill;* leaves cordate or reniform. *O. renifor'mis,* Hk.

Mountain rocks and streams, from N. Wales and Westmoreland to Orkney; ascends to near 4,000 ft.; S.W. Ireland; fl. July–Aug.—Perennial, glabrous, rather fleshy, acid. *Rootstock* tufted. *Stem* 6–18 in., stout, subsimple. *Leaves,* radical many, ½–¾ in. broad, long-petioled, rounded or retuse, rarely 3-lobed or subhastate; cauline solitary. *Panicle* slender, leafless, lax-fld.; pedicels slender, jointed at the middle, top thickened. *Outer sepals* spreading or reflexed; inner ₁₁₀ in., spathulate, 3–5-nerved. *Fruit* ⅛ in. diam.; wing orbicular-cordate, membranous, veined, top notched.—An excellent pot-herb and antiscorbutic.

ORDER LXIII. ARISTOLOCHIA'CEÆ.

Herbs or shrubs, often climbing. *Leaves* alternate, entire or lobed, exstipulate. *Flowers* 2-sexual, solitary spiked or racemed, regular or irregular. *Perianth* superior, tubular, campanulate or trumpet-shaped; limb 3-lobed or 1-lipped, valvate in bud. *Stamens* 6 or 12, rarely 5; anthers subsessile, free or adnate to the style, cells introrse or extrorse. *Ovary* 4–6-celled; styles 6, inner surface stigmatic; ovules very many, anatropous, 2-seriate in the inner angles of the cells. *Fruit* a 4–6-valved septicidal capsule, or a berry. *Seeds* horizontal flattened or boat-shaped, raphe thickened, albumen copious fleshy or horny; embryo minute, basilar, cotyledons short, radicle usually next the hilum.—DISTRIB. Chiefly trop.; genera 5; species 200.—AFFINITIES. With *Nepenthaceæ* and *Rafflesiaceæ.*—PROPERTIES. Bitter, acrid, sometimes aromatic.

Calyx campanulate, regularly 3-cleft. Stamens 12...................1. Asarum.
Calyx tubular, mouth oblique. Stamens 6 ....1*. *Aristolochia.*

1. AS'ARUM, *Tournef.* ASARABACCA.

Perennial herbs. *Rootstock* stout, branched, woody. *Leaves* radical. *Flower* solitary, peduncled, terminal, lurid purple, proterogynous. *Perianth* campanulate, regular, persistent, 3-lobed. *Stamens* 12; anthers bursting outwards, connective produced. *Ovary* inferior or ½-inferior, 6-celled; styles 6, tubular, grooved or 2-fid. *Fruit* coriaceous, bursting irregularly. *Seeds* boat-shaped, wrinkled on the convex face, with a median winged or fleshy raphe on the other.—DISTRIB. Europe, N. Asia, Himalaya, N. America; species 13.—ETYM. doubtful.

A. europæ'um, *L.*; pubescent, leaves evergreen reniform.
Copses in Wilts, Hereford, Bucks, York, Denbigh, Lancaster; (a denizen, *Wats.*); fl. May.—*Rootstock* creeping, fleshy; branches and stems short, sending up annually a pair of leaves and 2 large scales. *Leaves* 2–3 in. diam., dark green, petiole 3–5 in. *Scapes* from between the leaves, very short, woolly. *Perianth* ½ in., greenish-purple, lobes incurved. *Filaments* subulate, alternate longer; connective with a long subulate tip. *Styles* recurved, stigmas projecting between the anthers. *Fruit* globose.—DISTRIB. From Belgium southd., W. Siberia.—Root cathartic, emetic, and sternutatory.

1*. *ARISTOLO'CHIA, Tournef.* BIRTHWORT.

Shrubs or perennial herbs, often twining. *Leaves* cauline; petioles with dilated bases, having in their axils solitary or racemose proterogynous flowers and often the stipule-like leaf of an undeveloped bud. *Perianth* coloured, tube inflated at the base, then contracted, hairy inside; limb dilated, obliquely 1–2-lipped. *Anthers* 6, rarely 5 or more, adnate in a whorl to the very stout short 6-lobed style. *Capsule* septicidally 6-valved. —DISTRIB. Chiefly trop.; species 160.—ETYM. The old Greek name.

A. CLEMATI'TIS, *L.*; glabrous, flowers clustered, lip narrow acute.
Ruins, &c., from York southd., rare, not indigenous; fl June–Sept.—*Rootstock* creeping, woody. *Stems* many, erect, simple, angled. *Leaves* 3–6 in. diam., broadly cordate, obtuse, apiculate, reticulate, glaucous beneath; auricles rounded, incurved, almost closing the deep sinus. *Flowers* 4–8 in a cluster; pedicel very short. *Ovary* fusiform. *Calyx* 1 in., yellow; tube slender, curved, base globose; lip ½ in., oblong or ovate; throat dilated. *Capsule* ¾–1 in., pyriform; peduncle decurved. *Seeds* suborbicular, much compressed, granulate, deeply excavated on the ventral face.—DISTRIB. From Denmark southd., W. Asia.

Order LXIV. **THYMELÆA'CEÆ.**

Herbs, shrubs, or trees ; juice acrid ; inner bark tenacious. *Leaves* alternate or opposite, quite entire, exstipulate. *Flowers* 2-sexual (rarely polygamous), solitary fascicled cymose or capitate. *Perianth* inferior, throat naked or bearing scales glands or staminodes ; lobes 4–5, imbricate in bud. *Stamens* 2, 4, 8, or 10, adnate in 1–2 series to and included within the perianth-tube, when equalling its lobes alternate with them. *Disk* 0 or of 4–8 hypogynous scales or glands. *Ovary* free, 1- rarely 2-celled ; style terminal or lateral, stigma capitate ; ovules 1–3, pendulous, anatropous. *Fruit* a drupe or berry (rarely capsular). *Seed* pendulous, testa thin or crustaceous, albumen scanty or 0 ; embryo straight, cotyledons plano-convex, radicle short superior.—Distrib. Temp. and trop. regions ; genera 40 ; species about 300.—Affinities. With *Elæagnaceæ*, *Proteaceæ*, and *Santalaceæ*.—Properties acrid.

1. DAPH'NE, *L.*

Shrubs, rarely tall. *Leaves* usually alternate and persistent. *Flowers* odorous, honeyed. *Perianth* tubular ; lobes 4, spreading ; throat naked. *Stamens* 8, subsessile, 2-seriate ; anthers fixed by the back. *Style* subterminal, short or 0. *Fruit* coriaceous or fleshy. *Testa* crustaceous. Distrib. Europe, N. Africa, temp. Asia ; species about 50.—Etym. doubtful.

1. **D. Laure'ola,** *L.* ; leaves evergreen, flowers fascicled in the upper leaf-axils green glabrous. *Spurge Laurel.*

Copses and hedgebanks in stiff soils, from Durham to Devon and Kent ; Channel Islands ; fl. Jan.–April.—Shrub 1–3 ft., leafless below, branches few. *Leaves* 2–5 in., very coriaceous, obovate-lanceolate, acute, subsessile. *Cymes* few-fld. ; bracts oblong, deciduous ; flowers ½ in., inclined, males and 2-sexual intermixed. *Calyx-lobes* ½ the length of the tube. *Fruit* ½ in., ovoid, black.—Distrib. From Belgium southd. (excl. Russia and Greece), N. Africa, W. Asia.—Berry very poisonous.

2. **D. Meze'reum,** *L.* ; leaves deciduous, flowers appearing before the leaves clustered on the branches pink silky. *Mezereon.*

Copses and woods, perhaps native in the S. ; (an alien or denizen, *Wats.*) ; fl. Feb.–April.—Shrub 2–4 ft. ; branches few, erect. *Leaves* 2–3 in., obovate- or spathulate-lanceolate, acute, membranous, petioled. *Flowers* usually 3-nate, subsessile in the axils of the last year's leaves, very fragrant, rarely white ; bracts small. *Perianth* ⅓ in. diam., tube as long as the lobes. *Fruit* ½ in., bright red, ovoid.—Distrib. Europe (excl. Greece), Siberia.—Acrid and poisonous ; leaves used as a vesicant ; berries cathartic.

ORDER LXV. ELÆAGNA'CEÆ.

Shrubs or trees, with copious silvery or brown scales; buds naked. *Leaves* alternate or opposite, quite entire, exstipulate. *Flowers* small, regular, 1–2-sexual, axillary, fascicled or cymose, white or yellow. *Perianth* in the 2-sexual and female fl. tubular, 2–6-cleft, lobes imbricate or valvate in bud; in male fl. of 2 or 4 sepals free or connate below. *Disk* 0, or lining the calyx-tube. *Stamens* adnate to the calyx-tube, in the male fl. twice as many as the lobes, in the 2-sexual as many as and opposite the lobes; anthers fixed by the back or base. *Ovary* free, sessile, enclosed in the thickened calyx-base, 1-celled; style filiform, stigma lateral; ovule 1, basal, erect, anatropous. *Fruit* indehiscent, enclosed in the calyx-tube. *Seed* ascending, testa thick or thin, albumen 0 or scanty; embryo straight axile, cotyledons thick, radicle inferior.—DISTRIB. N. temp. and trop. zones; genera 3; species 16.—AFFINITIES. With *Thymeleaceæ*.—PROPERTIES unimportant.

1. HIPPOPH'AE, *L.* SEA BUCKTHORN.

A shining silvery willow-like diœcious shrub. *Leaves* alternate. MALE fl. in axillary clusters. *Sepals* 2. *Stamens* 4. FEM. fl. solitary. *Calyx* tubular, minutely 2-lobed. *Fruit* a membranous utricle enclosed in the succulent calyx-tube. *Seed* oblong, grooved on one side, testa crustaceous shining, albumen a thin fleshy layer; embryo amygdaloid.—DISTRIB. Europe, N. and Central Asia, Himalaya.—ETYM. doubtful.

H. rhamnoi'des, *L.*; leaves obovate at length lanceolate.

Sandy sea-shores, York to Kent and Sussex, not common; naturalized in Scotland and Ireland.—Shrub 1–8 ft.; branches slender and subpendulous, or short and spinescent. *Leaves* ¼–2 in., lengthening after flowering to 3 in., dull green above, silvery beneath. *Flowers* on the old wood; male minute; sepals broadly oblong; filaments short; anthers yellow. *Fruit* ⅓ in. diam., globose or oblong, orange-yellow.

ORDER LXVI. LORANTHA'CEÆ.

Evergreen parasitic shrubs. *Stem* often jointed. *Leaves* usually opposite, coriaceous, exstipulate. *Flowers* 1–2-sexual. *Sepals* thick, 4, 6, or 8, superior, free or united into a tubular calyx, lobes valvate in bud. *Stamens*, one adnate to each calyx-lobe; anther 1–2 celled opening by slits, or many-celled and opening by many pores. *Disk* annular, epigynous or 0. *Ovary* inferior, 1-celled; style simple or 0, stigma simple; ovule 1, reduced to a nucleus or to an embryo sac, adnate to the substance of the ovary. *Berry* 1-seeded. *Seed* erect, testa thin, albumen copious fleshy; embryos 1 or more, cotyledons thin or plano-convex, radicle superior.—DISTRIB. Trop., temp.; genera 13; species 500.—AFFINITIES. Very near *Santalaceæ*.—PROPERTIES unimportant.

1. VIS'CUM, *L.* Mistletoe.

Leaves opposite whorled or 0. *Flowers* diœcious, small, green, spiked or clustered in the forks or internodes of the branches. *Sepals* 4, triangular. *Anthers* sessile, cells many, opening by pores. *Stigma* sessile.— Distrib. Of the Order ; species about 30.—Etym. ἰξός, or βισκός, of the Greeks.

V. al'bum, *L.* ; leaves obovate-lanceolate obtuse 5-7-nerved.

On various trees, most rare on the oak, from York and Denbigh to Devon and Kent ; fl. March–May.—*Shrub* 1-4 ft., yellow green, glabrous; branches terete, dichotomous, knotted. *Leaves* 1-3 in., opposite or in whorls of 3. *Flowers* 3-nate, inconspicuous, green, 2-bracteate rarely monœcious. *Berry* white, nearly ½ in. diam., ovoid or globose, viscid. *Embryos* 1-3, green ; when 2, often united by the cotyledons.—Distrib. Europe, N. Asia.

Order LXVII. SANTALA'CEÆ.

Herbs, shrubs, or trees, usually parasitic on roots. *Leaves* mostly alternate, quite entire, exstipulate. *Flowers* 1-2-sexual, small or minute, solitary or cymose, 2-bracteolate. *Calyx* inferior or becoming adherent to the ovary ; lobes 3-5, valvate in bud, often with a tuft of hairs on their face. *Stamens* opposite and adnate to the calyx-lobes, filaments short ; anthers fixed by the base or back. *Disk* epigynous, often dilated and lobed. *Ovary* 1-celled ; style short, stigmas 1-5 ; ovules 2-5, reduced to a naked nucleus pendulous from a basal erect column. *Fruit* indehiscent, 1-celled, 1-seeded. *Seeds* adhering to the placenta, and often to the pericarp, albumen fleshy ; embryo straight axile, cotyledons plano-convex, radicle superior.—Distrib. All regions ; genera 28 ; species 220.—Affinities. With *Loranthaceæ, Olacineæ,* and *Corneæ.*—Properties unimportant, except the fragrant wood of *Santalum.*

1. THE'SIUM, *L.* Bastard Toad-Flax.

Slender, herbaceous, perennial root-parasites. *Leaves* alternate, narrow, decurrent 1-3-nerved. *Flowers* minute, green, solitary and axillary or in 2-chotomous cymes, 2-sexual. *Calyx-tube* short or long, limb 5- rarely 4-lobed, persistent ; lobes with a tuft of hairs on the face. *Ovary* inferior ; style short, stigma capitate ; ovules 3. *Fruit* ribbed.—Distrib. Europe, Asia, and Africa ; species about 100.—Etym. obscure.

T. linophyl'lum, *L.* ; stems diffuse, leaves 1-nerved, pedicels scabrid. *T. humifu'sum,* DC. ; *T. divarica'tum,* var. *ang'licum,* A.DC.

Dry chalky pastures, from Norfolk and Gloster to Cornwall and Sussex ; Channel Islands ; fl. May–July.—*Rootstock* woody, yellow ; roots fibrous, attached to those of various plants. *Stems* many, 6-18 in., leafy, prostrate. *Leaves* ½-1½ in., linear-lanceolate, acute or obtuse. *Flowers* ⅙ in. diam., racemed or

fascicled, pedicelled, white inside. *Calyx* funnel-shaped, lobes incurved in
fruit toothed. *Fruit* ⅓ in., green, ovoid, contracted into the short stout
pedicel.—Distrib. From Belgium southd., N. Africa, W Asia.

Order LXVIII. EUPHORBIA'CEÆ.

Herbs, shrubs, or trees, juice often milky. *Leaves* usually alternate,
simple, often stipulate. *Flowers* small or minute, usually 1-sexual, bracteate or involucrate. *Perianth* 0, or sepals 2 or more.—Male. *Stamens*
1 or more ; anthers didymous. *Ovary* rudimentary or 0.—Female. *Ovary*
2–3-lobed, 2–3-celled ; styles 2–3, stigmas entire or lobed. *Ovules* 1–2,
collateral, pendulous from the top of each cell, funicle dilated over the
micropyle. *Capsule* 2–3-lobed and -celled, cells 1–2-seeded. *Seeds* pendulous, testa usually crustaceous, funicle often swollen at the top (seeds
carunculate), albumen copious fleshy ; embyro axile, radicle superior.—
Distrib. All climates except Arctic ; genera 197 ; species about 3,000.—
Affinities. Close with *Malvaceæ* and *Urticaceæ.*—Properties. Usually
acrid, but too numerous to specify.—The above diagnosis applies to the
British genera.

Tribe I. **EUPHORBIEÆ.** *Involucre* calyciform with many male monan-
    drous flowers surrounding one female. *Perianth* minute or 0..1. Euphorbia.

Tribe II. **BUXEÆ.** *Flowers* distinct. *Stamens* opposite the sepals. *Ovules*
    2 in each cell ....2. Buxus.

Tribe III. **CROTONEÆ.** *Flowers* distinct. *Stamens,* outer or all opposite
    the sepals. *Ovules* solitary in each cell ....3. Mercurialis.

1. EUPHOR'BIA, *L.* Spurge.

Herbs (the British species). *Inflorescence* of many male and one female
flower in a 4–5-lobed involucre (perianth of some) ; lobes with thick
glands at the sinuses.—Male fl. a pedicelled stamen ; anther didymous.
—Female fl. *Ovary* on a lengthening pedicel, inclined or pendulous ;
stigmas 2-fid. *Capsule* 3-lobed, 3-valved, valves with a coriaceous exocarp
separable from a hard 2-valved endocarp. Distrib. Of the Order ;
species 600.—Etym. The old Greek name.

Section 1. *Leaves* exstipulate. *Branches* (or stem, if simple) terminated by umbels of forked branchlets (rays) subtended by a whorl of
leaves ; rays 2-bracteate at the forks.

* *Leaves* alternate. *Umbels* 5- rarely 3–6-rayed. *Glands of involucre*
    transversely oblong reniform or orbicular, not cuspidate.

1. **E. Heliosco'pia,** *L.* ; annual, glabrous, rarely hairy, leaves narrowly obovate serrate above the middle, upper bracts broadly ovate-cordate,
capsule smooth, seeds deeply pitted. *Sun Spurge,*

Fields and waste places, N. to Shetland; ascends to 1,000 ft. in the Highlands; Ireland; Channel Islands; fl. June–Oct.—Subglaucous. *Stem* 6–18
in., simple or 3-fid below. *Leaves* 1–2 in., subpetioled, sometimes cuneate,
tip rounded, membranous, lower smaller. *Involucral glands* orbicular,
yellow. *Capsule* ⅙ in. *Seeds* brown.—DISTRIB. From Belgium southd.,
N. Africa, N. and W. Asia, India; introd. in N. America.

2. **E. platyphyl'los,** *L.* ; annual, glabrous or hairy, leaves linear-
oblong or obovate-lanceolate acute serrulate above the middle, bracts
cordate, capsule warted, seeds smooth.

Fields and waste places, rare, from York and Gloster southd.; fl. July–Oct.—
Stem ½–3 ft., usually simple, stout, erect; branches numerous, alternate,
slender, ascending. *Leaves* ½–1½ in., sessile, spreading or reflexed. *Bracts*
short, broad, apiculate. *Involucral glands* suborbicular.—DISTRIB. From
Belgium southd., N. Africa, W. Asia.; introd. in N. America.
E. PLATYPHYL'LOS proper; bracts ⅜–¾ in., capsule ⅙ in. long, warts hemispherical, seeds olive-brown. *E. stric'ta,* Sm., not L.
Sub-sp. E. STRIC'TA, *L.*; bracts ¼–⅜ in., capsule $\frac{1}{12}$ in., warts conical, seeds
oblong smaller red-brown.—Woods on limestone in Gloster and Monmouth.

3. **E. hiber'na,** *L.* ; perennial, pubescent, leaves elliptic- or lanceolate-
oblong quite entire, upper cordate, bracts ovate-cordate, capsule furrowed
and warted, seeds smooth.

Copses and hedges, N. Devon; S. and W. Ireland, rare; fl. May–June.—*Rootstock* stout. *Stems* 1–2 ft., several, subsimple, leafy. *Leaves* 2–4 in., 1–1¼
in. broad, sessile, obtuse or notched at the tip, thin. *Bracts* broad, upper
rounded at the base. *Involucral glands* reniform. *Capsule* ¼ in., subglobose,
valves not keeled, warts cylindric. *Seeds* broad, pale brown.—DISTRIB.
W. France, N. Spain.—Used in Ireland to poison fish.

4. **E. pilo'sa,** *L.* ; perennial, hairy, leaves oblong-lanceolate tips serrulate, bracts elliptic obtuse, capsule glabrous or hairy smooth or minutely
warted, seeds smooth. *E. palus'tris,* Forst., not L.

Near Bath, in shaded places; (alien or denizen, *Wats.*); fl. May–June. *Rootstock* stout. *Stems* tall, leafy, branched above. *Leaves* 3–4 in., membranous.
Bracts yellow. *Rays* of umbel 4–6, 3-fid, then 2-fid. *Involucre* glabrous or
hairy, glands transversely oblong. *Seeds* obovoid.—DISTRIB. Mid. and S.
Europe to W. Siberia and the Caucasus.

** *Leaves* alternate. *Umbels* 3- or many-rayed. *Glands of involucre*
reniform or lunate with cuspidate tips.

† *Bracts connate at the base.*

5. **E. amygdaloi'des,** *L.* ; perennial, hairy, leaves obovate-lanceolate
quite entire, capsules glabrous minutely dotted, seeds smooth grey.

Woods, copses, &c., from Northumbd. southd., local; Bandon and Donegal in
Ireland; Channel Islands; fl. March–May.—*Rootstock* woody. *Stems* 6–12
in., erect, very stout, leafy, barren the first year, elongating the following
to 2 ft., then throwing out slender branches. *Leaves* 2–3 in., obtuse or
acute, lower petioled, upper sessile often oblong. *Rays* 5–10; bracts

connate into an orbicular limb ¾–1 in. diam., yellow. *Involucres* broad, pedicels slender, cusps of glands converging. *Capsule* with rounded valves. *Seed* subglobose, acute.—DISTRIB. From Holland southd., W. Asia.

†† *Bracts free at the base. All glabrous.*

6. **E. Pep'lus,** *L.* ; annual, leaves orbicular-obovate quite entire, bracts ovate, capsules small, valves keeled, seeds pitted whitish.

Waste places, N. to Orkney; ascends to 1,200 ft. in Derby; Ireland; Channel Islands; fl. July–Nov.—*Stems* 6–10 in., simple or 3-chotomous below. *Leaves* ½–¾ in., thin, petiole short slender. *Rays* 3, repeatedly forked. *Involucres* small; cusps of glands slender, curved. *Capsule-valves* with 2 keels on the back. *Seeds* 3-gonous, dorsally deeply pitted and keeled, ventrally 2-sulcate. —DISTRIB. Europe, N. Africa, N. and W. Asia, N.W. India; introd. in N. America.

7. **E. exig'ua,** *L.* ; annual, leaves linear-lanceolate quite entire obtuse or acute, bracts cordate at the base, capsules rough on the back of the valves, seeds pale deeply pitted.

Fields, &c., from Banff and the Clyde southd.; Scotland, rare; Ireland, local; Channel Islands; (a colonist, *Wats.*); fl. July–Oct.— Very variable. Usually excessively branched from the base; branches 6–15 in., erect and strict, or prostrate and ascending. *Leaves* ⅓–1 in., broadest above or below the middle, sometimes truncate and apiculate. *Rays* 3–5, often forked; bracts often toothed at the base. *Involucres* small, subsessile; tips of glands obtuse. *Capsules* small. *Seeds* obtusely 3-gonous, keeled, grey.—DISTRIB. Europe, N. Africa, W. Asia, N.W. India.

8. **E. portland'ica,** *L.* ; perennial, leaves coriaceous obovate or oblong-obovate quite entire, bracts deltoid- or reniform-cordate, capsules slightly rough on the back of the valves, seeds opaque brown pitted.

Sea-shore, from Wigton southd. to Hants, rare (absent on the E. coast); Ireland; Channel Islands; fl. May–Aug.—*Rootstock* cylindric, woody, tortuous. *Stems* 6–18 in., very many, tufted, naked and scarred below, branched, leafy above. *Leaves* ½–⅔ in., spreading, acute or apiculate. *Rays* 3–5; bracts ⅓–½ in. diam., broader than long, often keeled and cuspidate. *Capsule-valves* faintly keeled, granulate. *Seeds* with shallow pits, 'cuticle brown, caruncle large.—DISTRIB. W. France, Spain, Portugal.

9. **E. Para'lias,** *L.* ; perennial, leaves imbricate coriaceous quite entire, lower linear-obovate or -oblong, upper ovate, bracts broadly cordate, capsules leathery wrinkled, seeds minutely dotted whitish.

Sandy shores, from Cumberland and Suffolk southd.; local in Ireland; Channel Islands; fl. July–Oct.—Bushy, glaucous, often reddish. *Rootstock* woody. *Stems* 6–18 in., many, stout, erect or ascending, naked and tubercled below. *Leaves* ½–1 in., very thick, sessile, obtuse, concave, nerveless. *Rays* 5–8, short, stout, forked once or twice; bracts variable, ¼–¾ in. diam., sometimes broader than long. *Involucres* sessile or pedicelled, cusps of glands short. *Capsules* deeply lobed, valves very rugose, with a dorsal furrow. *Seeds* ovoid, caruncle minute.—DISTRIB. From Belgium southd. and eastd.

10. **E. E'sula,** *L.* ; perennial, leaves linear or oblong-lanceolate, bracts broadly cordate mucronate, capsule granulate, seeds smooth ovoid brown.

Woods and fields, native in Jersey; naturalized in Forfar, Edinburgh, and Alnwick; fl. July.—*Rootstock* creeping. *Stem* 1–2 ft., erect, slender, naked below, simple or with flowerless side-branches. *Leaves* 1–1½ in., sessile, spreading, acute or obtuse, sometimes denticulate, thin, 1-nerved. *Rays* 10–20, long, slender, forked only at the tips; bracts ½–¾ in. diam., reniform-cordate, acute or obtuse. *Involucres* small, long-pedicelled; glands with short straight cusps. *Capsule* small, valves with a dorsal furrow. *Seeds* with a small caruncle.—DISTRIB. Europe, N. and W. Asia; introd. in N. America.

E. CYPARIS'SIAS, *L.* ; perennial, leaves narrow-linear quite entire, bracts cordate obtuse, capsule granulate, seeds smooth globose pale.

Woods and plantations, from Cumberland southd.; fl. June–July.—Habit of *E. E'sula*, but rather glaucous; rootstock creeping and stoloniferous; more leafy; leaves narrower; bracts smaller; and seeds almost white.—DISTRIB. Europe; introd. in N. America.

*** *Leaves* opposite. *Umbels* 3–4-rayed. *Glands of involucre* lunate, cuspidate.

11. **E. Lath'yris,** *L.* ; biennial, leaves decussate linear-oblong broader at the base obtuse, bracts cordate at the base, capsule smooth, seeds ridged and wrinkled dusky brown. *Caper Spurge.*

Copses and woods, native ? in Somerset and Sussex, naturalized else-where and in the Channel Islands; (an alien, *Wats.*); fl. June–July.— Glabrous, glaucous. *Stem* stout, erect, short, leafy the first year, during the next elongating to 3–4 ft., and flowering. *Leaves* 2–8 in., sessile, spreading, 1-nerved, tip rounded apiculate. *Rays* stout, unequal, irregularly forked; bracts 1–3 in., ovate-lanceolate, acute. *Involucres* large, sessile, cusps of glands suberect. *Capsule* ⅓ in. diam. *Seed* ⅛ in., broadly oblong, obliquely truncate, caruncle large.—DISTRIB. S. Europe; introd. in N. America.

SECTION 2. *Stems* prostrate, dichotomously branched. *Leaves* opposite, stipulate. *Involucres* axillary or in the forks, solitary.

12. **E. Pep'lis,** *L.* ; annual, glabrous, leaves dimidiate-cordate subentire.

Sandy shores, S. Wales and Cornwall to Hants, very rare; Waterford; Channel Islands; fl. July–Sept.—Procumbent, glaucous, purplish. *Stems* many, 6–12 in., spreading from the root. *Leaves* ½ in., coriaceous, shortly petioled, obtuse or retuse, base auricled on one side truncate on the other; stipules ovate, 2-fid. *Involucres* shortly pedicelled, glands oblong. *Capsule* ⅛ in., valves smooth, keeled, glabrous. *Seeds* white, not caruncled.—DISTRIB. Atlantic and Mediterranean shores, from France southd.; salt tracts of Asia.

2. BUX'US, *Tournef.* Box.

Evergreen shrubs. *Leaves* opposite, exstipulate. *Flowers* monœcious, in axillary fascicles or spikes, green, 4-bracteolate, uppermost female. —MALE. *Sepals* 4, 2 outer imbricate in bud. *Stamens* 4, hypogynous, opposite the sepals, filaments stout fleshy ; anthers introrse. *Ovary* rudimentary.—FEMALE. *Sepals* 4-12, often imbricate in threes. *Ovary* 3-celled, top 3-lobed ; styles 3, excentric, spreading, persistent, grooved and stigmatic on the inner face ; ovules suspended in pairs in each cell, anatropous, raphe dorsal. *Capsule* coriaceous, 3-celled, loculicidally 3-valved ; cells 1-2-seeded. *Seeds* pendulous, testa crustaceous shining, base thickened, albumen fleshy ; embryo axile, curved, radicle superior.— DISTRIB. Europe, Africa, Madagascar, Asia, W. Indies ; species 17.— ETYM. πίξος of the Greeks.

B. sempervi'rens, *L.* ; leaves oblong, flowers crowded sessile.

Chalk hills, Kent, Surrey, Bucks, Gloster, (indigenous), naturalized elsewhere fl. April–May.—A shrub or small tree, 8–12 ft., branches erect or drooping. young pubescent, wood close-grained. *Leaves* ½–1 in., obtuse or retuse. *Spikes* small ; bracts and sepals obtuse ; flowers whitish. *Stamens* much exserted ; anthers didymous. *Ovary* globose. *Capsule* ½ in., ovoid, 3-horned, wrinkled. *Seeds* black.—DISTRIB. From Belgium southd., N. Africa, N. and W. Asia, W. Himalaya.—Bitter and poisonous ; wood used for engraving upon.

3. MERCURIA'LIS, *Tournef.* DOG'S MERCURY.

Erect herbs. *Leaves* opposite, petioled, serrate, stipulate. *Flowers* mon- di-œcious, minute ; males in interrupted axillary spikes or racemes ; females clustered, spiked or racemose. *Sepals* 3, valvate. *Disk* in the female of 2 elongate glands, alternating with the carpels. *Stamens* 8–20, on a central disk, filaments slender erect ; anther-cells pendulous from a sub-globose connective, extrorse. *Ovary* 2-celled ; styles long, subulate ; ovule solitary, pendulous. *Capsule* didymous, outer coat separating from the cartilaginous 2-valved inner. *Seeds* pitted or wrinkled, funicle thickened ; cotyledons broad.—DISTRIB. Europe, N. Africa, N. Asia ; species 6.—ETYM. unknown.

1. M. peren'nis, *L.* ; perennial, hairy, diœcious, stem simple.

Shady places, N. to Orkney ; ascends to 1,700 ft. in the Highlands ; Ireland ; Channel Islands ; fl. March–April.—*Rootstock* slender, creeping. *Stem* 6–18 in., solitary, erect, terminal. *Leaves* larger upwards, upper 2–3 in., shortly petioled, ovate or elliptic-lanceolate, broader and subsessile in *M. ova'ta,* Steud., crenate-serrate, green, often blue when dry ; stipules minute. *Male racemes* very slender, long-peduncled ; flowers pedicelled ; sepals acute. *Female spikes* or racemes shorter, 1–3-fld. ; styles long, recurved, stigmatic all over the front. *Capsule* ⅓ in. diam., hispid. *Seeds* grey, cuticle white.—DISTRIB. Europe, N. Africa.

2. M. an'nua, *L.* ; annual, nearly glabrous, stem branched.

Fields, gardens, in England, a casual in Scotland ; rare in Ireland ; (a colonist ? *Wats.*) ; fl. July–Oct.—Diœcious, or monœcious (*M. ambigua,* L. fil.). *Stem* 6–18 in. *Leaves* membranous, shortly petioled, ovate or lanceolate, narrowest in the female, acute, base rounded or cordate, shining, crenate-serrate, ciliate. *Female* clusters with sometimes male flowers intermixed ; styles diverging, stigmatic on the sides in front. *Capsule* small, tubercled, hispid. *Seeds* brown, reticulate.—Distrib. Europe, N. Africa, W. Asia.—Leaves boiled and eaten as a pot-herb.

Order LXIX. URTICA'CEÆ.

Herbs, shrubs, or trees. *Leaves* opposite or alternate usually stipulate. *Flowers* 1- rarely 2-sexual, small, green. *Perianth* of male 3–8-lobed or 3–5-partite, of female tubular or 3–5-cleft, or of a scale-like sepal. *Stamens* as many and opposite the perianth-lobes, filaments straight with erect anthers, or inflexed in bud with reversed anthers. *Ovary* sessile, 1- rarely 2-celled ; style 1 or 0 with a capitate stigma, or styles 2 papillose ; ovules solitary in the cells, erect and orthotropous, or pendulous and anatropous. *Fruit* indehiscent. *Seed* pendulous or erect, albumen 0 or fleshy, radicle superior. —Distrib. All climates ; genera 108 ; species over 1,500.— Affinities. With *Malva'ceæ* and *Euphorbia'ceæ.*—Properties. Very various.

Tribe I. **ULME'Æ.** *Flowers* usually 2-sexual. *Perianth* 3–8-lobed or -partite. *Filaments* slender, straight in bud. *Ovary* 1–2-celled, ovules pendulous. *Albumen* 0; embryo straight.—Trees.

Samara stalked, winged. *Cotyledons* flat...................................1. Ulmus.

Tribe II. **URTICE'Æ.** *Flowers* 1-sexual. *Perianth* of male 4–5-partite, of female tubular or 4–5-cleft. *Filaments* inflexed in bud with reversed anthers. *Ovary* 1-celled; style simple or 0; ovule erect, orthotropous. *Albumen* fleshy or 0; embryo straight.—Herbs or shrubs.

Leaves opposite, with stinging hairs ....2. Urtica. Leaves alternate, with simple hairs ....3. Parietaria.

Tribe III. **CANNABINE'Æ.** *Flowers* diœ:ious, males panicled, females clustered or spicate. *Perianth* of male 5-partite, of female 1 scale-like sepal. *Filaments* straight in bud. *Ovary* 1-celled ; styles 2, ovule pendulous. *Albumen* fleshy ; embryo curved.

Stem twining. Embryo spiral ....4. Humulus.

1. UL'MUS, *L.* Elm.

Trees or shrubs ; juice watery. *Leaves* alternate, simple, distichous, oblique, scabrid ; stipules caducous. *Flowers* 2-sexual, fascicled, lateral. *Perianth* campanulate, 5- rarely 4- 8- or 9-fid, imbricate in bud, persistent. *Stamens* usually 5, filaments adnate to the perianth-tube ; anthers extrorse. *Ovary* free, 1–2-celled ; styles 2, subulate, stigmatic on the inner face ; ovule 1 in each cell, pendulous, anatropous. *Fruit* a 1-seeded samara. *Seed* pendulous, testa thin, albumen 0 ; embryo straight, cotyledons large flat or folded, radicle superior.—Distrib. N. temp. zone ; species 16. —Properties. Bitter, mucilaginous, astringent.

U. monta'na, *Sm.* ; seed in the centre of the oblong or suborbicular samara.   *U. campes'tris,* L. herb. ;   *U. ma'jor,* Sm.   *Scotch, Wych, or Mountain Elm.*

Woods, N. to Sutherland, indigenous and naturalized; Ireland; Channel Islands; ascends to 1,300 ft. in Yorkshire; fl. March–April.—A large tree, 80–120 ft., trunk attaining 50 ft. in girth; branches long, spreading, bark corky or not; twigs pubescent. *Root* sending up suckers, chiefly when cut. *Leaves* 3–6 in., often 3 in. diam., ovate-oblong, cuspidate, doubly and trebly serrate, base unequally rounded or cordate. *Perianth* ¼ in., ciliate, lobes obtuse. *Stamens* 4–6; anthers purple. *Samara* ¾–1¼ in., very variable in breadth and the depth of the notch.—DISTRIB. Europe, Siberia.

U. CÁMPES'TRIS, *Sm.* ; seed above the centre of the obovate or oblong samara. *U. carpinifo'lia,* Lindl. *Common Elm.*

Woods and hedgerows, rarer in Scotland; ascends to 1,500 ft. in Derby; Ireland; Channel Islands; a denizen, never seeding; fl. March–May.—A very large tree, 125 ft., trunk attaining 20 ft. in girth; bark rugged. *Root* sending up abundant suckers. *Branches* spreading (suberect in *U. stric'ta,* Lindl.); twigs often corky. *Leaves* smaller than in *U. monta'na,* 2–3 in., less cuspidate, often narrow at the base (scabrid above and pubescent beneath in *U. subero'sa,* Ehrh., nearly glabrous in *U. gla'bra,* Mill.). *Perianth* smaller. *Stamens* often 4. *Samara* ½–¾ in., usually obovate.—DISTRIB. Mid. and S. Europe, N. Africa, Siberia.—Many vars. of this and the preceding are described, differing in habit and foliage, but they offer no constant characters.

2. URTI'CA, *Tournef.* NETTLE.

Herbs, rarely shrubs, with stinging hairs and tenacious inner bark. *Leaves* opposite ; stipules 2 on each side. *Flowers* mon- di-œcious in bracteate clusters, ebracteolate ; pedicel of male jointed. *Perianth* 4-partite ; segments imbricate in bud, persistent, of male concave, of female flat unequal. *Stamens* 4 ; anthers reniform. *Stigma* subsessile, penicillate. *Fruit* minute, compressed.—DISTRIB. Temp. and trop. regions ; species 30.—ETYM. *uro,* from the *burning* pain of its stings.

1. **U. u'rens,** *L.* ; annual, glabrous except for the stinging hairs, leaves ovate-oblong coarsely serrate, panicles 2-sexual.

Fields and waste places, N. to Shetland ; ascends to 1,600 ft. in Northumbd. ; Ireland ; Channel Islands; fl. June–Sept.—*Stem* 1–2 ft., erect, branched. *Leaves* 1–2 in., petioled, teeth few, terminal oblong. *Spikes* ¼–1 in.; in pairs. *Flowers* few in a cluster; pedicels long or short.—DISTRIB. Europe (Arctic), N. Africa, N. and W. Asia, Himalaya ; introd. in N. America, &c.

2. **U. dioi'ca,** *L.* ; perennial, pubescent, leaves ovate-cordate or lanceolate, deeply serrate, panicles usually 1-sexual.

Hedgebanks, &c., N. to Shetland ; ascends to 2,500 ft. in the Highlands; fl. June–Sept.—*Rootstock* creeping, stoloniferous. *Stem* 2–4 ft., simple or branched. *Leaves* 2–4 in., petiole long or short, nerves impressed ; stipules linear-oblong. *Panicles* 1-3 in., in pairs, males lax- females dense-fld., recurved.—DISTRIB. N. temp. regions (Arctic), S. Africa, Andes.—The young leaves are a good pot-herb, and yield a green dye.

3. **U. pilulif'era,** *L.* ; annual, glabrous except for the stinging hairs, leaves ovate or cordate entire or toothed, female flowers capitate.

Waste places, E. England, chiefly near the sea ; (an alien, *Wats.*); fl. June–Aug.—*Stem* 1-2 ft., simple or branched. *Leaves* 1-3 in., long-petioled ; stipules ovate. *Male spikes* panicled, peduncles very slender elongate. *Female heads* ½ in. diam.; peduncles ½ in., stout. *Flowers* much larger than in the preceding species.—DISTRIB. Europe, N. Africa, W. Asia ; introd. in N. America.—The most virulent British nettle.

U. pilulif'era proper ; leaves deeply serrate.—VAR. *U. Dodar'tii,* L. ; leaves entire or nearly so.

3. PARIETA'RIA, *Tournef.* PELLITORY.

Herbs, rarely shrubs. *Leaves* alternate, quite entire, exstipulate. *Flowers* clustered or cymose, polygamous, 1-3-bracteate, proterogynous. *Perianth* of male 4-partite, valvate in bud ; of female tubular, 4-fid. *Stamens* 4. *Style* long or short, stigma papillose. *Fruit* minute, included in the enlarged calyx.—DISTRIB. Temp. and trop. regions ; species 8.— ETYM. *paries,* from growing on *walls*.

P. officina'lis, *L.* ; leaves triple-nerved. *P. diffu'sa,* Koch.

Old walls, hedgebanks, &c., from Ross southd.; Ireland ; Channel Islands ; fl. June–Oct.—Perennial, pubescent with curled hairs. *Rootstock* short, woody. *Stems* 1-2 ft., tufted, erect, or decumbent, terete ; branches slender, leafy. *Leaves* ½ 4 in., elliptic-lanceolate or ovate, obtuse or acute, petiole slender. *Flowers* in axillary clusters, mostly 2-sexual, in a 3-6-lobed few-fld. involucre. *Calyx* elongate and tubular after flowering.—DISTRIB. Europe, N. Africa, W. Asia.

4. HU'MULUS, *L.* HOP.

Perennial, twining herbs ; juice watery. *Leaves* opposite, lobed ; stipules connate. *Flowers* minute, diœcious.—MALE panicled. *Sepals* 5, free, imbricate in bud. *Stamens* 5, adnate to the base of the sepals ; anthers oblong, basifixed, slits subterminal.—FEMALES in pairs in the axils of the bracts of a dense spike which forms a catkin-like head in fruit, bracteate and bracteolate. *Sepal* 1, membranous, bract-like. *Ovary* free, compressed ; styles 2, subulate, stigmatic all over ; ovule 1, pendulous, campylotropous, micropyle superior. *Fruit* dry, indehiscent, enclosed in the sepal. *Seed* pendulous, testa coriaceous, albumen 0 ; embryo a flat helix. —DISTRIB. N. temp. and trop. regions ; species 2.—ETYM. doubtful.

H. Lu'pulus, *L.* ; bracts of fruit much enlarged scarious.

Hedges and copses, from York southd. ; Channel Islands ; naturalized N. to Renfrew and Elgin, and in Ireland ; ascends to 1,000 ft. in the Highlands fl. July–Aug.—*Rootstock* stout, branched. *Stems* tall, scabrid, almost prickly very tough; branchlets glabrate. *Leaves* 3–4 in. diam., cordate, petioled, uppermost ovate, the rest palmately 3–5-lobed to the middle; lobes ovate, acutely toothed. *Male fl.* ¼ in. diam., in panicles 3–5 in.; *female heads* ½ in. diam., on curved peduncles ½–1 in.; stigmas purple. *Fruiting heads* 1½ in., broadly ovoid or subglobose, yellow; scales orbicular, covered with resinous glands at the base, as are the bracteoles and fruit.—DISTRIB. Temp. Europe, Asia, N. America.—Heads of fruit used in brewing; the young blanched foliage is a good pot-herb.

ORDER LXX. **MYRICA'CEÆ.**

Shrubs or trees, often with a glandular wax-secreting pubescence; buds scaly. *Leaves* alternate, simple, exstipulate (with one exception). *Flowers* in simple or compound spikes, usually bracteate, 1-sexual, perianth 0.—MALE fl. *Stamens* 2–16, filaments adnate to the base of the bract, free or connate; anthers basifixed, extrorse. FEMALE fl. 2–4-nate and usually 2–4-bracteolate. *Ovary* sessile, 1-celled; styles 2, lateral, filiform, stigmatic all over; ovule 1, basal, orthotropous. *Drupe* papillose, sometimes 2-winged from being adnate to the enlarged bracteoles, stone 1-seeded. *Seed* erect, testa thin, albumen 0; cotyledons plano-convex, radicle superior.—DISTRIB. Temp. and trop. Asia, S. Africa, N. America; genus 1; species 35.—AFFINITIES. Close with *Juglandeæ.*—PROPERTIES. Yield wax, resin, benzoic acid, and tannin.

1. MYRI'CA, *L.* SWEET-GALE, BOG-MYRTLE.

Character of the Order.—ETYM. The Greek name.

M. Ga'le, *L.* ; leaves narrowly cuneate-obovate or -lanceolate serrate towards the tip, stamens usually 4, ovary 2-bracteolate.

Bogs and moors, Caithness to Cornwall and Sussex, ascends to 1,800 ft. in the Highlands; Ireland; fl. May–July.—*Shrub* 2–3 ft., twiggy, suberect, resinous, fragrant, flowering before leafing. *Leaves* 2–3 in., rarely quite entire, obtuse or acute, very shortly petioled, often pubescent beneath. *Male spikes* ½–1 in., racemose, crowded, erect; bracts broadly ovate, concave, anthers red; female ¼ in., styles red. *Drupe* minute, lenticular, adnate to the persistent bracts.—DISTRIB. W. and N.W. Europe, N. Asia, N. America.— Much used in cottage-practice and for tea-making.

ORDER LXXI. **CUPULIF'ERÆ.**

Trees or shrubs. *Leaves* alternate, stipulate. *Flowers* monœcious, anemophilous.—MALE fl. solitary, crowded, or in spikes, bracteate. *Sepals* 1–5 or more, unequal, or 0. *Stamens* 2–20, on a disk or adnate to the bases of the sepals; filaments free or connate; anthers introrse, 2-celled.—FEMALE fl.

Calyx adnate to the ovary, or 0. *Ovary* inferior, after fertilization more or less completely 2-3- (rarely 4-6-) celled ; styles as many, stigmatose above and within ; ovules 1, or 2 collateral, erect or pendulous, anatropous. *Fruit* indehiscent, 1- rarely 2-seeded, seated on or enclosed in the hardened or accrescent bracts. *Seed* large, testa thin often adherent to the pericarp, albumen 0 ; cotyledons thick, fleshy or farinaceous, often grooved or folded ; radicle short, superior.—Distrib. N. hemisphere, from N. Africa, N. India, the Malay Is. and Darien northwards, Mts. of South Australia, N. Zealand, Chili ; genera 10 ; species about 400.—Affinities. With *Juglandeæ.*—Properties. Yield tannin, many good woods, and esculent embryos.

Tribe I. **BETULE'Æ.** Male fl. *Spikes* pendulous. *Sepals* 4 or fewer. *Stamens* 2-4. Female fl. 2-3 under each side of a catkin-like spike. *Perianth* 0. *Ovary* 2-celled, cells 1-ovuled ; styles 2. *Fruit* small, compressed, covered by the scales of the spike.

Stamens 2 ; scales of female spike thin, deciduous, 3-fid ....1. Betula.
Stamens 4 ; scales of female spike persistent, woody ....2. Alnus.

Tribe II. **QUERCINE'Æ.** Male fl. *Calyx* 4-10-lobed or -partite. *Filaments* simple ; anther-cells connate. Female fl. 1-3 in an involucre of many bracteoles which enlarges in fruit. *Ovary* 3-7-celled ; ovules 2 in each cell. *Fruit* seated in a cupular involucre.

Male catkins slender. Styles 3, short....................................3. Quercus.
Male catkins globose. Styles 3, filiform ....4. Fagus.

Tribe III. **CORYLE'Æ.** Male fl. *Catkins* pendulous. *Perianth* 0. *Stamens* included between 2 bracteoles ; anther-cells separate or connate, hairy at the tip. Female fl. in pairs ; bracts enlarging in fruit. *Ovary* imperfectly 2-celled ; ovules 2, pendulous from one placenta only. *Fruit* enclosed in the coriaceous bracts.

Female spike minute, with few brown scales..........................5. Corylus.
Female spike large, with many leafy scales.........................6. Carpinus.

1. BE'TULA, *Tournef.* Birch.

Trees or shrubs. *Flowers* monœcious.—Male *catkin. Scales* peltate, with 3-bibracteolate flowers. *Sepals* 1-4. *Stamens* 2, filaments forked, separating the anther-cells.—Female. *Bracts* imbricate, usually 3-lobed, 2-3-fld. ; bracteoles 0. *Perianth* 0. *Ovary* compressed, 2-celled ; styles 2, slender, stigmas terminal. *Ovules* 1 in each cell. *Fruit* lenticular, winged or margined, 1-seeded. *Cotyledons* flat.—Distrib. N. temp. and Arctic regions, Mexico to Peru ; species about 25.—Etym. The Latin name.

1. **B. al'ba,** *L.* ; a tree, leaves long-petioled deltoid rhomboid or ovate acute doubly serrate, fruit broadly winged.

Woods and copses, N. to Orkney ; ascends to 2,500 ft. in the Highlands ; Ireland ; Channel Islands ; fl. April–May.—A short-lived tree, 40–50 rarely

80 ft., trunk 8–10 in. diam.; bark flaking, silvery white; branches often
weeping (*B. pen'dula,* Wahlb.). *Leaves* 1–3 in., sometimes pubescent, rather
coriaceous, resinous or glandular when young; petiole slender; stipules
broad. *Male catkins* ½–2 in., pendulous; sepal 1; female spike solitary,
shorter, suberect. *Scales of fruit* cuneate, brown, 3-lobed to the middle.
Fruit orbicular, wing notched.—DISTRIB. Europe (Arctic), N. Asia, N.
America (a variety).—Bark used in tanning, and yields a fragrant oil; juice
sugary in spring, and a wine is made from it; wood durable.
B. ALBA proper; leaves truncate at the base, lateral lobes of fruiting bracts
spreading. *B. verruco'sa,* Ehrh.
Sub-sp. B. GLUTINO'SA, *Fries;* sometimes bushy, leaves rhomboid-ovate, lateral
lobes of fruiting bracts erect.—VAR. *denuda'ta,* leaves glabrous resinous.—
VAR. *pubes'cens,* leaves and twigs pubescent.

2. **B. na'na,** *L.* ; a bush, leaves short-petioled orbicular crenate, fruit
very narrowly winged.

Mts. of Northumbd., Peebles, and from Perth to Sutherland, local; ascends
to 2,700 ft.; fl. May.—Bush 1–3 ft. *Leaves* ¼–½ in., glabrous, dark green.
Catkins ¼ the size of those of *B. alba.* *Bracts* of fruiting catkins broadly
obcuneate, with 3 rounded lobes.—DISTRIB. Arctic and Alpine N.W. and
W. Europe, N. Asia, N. America.

2. AL'NUS, *Tournef.* ALDER.

Trees or shrubs.—MALE fl. in *catkins.* *Scales* peltate, 3-fld. *Sepals*
and *stamens* 3–5 ; anthers 2-celled.—FEMALE fl. in broadly ovoid spikes ;
scales fleshy, 2–3-fld., each with 2–3 bracteoles or sepals adnate to the
bract. *Fruiting spike* woody ; fruit compressed, winged or not, 1-celled,
1-seeded. *Cotyledons* flat.—DISTRIB. Europe, Asia from the Himalaya
northd., N. America, Andes ; species 14.—ETYM. The Latin name.

A. glutino'sa, *Gœrtn.* ; leaves obovate- or orbicular-cuneate green on
both sides, female spikes racemose.

River-banks, marshes, &c., N. to Caithness ; ascends to 1,600 ft. in the Highlands ; Ireland ; Channel Islands ; fl. March–April.—A bush or tree, 20–40
very rarely 70 ft. ; trunk 1–2 ft. diam. ; bark black ; wood white when alive,
red when cut, then pale pink. *Leaves* 2–4 in., shortly petioled, glutinous
and hairy when young, sinuate and serrulate, laciniate in var. *inci'sa* ; stipules ovate. *Catkins* appearing before the leaves; male 2–4 in., bracts
orbicular red ; female spikes ½–1 in., terminal, racemose, obtuse, bracts red-
brown, woody. *Fruit* ⅒ in., pale, hardly winged.—DISTRIB. Europe, N.
Africa, W. and N. Asia.—Wood soft, durable.

3. QUER'CUS, *Tournef.* OAK.

Trees ; buds scaly ; hairs often stellate. *Leaves* evergreen or deciduous ;
stipules deciduous.—MALE fl. in *catkins,* with usually a caducous bract at
the base. *Calyx* 4–7-lobed. *Stamens* indefinite (10 in the British sp.),
filaments slender exserted.—FEMALE fl. spicate, enclosed in imbricating
bracts. *Calyx-limb* 3–8-lobed. *Ovary* 3-celled ; styles 3 ; ovules 2 in

each cell. *Fruit* terete, 1-seeded, seated in a cupule of imbricating scales. *Seed* with the remains of the septa and undeveloped ovules attached to its upper or lower part ; cotyledons included in the pericarp in germination. —Distrib. Of the Order, excl. the S. temp. hemisphere ; species about 300.—Etym. The Latin name.

Q. Ro'bur, *L.* ; leaves deciduous oblong-obovate sinuate-lobed.

Woods, &c., from Sutherland southd.; ascends to 1,350 ft. in the Highlands ; Ireland ; Channel Islands ; fl. April–May.—*Trunk* 60 to 100 ft., 70 ft. in girth (Cowthorpe oak, Yorkshire) ; bark rugged ; branches tortuous. *Leaves* 3–6 in., base narrowed rounded or cordate, young pubescent beneath, plaited in bud. *Male catkins* 1–3 in., appearing with the leaves, pendulous ; bracts linear ; flowers small, in distant clusters ; sepals hairy. *Acorns* ¾–1 in., 2–3 together ; cupule with many appressed triangular obtuse imbricating scales.—Distrib. From the Atlas, Taurus, and Syria, almost to the Arctic circle.—The following varieties are very inconstant.

Q. sessil'iflo'ra, Salisb. ; leaves petioled, peduncles very short.—*Q. peduncula'ta,* Ehrh. ; leaves sessile, peduncles long.—*Q. interme'dia,* D. Don. ; leaves downy beneath, petioles and peduncles short.

4. FA'GUS, *Tournef.* BEECH.

Trees. *Leaves* deciduous or evergreen ; stipules caducous.—Male fl. in long-peduncled heads ; bracts small or 0. *Calyx* 4–7-lobed. *Stamens* 8-40, filaments slender, exserted ; anthers oblong.—Female fl. 2–4, in a 4-partite involucre of imbricating bracts. *Calyx-limb* 4–5-toothed. *Ovary* 3-gonous, 3-celled ; styles 3, linear ; ovules 2 in each cell, pendulous from the top. *Fruits* usually 2 together, 1- rarely 2–3-seeded, compressed 3-gonous or 2–3-winged, enclosed in the hardened or coriaceous scaly involucre. *Seed* pendulous, crowned with the undeveloped ovules ; cotyledons thin, plaited, leafy after germination.—Distrib. Temp. Europe, N. Asia (excl. the Himalaya), N. and S. America, S. Australia, New Zealand ; species 15.—Etym. from φάγω, from the *eatable* seeds.

F. sylvat'ica, *L.* ; leaves deciduous oblong-ovate obscurely toothed.

Wcods in England, especially on chalk and limestone ; ascends to 1,200 ft. in Derby ; planted in Scotland and Ireland ; fl. April–May.—*Trunk* 118 ft. (King's beech, Ashbridge), and 29 ft. girth (Bicton, Devon) ; head 352 ft. diam. (Knowle beech) ; bark smooth, white ; branches horizontal ; buds acute. *Leaves* 2–3 in., shortly petioled, acuminate, silky when young ; plaited parallel to the nerves in bud ; stipules scarious. *Male fl.* capitate, pendulous ; peduncle 1–2 in. ; anthers yellow ; female on shorter peduncles. *Cupule* ¾ in., 4-cleft, segments bristly. *Fruit* 3-quetrous, smooth.—Distrib. A triangular area between Norway, Asia Minor, and Spain ; Japan.—Wood used for tools, carpentry, and fuel ; fruit yields oil.

5. COR'YLUS, *Tournef.* HAZEL.

Shrubs or trees. *Leaves* deciduous, plaited in bud. Male *catkins* slender, pendulous ; bracts cuneate, with 2 bracteoles above their inner base. *Perianth* 0. *Stamens* 4-8, filaments short; anther-cells separate, tips hairy. FEMALE fl. sessile in pairs in the upper bracts of a minute head, each enclosed in a 3-partite bracteole which enlarges after flowering. *Calyx-limb* unequally toothed. *Ovary* 2-celled; styles filiform; ovules 1 in each cell, pendulous. *Fruit* woody, 1-celled, 1-seeded, enclosed in the greatly enlarged coriaceous more or less cut bract and bracteoles. *Cotyledons* thick, plano-convex, included in the nut in germination.— DISTRIB. Temp. N. hemisphere; species 7.—ETYM. κόρυς, from the *cap*-like form of the involucre.

C. Avella′na, *L.* ; leaves orbicular-cordate doubly serrate cuspidate.

Copses and hedges, N. to Orkney; ascends to nearly 1,900 ft. in the Highlands; Ireland; Channel Islands; fl. Feb.-March.—A glandular, hispid and pubescent shrub; rarely a tree 30 ft, with trunk 3 ft. girth at the ground (Eastwell Park, Kent). *Leaves* 2-4 in., distichous, base unequal, plaited parallel to the midrib in bud; petiole short; stipules oblong, obtuse. *Male catkins* 1-2 in., 2-4 in a raceme, female heads subsessile. *Fruit* on an elongated branch, ½-¾ in., clustered, woody; involucre palmately lobed and cut, unarmed.—DISTRIB. Europe, N. Africa, temp. Asia.—Wood very elastic. Nuts yield abundance of bland oil.

6. CARPI′NUS, *Linn.* HORNBEAM.

Trees. *Leaves* deciduous.—MALE. *Catkins* lateral; bracts ovate, acute. *Stamens* 3-12 in the axil of the bract, filaments slender forked; anther-cells separate, stipitate, tips hairy.—FEMALE. *Spikes* terminal, erect in flower, pendulous in fruit; bracts ovate-lanceolate, caducous; flowers in pairs, each in a lobed bracteole which enlarges after flowering. *Calyx-limb* toothed. *Ovary* strongly nerved, 2-celled; styles 2, filiform; ovules 1 in each cell, pendulous. *Fruit* almost woody, nerved, 1-celled, 1-seeded, enclosed in the large leafy lobed bracteole. *Cotyledons* fleshy.—DISTRIB. N. temp. zone; species 9.—ETYM. The Latin name.

C. Bet′ulus, *L.* ; leaves elliptic-ovate doubly serrate hairy beneath, female bracts 3-lobed 3-nerved mid-lobe much the longest.

From N. Wales, Stafford and Norfolk to Devon and Kent; planted N. of this and in Ireland; Channel Islands; fl. May.—A small tree, sometimes 70 ft.; with the trunk (usually flattened) 10 ft. in girth; bark smooth, light-grey, wood close, white, heavy. *Leaves* 2-3 in., subdistichous, acute or acuminate, shortly petioled, plaited parallel to the nerves in bud; stipules large, linear-oblong. *Male catkins* 1-2 in., pendulous; bracts ovate-lanceolate, acute. *Female* 2-4 in., pendulous in fruit, cylindric; bracteole 1-1½ in., entire or toothed. *Fruit* ¼ in., green, 7-11-nerved.—DISTRIB. From Gothland southd., W. Asia.—Wood the best fuel, very tough and difficult to work.

ORDER LXXII. **SALICI'NEÆ.**

Trees or shrubs. *Leaves* alternate, simple, deciduous, stipulate. *Flowers* dioeciós, in catkins which usually precede the leaves. *Perianth* 0. *Disk* annular urceolate or glandular.—MALE. *Stamens* 2 or more, inserted under the disk, filaments free or connate ; anthers basifixed, introrse.— FEMALE. *Ovary* sessile or pedicelled, 1-celled ; styles 2, short, stigmas 2 entire or 2-4-lobed ; ovules many on 2 parietal placentas, ascending, anatropous, raphe dorsal. *Capsule* 1-celled, loculicidal ; valves 2, rolling back, many-seeded. *Seeds* minute, testa membranous, funicle short, with a pencil of silky hairs that conceals the seed, albumen 0 ; embryo straight, cotyledons plano-convex, radicle inferior.—DISTRIB. Arctic and N. temp. zones, rare in the tropics and S. ; absent from Australia and the Pacific ; genera 2 ; species about 180.—AFFINITIES. Very obscure.—PROPERTIES. Bitter, astringent, febrifuge, aromatic.

Leaves broad. Catkins drooping, scales cut.............................1. Populus.
Leaves usually narrow. Catkins usually erect, scales entire..........2. Salix.

1. PO'PULUS, *Tournef* POPLAR.

Catkins drooping ; scales crenate lobed or cut. *Disk* oblique, cupular *Stamens* 4-30, filaments free. *Stigmas* slender, 2-4-cleft.—DISTRIB. N. temp. regions ; species 18.—ETYM. The Latin name.

SECTION 1. **Leu'ce.** Young shoots pubescent. *Fruiting catkins* dense scales ciliate. *Stamens* 4-12. *Stigmas* 2-4-lobed.

1. **P. al'ba,** *L.* ; buds not viscid, leaves of shoots more or less lobed, of branches broadly ovate cordate sinuate white and cottony beneath.

Moist woods, river-banks, &c., from Elgin and the Clyde southd. ; Ireland Channel Islands ; fl. March-April.—A large tree, 60-100 ft. ; bark grey, smooth ; wood white ; branches spreading, buds cottony ; suckers many, with large deltoid-ovate lobed and toothed leaves 2-4 in. diam. *Leaves* on the branches 1-3 in., glabrous in age ; petiole very long, slender, compressed *Catkins* 2-4 in., cylindric ; female shorter. *Stamens* 6-10 ; anthers purple. *Capsules* ¼ in., narrow ovoid.—DISTRIB. From Gothland southd., N. Africa, N. and W. Asia, N.W. Himalaya.—Wood light, does not burn easily.

P. AL'BA proper ; leaves of the suckers lobed, of the branches white and cottony beneath, stigmas usually 2-4 linear. *White Poplar, Abele.*— A doubtful native, cultivated as far N. as Forfar, but does not flower in Scotland ?

Sub-sp. P. CANES'CENS, *Sm.* ; leaves of the suckers angled and toothed, of the branches hoary beneath or glabrous, stigmas 2-4 rarely 2 each 4-cleft. *Grey Poplar.*—A supposed hybrid with *tremula,* indigenous in S.E. England. Wood said to be superior to that of *P. alba* proper. I have never seen stigmas like those figured in *Engl. Bot.* (? copied from Reichenbach).

2. **P. trem'ula,** *L.* ; buds not viscid, leaves of shoots cordate acute entire, of branches suborbicular-ovate sinuate-serrate with incurved teeth glabrous or silky beneath. *Aspen.*

B B

Copses, &c., N. to Orkney, indigenous, more often planted; ascends to 1,600. ft. in Yorkshire; Ireland; Channel Islands; fl. March–April.— Erect, 40–80 ft., short-lived. *Bark* grey, wood white; suckers many, pubescent; branches spreading; buds pubescent. *Leaves* 1–4 in., old obtuse, young acute, cottony beneath; petiole very long, slender, glabrous, compressed. *Catkins* 2–3 in., cylindric; scales laciniate. VARS. *villo'sa* and *gla'bra*, have respectively villous and subglabrous foliage.—DISTRIB. Europe (Arctic), N. Africa, N. Asia.—Wood indifferent.

SECTION 2. **Aigei'ros.** Young shoots glabrous. *Fruiting catkins* lax; scales subglabrous. *Stamens* 8–30. *Stigmas* 2-fid, short, cuneate.

P. NI'GRA, *L.*; buds viscid, leaves rhombic deltoid or suborbicular finely crenate-serrate at length glabrous. *Black Poplar.*

Moist places, river-banks, &c., not indigenous; fl. April.—Erect, 50–60 ft., short-lived, growth rapid. *Bark* grey; wood soft, white; branches spreading. *Leaves* 1–4 in., angles rounded, acuminate, young silky beneath and ciliate; petiole slender, compressed. *Male catkin* 2–3 in., cylindric; female shorter, ascending, peduncle curved in fruit; scales shortly cut. *Stamens* 12–20; anthers purple. *Capsules* ¼ in., ovoid, pedicelled, recurved.—DISTRIB. Europe, N. Asia.—Wood light, much used for carving, charcoal, &c.; bark for tanning.

2. SA'LIX, *Tournef.* WILLOW.

Trees or shrubs. *Leaves* quite entire or serrate; stipules persistent or deciduous. *Stamens* 2 or more, filaments free or connate. *Catkins* usually erect; scales entire. *Disk* of 1–2 distinct glands. *Stigmas* entire or 2-fid. —DISTRIB. Of the Order; species 160.—ETYM. The Latin name.—For the species of this troublesome genus I have followed Andersson (in *D.C. Prodr.* XVI. part 2), and for the vars., principally Syme.

SECTION 1. *Catkins* on short peduncles that bear fully developed leaves; scales pale, persistent or deciduous. *Filaments* hairy below, all free. *Capsule* glabrous in the British species.

* *Stamens 3 or more, free. Petiole glandular at the top. Capsule pedicelled.*

1. **S. trian'dra,** *L.*; leaves linear- or oblong-lanceolate acuminate glandular-serrate glabrous paler or glaucous beneath, disk of the male flower 2-glandular, stamens 3. *Almond-leaved* or *French Willow.*

River-banks and osier-grounds from Perth southd.; doubtfully native of Scotland and Ireland; fl. April–June.—A tree, 20 ft., bark flaking. *Leaves* 2–4 in., base broad or narrow; stipules large, ½-cordate. *Catkins* 1–2 in., appearing with the leaves, slender, female narrower; scales nearly glabrous. *Capsule* small, terete or furrowed, glabrous; style thick, short.—DISTRIB. Europe (Arctic), N. Asia.

S. trian'dra proper (*S. amygdali'na,* L.); leaves more linear narrow glaucous beneath.—VAR. *S. Hoffmannia'na,* Sm.; leaves broader at the base green beneath.—*S. contor'ta,* Crowe, is a var. cultivated in Sussex.—*S. undula'ta,* Ehrh. (*S. lanceola'ta,* Sm.), distinguished from *S. trian'dra* by its shaggy scales and distinctly developed style, is commonly cultivated for basketwork, but is not indigenous. Andersson supposes it to be a smooth-fruited hybrid between *triandra* or *alba*, and *viminalis.—S. Trevira'ni,* Spr., is another hybrid with *viminalis,* found in Staffordshire.

2. **S. pentan'dra,** *L.* ; leaves elliptic or ovate- or obovate-lanceola'e acuminate glandular-serrulate viscid shining paler beneath, disk of male and female flowers 2-glandular, stamens 5 (4–12). *Bay-leaved Willow.*

River-banks and wet places, from N. Wales, Worcester, and York to Argyll and Banff ; planted S. of this ; ascends to 1,300 ft. in Northumbd.; rare in Ireland; fl. May–June.—A glabrous shrub, 6–8 ft., or tree, 20 ft.; dark brown. *Leaves* 1–4 in., fragrant, reticulate beneath ; stipules ovate oblong or 0. *Catkins,* male 1–2 in., erect then pendulous, scales oblong ; female shorter. *Capsule* glabrous ; style short.—DISTRIB. Europe (Arctic), excl. Greece and Turkey, W. and N. Asia.—The latest-flowering willow.—*S. cuspida'ta,* Schultz (*Meyeria'na,* Willd.), found in Shropshire, is probably a hybrid with *frag'ilis.*

** *Stamens 2, rarely more, free. Petiole glandular or not at the top.*

3. **S. frag'ilis,** *L.* ; leaves lanceolate long-acuminate glabrous glandular-serrate pale or glaucous beneath, young hairy, stipules ½-cordate deciduous, capsule pedicelled. *Crack Willow, Withy.*

Marshy ground, from Ross southd.; ascends to 1,300 ft. in Northumbd.; ? native in Scotland or Ireland; Channel Islands; fl. April–May.—A tree, 80–90 ft., trunk sometimes 20 ft. in girth ; branches spreading obliquely from the trunk ; twigs easily detached, smooth, polished. *Leaves* 3–6 in., petiole often glandular at the top. *Catkins* usually spreading, stout, male 1–2 in., female slender, often longer ; scales linear-lanceolate; disk 2-glandular. *Stamens* 2, rarely 3–5. *Capsule* glabrous ; style short.—DISTRIB. Europe, N. and W. Asia ; introd. in America.—Andersson doubts this being indigenous except in S.W. Asia.

S. frag'ilis proper ; twigs yellow-brown, leaves elliptic-lanceolate.—VAR. *S. decip'iens,* Hoffm.; twigs orange or crimson, leaves smaller, style longer.

S. Russellia'na, Sm. (*Bedford Willow*), a tree, 50 ft., trunk 12 ft. in girth, is considered a hybrid between *fragilis* and *alba,* and referred to *S. vir'idis,* Fries. Mr. Baker says it is a synonym of *S. frag'ilis ;* and that Fries' *S viridis* is not a British plant.

4. **S. al'ba,** *L.* ; leaves narrowly lanceolate long-acuminate silky on both surfaces (except when old) glandular-serrate, stipules ovate-lanceolate deciduous, capsule subsessile. *White Willow.*

Marshy ground, always? planted, from Sutherland southd.; Ireland ; Channel Islands; fl. May.—A large tree, 80 ft., trunk 20 ft. in girth ; bark fissured : twigs not easily detached, silky. *Leaves* 2–4 in., glabrous when old, petiole eglandular. *Catkins* slender, lax, erect, scales linear. *Capsule* glabrous, style very short —DISTRIB. Europe, N. Africa, N. and W. Asia, N.W. India.—Timber most useful for carpentry and fuel ; bark for tanning.

S. al'ba proper ; twigs olive, old leaves silky on both surfaces.—VAR. *S. ceru'lea,* Sm. ; twigs olive, old leaves glabrous glaucous beneath.—VAR. *S. vitelli'na*

L.; twigs yellow or reddish, old leaves glabrous above, scales of catkins longer. *Golden Willow.*

SECTION 2. *Catkins* on leafy or bracteate peduncles; scales persistent, discoloured at the tip (except *S. reticulata*). *Stamens* 2, filaments free. *Disk* 1-glandular. *Capsule* tomentose or silky, rarely glabrous.

  * *Capsule with a slender pedicel; style very short or* 0.

5. **S. Capre'a,** *L.*; leaves elliptic or oblong-obovate or -lanceolate acute or acuminate crenate reticulate on both surfaces tomentose beneath, stipules ½ reniform, catkins silky, male ovoid-oblong, female elongate at length nodding, scales hairy, tip black. *Common Sallow, Goat Willow.*

Copses, pastures, &c., by streams, from Argyll and Inverness southd.; ascends to 2,000 ft. in the Highlands; Ireland; Channel Islands; fl. April–May.— A grey tree or large shrub. *Leaves* 2–4 in., dark green above, cuspidate, margins narrowly recurved. *Catkins* short, preceding the leaves, sessile, bracteate; male 1 in., very stout, female lengthening to 3 in. *Capsule* ½ in., silky; pedicel very slender.—DISTRIB. Europe, N. and W. Asia, Himalaya. —The earliest-flowering British willow. The twigs with catkins gathered at Easter, are called Palm-branches. Andersson points out the impossibility of distinguishing this from *S. cine'rea*, L.
S. CAPRE'A proper; buds and twigs glabrous or puberulous, leaves usually broad glabrous and dull green above undulate crenate-serrate or subentire, stipules long or 0 —*S. sphacela'ta*, Sm., is a subalpine form, without stipules and with subentire leaves.
Sub-sp. S. CINE'REA, *L.*; buds and twigs tomentose, leaves smaller narrower from elliptic-oblong to oblanceolate margins undulate pubescent above, male catkins less stout opening later, anthers pale yellow, capsule smaller (filaments hairy at the base, *Syme*).—*S. aquat'ica*, Sm., with leaves more obovate glaucous, hairs beneath white, stipules large, and *S. oleifo'lia*, Sm., with leaves narrow rigid glaucous, hairs beneath red-brown, stipules small, are slight varieties.—Andersson correctly refers *S. cine'rea*, Sm., in *Engl. Bot.*, t. 1897, to this; but also quotes that plate and name under *daphnoi'des*, Vill., a very different plant.

6. **S. auri'ta,** *L.*; leaves obovate-oblong rarely oblanceolate crenate much wrinkled pubescent and reticulate beneath, stipules reniform, catkins short dense-fl., male ovoid, female cylindric.

Moist copses, heaths, &c., N. to Shetland; ascends to 2,000 ft. in the Highlands; Ireland; Channel Islands; fl. April–May.—A small bush, 2–4 ft., with straggling branches; probably a form of *S. Caprea*, being so closely allied to sub-sp. *cine'rea* that it is chiefly distinguishable by its smaller size, reddish twigs, leaves rarely 2 in., very much wrinkled, young reddish and crisped, often petioled, large stipules, shorter catkins, ½–¾ in., and narrower more tomentose capsule.—DISTRIB. Europe, N. and W. Asia.

7. **S. re'pens,** *L.*; leaves small oblong or linear-lanceolate obtuse or acute, margins recurved entire or serrulate shining and reticulate above, silky or glaucous beneath, stipules 0 or lanceolate, catkins cylindric-oblong, scales spathulate, anthers at length black. *S. fœ'tida*, Sm.

Heaths, commons, &c., from Shetland southd.; ascends to 2,500 ft. in the Highlands; Ireland; Channel Islands; fl. April–May.—A small, straggling bush; branches slender, elongate, erect or decumbent; buds silky. *Leaves* excessively variable, ½–1½ in., young always silvery silky. *Catkins* preceding or appearing with the leaves, erect, short, sessile, rarely on lengthening leafy peduncles ½–1 in.; bracts leafy; scales yellow-green or purple, silky, always dark at the tip; anthers yellow till the pollen is shed. *Capsule* pedicelled, glabrous or silky.—DISTRIB. Europe, Siberia.—The following are the chief British forms.

S. re'pens proper; stem decumbent below, flowering branches erect or ascending, leaves appearing with the flowers elliptic-oblong quite entire silky beneath tip straight, stipules 0, capsule glabrous.—*S. fus'ca*, L.; stem suberect, branches spreading, leaves elliptic-oblong faintly serrate tip straight; the rest as in *repens.*—*S. prostra'ta*, Sm.; prostrate, branches many slender, leaves appearing after the flowers faintly serrate puberulous above glaucous and silky beneath, stipules minute or 0, capsule silky.—*S. ascen'dens*, Sm.; stem decumbent, branches ascending, leaves as in *prostra'ta* but appearing with the flowers more silky beneath and tips recurved, stipules ovate or lanceolate or 0, capsule silky at length glabrous.—*S. parvifo'lia*, Sm.; stem as in *prostra'ta*, leaves and capsules as in *ascendens*, stipules small ovate or 0. —*S. argen'tea*, Sm. (*arena'ria*, L. partly); stem and simple slender branches erect, leaves appearing with the flowers elliptic-ovate quite entire densely silky and silvery especially beneath, tip recurved, capsule silky. A large form, growing in sandy places.—*S. rosmarinifo'lia*, L.; leaves 2–3 in. linear or linear-lanceolate faintly glandular-serrate or entire glabrous or silky beneath, stipules ovate or lanceolate, catkins short sessile dense, scales black, capsule tomentose. Said to have been found in the last century by Sherard in bogs in Scotland.

S. incuba'cea, L. (? *S. ambig'ua*, Ehrh.; *S. re'pens*, var. *incuba'cea*, Syme), is referred to *re'pens* by Wimmer, and to a hybrid between this and *auri'ta* by Andersson; it has reticulate leaves with recurved margins and large stipules.—*S. spathula'ta*, Willd., is referred by Andersson to a hairy form of it.

** *Capsule with a slender pedicel; style distinct.*

8. **S. ni'gricans**, *Sm.*; leaves thin ovate-oblong cordate or lanceolate subacute reticulate above, stipules 0 or ½-cordate, catkins sessile or on short leafy peduncles ovoid or cylindric, scales linear-oblong hairy.

Rocks and banks of streams, from York to Argyll and Aberdeen; ascends to 2,300 ft. in the Highlands; rare in Ireland; fl. May–June.—A procumbent shrub or tree, 10 ft. *Leaves* entire or serrate, black when dry. *Catkins* ½–1 in., appearing before or after the leaves, scales acute or obtuse. *Filaments* hairy at the base. *Capsule* narrowly conical, glabrous or tomentose; pedicel and styles slender.—DISTRIB. Alps of N. and Mid. Europe. Andersson describes this as the most variable of the genus. It is perhaps a form of *S. phylicifo'lia*, which presents a parallel series of variations. Syme sums its differences in the thinner more reticulate darker leaves, more or less glaucous beneath, blackening more when dry, and more permanently hairy; more pubescent twigs; large stipules; more glabrous capsule; shorter style and longer pedicel.

Mr. Baker's experience is:—*phylicifo'lia*, twigs bright chestnut, rarely hairy; leaves firmer brighter coloured above, glaucous (very rarely not), quite

glabrous beneath ; stipules of the leaf-shoots smaller and more deciduous ;
plant drying without turning black :—*ni'gricans*, twigs much shorter, dull-
coloured, pubescent (like *ciné'rea*) ; leaves softer usually grey-pubescent and
much less if at all glaucous beneath ; stipules of barren shoots larger, more
persistent ; style and pedicel the same in both, and ovary similarly variable
in silkiness; plant turning black when dried. Both, when growing with
Caprea and *cinerea*, flower a little later (through May into June), and are
mostly plants of subalpine valleys.
The erect varieties with broad leaves 1–4 in., more or less glaucous beneath,
and silky capsules are :—True *ni'gricans ; cotinifo'lia*, Sm. ; and *Forsteria'na*,
Sm.—*S. rupes'tris*, Sm., is a trailing variety with small broad leaves.- *S.
Andersonia'na*, Sm. ; *damasce'na*, Forbes, and *petræ'a*, G. Anders., are erect
shrubs with glabrous capsules.—*S. hir'ta*, Sm., is a subarborescent form (male
only) with silky twigs and leaves densely pubescent beneath.—*S. floribun'da*,
Forbes (*tenuifo'lia*, Sm. ; *bi'color*, Hook.), is a doubtful plant.

9. **S. phylicifo'lia**, *L.* ; leaves ovate-oblong or elliptic-lanceolate quite
glabrous shining above glaucous beneath, stipules 0 or very small, catkins
sessile bracteate, scales linear-oblong black acute.   *Tea-leaved Willow.*

Rocks and mountain streams, from York and Lancashire to Orkney ; ascends
to 2,000 ft. in Yorks. and Perth ; rare in Ireland; fl. April-May.—A
handsome large bush or small tree (10 ft.). When fully developed conspicuous from its spreading shining chestnut or reddish branches, and
glistening green and glaucous foliage. Andersson distinguishes it from *S.
ni'gricans* by the less unequal thicker leaves, which blacken less in drying,
and have no minute white dots; capsule larger, with longer beaks; he
adds that it is the earliest flowerer of the genus (in England *S. Capre'a*
flowers first) ; (Syme says it is among the latest).—DISTRIB. Almost the same
as *S. ni'gricans*.
The British forms are, 1st, the erect with silky capsules, *S. Davallia'na*, Sm. ;
Weigelia'na, Willd. (*Wulfenia'na*, Sm.) ; *ni'tens*, G. Anders. ; *Crowea'na*, Sm. ;
Dicksonia'na, Sm. (*myrtilloi'des*, Sm., not L.) ; and *tenu'ior*, Borr.—2d, erect
with glabrous or nearly glabrous capsules, *laxiflo'ra*, G. Anders. ; *propin'qua*,
Borr. ; *tetrap'la*, Walker ; *Borreria'na*, Sm.; *phillyreæfo'lia*, Borr ; *tenuifo'lia*,
Borr.—3d, a more or less decumbent rooting form, with silky capsules, *S.
radi'cans*, Sm. (*phylicifo'lia*, Sm.).
S. lauri'na, Sm. (*laxiflo'ra*, Borr.; *bi'color*, Sm.), a small handsome tree, 20–30
ft., found in various parts of England and Ireland, is a hybrid between *S.
phylicifo'lia* and *Capre'a*.

*** *Capsule with a short pedicel or* 0. *Style slender.*

S. DAPHNOI'DES, *Vill.* ; arboreous, buds large, leaves narrow oblong or
linear-lanceolate acuminate acutely serrate shining above glaucous beneath,
stipules ½-cordate acute, catkins stout sessile, scales black acute very villous, capsules glabrous.   *S. acutifo'lia*, Willd.   *Violet Willow.*

Great Ayton, Yorkshire, not indigenous; fl. April.—A small tree, 10–12 ft.,
twigs violet. *Leaves* 3–6 in., very acuminate, with persistent glaucous
bloom. *Catkins* clothed with silky hairs, appearing before the leaves ;
scales black-pointed ; anthers yellow.—DISTRIB. S. Scandinavia across
Europe and Asia to the Amur.

10. **S. vimina'lis,** *L.* ; leaves linear-lanceolate acuminate reticulate above silvery silky beneath, margin revolute quite entire, stipules linear-lanceolate, catkins sessile, bracts small or 0, capsules tomentose. *Osier.*

Wet places and osier-beds, from Elgin and Argyll southd.; Ireland, native? ; fl. April–June.—A shrub or small tree, 30 ft.; branches long, straight, young silky, old polished, leafy. *Leaves* 4–10 in., narrowed into the petiole, margins waved. *Catkins* ¾–1 in., mature long before the leaves, golden-yellow; scales oblong, brown; gland slender. *Capsule* shortly pedicelled, base broad, white; stigmas rarely 2-fid.—DISTRIB. Russia, N. Asia, Soongaria; cult. throughout Europe; introd. in N. America.

S. stipula'ris, Sm., is a 'supposed hybrid, probably with *Caprea* or *cinerea*, cultivated in Essex and Suffolk; Channel Islands; it has broader more undulating leaves, large stipules, and a shortly pedicelled capsule.—In Britain this and *Smithia'na* shade off into *viminalis*, and keep quite distinct from *cinerea* (Baker).

S. Smithia'na, Willd. (*S. mollis'sima,* Sm.); this Andersson regards as an undoubted and excessively variable hybrid between *viminalis* and *Caprea;* it is very common in osier-grounds of England and Ireland, and found in Scotland; the leaves are not so silvery beneath, usually dull and hoary; capsule long pedicelled.—*S. acumina'ta,* Sm., *ferrugin'ea,* G. Anders., *S. holoseric'ea,* Willd., and *rugo'sa,* Leefe (*holoseric'ea,* Hook. and Arn.), are varieties or hybrids with *S. cinerea.*

11. **S. lana'ta,** *L.* ; leaves broadly ovate or oblong-lanceolate acute woolly entire cottony beneath with raised reticulate veins, stipules large ½-cordate glandular-serrate, catkins terminal sessile, scales obtuse clothed with long golden hairs, capsules glabrous.

Alpine cliffs and rills, Perth, Fo:far, Inverness, and Sutherland; alt. 2,000–2,500 ft.; fl. May–June.—A small shrub, 2–3 ft., branches tortuous, twigs tomentose; buds large, black, hirsute. *Leaves* 1–3 in., coriaceous, petiole very short. *Catkins* appearing with the leaves, crowded, stout, sometimes ashy white as in *S. Lappo'num,* especially in age; male 1–2 in., female 2–4 in.; scales black, oblong. *Stamens* glabrous. *Capsule* shortly pedicelled; style very slender, stigmas filiform notched or 2-fid.—DISTRIB. Arctic and Alpine Scandinavia, Arctic Asia, Altai Mts., Greenland.

12. **S. Lappo'num,** *L.* ; dwarf, leaves elliptic or obovate-lanceolate acuminate reticulate silky and villous above cottony beneath with straight raised veins, margins recurved, stipules 0 or small, catkins subsessile bracteate, scales acute clothed with long white hairs, capsules tomentose.

Alpine rocks, Edinburgh and Argyll to Sutherland; alt. 2,000–2,700 ft.; fl June–July.—Shrubby, 2–3 ft., erect or decumbent; branches stout, brown, buds woolly. *Leaves* 1–2 in., very variable, quite entire or sinuate-serrate, dull green above; petiole rather long, base dilated. *Catkins* preceding the leaves; male ovoid, scales black, anthers yellow; female longer, 1–3 in. *Capsule* conical, subsessile; style very long, stigmas filiform cleft.—DISTRIB. Scandinavia to the Arctic circle, Mts. of France and N. Italy, Siberia, N. America.—The Edinburgh specimens seen by me are flowerless; this locality is anomalous for so alpine a plant.

The following are slight varieties:—*S. arena'ria,* L. (partly); leaves downy
above woolly beneath, style equalling the capsule.—*S. Stuartia'na,* Sm.;
leaves woolly above silky and cottony beneath, style equalling the capsule.—*S. glau'ca,* Sm. (not L.); leaves snow-white and woolly beneath, style
much shorter than the capsule. (I should doubt this being the same
species.)

13. **S. Myrsini'tes,** *L.*; dwarf, leaves small rigid ovate obovate or
lanceolate glandular-serrate shining and reticulate on both surfaces, stipules
0 or lanceolate, catkins on leafy peduncles, scales spathulate blackish,
capsules hairy. *S. retu'sa,* Dickson (Andersson).

Alpine rocks and rivulets in Mid. Scotland; alt. 1,000–2,700 ft; fl. June–
July.—A small rigid suberect or creeping shrub, young parts clothed with
silky deciduous hairs. *Leaves* usually ½–1 in., very variable, dark green
and glossy; petiole very short; stipules ovate-lanceolate, serrate. *Catkins*
½–1 in., appearing with or after the leaves, on stout peduncles often as long,
oblong, male ovoid; scales pilose; disk large; anthers at length black.
Capsule hairy or pubescent, distinctly pedicelled; style long, stigmas thick.
—Distrib. Alps of Scandinavia (Arctic), Mid. Europe, Siberia, N. America.
The following are British varieties: *S. procum'bens,* Forbes (*S. læ'vis,* Hook.);
leaves broad subacute faintly serrate.—Var. *arbutifo'lia,* Syme; leaves
narrow acute or acuminate very faintly serrate.—Var. *serra'ta,* Syme;
leaves ovate acute serrate.

S. Graha'mi, Borr. MS. (Baker in *Seem. Journ. Bot.* 1867, 157, t. 66), is only
known from female specimens cultivated in the Edinburgh Bot. Garden,
said to have been brought by Prof. Graham from Frouvyn in Sutherland. It
appears to me to be a form of *S. Myrsini'tes,* with smaller catkins, paler scales,
and a perfectly glabrous capsule with a rather long very silky pedicel; and
not allied to *S. pola'ris* or *herba'cea.* Syme suspects it to be a hybrid
between *herbacea* and *nigricans* or *phylicifolia*; and Nyman, a sub-sp. of
S. retusa, L. The *Engl. Bot.* figures of the ovary and scale are very in-
correct. A similar plant occurs in Muckish Mt., Donegal.

14. **S. Sadle'ri,** *Syme;* dwarf, leaves small short-petioled broadly ovate
or ovate-cordate subacute entire smooth and cottony above reticulated
and glabrous beneath, stipules 0, catkins terminating leafy branches,
scales oblong obtuse, capsule glabrous.

Rocky ledges of Glen Callater, alt. 2,500 ft.; frt. Aug.—Prostrate branches
few, tortuous; bark shining, red-brown; shoots woolly. *Leaves* few, ½–1 in.,
firm, dark green, young cottony beneath; petiole ⅛ in. *Catkins* ½–¾ in.,
cylindric, many-fld.; peduncle woolly, ¼–⅜ in., leafless; scales woolly at
length dark brown, shorter than the capsules which are ⅙ in., glabrous,
conical-ovoid; pedicels and persistent styles slender.—Only two plants
hitherto seen, both in ripe fruit. I am indebted to Dr. Balfour for the
loan of one here described. Nyman regards it as a sub-sp. of *S. lana'ta,* or
hybrid with this and *reticula'ta.*

15. **S. Arbus'cula,** *L.*; leaves ovate-lanceolate or obovate acuminate
serrulate shining above pale or glaucous beneath, catkins lateral on bracteate peduncles, scales obtuse hirsute, capsules sessile tomentose.

Rocks, Dumfries, Argyll, Perth, Forfar, Aberdeen, alt. 1,000–2,400 ft.; fl. June–July.—A small, rigid, decumbent, rooting shrub; twigs yellow, pubescent, then brown. *Leaves* ½–1½ in., very variable, at first silky beneath. *Catkins* ½–1 in., females often long-peduncled; scales obovate or rounded, reddish, very pubescent. *Capsules* conical, reddish, base embraced by the scale; style long, deeply cleft, stigmas thick notched.—DISTRIB. Alps of Mid. and N. Europe (Arctic), N. and Central Asia, Greenland.—Intermediate between *S. phylicifo'lia* and *Myrsini'tes.*
British forms are: *S. carina'ta*, Sm.; suberect, leaves folded and recurved.—*S. fœ'tida*, Schl. (*prunifo'lia*, Sm.); decumbent or ascending, leaves flat.—*S. venulo'sa*, Sm.; decumbent, leaves reticulate on both surfaces.—*S. vaccinii-fo'lia*, Walker (*liv'ida*, Sm., not Wahl); decumbent, leaves silky beneath.

**** *Capsule sessile or subsessile. Style short or 0.*

16. **S. herba'cea,** *L.* ; very dwarf, branches buried, leaves oblong or orbicular obtuse or retuse shining reticulate serrate, catkins on 2-leaved peduncles oblong few-fld., scales concave glabrous or pubescent.

Loftiest Welsh, N. English, Scotch and Irish Alps, ascending to 4,300 ft.; fl. June.—*Stem* and *branches* spreading under the turf, sending up short flowering few-leaved twigs; bud-scales persistent, brown. *Leaves* ¼–½ in., shortly petioled, curled; stipules minute ovate or 0. *Catkins* ⅛–¼ in., flowering after the leaves, subterminal, shortly peduncled, 4–10-fld.; scales obovate, obtuse; anthers yellow-brown or purple. *Capsule* rarely pubescent, subsessile; style rather short.—DISTRIB. Arctic and Alpine Europe, N. Asia, N. America.—The smallest British shrub.

17. **S. reticula'ta,** *L.* ; dwarf, branches buried, leaves orbicular-oblong or obovate strongly reticulate on both surfaces green above glaucous beneath, catkins on very long leafy peduncles, style very short.

Lofty Mts. of Perth, Forfar, Aberdeen, Inverness, and Sutherland; alt. 2,000–3,200 ft.; fl. July–Aug.—*Stem* 1–2 ft., procumbent, short, woody; branches tortuous, sparingly leafy. *Leaves* sometimes cuneate, obtuse or retuse, margin entire or waved, young hairy, older glabrous and rugose above, usually hoary beneath, stipules 0. *Catkins* ½–1 in., subterminal, flowering after the leaves, oblong, many-fld.; scales obovate, purplish or yellow, of one colour; anthers purplish; disk a laciniate cup. *Capsule* sessile, hoary, obtuse; stigmas notched.—DISTRIB. As *S. herba'cea*, reaching the limits of Arctic vegetation.

SECTION 3. **Syn'andræ,** *Anderss.* *Catkins* appearing before the leaves, on short bracteate or leafy peduncles; scales tipped with dark colour. *Stamens* 2, filaments more or less combined.

18. **S. purpu'rea,** *L.* ; shrubby, leaves often subopposite thin linear-lanceolate serrulate glabrous, stipules ½-ovate or 0, catkins subsessile, scales small, capsule subsessile, styles very short. *Purple Osier.*

River-banks and osier-beds, from Banff and Isla to Devon and Kent; native? in Ireland; fl. March–April.—An erect or decumbent shrub, 5–10 ft.; twigs slender, tough; bark red or purple. *Leaves* 3–6 in., sparingly hairy when young, shortly petioled, broadest about or beyond the middle, glaucous but most so beneath, black in drying. *Catkins* ¾-1½ in., opposite or alternate, erect, then spreading or recurved, cylindric; scales purple-black above, hairy or woolly; filaments hairy at the base; anthers red, then black. *Capsule* broadly ovoid, obtuse; stigmas entire or cleft.—DISTRIB. Europe, N. Africa, N. and W. Asia, India; introd. in N. America.—Bark very bitter; used for basket-making, but not so commonly as *S. vimina'lis.*—The best-marked British varieties are:—*S. Woolgaria'na*, Borr., and *ramulo'sa*, Borr.; erect, branches yellowish, stigmas notched. *S. Lambertia'na*, Sm.; erect, leaves broader above, branches purplish glaucous, stigmas subsessile short thick.

S. ru'bra, Huds., a common osier-bed shrub, or tree 10 ft., is a hybrid with *viminalis ;* the leaves are silky beneath, filaments usually more or less free, and style longer; it is very variable, and includes:—*S. ru'bra* proper; filaments connate at the base only; *S. Forbya'na*, Sm.; filaments united to the top; *S. Hel'ix*, L.; leaves often subopposite filaments united to the top (bears fascicles of diseased leaves, owing to the puncture of a Cynips, hence called *Rose Willow*).

S. Donia'na, Sm., a native of dry places in Mid. and S. Germany, is interme-diate between *re'pens* and *purpu'rea* (Andersson), was stated by G. Don. to be a native of Forfarshire, no doubt erroneously ; it may be known from *purpurea* by its yellow anthers, pedicelled capsule, and filaments more or less connate at the base only.

S. Pontedera'na, Schleich., gathered by the Rev. J. E. Leefe near Rothbury, Northumbd., is probably a hybrid with *cine'rea* (Baker).

ORDER LXXIII.   **CERATOPHYL'LEÆ.**

A submerged, branched, slender, fragile herb; stems cylindric. *Leaves* whorled, sessile, exstipulate, 2-chotomously cut into linear-toothed lobes. *Flowers* solitary, axillary, minute, monœcious, enclosed in an 8–12-partite persistent involucre with subulate lobes.   *Perianth* 0.—MALE.   *Anthers* many, crowded, sessile, oblong; cells linear, sunk in a fleshy 2-cuspidate connective, bursting irregularly.—FEMALE.   *Ovary* oblong, 1-celled; style terminal, subulate, persistent, stigma unilateral papillose; ovule 1, pendulous from the top of the cell, orthotropous. *Fruit* coriaceous, in-dehiscent, base tubercled winged or spurred. *Seed* pendulous, testa membranous, hilum thickened, albumen 0 ; cotyledons 2-fid ovoid thick, plumule large many-leaved, radicle very short.—DISTRIB. Europe, Asia, Africa, N. America (Arctic).—AFFINITIES doubtful.—PROPERTIES un-known.—ETYM. κέρας and φύλλον, from the *horn*-like *leaf*-lobes.

1. CERATOPHYL'LUM, L. HORNWORT.

C. demer'sum, *L.* ; leaves remotely serrate.

Ponds and ditches from Forfar southd.; E. Scotland only; rare in Ireland : Channel Islands; fl. July-Sept.—*Stems* 8 in.-3 ft., densely leafy. *Leaves* 1 in., segments spreading, subulate, dark green. *Flowers* found in shallow water only, very inconspicuous. *Fruit* ¼ in., tipped with the slender curved style, very variable.

C. DEMER'SUM proper; fruit smooth, spurs 2 subulate. *C. apicula'tum,* Cham.

Sub-sp. C. SUBMER'SUM, *L.*; fruit not spurred, covered when mature with cylindric projections.—S.E. England, rare.

SUB-CLASS II. GYMNOSPERM'Æ.

ORDER LXXIV. **CONIF'ERÆ.**

Trees or shrubs ; wood without ducts (except in the first year) ; wood-cells studded with disks. *Leaves* usually alternate, rigid, linear or subulate, solitary, or fascicled in membranous sheaths. *Flowers* monœcious or diœcious, anemophilous ; males in deciduous catkins ; females in cones or solitary ; perianth 0.—MALE of many 1- or more-celled anthers seated on the scales of the catkin, filaments 0 or connate.—FEMALE of 1 or more sessile naked orthotropous or anatropous ovules seated on an open carpellary leaf (bracts of some), which is free or adnate to the scale of the cone. *Seeds* often winged, testa thin or thick, albumen densely fleshy ; embryo axile, straight, cotyledons 2 or more, radicle terete often attached to a crumpled thread (suspensor).—DISTRIB. Especially cold regions ; very rare in trop. Africa and America ; genera 33 ; species about 300.—AFFINITIES. With *Cycadeæ* and *Gnetaceæ.*—PROPERTIES. Yield terebinthine, succinic acid, pitch, tar, turpentine, valuable woods, and a few edible seeds.

TRIBE I. **ABIETI'NEÆ.** *Flowers* monœcious. *Cones* usually large, conical ; scales more or less woody. *Pollen* curved.

Cone woody, scales persistent..1. Pinus.

TRIBE II. **CUPRESSI'NEÆ.** *Flowers* mono- di-œcious. *Cones* usually globose or short; scales woody or fleshy, persistent. *Pollen* globose.

Cone fleshy, globose; scales at length connate......................2. Juniperus.

TRIBE III. **TAXI'NEÆ.** *Flowers* diœcious. Cones much reduced ; scales small, thin or coriaceous, the upper with 1 ovule. *Seed* hard, with a fleshy coat, or seated in a fleshy cup. *Pollen* globose.

Seed solitary, seated in a fleshy cup.......................................3. Taxus.

1. PI'NUS, *L.* PINE.

Trees ; branches more or less whorled. *Leaves* evergreen, in clusters of 2, 3, or 5. *Male catkins* spicate, ovoid or oblong of many 2-celled anthers spirally arranged ; pollen-grains curved, 2-globose. *Fruit,* a cone, usually ripening in the second year ; scales woody. *Ovules* 2, inverted, adnate to the scale ; cotyledons 3 or more, linear ; radicle inferior.—DISTRIB. N. hemisphere, from Mexico and Borneo to the Arctic circle ; species about 70.—ÉTYM. The classical name.

P. sylves'tris, *L.* ; leaves in pairs, cones ovoid young recurved, seeds winged. *Scotch Fir.*

In a few spots, York to Sutherland; ascends to 2,200 ft.; Ireland; once native of many parts of Britain; planted elsewhere; fl. May–June.—A tree, 50–100 ft., trunk attaining 12 ft. girth; wood red or white; bark red-brown, rough. *Leaves* 2–3 in., falling in the 3d year, acicular, acute, grooved above, convex and glaucous beneath, minutely serrulate, sheath fimbriate. *Male catkins* ¼ in., yellow; connective produced. *Cones* 1–2 in., 1–3 together, acute; scales few, ends rhomboid with a transverse keel and deciduous point. *Seed* ½ in., wing cuneate, much exceeding the nucleus.— DISTRIB. Europe, N. Asia.—Yields tar, pitch, rosin, turpentine, and deals.

2. JUNIP'ERUS, *L.* JUNIPER.

Trees or bushes; heart-wood red, odorous. *Leaves* opposite or whorled in threes, all subulate, or on the young shoots subulate, on the old scale-like and appressed. *Male catkins* solitary or crowded, of many 2–6-celled anthers. *Cone* ripening the 2d year, small, globose, baccate, of 4–6 decussate or whorled confluent fleshy scales, the upper and lower flowerless. *Ovules* 1–2 under each scale, erect. *Seeds* 1–8, enclosed in the fleshy confluent scales, free or connate, testa various; cotyledons 2–5, oblong, radicle superior.—DISTRIB. Temp. and cold N. hemisphere; species 27.—ETYM. The classical name.

J. commu'nis, *L.* ; leaves whorled in threes subulate pungent glaucous above, margins and midrib thickened.

Open hill-sides, N. to Shetland; ascends to 2,400 ft.; Ireland; fl. May–June. —Shrubby, 1–5 ft., rarely subarboreous (10–20 ft., with trunk 5 ft. in girth). *Bark* flaking, fibrous, red-brown. *Leaves* ½–1 in., crowded; lower shorter, oblong-lanceolate, concave. *Cone* ¼–⅓ in. diam., very fleshy, blue-black, glaucous, with scarious empty scales at its base.—DISTRIB. Europe, N. Africa, N. and Mid. Asia, N. America, Arctic regions.—A diuretic.

J. commu'nis proper; leaves spreading straight subulate.—VAR. *J. na'na,* Willd.; leaves shorter broader imbricate incurved. Mountains from N. Wales and Westmoreland to Shetland, ascends to 2,700 ft.

3. TAX'US, *Tournef.* YEW.

A tree or shrub; wood very tough, heart-wood red; wood-cells with a spiral thickening within. *Leaves* linear, 2-farious; petiole very short, with a half twist. *Male catkin* peltate, subglobose with 5–8 anthers surrounded at the base by imbricate scarious empty scales.—*Female* of a few minute scales, and 1 terminal erect ovule seated on a fleshy disk, which enlarges into a red fleshy cup containing the seed. *Seed* ovoid, subcompressed, testa bony; cotyledons 2, short, radicle superior.—DISTRIB. N. temp. regions to the Arctic circle.—ETYM. possibly τόξον, from the wood being used for *bows*.

T. bacca'ta, *L.* ; leaves linear more or less falcate acute.

Hill-sides and woods, from Perth and Argyll to Somerset and Kent; ascends to 1,500 ft. in Northumbd.; Ireland; fl. March.—*Tree* 15-50 ft., in England; loftier in India; sometimes 27 ft. in girth, channelled; bark thin, flaking; branches spreading. *Leaves* ⅓-1½ in., coriaceous, shining above, paler beneath. *Male catkins* ¼ in., yellow; female minute; scales green. *Fruit* ⅓ in.; cup red, mucilaginous; seed olive-green, punctulate.—I believe also the 6 supposed species. of this genus to be forms of one. *T. fastigia'ta,* Lindl. (Irish or Florence-court yew), is a fastigiate variety.

Class II. MONOCOTYLEDONES.

Order LXXV. HYDROCHARI'DEÆ.

Aquatic herbs. *Leaves* aërial floating or submerged, opposite or whorled, convolute in bud. *Flowers* usually diœcious, buds inclosed in one or more spathaceous bracts.—MALE. *Perianth* of 6 segments in 2 series (rarely 0), 3 inner often petaloid, imbricate or valvate in bud. *Stamens* on the base of the segments, 3 and opposite the sepals, or more and in 2 or more series, filaments free or connate below; anthers adnate. *Ovary* rudimentary.—FEMALE. *Perianth* superior, 6-partite. *Staminodes* various, sometimes antheriferous. *Ovary* 1- or 3-6-celled; styles 3 or 6, 2-fid, free or connate below, stigmas decurrent; ovules many, ascending. *Fruit* usually baccate, submerged, 1-6-celled. *Seeds* many or few, on projecting placentas, testa firm, albumen 0; embryo straight, plumule lateral, radicle next the hilum.—DISTRIB. All climates; genera 14; species about 40.—AFFINITIES. None.—PROPERTIES unimportant.

Leaves orbicular, floating ....1. Hydrocharis.
Leaves ensiform, serrate, submerged....................................2. Stratiotes.
Leaves linear opposite or whorled, submerged ....2*. Elodea.

1. HYDROCH'ARIS, *L.* FROG-BIT.

A floating herb. *Leaves* orbicular, quite entire.—MALE fl. 2-3, sub-umbelled; spathe peduncled, 2-leaved. *Sepals* 3, herbaceous. *Petals* 3, white membranous. *Stamens* 12, 3-6 without anthers, filaments connate below, forked, fleshy; anthers basifixed, cells separate, both on one fork. *Pistillodes* 3.—FEMALE solitary, long-peduncled; spathe 1, radical. *Sepals* of the male. *Petals* with a fleshy gland at the base. *Staminodes* 6, filiform, in pairs opposite the sepals. *Ovary* ovoid, 6-celled; styles 6, short, connate at the base; ovules many, inserted on the septa, orthotropous. *Fruit* fleshy, indehiscent. *Seeds* few, ovoid, immersed in mucus, testa lax papillose; embryo ovoid.—DISTRIB. Europe, N. Asia.—ETYM. ὕδωρ, *water,* and χάρις, *elegance.*

H. **Mor'sus-Ra'næ,** *L.* ; leaves orbicular-reniform, flowers white.

Ponds and ditches, from Durham to Devon and Kent; local in Ireland; fl.
July–Aug.—Stoloniferous, roots fibrous and bulbiferous. *Leaves* 1–1½ in.
diam., deep green above, reddish beneath. *Flowers* erect, ¾–1 in. diam.;
sepals small, oblong; petals broadly obovate, crumpled. *Fruit* I have not
seen.—The cells of the testa swell in water and emit a spiral thread.

2. STRATIO′TES, *L.* WATER SOLDIER.

A stoloniferous submerged herb. *Leaves* all radical. *Flowers* subdiœcious, submerged before flowering and when fruiting, floating when in
flower, honeyed. *Inflorescence* and *perianth* of *Hydrocharis*, but female
flower sessile on a 2-spathed peduncle. —MALE fl. *Stamens* many, filaments
subulate, 12 antheriferous; anthers linear.—FEMALE. *Staminodes* many,
a few antheriferous. *Ovary* compressed with a narrow neck, and fruit as
in *Hydrocharis;* ovules anatropous.—DISTRIB. Europe (excl. Greece),
Siberia.—ETYM. στρατιώτηs, a *soldier*, from the sword like foliage.

S. aloi′des, leaves tapering spinous-serrate.

Ponds and ditches in E. of England from Northumbd. to Suffolk and
Northampton, and in Lancashire and Cheshire; naturalized in E. Scotland
and Ireland; fl. June–Aug.—*Leaves* 6–18 in., spreading, base ¼–½ in. diam.,
rigid, brittle, deep green, many-nerved. *Peduncles* axillary, short, stout.
Flowers 1½ in. diam. *Fruit* decurved at right angles to the peduncle,
flagon-shaped, 6-gonous, green; carpels separating. *Seeds* with a mucous
coat.

2*. ELO′DEA, *Michx.* WATER-THYME.

Stem slender, submerged, elongate, branched. *Leaves* in whorls of 3
(rarely 4), lower opposite. *Flowers* subdiœcious, axillary, solitary, sessile;
spathe tubular, slender, 2-lipped. *Sepals* and *petals* 3 each, small, green.
—MALE. *Stamens* 3–9, filaments short or 0; anthers oblong.—FEMALE.
Tube of perianth long, slender. *Staminodes* 3 or 6, sometimes antheriferous. *Ovary* 1-celled; style slender, adnate to the perianth-tube,
stigmas 3; ovules few, orthotropous, placentas 3 parietal. *Berry* oblong,
1-celled, few-seeded.—DISTRIB. Temp. and trop. climates: species 8.—
ETYM. ἑλώδηs, growing in watery places.

E. CANADEN′SIS, *Michx.*; leaves linear- or lanceolate-oblong serrulate.
Anach′aris Alsinas′trum, Bab.; *Udo′ra canaden′sis,* Nutt.

Ponds, ditches, and streams, from Aberdeen to Cornwall and Kent; introduced
from America into County Down about 1836, and into England about 1841;
fl. May–Oct.—Dark green, pellucid. *Stem* 1–4 ft., brittle, terete, rooting at
the nodes. *Leaves* ¼–½ in., sessile, acute, margined. *Flowers* floating, ⅛ in.
diam., greenish-purple, tube of female 4–8 in., capillary. *Sepals* boat-
shaped. *Petals* recurved. *Stigmas* long, terete, notched. *Male* hitherto
found near Edinburgh only.—DISTRIB. N. America.

ORDER LXXVI. **ORCHID'EÆ.**

Terrestrial herbs, roots fascicled or tuberous (many exotics are epiphytes, with pseudo-bulbs). *Leaves* sheathing at the base, or scales. *Flowers* solitary, spiked, racemed or panicled. *Perianth* superior, irregular, of 6 coloured segments; 3 outer (*sepals*) nearly similar; inner lateral similar (*petals*); third inner (inferior by torsion of the pedicel) (*lip*) dissimilar, usually larger, often spurred. *Stamens* confluent with the style into an unsymmetrical column; anthers 1 superposed to the outer sepal (in *Cypripedium* 2 opposite the petals), 2-celled, persistent or deciduous; pollen in 2, 4, or 8 pyriform usually pedicelled masses (*pollinia*), pedicels (*audicles*) terminating in a gland; glands exposed or in 1 or 2 pouches. *Ovary* usually long and twisted, 3-gonous, 1-celled; style often terminating in a beak (*rostellum*) at the base of the anther or between its cells; stigma a viscid surface (of three confluent stigmas) facing the lip, beneath the rostellum; ovules very many, anatropous, placentas 3 parietal. *Capsule* 3-valved, valves separating from a framework that bears the placentas. *Seeds* innumerable, very minute, fusiform, testa very lax reticulate, albumen 0; embryo fleshy.—DISTRIB. All climates and situations except the very cold maritime, and aquatic; genera 334; species 5,000.—AFFINITIES. With *Irideæ.*—PROPERTIES. A few are aromatic (*Vanilla*); the tubers of *Orchis* are nutritive, and yield starch and salep.

TRIBE I. **EPIDENDRE'Æ.** *Anther* a 2-celled cap, hinged upon the column : pollen-masses 4, waxy, free or connate, not attached to the rostellum.

Leafy herb. Lip superior. Column short ....1. Malaxis.
Leafy herb. Lip superior or inferior. Column slender ....2. Liparis.
Leafless brown saprophytes. Lip inferior ....3. Corallorhiza.

TRIBE II. **NEOTTIE'Æ.** *Anther* a 2-celled deciduous cap, hinged on to the top or back of the column; pollen-masses 2 or 4, granular or powdery, grains free or united by an elastic web.

* *Anther hinged on the back of the column; rostellum beaked.*

A leafless brown saprophyte. Pollen powdery ....4 Neottia.
Leaves 2 subopposite. Lip free. Pollen powdery....................5. Listera.
Leaves several. Lip free. Pollen-grains coherent ....6. Goodyera.
Leaves several. Lip adnate to the base of the column. Pollen powdery.
7. Spiranthes.

** *Anther hinged on the top of the column; rostellum very short.*

A leafless brown saprophyte. Pollen granular. ....8. Epipogium.
Stem leafy. Flowers racemed; ovary straight.....................9. Epipactis.
Stem leafy. Flowers spiked; ovary twisted ....10. Cephalanthera.

TRIBE III. **PHRYDE'Æ.** *Anther* 1, confluent with the column, 2-celled, erect; pollen-masses 2, granular, grains united by an elastic web.

* *Glands of stalks of pollen masses in pouches of the rostellum.*

Spur long; both glands in one pouch ....11. Orchis.
Spur 0; both glands in one pouch....... .... 12. Aceras.
Spur 0; glands in separate pouches........... ....13. Ophrys.

** *Glands of stalks of pollen masses naked, not in pouches of the rostellum.*

Spur 0 .... 14. Herminium.
Spur long or short................. ....15. Habenaria.

TRIBE IV. **CYPRIPEDIE′Æ.** *Anthers* 2; rostellum prolonged into a
shield between the anthers; pollen powdery.

Lip large, saccate ... .16. Cypripedium.

1. MALAX′IS, *Sw.*

A small green subsucculent herb. *Leaves* few, broad, short. *Flowers*
minute, racemose, green. *Sepals* and very small *petals* spreading. *Lip*
superior, minute, entire, concave. *Anther* hinged on to the top of the
minute column, persistent; pollen-masses 4, waxy, attached to one gland.
Stigma depressed; rostellum minute.—DISTRIB. Europe (Arctic) N. of
the Alps, N. Asia.—ETYM. μάλαξις, in allusion to its softness.

M. paludo′sa, *Sw.*; leaves obovate obtuse. *Bog Orchis.*

Sphagnum swamps and bogs, from Devon and Kent to Sutherland, rare,
and easily overlooked; ascends to 1,500 ft. in N. England; local in
Ireland; fl. July–Sept.—*Stem* 1–4 in., swollen and sheathed with white
scales at the base, forming a new plant at the side of the old, angled above.
Leaves few, fringed with cellular bulbils that develop new plants. *Raceme*
elongate, many-fld.; bracts minute; pedicel twisted. *Flowers* ⅛ in., yellow-
green; sepals ovate; petals linear-oblong, recurved; lip about equalling
the petals, erect, acute, embracing the column at its base.

2. LI′PARIS, *Rich.*

Habit of *Malaxis,* but usually 2-leaved. *Sepals* and *petals* spreading,
linear. *Lip* inferior or superior, broader than the sepals, entire. *Anther*
terminal on the slender column, deciduous; pollen-masses 4, waxy, glands
evanescent. *Stigma* small, depressed; rostellum minute.—DISTRIB.
Temp. and trop. regions; species 100.—ETYM. λιπαρός, from its *greasy*
texture.

L. Loesel′ii, *Rich.*; leaves elliptic-lanceolate acute keeled petioled, lip
oblong-obovate mucronate. *Sturmia,* Reichb. *Fen Orchis.*

Spongy bogs, Norfolk, Suffolk, Hunts, Cambridge; fl. July.—*Stem* 4–8 in.,
swollen and sheathed with white scales at the base, 3-gonous and leafless
above. *Leaves* 1–3 in. *Flowers* few, ½ in., ascending, pale yellow-green;
bracts mostly small.—DISTRIB. Europe N. of the Alps, Italy, Turkey.

3. CORALLORHI′ZA, *Haller.*   CORAL-ROOT.

Brown leafless saprophytes.   *Root* of branched, fleshy, interlaced fibres.
Stem with sheathing scales.   *Flowers* few, small, subracemose.   *Upper sepal*
and *petals* connivent, lateral sepals spreading.   *Lip* deflexed, short, lateral
lobes small; spur minute, adnate to the ovary.   *Anther* terminal on the
short column, deciduous, 2-celled ; pollen-masses 4, subglobose, granular,
free.   *Stigma* discoid ; rostellum inconspicuous.—DISTRIB.   N. temp.
regions ; species 12.—ETYM. κοράλλιον and ῥίζα, *coral root.*

C. inna′ta, *Br.* ; sepals linear-lanceolate, lip oblong.

Boggy or sandy woods and copses in E. Scotland, from Ross to Berwick, very
rare ; fl. July-Aug.—*Stem* 6–10 in., slender ; sheaths lax, red-brown.
Raceme 4–8-fld., pedicels very short ; bracts minute.   *Flowers* horizontal ;
perianth ¼ in. ; sepals ovate-lanceolate, olive-green, lateral deflexed, and
petals narrower ; lip whitish with small purple tubercles.—DISTRIB.
Arctic, N. and Mid. Europe, N. Asia, N. America.

4. NEOT′TIA, *L.*   BIRDS′-NEST ORCHIS.

Leafless brown saprophytes, stem with sheathing scales.   *Flowers*
racemed.   *Sepals* and *petals* incurved.   *Lip* decurved, base saccate, apex
with two straight or spreading lobes.   *Anther* hinged on to the back of
the slender free column, 2-celled ; pollen-masses 2, powdery, glands connate.
Stigma prominent ; rostellum tongue-shaped.—DISTRIB.   N. temp. Europe
and Asia ; species 3.—ETYM. νεοττιά, a bird's nest, in allusion to the
curious roots.

N. Ni′dus-avis, *L.* ; glabrous, lobes of lip spreading.

Dark woods, especially beech, from Banff and Argyll southd. ; Ireland ; fl.
June–July.—*Root* a mass of succulent, stout, interlaced fibres.   *Stem* robust,
1–1½ ft., dirty-brown.   *Flowers* ⅓ in., grey-brown, bracts short.—DISTRIB.
Europe, W. Siberia.

5. LIS′TERA, *Br.*

Root of fleshy fibres.   *Leaves* 2, subopposite.   *Flowers* racemed, green.
Sepals and *petals* spreading.   *Lip* deflexed, entire, lateral lobes 0 or
minute ; spur 0.   *Anther* hinged on to the back of the column, 2-celled ;
pollen-masses 2, powdery, glands connate.   *Stigma* prominent ; rostellum
tongue-shaped.—DISTRIB. N. temp. and cold regions ; species 10.—ETYM.
Dr. Martin Lister, a British naturalist.

1. **L. ova′ta,** *Br.* ; stem tall terete pubescent above, leaves broadly
elliptic, lip 2-fid without lateral lobes.   *Tway-blade.*

Woods and pastures, N. to Sutherland ; ascends to near 1,900 ft. in N. England ; Ireland ; Channel Islands ; fl. May–July.—*Stem* 1–2 ft., solitary,
stout.   *Leaves* 3–8 in., ribbed.   *Raceme* elongate ; bracts minute.   *Flowers*
½ in. ; sepals deep green, ovate, subacute ; petals yellow-green ; lip the
same, base slightly saccate, apiculate between the terminal lobes.   *Rostellum*

C C

emitting when touched a viscid fluid that attaches the pollen to foreign bodies.—DISTRIB. Europe (Arctic), Siberia.

2. **L. corda'ta,** *Br.* ; small, glabrous, stem angled fragile, leaves ovate-cordate, lip 2-fid and with 2 small basal linear lobes. *Lesser Tway-blade.*
Mountain woods and moors, especially under heather, from Hants and Devon to Shetland ; ascends to 2,700 ft. in the Highlands ; rare in Ireland ; fl. July-Sept.—*Stem* 4–8 in., brownish-green. *Leaves* ½–1 in., sessile, membranous, acute. *Raceme* lax, few-fld. ; bracts minute. *Flowers* ⅛ in. ; sepals and petals olive-brown, obtuse ; lip dirty yellow-green, terminal lobes linear. —DISTRIB. Arctic and Alpine Europe, N. Asia, N. America.

6. GOODYE'RA, *Br.*

Rootstock creeping. *Leaves* usually ovate and petioled. *Flowers* small, spiked, in spiral series.· *Upper sepal* and *petals* ascending, free or connate ; lateral sepals deflexed, embracing the base of the lip. *Lip* decurved, entire, base saccate. *Anther* hinged on to the back of the column, 2-celled ; pollen-masses of loosely cohering grains, sessile on one oblong gland. *Stigma* discoid ; rostellum beaked, finally 2-fid.—DISTRIB. N. temp. regions ; species 25.—ETYM. *John Goodyer,* an English botanist.

G. re'pens, *Br.* ; leaves ovate acute, nerves reticulate.
Fir woods in E. Scotland, from Cumberland, Berwick, and Ayr to Ross, rare ; fl. July-Aug.—*Rootstock* slender, matted, widely creeping. *Stem* 4–8 in , slender, and spike glandular-pubescent. *Leaves* ½–1 in., acute, dark green, pubescent beneath. *Spike* slender ; bracts subulate-lanceolate, longer than the ovary. *Flowers* cream-white; perianth ¼ in.—DISTRIB. Mid and N. Europe (Arctic), Siberia, Himalaya, N. America.

7. SPIRAN'THES, *Rich.* LADY'S-TRESSES.

Root of tubers or stout fibres. *Stem* leafy. *Spike* of small flowers in 1–3 spirally-twisted rows. *Sepals* and *petals* similar, suberect ; sepals gibbous at the base ; upper adnate to the petals, forming a tube round the lip. *Lip* embracing and adnate to the base of the column, tip entire, disk 2-tubercled. *Anther* hinged on to the back of the column, 2-celled ; pollen masses 4, powdery, sessile on one linear gland. *Stigma* discoid ; rostellum beaked, finally 2-fid.—DISTRIB. Trop. and temp. regions ; species 80.—ETYM. σπεῖρα and ἄνθος, from the *twisted inflorescence.*

1. **S. autumna'lis,** *Rich.* ; tubers 2–3 ovoid, flowering-stem sheathed distinct from the root-leaves, spike slender, flowers in 1 series.
Dry pastures from Westmoreland and York southd. ; S. and Mid. Ireland ; Channel Islands ; fl. Aug.-Sept.—*Stem* 4–8 in., slender, upper part and spike pubescent. *Leaves* 1 in., in lateral rosettes, ovate, acute, appearing after the flowers. *Flowers* ⅛ in., sheathed by the cucullate cuspidate bracts, fragrant, white; lip channelled at the base, tip exserted crenate.—DISTRIB. From Denmark southd., N. Africa.

2. S. æstiva'lis, *Rich.* ; tubers several cylindric, radical leaves on the flowering stem linear, spike slender many-fld., flowers in 1 series.

Bogs; Wyre Forest, Worcester, and New Forest, Hants; Channel Islands; fl. July–Aug.—*Stem* 6–18 in., glabrous. *Leaves* 2–6 in., narrowed below. *Spike* slightly pubescent. *Flowers* and bracts as in *S. autumna'lis,* but rather larger.—DISTRIB. W. Europe from Belgium southd., N. Africa.

3. S. Romanzovia'na, *Cham.* ; tubers several cylindric, radical leaves on the flowering stem narrow obovate-lanceolate, spike stout, flowers in 3 series. *S. cer'nua,* Hook, *f.,* not Rich. ; *S. gemmip'ara,* Lindl.

Meadows, Bantry Bay, Co. Cork; fl. Aug.–Sept.—*Stem* 6–10 in., stout, glabrous, leafy throughout. *Leaves,* lowest 3–6 in. *Spike* 2–3 in., glandular-pubescent ; bracts sheathing the base of the ovary, subulate-lanceolate. *Flowers* white, much larger and broader than in the preceding species; lip tongue-shaped, contracted below the crenate recurved tip, tubercles at the base smooth and shining.—DISTRIB. Kamtschatka.—A. Gray correctly refers *S. gemmip'ara* to *Romanzovia'na.*

8. EPIPO'GUM, *Gmelin.*

Leafless saprophytes. *Root* of fleshy branched fibres. *Flowers* racemed. *Sepals* and *petals* rather spreading. *Lip* superior, ovate, 3-lobed, disk with rows of glands; spur short, stout. *Anther* terminal, deciduous ; pollen-masses 2, in cavities of the cylindric column, stalked, glands connate. *Stigma* prominent, horseshoe like ; rostellum 0. *Ovary* not twisted.—DISTRIB. Europe N. of the Alps, N. Italy, N. Asia, Himalaya ; species 2.—ETYM. ἐπί and πώγων, from the *lip* being uppermost.

E. Gmeli'ni, *Rich.* ; flowers pale yellow. *E. aphyl'lum,* Sw.

Amongst decayed leaves, Herefordshire, most rare; fl. Aug.—*Stem* 4–8 in., tumid above the base, pale yellow-brown, with 1 or 2 appressed sheaths. *Bracts* as long as the pedicels. *Flowers* 2–6, shortly pedicelled, 1 in. ; ovary broad, short ; sepals and petals narrow-lanceolate, subequal, margins involute ; lip recurved, lateral lobes small, middle whitish with red glands ; spur obtuse.

9. EPIPAC'TIS, *Rich.* HELLEBORINE.

Rootstock creeping. *Stem* leafy. *Flowers* racemed ; ovary straight ; pedicel twisted. *Sepals* and *petals* conniving or spreading. *Lip* much contracted in the middle, basal lobe concave, terminal entire with 2 basal tubercles. *Anther* sessile, hinged on the top of the column ; pollen-masses 2, powdery, glands connate. *Stigma* prominent ; rostellum short, erect. *Capsule* pendulous.—DISTRIB. Europe, N. Asia, Himalaya ; species about 8.—ETYM. The classical name for this or another plant.

1. E. latifo'lia, *Sw.* ; leaves orbicular ovate-lanceolate or oblong, bracts mostly exceeding the flowers, basal lobe of lip with rounded margins, terminal broadly ovate, ovary broadly pyriform. *E. Hel'leborine,* Crantz.

Woods, from Ross southd. ; Ireland ; fl. July–Aug.—*Stem* 1–3 ft., pubescent above. *Leaves* variable, ribbed. *Raceme* many-fld., bracts green. *Flowers* subsecund, ½–⅔ in. diam., green, variously marked with yellow white or purple; sepals broadly ovate; petals ovate-lanceolate; lip variable in form and colour, as long as the sepals or shorter, terminal lobe with thick ned ridges on the disk.—DISTRIB. Europe (Arctic), N. Africa, Siberia, Himalaya.—I am indebted to Mr. Baker for the diagnoses of the following subspecies, which coincide with Syme's, and appear to embrace the prevalent forms; they do not however precisely accord with those of other countries, nor do materials from different parts of England give quite the same results.

Sub-sp. LATIFU'LIA proper; stems 2–3 ft. not tufted, lower leaves 4–5 by 2–3 in., sepals ¼–⅓ in. ovate-oblong, tip of lip broader than long obscurely pointed. *E. viridiflo'ra*, Hoffm.; *E. Hel'leborine*, var. *var'ians*, Crantz.— Common. (Also found in one spot in E. U. States.)

Sub-sp. E. PURPURA'TA, *Sm.*; stems 2–3 ft. often tufted, lower leaves 3–4 by 1½–2 in., sepals oblong-lanceolate ¼–½ in. more pointed, tip of lip as broad as long subdeltoid. Flowers usually tinted violet-purple, except var. *E. me'dia*, Fries, which is also less robust. *E. viola'cea*, Bor.—S. of England rare.

Sub-sp. E. ATRO-RU'BENS, *Hoffm.*; dwarfer, lower leaves 1½–2 by 1 in. ovate acute, tip of lip broader than long rounded obscurely cuspidate. Flowers a month earlier, reddish brown. *E. ova'lis*, Bab.; *E. rubigino'sa*, Crantz.— Limestone cliffs, Orme's Head, Yorkshire (ascends to 1,200 ft.), Sutherland, &c.

2 **E. palus'tris,** *Sw.*; leaves lanceolate, bracts mostly short, basal lobe of lip angular terminal obtuse crenate, ovary narrowly pyriform.

Marshy places, from Fife and Perth southd., local; rare in Ireland; Channel Islands; fl. July.—*Stem* 8–18 in., slender, wiry, pubescent above. *Leaves* acute, upper acuminate. *Flowers* few, ⅓–⅔ in. broad; sepals and petals ovate, subacute, green striped with purple; lip white, streaked with red, terminal lobe tubercled towards the base.—DISTRIB. Europe, Siberia.

10. CEPHALAN'THERA, *Rich.*

Rootstock creeping. *Stem* leafy. *Leaves* subdistichous. *Spikes* few-fld.; ovary twisted. *Flowers* suberect, sepals and petals incurved. *Lip* decurved, constricted in the middle, basal lobe saccate, terminal not tubercled, disk crested. *Anther* hinged on the contracted top of the column; pollenmasses 2, powdery, glands connate. *Stigma* prominent; rostellum 0. *Capsule* erect.—DISTRIB. Europe, N. Asia, Himalaya; species 4.— ETYM. κεφαλή and ἄνθηρα, from the position of the *anther*.

1. **C. pal'lens,** *Rich.*; leaves ovate-oblong, lower bracts large much exceeding the almost glabrous ovary, flowers white. *C. grandiflo'ra*, S. F. Gray. *White Hel'leborine.*

Woods and copses chiefly in chalky districts, from Cumberland to Somerset and Kent, rare; fl. May–June.—*Stems* tufted, 1–2 ft. *Leaves* 3–6 in., upper narrower. *Flowers* ¾ in. distant, suberect, cream-white; sepals and petals ovate-oblong, obtuse; terminal lobe of lip orbicular, erect, yellow.—DISTRIB. From Denmark southd.

2. **C. ensifo'lia,** *Rich.*; leaves ovate or lanceolate, bracts of upper flowers much smaller than the almost glabrous ovaries, flowers white.

Woods and copses from Mull and Perth to Dorset and Sussex, local; rare in
Ireland; fl. May–June.—*Stems* 1-2 ft., subsolitary, slender. *Upper bracts*
minute. *Leaves* usually longer than in *C. grandiflo'ra*, flowers whiter and
narrower, sepals more acute.—DISTRIB. Europe, W. Asia.

3. **C. ru'bra,** *Rich.* ; leaves lanceolate, bracts exceeding the glandular
pubescent ovaries, flowers rose-purple. *Red Helleborine.*

Woods and copses on limestone, Gloster and Somerset, very rare; fl. June–
July.—*Stem* 6-18 in., slender. *Flowers* few or many; sepals and petals
acuminate; lip white, terminal lobe ovate-lanceolate. *Column* slender.—
DISTRIB. From Gothland southd., W. Siberia.

11. OR'CHIS, *L.*

Tubers globose ovoid or palmate. *Leaves* chiefly radical, sheathing.
Flowers spiked. *Sepals* and petals ascending, connivent or the lateral sepals
spreading. *Lip* spurred, decurved or deflexed, spur not honeyed. *Anther*
confluent with the column, cells diverging at the base; pollen-masses decurving after removal, glands in one 2-lobed pouch ; rostellum projecting
between the lobes of the pouch.—DISTRIB. Europe, N. and W. Asia,
Himalaya, rare in N. America ; species about 80.—ETYM. The old Greek
name.

SECTION 1. **Or'chis** proper. *Lobes* of lip not spirally coiled. *Pollen-
glands* not connate.

* Lateral sepals spreading or reflexed.

1. **O. mas'cula,** *L.* ; tubers ovoid, leaves usually spotted, spike lax,
bracts 1-nerved coloured, lip 3-lobed, spur longer than the ovary. *Purple
Orchis.*

Copses and pastures, N. to Shetland ; ascends to 1,500 ft. in the Lake district; Ireland; Channel Islands ; fl. April–June.—*Stem* 6-18 in. *Leaves*
narrow-oblong, obtuse, spots purple-black. *Bracts* equalling the ovary.
Flower red-purple, rarely white; sepals ½ in., acute or obtuse ; lip as broad
as long, margins recurved, spotted with purple, mid-lobe longest crenate,
tip notched; spur stout, obtuse, variable in direction.—DISTRIB. Europe,
N. Africa, W. Siberia.—Yields salep.

2. **O. laxiflo'ra,** *Lamk.* ; tubers globose, leaves lanceolate, spike lax,
bracts 3-5-nerved coloured, lip 2-3-lobed, spur ½ as long as the ovary.

Wet meadows, Channel Islands ; ballast heaps, Hartlepool ; fl. May–June.—
Stem 1-3 ft., grooved. *Leaves* cauline and radical, acuminate, not spotted.
Bracts as long as the slender ovary. *Flowers* 1 in. from dorsal sepal to tip
of lip, bright red-purple; sepals and petals obtuse; lip as broad as long,
sides reflexed, spotted, lateral lobes very large, crenulate, mid-lobe shorter
or 0 ; spur stout, obtuse, variable in direction.—DISTRIB. From Belgium
southd.

3. **O. latifo'lia,** *L.* ; tubers palmate, leaves usually spotted, spike
dense, bracts 3-nerved green, lip obscurely 3-lobed, spur usually shorter
than the ovary. *O. palma'ta,* Syme. *Marsh Orchis.*

Moist meadows, &c., N. to Shetland ; ascends to near 1,600 ft. in Northumbd. ; Ireland ; Channel Islands ; fl. May–July.—*Stem* 1–3 ft., usually tubular, leafy upwards. *Leaves* oblong or lanceolate. *Bracts* mostly exceeding the flowers. *Flowers* ⅔ in. from dorsal sepal to tip of lip, dull purple ; sepals and petals obtuse or acute ; lip spotted with purple, margins recurved, mid-lobe narrowest ; spur nearly straight or decurved.—DISTRIB. Europe, N. Africa? N. Asia, Himalaya.

O. LATIFJ'LIA proper ; leaves oblong, tip flat, lip spotted. *O. maja'lis,* Wats. Fl. May–June.

Sub-sp. O. INCARNA'TA, *L.* ; leaves lanceolate acute unspotted, tip concave, base broader, flowers larger. Wilts, Hants, Cornwall, Cork ; fl. June–July.—VAR. *angustifo'lia,* Bab., is a narrow-leaved form.

4. **O. macula'ta,** *L.* ; tubers palmate, leaves usually spotted, spike oblong-pyramidal dense, bracts 3-nerved green, lip deeply 3-lobed, spur equalling the ovary or shorter. *Spotted Orchis.*

Moist places, N. to Shetland ; ascends to 3,000 ft. in the Highlands ; Ireland ; Channel Islands ; fl. May–July.—*Stem* 6–18 in., slender, leafy upwards. *Leaves* narrow, oblong-lanceolate, acute or obtuse. *Bracts* subulate, about equalling the ovary. *Flowers* ½ in. from the dorsal sepal to the tip of the lip, very pale purple or white, spotted, rarely white ; lip as broad as long, margins recurved, mid-lobe narrower and about as long as the lateral, which are toothed ; spur straight.—DISTRIB. Europe (Arctic) (excl. Greece), N. and W. Asia.

** *Lateral sepals arching and forming a hood with the dorsal and the petals.*

5. **O. Mo'rio,** *L.* ; tubers globose, spike lax, bracts 1-nerved coloured, lip 3-lobed, spur ascending equalling the ovary. *Green-winged Orchis.*

Meadows, &c., from Northumbd. southd. ; Ireland ; fl. May–June.—*Stem* 6–12 in. *Leaves* rather small and narrow, unspotted. *Bracts* about equalling the ovary. *Flowers* ⅔ in. from the dorsal sepal to tip of lip, dingy purple ; sepals obtuse, veins green ; lip spotted, lateral lobes broad crenate, middle about as long ; spur nearly straight, obtuse.—DISTRIB. From Gothland southd., N. and W. Asia.

6. **O. ustula'ta,** *L.* ; short, tubers ovoid, spike dense, bracts 1-nerved green, sepals and petals very dark white-spotted, lip 3-lobed, spur ¼ as long as the ovary. *Dark-winged* or *Dwarf Orchis.*

Dry pastures, from Northumbd. to Devon and Kent, local ; fl. May–June.—*Stem* 3–10 in. *Leaves* narrow-oblong, acute, unspotted. *Bracts* variable, scarious. *Flowers* ⅔ in. from the dorsal sepal to tip of lip, eventually white ; sepals and petals dark-purple and green ; lip with raised purple spots, lateral and segments of 2-fid mid-lobe nearly equal ; spur decurved, obtuse.—DISTRIB. From Gothland southd. (excl. Greece), W. Siberia.

7. **O. purpu'rea,** *Huds.* ; tall, stout, tubers ovoid, spike dense, bracts 1-nerved, sepals obtuse green and purple, lip 3-lobed, lateral lobes narrow, mid-lobe obcordate crenulate, segments broad flat crenulate, spur ½ as long as the ovary. *O. milita'ris,* Sm. ; *O. fus'ca,* Jacq.

Downs and copses in chalk soils, Kent and Sussex; fl. May.—*Stem* 1-3 ft.,
stout. *Leaves* oblong, 3-5 in., obtuse. *Spikes* usually large, many-fld.
Flowers ¾ in. from the dorsal sepal to tip of lip; sepals and petals
hardly acute, green and purple outside, paler inside, spotted; lip pale rosy,
spotted with purple, with a notch in the sinus of the mid-lobe; spur
decurved.—DISTRIB. From Denmark southd. (excl. Greece).

8. **O. milita′ris,** *L.* ; tubers ovoid, spike oblong dense, bracts 1-nerved,
sepals acuminate and petals pale purple or white, lip 3-lobed, lateral lobes
narrow, middle 2-fid with narrow upcurved segments and a tooth in the
sinus, spur ½ as long as the ovary.

Woods and chalk downs, Oxford, Berks, Herts, Bucks, Kent; fl. May–June.—
Stem 1-1½ ft. *Leaves* large, oblong, obtuse, concave, unspotted. *Bracts*
very short. *Flowers* about 1 in. from dorsal sepal to tip of lip, bright or
pale purple; lip pale, dotted with raised rough points, lateral lobes linear,
segments of mid-lobe rather broader; spur decurved, obtuse.—DISTRIB.
From Gothland southd. (excl. Greece), N. Africa (?), N. Asia.
O. MILITA′RIS proper; lateral lobes of pale purple lip narrow veined, mid-lobe
deeply 2-fid, segments broader than the lateral lobes.
Sub-sp. O. SIM′IA, *Lamk.*; more slender, lobes of crimson lip and segments
of the mid-lobe very narrow. *O. tephrosan′thos*, Vill.

SECTION 2. **Anacamp′tis,** *Rich.* (gen.). *Lip* ascending, lobes broad,
not spirally coiled ; spur very long. *Pollen-glands* connate. *Stigmatic
surfaces* distinct. *Rostellum* overhanging the mouth of the spur.

9. **O. pyramida′lis,** *L.* ; tubers globose, leaves acuminate, spike pyramidal, bracts 1-3-nerved coloured, lateral sepals spreading, lip 3-lobed
2-tubercled at the base, spur longer than the ovary.

Pastures, &c., from Wigton and Berwick southd.; Mid. Ireland; fl. June–
Aug.—*Stem* 6-24 in., slender. *Leaves* chiefly radical, lanceolate. *Spike*
pyramidal, then oblong. *Bracts* as long as the ovary. *Flowers* rosy, rarely
white; sepals and petals obtuse; lip broader than long, lobes subentire,
variable in shape and relative size.—DISTRIB. From the Baltic southd.,
N. Africa.—Scent of flowers peculiar.

SECTION 3. **Loroglos′sum,** *Rich.* (gen.) *Lobes of lip* very long,
spirally coiled in bud ; spur very short. *Pollen-glands* connate.

10. **O. hirci′na,** *L.* ; tubers ovoid, stem tall, spike long, bracts very
long ribbed green, lateral sepals conniving, mid-lobe of lip strap-shaped.
Lizard Orchis.

Copses in E. Suffolk and Kent, extremely rare ; fl. July-Aug.—*Stem* 1-5 ft.
Leaves chiefly radical, oblong, obtuse. *Spike* 6-17 in. *Bracts* 1-2 in.,
much exceeding the flowers. *Flowers* large; sepals and petals forming a
green hood ½ in. long; lip 1½ in., white, purple-spotted at the base, lateral
lobes slender, mid-lobe about 1 in. by ⅛ broad, green.—DISTRIB. From
Belgium southd., N. Africa.—Odour detestable, hircine.

12. A'CERAS, *Br.*   MAN ORCHIS.

Tubers ovoid. *Sepals* and *petals* forming a hood. *Lip* elongate, 4-lobed ; spur 0. *Anther* confluent with the column ; cells parallel ; pollen-masses 2, decurving after removal ; glands connate in one pouch. *Stigma* depressed ; rostellum obsolete.—DISTRIB.   Europe, N. Asia, Himalaya ; species 11.—ETYM. *ă, privative,* and κέρας, *spur.*

A. **anthropoph'ora,** *Br.* ; perianth green, lobes of lip linear.

Pastures and copses in chalky soil in E. England, from York to Kent and Sussex, scarce ; fl. June–July.—*Stem* 8–16 in. *Leaves* oblong-lanceolate, lower obtuse, upper acute. *Spike* lax-fld., narrow ; bracts small. *Flowers* ⅔–¾ in. ; sepals and petals often edged with red ; lip perpendicular, yellow, edges red, narrow, with 2 lateral and 2 terminal lobes, all similar and linear.—DISTRIB.   From Belgium southd., N. Africa.

13. O'PHRYS, *L.*

Tubers ovoid. *Perianth* spreading. *Petals* small. *Lip* perpendicular, usually convex, velvety ; spur 0. *Anther* capping the column, arched forwards, often beaked, cells parallel ; pollen-masses 2, glands in separate pouches. *Stigma* a depressed disk ; rostellum 0. *Ovary* not twisted.— DISTRIB.   Europe, N. Africa, W. Asia ; species about 30.—ETYM. ὀφρύς, an *eyebrow,* from the markings of the lip.

1. **O. apif'era,** *Huds.* ; sepals pink or white inside, lip broad convex 3-lobed with a terminal appendage, anther-beak hooked. *Bee Orchis.*

Copses and fields in chalk and limestone districts, from Durham and Lancaster southd.; ascends to 1,000 ft. in W. England ; S. and Mid. Ireland ; Channel Islands ; fl. June–July.—*Stem* 6–18 in. *Leaves* short, oblong. *Spike* 3–6-fld. ; bracts large, leafy. *Flowers* 1–1¼ in. ; sepals ovate ; petals small, downy, linear-oblong, obtuse ; lip brown-purple, lateral lobes tubercled at the base, disk spotted with orange yellow. *Pollen-gland* persistent on the column, head falling over on the stigma, and fertilising the ovules.—DISTRIB.   From Belgium southd., N. Africa.

O. APIF'ERA proper ; petals linear, lip equalling the sepals deeply 3-lobed, appendage recurved, or acute and triangular in *O. Trol'lii,* Heg.

Sub-sp. O. ARACHNI'TES, *Hoffm.* ; petals subdeltoid-ovate, lip longer than the sepals, appendage straight or incurved.—Kent, Surrey.—Pollen said to be stiff and not falling over the stigma.

2. **O. aranif'era,** *Huds.* ; sepals yellow-green inside, petals oblong, lip broad convex without an appendage, anther-beak not hooked. *Spider Orchis.*

Copses and downs in chalk and limestone, rare, from Northampton and Suffolk to Dorset and Kent ; fl. April–May.—Habit of *O. apif'era.* Lip brown with various glabrous markings.—DISTRIB.   From France southd.

O. *aranif'era* proper ; petals almost glabrous, lip usually lobed at the margin.— VAR. *O. fucif'era,* Smith ; petals downy within, lip rarely lobed, its tubercles less prominent.

3. O. muscif'era, *Huds.* ; sepals yellow-green, petals narrow linear, lip narrow nearly flat, anther not. beaked. *Fly Orchis.*

Copses and downs on chalk and limestone, from Durham and Westmoreland to Kent and Somerset; Mid. Ireland, very rare; fl. May–July.—*Stem* slender, 10–18 in. *Leaves* few, linear-oblong. *Flowers* distant, ¾ in. ; petals and lip bright red-brown; lip with a blue patch, sometimes edged with yellow, lateral lobes reflexed, terminal 2-fid.—Distrib. From Norway southd. (excl. Greece).

14. HERMIN'IUM, *Br.* Musk Orchis.

Tubers ovoid. *Leaves* 2 or few. *Perianth-segments* incurved. *Lip* 3-lobed ; spur 0. *Column* with short lateral arms. *Anther* confluent with the column, cells diverging below ; pollen-masses 2, subsessile, glands large exposed. *Stigma* discoid ; rostellum 0.—Distrib. Europe, temp. Asia ; species 4.—Etym. ἑρμίν, the *foot of a bed-post*, from the shape of the tubers.

H. Monor'chis, *Br.* ; flowers minute subsecund green.

Chalk downs, from Norfolk, Cambridge, and Gloster to Somerset and Kent ; fl. June–July.—*Tubers* at the end of fleshy fibres. *Stem* 4–10 in., slender. *Leaves*, radical 2, narrow-oblong, acute ; cauline solitary. *Spike* slender, rather lax ; bracts green, as long as the ovary. *Flowers* ⅛ in., not honeyed, musky at night ; sepals broad ; petals narrower, longer, obscurely lobed at the side ; lip narrow, 3-lobed, base saccate,. mid-lobe entire narrow.— Distrib. Europe (Arctic) (excl. Spain, Greece), Siberia, Himalaya.

15. HABENA'RIA, *Br.*

Habit of *Orchis.* *Tubers* 2, ovoid, entire or lobed, or of several fleshy fibres. *Lip* spurred, decurved or deflexed. *Anther* confluent with the column ; cells parallel or diverging ; pollen-masses decurving after removal ; glands exposed (or partially concealed in *H. viridis* and *intacta*) ; rostellum produced or not ; stigma 2-lobed or depressed.—Distrib. N. temp. and trop. regions ; species 400.—Etym. doubtful.

The minute modifications of the rostellum, &c., by which the genera here united by Bentham (and most of them also by A. Gray) were characterized, do not hold good for numerous exotic species of the genus.

Section 1. **Gymnade'nia,** *Br.* (gen.). *Tubers* 2, lobed. *Sepals* spreading. *Spur* long or short. *Anther-cells* parallel ; pollen-glands remote, linear ; rostellum produced between the glands. *Stigmas* lateral, large, tumid.

1. H. conops'ea, *Benth.* ; flowers purple, lip obtusely 3-lobed, spur very slender. *G. conops'ea,* Br. *Fragrant Orchis.*

Dry pastures, N. to Shetland ; ascends to 2,000 ft. in the Highlands ; Ireland ; fl. June–Aug.—*Stem* 6–18 in. *Leaves* oblong-lanceolate, keeled, acute. *Spike* dense or lax, narrow. *Bracts* as long as the ovary, green, 3-nerved.

Flowers bright rose-red or purple, broader than long, $\frac{1}{2}$ in. diam.; sepals and petals obtuse; lip broad, lobes 3, subequal, rounded; spur flexuous.—DISTRIB. Europe (Arctic), N. and W. Asia.—Very fragrant.

SECTION 2. **Ti'nea,** *Bivoni* (gen.) (*Neotin'ea,* Rchb. *f.,* gen.). *Tubers* 2, ovoid, entire. *Sepals* conniving into a hood. *Spur* very short. *Anther-cells* parallel; pollen-glands globose, remote, partially concealed by an ascending process of the rostellum. *Stigmas* lateral, large, reniform.

2. **H. intac'ta,** *Benth.* ; leaves oblong, often spotted, flowers pink, lip 3-lobed, spur subglobose. *N. intac'ta,* Rchb. *f.* ; *T. cylindra'cea,* Biv.

Limestone pastures, Mayo and Galway; fl. June.—Habit of *Gymnade'nia al'bida,* but smaller, 4–10 in. *Spike* dense-fld., sometimes twisted; bracts shorter than the ovary, 1-nerved. *Flowers* $\frac{1}{3}$ in., pink or purplish; sepals darker; petals acute; lip projecting, lateral lobes short linear, middle entire or lobed.—DISTRIB. France, S. Europe, N. Africa, Asia Minor.— This little plant has been referred to 7 genera.

SECTION 3. **Leucor'chis,** E. H. F. Mey. (gen.) (*Bicchia,* Parl., gen.). *Root-fibres* many, fleshy. *Sepals* conniving into a hood. *Spur* short. *Anther-cells* parallel; pollen-glands remote, orbicular; rostellum prominent between the glands. *Stigma* depressed.

3. **H. al'bida,** *Br.* ; flowers white, lip acutely 3-lobed. *Gymnade'nia, al'bida,* Rich.

Hilly pastures, Sussex, Wales, and from York and Lancaster to Shetland; ascends to near 1,900 ft. in the Highlands; W. and N. Ireland; fl. June–Aug.—*Stem* 6–12 in. *Leaves* small, obtuse, upper acute. *Spike* narrow, dense; bracts green, equalling the ovary. *Flowers* $\frac{1}{5}$ in., subsecund, sweet-scented; ovary short; sepals and petals obtuse; lip small, projecting; lobes triangular.—DISTRIB. Europe (excl. Greece, Turkey), W. Siberia, Greenland.—This plant has been placed under 6 genera.

SECTION 4. **Coeloglos'sum,** *Hartm.* (gen.). *Tubers* 2, lobed. *Sepals* conniving into a hood. *Spur* very short. *Anther-cells* parallel, remote; pollen-glands oblong, partially concealed by a small pouch, rostellum 2-fid. *Stigma* depressed.

4. **H. vir'idis,** *Br.* ; leaves several, flowers green, lip linear-oblong 2-fid. *Frog Orchis.*

Hilly meadows, Shetland to Devon and Kent; ascends to 2,500 ft. in the Highlands; Ireland; fl. June–Aug.—Tubers ovoid, often lobed. *Stem* 3–12 in. *Leaves* narrow-oblong, obtuse, smaller upwards. *Spike* lax; bracts green, exceeding the ovaries. *Flowers* $\frac{1}{4}$–$\frac{1}{3}$ in.; hood hemispheric, petals and sepals striped with dark red; lip paler, browner. *Anther-cells* diverging, rostellum 0.—DISTRIB. Europe (Arctic) (excl. Greece), N. Asia, N. America.

SECTION 5. **Platan'thera,** *Rich.* (gen.). *Tubers* 2, lobed. *Sepals* spreading. *Spur* long. *Anther-cells* parallel or diverging; pollen-glands remote, orbicular or oblong; rostellum not produced. *Stigma* depressed.

5. H. bifo'lia, *Br.* ; leaves 2, flowers whitish, lip linear-oblong entire, spur twice as long as the ovary. *Butterfly Orchis.*

Wet meadows, woods, and heaths, N. to Ross; ascends to 1,500 ft. in N. England; Ireland; Channel Islands; fl. June–Aug.—*Tubers* 2, ovoid. *Stem* 6–18 in. *Leaves* rarely 3, lower 3–6 in., ovate or oblong, obtuse, narrowed at the base; upper small, lanceolate. *Spike* 4–6 in., lax-fld.; bracts equalling or exceeding the ovary, green. *Flowers* 1 in. from upper sepal to tip of lip, tinged with green or yellow; sepals subacute, dorsal broad, lateral large; petals small; lip obtuse.—DISTRIB. Europe, N. Asia (Arctic).—Very fragrant. The following sub-species are, according to Darwin, distinct, and require different species of moths to fertilize them. They vary in the position and distances of their anther-cells, but intermediates occur.

H. BIFO'LIA proper; lateral sepals narrow, spur slender spreading, anther-cells parallel, caudicle short, gland oblong. *Platan'thera solstitia'lis,* Bœnn. Fl. June–July.

Sub-sp. H. CHLORAN'THA, *Bab.*; flowers usually larger, lateral sepals broader, spur stout decurved more clavate, anther-cells more distant diverging, caudicle longer attached by a short drum-like pedicel to the orbicular gland; fl. July–Aug.

16. CYPRIPE'DIUM, *I.* LADY'S SLIPPER.

Rootstock creeping. *Stem* leafy at the base or upwards. *Sepals* and *petals* spreading. *Lip* large, inflated. *Column* curved over and nearly closing the small orifice of the lip, bearing a terminal dilated lobe (deformed stamen), on each side of the base of which is a short antheriferous arm, and below it a discoid pedicelled stigma. *Anthers* 2, partially 2-celled; pollen viscid, granular; rostellum 0. *Ovary* straight.—DISTRIB. Trop. and temp. regions; species 40.—ETYM. Κύπρις and πόδιον, *Venus' Slipper.*

C. Cal'ceolus, *L.*; bracts foliaceous, lip obovoid.

Woods in limestone districts, Durham and York, very rare; fl. May.— Pubescent. *Stem* 6–18 in. *Leaves* oblong, acuminate, ribbed. *Flowers* 1–2, odorous; bracts foliaceous; sepals red-brown, upper 1–1½ in., erect, ovate-lanceolate, acuminate; lateral narrower, usually connate, placed under the lip; petals 1½ in., linear; lip as long, pale yellow, obovoid, with a rounded upturned end.—DISTRIB. Europe (Arctic) (excl. Turkey), N. Asia.

ORDER LXXVII. IRIDE'Æ.

Perennial herbs; rootstock tuberous, bulbous, or creeping, or a corm. *Leaves* often equitant and ensiform. *Flowers* regular or not, 2-bracteate. *Perianth* superior, petaloid, of 6 imbricate segments in 2 series, often twisted and persistent after flowering. *Stamens* 3, epigynous or inserted on the outer perianth-segments; anthers usually narrow, extrorse.

Ovary 3-celled ; style simple, stigmas 3 often dilated, simple or divided ; ovules very many, in the inner angles of the cells, anatropous. *Capsule* 3-gonous, 3-celled, loculicidally 3-valved. *Seeds* many, testa coriaceous or thin, albumen horny or fleshy ; embryo terete, short, cylindric.— DISTRIB. Chiefly extra-tropical ; genera about 57 ; species 700.—AF-FINITIES. With *Amaryllideæ* and *Orchideæ.*—PROPERTIES. Purgative and diuretic.

Segments of perianth nearly equal.
    Perianth-tube short. Stigmas 3, 2-partite ....1. Romulea.
    Perianth-tube short. Stigmas 3, entire ....2. Sisyrinchium.
    Perianth-tube long ....2*. Crocus.
Segments of perianth unequal.
    Perianth regular ....3. Iris.
    Perianth irregular...4. Gladiolus.

1. ROM'ULEA, *Maratti.*

Corm sheathed. *Leaves* radical, slender, linear. *Scape* simple or branched. *Perianth* regular ; tube very short ; segments equal, suberect, tips recurved. *Stamens* on the throat of the perianth, filaments free hairy ; anthers basifixed. *Ovary* short, 3-gonous ; stigmas linear 2-cleft. *Capsule* ovoid, 3-lobed. *Seeds* subglobose, testa coriaceous.—DISTRIB. Chiefly S. Africa ; species 54.—ETYM. *Romulus,* the founder of Rome.

R. Colum'næ, *Seb.* and *Maur.* ; leaves·wiry. *Trichoné'ma Colum'næ,* Reichb. ; *T. Bulboco'dium,* Sm.

Sandy pastures, Dawlish ; Channel Islands ; fl. March–May.—*Corm* size of a pea, ovoid ; sheaths brown, shining. *Leaves* 2–4 in., recurved, subcylindric, grooved above. *Scape* very short, 1–3-fld.; spathe longer than the perianth-tube ; pedicels curved in fruit. *Perianth-segments* ½–¾ in., subacute, green-ish outside, whitish inside, veins purple, claw yellow. *Capsule* small.— DISTRIB. W. France, S. Europe, N. Africa.

2. SISYRIN'CHIUM, *L.* BLUE-EYED GRASS.

Root of rigid fibres. *Leaves* radical, linear, equitant. *Scapes* usually 2-edged. *Flowers* umbellate. *Perianth* regular, tube very short ; seg-ments equal, spreading or suberect. *Stamens* on the throat of the perianth, free or connate at the base ; anthers basifixed. *Ovary* short, 3-gonous ; style short, stigmas 3 filiform involute. *Capsule* subglobose, coriaceous. *Seeds* subglobose or angled, testa hard.--DISTRIB. N. and S. America ; species 50.—ETYM. obscure.

S. angustifo'lium, *Miller ;* bracts suberect lanceolate. *S. Bermu-dia'na,* var. *a,* L.

Bogs, Galway and Kerry ; fl. July–Aug. —*Leaves* 3–5 in., ¼ in. broad, ensiform. *Scape* 6–18 in., flattened, wing narrow. *Bracts* shorter than the 1–4 flowers, which are ⅔ in. diam. *Perianth-segments* blue inside, oblong, retuse, caudate.

—Distrib. Arctic and temp. N. America.—Mr. Wynn assures me that this plant is truly wild in Kerry. It differs entirely from the Bermudian plant.

2*. *CRO'CUS, L.*

Corm with sheathing fibrous coats. *Stem* 0. *Leaves* radical, surrounded by scarious sheaths, narrow-linear, channelled, white beneath, margins recurved. *Flowers* solitary or fascicled, subsessile, honeyed. *Perianth* large, tube very long; segments equal, narrow-oblong, concave. *Stamens* on the bases of the outer segments, filaments free; anthers basifixed. *Ovary* subterranean, hidden amongst the leaf-bases, ovoid; style filiform, stigmas 3 cuneate dilated or laciniate. *Capsule* on a long slender pedicel, fusiform. *Seeds* globose, testa thick.—Distrib. Europe, N. Africa, N. and W. Asia; species 70.—Etym. The old Greek name.

C. NUDIFLO'RUS, *Sm.*; flowers solitary autumnal, stigmas multifid. *C. specio'sus,* Hook.

Meadows, Midland counties, local, Salop and Warwick to York and Lancashire; (native? *Wats.*); fl. Sept.-Oct.—*Corm* subglobose, clothed with rich brown coats of parallel fibres. *Leaves* vernal. *Perianth-lobes* 2 in., bright purple. *Anthers* pale orange-yellow. *Stigmas* orange, their segments truncate and crenate. *Seeds* as in *C. ver'nus.*—Distrib. S. France, Spain.

C. VER'NUS, *All.*; flowers few vernal, stigmas toothed. *Purple Crocus.*

Naturalized in meadows, Nottingham, Suffolk, Middlesex; fl. March-April.— *Corm* broad, depressed; sheaths of reticulate fibres, much torn, dirty brown. *Perianth-lobes* 1-2 in., purple or white. *Anthers* pale bright-yellow. *Stigmas* deep orange. *Capsule* ½-¾ in. *Seeds* reddish, small.—Distrib. Mid. and S. Europe.

3. *I'RIS, L.*

Rootstock tuberous or creeping. *Leaves* chiefly radical, equitant, ensiform. *Scape* compressed; spathes terminal with scarious borders. *Perianth-tube* short, rarely long; sepals large, stipitate, reflexed, stipes channelled; petals smaller, suberect, stipitate, margins of stipes involute. *Stamens* inserted on the base of the sepals, filaments free; anthers basifixed. *Ovary* 3-gonous; style stout, stigmas 3 very broad petaloid arching over the stamens 2-fid and with a transverse lamella, stigmatic surface a point below the lamella. *Capsule* coriaceous, 3-gonous, 3-ribbed. *Seeds* many, flat or globose, testa coriaceous hard or thick and fleshy.— Distrib. N. temp. regions; species 100.—Etym. The Greek name, from the hues of the flower.

1. **I. Pseud-a'corus,** *L.*; flowers yellow, petals ¾ shorter than the sepals. *Yellow Flag.*

River-banks, ditches, &c., N. to Shetland; Ireland; Channel Islands; fl. May-Aug.—*Rootstock* creeping, stout. *Leaves* 2-4 ft., ½-1 in. broad. *Scape* 2-4 ft., leafy, often branched at the top; pedicel about as long as the ovary;

spathes 2 in., acute or obtuse. *Flowers* 3-4 in. diam., variable in colour, (pale in *I. Bastar'di,* Bor.) and form of the segments; tube cylindric; sepals often purple-veined, with an orange spot near the base; petals spathulate. *Stigmas* yellow. *Seeds* much vertically compressed, faces flat, testa hard.— Distrib. Europe, N. Africa, Siberia.—Rootstock acrid.

I. Pseud-a'corus proper; sepals clear yellow, claw short, blade broad, stigmas long narrow.—Var. *I. acorifor'mis,* Bor.; sepals with a dark yellow obicular blade and long greenish purple-veined claw, stigmas short broad.— Common.

2. I. fœtidis'sima, *L.* ; flowers generally purple, petals ¼ shorter than the sepals. *Fœtid Iris, Roast-beef plant.*

Copses and hedgebanks, chiefly on limestone, from Durham southd.; naturalized in Scotland and Ireland; Channel Islands; fl. May–July.—*Rootstock* stout, creeping. *Leaves* dark green, flaccid. *Scape* 1-2 ft., leafy; pedicels longer than the ovary; spathes 3-4 in., acuminate. *Flowers* 3 in. diam.; sepals obovate-lanceolate, blue-purple, rarely yellow; petals and stigmas spathulate, yellow. *Capsule* 2-3 in., clavate. *Seeds* globose, testa fleshy orange-red.—Distrib. W. Europe to Italy.

4. GLADI'OLUS, *L.*

Corms with reticulate fibrous coats. *Leaves* equitant, ensiform. *Scape* tall. *Flowers* secund, spiked, inclined or horizontal. *Perianth* sub-2-labiate; tube short, curved; segments obovate. *Stamens* ascending, inserted on the perianth-tube; anthers linear, versatile. *Ovary* ovoid; style filiform, stigmas 3 broad undivided. *Capsule* coriaceous. *Seeds* compressed and winged, or globose, testa fleshy.—Distrib. Europe, W. Asia, chiefly S. African; species 90.—Etym. *gladiolus,* a *little sword.*

G. commu'nis, *L.* ; var. *illyr'icus,* Koch (sp.); leaves glaucous, spathes subequal, perianth campanulate.

Open grounds, New Forest and I. of Wight, rare; fl. June–July. *Corm* size of a hazel-nut, with many bulbils at its base. *Leaves* 6-10 in., ⅓-½ in. diam., glaucous, acuminate. *Scape* 2-3 ft., leafy. *Spike* 4-8-fld.; spathes subequal, lanceolate, acuminate. *Perianth* 1-1½ in., curved, crimson-purple; 3 upper segments spathulate; 3 lower more obovate, paler with strong red-purple veins. *Stigmas* spathulate, margins involute after flowering. *Capsule* ½ in., clavate. *Seeds* narrowly winged.—Distrib. Europe, from W. France southd. and eastd., N. Africa.—Var. *illyr'icus* differs from *commu'nis* in the more slender fibres of the corm sheath; broader stigmas not papillose on the margin throughout their length; and narrower wing of the seed.

Order LXXVIII. **AMARYLLIDE'Æ.**

Rootstock bulbous. *Leaves* radical. *Scape* naked. *Flowers* bracteate. *Perianth* superior, regular or irregular, coloured, of 6 lobes or segments in 2 series, with sometimes a crown at the mouth of the tube. *Stamens* 6,

on the perianth-tube or bases of the segments (rarely epigynous), filaments free or connate ; anthers versatile, linear or oblong, bursting inwards
or by terminal pores. *Ovary* ovoid or globose, 3-celled ; style filiform or
columnar, stigmas 1 or 3 ; ovules many, in 2 series, in the inner angles of
the cells, anatropous. *Fruit* usually capsular, rarely fleshy, 3-celled,
loculicidally 3-valved, cells 1- or more-seeded. *Seeds* turgid or compressed,
testa various, albumen fleshy ; embryo straight, axile, terete.—Distrib.
Temp. and trop. ; genera about 64 ; species 650.—Affinities. Close
with *Irideæ* and *Liliaceæ.*—Properties. Emetic, narcotic, and poisonous.
Agave yields textiles, and a fermentable liquor (pulque).

Mouth of perianth with a circular crown.............................1. Narcissus.
Crown 0. Outer perianth-segments largest.........................2. Galanthus.
Crown 0. Perianth-segments equal ....,.......3. Leucojum.

1. NARCIS'SUS, *L.*

Scape compressed. *Leaves* narrow, linear. *Flowers* solitary or umbellate, large, white or yellow, drooping or inclined ; spathe membranous.
Perianth tubular below ; segments spreading, mouth surmounted by a
circular crown. *Stamens* inserted in the tube, included within the crown,
filaments free or adnate to the tube ; anthers versatile. *Ovary* 3-gonous ;
style filiform, stigma obtuse. *Capsule* coriaceous. *Seeds* globose, testa
smooth, rough when dry.—Distrib. Europe, N. and W. Asia ; species
about 20.—Etym. mythological.

N. Pseudo-narcis'sus, *L.* ; leaves nearly flat, flower solitary yellow
campanulate, crown campanulate as long as the perianth-segments, margin
crisped obscurely 6-lobed. *Daffodil, Lent Lily.*

Copses and pastures throughout England, local ; naturalized in Scotland and
Ireland ; fl. March–April.—*Bulb* 1 in., outer scales membranous. *Leaves*
glaucous, obtuse. *Scape* 6–10 in. *Flower* primrose-yellow, 2 in. ; pedicel
short. *Perianth-lobes* acute. *Capsule* turbinate.—Distrib. From Gothland
southd. (excl. Greece, Turkey).—Of the following forms the first only is
indigenous.
N. Pseudo-narcis'sus proper ; leaves slightly glaucous, perianth-segments
oblong-lanceolate sulphur-yellow, crown lemon-yellow obscurely 6-lobed.—
Var. *N. lobula'ris,* Haw. (*N. Bromfield'ii,* Syme) ; perianth and corona both
lemon-yellow, corona distinctly 6-lobed. The Tenby Daffodil (*N. camb'ricus,*
Haw.) scarcely differs.—Var. *N. ma'jor,* L. ; more robust, perianth-segments
broader and corona lemon-yellow, crown with 6 rounded lobes. The great
Spanish Daffodil.

N. biflo'rus, *Curt.* ; leaves keeled, flowers 1–3 salver-shaped, crown
short concave membranous.

Naturalized in sandy fields ; fl. April–May.—*Bulb* 1–1½ in., outer scales
membranous. *Leaves* very long, 10–18 in., hardly glaucous, obtuse. *Scape*
as long, acutely 2-edged. *Perianth* 1½ in. diam., white or pale straw-
coloured, pedicel slender ; tube 1 in., slender ; segments broadly ovate,
obtuse ; crown pale yellow.—Distrib. W. Europe.

2. GALAN'THUS, *L.* SNOWDROP.

Leaves 2, linear. *Scape* compressed. *Flowers* solitary, pendulous, white ; spathe membranous. *Perianth* campanulate ; sepals spreading ; petals small, erect, notched, with 2 green honeyed grooves. *Stamens* 6, epigynous ; anthers pointed, connivent, slits 2 terminal. *Ovary* ovoid ; style subulate, stigma simple. *Capsule* ovoid, herbaceous. *Seeds* few, subglobose, testa soft white.—DISTRIB. Europe ; species 3.—ETYM. γάλα, *milk,* and ἄνθος, *flower.*

G. niva'lis, *L.* ; leaves glaucous keeled.

Meadows and copses, wild ? in Hereford and Denbigh ; naturalized elsewhere in England, Scotland, and Ireland ; fl. Jan.-March.—*Bulb* ½ in., ovoid. *Leaves* 6–10 in., obtuse. *Scape* longer, prostrate in fruit ; spathe 2-fid, 2-nerved ; pedicel slender. *Flower* 1 in. ; sepals obovate, concave ; petals white.—DISTRIB. From Holland southd. (excl. Greece), W. Asia.

3. LEUCO'JUM, *L.* SNOW-FLAKE.

Characters of *Galanthus,* but leaves numerous ; spathes 2, free or connate, 1–6 fld. ; sepals and petals subequal ; anthers obtuse, opening by slits ; style clavate ; testa crustaceous fleshy.—DISTRIB. Chiefly European ; species 9.—ETYM. λευκίς, *white,* and ἴον, a *violet.*

1. L. æsti'vum, *L.* ; leaves hibernal, flowers æstival, scape 2–6-fld.

Wet meadows and osier holes in S.E. England ; Suffolk to Oxford and Kent to Dorset ; (a denizen, *Wats.*) ; fl. May.—*Bulb* 1 in. *Leaves* 12–18 in., obtuse, subglaucous. *Scape* as long, prostrate in fruit, 2-edged ; tip of spathe entire, green. *Flowers* drooping, buds erect. *Sepals* ¾ in., ovate, white, tips green. *Fruit* turbinate. *Seeds* not caruncled.—DISTRIB. From Denmark southd.

2. L. vernum, *L.* ; leaves and flowers vernal, scape 1–2-fld.

Copses, Dorset ; (an alien or denizen, *Wats.*) ; fl. March–April.—Much smaller than *L. æsti'vum ;* leaves subdistichous ; scape less winged ; spathe 2-fid at the tip ; flowers about as large ; ovary more globose ; seeds caruncled.— DISTRIB. From Belgium southd. (excl. Greece, Russia).

ORDER LXXIX. DIOSCORE'Æ.

Rootstock often tuberous. *Stem* twining to the left, leafy. *Leaves* alternate, veins reticulate. *Flowers* inconspicuous, 1-sexual, in axillary panicles or racemes. *Perianth* herbaceous, superior in the female flower ; segments 6, in two series, regular, persistent. *Stamens* 6, inserted on the perianth-segments, free ; anthers introrse. *Ovary* 3-celled ; styles 3, short, stigma entire or lobed ; ovules 2, collateral or superposed, anatropous. *Fruit* 3-angled, 3-celled, indehiscent or loculicidally 3-valved, rarely a berry or 1-celled. *Seeds* winged, compressed or globose, albumen dense ; embryo small.—DISTRIB. Chiefly trop. ; genera 6 ; species about 100.—AFFINITIES. With *Smilaceæ.*—PROPERTIES. The acrid yam tubers are nutritious when cultivated or boiled.

1. TAMUS, *L.* BLACK BRYONY.

Perianth campanulate. *Stigmas* 2-lobed. *Berry* imperfectly 3-celled, few-seeded. *Seeds* globose.—DISTRIB. Europe, Mediterranean ; species 2.—ETYM. doubtful.

T. commu'nis, *L.* ; leaves ovate-cordate acuminate.

Copses and hedges from Cumberland southd.; Channel Islands; fl. May– June.—*Rootstock* ovoid, black, fleshy, subterranean. *Stem* many feet long, very slender, angular, branched. *Leaves* 2–3 in., long-petioled, obscurely laterally lobed, 5–7-nerved, tip setaceous ; stipules reflexed. *Flowers* ⅙ in. diam.; males solitary or fascicled in slender racemes which are branched at the base ; female racemes 1 in., shorter, recurved, few-fld.; bracts minute. *Berry* ½ in., oblong, red.—DISTRIB. From Belgium southd., N. Africa, W. Asia.

ORDER LXXX. LILIA'CEÆ.

Root fibrous ; rootstock bulbous or creeping. *Stem* rarely shrubby or arborescent. *Flowers* 2- rarely 1-sexual. *Perianth* herbaceous, peta-loid, inferior ; segments 6 in 2 series, rarely 4, 8, or 10, free or connate, imbricate (rarely valvate) in bud. *Stamens* 6 (3 in *Ruscus*) hypogynous or inserted on the perianth, filaments long or short ; anthers oblong or linear. *Ovary* 3-celled ; styles 1 or 3, rarely 0, stigma simple or 3-lobed ; ovules 2 or more in the inner angle of each cell, anatropous. *Fruit* a 3- rarely 1–2-celled capsule or berry. *Seeds* 1 or more in each cell, albumen horny or fleshy ; embryo small, terete, radicle next to or far from the hilum.—DISTRIB. All climates ; genera 187 ; species about 2,500.—AFFINITIES. With *Junceæ.*—PROPERTIES various.

SERIES A. *Rootstock* not bulbous. *Anthers* bursting inwards. *Fruit* a berry.

TRIBE I. **ASPARAGEÆ.** *Stem* rigid, branched or climbing. *Leaves* minute, scale-like, with leaf-like branchlets (cladodes) in their axils. *Ovules* orthotropous or hemi-anatropous.

Flowers on the cladodes diœcious. Stamens 3, filaments connate...1. Ruscus. Flowers axillary. Stamens 6, filaments distinct ....2. Asparagus.

TRIBE II. **POLYGONATEÆ.** *Stem* herbaceous, leafy. *Flowers* axillary or terminal. *Ovules* anatropous.

Flowers axillary; perianth tubular 6-cleft ....3. Polygonatum. Flowers in terminal racemes ; perianth-segments 4 ....4. Maianthemum.

D D

Tribe III. **CONVALLARIEÆ.** *Leaves* radical. *Flowers* on a lateral naked scape, racemed. *Ovules* anatropous ....5. Convallaria.

Series B. *Rootstock* not bulbous. *Leaves* radical. *Flowers* racemed or panicled. *Anthers* bursting inwards. *Fruit* a loculicidal capsule.

Tribe IV. **ASPHODELEÆ**..6. Simethis.

Series C. *Roctstock* bulbous. *Anthers* bursting inwards. *Fruit* a loculicidal capsule.

Tribe V. **ALLIEÆ.** *Leaves* radical. *Flowers* umbelled or capitate, on a naked terminal scape; heads or umbels enclosed at first in a 2-leaved membranous involucre ....7. Allium.

Tribe VI. **SCILLEÆ.** *Leaves* radical. *Flowers* 1-bracteate, racemed on a terminal naked scape.
Perianth globose; mouth constricted 6-cleft ....8. Muscari.
Perianth of 6 blue or red segments ....9. Scilla.
Perianth of 6 white segments ....10. Ornithogalum.

Tribe VII. **TULIPEÆ.** *Leaves* radical and cauline. *Flowers* few, solitary or loosely racemed or whorled; perianth-segments 6, free.
Flowers few, large, nodding. Nectary 0 or obscure. Anthers versatile.
                                                 10*. Lilium.
Flowers large, nodding. Nectary oblong. Anthers erect......11. Fritillaria.
Flowers large, subsolitary. Nectary 0. Anthers erect ....12. Tulipa.
Flowers few, small (yellow). Nectary 0. Anthers erect............13. Gagea.
Flowers few, small (white). Nectary transverse. Anthers erect...14. Lloydia.

Series D. *Rootstock* various. *Anthers* usually bursting laterally or outwards. (*Melanthaceæ.*)

Tribe VIII. **COLCHICEÆ.** *Rootstock* a corm. *Leaves* radical. *Scape* very short, subterranean, 1–3-fld. *Perianth* with a very long slender tube ....15. Colchicum.

Tribe IX. **NARTHECIEÆ.** *Rootstock* short or creeping. *Leaves* radical. *Scape* erect. *Perianth* 6-cleft, usually persistent. *Fruit* capsular.
Style very short, stigma small. Capsule loculicidal............16. Narthecium.
Style 0; stigmas 3 short. Capsule septicidal ....17. Tofieldia.

Tribe X. **MEDEOLEÆ.** *Rootstock* stout, creeping; stem simple. *Leaves* radical or cauline and opposite, or whorled. *Flowers* terminal, solitary or umbelled. *Fruit* a berry...18. Paris.

1. RUS'CUS, *L.* Butcher's Broom.

Evergreen, subdiœcious shrubs. *Rootstock* stout, creeping. *Leaves* minute scales, bearing in the axils leaf-like branches ("cladodes"). *Flowers* minute, on the face or margin of the cladode. *Perianth* herbaceous, persistent; segments 6, spreading, inner smaller, all partially valvate in bud. *Stamens* 3, filaments connate in a short stout column;

anthers sessile, cells diverging below. *Ovary* enclosed in a fleshy cup
(staminal) 3-celled ; style short, stigma discoid ; ovules few. *Berry* usually
1-celled. *Seeds* solitary globose, or 2 plano-convex, testa thin, adherent,
albumen horny ; embryo minute, lateral, radicle far from the hilum.—
DISTRIB. Temp. Europe, W. Asia, N. Africa ; species 2–3.—ETYM.
obscure.

R. aculea'tus, *L.* ; flowers 1–2 subsessile on the ovate spinescent
cladodes.

Copses and woods, from Norfolk, Leicester, and S. Wales southd., rare ;
naturalized in Scotland and Ireland ; Channel Islands ; fl. Feb.–April.—
Stems 10–24 in., tufted, branched, erect, stout, angled, young shoots scaly.
Cladodes ½–1½ in., twisted at the base. *Flowers* ⅛ in. diam., bracteate and
bracteolate, males on narrower cladodes. *Berry* ½ in. diam., bright red,
rarely yellow.—DISTRIB. From France southd., N. Africa, W. Asia.

2. ASPAR'AGUS, *L.* ASPARAGUS.

Rootstock stout, creeping. *Stem* slender, branched, terete or angled
(sometimes spiny and climbing). *Leaves* minute scales, bearing in their
axils fascicles of needle-like branches ("cladodes)." *Flowers* small,
1–2-sexual, pendulous, axillary, honeyed ; pedicel jointed. *Perianth*
campanulate, segments connate at the base. *Stamens* on the base of the
segments ; anthers oblong. *Ovary* 3-gonous ; styles combined, stigmas
3 ; cells 2- or several-ovuled. *Berry* globose. *Seeds* 3–6, testa black
brittle ; embryo dorsal, clavate, radicle far from the hilum.—DISTRIB.
Temp. and trop. Asia, Africa ; species 100.—ETYM. The old Greek
name.

A. officina'lis, *L.* ; stems annual suberect terete flexuous, branches
slender.

Coasts of Wales, Cornwall, and Dorset, rare ; naturalized elsewhere ; Tramore,
Ireland ; Channel Islands ; fl. June–Aug.—*Rootstock* 1–2 ft., prostrate ;
young shoots scaly below, scales triangular. *Cladodes* ½–2 in. *Flowers* 1–2,
axillary, 1-sexual, dirty white, or yellow with red veins, males the largest ;
pedicel as long, jointed at the middle. *Berry* ¼ in. diam., red.—DISTRIB.
From Sweden southd. (excl. Greece), N. Africa, Siberia ; introd. in N.
America.—Diuretic. Cultivated since the Roman period.

3. POLYGONA'TUM, *Tournef.*

Rootstock creeping. *Stem* leafy. *Leaves* alternate opposite or whorled.
Flowers axillary, solitary or racemed, pendulous, white green or purplish,
honeyed, homogamous, ebracteate. *Perianth* tubular-campanulate, mouth
6-cleft, outer lobes subvalvate with replicate edges. *Stamens* on the
middle of the tube, included. *Ovary* and *fruit* of *Convallaria.*—DISTRIB.
Europe, N. Asia, Himalaya, N. America ; species about 23.—ETYM. πολύς
and γόνυ, alluding to the *many* nodes (*knees*).

1. P. verticilla'tum, *All.* ; stem angled, leaves whorled narrow-lanceolate, perianth constricted in the middle, filaments papillose.

D D 2

Wooded banks and glens, very rare, Northumbd., Perth and Forfar; fl. June–
July.—*Stem* 2–3 ft. *Leaves* 3–5 in., sessile, 3–6 in a whorl (rarely 1 or 2),
flaccid, margins and veins beneath ciliolate. *Peduncles* ¼–¾ in., 1–3-fld.
Perianth ¼ in., greenish. *Berry* ¼ in. diam.—DISTRIB. Europe (Arctic)
(excl. Greece), Siberia, Himalaya.

2. **P. multiflo′rum,** *All.* ; stem terete, leaves alternate subbifarious or
secund oblong ½-amplexicaul, perianth constricted in the middle, filaments
pubescent. *Solomon's Seal.*

Woods, rare, from Northumbd. to Kent and Devon (excl. Wales); naturalized
in Scotland and Ireland; fl. May–June.—*Stem* 2–3 ft., naked below, arched.
Leaves 3–5 in., very shortly petioled, acute or obtuse. *Peduncles* 2–5-fld.
Perianth ⅔ in., greenish white. *Berry* ⅓ in. diam., blue-black.—DISTRIB.
Europe, Siberia, Dahuria.

3. **P. officina′le,** *All.* ; stem angled, leaves alternate oblong subbifarious
½-amplexicaul, perianth cylindric, filaments glabrous.

Wooded limestone cliffs, rare, from the Border to Somerset and Dorset; fl.
May–June.—Smaller than *P. multiflo′rum*, with more leathery leaves and
usually solitary larger flowers, perianth cylindric with broader lobes. *Stem*
6–12 in., arched. *Leaves* 3–4 in., subacute. *Peduncles* rarely 3-fld. *Perianth*
1 in., greenish-white. *Berry* ¼ in. diam., blue-black.—DISTRIB. Europe,
N. Asia.

4. MAIANTHEMUM, *Wigg.*

Rootstock slender, creeping. *Stem* erect, leafy. *Leaves* alternate.
Flowers white, terminal, racemed. *Perianth* of 4 free segments in 1 series,
or 6 in 2 series, deciduous. *Stamens* 4 or 6, on the bases of the segments.
Ovary 2–3-celled ; style short, simple, stigma obscurely 2–3-lobed ; cells
1–2-ovuled. *Berry* and *seeds* as in *Convallaria.*—DISTRIB. Europe
(Arctic) (excl. Turkey), N. Asia, N. America.—ETYM. μαῖος and ἄνθεμον,
May-flowerer.

M. Convallaria, *Roth ; M. bifolium,* DC. ; *Smilacina bifolia,* Desf.

Woods, very rare, wild in Yorkshire, and probably in Lancashire and Bedford ;
fl. May–June.—Glabrous or pubescent. *Stem* 4–8 in., flexuous. *Radical
leaves* 1½–2½ in., cordate, acute, base deeply 2-lobed, many-nerved, long-
petioled ; cauline 2–3, short petioled or upper sessile. *Raceme* 1–2 in.,
8–10-fld. ; pedicels slender, solitary or 2–3 ; bracts minute. *Flowers* ⅛ in.
diam., 4-merous, suberect, fragrant. *Berry* ¼ in. diam., apiculate, white,
dotted.

5. CONVALLA′RIA, *L.* LILY OF THE VALLEY.

Rootstock creeping. *Stem* 0. *Leaves* 2–3, sheathed at the base. *Scape*
slender ; flowers racemose, homogamous. *Perianth* shortly campanulate,
mouth 6-cleft, lobes recurved. *Stamens* on the base of the perianth,
included ; anthers subsagittate. *Ovary* ovoid, terete ; style simple,
stigma 3-gonous ; cells 4–8-ovuled. *Berry* globose. *Seeds* 2–3, subglobose,
testa thin white adherent, albumen horny ; embryo dorsal, radicle far

from the hilum.—DISTRIB. Europe (excl. Greece), N. Asia.—ETYM. *convallis*, a *valley*.

C. maja'lis, *L.* ; leaves ovate-lanceolate petioled.
Woods, from Caithness to Kent and Devon, not common ; ascends to 1,000 ft. in N. England ; naturalized in Scotland and Ireland ; fl. May–June. *- Leaves* 6–8 in.; petiole long, slender, sheathing. *Scape* 6–10 in., angular ; bracts membranous ; raceme 6–12-fld. ; pedicels curved. *Flowers* ¼ in. diam., white or rose, odorous, drooping, subglobose. *Berry* red.

6. SIME'THIS, *Kunth.*

A slender herb. *Root* of fascicled fibres. *Leaves* radical, grassy. *Scape* panicled, bracteate. *Flowers* jointed on the pedicel. *Perianth* spreading, deciduous ; segments 6, almost free. *Stamens* 6, on the base of the segments, filaments woolly below ; anthers oblong. *Ovary* subglobose, 3-celled ; style slender, stigma a point ; ovules 2 in each cell, superposed. *Capsule* loculicidally 3-valved. *Seeds* 6, subglobose, arillate, testa black crustaceous shining ; embryo long, radicle near the hilum.—DISTRIB. S.W. France, Spain, Italy, N. Africa.—ETYM. Classical.

S. bi'color, *Kunth.* *S. planifo'lia,* Woods.
Fir woods, Dorset (extinct) ; Derrynane, Ireland ; (an alien or denizen, *Wats.*) ; fl. June.—*Root-fibres* stout. *Leaves* 6–18 in., ¼ in. diam., linear, acuminate, recurved, surrounded at the base with torn fibrous brown sheaths, flat or concave. *Scape* as long, dichotomously branched ; bracts slender ; pedicels rigid, slender. *Flowers* ¾ in. diam., corymbose ; segments oblong, obtuse, concave, purple on the back, white inside. *Capsule* obtusely angled ¼ in. diam.

7. AL'LIUM, *L.*

Fœtid, pungent herbs. *Bulb* coated. *Leaves* radical. *Flowers* capitate or umbelled ; spathes 1–2, membranous. *Perianth-segments* 6, free, spreading or campanulate. *Stamens* 6, hypogynous or on the base of the segments, filaments free or connate below ; anthers oblong. *Ovary* 3-gonous ; style filiform, simple or 3-cleft, stigmas simple ; ovules few in each cell. *Capsule* membranous, top depressed, 3-lobed, loculicidally 3-valved. *Seeds* 1–2 at the base of each cell, turgid or compressed, testa black ; embryo curved, excentric, radicle next the hilum.—DISTRIB. N. temp. regions ; species 250.—ETYM. Latin for Garlic.

SECTION 1. **Por'rum.** *Leaves* sheathing the scape to the middle. *Perianth-segments* erect. *Three outer filaments* broader, tips 3-fid ; lateral cusps subulate, about as long as the antheriferous.

* *Leaves fistular.*

1. **A. vinea'le,** *L.* ; leaves flattened or grooved above, spathe solitary short with a long beak, head globose usually with bulbils. *Crow Garlic.*

Pastures and waste dry places, from the Clyde and Aberdeen southd., not frequent; S. and E. Ireland; Channel Islands; fl. June–July.—*Bulb* small. *Leaves* 8–24 in., strict. *Scape* longer, cylindric. *Flowers* ⅓ in., green or pink, sometimes replaced by bulbils (*A. compac'tum*, Thuill.); pedicel slender, tip thickened. *Filaments* exposed. *Bulbils* ¼ in., green or purplish. —Distrib. Europe (excl. Greece), Canaries; introd. in N. America.

2. **A. sphæroceph'alum**, *L.*; leaves terete or flattened or grooved above, spathes 2 shortly beaked, head globose dense-fld.

St. Vincent's Rocks, Bristol; Channel Islands; fl. June–Aug.—Habit of *A. vineale*, but ribs of leaf rough when young, and heads dense, globose, red-purple, without bulbils.—Distrib. From Belgium southd. (excl. Greece).

*** Leaves not fistular, glaucous.*

3. **A. Scorodopra'sum**, *L.*; leaves flat keeled, edges scabrid, sheaths 2-edged, spathes 2 shortly beaked, head with bulbils, anthers not exserted. *A. arena'rium*, Sm. *Sand Leek.*

Dry pastures and copses, rare, York and Lancaster to Fife and Perth; Ireland; fl. May–Aug.—*Bulb* ovoid, with small stalked bulbils. *Leaves* 6–8 in., ⅓–⅔ in. broad. *Scape* slender. *Head* lax-fld.; bulbils purple; pedicels slender. *Perianth* ⅓ in., segments red-purple, margins white, keel of outer scabrid.—Distrib. Europe (excl. Spain, Greece).

3*. A. Ampelop'rasum, *L.*; leaves distichous folded, edges scabrid, sheaths cylindric, spathe 1 with a compressed long beak, head often with bulbils, anthers exserted. *Wild Leek.*

Rocky banks, naturalized; fl. July–Aug.—*Bulb* large, with often stalked bulbils. *Leaves* 1–2 ft., 1–1½ in. broad. *Scape* 3–6 ft., very stout. *Heads* globose, 3–4 in. diam., very many-fld., pedicels unequal. *Perianth* ⅓ in., white or greenish.—Distrib. Switzerland, Europe S. of the Alps, W. Asia.
A. Ampelop'rasum proper (*A. holmen'se*, Mill.); head compact, bulbils 0. Steep holmes in the Severn.—Var. *bulbif'erum*, Syme; head compact, bulbils few. Guernsey.—Var. *A. Babingto'nii*, Borrer (*A. Halle'ri*, Bab.); flowers few, bulbils very many, pedicels sometimes proliferous. Dorset and Cornwall,? wild; Roundstone and Great Aran Is., Ireland.

Section 2. **Codonop'rasum.** *Perianth-segments* erect or spreading *Filaments* all simple or obscurely 3-fid.

4. **A. Schœnop'rasum**, *L.*; leaves fistular, head dense-fld. without bulbils, spathes 2 shortly beaked, stamens included. *Chives.*

Rocky pastures, very rare, Northumbd., Lancashire, Brecon, Cornwall; fl. June–July.—*Bulbs* narrow, small, tufted on short rootstocks. *Leaves* 4–10 in., few, terete or grooved above. *Scapes* 6–14 in., stout or slender, hollow. *Perianth* ⅓–½ in., campanulate, pale purple; pedicels short. *Stamens* connate at the very base. *Capsule* globose, small.—Distrib. Mid. Europe (excl. Turkey), N. Asia, Himalaya, N. America.

A. Schœnop'rasum proper; leaves straight, ribs smooth or scaberulous, perianth-segments gradually acuminate. N. of England.—Var. *A. sibir'i-cum*, L.; larger, leaves recurved, ribs more scabrid, perianth-segments abruptly-acuminate. *A. arena'rium*, Sm. in *Engl. Bot.* Kynance Cove.

5. **A. olera'ceum,** *L.* ; leaves nearly flat or ½-terete sheathing the cylindric scape to the middle, head lax-fld. with bulbils, spathes 2, beaks slender unequal, stamens equalling the perianth. *Field Garlic.*

Borders of fields, &c., rare, E. Scotland, Forfar and Perth to Berwick, and southd. to Devon and Kent; fl. July.—*Bulb* small. *Leaves* very slender, variable in breadth and thickness, flat towards the tip, with many striæ, ribs rough. *Scapes* 10-18 in., very slender. *Pedicels* flexuous, spreading; 2-4 in. *Perianth* campanulate, segments obtuse, pale, olive-green pink or brownish. *Stamens* included, filaments shortly connate, subulate.—Distrib. Europe (excl. Greece), W. Siberia, Himalaya.

A. olera'ceum proper; leaves narrow ½-terete subfistular grooved above. Devon, Somerset, Gloucester.—Var. *A. complana'tum*, Bor. (*A. carina'tum*, Sm. not L.); leaves broader linear almost solid nearly flat. Yorkshire and Northumbd.

A. carina'tum, *L.* ; leaves linear channelled sheathing the cylindric scape to the middle, head with bulbils, spathes 2, beaks long slender very unequal, stamens at length twice as long as the perianth.

Nottingham, Newark, Lincoln, Edinburgh, and Perth ; naturalized ; fl. Aug.— Very similar to *A. olera'ceum*, but at once distinguished by the long filaments. *Leaves* flat towards the tip, with 3-5 striæ. *Flowers* bright rose-pink, proterandrous.—Distrib. Europe (excl. Spain, Greece, Turkey).

6. **A. trique'trum,** *L.* ; leaves linear sharply keeled sheathing the base of the triquetrous scape, head lax-fld. without bulbils, spathes 2 lanceolate.

Hedgebanks and meadows, Cornwall, Guernsey ; fl. April-June.—*Bulb* ovoid, rather small. *Leaves* 5-7 in., ¼-⅓ in. broad, recurved or revolute. *Scapes* 10-18 in., rather stout, bending over in fruit. *Flowers* secund, drooping or inclined ; perianth ⅔ in., white, campanulate; segments linear-oblong; pedicels curved, tip clavate. *Stamens* short, free, included, filaments slender. *Stigmas* 3, distinct, filiform.—Distrib. S. France, Spain, Italy.

7. **A. ursi'num,** *L.*; leaves ovate-lanceolate sheathing the base of the 3-gonous scape, spathes 2 ovate acuminate. *Ramsons.*

Woods, hedgebanks, &c., from Skye and Ross southd.; ascends to 1,200 ft. in Yorkshire ; Ireland ; fl. April-June.—*Bulb* narrow, compressed, on a short rootstock, outer coat fibrous. Leaves 4-8 in., vernal, acuminate; petiole 2-4 in. *Scape* 6-18 in. *Umbel* regular, flat-topped; spathes 1 in.; pedicels strict, 1½-2 in.; bulbils 0. *Flowers* honeyed, proterandrous, white; segments spreading, lanceolate, acute. *Stamens* shorter than the segments, filaments free slender; anthers dehiscing in succession, the inner first. *Stigma* minute. *Capsule* turbinate.—Distrib. Europe (excl. Greece), N. Asia.

8. MUSCA'RI, *Tourn.* GRAPE HYACINTH.

Perianth globose, mouth 6-fid. *Stamens* on the middle of the tube, included, filaments very short; anthers short. *Ovary* ovoid, deeply 3-

lobed ; style short, stigma simple ; cells few-ovuled. *Capsule* 3-quetrous, loculicidally 3-valved. *Seeds* 6 or fewer, as in *Scilla.*—DISTRIB. Europe, W. Asia ; species about 40.—ETYM. from the *musky* scent.

M. racemo′sum, *Miller ;* leaves slender prostrate flexuous.

Sandy pastures, Norfolk and Suffolk, Cambridge; fl. May.—*Bulb* small, with bulbils at the base. *Leaves* 6–10 in., ½-terete, grooved above. *Scape* short. *Raceme* short, cylindric, many-fld.; rachis dilating after flowering; pedicels slender, lengthening in fruit; bracts minute. *Flowers* ⅛ in. diam., dark blue, upper imperfect. *Capsule* ⅛ in.—DISTRIB. From Belgium southd., N. Africa.

9. SCIL′LA, *L.* SQUILL.

Bulb coated. *Leaves* radical, linear. *Flowers* usually racemose, blue, rarely purple or white. *Perianth* deciduous ; segments 6, spreading or conniving, free or nearly so ; nectary 0. *Stamens* 6, on the base of the segments or above it, filaments flattened ; anthers oblong. *Ovary* ovoid ; style filiform, stigma minute ; ovules 4 or more in each cell. *Capsule* 3-angled, loculicidally 3-valved. *Seeds* many, testa black, albumen fleshy ; embryo terete, radicle next the hilum.—DISTRIB. Chiefly Europe and W. Asia ; species about 80.—ETYM. Classical.

SECTION 1. *Perianth-segments* free, spreading ; stamens inserted on their bases ; anthers purple. *Seeds* angular.

1. S. ver′na, *Huds.* ; leaves vernal preceding the subcorymbose flowers, bracts as long as the pedicels or longer.

Rocky pastures rare, W. coast of England and Wales from Flint to Devon ; Scotland from Ayr and Berwick to Shetland ; E. and N.E. Ireland, very rare ; fl. April–May.—*Bulb* as large as a hazel-nut. *Leaves* 3–10 in., ⅛–⅙ in. broad, recurved, concave. *Scapes* 1–2, shorter than the leaves. *Flowers* ½ in. diam., bright-blue, fragrant ; lower pedicels ½ in. *Capsule* ⅛ in. diam.—DISTRIB. Coasts of Norway, France, Spain.

2. S. autumna′lis, *L.* ; leaves autumnal narrow succeeding the shortly racemose flowers, bracts 0.

Rocks and pastures from Gloster and Middlesex to Cornwall and Kent ; Channel Islands ; fl. July–Sept.—*Bulb* ¾–1½ in. diam. *Leaves* 3–6 in., ½-terete, grooved above. *Scapes* several, equalling the leaves. *Flowers* ½ in. diam., reddish purple ; pedicels ascending or spreading. *Capsule* small. —DISTRIB. W. and S. Europe, N. Africa, Crimea.

SECTION 2. *Perianth* campanulate, segments connate at the base ; stamens inserted below their middle ; anthers yellow. *Seeds* subglobose.

3. S. nu′tans, *Sm.* ; leaves and flowers vernal, bracts in pairs. *Ag′raphis nu′tans,* Link ; *Endym′ion nu′tans,* Dumort ; *Hyacin′thus nonscrip′tus,* L. *Bluebell, Wild Hyacinth.*

Woods, banks, &c., from Caithness southd.; ascends to 1,500 ft. in the Lake District; Ireland ; Channel Islands ; fl. April–June.—*Bulb* ¾–1 in. diam.

Leaves 10–18 in., ½ in. broad, subacute, concave. *Scape* solitary, tall, stout. *Raceme* 6–12-fld.; bracts linear, membranous. *Flowers* 1 in., blue purple white or pink, drooping; pedicel short, curved, erect in bud and fruit. *Capsule* subglobose.—DISTRIB. W. Europe from Belgium southd to Italy,

10. ORNITHO'GALUM, *L.*

Bulb coated. *Leaves* all radical, linear. *Flowers* racemose or corymbose, white. *Perianth* spreading, persistent; segments 6, free, with a basal nectariferous gland. *Stamens* 6, hypogynous, filaments flattened; anthers versatile, linear-oblong. *Ovary* 3-quetrous, with 3 glands on the top; style 3-gonous, stigma obtuse; cells many-ovuled. *Capsule* grooved, loculicidally 3-valved. *Seeds* terete or angled, testa black, rough when dry; embryo cylindric, radicle next the hilum.—DISTRIB. Europe, N. and W. Asia, N. Africa; species about 70.—ETYM. The classical name.

O. pyrena'icum, *L.* ; raceme many-fld., pedicels spreading, filaments much dilated to above the middle.

Woods and copses, local, Somerset, Wilts, Bedford, Berks, Sussex; fl. June–July.—*Bulb* 2 in., ovoid. *Leaves* 1–2 ft., ¼–½ in. diam., vernal, concave, glaucous, soon withering at the flowering season. *Scape* tall, stout. *Raceme* 4–8 in., very many-fld.; bracts subulate. *Perianth* 1 in. diam.; segments narrow-oblong, green, margins white inside. *Capsule* ½ in.—DISTRIB. From Belgium southd.—Young shoots eaten, sold at Bath as French Asparagus.

O. UMBELLA'TUM, *L.* ; corymbs 6–10-fld., pedicels suberect, filaments broadly subulate. *Star of Bethlehem.*

Copses, meadows, &c., naturalized; fl. May–June.—*Bulb* 1 in., with many bulbils. *Leaves* 6–8 in., ¼ in. broad, concave, green with a white stripe. *Scape* tall; rachis of corymb elongate after flowering; bracts very long. *Perianth* 1–1½ in. diam.; segments linear-oblong, white with a dorsal green midrib. *Capsule* obovoid.—DISTRIB. From Holland southd.

O. NU'TANS, *L.* ; racemes few-fld. drooping, filaments broadly dilated upwards 3-fid.

Copses, &c., Midland and E. counties from Yorkshire and Durham to Hereford, naturalized; fl. April–May.—*Bulb* 2 in. *Leaves* 1–2 ft., ¼–½ in. broad. concave, glaucous with a white stripe. *Scape* as long; bracts long, slender; pedicels curved, ¼–½ in. *Perianth* 1–1½ in. broad; segments lanceolate, white with a dorsal green midrib. *Capsule* pendulous, broadly ovoid, green, fleshy.—DISTRIB. S. Europe.

10.* *LIL'IUM, L.*

Bulb of many imbricate fleshy scales. *Leaves* all cauline, not sheathing, alternate or whorled. *Flowers* few or many, large, erect or drooping, homogamous; perianth-segments free, caducous, erect below, recurved or revolute above; nectary median, elongate. *Stamens* hypogynous or on the base of the segments, filaments subulate; anthers fixed above the base in front, versatile, bursting inwards. *Ovary* 6-grooved ; style terete, stigma obtuse ; cells many-ovuled. *Capsule* erect and seeds as in *Fritillaria.*—DISTRIB. Temp. N. regions ; species 40.—ETYM. unknown.

L. MAR'TAGON, *L.* ; leaves petioled obovate-lanceolate whorled, upper linear, flowers erect racemose. *Purple Martagon Lily.*

Copses, Mickleham, Surrey, naturalized ; fl. Aug.–Sept.—*Bulb* large, scales white. *Stem* 2–3 ft., terete. *Leaves* 3–8 in., chiefly in a few whorls of 6–8, subacute. *Flowers* 1½ in. diam., drooping, odorous at night ; bracts ½–1 in., green, linear-lanceolate ; pedicels 1–5 in., stout, curved ; perianth-segments oblong, pale purple or white, with dark raised papillæ ; nectary with thick raised borders. *Anthers* red-brown. *Capsule* 1 in.—DISTRIB. Europe from Mid. France southd. and eastd.

11. FRITILLA'RIA, *L.*

Bulbs often clustered ; scales few, thick. *Stem* leafy, 1- or more-flowered. *Leaves* sessile, not sheathing. *Flowers* drooping ; perianth campanulate ; segments free, caducous, tip not recurved, nectariferous gland basal. *Stamens* on the very base of the segments, filaments subulate ; anthers oblong or linear, fixed above the base in front. *Ovary* long, 3-gonous ; style 3-grooved, stigmas 3 glandular inwards ; cells many-ovuled. *Capsule* erect, oblong, 3-gonous, loculicidally 3-valved above, margins of valves ciliate. *Seeds* many, 2-seriate, horizontal, vertically compressed, margined or winged, testa spongy pale ; embryo terete, radicle next to the hilum.—DISTRIB. N. temp. regions ; species 50.—ETYM. *fritillus,* a *dice-box,* from the chequered petals.

F. **Melea'gris,** *L.* ; leaves linear flat subacute. *Snake's Head.*

Moist meadows, rare, from Norfolk and Stafford to Somerset and Hants ; (a denizen, *Wats.*) ; fl. May.—Bulb small, of 2–3 turgid scales. *Leaves* 6–8 in., ¼–½ in. broad, cauline few, short. *Stem* 10–18 in. *Flower* solitary, rarely 2, 1½ in., segments narrow-oblong, tesselated with dull purple, rarely almost colourless ; nectary narrow. *Anthers* ½ in., yellow. *Seeds* close-packed.—DISTRIB. Europe (excl. Greece, Turkey), W. Asia.

12. TU'LIPA, *L.*

Bulbs of thick convolute scales. *Leaves* radical and cauline, lower sheathing. *Flowers* usually solitary, erect or inclined ; perianth campanulate ; segments free, tips recurved, nectary 0. · *Stamens* hypogynous, filaments short subulate glabrous or hairy below ; anthers fixed by the base, mobile, linear, bursting inwards. *Ovary* 3-gonous ; stigma sessile with 3 radiating lobes ; cells many-ovuled. *Capsule* erect, coriaceous, and seeds as in *Fritillaria.*—DISTRIB. Europe, N. and W. Asia ; species 50.—ETYM. *Tulipan,* a *turban,* in Persian.

T. **sylves'tris,** *L.* ; flowers bright yellow, perianth-segments elliptic-lanceolate. *Wild Tulip.*

Chalk pits, &c., wild in S.W. York, Norfolk, Suffolk, Somerset; naturalized elsewhere; fl. April–May.—*Bulb* small, ovoid, stoloniferous; scales chestnut-brown. *Leaves* few, 6–10 in., linear, ½–1 in. broad, glaucous. *Stem* 1–2 ft., terete, flexuous. *Flowers* 2 in., fragrant. *Filaments* woolly at the base. *Capsule* 1 in., acute above and below.—DISTRIB. From Holland southd.

13. GA'GEA, *Salisb.*

Bulb coated; coats few coriaceous. *Leaves* radical, linear. *Scapes* with leafy bracts. *Flowers* yellow, corymbose or umbellate, proterogynous; perianth-segments persistent, free spreading from an erect base; nectary 0. *Stamens* on the base of the segments, filaments flattened subulate; anthers linear, basifixed. *Ovary* 3-gonous or 3-quetrous; style 3-gonous, stigma obtuse; cells many-ovuled. *Capsule* membranous, loculicidally 3-valved. *Seeds* many, subglobose, pendulous, raphe thick, testa soft yellow; embryo cylindric, radicle next to the hilum.—DISTRIB. Europe. N. Asia; species 20.—ETYM. *Sir Thomas Gage*, a British botanist.

G. lu'tea, *Ker ;* radical leaf solitary ribbed, sheath slender, bracts 1–3. *G. fascicularis*, Salisb.; *Ornitho'galum lu'teum*, L. *Yellow Star of Bethlehem.*

Copses and pastures on the E. from Perth and Moray to Gloster, Sussex, and Somerset, local and rare; fl. March–May.—*Bulb* small, subglobose, with basal bulbils. *Radical leaf* linear, 6–18 in., ⅓–⅔ in. broad. *Scape* short; bracts 1–3, 2 sometimes opposite; pedicels 1–2 in. *Perianth* ½–¾ in., opens in forenoon only; segments linear-oblong, obtuse, inner narrowest, thin, yellow, back green.—DISTRIB. Europe (excl. Greece), N. Asia, Himalaya.

14. LLOY'DIA, *Salisb.*

A small slender herb. *Bulb* minute, thickly scaly. *Leaves* filiform. *Stem* leafy, slender, 1–2-fld. *Flowers* erect, honeyed, homogamous; perianth-segments persistent, free, spreading, equal, yellow or white, with a transverse cavity. *Stamens* 6, on the base of the segments, filaments filiform; anthers oblong, basifixed. *Ovary* 3-gonous; style filiform, stigma obtuse; cells many-ovuled. *Capsule* 3-quetrous, loculicidally 3-valved at the top. *Seeds* many, horizontal, 3-quetrous, testa black rugose; embryo minute, next the hilum.—DISTRIB. Europe, Asia, and N. America; species 2.— ETYM. *E. Lloyd*, an antiquary, its discoverer in Wales.

L. sero'tina, *Reichb ;* leaves 3-gonous incurved.
Rocky ledges of the Snowdon range, very rare; fl. June.—*Sheaths* of bulb very many and loose. *Leaves* 6–10 in., cauline shorter. *Stem* 2–8 in., terete. *Perianth-segments* ½ in., white (in England), obovate-oblong, veined with purple. *Seeds* red-brown till quite ripe.—DISTRIB. Arctic, Alps of Mid. Europe, Himalaya, N. America.—Flowers in the Himalaya vary from white to primrose or deeper yellow.

15. COL'CHICUM, *L.* MEADOW SAFFRON.

Corm coated. *Leaves* all radical, usually vernal. *Bracts* spathaceous. *Flowers* solitary or fascicled, subsessile on the corm, erect, crocus-like, honeyed, proterogynous; perianth-tube very long, slender, limb campanulate; segments 6, oblong. *Stamens* on the mouth of the perianth, included, filaments filiform; anthers oblong, fixed above the base in front, versatile, bursting inwards. *Ovary* deeply 3-grooved; styles 3, free, filiform, tips recurved, stigmatic inwards; cells few- or many-ovuled. *Capsule* membranous, 3-grooved to the axis, septicidally 3-valved at the top. *Seeds* subglobose, testa rugose, funicle fleshy; embryo minute, remote from the hilum.—DISTRIB. Europe, N. and E. Asia; species 30.—ETYM. Classical.

C. autumna'le, *L.*; leaves lanceolate, sheaths stout long large.

Meadows, from Westmoreland and Durham to Sussex and Somerset, local; naturalized in Scotland; Ireland; fl. Aug.–Oct.—*Corm* large, oblique, compressed; scales shining, chestnut. *Leaves* flat, 6–10 in., nerves close-set. *Flowers* appearing in succession; ovary subterranean; perianth-tube 2–6 in.; limb 1½ in., pale purple. *Anthers* ½ in., yellow. *Capsule* 1–2½ in., shortly peduncled, ellipsoid, acute at both ends, ripening in spring. *Seeds* many, small, pale brown.—DISTRIB. From Denmark southd.—Yields the famous drug. Flowers, when vernal, greenish and imperfect.

16. NARTHE'CIUM, *Mœhr.* BOG ASPHODEL.

Rigid herbs with the habit and leaves of *Tofieldia. Flowers* racemose, golden-yellow; perianth-segments subequal, spreading, persistent, erect in fruit. *Stamens* 6, 3 hypogynous and 3 on the base of the segments, filaments subulate villous; anthers linear, fixed by the back, bursting inwards. *Ovary* narrow-ovoid, 3-gonous, narrowed in the short style, stigma obtuse; cells many-ovuled. *Capsule* 3-gonous, narrow, pointed, loculicidally 3-valved above. *Seeds* many, terete, testa filiform at each end, attached to the inner angle of the cell by a hair-like funicle; embryo minute, next the hilum.—DISTRIB. Europe N. of the Alps and Pyrenees, N. Asia, N. America.—ETYM. ναρθήκιον, a *rod.*

N. ossif'ragum, *Huds.*; perianth-segments linear-oblong.

Bogs, N. to Shetland; ascends to near 3,200 ft. in the Highlands; Ireland; fl. July–Aug.—*Rootstock* wiry, long, slender. *Leaves* 6–12 in., rigid, strongly ribbed, acuminate. *Stem* leafless or with 1–2 short ½-amplexicaul leaves. *Raceme* 2–4 in.; bracts subulate; pedicels sometimes bracteolate. *Flowers* ½ in. diam.; segments ribbed and green on the back. *Filaments* white; anthers deep-orange. *Capsule* red, longer than the perianth.—DISTRIB. Of the genus.

17. TOFIELD'IA, *Huds.*

Rootstock short creeping. *Leaves* radical, equitant, ensiform. *Scape* slender; bracteoles 3, minute, connate. *Flowers* racemed, small, green, honeyed, homogamous ; perianth-segments persistent, spreading, free. *Stamens* 6, on the base of the segments, filaments filiform ; anthers short, fixed by the back, bursting inwards. *Ovary* ovoid, 3-gonous ; styles 3, conic, persistent, stigmas obtuse ; cells many-ovuled. *Follicles* 3, almost free, membranous, oblong, acute. *Seeds* many, minute, marginal on the valves, cymbiform, testa thin ; embryo minute, next the hilum.—DISTRIB. Arctic and Alpine Europe, Asia, and America ; species 14.—ETYM. *Tofield*, a Yorkshire botanist.

T. palus'tris, *Huds.* ; bracteoles scarious. *T. borea'lis*, Wahlnb.

Mountain rills and bogs, York, Durham, and from Argyll and Perth to Sutherland ; ascends to 2,400 ft. in the Highlands ; fl. July–Aug.—*Leaves* tufted, 2–3 in., 3–5-nerved. *Scape* 4–8 in., naked or 1–2-leaved, slender, terete. *Raceme* dense-fld., ½–1 in. ; pedicels short, bracteolate at the base. *Flowers* ¼ in. diam., pale-green ; segments linear-oblong, obtuse. *Capsule* ⅛ in. diam., subglobose.—DISTRIB. Of the genus.

18. PAR'IS, *L.* HERB PARIS.

Rootstock stout. *Stem* simple, with 1 whorl of usually 4–9 leaves. *Flowers* solitary, not honeyed, malodorous, proterogynous. *Sepals* 3–5, lanceolate. *Petals* as many, filiform. *Stamens* 8–12, subhypogynous ; anthers basifixed, cells narrow ; connective sometimes produced. *Ovary* subglobose, 4–5-lobed and -celled ; styles 4–5, free, stigmas decurrent ; cells 4- or more-ovuled. *Berry* indehiscent or loculicidally 4–5-valved. *Seeds* few or many, testa coriaceous or fleshy, albumen horny ; embryo terete, radicle next the hilum.—DISTRIB. Europe, N. Asia, Himalaya ; species 3–4.—ETYM. *par*, *equal*, from the 4-nary parts of *P. quadrifolia.*

P. quadrifo'lia, *L.* ; leaves ovate-oblong or obovate.

Woods, local, from Caithness to Kent and Somerset ; fl. May–June.—Glabrous. *Rootstock* white, creeping. *Stem* 6–12 in., terete, leafy at the top, sheath basal. *Leaves* 4 (rarely 3–8), 3–5 in., acute, 3–5-nerved. *Flower* 1½–1¾ in. diam., solitary ; peduncle ½–3 in., erect. *Sepals* green, acuminate. *Petals* as long, yellow. *Connective* much produced. *Berry* black, bursting irregularly. *Seeds* black, testa coriaceous.—DISTRIB. Europe (Arctic) (excl. Greece), N. and W. Asia.—Rootstock purgative.

ORDER LXXXI. JUNCE'Æ.

Perennial, rarely annual herbs. *Rootstock* usually creeping, scaly. *Stems* erect, usually simple, sometimes septate within, pith often thick, continuous or interrupted. *Leaves* slender, flat or terete, or reduced to sheathing scales. *Flowers* green or brown, in axillary or terminal cymes, regular, 2-sexual, bracteolate. *Perianth* inferior, scarious or coriaceous ; segments 6 in 2 series, free, subequal, persistent, imbricate in bud. *Stamens* 6, inserted on the bases of the segments, rarely 3 on the outer only, filaments flattened ; anthers basifixed, usually linear, bursting inwards. *Ovary* free, 1-3-celled ; style short or 0, stigmas 3 filiform papillose all over ; ovules 3, basilar, or many on 3 parietal or axile placentas, erect, anatropous. *Capsule* 1-3-celled, loculicidally 3-valved, 3-many-seeded. *Seeds* erect, testa membranous often lax, albumen dense ; embryo next the hilum, small.—DISTRIB. Chiefly temp. and Arctic ; genera 4-5 ; species about 130.—AFFINITIES. With *Liliaceæ.*—PROPERTIES unimportant.

Glabrous. Ovules many, parietal or axile.............................1. Juncus.
More or less pilose. Ovules 3, basal...2. Luzula.

1. JUN'CUS, *L.* RUSH.

Glabrous herbs. *Outer perianth-segments* keeled or midrib thickened. *Flowers* proterogynous, anemophilous. *Stamens* 6, rarely 3. *Ovary* 3-rarely 1-celled ; ovules many, placentas axile, rarely parietal. *Capsule* completely or incompletely 3-celled. *Seeds* many ; embryo minute.— DISTRIB. Of the Order ; species 100.—ETYM. *jungo,* from their use in *tying.*

SECTION 1. *Rootstock* perennial, usually creeping. *Stems* rarely septate within, terete, rarely compressed. *Leaves* solid, not septate within.

* *Leaves all reduced to sheaths. Cymes wholly lateral, many-fld. ; flowers not or rarely clustered. Testa not produced at either end.*

1. **J. effu'sus,** *L.* ; stems soft, pith . continuous, perianth-segments lanceolate exceeding the obovoid retuse capsule, stamens 3. *J. communis,* Meyer.

Moist places, N. to Shetland ; ascends to 2,400 ft. in the Highlands; Ireland ; Channel Islands ; fl. July-Aug.—*Tufts* circular, densely matted. *Stems* 1-3 ft., sometimes ¼ in. in diam., green, very finely striate. *Cymes* very compound, variable in form and size. *Perianth* ₁₀-⅛ in. *Stamens* rarely 6. *Seeds* minute, yellow-brown.—DISTRIB. Europe, N. Africa, temp. Asia and America, N. Zealand.
J. effu'sus proper ; cymes usually lax effuse, perianth olive-green, anthers oblong, capsule not mucronate.—VAR. *J. conglomera'tus,* L. ; cymes usually dense subglobose, perianth tinged with brown, anthers longer linear, capsule mucronate.

2. **J. glau'cus,** *Ehrh.* ; stems rigid glaucous striate, pith interrupted, perianth-segments narrow-lanceolate about equalling the ovoid mucronate capsule, stamens 6.

Wet, usually stiff soils, from Aberdeen and the Clyde southd. ; ascends to 1,200 ft. in N. England ; Ireland ; Channel Islands ; fl. July-Aug.—Habit of *J. commu'nis,* but not so tall, and stems deeply grooved. *Cymes* effuse, suberect. *Perianth* brown.—DISTRIB. Europe, N. Africa, Siberia.
J. diffu'sus, Hoppe (a hybrid between *glau'cus* and *effu'sus*) ; stems softer less glaucous and striate, pith continuous, cyme elongate, capsule more obovoid, seeds imperfect.

3. J. bal'ticus, *Willd.* ; stems rigid, pith continuous, perianth-segments ovate-lanceolate about equalling the ovoid abruptly mucronate capsule, stamens 3. *J. arc'ticus,* Hook., not Willd.

Sandy seashores, rarely by inland lakes, from Fife and Kincardine to Caithness; fl. July.—*Rootstock* creeping, not tufted. *Stems* few, 1–2 ft., $\frac{1}{18}$–$\frac{1}{10}$ in, diam., pale green, scarcely striate. *Cymes* small, corymbose, suberect, few-fld. *Perianth* about $\frac{1}{8}$ in., dark brown, midrib pale. *Anthers* much longer than the filaments.—DISTRIB. N. Germany to the Arctic circle, N. America.

4. J. filifor'mis, *L.* ; stems wiry pale green filiform, pith interrupted, cyme midway up the stem small, perianth-segments lanceolate exceeding the turbinate obtuse mucronate capsule, stamens 6.

Stony and gravelly margins of lakes, Westmoreland, Cumberland, Kincardine; fl. July–Aug.—*Rootstock* loosely tufted. *Stems* 4–8 in., very slender, faintly striate. *Sheaths* often with subulate tips. *Cymes* sessile; flowers few, crowded, very pale. *Anthers* shorter than the filaments. *Seeds* very minute.—DISTRIB. Europe N. of the Alps (Arctic), N. Asia, N. America.

** *Leaves all reduced to sheaths, or a few elongate and stem-like. Cymes lateral, flowers many clustered. Stamens 6. Testa produced at each end.*

5. J. acu'tus, *L.* ; stems rigid pungent, sheaths long shining, perianth-segments ovate-lanceolate, inner obtuse with a broad scarious margin half as long as the broadly ovoid mucronate capsule.

Sandy seashores from Norfolk and Carnarvon to Kent and Devon, rare ; S. and S.E. Ireland ; Channel Islands; fl. July–Aug.—*Tufts* circular. *Stems* 2–4 ft., stout, terete, hardly striate, many flowerless. *Cymes* corymbose, dense-fld., very large in fruit; bracts lanceolate-subulate, exceeding the flowers; flowers $\frac{1}{6}$–$\frac{1}{4}$ in. *Perianth* brown, inner segments retuse winged towards the tip. *Capsule* nearly $\frac{1}{4}$ in., turgid, hardly 3-gonous.—DISTRIB. Coasts from France to Turkey, N. Africa.

6. J. marit'imus, *Sm.* ; stems wiry pungent, sheaths short pale, perianth-segments lanceolate all acute and without scarious margins equalling the elliptic-oblong acuminate capsule.

Salt marshes from Isla and Elgin southd.; rare in Scotland; common in Ireland; fl. July–Aug.—Habit of *J. acu'tus*, but irregularly tufted ; stems less rigid and usually more slender ; cymes more interrupted, branches long erect; bracts not exceeding the very pale flowers.—DISTRIB. Coasts from Gothland to Turkey, N. Africa, W. Siberia, N. America.

*** *Leaves all terete compressed or channelled. Cymes terminal or lateral, 1–3- (rarely 6-) fld. Testa produced at each end.*

7. J. triglu'mis, *L.* ; stems terete, leaves radical subulate channelled, flowers usually 3 terminal equalling the membranous bracts, perianth-segments obtuse much shorter than the ellipsoid beaked capsule.

Alpine bogs, from Carnarvon and Durham to Shetland ; ascends to 3,000 ft. in the Highlands; fl. July–Aug.—Tufted, black; stolons 0. *Stems* 6–18 in., slender, wiry. *Leaves* very short, formed of 2 separate tubes, sheaths auricled. *Flowers* erect, $\frac{1}{6}$–$\frac{1}{5}$ in., pale red-brown in fruit.—DISTRIB. Arctic and Alpine Europe, N. Asia, Himalaya, N. America.

8. **J. biglu'mis,** *L.* ; stems $\frac{1}{2}$-terete, leaves radical subulate compressed, flowers usually 2 shorter than the bract, perianth-segments obtuse nearly equalling the turbinate retuse 3-lobed capsule.

Bogs, alt. 2,000–3,300 ft. on the alps of Perth, Argyll, Skye ; fl. July–Aug.—Not tufted, stoloniferous. *Stem* 2–6 in. *Leaves* shorter, septate within, sheaths not auricled. *Flowers* $\frac{1}{8}$ in., upper pedicelled, lower sessile, chestnut-brown. *Capsule* mucronate between the lobes.—DISTRIB. Arctic and subarctic Europe, Asia, Himalaya, N. America.

9. **J. casta'neus,** *L.* ; stems terete leafy, leaves $\frac{1}{2}$-terete fistular, cymes 1–3 lateral and terminal 2–6-flowered, perianth-segments acute half as long as the elliptic-oblong beaked capsule.

Alpine bogs, alt. 2,500–3,000 ft., very rare ; Perth, Forfar, Inverness, and Aberdeen ; fl. July–Aug.—Not tufted, stoloniferous. *Stem* 6–16 in. *Leaves* 2–3, variable in length, channelled above, sheaths not auricled, walls thin. *Flowers* $\frac{1}{8}$–$\frac{1}{6}$ in., bright brown. *Capsule* nearly $\frac{1}{4}$ in., 3-gonous. *Seeds* large for the genus.—DISTRIB. Arctic and Alpine N. and Mid. Europe, N. Asia, Himalaya, N. America.

10. **J. trif'idus,** *L.* ; stems wiry terete, leaves subulate short mostly reduced to sheaths, flowers 1–3 between 2 filiform bracts, perianth-segments acuminate shorter than the ovoid beaked capsule.

Alpine rocks, alt. 1,200–3,000 ft., from the Clyde, Perth, and Forfar to Shetland ; July–Aug.—Forms dense matted rigid masses of rootstocks and sheaths. *Stems* 3–10 in. *Upper leaf-sheath* with one short subulate leaf ; another leaf occurs on the stems beneath the inflorescence, with a tubular sheath, which has often scarious edges. *Bracts* 2–4 in. *Flowers* $\frac{1}{5}$ in., pale brown ; perianth-segments very narrow, margins pale. *Style* and stigmas long. *Seeds* large, appendages short.—DISTRIB. Arctic and Alpine Mid. Europe, N. and W. Asia, N. America.

**** *Leaves chiefly radical, flat or grooved above. Cymes terminal, 3–many-fld. Testa not produced at either end.*

11. **J. squarro'sus,** *L.* ; rigid, stems compressed, leaves subulate $\frac{1}{2}$-terete below channelled above, cymes terminal, clusters 2–3-fld., perianth-segments oblong obtuse equalling the obtuse mucronate capsule.

Moorlands, &c., N. to Shetland ; ascends to 3,200 ft. in the Highlands ; Ireland ; fl. June–July.—*Stems* densely tufted, 4–10 in., stout, solid, naked or with 1–2 leaves. *Leaves* 3–7 in., densely crowded, recurved from the broad keeled sheath, striate. *Cyme* irregularly corymbose, branches erect ; bracts broad, scarious, shorter than the flowers. *Flowers* $\frac{1}{5}$ in., pale. *Capsule* obtusely 3-gonous.—DISTRIB. N. Europe (excl. Greece), Siberia, Greenland.

12. **J. ten'uis,** *Willd.* ; stems wiry terete, leaves channelled, cymes terminal, bracts long, perianth-segments lanceolate acuminate larger than the ovoid obtuse or retuse pointed capsule.

Sandy moist soil, Hereford, very rare; fl. June–Aug.—*Rootstock* tufted. *Stem* 6–15 in., very slender, cylindric. *Leaves* chiefly radical, few, very slender, deeply striate; base dilated, membranous. *Cymes* shorter than the filiform erect bracts. *Flowers* sessile and pedicelled, ⅛ in. long, pale. *Style* very short. *Capsule* turgid. *Seeds* minute, ellipsoid, acute at both ends, minutely ribbed.—DISTRIB. W. France, Holland, Germany, W. United States.

13. **J. compres'sus,** *Jacq.* ; slender, stems 1-2-leaved subcompressed, leaves slender ½-terete channelled above, cymes terminal, flowers sub-solitary, perianth-segments oblong obtuse equalling or shorter than the ovoid obtuse mucronate capsule. *J. bulbo'sus,* L. (name given by error). Marshy places, N. to Shetland; Ireland; Channel Islands; fl. June–July.— *Rootstock* creeping. *Stems* 6–24 in., very slender, hollow. *Leaves* narrow, flaccid, suberect, equalling the stem or shorter. *Cymes* irregularly corymbose; branches suberect, slender, few-fld.; bracts small. *Flowers* ⅛–⅙ in. perianth-segments pale in the middle, margins broad scarious pale or dark brown or purple.—DISTRIB. Europe (excl. Greece), N. and W. Asia, N America.
J. COMPRES'SUS proper ; stems tufted, capsule broader almost obovoid shortly mucronate much longer than the perianth.—Rather rare.
Sub-sp. J. GERAR'DI, *Loisel ;* stems more remote, capsule narrower strongly mucronate not exceeding the perianth. *J. bott'nicus,* Wahl.; *J. cœno'sus* Bich.—Salt marshes. Also in N. U. States.

SECTION 2. *Rootstock* perennial, usually creeping. *Stems* solid. *Leaves* hollow and septate within. *Testa* not produced.

14. **J. obtusiflo'rus,** *Ehrh.* ; stem tall and leaves (few) erect terete, cymes lateral or subterminal in very compound corymbs, branches zigzag, perianth-segments obtuse equalling the ovoid mucronate capsule.
Marshy places, from the Clyde and Haddington southd.; rare in Scotland and Ireland ; fl. July–Aug.—*Rootstock* widely creeping. *Stems* not tufted, 2–3 ft., usually stout, soft, hardly striate, sheathed at the base. *Leaves* 1–2, like the stem. *Flowers* ⅛ in., sessile in dense peduncled or sessile clusters of 3–8, pale ; bracts small, obtuse, scarious.— DISTRIB. From Gothland southd. (except Greece, Turkey, and Russia), N. Africa.

15. **J. articula'tus,** *L.* ; stems slender and leaves slightly compressed, cymes lateral or subterminal compound corymbose, perianth-segments acuminate not exceeding the obovoid narrow acuminate capsule.
Bogs, especially in mountain districts, N. to Shetland ; ascends to near 3,500 ft. in the Highlands ; Ireland ; Channel Islands ; fl. June–Aug.—Very variable in habit, size, robustness, amount of foliage, and size and composition of the cyme ; the following sub-species express its principal modifications.—DISTRIB. Europe, N. Africa, N. and W. Asia, Himalaya.
J. ARTICULA'TUS proper (*J. acutiflo'rus,* Ehrh.); tall, leaves very conspicuously jointed when dry, flowers in dense distant sessile or peduncled clusters of 3–12 dark chestnut, bracts acuminate ½–⅔ as long as the flower, perianth-segments equalling the narrow acuminate capsule, stamens 6. *J. sylvat'icus,* Reichb.—Ascends to 1,200 ft. in the Lake District.

E E

Sub-sp. J. SUPI'NUS, *Mœnch*; rootstock sometimes tuberous, stems terete 3–10 in. often floating with flaccid straggling branches, joints very obscure, leaves slender, cymes terminal, branches few long suberect, bracts scarious acute sometimes equalling the flower, perianth-segments acute equalling the ovoid obtuse mucronate capsule, stamens 3.—Ascends to 3,500 ft.—VAR. *J. uligino'sus*, Sibth; erect.—VAR. *J. subverticilla'tus*, Wulf.; decumbent or floating.

Sub-sp. J. LAMPROCAR'PUS, *Ehrh.*; stem slightly compressed stout or slender, and leaves evidently septate when dry, cyme terminal, branches long suberect, perianth-segments shorter than the narrow beaked glossy capsule, inner obtuse, stamens 6.—Ascends to 2,400 ft. in the Highlands; also found in N. America.—Of *J. nigritel'lus*, D. Don (not Koch), of the Clova Mts., nothing satisfactory is known; garden specimens from Don himself in Borrer's Herbarium have compressed stems, and seem to be *lamprocarpus*.

SECTION 3. Annual. *Stem* hollow, septate within or not. *Testa* not produced.

16. **J. bufo'nius,** *L.* ; very pale, stems slender septate, upper part or cyme dichotomously branched, perianth-segments lanceolate much longer than the obtuse mucronate capsule, stamens 3 or 6.

Moist places, N. to Shetland; ascends to near 2,000 ft. in the Lake District; Ireland; Channel Islands; fl. June–Aug.—Very variable, densely aggregated, from the seedlings growing in masses. *Stem* 1–8 in., erect or ascending. *Leaves* few, setaceous, channelled above, not jointed, pale green, very narrow, sheaths short. *Cyme* occupying most of the stem; branches short or long, often flexuous; flowers ⅛–⅓ in., solitary and distant, or in clusters of 2–4, lateral open and hexandrous, terminal cleistogamous and 3-androus; bracts small, scarious, obtuse. *Flowers* usually secund, pale green. *Perianth-segments* very unequal, long-acuminate, sometimes much longer than the pale obovoid capsule.—DISTRIB. Europe, N. Africa, N. Asia, Himalaya, N. America, N. Zealand.

17. **J. capita'tus,** *Weigel ;* stems setaceous grooved not septate, heads terminal bracteate, perianth-segments elliptic-ovate acuminate awned longer than the broadly ovoid mucronate capsule, stamens 3.

Sands inundated in winter, W. Cornwall, Channel Islands; fl. May–July.—Very small, 1–4 in., tufted, reddish when dry. *Stems* strict. *Leaves* all radical, short, setaceous, channelled; sheaths short. *Heads* solitary, rarely 2–3, 2-6-fld ; outer bracts setaceous, about twice as long as the flowers, inner smaller, broadly ovate, aristate. *Perianth* ⅙ in., pale, tips recurved.—DISTRIB. Europe, N. Africa.

18. **J. pygmæ'us,** *Rich.* ; stems slender terete not septate, flowers few bracteate, perianth-segments linear-lanceolate acuminate not awned longer than the narrow acute capsule, stamens 3 or 6.

Damp places, Lizard and Kynance Downs; fl. May–June.—Very small, 1–2 in., tufted, pink when dry. *Stem* simple or once branched. *Leaves*, radical setaceous faintly jointed channelled, cauline solitary base auricled. *Flowers* 1–5, subsessile ; bracts ovate. *Perianth* ¼ in., segments membranous 3-nerved, with hyaline margins. *Anther* shorter than the filament. *Capsule*

oblong-lanceolate, obtuse, 3-gonous, pale.   *Seeds* many, filiform, ribbed.—
DISTRIB. Holland to Portugal and Greece.

2. LU'ZULA, *DC.* WOOD-RUSH.

Characters of *Juncus*, but always perennial, foliage more grass-like,
always more or less ciliate with long flexuous white hairs.   *Flowers* proterogynous, anemophilous.   *Anthers* usually longer than the filaments.
Ovary 1-celled ; ovules 3, subbasal, erect.   *Capsule* 1-celled, 3-valved,
3-seeded.   *Seeds* with a basal or terminal appendage.—DISTRIB.   All
temp. and cold regions ; species about 26.—ETYM. *luciola,* a *glowworm.*

1. **L. max'ima,** *DC.* ; tall, cymes very compound, flowers clustered,
perianth-segments awned hardly equalling the ovoid acute beaked capsule,
seeds tubercled at the tip.   *L. sylvat'ica,* Gaud.

Woods and heaths, especially uplands, N. to Shetland; ascends to 2,300 ft. in
  the Highlands; Ireland; Channel Islands; fl. May–June.—*Rootstock* short,
  tufted; stolons short.   *Stems* few, 1–2 ft.   *Leaves,* radical ½–1 ft., often ½
  in. broad, with scanty silky hairs, channelled ; cauline few, short.   *Cymes*
  large, branches often 3–4 in., spreading in fruit.   *Flowers* $\frac{1}{12}$ in., pale,
  usually 3–4 together ; bracteoles ovate, acute, scarious.—DISTRIB.   Europe
  (Arctic) (excl. Greece, Russia).

2. **L. verna'lis,** *DC.* ; slender, cymes lax, branches few reflexed in
fruit, flowers subsolitary, perianth-segments acuminate shorter than the
very broadly ovoid obtuse capsule, crest of seeds long curved terminal.
L. pilo'sa, Willd.

Shady places, N. to Shetland; ascends to near 1,900 ft. in the Highlands;
  Ireland; Channel Islands; fl. April–May.—*Rootstock* short, tufted; stolons
  slender.   *Stems* many, ½–1 ft.   *Leaves* about half as long as the stem, $\frac{1}{8}$–$\frac{1}{4}$
  in. broad, soft, sparingly hairy.   *Cymes* with capillary branches and pedicels.
  Flowers $\frac{1}{8}$–$\frac{1}{4}$ in., chestnut-brown, rarely in pairs ; bracteoles broad, short.
  Capsule very broad below, suddenly contracted to a conical top above the
  middle.—DISTRIB.   Europe (Arctic) (excl. Greece), N. Africa, temp. Asia,
  N. America.—*L. Borre'ri,* Bromf., is a hybrid? found in S. England and
  Wicklow, with an acute shorter capsule that ripens no seed.

3. **L. Fors'teri,** *DC.* ; characters of *L. pilosa,* but more slender, capsule acuminate, and seeds with a shorter terminal straight obtuse crest.

Shaded places in chalky soil, from S. Wales, Oxford, and Essex to Cornwall
  and Kent; Channel Islands; fl. April–June.—Habit of *L. pilo'sa,* from which
  luxuriant specimens can hardly be distinguished, except by the characters
  given above.—DISTRIB.   Mid. and S. Europe and N. Africa.

4. **L. campes'tris,** *Willd.* ; leaves very hairy, cymes short, flowers in
dense clusters, perianth-segments acuminate longer than the broad obovoid
obtuse apiculate capsule, seeds with a conical white basal appendage.

Heaths, meadows, and pastures, N. to Shetland, ascends to 3,200 ft. in the
  Highlands and Wales; Ireland; Channel Islands; fl. April–June.—*Rootstock* creeping, tufted.   *Stems* 4–12 in., stout or slender.   *Leaves* shorter

than the stem, usually copiously hairy. *Cymes* very variable; clusters subsessile, or on long drooping branches. *Flowers* ⅛ in., pale or dark; bracteoles ciliate. *Seeds* with no terminal crest.—DISTRIB. Europe (excl. Greece),
all temp. and cold regions.

L. campes'tris proper; usually short, clusters usually 3–4-fld., seeds subglobose.
—VAR. *L. erec'ta*, Desv. (*L. multiflo'ra*, Lej.; *L. conges'ta*, Lej.); larger,
stouter, cymes more contracted, flowers more in a cluster, perianth-segments
narrower, filaments longer, capsule narrow, seeds oblong with a shorter
basal appendage.—Most frequent on heaths.

5. **L. spica'ta,** *DC.* ; leaves slender, cymes drooping dense-fld.
spike-like, bracteoles silvery, perianth-segments awned exceeding the
broad ellipsoid apiculate capsule, seeds with an obscure white basal
appendage.

Mts. of N. Wales, Westmoreland, and from Perth and Stirling to Shetland ;
alt. 1,000–4,300 ft.; fl. July.—*Rootstock* densely tufted ; stolons short.
Stems 6–12 in., slender. *Leaves* much shorter than the stem, small, hairy on the
sheaths chiefly, coriaceous, recurved, narrow, channelled ; cauline few, short.
Cymes ¾–1½ in., shorter than the leafy bracts. *Flowers* ⅛ in., usually shorter
than the scarious transparent ciliate awned bracteoles. *Seeds* oblong.—
DISTRIB. Arctic and Alpine Europe, N. Asia, Himalaya, N. America.

6. **L. arcua'ta,** *Swartz ;* dwarf, leaves short, cymes umbelled, branches
few outer slender recurved, perianth-segments acuminate exceeding the
broadly ovoid apiculate capsule, seeds with an obscure basal appendage.

High alps of Aberdeen, Banff, Inverness, and Sutherland, alt. 3,000–4,300
ft.; fl. July.—*Rootstock* creeping, loosely tufted ; stolons slender. *Stems*
1–4 in., rather stout. *Leaves* coriaceous, sparingly hairy, narrow, recurved,
channelled. *Cymes* lax, outer branches 1–2 in., 1–3-fld. *Flowers* ⅛ in., 3–5
in a cluster, dark chestnut; bracteoles lanceolate, acute, not silvery. *Seeds*
small, oblong.—DISTRIB. Norway, Arctic regions.

ORDER LXXXII. **ERIOCAULONE'Æ.**

Perennial, scapigerous herbs. *Leaves* chiefly radical, often cellular,
sheaths narrow. *Flowers* minute, usually monœcious, in involucrate
heads, bracteate.—MALE. *Perianth* membranous or scarious, outer of 2–3
free segments ; inner a 2–3-lobed tube. *Stamens* 2–3, inserted on the tube
opposite its lobes, with sometimes alternate perfect or imperfect ones,
filaments inflexed in bud ; anthers fixed by the back, 2-celled, bursting
inwards. *Ovary* rudimentary.—FEMALE. *Perianth* inferior, persistent,
outer as in the male, inner of 2–3 petals or pencils of hairs. *Staminodes*
0. *Ovary* of 2–3 connate carpels ; style short, terminal, persistent, stigmas 2–3 slender ; ovules solitary and pendulous from the top of each
cell, orthotropous. *Capsule* membranous, 2–3-celled, loculicidally 2–3-
valved. *Seed* pendulous, testa coriaceous, epidermis hyaline splitting into
hairs, albumen floury ; embryo outside and at the base of the albumen

farthest from the hilum.—Distrib. Chiefly trop.; genera 6; species 325.—Affinities. With *Restiaceæ* and *Xyrideæ.*—Properties unimportant.

1. ERIOCAU'LON, *L.* Pipe-wort.

Male fl. chiefly in the centre of the head. *Outer perianth-segments* subspathulate. *Stamens* 4 or 6.—Distrib. Of the Order; species 100.—Etym. ἔριον and καυλός, from the *woolly scapes* of some species.

E. septangula're, *With.*; leaves subulate, scape 6–8-furrowed. Lakes in Skye and W. Ireland; fl. Aug.—*Rootstock* creeping; roots white cellular. *Stem* very short, leafy. *Leaves* 2–4 in., compressed laterally, green, translucent, septate. *Scape* 6–24 in., rarely more, twisted. *Head* ¼–¾ in. diam.; bracts lead-coloured, oblong-obovate, obtuse. *Flowers* 2-merous; outer segments dark, bearded at the tip, inner ciliate with a black spot towards the tip. *Anthers* dark. *Ovary* stalked.—Distrib. N. America.

Order LXXXIII. TYPHA'CEÆ.

Marsh or aquatic herbs. *Rootstock* creeping. *Leaves* narrow, linear, obtuse, bases sheathing. *Spathe* 0 or caducous. *Flowers* monœcious, in cylindric oblong or globose spikes or heads, the male heads uppermost. *Perianth* of persistent membranous scales or hairs.—Male. *Stamens* few or many, filaments slender; anthers basifixed, 2-celled, dehiscence lateral.—Female. *Ovary* sessile or stalked, 1- rarely 2-celled, contracted into a simple persistent style, stigmatose ventrally; ovule 1, pendulous from the top of the cell, anatropous. *Fruit* small, coriaceous, or a drupe. *Seed* pendulous, albumen fleshy or mealy; embryo straight, terete, radicle next the hilum.—Distrib. Temp. and trop.; genera 2; species about 16.—Affinities. With *Aroideæ* and *Pandaneæ* (screw-pines), of which *Sparganium* is almost a member.—Properties. The roots of *Typha* are farinaceous; and the pollen is made into cakes in Sind and New Zealand.

Flowers in globose heads..1. Sparganium.
Flowers in cylindric or oblong spikes...2. Typha.

1. SPARGA'NIUM, *L.* Bur-reed.

Heads globose, subtended by leafy bracts. *Perianth* of 3–6 spathulate membranous scales. *Stamens* 2–3, connective hardly produced at the tip. *Ovary* 1- rarely 2-celled. *Drupe* angled, small, 1–2-celled.—Distrib. Temp. and trop.; species about 6.—Etym. σπάργανον, a *band*, from the form of the leaf.

1. **S. ramo'sum,** *Huds.* ; erect, branched, leaves erect 3-quetrous at
the base keeled, stigma linear, drupes sessile broadly ovoid, beak short.

Ponds, ditches, and river-banks, N. to Shetland ; ascends to 1,200 ft. in Derby ;
Ireland ; Channel Islands ; fl. June–July.—*Stems* 1–4 ft. *Leaves* 2–5 ft.,
1 in. broad. *Male heads* olive-brown, deciduous, ½–¾ in. diam.; *female* as
large, 1 in. when in fruit; bracts linear. *Drupe* ¼ in.—DISTRIB. Europe,
N. Asia, N. Africa, N. America.

2. **S. sim'plex,** *Huds.* ; erect, simple, leaves erect (sometimes floating)
keeled 3-gonous below, heads racemose, stigma linear, drupe shortly stalked
fusiform, beak long.

Ponds, ditches, and river-banks, N. to Shetland ; Ireland ; fl. June–July.—
Stem 1–2 ft. *Male heads* yellow, sessile; female peduncled. *Drupe*
narrowed at both ends.—DISTRIB. Europe (excl. Greece), N. and W. Asia,
N. America.

3. **S. na'tans,** *L.* ; floating, simple, leaves flat at the base not keeled,
heads racemose, stigma tongue-shaped, drupe stalked, beak rather long.

Lakes, ditches, &c., from Shetland southd.; Ireland ; ascends to near 1,600 ft.
in the Highlands ; fl. July–Aug.—*Stem* 1–3 ft., suberect in flower, leafy,
flaccid, upper part floating. *Leaves* ⅛–¼ in. diam. *Heads* ¼ in. diam., female
peduncled.—DISTRIB. Europe, N. Africa, Siberia, W. Asia, N. America.—
Probably a form of *S. sim'plex*, as suggested by Bentham.

S. NATANS proper ; sheaths rather inflated, male heads several, drupe fusiform.
S. affi'ne, Schn.—Lakes, England, Scotland and Ireland, not common.
Sub-sp. S. MINIMUM, Fries ; more slender, sheaths not inflated, male heads
1–2, drupe more obovoid.—Caithness to Hants.

2. TY'PHA, *L.* REED-MACE, CAT'S-TAIL, CLUB-RUSH.

Spikes superposed, cylindric, with deciduous leafy bracts. *Perianth* of
2–3 extremely slender cellular hairs. *Stamens* several, monadelphous ;
connective produced. *Ovaries* stalked, many imperfect ; style very
slender, stigma unilateral narrow. *Fruit* minute, stalked, dehiscent
along the inner face. *Seed* cylindric, testa striate.—DISTRIB. Temp. and
trop. ; species 10.—ETYM. τῖφος, a *fen.*

1. **T. latifo'lia,** *L.* ; leaves ½–1½ in. broad subglaucous, spikes contiguous or nearly so, rachis naked. *T. me'dia*, DC.

Lakes, river-banks, &c., from Orkney (probably) southd.; Ireland ; Channel
Islands ; fl. July–Aug.—*Stem* 3–7 ft., terete. *Leaves* distichous, 3–6 ft.,
linear, obtuse, nearly flat. *Spikes* ½ to nearly 1 ft., 1 in. diam., dark brown,
silky from the copious filiform scales ; female ebracteate. *Stigma* lanceolate.
—DISTRIB. Europe (excl. Greece), N. Africa, N. and W. Asia, N. America.

2. **T. angustifo'lia,** *L.* ; leaves ¼–¾ in. broad dark-green not glaucous
convex beneath, spikes separate, rachis hairy.

Ditches and ponds, from Fife and Lanark southd.; E. Ireland, rare ; Channel
Islands ; fl. July.—Smaller in all its parts than *T. latifo'lia* ; leaves narrower,

channelled towards the base; spikes ½–¾ in. diam., separated by ½–1 in., female often interrupted, bracteate; female flowers bracteate, perianth-scales dilated towards the tip; stigmas broader.—Distrib. Europe (excl. Greece), N. Africa, N. Asia, India, N. America.

Order LXXXIV. AROIDE'Æ.

Herbs or shrubs. *Leaves* various. *Spathe* 1-leaved. *Flowers* on a spadix, 1- or 2-sexual; perianth 0, or hypogynous and polyphyllous. *Stamens* few or many; anthers 2-celled, dehiscing outwards, or by terminal pores. *Ovary* 1- or more-celled; style simple or 0, stigma capitate or discoid; ovules 1 or more, variously attached. *Berry* 1- or more-celled. *Seeds* 1 or more, albumen abundant fleshy or mealy, rarely 0; embryo various.—Distrib. Trop. and temp. regions; genera 100; species 1,000.—Affinities. With *Pandaneæ* and *Typhaceæ.*—Properties. Acrid and poisonous.

Flowers 1-sexual; perianth 0...1. Arum.
Flowers 2-sexual; perianth-segments free..........................2. Acorus.

1. A'RUM, *L.*

Rootstock tuberous. *Leaves* radical, nerves reticulate; petiole sheathing at the base. *Scape* terete. *Spathe* convolute, contracted above the base. *Spadix* contracted below the middle, terminated by a naked cylindric column, bearing from the base upwards, 1stly a crowd of naked sessile ovaries; 2dly pistillodes; 3dly a crowd of naked sessile anthers; 4thly staminodes. *Anthers* 2–4-celled, dehiscence terminal. *Ovary* 1-celled; stigma sessile; ovules few, basal, erect, orthotropous. *Berry* fleshy. *Seeds* few, testa coriaceous, albumen mealy; embryo short, radicle opposite the hilum.—Distrib. N. temp. and sub-trop. regions; species about 20.—Etym. doubtful.

1. **A. macula'tum,** *L.*; leaves vernal, spathe twice as long as the spadix. *Cuckoo-pint, Lords and Ladies.*
Woods and hedges, from Caithness southd.; ascends to 1,000 ft. in N. England; ? wild in Scotland; Ireland; Channel Islands; fl. April–May.—*Corms* annual, new produced at the base of the stem. *Leaves* 6–10 in., hastate-cordate; often spotted black, lobes acute or obtuse. *Scape* short, lengthened in fruit. *Spathe* 6–10 in., erect, yellow-green, edged and often spotted with purple, base persistent. *Spadix* above dull purple, rarely yellow. *Berries* ½ in. diam., crowded, scarlet, bursting the base of the spathe. *Seeds* 2–3, testa reticulate.—Distrib. From Gothland southd., N. Africa.—Corms yield Portland arrow-root.

2. **A. ital'icum,** *Miller;* leaves hibernal, spathe thrice as long as the spadix.

Cornwall to Sussex, very local; Channel Islands; fl. June.—A larger, stouter plant than *A. macula'tum,* leaves more triangular; spathe falling over at the top as soon as it expands ; spadix always yellow; pistillodes much longer; berries longer; seeds larger.—DISTRIB. W. France, S. Europe, N. Africa.

2. A'CORUS, *L.* SWEET-FLAG.

Rootstock creeping. *Leaves* radical, ensiform, equitant. *Scape* flattened. *Spathe* continuous with the scape, 2-edged. *Spadix* lateral, terete, narrowed upwards, covered with flowers. *Perianth-segments* 6, free, membranous, oblong, persistent. *Stamens* 6, on the base of the segments, filaments flattened ; anthers didymous. *Ovary* 2-3-celled ; stigma sessile, minute ; ovules many, pendulous from the top of the cell, orthotropous. *Berry* 6-gonous, full of mucus. *Seeds* 1–3, testa thin, albumen horny ; embryo green, cylindric, radicle next the hilum.—DISTRIB. Europe, temp. Asia and America ; species 2.—ETYM. Classical.

A. Cal'amus, *L.* ; midrib of leaf thick.

Ditches, ponds, &c., from York and Lancaster to Somerset and Sussex, rare ; naturalized in Scotland and Ireland ; fl. June–July. *Leaves* 3–6 ft., ⅜–1¼ in. diam., margins waved. *Scape* leaf-like, ½ in. diam. *Spathe* long. *Spadix* 3–4 in., ¾ in. thick, curved. *Perianth-segments* not longer than the ovary. *Fruit* obovoid, top pyramidal.—DISTRIB. Europe (excl. Greece), N. Asia, Himalaya, N. America.—Aromatic, stimulant, tonic. Supposed to have been introd. from India. I have never seen fruit.

ORDER LXXXV. **LEMNA'CEÆ.**

Minute annual floating green scale-like plants, rootless or with capillary simple roots, propagated by budding, and by autumnal hibernating bulbils, rarely by seed ; vascular tissue 0 or rudimentary ; roots tipped by a membranous sheath. *Flowers* rarely produced, most minute, 1–3 in a spathe, or naked. *Perianth* 0. *Stamens* 1–2 ; anthers 1-2-celled, dehiscence transverse ; pollen spherical. *Ovary* 1-celled ; style short, stigma truncate or funnel-shaped ; ovules 1–7, orthotropous anatropous or ½-anatropous. *Utricle* bottle-shaped. *Seeds* 1 or more, testa coriaceous, inner coat thickened and discoid over the radicle, albumen fleshy or 0 ; embryo axile, straight, stout, cylindric.—DISTRIB. All standing waters ; genera 2 ; species about 20.—AFFINITIES. With *Naiadaceæ* and *Aroideæ.* —PROPERTIES 0.

Frond with 1 or more capillary roots. Anthers 2-celled............1. Lemna. Frond rootless. Anthers 1-celled.......................................2. Wolffia.

1. LEM'NA, *L.* DUCKWEED.

Fronds with roots. *Flowers* in marginal clefts of the frond. *Stamens* 1–2, filaments slender ; anthers 2-celled, didymous ; pollen muricate. *Ovules* 1–7.—DISTRIB. All latitudes ; species 7.—ETYM. The Greek name.

SECTION 1. **Lem'na** proper. *Root* single. *Ovule* 1, $\frac{1}{2}$-anatropous. *Seed* horizontal ; albumen copious.

1. **L. mi'nor,** *L.* ; frond obovate or oblong slightly convex beneath. Still waters, N. to Orkney ; ascends to 1,200 ft. in Derby ; Ireland ; Channel Islands ; fl. July.—*Frond* $\frac{1}{8}$–$\frac{1}{6}$ in., young sessile on the old, soon disconnected, green above, paler beneath ; epidermal cells with flexuous walls. *Spathe* unequally 2-lipped. *Stamens* 2, developed successively (each a male flower). *Style* long.—DISTRIB. Almost ubiquitous.

2. **L. trisul'ca,** *L.* ; frond flat obovate-lanceolate, tip serrate, young hastate persistent. *Ivy-leaved Duckweed.* Still waters, from Lanark and Banff southd.; rare in Scotland ; local in Ireland ; Channel Islands ; fl. June–July.—*Frond* $\frac{1}{2}$–$\frac{3}{4}$ in., narrowed at the base, without epidermis, proliferous on one or both sides ; young fronds numerous, placed crosswise to the old. *Style* very short. *Testa* rough, grooved.—DISTRIB. Europe, Siberia.

SECTION 2. **Telmatopha'ce,** *Schleid.* (gen.). *Root* single. *Ovules* 2–7, erect, anatropous. *Seeds* erect ; albumen scanty or 0.

3. **L. gib'ba,** *L.* ; frond obovate or orbicular tumid beneath, stamens 2, utricle bursting transversely. Still waters, from Edinburgh and Lanark to Devon and Kent ; local in Ireland ; fl. June–Sept.—*Frond* $\frac{1}{6}$–$\frac{1}{4}$ in., opaque, pale green, young sessile ; cells beneath very large, epidermal with flexuous walls.—DISTRIB. Europe (excl. Greece), Siberia, N. Africa, America.

SECTION 3. **Spirode'la,** *Schleid.* (gen.). *Roots* many. *Ovules* 2, erect, anatropous. *Fruit* unknown.

4. **L. polyrhi'za,** *L.* ; frond broadly obovate plano-convex 7-nerved. Ponds and ditches, from Lanark and Edinburgh to Devon and Kent ; local (if native) in Scotland ; Mid. Ireland.—*Frond* $\frac{1}{4}$–$\frac{1}{2}$ in., dark green above, purple beneath, tracheæ abundant ; epidermal cells with flexuous walls. *Spathe* 2-lipped. *Stamens* 2.—DISTRIB. Europe (excl. Greece), Siberia, Madeira, N. America.—Flower unknown in Britain.

2. WOLFF'IA, *Hork.*

Fronds like grains of sand, rootless, oblong or subglobose, flattened above, proliferous, cleft near the base. *Flowers* bursting through the upper surface of the fronds. *Spathe* 0. *Anthers* sessile, 1-celled ; pollen smooth. *Ovary* globose ; style short, stigma depressed ; ovule 1, erect, orthotropous. *Utricle* spherical, indehiscent. *Seed* oblique : albumen fleshy, scanty.—DISTRIB. Europe local, Africa, America ; species 12.— ETYM. *J. F. Wolff*, a writer on *Lemna.*

W. arrhi'za, *Wimm.* ; frond loosely cellular beneath. *Lem'na Michel'ii,* Schleid.

Ponds, Essex, Middlesex, Kent, Surrey, probably common elsewhere.—The smallest known flowering plant. *Frond* $\frac{1}{20}$ in. long, $\frac{1}{40}$ in. broad, young solitary at the base of the old, soon detached ; epidermal cells with straight walls. *Flowers* described from African specimens.—DISTRIB. Europe local, from Holland southd.

ORDER LXXXVI. **ALISMA'CEÆ.**

Marsh or aquatic scapigerous herbs. *Leaves* chiefly radical, erect or floating ; petiole sheathing. *Flowers* usually 2-sexual, usually panicled or umbellate. *Perianth* inferior of 6 segments in two series. *Stamens* 6, 9, or more, hypogynous, filaments free ; anthers oblong. *Ovary* of 3, 6, or more carpels, free or subconnate ; styles short or 0, stigma terminal simple or feathery ; ovules 1 or more in each cell, anatropous or campylotropous. *Fruit* of indehiscent coriaceous carpels or follicles. *Seeds* 1 or more in each carpel, testa coriaceous or membranous, albumen 0 : embryo straight or hooked, radicular end thick, next the hilum.—DISTRIB. All climates ; genera 10 ; species about 50.—AFFINITIES. With *Naiadaceæ.* —PROPERTIES unimportant.

TRIBE I. **ALISME'Æ.** *Sepals* green. *Petals* usually large, fugacious. *Ripe carpels* indehiscent. *Ovules* solitary, basal, or numerous in the inner angle of the carpel.

Flowers bisexual, whorled ; carpels free. Leaves erect...............1. Alisma.
Flowers bisexual, subsolitary ; carpels free. Leaves floating.......2. Elisma.
Flowers bisexual, whorled ; carpels connate. Leaves erect..3. Damasonium.
Flowers unisexual, whorled ; carpels free. Leaves erect.........4. Sagittaria.

TRIBE II. **BUTOME'Æ.** *Petals* and *sepals* similar. *Ripe carpels* dehiscent. *Ovules* numerous, on branching parietal placentas...............5. Butomus.

1. **ALIS'MA,** *L.*

Roots fibrous. *Leaves* erect. *Flowers* umbelled or whorled, bracteate, honeyed. *Sepals* herbaceous. *Petals* deciduous, membranous, involute in bud. *Stamens* 6, filaments filiform ; anthers versatile, introrse, *Carpels* many, free ; style ventral or terminal, short, stigma simple ; ovules solitary, erect, campylotropous. *Ripe carpels* turgid or compressed, ribbed, keeled, or grooved. *Seed* erect, testa very thin, raphe ventral ; embryo hooked.—DISTRIB. Temp. and trop. ; species 10.—ETYM. doubtful.

1. **A. Planta'go**, *L.* ; leaves erect, flowers panicled, carpels in one whorl laterally compressed, styles ventral. *Water Plantain.*

Ditches, edges of streams, &c., from Ross southd., rare in the N.; Ireland; Channel Islands ; fl. June–Aug.—*Base of stem* swollen, fleshy. *Leaves* 6–8 in., petioled, ovate-lanceolate, base acute obtuse or subcordate, 5–7-ribbed, young submerged or floating. *Scape* 1–3 ft.; pedicels 1–1½ in. *Flowers* ⅔ in. diam., homogamous. *Petals* pink or rose, claw yellow. *Carpels* 20–30. —DISTRIB. Arctic and N. temp. regions, Himalaya, Australia.—Juice acrid.

A. Planta'go proper; leaves subcordate, sepals oblong, styles twice as long as the ovary.—VAR. *A. lanceola'tum*, With.; leaves lanceolate, sepals ovate, styles as long as the ovary.

2. **A. ranunculoi'des**, *L.* ; leaves erect linear-lanceolate, flowers umbelled or whorled, carpels capitate 4–5-ribbed turgid, styles terminal.

Bogs and ditches, from Ross southd.; Ireland; Channel Islands; fl. May–Sept.—Tufted. *Leaves* petioled, blade 2–3 in., 3-ribbed; the first developed submerged, pellucid, sometimes floating. *Scape* 6–18 in.; pedicel 1–3 in. *Flowers* as in *A. Planta'go*, pale purplish. *Carpels* ovoid, apiculate.—DISTRIB. From Gothland southd. (excl. Greece, Turkey), N. Africa.

A. ranunculoi'des proper; erect or suberect.—VAR. *A. ré'pens*, Davies; stem procumbent geniculate rooting. Lakes, N. Wales and Ireland.

2. ELIS'MA, *Buchenau*.

Roots fibrous. *Leaves* floating. *Flowers* subsolitary from the nodes, pedicelled, bracteate. *Sepals, petals,* and *stamens* of *Alisma. Carpels* 10–12; style terminal, short, stigma simple ; ovules solitary, erect, anatropous, raphe dorsal. *Ripe carpels* oblong-ovoid, many-ribbed. *Seed* erect, testa thin ; embryo hooked.—DISTRIB. N. and W. Europe, local. —ETYM. A variation from *Alisma.*

E. na'tans, *Buch. Alisma na'tans,* L.

Lakes, very rare, Ayr and Wigton; W. England, from Cumberland to Hereford and Wales; W. Ireland, rare; fl. July–Aug. *Radical leaves* 2–8 in., submerged, subulate-lanceolate, pellucid. *Stem* floating and rooting, giving off long-petioled, oblong, floating leaves, ½–1 in., and 1–5 slender erect 1-fld. peduncles, 2–3 in., the uppermost rarely umbellate. *Flowers* ½ in. diam., petals white, claw yellow. *Ripe carpels* beaked.

3. DAMASO'NIUM, *Juss.* STAR-FRUIT.

Habit and inflorescence of *Alisma*, but carpels 6–10, connate at the base, spreading horizontally, 2- or more-ovuled. *Fruit* of as many stellately spreading 1- or more-seeded long-beaked carpels. *Seeds*, lower erect, upper horizontal, testa membranous rugose ; embryo hooked.—DISTRIB. Europe, California, Australia ; species 4.—ETYM. obscure.

D. stella'tum, *Pers.* ; leaves narrow oblong 3–5-nerved, base cordate. *Actinocar'pus Damaso'nium,* Br.

Gravelly ditches and pools, rare, from Salop and Suffolk to Kent and Hants;
fl. May–July.—*Leaves* many, 2 in., obtuse, floating or emersed; petiole stout,
2–5 in. *Scape* 4–6 in., stout, with usually 2 whorls of flowers; pedicels 1 in.
Flowers ¼ in. diam.; petals caducous. *Fruit* ⅔ in. diam., carpels dehiscing
ventrally.—DISTRIB. France, Spain, Italy, N. Africa.

4. SAGITTA'RIA, *L.*

Habit and inflorescence of *Alis'ma,* but flowers 1-sexual; stamens
numerous ; anthers basifixed, dehiscence lateral.—DISTRIB. Temp. and
trop. ; species about 15.—ETYM. *sagitta,* an *arrow.*

S. sagittifo'lia, *L.* ; leaves hastate obtuse or acute. *Arrow-head.*

Ditches, canals, &c., from Cumberland to Kent and Devon; naturalized in
Scotland ; local in Ireland ; fl. July–Sept.—*Stem* swollen at the base, stoloniferous ; stolons producing globose winter tubers ½ in. diam. *Leaves* 2–8 in.,
erect, lobes long more or less diverging acuminate, the first developed submerged, pellucid, linear ; petiole 8–18 in., stout, 3-gonous. *Scape* 6–18 in.,
with 3–5 distant whorls of 3–5 flowers each ; bracts short, obtuse, membranous ; lower whorls female, pedicels short; upper male with longer
pedicels. *Flowers* ½ in. diam., males larger. *Petals* white, caducous, claw
purple. *Anthers* purple. *Ripe carpels* numerous, much compressed laterally,
obliquely obovate, apiculate, wings broad thick, cell small. *Seeds* as in
Alis'ma.—DISTRIB. Europe (Arctic) (excl. Greece), N. Asia, N.W. India.

5. BU'TOMUS, *L.* FLOWERING RUSH.

A tall marsh herb. *Rootstock* creeping. *Leaves* slender, erect. *Scape*
naked. *Flowers* in a bracteate umbel. *Perianth-segments* 6, oblong,
spreading, subequal, all coloured. *Stamens* 9, hypogynous, 6 in pairs
opposite the outer segments, 3 opposite the inner, filaments subulate ;
anthers basifixed, oblong, introrse. *Carpels* 6, beaked, connate below ;
styles short, stigmas sessile ; ovules covering the walls of the carpel,
anatropous, ascending. *Follicles* 6, beaked, coriaceous, turgid. *Seeds*
many, minute, ascending, testa thin furrowed ; embryo straight.—DISTRIB.
Europe, N. and W. Asia, N.W. India.—ETYM. obscure.

B. umbella'tus, *L.* ; leaves long slender 3-quetrous, scape terete.

Ditches and river-sides from York and Durham southd. ; naturalized in Scotland ; rare in Ireland ; fl. June–July.—*Rootstock* stout. *Leaves* 3–4 ft., base
sheathing, twisted. *Flowers* many, 1 in. diam., rose-red, proterandrous ;
pedicels 2–4 in. *Anthers* and *carpels* red.

ORDER LXXXVII. NAIADA'CEÆ.

Marsh- or water-herbs. *Rootstock* usually creeping. *Stems* elongate
(rarely 0), jointed, branched, slender. *Leaves* often floating, alternate
or distichous, rarely opposite, sheathing at the base ; stipules 0

or sheathing and inserted within the petiolar sheath. *Flowers* inconspicuous, 1–2-sexual, green. *Perianth* 0, or tubular, or cup-shaped, or of
3–4 inferior valvate segments. *Stamens* hypogynous; anthers 1–2-celled.
Ovary of 1–4 carpels; style 1, stigma various; ovules one in each carpel,
rarely more, erect or pendulous. *Fruit* of one or more utricles achenes or
drupes. *Seed* solitary, testa membranous, albumen 0; embryo straight or
curved, radicular end very large.—DISTRIB. All climates; genera 16;
species 120.—AFFINITIES. With *Alismaceæ.*—PROPERTIES unimportant.

TRIBE I. **JUNCAGIN'EÆ.** Erect marsh herbs, with rush-like leaves.
Flowers spiked or racemed, 1–2-sexual. *Perianth-segments* 3 or 6, herbaceous
Stamens 6. *Carpels* 3 or more; ovules basilar, erect, anatropous. *Embryo*
straight.

Flowers ebracteate. Anthers subsessile, short......................1. Triglochin.
Flowers bracteate. Filaments and anthers long...............2. Scheuchzeria.

TRIBE II. **POTAME'Æ.** Aquatic herbs. *Flowers* spiked, 2-sexual. *Perianth-
segments* 4, herbaceous, or 0. *Stamens* 2 or 4. *Carpels* 4; ovules solitary,
axile, campylotropous. *Embryo* curved.

Perianth-segments 4. Achenes sessile............................3. Potamogeton.
Perianth 0. Achenes stipitate..4. Ruppia.

TRIBE III. **ZANNICHELLIE'Æ.** Aquatic herbs. *Flowers* axillary, 1-sexual.
Perianth 0 or hyaline. *Stamen* 1. *Carpels* 2–9; ovules solitary, pendulous, orthotropous. *Embryo* induplicate or involute.

Perianth 0...5. Zannichellia.

TRIBE IV. **ZOSTERE'Æ.** Marine herbs. *Flowers* sessile on a flattened
spadix, 1-sexual. *Perianth* 0. *Anthers* sessile, 1-celled, pollen confervoid.
Carpels solitary, sessile; ovules solitary, pendulous, orthotropous. *Embryo*
straight with the cotyledonary end in a slit.

Flowers monœcious...6. Zostera.

TRIBE V. **NAIADE'Æ.** Aquatic herbs. *Flowers* axillary, solitary or
crowded. *Perianth* hyaline. *Anther* 1. *Carpels* solitary; ovules solitary, basilar, anatropous. *Embryo* straight.

Perianth tubular.......................................,...............................7. Naias.

1. TRIGLO'CHIN, *L.*

Roots fibrous. *Leaves* erect, very narrow. *Flowers* racemed, small, green,
ebracteate, anemophilous, proterogynous. *Perianth-segments* 6, all similar,
cucullate, subequal, deciduous. *Stamens* 6, on the base of the perianth-
segments, filaments very short; anthers broad, extrorse. *Ovary* 6-celled;
stigmas 3 or 6, feathery; ovules one in each cell or the alternate cells
empty, anatropous. *Fruit* of 3–6 1-seeded coriaceous carpels, separating
from a central axis and dehiscing in front. *Seeds* erect, terete, testa coriaceous.—DISTRIB. Temp. regions, many Australian; species 12.—ETYM.
τρεῖς and γλωχίν, from the 3-pointed carpels.

430 *NAIADACEÆ.* [Triglochin

1. **T. palus'tre,** *L.* ; leaves filiform ½-terete throughout, fruit clavate, carpels 3 slender long attached to the axis by a point.

Marshes and wet meadows, N. to Shetland; ascends to 2,000 ft. in the Highlands; Ireland; Channel Islands; fl. June–Aug.—*Stem* swollen at the base, stoloniferous. *Leaves* 2–12 in., flaccid, upper surface faintly grooved. *Scape* slender. *Raceme* elongating after flowering, pedicels short. *Perianth* purple-edged. *Anthers* purple. *Fruit* appressed to the scape. *Carpels* terete at the back, narrowed below, axis 3-quetrous.—Distrib. Europe (Arctic) (excl. Greece), N. Africa, N. Asia, N.W. India, N. America.

2. **T. marit'imum,** *L.* ; stems tufted thickened at the base, leaves slightly flattened at the tip, fruit oblong of 6 separable carpels.

Salt marshes, N. to Shetland; Ireland; Channel Islands; fl. May–Sept.—Larger and stouter than *T. palus'tre;* scape curved; raceme longer, not dense-fld., nor so long in fruit; flowers larger; fruit not appressed to the scape, and carpels grooved at the back.—Distrib. Coasts of Europe (Arctic) (excl. Greece, Turkey), N. Africa, salt districts of Asia, N. America.

2. SCHEUCHZE'RIA, *L.*

A small marsh herb. *Rootstock* creeping. *Leaves* erect, slender. *Scape* leafy; flowers racemose, bracteate. *Perianth-segments* 6, herbaceous, reflexed, connate at the base, persistent. *Stamens* 6, hypogynous, filaments short; anthers long, narrow, adnate to the filament, extrorse. *Ovary* of 3 carpels connate at the base, stigma sessile; ovules 2–3 in each carpel, basilar, anatropous. *Fruit* of 2–3 inflated spreading follicles. *Seeds* 1–2, erect, ovoid, testa coriaceous.—Distrib. N. and Mid. Europe (Arctic), N. Asia, Rocky Mts.—Etym. Two *Scheuchzers,* Swiss botanists.

S. palus'tris, *L.* ; leaves ½-terete, sheaths turgid.

Marshes, rare, Salop, Notts, Chester, York, Perth; fl. July.—*Rootstock* long, slender, clothed with old leaf-sheaths. *Leaves* 6–10 in., with dilated brown sheaths and a pore at the tip. *Scape* stout, curved. *Raceme* lax, few-fld.; pedicels much lengthened in fruit. *Perianth-segments* linear-oblong. *Ripe carpels* large for the plant.

3. POTAMOGE'TON, *L.*

Aquatic herbs. *Leaves* submerged and translucent, or floating and opaque, alternate or opposite; stipules connate, membranous or 0. *Flowers* 2-sexual, in axillary or terminal spikes, proterogynous. *Perianth-segments* 4, small, herbaceous, clawed, persistent, valvate in bud. *Stamens* 4; anthers subsessile on the claw, 2-celled, extrorse. *Carpels* 4 (rarely 1) free, sessile, stigma subsessile; ovules solitary, ascending, campylotropous. *Drupelets* 4 (rarely 1), small, green. *Seed* curved round a lateral process from the cell, testa membranous; radicular end of embryo inferior, large, cotyledonary narrowed hooked or involute, plumule immersed.—Distrib. Temp. regions, more rare in trop.; species about 50.—Etym. ποταμός, a river, and γείτων, a neighbour.

Mr. A. Bennett, F.L.S., of Croydon, has given me the benefit of his unrivalled knowledge of this difficult genus by revising my MS. for this edition.

Section 1. *Leaves alternate, or the upper opposite, oblong obovate or lanceolate, not truly linear, margins involute in bud ; stipules free.*

* *Flowering-stem without barren branches below ; upper leaves oblong floating, lower (rarely all) submerged ; peduncles axillary, many- and dense-fld.*

1. **P. na′tans,** *L.* ; leaves long-petioled, submerged 0 or reduced to phyllodes, floating elliptic to lanceolate coriaceous, stipules very long acuminate, peduncle stout, spike dense-fld., dry drupelets large keeled dorsally, beak short.

Lakes and ponds, N. to Shetland ; Ireland ; Channel Islands ; fl. June–Sept. —*Stem* terete. *Floating leaves* 2–6 in., mostly alternate, blade very shortly decurrent on the petiole, with the margins minutely incurved or auricled ; submerged 6–12 in., with very rarely a limb. *Drupelets* ⅛ in., ventral margin convex, dorsal ½-circular.—Distrib. Europe, N. Asia, India, Africa, N. America, Australia.—*P. polygonifo′lius,* var. *linea′ris,* Syme, with submerged leaves 12–16 by ⅛–½ in., from Killarney, is probably referable here, but the fruit is unknown.

2. **P. polygonifo′lius,** *Pourr.* ; leaves long-petioled, submerged narrowly lanceolate, phyllodes 0, floating obovate to narrowly lanceolate rather membranous, peduncles and spike slender, dry drupelets very small not keeled, beak very short. *P. oblon′gus,* Viv.

Lakes and pools on heaths, &c., N. to Shetland ; ascends to 1,600 ft. in the Lake District ; Ireland, rare ; fl. June–Sept.—More membranous than *P. na′tans,* with well-developed submerged leaves, blade often very decurrent on the petiole, base acute rounded or cordate ; upper opposite. *Drupelets* ¹¹⁄₁₀ in., red, margins rounded.—Distrib. Europe, Asia, Canada?, N. Zealand. —Var. *flu′itans,* Syme, is a deep-water form with very long submerged leaves and subcoriaceous floating ones.

3. **P. plantagineus,** *Du Croz;* leaves long-petioled chiefly submerged all membranous translucent from orbicular and cordate to elliptic-lanceolate upper opposite, stipules short broad obtuse, peduncle very slender, spike dense-fld., dry drupelets very small rounded not keeled green, beak obsolete.

Ditches, fens, &c., from the Clyde and Haddington to Somerset and Kent Ireland ; Channel Islands ; fl. June–Sept.—*Leaves* broader, more membranous and translucent with more slender petioles than *P. polygonifo′lius* and *na′tans,* and with very different stipules. *Drupelets* ¹⁄₁₀ in. long.

4. **P. rufes′cens,** *Schrad.* ; leaves short-petioled translucent, lower submerged linear-lanceolate many-nerved, upper floating or erect broader subcoriaceous, petiole short, stipules large, peduncle stout, spike stout, dry drupelets ovoid acuminate red. *P. flu′itans,* Sm., not Schrad.

Ponds, canals, and ditches, from Caithness southd.; ascends to 3,000 ft. in Perthshire ; Ireland ; fl. July–Sept.—*Stem* 1–4 ft., terete. *Leaves* reddish, quite entire, upper 2–7 in., obovate or oblong-lanceolate, obtuse, lower sessile; stipules very variable, obtuse, not winged. *Peduncles* 2–7 in. *Drupelets* ventrally convex, dorsally ½-circular, keeled.—Distrib. Europe (Arctic), N. Asia, N.W. India, N. America.—Var. *P. spathula'tus,* Koch and Ziz, almost connects this with *P. polygonifo'lius.*—Perthshire.

** *Flowering-stem with copious barren branches below; upper leaves usually floating, broader than the lower submerged ones; peduncles axillary and terminal, many- and dense-fld.*

5. **P. heterophyl'lus,** *Schreb.* ; stem slender, submerged leaves linear-lanceolate, stipules small, peduncle stout thickened upwards, dry drupelets small 3-keeled, beak short.

Pools and lakes, from Shetland to Hants and Kent; ascends to 2,800 ft. in Perthshire; Ireland, rare; fl. June–Sept.— *Stem* 2–4 ft., terete, green or reddish, much branched below. *Submerged leaves* 1–7 in., sessile, flaccid, acuminate or cuspidate, floating 0 or similar or oblong petioled and coriaceous. *Peduncles* axillary and terminal. *Drupelets* ₁⁄₁₂ in., ventrally nearly straight, dorsally ½-circular.—Distrib. Europe (Arctic), N. Asia, N. America.

P. heterophyl'lus proper; leaves not amplexicaul almost flat, upper opposite coriaceous floating. *P. gramin'eus,* Fries, Koch.

Sub-sp. P. ni'tens, *Weber;* much branched, leaves usually all submerged alternate recurved undulate shining, peduncles less thickened upwards, spike shorter, drupelets smaller.—Aberdeen to Northumbd., Anglesea, Surrey; Ireland, rare.

6. **P. lanceola'tus,** *Sm.* ; stem filiform fragile, leaves sessile straight linear- or oblong-lanceolate obtuse entire translucent nerves fenestrate, upper opposite petioled, stipules slender acuminate, peduncles short not thickened upwards, spike very short, dry drupelets ventrally 3-toothed.

Rivers, Anglesea, Cambridge ; Co. Down; fl. July–Sept.—Resembles slender forms of *P. heterophyl'lus,* but leaves darker green, very obtuse, fenestrate all over; peduncles very slender, and spike shortly ovoid. *Drupelets* ⅛ in., nearly straight dorsally, much rounded ventrally, with a ventral and 2 lateral teeth, beak ventral.

7. **P. lonchites,** *Tuckerm.* ; stem stout, submerged leaves alternate very long straight strap-shaped 7–9-ribbed quite entire translucent, floating 0 or opposite oblong-lanceolate long-petioled, peduncle slightly thickened upwards, spike short, dry drupelets obovoid dorsally 3-keeled, beak short.

Ireland, River Boyne; fl. June–Sept.—*Stems* much longer and less branched than in *P. heterophyl'lus,* and submerged leaves much longer, drupelets larger more like those of *P. flu'itans,* cotyledonary end of embryo involute. —Distrib. Canada and California to Mexico and Florida.—Fruit described from American specimens.

8. **P. lu'cens,** *L.* ; stem stout, leaves large subsessile all translucent linear- or oblong-lanceolate cuspidate undulate serrulate upper opposite

often floating, stipules large lŏng 2-winged or keeled, peduncle robust thickened upwards, spike stout, dry drupelets small turgid, beak short.

Lakes, ponds, and streams, from Banff and Argyll to Devon and Kent; Ireland; fl. June–Sept.—The largest British species, very lucid and glistening, pale green. *Stem* 3–6 ft. *Leaves* 4–10 in., mostly submerged, very variable, many-nerved, mostly serrulate towards the long or short tip; upper broader rarely floating or coriaceous. *Peduncles* very variable in length; spike usually dense-fld. *Drupelets* 1–10 in., convex on both faces, beak obtuse.—DISTRIB. Europe, N. Africa, N. Asia, N.W. India, N. America, Australia.

P. LU'CENS proper; leaves shortly petioled, strongly serrulate apiculate or mucronate, peduncles usually short.—VAR. *P. acumina'tus,* Schum. (*P. longifo'lius,* Gay); leaves very narrow tapering to both ends, peduncles more than twice as long as the spike.

Sub-sp. P. ZIZ'II, *Roth;* much branched, lower leaves often recurved, upper cuneate at the base, floating obovate or oblong subcoriaceous, peduncles very long, spike 1–2 in., drupelets more rounded.—Forfar and Perth to Hants and Surrey.—The Lough Corrib plant, of which only one specimen is known, and which was referred by Babington to *P. longifolius,* Gay, differs in the narrower entire leaves and small scattered flowers.

Sub-sp. DECIP'IENS, *Nolte;* leaves sessile very variable suborbicular to oblong-lanceolate mucronate subentire undulate, stipules short scarcely winged, peduncles stouter, spike denser, drupelets smaller.

*** *Flowering-stem with barren branches below; leaves uniform, ½- or wholly-amplexicaul, all submerged, oblong or ovate-oblong, upper opposite.*

9 **P. Griffith'ii,** *A. Bennett;* stem terete, lower leaves subamplexicaul, strap-shaped, tip concave, upper long-petioled oblanceolate tapering into the petiole, stipules long narrow obtuse, peduncles slender, spike dense-fld.

Wales, near Aber; fl. summer.—*Stem* branched. *Lower leaves* 7–12 in., 11-nerved, with 4–5 fainter nerves near the midrib; upper 13–17-nerved. *Peduncles* shorter than the upper leaves. Young *drupelets* ovoid, beak terminal from the ventral face.—Habit between *prælon'gus* and *rufes'cens.*

10. **P. prælon'gus,** *Wulfen.;* stem terete robust, leaves ½-amplexicaul oblong obtuse entire 3-nerved tip usually concave, stipules large obtuse not winged, peduncles very long stout, spike dense-fld., dry drupelets large strongly acutely keeled.

Lakes and deep rivers, chiefly in the E.; Caithness to Essex and Salop; ascends to 2,800 ft. in Perthshire; rare in Ireland; fl. May–July.—About as large as *P. lu'cens,* but barren branches few. *Stem* greenish-white. *Leaves* 3–10 by 1–1½ in., linear-oblong, midrib dilated, upper opposite, nerves numerous and close. *Peduncles* hardly thickened upwards. *Drupelets* twice as large as in *P. lucens.*—DISTRIB. Europe (Arctic), Himalaya, America.

11. **P. salicifo'lius,** *Wolfg.;* stem slender terete, leaves all translucent, submerged ½-amplexicaul lanceolate acute or subacute entire, stipules obtuse, peduncles not enlarged upwards, spike short, dry drupelets compressed rounded obtuse. *P. lithuan'icus,* Gorski.

F F

Rivers near Hereford ; fl. July.—Habit between *P. prælon'gus* and *rufes'cens*, sparingly branched. *Leaves* 4–6 in., 3-nerved with many secondary nerves. *Peduncles* 2–4 in. ; spike ¾–1½ in. *Drupelets* not seen in British examples. —DISTRIB. Sweden, Lithuania.

12. **P. perfolia'tus,** *L.* ; stem stout terete, leaves amplexicaul ovate-cordate obtuse or subacute entire 5–9-nerved tip flat, stipules small subacute caducous, peduncles short stout, spike dense-fld., dry drupelets hardly keeled, beak short.

Ponds, lakes, and streams, N. to Shetland ; ascends to 1,500 ft. in Wales ; fl. June–Sept.—*Branches* dichotomous. *Leaves* 1–4 in., upper opposite, translucent. *Peduncles* terminal and in the forks, not thickened upwards. *Perianth-segments* long-clawed. *Drupelets* ${}_{10}^{1}$ in., compressed.—DISTRIB. Europe, N. Asia, N.W. India, W. Africa, America, Australia.—Forms approaching var. *lanceola'tus*, Robbins (A. Gray, Man. Bot. N. U. States), are found in Scotland.—VAR. *Jackso'ni*, Lees, from Yorkshire.

13. **P. cris'pus,** *L.* ; stem slender compressed, leaves distichous ½-amplexicaul oblong acute or obtuse crisped serrulate 3-nerved, stipules small obtuse caducous, peduncles curved tapering upwards, spikes few-fld., dry drupelets acuminate, beak long.

Ponds, ditches, &c., from Orkney southd.; Ireland ; Channel Islands ; fl. July–Aug.—Dichotomously branched. *Leaves* 1½–3 in., close-set, spreading and recurved, margins rarely flat (*P. serratus*, Huds.). *Peduncles* stout or slender ; flowers 6–8, lax, very small. *Drupelets* ¼ in., obliquely ovoid, compressed.—DISTRIB. Europe, N. Asia, India, Japan, N. America, Australia.

SECTION 2. *Leaves* all subopposite, margins involute in bud ; stipules 0.

14. **P. den'sus,** *L.* ; stem slender brittle, leaves subopposite distichous ovate-cordate serrulate translucent 3–5-nerved, stipules 0 except on the upper pair of leaves, peduncles very short, spikes few-fld., dry drupelets suborbicular sharply keeled, beak short recurved.

Ponds and streams, from the Forth of Clyde to Somerset and Kent ; ascends to 1,000 ft. in the Lake District ; very local in Scotland and Ireland ; fl. July–Sept.—Dichotomously branched. *Leaves* ½–1 in., close-set, recurved, acute or acuminate, keeled, the nodes of the subopposite pairs being confluent, but their bases overlap. *Peduncles* rarely longer than the leaves, at length recurved ; spike laxly 3–6-fld. *Drupelets* compressed, pericarp thin ; cotyledonary end of embryo circumvolute.—DISTRIB. From Denmark southd., Himalaya, N. Africa, America.

SECTION 3. *Leaves* alternate or the upper opposite, all similar, ligulate (not broader in the middle), margins flat in bud ; stipules free.

15. **P. zosterifo'lius,** *Schum.* ; stem compressed winged, leaves ½-amplexicaul linear abruptly acuminate 3- rarely 5-nerved, stipules large acuminate, peduncle much longer than the spike, drupelets subreniform 3-ribbed. *P. cuspidatus*, Sm. ; ? *P. compres'sus*, L.

Forfar to York and Essex, local; Co. Down; fl. July-Aug.—*Stem* broad, internodes long. *Leaves* 4–10 in., often ¼ in. broad, with 3 strong and many slender nerves. *Peduncles* 2–3 in.; spike many-fld. *Drupelets* ⅛ in., slightly compressed; beak terminal.—DISTRIB. Europe, N. Asia, N. America.—The *P. compres'sus,* L., being a doubtful plant, the name *zosterifo'lius* is adopted by most authors.

16. **P. acutifo'lius,** *Link ;* stem compressed, leaves ½-amplexicaul linear finely acuminate 3-nerved, stipules lanceolate acute, peduncle rarely equalling the very short spike, dry drupelets convex with a strong tooth near the base ventrally, beak recurved.

Lakes and ditches, Yorkshire to Kent and Dorset, rare; fl. June–Aug.—*Stem* narrower than in *P. zosterifolius. Leaves* 2–6 in., with 3 strong and many slender nerves. *Peduncles* ½– (rarely) 1 in.; spike about ¼ in. *Drupelets* ⅛ in., compressed; beak ventral.—DISTRIB. Europe, Australia.

17. **P. obtusifo'lius,** *Mert.* and *Koch ;* stem slender compressed, leaves sessile linear subacute or obtuse 3-nerved, stipules very obtuse, peduncles very short, spike small ovoid, dry drupelets broad keeled, beak straight. *P. gramin'eus,* Sm., not L.

Pools and ditches from Aberdeen to Hants and Kent; Ireland, rare; fl. July–Aug.—*Stem* flexuous, much branched, 4-gonous, not winged. *Leaves* 2–6 by ⅛–⅓ in., dark green, rather opaque, with no visible nerves between the 3 principal ones; stipules ½ in., tip broad. *Peduncles* much shorter than the leaves, terminal or in the forks, equalling or exceeding the densely few-fld. spike. *Drupelets* compressed, broadly obliquely ovoid.—DISTRIB. Europe, W. Siberia, N. America.

18. **P. pusil'lus,** *L.* ; stem filiform, leaves ½-amplexicaul narrowly linear acute acuminate or subacute 1–3- (rarely 5–7-) nerved, stipules small acute, peduncles slender, spike few-fld., dry drupelets small turgid obtusely keeled, beak stout.

Rivers, ponds, and ditches, from Orkney southd.; ascends to 1,000 ft. in the Highlands; Ireland; fl. July-Aug.—*Stem* often much branched, rarely slightly compressed. *Leaves* ⅓–3 by ₁₀–₁₀ in., rarely acuminate, green, rather opaque. *Peduncle* usually much longer than the 6–10-fld. spike; flowers minute. *Drupelets* obliquely ovoid; beak subterminal. DISTRIB. Europe (Arctic), N. Africa, N. Asia, N. America.

P. PUSIL'LUS proper; leaves acute or subacute, 1–3-nerved.—VAR. *tenuis'sima,* Koch (*P. grac'ilis,* Bab., not Fries); smaller, leaves acuminate more spreading 1-nerved.—VAR. *rig'ida,* A. Benn.; rigid, fragile, lateral nerves of leaf faint or 0, stipules long, spike ½ in., drupelets obscurely keeled. —Orkney.

Sub-sp. P. FRIES'II, Rupr.; stem compressed, leaves often fascicled broader 5- rarely 7-nerved, peduncles 1–2 in. compressed, spike ½–¾ in., interrupted, dry drupelets larger, beak shorter. *P mucrona'tus,* Schrad.; *P. compres'sus,* Sm.—Orkney to Dorset, local; Sussex and Hants; Co. Down.

Sub-sp. P. STURROCK'II, A. Benn.; stem filiform, leaves 2–3 in. subobtuse 3–5- nerved bright-green pellucid, peduncles very slender 2–4 in., spike ⅛–½ in., dry drupelets much smaller, beak short.—Forfar and Perth.—A very elegant and delicate plant.

19. **P. trichoi'des,** *Cham.* and *Schl.* ; stem capillary, leaves ½-amplexicaul setaceous 1- (rarely 3-) nerved, stipules slender acute, peduncles
filiform much longer than the few-fld. spike, dry drupelets solitary, beak
short.

Muddy ponds and ditches, Norfolk, E. Suffolk; W. Ireland; fl. Aug.–Oct.—
Stem repeatedly dichotomously and divaricatingly branched. *Leaves* 1–2½
in., spreading, acuminate, dark green, rigid. *Peduncles* curved, longer than
the leaves; flowers 3–6, very minute, monogynous. *Drupelets* $\frac{1}{10}$ in., obliquely
ovoid, compressed, dorsally more or less tubercled, ventrally toothed near
the base.—DISTRIB. From Sweden southd. and eastd., N. Africa.

SECTION 4. *Leaves* all similar, submerged, linear, margins flat in bud;
stipules adnate with the leaf-base into a sheath.

20. **P. pectina'tus,** *L.* ; stem filiform, densely distichously branched,
leaves very long linear or filiform acuminate 1–3-ribbed, peduncles long
slender, flowers few remote, dry drupelets large turgid, beak short.

Fresh and brackish ditches, &c., from Orkney southd.; Ireland; Channel
Islands; fl. June–Aug.—Forms dense masses. *Root* a small tuber. *Leaves*
3–8 by $\frac{1}{20}$–$\frac{1}{4}$ in., lower 5-nerved; stipular sheath often 1–1½ in', tip free.
Peduncles not thickened upwards; flowers interruptedly whorled. *Drupe-
lets* very large for the plant, $\frac{1}{6}$ in., dimidiate-obovoid, obscurely keeled
dorsally, ventrally slightly convex.—DISTRIB. Europe (Arctic), N. Asia,
India, Africa, N. America, Australia.

P. PECTINA'TUS proper; upper leaves 1-nerved, channelled bifistular, lower
flat 3-ribbed, lateral ribs of dry drupelets conspicuous. Common. *P.
mari'nus,* Huds., not L., is a var. with stems naked below.
Sub-sp. P. FLABELLA'TUS, Bab.; upper leaves 1–3-nerved, lower flat 3–5-nerved,
lateral ribs of dry drupelets obscure. *P. juncifo'lius,* Kerner.—England,
Scotland (very rare), Ireland.

21. **P. filifor'mis,** *Nolte* ; stem filiform, branches short, leaves capillary
1-nerved, peduncles longer than the leaves very slender, flowers in distant whorls, dry drupelets small, beak very short.

Lakes and ditches, Shetland to Berwick, local; Anglesea; Ireland, rare; fl
July–Sept.— *Stem* branched below, simple above. *Leaves* all capillary 1-
nerved and channelled. *Flowers* in dense whorls. *Drupelets* scarcely
keeled, beak terminal.—DISTRIB. Europe (Arctic), N. Asia, N. India, N.
America, Australia.

4. RUP'PIA, *L.*

Slender brackish-water herbs. *Leaves* alternate or subopposite, submerged, filiform, with stipuliform sheaths. *Flowers* minute, 2-sexual,
proterandrous, usually 2 on opposite sides of a filiform rachis; peduncle
ebracteate, terminal, but apparently lateral from being pushed aside by an
axillary shoot from the last leaf, elongate after flowering. *Perianth* 0.
Stamens 2, an upper and a lower, filaments short broad; anthers attached
by the back; cells reniform distant, dehiscence vertical; pollen a curved
tube with 1 median and 2 terminal nuclei. *Carpels* 4, sessile; **stigmas**

sessile peltate; ovules solitary, pendulous. *Drupelets* stipitate, ovoid,
carried up on the greatly lengthened usually spirally coiled peduncle.
Seed pendulous, testa thin; radicle large ovoid, cotyledonary end small
hooked subterminal.—DISTRIB. Temp. and trop. regions; species 1 or
more.—ETYM. *H. B. Ruppius,* a botanical author.

R. marit'ima, *L.* ; leaves opposite and alternate.

Brackish ditches, &c., N. to Shetland; rare in W. Scotland; Ireland;
Channel Islands; fl. July-Sept.—Habit of *Potamoge'ton pectina'tus. Stem*
filiform, much branched, 2 ft. and upwards. *Leaves* 1–3 in.; the first leaf of
each axillary shoot is opposite to a narrow obtuse or notched cellular scale
½ in. long arising from the base at the side next the axis. *Peduncle* short
and straight in flower, fruiting 5–6 in. *Drupelets* ⅛–⅙ in., green, beaked,
pedicel 1–2 in.—DISTRIB. All shores temp. and trop.

R. MARIT'IMA proper; sheaths inflated, fruiting peduncles spirally coiled,
anther-cells oblong, drupelets nearly straight, beak short. *R. spira'lis,*
Hartm.—Orkney to Somerset; rare in Ireland.

Sub-sp. R. ROSTELLA'TA, *Koch;* sheaths appressed, fruiting peduncles short
flexuous, anther-cells subglobose, drupe gibbous, beak longer. Common;
flowers earlier.—VAR. *na'na,* Syme, has creeping stems buried in the mud,
and very short pedicels. Orkney.

5. ZANNICHEL'LIA, *L.* HORNED PONDWEED.

Slender water-plants. *Leaves* submerged, usually opposite, linear;
stipules adherent to the sheathing leaf-base. *Flowers* minute 1- or 2-
sexual, solitary or in pairs, axillary, situated in a cup-shaped sheath. *Stamen* 1; anther 2-celled, sagittate, cells adnate to the slender filament,
dehiscence lateral, connective excurrent; pollen globose. *Carpels* 4–6,
styles long or short, stigma peltate persistent; ovule pendulous, orthotropous. *Drupelets* 4–5, sessile or stalked, oblong, curved. *Seed* pendulous,
testa membranous; radicle large, clavate; cotyledonary end slender, twice
folded.—DISTRIB. Temp. and trop. regions; species 1 or several.—ETYM.
Zannichelli, a Venetian botanist.

Z. palus'tris, *L.* ; achenes curved, keel smooth tubercled or crenulate.

Fresh and brackish ditches and pools, N. to Orkney; rare in W. Scotland;
Ireland; Channel Islands; fl. May-Aug.—Annual. *Stem* 3–6 in., filiform or
setaceous; branches divaricate. *Leaves* 1–3 in., opposite, subwhorled, from
capillary. *Flowers* sessile or shortly pedicelled. *Achenes* about ¹₁₂ in.,
narrow.—DISTRIB. Europe, N. Africa, Siberia, India.—The following forms
are defined by Mr. Baker.

Z. PALUS'TRIS proper; drupelets 2–4 sessile, back rarely crenulate, style half
as long, stigma small and crenulate, filament ½–1 in., anther 4-celled.—
Z. macroste'mon, Gay.

Sub-sp. Z. BRACHYSTE'MON, *Gay;* drupelets 2–4 nearly sessile, back crenate,
style about half as long, stigma large crenulate, filament ⅛–¼ in., anther
2-celled. *Z. palus'tris,* E. B.; *Z. ma'jor,* Bonn.—Common.

Sub-sp. Z. PEDUNCULA'TA, *Reichb.*; drupelets pedicelled, back strongly muricate,
style about as long, stigma large crenulate, filament ¼–½ in., anther 2-celled.
Z. pedicella'ta, E. B.

Sub-sp. Z. POLYCAR'PA, *Nolte ;* drupelets often 5–6 subsessile, back cylindric, style $\frac{1}{2}-\frac{1}{4}$ as long, stigma large repanded, filament $\frac{1}{10}-\frac{1}{8}$ in., anther 2-celled. —VAR. *tenuis'sima*, Fr.; very slender.

6. ZOSTE'RA, *L.* GRASSWRACK.

Grass-like marine plants ; rootstocks matted, creeping. *Stem* compressed. *Leaves* distichous, sheathing, long, linear ; stipules adherent to the sheathing leaf-base. *Flowers* in 2 parallel series of alternating anthers and carpels on one surface of a linear membranous peduncled spadix, which is enclosed in a sheathing leaf-like spathe. *Perianth* 0. *Anthers* sessile, 1-celled, dehiscence longitudinal ; pollen of slender tubes. *Carpels* ovoid, fixed laterally ; style subulate persistent, stigmas 2 capillary exserted from the spathe, deciduous ; ovule pendulous, orthotropous. *Utricle* dehiscent. *Seed* ovoid, testa tough ; embryo large, oblong, deeply grooved, cotyledonary end sigmoid, slender, sunk in the groove, plumule immersed.— DISTRIB. Various coasts ; species 4.—ETYM. ζωστήρ, a *riband.*

1. **Z. mari'na,** *L.* ; spathe dilated above the peduncle, its blade long. spadix many-fld. margin entire.

Muddy and sandy estuaries near low-water mark, N. to Shetland ; Ireland ; Channel Islands ; fl. July–Sept.—*Rootstock* slender, rather fleshy. *Leaves* 1–3 ft., $\frac{1}{8}-\frac{1}{2}$ in. broad, obtuse, bright green, opaque, 1–7-nerved. *Spathe*, including the leafy portion, 6–10 in., varying in breadth as the foliage does. *Spadix* 1–3 in. *Flowers* green, usually in series of 2 anthers (perhaps one 2-celled anther) and an ovary. *Fruit* $\frac{1}{8}$ in. furrowed.—DISTRIB. Most temp. coasts (Arctic).

VAR. *mari'na* proper ; leaves $\frac{1}{4}-\frac{1}{3}$ in. broad 3–7-ribbed.—VAR. *angustifo'lia*, Syme ; leaves $\frac{1}{8}-\frac{1}{4}$ in. broad 1–3-nerved. Orkney, &c.

2. **Z. na'na,** *Roth ;* spathe dilated above the peduncle, its blade short, spadix few-fld., margin with inflexed membranous appendages.

Estuaries, rare, from Forfar and Argyll to Sussex and Cornwall ; Dublin Bay ; fl. July–Sept.—Similar to *Z. mari'na*, var. *angustifo'lia*, but leaves not more than 6 in. by $\frac{1}{8}-\frac{1}{6}$ broad ; fruit shorter and very obscurely striate.—DISTRIB. Atlantic coast of Europe, N. Africa.

7. NA'IAS, *L.*

Very slender, submerged, fresh-water herbs. *Leaves* linear, opposite fascicled or whorled ; stipules adnate to the leaf-base. *Flowers* 1-sexual, solitary or crowded.—MALE. A solitary 1–4-celled anther enclosed in 2 sheaths, the outer sheath toothed ; pollen large, globose.—FEMALE. A solitary sessile carpel, naked or enclosed in sheaths ; style short, stigmas 2–4 persistent subulate ; ovule basal, erect, anatropous. *Drupe* small ; epicarp thin, separable. *Seed* ovoid, testa thin ; embryo straight, oblong, radicular end largest.—DISTRIB. Various climates ; species 10.—ETYM. ναϊάς, a *water-nymph.*

1. **N. flex'ilis,** *Rostkov.* ; leaves very slender, subentire.

Lakes, Perthshire, Skye, and Connemara; fl. Aug.-Sept.—*Stem* filiform, branched, brittle. *Leaves* ½-1 in., opposite or 3 in a whorl, linear, quite entire or remotely serrulate; sheaths ciliate. *Flowers* 2-3 or solitary. *Drupe* ⅛ in.—Distrib. Europe, Asia, America.

2. **N. mari'na,** *L.* ; leaves strongly spinular-serrate. *N. major,* All.

Hickling Broad, Norfolk ; fl. July.—*Stem* 3-9 in., sparingly branched, toothed here and there. *Leaves* opposite and 3-nate, ½-2 in. long ; sheath entire. *Flowers* diœcious, solitary. *Drupe* ¼ in., nearly ellipsoid, purplish.—Distrib. Temp. and trop. regions of the Old World.

Order LXXXVIII. **CYPERA'CEÆ.**

Grassy or rush-like herbs, usually perennial. *Stem* solid, often 3-gonous. *Leaves* with closed sheaths. *Flowers* 1-2-sexual, in the axils of small bracts (glumes), which are arranged in terete angled or compressed spikelets, beyond which the anthers and styles project. *Glumes* concave, often rigid, distichous or inserted all round the rachis, lower of each spikelet often empty. *Perianth* 0, or of 3-6 or more hypogynous scales or bristles. *Stamens* 1-6, hypogynous, filaments linear flat ; anthers basifixed, linear, dehiscing inwards, 2-celled, often with a claw at the tip. *Ovary* 1-celled (in *Carex* enclosed in a coriaceous utricle, *perigynium*) formed of 1 folded or 2 connate bracteoles ; style 1, stigmas 2-3 filiform papillose all over ; ovule solitary, erect, anatropous. *Fruit* small, indehiscent, compressed or 3-gonous (in *Carex* enclosed in the perigynium). *Seed* erect, testa membranous, albumen floury ; embryo minute, lenticular, at the base and outside of the albumen.—Distrib. All climates ; genera 61 ; species 2,200.—Affinities obscure.—Properties 0.

Tribe I. **SCIRPE'Æ.** *Spikelets* many-fld. *Flowers* 2-sexual. *Perianth* 0, or of scales or bristles.

Spikelets compressed ; glumes distichous, deciduous ....1. Cyperus.
Spikelet solitary, terete, terminal. Bristles 3-8 included......2. Heleocharis.
Spikelets usually clustered and lateral. Bristles 0, or 3-8 included..3. Scirpus.
Spikelets solitary or clustered, terete, terminal. Bristles very
    long flexuous exserted..4. Eriophorum.

Tribe II. **RHYNCHOSPORE'Æ.** *Spikelets* 1- or few-fld., terete or compressed. *Flowers* upper or all bisexual. *Perianth* 0, or of bristles.

Spikelets terete. Bristles slender or 0. Nut beaked ......5. Rhynchospora.
Spikelets compressed ; glumes distichous. Bristles various or 0.
    Nut not beaked ....6. Schœnus.
Spikelets terete. Bristles 0. Nut obtuse.............................7. Cladium.

TRIBE III. **SCLERIE'Æ.** *Spikelets* 1–2-fld., terete or compressed. *Flowers* unisexual. Bristles 0..8. Kobresia.

TRIBE IV. **CARICE'Æ.** *Spikelets* many-fld., terete, glumes most or all flower-bearing. *Flowers* unisexual; male naked; female enclosed in a perigynium..9. Carex.

1. CYPE'RUS, *L.*

Perennial, rarely annual, rushy or grass-like herbs of various habit. *Spikelets* linear, compressed, in lateral or terminal usually bracteate'heads, or branched umbels, or panicles. *Glumes* many, distichous, concave, keeled, deciduous, all or most flower-bearing. *Flowers* 2-sexual. *Bristles* 0. *Stamens* 1–3. *Styles* deciduous, not tumid at the base, stigmas 2–3. *Nut* 3-gonous or compressed.—DISTRIB. All climates but cold ; species about 700.—ETYM. The old Greek name.

1. **C. lon'gus,** *L.* ; perennial, tall, cyme umbellate, glumes erect red-brown. *Galingale.*

Marshes, very rare, Pembroke, and from Kent to Cornwall ; Channel Islands ; fl. Aug.-Sept.—*Rootstock* stout, creeping. *Stems* 2–3 ft., stout, erect, 3-quetrous, leafy at the base. *Leaves* few, flat, keeled ; margins hardly scaberulous. *Rays* many, 3–6 in., slender, again umbellate. *Bracts* leaf-like, far exceeding the rays. *Spikelets* ½–¾ in., 4–8, linear, curved, distichously crowded. *Glumes* lanceolate, midrib green scabrid. *Nut* 3-quetrous, pale.—DISTRIB. From France and Germany southd., N. Africa.—Rootstock aromatic, formerly used as a medicine.

2. **C. fus'cus,** *L.* ; annual, dwarf, spikelets corymbose or capitate, glumes at length spreading green or pale brown.

Ditches and wet meadows, very rare, Surrey (Chelsea naturalised, now extinct) ; Channel Islands ; fl. Aug.-Sept.—*Stems* 3–10 in., many ascending from a fibrous root, 3-quetrous. *Leaves* flat, spreading, grass-like. *Rays* few, short, usually simple. *Bracts* 3, leaf-like, unequal, broad at the base. curved, spreading. *Spikelets* ⅓–½ in., crowded, slender. *Glumes* many, oblong-ovate, subacute ; midrib broad or narrow, smooth, green. *Nut* minute, white.—DISTRIB. From Gothland southd., N. Africa, N. and W. Asia.

2. HELEO'CHARIS, *Br.*

Tufted, erect, usually perennial glabrous herbs. *Stems* slender, sheathed at the base. *Spikelets* solitary, terminal, erect, terete angled or compressed. *Glumes* many, imbricate all round the rachis ; lower 1–2, if any, flowerless. *Flowers* 2-sexual. *Bristles* 3–6, not longer than the glumes. *Stamens* 3. *Style* deciduous, articulate with the top of the fruit, stigmas 2–3. *Fruit* compressed or 3-gonous, tipped with the style-base.—DISTRIB. All climates, especially temp. ; species about 80.—ETYM. ἕλος and χαίρω, from *delighting* in *marshes.*

1. **H. palus'tris,** *Br.* ; rootstock stout creeping branched with many tufts of leaves and stems, lowest glume broadest, bristles 4–6, nut compressed, stigmas 2.

Marshes, lake borders, and ditches, N. to Sutherland ; ascends to 1,200 ft. in Yorkshire; Ireland; Channel Islands; fl. June–July.—*Rootstock* elongate, black. *Stems* 6–18 in., stout or slender, slightly compressed. *Sheaths* 2, brown, truncate, leafless. *Spike* ¼–½ in., terete, narrow-ovoid, red-brown. *Glumes* lanceolate, subacute; lowest much shortest, obtuse. *Anthers* apiculate. *Nut* compressed, obovate, striate, top triangular.—DISTRIB. Europe (Arctic), N. Africa, N. Asia, N. India, N. America.

H. PALUS'TRIS proper ; glumes dark, keel green, edges pale, lowest suborbicular half-embracing the base of the spikelet, nut faintly striate.

Sub-sp. H. UNIGLU'MIS, *Link;* glumes brown, edges narrow pale, lowest ovate almost embracing the base of the spikelet. Less common.—*H. Watso'ni.* Bab., from Argyll and Wicklow, is a short, more rigid form with dark brown glumes, lowest embracing the base of the spikelet, and nut more evidently punctate in lines longer than the bristles.

2. **H. multicau'lis,** *Sm.* ; rootstock short with one tuft of leaves and stems, lowest glume largest, bristles 5–6, fruit 3-gonous, stigmas 3.

Marshes, pools, &c., chiefly on moorlands, N. to Orkney; Ireland; Channel Islands; fl. July–Aug.—Similar to *H. palus'tris,* but differs in habit, in the obliquely truncate leaf-sheaths, and usually blunter glumes with narrower margins. *Nut* hardly striate.—DISTRIB. N. Europe, N. America.

3. **H. acicula'ris,** *Sm.* ; rootstock stoloniferous, stems setaceous obtusely 4-gonous grooved, spikelets minute, glumes ovate obtuse, lowest broadest, bristles 1–3, nut 3-gonous ribbed, stigmas 3.

Sandy edges of lakes and pools, from Forfar and the Clyde to Surrey and Cornwall, rare in Scotland; W. Ireland; Channel Islands; fl. July–Aug.— *Stolons* capillary. *Stems* many, 2–8 in., extremely slender. *Sheaths* membranous, acute. *Spikelets* ⅛–¼ in., compressed, red-brown. *Nut* very minute, pale, top subglobose.—DISTRIB. N. and Mid. Europe (Arctic), N.W. India, N. America.

3. SCIR'PUS, *L.*

Leafy or leafless, usually tall, marsh- or water-plants ; rootstock creeping. *Spikelets* several, in terminal or lateral cymes heads or clusters, or solitary, terete or compressed. *Glumes* imbricate all round the rachis or distichous, all but the 1–2 lowest flower-bearing. *Flowers* 2-sexual. *Bristles* 1–6, included, or 0. *Stamens* 3. *Style* 2-3-cleft, not swollen at the base, deciduous. *Nut* compressed or 3-gonous, top not swollen.—DISTRIB. All climates ; species about 300.—ETYM. The old Latin name.

SECTION 1. *Spikelets* large, lateral or terminal, cymose or clustered and sessile ; glumes numerous. *Bristles* 1–6.

* Stem leafless or nearly so. Cymes leafless.

1. **S. lacus'tris,** *L.* ; stems terete usually leafless, cymes terminal branched longer than the bracts, branches stout, spikelets solitary or clustered, glumes obtusely 2-lobed mucronate ciliate. *Bulrush.*

Lakes, ditches, and marshes, N. to Shetland; Ireland; Channel Islands; fl. July–Aug.—*Stems* 1–8 ft., spongy, as thick as the thumb or less, base sheathed. *Leaves* 0, or short flat and keeled in still water, or long and strap-shaped in streams. *Cyme* lateral at first, then expanding and over-topping the stem; branches few, $\frac{1}{2}$–3 in., strict, stout. *Spikelets* 1–6, $\frac{1}{4}$–$\frac{1}{2}$ in., sessile, cylindric in flower, ovoid in fruit, red-brown. *Glumes* mucronate or awned. *Bristles* 4–6, shorter than the broad nut.—DISTRIB. Arctic, temp. and trop. regions.
S. LACUS'TRIS proper: stems terete green, leaves often floating, glumes glabrous, anther-tips ciliate, nut 3-gonous.
Sub-sp. S. TABERNÆMONTA'NI, *Gmel.*; stems terete glaucous, glumes scabrid, anther-tips glabrous, nut compressed. *S. glau'cus,* Sm.—Usually near the sea, from Forfar and Dumbarton southd.; Ireland; Channel Islands.
Sub-sp. S. CARINA'TUS, *Sm.*; stems green obtusely 3-gonous above, glumes smooth, anther-tips glabrous, nut compressed.—Tidal rivers, Middlesex, Kent, Cornwall.

2. **S. tri'queter,** *L.* ; stem 3-quetrous usually leafless, cymes lateral, branches short stout, spikelets solitary or few elongate ovoid, glumes obtusely 2-lobed mucronate.

Muddy tidal rivers; Middlesex and Sussex to Cornwall; fl. Oct.—*Rootstock* slender. *Stems* 1–3 ft., strict, spongy, concave on one side. *Leaves* 0, or very short on the upper sheath. *Spikelets* in sometimes sessile clusters. *Glumes* obovate, brown, smooth. *Anther-tip* glabrous. *Bristles* 2–6. *Nut* obovoid, 3-gonous, smooth, shining.—DISTRIB. From Denmark southd. (excl. Spain, Greece, Turkey), E. Asia, Africa, America, Australia.

3. **S. pun'gens,** *Vahl ;* stems 3-quetrous, leaves 2–3 linear, spikelets lateral sessile, glumes ovate acutely 2-lobed mucronate. *S. Roth'ii,* Hoppe.

Sandy banks of St. Ouen's Pond, Jersey ; fl. June–July.—Habit of *S. tri'queter,* but smaller, more slender, 6–18 in.; leaves 4–6 in., channelled, keeled; spikelets $\frac{1}{2}$ in., sessile; glumes red-brown; anther-tips acute; bristles 1–2; nut obovoid, pale.—DISTRIB. W. Europe, from Denmark southd.

** Stem 3-gonous, leafy. Cymes terminal, leafy.

4. **S. marit'imus,** *L.* ; leaves channelled, cymes corymbose, spikelets few cylindric red-brown, glumes acutely 2-lobed awned.

Salt marshes from Ross and Skye southd.; Ireland; Channel Islands; fl. July–Aug.—*Rootstock* often tuberous. *Stems* 1–3 ft., tufted, rigid, leafy below. *Leaves* often $\frac{1}{2}$ in. broad, elongate, keeled, dark green. *Spikelets* $\frac{1}{2}$–1 in., brown, peduncled or sessile; bracts $\frac{1}{2}$–1 ft.; glumes glabrous or pubescent. *Bristles* 1–6. *Nut* compressed or 3-gonous, truncate, dotted, shining.—DISTRIB. Europe (Arctic), N. Africa, W. Siberia, N.W. India, N. America.

5. **S. sylvat'icus,** *L.* ; leaves flat, cymes effuse, branches many divaricate, spikelets small ovoid green, glumes entire obtuse.

Moist shaded places, from Argyll and Banff southd.; Ireland; fl. July.— *Stems* solitary, 1–3 ft., stout, leafy. *Leaves* large, sometimes ¾ in. broad, keeled. *Cymes* 2–5 in. diam.; branches slender; spikelets ⅛–⅙ in., 3–5 in a cluster. *Glumes* ribbed. *Bristles* 6, barbed. *Nut* obovoid, mucronate, punctulate.—Distrib. Europe (excl. Greece), N. Asia, temp. N. America.

Section 2. *Spikelets* small, lateral, sessile, fascicled, rarely solitary ; glumes few or many. *Bristles* 0. (*Isolepis,* Br.)

6. **S. seta'ceus,** *L.* ; stems filiform, leaves 1–2 narrow channelled, spikelets 1–3 lateral, nut obovoid 3-gonous ribbed and striate.

Gravelly and sandy damp places, N. to Orkney ; ascends to 1,500 ft. in Yorkshire ; Ireland ; Channel Islands ; fl. July–Aug.—*Stems* 3–6 in., tufted, terete, rigid. *Leaves* short, setaceous. *Spikelets* usually 2–3, ⅛–¼ in., ovoid. *Glumes* ovate, obtuse, green and brown. *Stamens* 2–3. *Stigmas* 3. *Nuts* broad, brown.—Distrib. Europe, N. Africa (?), Siberia.

7. **S. Sa'vii,** *Seb.* and *Maur.* ; stems filiform, leaves 1–2 narrow channelled, spikelets 1–3 subterminal, fruit subglobose 3-gonous not furrowed.

Wet bogs, W. Scotland, from Isla southd.; Lancashire; Wales and Suffolk to Hants and Cornwall; Ireland; Channel Islands; fl. July.—Very similar to *S. seta'cea,* but larger, paler, often 10 in., with longer leaves ; fruit paler, shining, dotted in lines. *Spikelet* sometimes solitary (*S. pyg'mæa,* Kunth).— Distrib. W. France, S. Europe, N. Africa.

8. **S. flu'itans,** *L.* ; floating, leafy, spikelet terminal solitary, nut obovate compressed mucronate. *Eleogi'ton fluitans,* Link.

Marshes and pools, from Orkney southd.; ascends to 1,200 ft. in Yorkshire ; Ireland ; Channel Islands ; fl. June–July.—*Stems* 6–18 in., compressed, slender, branched. *Leaves* 1–2 in., linear, very slender. *Spikelets* ⅛–⅙ in., narrow-ovoid, pale. *Glumes* 4–8, oblong. *Stigmas* 2. *Nut* plano-convex, pale, smooth, tipped with the base of the style.—Distrib. W. Europe, from Gothland southd., Azores.

9. **S. Holoschœ'nus,** *L.* ; stems tall terete stout, leaves few erect subulate rigid channelled, spikelets in compact globose cymose heads, fruit subglobose mucronate transversely wrinkled.

Sandy sea-coasts, N. Devon ; Channel Islands ; fl. Sept.—*Rootstock* creeping, stout. *Stems* 2–3 ft., as thick as a crowquill, tufted, margins of sheaths united by reticulate fibres. *Leaves* on the upper sheaths only, shorter than the stem, ½-terete, margins rough. *Heads* ¼–½ in. diam., upper subsessile ; branches of cyme 1–3 in., ½-terete, very stout. *Spikelets* minute. *Glumes* obovate, notched, mucronate, ciliate. *Stigmas* 3.—Distrib. From Belgium southd., N. Africa, Siberia.

Section 3. *Spikelets* small, terminal, solitary ; glumes few or many. *Bristles* 3–8. (*Bœo'thryon,* Ehrh.)

10. **S. cæspito'sus,** *L.* ; rootstock and leaves very short, 2 lowest glumes fertile equalling or exceeding the spike mucronate or awned, bristles 4–6 smooth, obovoid 3-quetrous acuminate. *Eleocharis,* Link.

Heaths and moors, N. to Shetland ; ascends to 3,500 ft.; Ireland ; fl. June–July.—*Stolons* 0. *Stems* 6–12 in., very densely tufted, wiry, grooved. *Sheaths,* lower split, large, stout, rigid, shining ; upper slender, with an erect short subulate blade. *Spikelets* ⅓–⅔ in., erect, chestnut-brown, shining. *Glumes* few, rigid, lowest flowering with usually a long green point. *Anthers* long, exserted, mucronate. *Stigmas* 3, very long. *Nut* brown.— DISTRIB. N. and Mid. Europe (Arctic), Siberia, N. America.

11. **S. pauciflo'rus,** *Lightf.* ; rootstock creeping, stolons long, upper sheaths truncate, lowest glume fertile obtuse not equalling the spikelet, bristles 3–6 barbed, nut obovoid 3-gonous. *Eleocharis,* Link.

Moorlands, N. to Caithness ; ascends to 2,100 ft. in Yorkshire ; Ireland ; Channel Islands ; fl. July–Aug.—Similar to *S. cæspito'sus,* but leafless, smaller ; lowest glume obtuse, with the rib not produced to the top ; and anthers not apiculate. *Nut* pale, minutely striate ; beak slightly contracted at the base, representing the tumid top of the nut of *Heleo'charis.*—DISTRIB. Europe (Arctic) (excl. Greece, Turkey), N. and W. Asia, N. America.

12. **S. par'vulus,** *R.* and *S.* ; minute, rootstock creeping, sheaths hyaline, leaves setaceous, spikelet minute, glumes pale lowest flowerless obtuse not exceeding the spikelet, bristles 4–8 barbed, fruit obovoid 3-gonous. *Eleocharis,* Hook.

Sandy seashores, Devon, Dorset, Hants, Wicklow ; fl. July.—*Rootstock* elongate, capillary, with distant tufts of a few soft stems and leaves, and small tubers. *Stems* 1–2 in., grooved ; sheath very inconspicuous, owing to its extreme tenuity. *Leaves* like the stem, slightly dilated at the base, recurved, subulate, channelled. *Spikelets* ₁¹₀ in., pale. *Glumes* membranous, obtuse. *Nut* pale.—DISTRIB. From Norway southd. (excl. Spain, Greece, Turkey).

SECTION 4. *Spikelets* small, in a terminal erect bracteate distichous spike ; glumes few. *Bristles* 3–6. (*Blys'mus,* Panz.)

13. **S. Cari'cis,** *Retz ;* leaves flat, edges rough, lower glume ribbed much shorter than the spikelet. *S. carici'nus,* Schrad. ; *S. planifo'lius,* Hull ; *Blys'mus compres'sus,* Panz.

Wet pastures and marshes, from the Forth and Clyde to Somerset and Kent ; ascends to 1,500 ft. in Northumbd. ; fl. June–July. *Rootstock* elongate. *Stems* 4–10 in., solitary, sheathed at the base. *Leaves* shorter than the stem, grass-like, keeled. *Bracts* leafy, long or short. *Spikes* ¾–1½ in. *Spikelets* many, ¼ in., pale brown. *Bristles* barbed. *Fruit* pale.—DISTRIB. Europe (excl. Spain, Greece), Siberia.

14. **S. ru'fus,** *Wahlb.* ; leaves ½-terete smooth, lowest glume not ribbed equalling the spikelet. *Blys'mus ru'fus,* Link.

Wet pastures, especially near the sea, from N. Wales and Lincoln to Shetland ; N. Ireland ; fl. July.—*Rootstock* creeping. *Stems* tufted, 3–12 in.

Leaves short, channelled. *Spikes* ½-⅔ in., chestnut-brown. *Spikelets* few, short. *Bristles* short, rough, very deciduous. *Fruit* brown.—Distrib. N.W. Europe (Arctic), N. Asia, Himalaya.

4. ERIOPH'ORUM, *L.* Cotton-Grass.

Perennial, tufted herbs. *Spikelets* terminal or lateral, solitary or cymose, terete. *Glumes* imbricate all round the rachis, all but 2-3 lowest flower-bearing. *Flowers* 2-sexual. *Bristles* 4-6 or very many, capillary, flat, at length greatly exceeding the spikelets. *Stamens* 1-3. *Style* deciduous, base not tumid, stigmas 2-3. *Nut* 3-gonous or compressed.—Distrib. N. temp. and Arctic regions; species about 12.—Etym. ἔριον and φορά, from the *cottony* heads.

* *Spikelet solitary, terminal, ebracteate.*

1. **E. vagina'tum,** *L.* ; stems glabrous, leaves filiform 3-quetrous, spikelet ovoid many-fld., bristles very many.

Boggy moors, Shetland to Sussex and Cornwall; ascends to near 3,000 ft. in the Highlands; Ireland; fl. April–May.—*Rootstock* short. *Stems* many, tufted, 6–10 in., longer in fruit, terete below, 3-gonous above, with 1-2 inflated leafless sheaths above the middle. *Leaves* very short. *Spikelet* ¾-1 in., erect. *Glumes* hyaline, broadly ovate, olive-green. *Nut* obovoid, obtuse, mucronate, compressed.—Distrib. N. and Mid. Europe (Arctic), N. and W. Asia, N.W. Tibet, N. America.

2. **E. alpi'num,** *L.* ; stems rough 3-gonous, leaves setaceous, spikelet narrow-oblong few-fld., bristles 4-6 crumpled.

Spongy bogs, Forfar (extinct); fl. June.—*Rootstock* creeping, producing a series of stems and leaves. *Stems* 6–10 in., very slender, rigid. *Leaves* short, rough, channelled, keeled. *Spikes* ¼ in., erect, oblong-lanceolate. *Glumes* yellow-brown. *Nut* minute, obovoid, trigonous.—Distrib. Arctic and Alpine Europe, N. Asia, N. America.

** *Spikelets in lateral corymbiform cymes, drooping, bracteate.*

3. **E. polysta'chion,** *L.* ; stems rigid obtusely 3-gonous, leaves smooth flat, tip 3-gonous, bracts 2-3, spikelets 4-12, nut obovoid mucronate.

Bogs, Shetland to Cornwall and Sussex; ascends to near 3,500 ft. in the Highlands; Ireland; Channel Islands; fl. May–June.—*Rootstock* stout. *Stems* 6–18 in., stout or slender, smooth, leafy. *Leaves* chiefly radical, variable in breadth. *Heads* rarely solitary, very variable in number, size, and length of peduncles. *Glumes* ⅛-¼ in., lead-coloured, oblong- or ovate-lanceolate, membranous. *Bristles* when fully grown 1½-2 in.—Distrib. Europe (Arctic) (excl. Greece), N. Asia, N. America.

E. polysta'chion proper; rootstock long, stems not tufted solid, leaves channelled 3-gonous above the middle, branches of cyme smooth, glumes ovate, margins broad scarious. *E. angustifo'lium,* Roth.; *E. grac'ile,* Sm., not Koch.

Sub-sp. E. latifo'lium, *Hoppe;* rootstock short, stems tufted slender trique-trous hollow, leaves flat tip short 3-gonous, branches of cyme scaberulous.

glumes lanceolate with very narrow scarious margins. *E. pubes'cens,* SM.—
Local ; ascends to 1,500 ft. in N. England ; rare in Ireland.

4. E. grac'ile, *Koch ;* stems very slender 3-gonous, leaves very narrow
3-quetrous throughout channelled, bracts 1–2 small, spikelets 3–6, fruit
narrowly obovate-lanceolate obtuse not mucronate.

Bogs, very rare, Yorkshire, Surrey, Hants ; fl. June–July.—A doubtful species
somewhat intermediate between *polysta'chion* and *latifo'lium.* Stem very
slender, 1–2 ft., leaves short, exceedingly narrow ; peduncles scabrid ;
heads small ; glumes broad, obtuse, distinctly ribbed, brown, without scarious margins ; nut very narrow.—DISTRIB. N. and Mid. Europe (Arctic),
Siberia, N. America.

5. RHYNCHO'SPORA, *Vahl.*

Perennial, tufted, leafy sedges. *Spikelets* terete, in axillary and terminal
corymbs or panicles. *Glumes* imbricate all round the rachis, 1–2 only
flower-bearing. *Flowers* 2-sexual, or the upper 1-sexual. *Bristles* 6 or more,
rarely 0. *Style-base* tumid, hardened, persistent, stigmas 3. *Nut* compressed or 3-gonous, tipped by a tumid tuberele.—DISTRIB. Temp. and
trop. ; species about 200.—ETYM. ῥύγχος and σπορά, from the *beaked
fruit.*

1. R. al'ba, *Vahl ;* spikelets pale 1-fld., bristles many barbed, stamens
2, fruit obovoid contracted below equalling the tumid tubercle.

Spongy bogs and wet meadows, Shetland to Cornwall and Sussex ; Ireland ;
fl. June–July.—*Rootstock* short. *Stems* 6–18 in., very slender, 3-gonous
above. *Leaves* subsetaceous, very narrow, channelled. *Corymbs* small, $\frac{1}{4}$–$\frac{3}{4}$
in. diam., terminal and axillary, long peduncled, flat-topped ; bracts leafy.
Spikelets $\frac{1}{4}$ in., crowded, white or pale brown. *Glumes* oblong-lanceolate,
acuminate, keeled, membranous.—DISTRIB. N. and Mid. Europe (Arctic),
Siberia, N. America.

2. R. fus'ca, *R.* and *S. ;* spikelets dark brown, bristles 6 barbed
upwards, stamens 3, fruit obovoid equalling the triangular serrulate
tubercle.

Bogs in Glamorgan, Somerset, Dorset, Hants, and Surrey, rare ; S.W. Ireland ;
fl. July–Aug.—Similar to *R. al'ba,* but more slender and rootstock elongate.
—DISTRIB. N. Europe, N.E. America.

6. SCHŒ'NUS, *L.*

Perennial, often leafless, rigid, rush-like herbs. *Spikelets* in compressed
terminal bracteate heads. *Glumes* subdistichous, rigid, the upper only
flower-bearing. *Flowers* 1–4, 2-sexual. *Bristles* 1–6. *Stamens* 3. *Style*
deciduous, base not tumid, stigmas 3. *Nut* 3-gonous, obtuse or mucronate.
—DISTRIB. Chiefly temp. ; species about 60.—ETYM. σχοῖνος, from the
use of some species as *cordage.*

S. ni'gricans, *L.* ; stems terete, spike obovoid usually much shorter than the bract.

Bogs and wet moors, rather local, Shetland to Cornwall and Surrey; ascends to 1,000 ft. in the Highlands; Ireland; Channel Islands; fl. June–July.—*Rootstock* short, stout, branched. *Stems* 6–24 in., in dense hard tufts of matted sheaths and leaves, terete, wiry, leafless above; sheaths copious, red brown or black, shining. *Leaves* wiry, terete, margins convolute. *Spikes* ½–¾ in., dark red-brown, shining; bract setaceous. *Spikelets* 4–10, erect, linear-oblong. *Glumes* irregularly distichous, oblong-lanceolate, subacute; keel scaberulous. *Bristles* barbed upwards. *Nut* small, ovoid, white.—Distrib. Europe, N. Africa, Siberia.

7. CLA'DIUM, *P. Brown.*

Coarse, harsh, perennial, usually tall, grassy herbs. *Spikelets* terete, usually panicled or cymose. *Glumes* few, concave, imbricate all round the rachis, 1–3 only flower-bearing. *Flowers* 2-sexual, or the lower male. *Bristles* 0. *Stamens* 2-3. *Style* deciduous, tumid but not jointed at the base, stigmas 2-3. *Nut* globose ovoid or 3-gonous, mucronate or beaked, pericarp thick corky, endocarp hard.—Distrib. All climates except very cold; species about 20.—Etym. κλάδος, a *twig.*

C. Maris'cus, *Br.* ; stems terete, spikelets clustered on the branches of many crowded compound cymes. *C. german'icum,* Schrad. (an older name).

Bogs and marshes, local in England, from the Border southd.; Sutherland, Wigton, and Berwick in Scotland; Ireland; fl. July–Aug.—*Rootstock* stout, creeping. *Stems* 2–5 ft., stout, erect, terete or obscurely 3-gonous, very leafy. *Leaves* 2–4 ft., ½ in. diam., rigid, glaucous, channelled, keeled, margins serrulate, points very long. *Cymes* axillary and terminal, corymbose; branches 1–3 in., erect or recurved, bracts setaceous. *Spikelets* ¼ in., crowded in pedicelled heads ½ in. diam. *Glumes* 5–6, obtuse, pale-brown, lower short, upper oblong-lanceolate flowering. *Flowers* about 2, one fertile. *Stamens* usually 2 ; anthers apiculate. *Nut* small, ovoid, 3-gonous, beaked, brown.—Distrib. From Gothland southd., N. Africa, Siberia.

8. KOBRE'SIA, *Willd.*

Small perennial sedges. *Leaves* rigid, keeled. *Spikelets* in a terminal compressed ovoid spike, few-fld., some male, others male (upper) and female. *Glumes* 2-3, imbricate all round the rachis, lowest flowerless. *Bristles* 0.—Male fl. *Stamens* 3.—Female fl. at the base of a convolute bract. *Ovary* 3-gonous; style-base simple, stigmas 3. *Nut* 3-gonous.—Distrib. Arctic and Alpine Europe, N. Asia, Himalaya; species 8.—Etym. *De Kobres,* a German patron of botany.

K. carici'na, *Willd.* ; lower spikelets with one female flower.

Upland moors, York, Durham, Westmoreland, Argyll, and Perth (ascends to about 2,500 ft.); fl. Aug.—*Rootstock* short. *Stems* 4–9 in., densely tufted, rigid, terete, leafy at the base only. *Leaves* 2–5 in., wiry, recurved, grooved, margins convolute. *Spike* ½–1 in., narrow. *Spikelets* subdistichous, sessile, cylindric; bracts small, lowest with a rigid serrulate point. *Glumes* ¼ in., rigid, ovate-oblong, obtuse, pale brown, basal always empty, second of the lower spikelets female, the third male; in the upper spikelets both flowers are male. *Nut* as long as the glumes, linear, beaked, pale.—DISTRIB. Arctic and Alpine Europe, Greenland, Rocky Mts.

9. CA'REX, *L.* SEDGE.

Perennial grass-like herbs. *Stems* usually leafy. *Spikelets* 1-2-sexual, rarely diœcious, terete, solitary or in heads spikes racemes or panicles, all 2-sexual or lower female with often a few male fl. at the base or top, and upper male with often a few female at the top or base. *Glumes* imbricate all round the rachis, persistent or deciduous.—MALE fl. *Stamens* 2-3, without perianth or bristles.—FEMALE fl. a compressed or 3-quetrous ovary, included in an urceolate 2-toothed sac, from which the 2-3 stigmas project. *Nut* minute, coriaceous, compressed or 3-gonous, included in the sac (*perigynium*).—DISTRIB. All climates, rare in trop., abundant in Arctic and cold ; species about 500.—ETYM. κείρω, from the *cutting* foliage. —Sometimes a slender rudimentary rachis (racheola) occurs in the perigynium, at the base of the ovary.

SECTION 1. *Spikelet* solitary, terminal.

1. **C. paucifio'ra**, *Lightf.* ; monœcious, leaves involute, spikelet ebracteate, glumes 4–6, upper male, perigynia reflexed, stigmas 3.

Moorland bogs, York to Caithness ; ascends to 2,700 ft. in the Highlands; fl. June–July.—*Rootstock* slender, creeping, stoloniferous. *Stems* 3-12 in., very slender, 3-gonous, smooth. *Leaves* setaceous. *Spikelets* ¼ in. *Glumes* ⅛–¼ in., rather distant, oblong-lanceolate, acute, pale, shorter than the fusiform pale beaked perigynia. *Nut* 3-quetrous, pale.—DISTRIB. Europe, chiefly Alpine and Arctic, N. America.

2. **C. pulica'ris**, *L.* ; monœcious, leaves involute, spikelet ebracteate, glumes 6–12, upper male, perigynia reflexed, stigmas 2.

Bogs, N. to Shetland ; ascends to 2,700 ft. in the Highlands ; Ireland ; Channel Islands ; fl. May–June.—*Rootstock* tufted; stolons 0 or short. *Stems* 3-8 in., smooth, rather rigid, terete, grooved. *Leaves* setaceous, often exceeding the stem. *Spikelets* ½–1 in. *Glumes* ⅛ in., lower sometimes remote, ovate-oblong, subacute, about ½ as long as the ovate-lanceolate stalked compressed beaked pale perigynia. *Racheola* linear, sometimes floriferous. *Fruit* oblong, plano-convex, grey.—DISTRIB. N. and Alpine Europe, N. Asia.

3. **C. Davallia'na**, *Sm.* ; diœcious, leaves filiform flat rough, spikelet ebracteate, glumes numerous. perigynia reflexed, stigmas 2.

Near Bath (extinct); fl. June.—*Rootstock* tufted, stolons 0. *Stems* 6-18 in., very slender, strict, rough. *Spikelet* ⅓–⅔ in. long, cylindric-oblong ; male very narrow, glumes linear-oblong ; female glumes ovate, acuminate, pale chestnut-brown, persistent, equalling the ovoid-lanceolate rather decurved sessile beaked perigynia.—DISTRIB. From Holland and the Pyrenees E. to Austria and S. Russia.

4. **C. dioi'ca,** *L.* ; leaves setaceous smooth, spikelet ebracteate diœcious or male below, glumes many, perigynia erect or spreading, stigmas 2.

Bogs and moorlands, Shetland to Somerset and Sussex; ascends to nearly 2,900 ft. in the Highlands; rarer in Ireland; fl. May–June.—*Rootstock* creeping, stoloniferous. *Stems* 6–12 in., terete, striate, wiry. *Spikelets :* male ⅓–⅔ in., very narrow, cylindric, glumes oblong-lanceolate, pale ; *female* ⅓–½ in., ovoid, glumes ovate, brown, deciduous, shorter than the sessile ovoid beaked ribbed plano-convex perigynia. *Nuts* lenticular, chestnut-brown.—DISTRIB. Europe (Arctic), Siberia, N. America.

5. **C. rupes'tris,** *All.* ; leaves channelled, bract subulate or 0, glumes few, upper male, perigynia erect, stigmas 3.

Ledges of alpine rocks, alt. 2,000–2,500 ft.; Perth, Forfar, Aberdeen, and Sutherland ; fl. July.—*Rootstock* creeping, tufted, stoloniferous. *Stems* 4–6 in., wiry, 3-gonous, rather rough. *Leaves* curved, rigid, margins recurved, tip wavy, rough. *Spikelet* ½–1 in., narrow, dark-brown, shining. *Glumes,* female few, broadly ovate, obtuse or mucronate, rather shorter than the elliptic, smooth, pale, abruptly beaked perigynia. *Nut* 3-quetrous, brown. —DISTRIB. Arctic and Alpine N. and Mid. Europe, N. Asia, N. America.

SECTION 2. *Spikelets* short, sessile, most or all 2-sexual, all similar. *Bracts* 0, or setaceous or leafy, never sheathing. *Stigmas* 2.

* *Spikelets in simple spikes or heads, male usually at the top only (spike sometimes compound at the base in* 8, dis'ticha; *see also* 12, murica'ta). *Rootstock creeping.*

6. **C. incur'va,** *Lightf.* ; stems short, spikelets capitate, bract 0, perigynia longer than the ovate obtuse glumes, beak abrupt short smooth.

Sandy shores, from Holy Isle to Shetland, rare; fl. June–July.—*Rootstock* very long, creeping. *Stems* stout, decurved, subterete, smooth, leafless. *Leaves* spreading and recurved, margins involute. *Spike* ½–¾ in., broadly ovoid. *Perigynia* elliptic-ovoid, turgid, spreading, pale, much larger than the pale brown glumes. *Nut* ferruginous, lenticular.—DISTRIB. Arctic and Alpine Europe, Siberia, N.W. India, N. America, S. Chili.

7. **C. divi'sa,** *Huds.* ; slender, spikelets short crowded bracteate, perigynia ovoid equalling the ovate cuspidate glumes, beak 2-fid serrulate.

Marshes near the sea, from York, N. Wales, and Lincoln southd.; Dublin; fl. May–June.—*Rootstock* stout. *Stems* 1–3 ft., very slender, leafy, 3-gonous, scabrid above ; basal sheaths leafless. *Leaves* long, very narrow, flexuous, margins involute. *Spikes* ½–1½ in., narrow, interrupted, pale brown ; bracts setaceous or filiform. *Glumes* membranous. *Perigynia* not margined. *Nut* brown, plano-convex, orbicular.—DISTRIB. From France and Germany southd., N. and S. Africa, W. Siberia, N.W. India.

8. **C. dis'ticha,** *Huds.* ; stems long, spikelets in an elongate head, bracts small, perigynia stipitate elliptic-ovoid ribbed exceeding the acuminate glumes, wing and 2-fid beak serrulate. *C. interme'dia,* Good.

Wet meadows and marshes, from Lanark and Aberdeen southd.; ascends to
1,200 ft. in the Highlands; Ireland; fl. June.—*Rootstook* creeping. *Stems*
1–3 ft., stouter than in *C. divi'sa*, leafy, 3-gonous, scaberulous above. *Leaves*
⅛–¼ in. broad, flat. *Spike* ½–3 in., subdistichous, pale brown, sometimes
compound at the base, the upper and lower spikelets usually wholly male.
intermediate chiefly male; bract never large and leafy. *Nut* ovoid, ferruginous.—DISTRIB. Europe, N. Asia, N. America.

9. **C. arena'ria,** *L.* ; stems short, spike oblong, bracts setaceous, perigynia substipitate elliptic-ovoid ribbed winged exceeding the subaristate
glumes, wing broad and 2-fid beak serrulate.

Sandy sea-shores, N. to Shetland, and inland in Surrey, Norfolk, and Suffolk;
Ireland; Channel Islands; fl. June.—Very near *C. disticha*, but habit
different, shorter; leaves rigid, curved; glumes more mucronate, and fruit
with a broad coriaceous wing. *Rootstock* very long, stout, branched, binding
the sands. *Stems* 8–12 in., stout, curved, 3-quetrous, scabrid above. *Leaves*
chiefly radical, stiff, margins involute. *Spike* 1–3 in., compressed, pale
brown. *Spikelets* many, lower female, upper usually male, intermediate 2-
sexual. *Nut* plano-convex, chestnut, shining.—DISTRIB. Europe, Siberia.
VAR. *C. liger'ica*, Gay, is a more slender form with female spikelets at the
top.—Scilly Islands. N. and W. Europe.

** *Spikelets male at the top only, in compound heads spikes or panicles (rarely
simply spiked or capitate in* 12, murica'ta; *see also* 8, dis'ticha).

10. **C. panicula'ta,** *L.* ; stout, rootstock very short, spikelets many in
a broad or narrow elongate panicle, perigynia ovoid many-nerved below
broadly 3-gonous equalling the pale margined ovate subaristate glumes,
beak narrow 2-toothed serrulate.

Wet copses and marshes, from Orkney southd.; Ireland; Channel Islands; fl.
June–July.—*Rootstocks* densely matted, forming tussocks 2–4 ft. diam.
Stems 1–4 ft., leafy, stout, 3-quetrous, scaberulous above. *Leaves* harsh,
long, narrow, flat. *Panicle* 2–6 in., very variable, rarely reduced to a simple
spike, pale brown. *Branches* short or long; bracts 0 or setaceous. *Spikelets*
crowded, pale brown. *Perigynia* truncate or cordate below, opaque, narrowed
into the long beak. *Nut* ovoid, base narrowed, biconvex, base of style
tumid.—DISTRIB. From Sweden southd., Canaries, W. Siberia.
C. PANICULA'TA proper; panicle usually broad.
Sub-sp. C. PARADOX'A, *Willd.*; stem clothed below with black erect nerves of
old sheaths, panicles smaller laxer, beak of perigynia narrower split to the
base with overlapping margins, nerves as in *C. panicula'ta.*—Bogs, very
rare; Middlesex, Norfolk, York, Westmeath.—DISTRIB. Europe (Arctic),
Siberia.

11. **C. teretius'cula,** *Good.* ; slender, rootstock obliquely creeping,
panicle narrow spike-like, spikelets few, perigynia spreading ovoid turgid
few-nerved exceeding the ovate acuminate glumes, beak long rough 2-
toothed.

Bogs and meadows, from Caithness to Devon and Kent; Ireland; fl. June.—
Near *C. panicula'ta*, but forming scattered tufts (not tussocks), rootstock
creeping; stems wiry, much more slender; spike shorter, broader, more

simple; perigynia brown, shining, ribbed only at the back, where 2–4 main ribs diverge from the base and then converge under the beak. *Nut* obpyriform, biconvex.—DISTRIB. Europe (Arctic) (excl. Greece), Canaries, Himalaya, N. America, N. Zealand.—*C. Ehrhartia'na*, Hoppe (*C. pseudo-paradox'a*, Gibs.), with more numerous stems and a larger more interrupted spike, found near Manchester and Birmingham, seems hardly a variety.

12. **C. murica'ta,** *L.* ; slender, spikelets few in spikes or slender panicles, bracts 0 or setaceous, perigynia spreading elliptic-ovoid smooth longer than the acuminate glumes, beak broad serrulate 2-fid. *C. spica'ta,* Huds.

Marshes, copses, &c., from Caithness, Elgin, and the Clyde southd.; Ireland; Channel Islands ; fl. May–June.—Densely tufted, stolons short. *Stems* 1–2 ft., wiry, 3-gonous, scaberulous above. *Leaves* shorter than the stem, narrow, flat. *Spike* very variable. *Spikelets* green, squarrose. *Perigynia* narrowed below, sessile, faintly ribbed, beak deeply 2-fid. *Nut* brown, plano-convex, base of style clavate.—DISTRIB. Europe, N. Africa, N. Asia, Himalaya, N. America.

C. MURICA'TA proper; stems erect, spike more continuous, glumes brown with green keels, beak serrulate.—Drier places.—A var. *pseudo-divul'sa* is recorded from Malvern Links; it is intermediate between *murica'ta* and *divul'sa*, the spike being interrupted below. It grows with *murica'ta* in W. Suffolk.

Sub-sp. C. DIVUL'SA, *Good.*; more slender, stems curved, spike elongate much interrupted, perigynia less spreading narrower and glumes paler, beak less serrulate.—Moist places, from York and N. Wales southd., Channel Islands.

13. **C. vulpi'na,** *L.* ; stout, leaves broad flat, spike subcylindric bracteate, perigynia spreading ovoid truncate below obscurely ribbed equalling the ovate awned pale brown glumes, beak long serrulate 2-fid.

Marshes, copses, and saline ditches, from Elgin and the Hebrides southd.; Ireland; Channel Islands; fl. June.—*Rootstock* tufted, stoloniferous. *Stems* many, 3-quetrous, leafy, angles scabrid, faces convex. *Leaves* ¼–½ in. *Spikes* 1–3 in., variable, squarrose; bracts setaceous, spreading, conspicuous. *Glumes* pale brown, midrib green, awn short scabrid. *Perigynia* compressed, ribs variable in stoutness. *Nut* ovoid, brown.—DISTRIB. Europe, N. Africa, Siberia, N. America.

*** *Spikelets male at the base, rarely at the top also, distant, alternate (rarely close in* 14, *echina'ta). Bracts often long and foliaceous.*

14. **C. echina'ta,** *Murr.*; slender, leaves narrow, spikelets 3–5 subglobose squarrose, bracts small, perigynia broadly ovoid plano-convex striate exceeding the ovate acute glumes, beak broad notched scabrid. *C. stellula'ta,* Good.

Moors and bogs, N. to Shetland; ascends to near 3,000 ft. in the Highlands; Ireland; Channel Islands; fl. May–June.—*Rootstock* densely tufted. *Stems* 6–18 in., 3-gonous, minutely scabrid above, stolons 0. *Leaves* channelled, shorter than the stem. *Spikes* ½–1 in., usually very pale; upper spikelets more slender. *Glumes* green or with pale brown sides. *Perigynia* olive-green, sessile, base rounded, lateral ribs stout, facial faint. *Nut* lenticular, pale olive.—DISTRIB. Europe (Arctic), N. Africa, N. Asia, N. America. —VAR. *C. Gry'pos*, Schk.; glumes with two broad red-brown bands. Perth.

15. **C. remo'ta,** *L.* ; slender, leaves narrow, bracts long, spikelets oblong, perigynia erect narrow-ovoid plano-convex striate exceeding the
oblong-ovate acuminate glumes, beak broad and margins above serrulate.

Copses, from Argyll and Ross southd. ; ascends to 1,000 ft. ; Ireland ; Channel Islands ; fl. June.—Tufted ; branches spreading *Stem* 1–2 ft., inclined,
3-gonous. *Leaves* ₁₂–⅔ in. broad, equalling the stems, flat. *Spike* 2–5 in.,
rachis scabrid ; lower bracts often very long. *Spikelets* 3–9. *Glumes* pale
brown, midrib broad green, margins white scarious. *Perigynia* pale,
narrowed below. *Nut* plano-convex, narrowed.—DISTRIB. Europe, N.
Africa, N. Asia, Himalaya, N. America.—*C. tenel'la,* Sm., not Schk., is a
starved form.—*C. axilla'ris,* Good. (? a hybrid with *murica'ta*), has taller
stouter 3-quetrous stems, shorter bracts, larger spikelets, the lower crowded
and compound, and broader rigid glumes. Marshes, from York southd. ;
Ireland.—*C. Bœnninghausenia'na,* Weihe (? a hybrid with *panicula'ta*), has
slender 3-quetrous scabrid stems, long spikes, no bracts, small spikelets, pale
membranous glumes, and perigynia serrulate below the middle. Marshes,
from Banff southd.

**** *Spikelets male at the base, or both at the top and base, in a compact head or
spike. Bracts 0 or subulate* (except *C. lepori'na*).

16. **C. lepori'na,** *L.* ; stout, leaves narrow, spike lobed compact, perigynia elliptic-ovoid plano-convex striate equalling the lanceolate acute
glumes, margins winged and long 2-fid beak serrulate. *C. ova'lis,* Good.

Wet places, N. to Shetland ; ascends to 1,700 ft. in the Lake District ; Ireland ;
Channel Islands ; fl. June.—*Rootstock* tufted ; stolons 0. *Stems* 6–12 in.,
3-quetrous, slightly scabrid above. *Leaves* shorter than the stem, flat ₁₀–⅛
in. broad, points fine. *Spike* ¾–1 in., often as broad, pale brown, compressed ;
spikelets male at the base only ; bracts 0, or subulate (*C. argyrolo'chin,*
Lond. Cat. 1867, not Hornem.). *Glumes* pale brown, midrib green, edges
whitish, acuminate or cuspidate. *Perigynia* sessile. *Nut* stipitate, oblong,
lenticular, shining.—DISTRIB. Europe (Arctic), N. and W. Asia, Rocky Mts.

17. **C. canes'cens,** *L.* ; slender, leaves narrow, spike interrupted,
bract 0 or minute, perigynia elliptic-ovoid plano-convex ribbed equalling
the ovate acute pale glumes, beak short 2-fid serrulate. *C. cur'ta,* Good.

Bogs and marshes, from Isla and Elgin southd. ; ascends to nearly 2,200 ft. in
the Highlands ; Ireland ; fl. June–July.—*Rootstock* tufted ; stolons 0. *Stems*
12–18 in., 3-quetrous, longer than the leaves, hardly scabrid above. *Leaves*
flat, ₁₀–¾ in. broad. *Spike* ½–1½ in., slender ; spikelets 3–8, ovoid, male at
the base only ; bracts 0, rarely 1, subulate. *Glumes* very pale, edges broad
scarious. *Perigynia* erect, pale olive, not winged, ribs slender, beak obscure.
Nut ovoid or obovoid, lenticular, pale.—DISTRIB. Europe (Arctic) (excl.
Spain, Greece, Turkey), N. Asia, N. America, S. Chili.
VAR. *C. alpic'ola,* Wahl. (*C. vit'ilis,* Fries ; *C. Persoo'nii,* Sieb.), has spikelets
fewer, few-fld., glumes browner, beak deeply 2-fid in fruit.—N. Wales,
York to Ross, rare, ascends to 3,600 ft.

18. **C. lagopi'na,** *Wahl.* ; stem wiry, leaves flat, spikelets 2–4 small
crowded, bracts minute, perigynia elliptic-ovoid biconvex ribbed exceeding
the ovate acute glumes, beak short 2-fid. *C. lepori'na,* L. in part.

Aberdeen Alps, alt. 3,600. ft., very rare; fl. Aug.—*Rootstock* tufted; stolons short. *Stems* 6–10 in., 3-quetrous, often curved below, smooth or scabrid above. *Leaves* $\frac{1}{10}$–$\frac{1}{8}$ in. broad. *Spikelets* $\frac{1}{4}$ in. ovoid. *Glumes* few, broad, dark brown, midrib green, edges white. *Perigynia* red-brown. *Fruit* obovoid.—DISTRIB. Arctic and Alpine Europe, N. Asia, N. America.

19. C. elonga'ta, *L.* ; slender, leaves narrow, spike slender, spikelets many close, perigynia elliptic-lanceolate biconvex ribbed much exceeding the ovate acute glumes, beak subulate entire decurved.

Wet copses and marshes, from York to Sussex and Kent; Lough Neagh; fl. June.—*Rootstock* tufted, matted; stolons 0. *Stems* very many, 1–2 ft., 3-quetrous, scabrid, graceful, leafy. *Leaves* longer than the stems, flaccid, flat, $\frac{1}{10}$–$\frac{1}{6}$ in. broad. *Spike* 1–2½ in.; bracts 0. *Spikelets* erect or spreading, lower rarely distant. *Glumes* red-brown, midrib green, edges white. *Perigynia* pale, sessile, spreading, strongly ribbed, gradually narrowed into the beak. *Fruit* oblong, obtuse, plano-convex.—DISTRIB. Europe (Arctic), N. Asia, N.W. America.

SECTION 3. *Spikelets* 2–6, short, ovoid, approximate, all subsimilar, 2–4 lower wholly female, upper male at the base only (rarely at the top or throughout). *Lowest bract* foliaceous, sheath very short or 0. *Stigmas* 3.

20. C. Buxbaum'ii, *Wahl.* ; leaves narrow, sheath-edges filamentous, spikelets 3–5 subsessile, perigynia green ellipsoid nerved larger than the lanceolate dark glume, beak 0. *C. polyg'ama,* Schk.

Stony banks, Lough Neagh; fl. July.—*Rootstock* short, creeping, stoloniferous. *Stems* 1–2 ft., rigid, 3-quetrous, leafy below. *Leaves* flat, $\frac{1}{8}$–$\frac{1}{4}$ in. broad ; sheaths red-brown, rigid. *Lower bracts* leafy, often exceeding the spike. *Spikelets* $\frac{1}{3}$–$\frac{3}{4}$ in., obtuse, conspicuous from the small dark glumes and large broad imbricating perigynia. *Glumes* rounded, mucronate acuminate or awned, dark red-brown, midrib green. *Perigynia* plano-convex, mouth 2-fid, ribs slender. *Nut* obovoid, 3-quetrous, brown, covered with white dots.—DISTRIB. Arctic and Alpine Europe, N. Asia, N. America, Australian Alps.

21. C. alpi'na, *Swartz ;* leaves short flat, spikelets 2–4 sessile subglobose, perigynia ovoid nerveless much larger than the ovate acute blackish glumes, beak very short notched scabrid. *C. Vahl'ii,* Schk.

Rocky mts., Aberdeen, Forfar, alt. 2,400–2,600 ft.; fl. Aug.—*Rootstock* with short stolons. *Stem* 6–18 in., subsolitary, rigid, 3-quetrous, smooth or scaberulous above. *Leaves* short, recurved $\frac{1}{8}$–$\frac{1}{6}$ in. broad. *Spikelets* $\frac{1}{4}$–$\frac{1}{3}$ in., close, lateral subhorizontal black ; bract slender. *Glumes* crowded, subacute, broad, without a green midrib. *Perigynia* broad, yellow-brown, scaberulous. *Nut* ellipsoid or ovoid, 3-quetrous, pale.—DISTRIB. Arctic and Alpine Europe, N. Asia, Himalaya, N. America.

22. C. atra'ta, *L.* ; leaves broad, spikelets 4–6 pedicelled inclined subcylindric, perigynia suborbicular 3-gonous compressed smooth larger than the ovate subacute blackish glumes, beak short slender smooth.

Ledges of Alpine rocks, N. Wales, Westmoreland, Dumfries, and Mid. Scotland, alt. 2,400–3,700 ft.; fl. July–Aug.—*Rootstock* tufted, small, stoloniferous. *Stem* 10–28 in., inclined, 3-gonous, hardly scabrid above. *Leaves* rather large, ¼–½ in. broad, flat, keeled. *Spikelets* ½–¾ in., rarely remote, uppermost sometimes wholly male; bracts usually exceeding the spikelets, sheaths very short. *Glumes* erect, imbricate. *Fruit* 3-quetrous, elliptic.—DISTRIB. Arctic and Alpine Europe, N. Asia, Himalaya, N. America.—Diandrous and 2-sexual flowers occur (*Boott.*).

SECTION 4. *Spikelets* mainly 1-sexual, lower all or chiefly female; upper different-looking, usually more slender, all or chiefly male.

* *Stigmas* 2 (rarely 3 in 27, *aquatilis*, and 28, *Goodenovii*). *Bracts* equalling the female spike, sheaths 0. (See also 56, *vesicaria*, sub-sp. *saxatilis*, and 59, *paludosa*.

† *Leaves with revolute or recurved margins (best seen when dry).*

23. **C. rig'ida,** *Good.* ; stout, leaves broad short recurved, sheaths all leafing, edges not filamentous, spikelets erect short, perigynia obovoid lenticular green equalling the obtuse dark glumes, beak very short smooth.

Stony mts., from N. Wales and York to Shetland; ascends to 4,300 ft.; W. and N. Ireland; fl. June–Aug.—*Rootstock* creeping, tufted or not. *Stems* rigid, 3-quetrous, usually curved, nearly smooth. *Leaves* many, stiff, keeled, ⅛–¼ in. broad. *Bracts* never much exceeding the stem. *Spikelets* very variable, subsessile; male cylindric, clavate or fusiform, sometimes female below; females 3–5, ¼–¾ in., sometimes pedicelled, obtuse, cylindric or oblong, sometimes male at the top. *Glumes* dark, midrib green, edges narrow pale. *Perigynia* broad, smooth, substipitate.—DISTRIB. Arctic and Alpine Europe, N. Asia, Himalaya, N. America.

24. **C. acu'ta,** *L.* ; rigid, leaves long broad, sheath-edges not filamentous, spikelets many long, perigynia compressed green broader than the dark narrow glumes, beak very short entire. *C. grac'ilis*, Curtis.

Wet places, from the Clyde and Berwick to Kent and Somerset; Ireland; fl. May–June.—*Rootstock* tufted, stoloniferous or not. *Stems* 2–3 ft., 3-quetrous, smooth, or scaberulous. *Leaves* equalling the stem, ⅛–⅓ in. broad, flaccid, lowest sheaths sometimes leafless. *Spikelets* 1–4 in., many, inclined, subsessile, sometimes long-pedicelled; males 2–4, stout or slender; females 3–5. *Glumes* obtuse acute or cuspidate, midrib green. *Perigynia* elliptic or suborbicular, smooth, substipitate, variable in size and breadth; ribs 3–5, faint, beak smooth. *Nut* plano-convex, orbicular or obovoid.—DISTRIB. Europe (Arctic), N. and W. Asia, E. and W. N. America.—Perigynia sometimes antheriferous.—*C. tricosta'ta*, Fries, is a slight variety with more orbicular and stronger ribbed perigynia, found N. Ireland.

25. **C. stric'ta,** *Good.* ; slender, leaves long erect flat, sheath-edges filamentous, lower leafless, spikelets sessile, perigynia orbicular-ovoid compressed green nerved much larger than the oblong obtuse dark glumes, beak very short notched. *C. cæspito'sa*, Gay, not L.

Marshy places, local, from the Clyde to Kent and Dorset; Ireland; fl. May–June.—*Rootstock* in large dense tufts. *Stems* 1–3 ft., 3-quetrous, slightly scabrid above. *Leaves* ⅛-¼ in. diam., long, flaccid; leafless sheaths long, strict, red-brown, shining. *Bracts* 0, or variable. *Spikelets* ¾-2 in., erect, cylindric; males 1-2, slender; females 1-3, sometimes male above, stout. *Glumes* in about 8 rows, pitchy, midrib green; of the male very narrow, subacute. *Nut* oblong.—DISTRIB. Europe, N. Africa, N. America.

†† *Leaves with incurved or involute margins (best seen when dry).*

26. **C. triner'vis,** *Degl.* ; rigid, leaves narrow, female spikelets 2-3 sessile oblong or cylindric, perigynia oblong compressed smooth 3-5-nerved, dotted equalling the dark obtuse glumes, beak very short.

Wet sandy places on the Norfolk coast; fl. July–Aug.—A stout short species with long scaly rootstocks and stolons; roots very stout. *Stems* 6-10 in., erect or curved, smooth, 3-gonous. *Leaves* equalling or exceeding the stem, smooth, keeled; sheaths not fibrous. *Spikelets* ½-1 in., close; lower bract slender, stiff, exceeding the spikelet, not sheathing. *Glumes* brown, midrib green. *Perigynia* stipitate, yellow, rarely orbicular. *Fruit* lenticular, brown, punctate.—DISTRIB. Shores of N.W. Europe.

27. **C. aquat'ilis,** *Wahl.* ; stem stiff, leaves long erect flat, spikelets 3-6 cylindric, perigynia suborbicular compressed pale smooth nerveless rather larger than the dark obtuse glumes, beak short smooth or 0.

Bogs and marshes in Scotland, from the Border to Caithness, rare; ascends to 3,300 ft.; fl. June–Aug.—*Rootstock* tufted, creeping, stoloniferous. *Stems* 10-24 in., 3-gonous, stout, polished, leafy below. *Leaves* ⅛-⅓ in. broad; sheaths all leafing, not filamentous. *Spikelets* 1-2½ in., slender, lower usually pedicelled; males 1-3, more slender and pale; bracts long, leafy. *Glumes* oblong or ovate, midrib bright green, margins not pale. *Perigynia* much compressed, very pale brown. *Fruit* variable, ovoid obovoid or orbicular, lenticular or 3-quetrous.—DISTRIB. Scandinavia (Arctic), Greenland, N. America.

C. aquat'ilis proper (including var. *mi'nor*, Boott); stem scaberulous above, bracts long, spikelets large pale. High mts.—VAR. *Watso'ni*, Syme; stems shorter smooth, leaves narrower, bracts shorter, spikelet smaller, glumes darker purple brown. The lowland form.

28. **C. Goodeno'vii,** *Gay;* leaves very narrow erect, spikelets 3-5 sub-sessile short, perigynia suborbicular plano-convex nerved below larger than the obtuse dark glumes, beak very short terete smooth. *C. cæspito'sa,* Sm. ; *C. vulga'ris,* Fries.

Marshes and wet meadows, N to Shetland; ascends to nearly 3,000 ft.; Ireland; Channel Islands; fl. May–July.—*Rootstock* tufted or creeping, sometimes extensively. *Stems* 6-24 in., 3-quetrous, scaberulous above, rigid, short and curved or long slender and erect. *Leaves* sometimes very slender, ¹⁄₁₂-¹⁄₁₀ in. broad, at others short, recurved, ⅛ in. broad. *Spikelets* ¼-1 in., erect, close or rather distant. *Glumes* imbricate, concolorous, or midrib green. *Perigynia* usually obtuse, green or olive, sometimes pitchy, much compressed, nerves slender, sometimes confined to the base. *Nut* orbicular, lenticular, rarely 3-quetrous.—DISTRIB. Europe (Arctic), N. America.

C. Gibso'ni, Bab.; glumes ⅓ shorter than the more elongate acute fruit

Yorkshile, extinct. Its author suggests its being an abnormal *Goodeno'vii.*

** *Stigmas* 3. *Beak of perigynium* short or 0; mouth truncate, entire or obliquely notched (rather long in 34, *vaginata,* and 35, *capillaris*). (See also 27 *aquatilis,* 28 *Goodenovii,* 46 *extensa,* 55 *strigosa,* 56 *vesicaria,* subsp. *saxatilis,* 59 *paludosa,* and 60 *riparia.*)

† *Perigynia glabrous.*

29. **C. limo'sa,** *L.* ; stems filiform, leaves narrow glaucous, bracts short, female spikelets 1–2 drooping short, pedicels capillary, perigynia ellipsoid green equalling the broad cuspidate pale glumes, beak very short entire.

Spongy bogs, local, Caithness to Dorset and Hants; Ireland; fl. June.—*Rootstock* slender, creeping. *Stems* 6–12 in., 3-quetrous. *Leaves* as long, $\frac{1}{12}$ in. broad, recurved, margins scabrous. *Male spikelet* erect, slender, sometimes female at the top; *female* 1–3, $\frac{1}{3}$ in., about equalling the pedicels, cylindric or oblong; bracts scarcely sheathing. *Glumes* few, large, lax, midrib greenish ribbed, edges brown. *Perigynia* sub-3-gonous, cuspidate, glaucous. *Nut* oval, 3-gonous.—DISTRIB. Europe (Arctic) (excl. Greece, Turkey), N. and W. Asia, N. America.

C. LIMO'SA proper; stem scabrid above, leaves concave, edges rough. glumes with green midrib, perigynia compressed, strongly ribbed.—England, Scotland, Ireland.

Sub-sp. C. IRRIG'UA, *Hoppe;* stems almost smooth, leaves flat shorter scarcely glaucous, glumes larger, points elongate, midrib not green, perigynia more turgid.—Argyll to Dorset.

30. **C. rariflo'ra,** *Sm.* ; stems wiry, leaves narrow erect, bracts short, female spikelets 2–3 pendulous few-fld., perigynia pedicelled elliptic or obovoid green embraced by the obtuse shining glume, beak short entire.

Alpine bogs, Forfar, Aberdeen, Banff, Inverness, alt. 2,400–3,000 ft.; fl. June–July.—*Rootstock* creeping. *Stems* 3-gonous, 6–12 in., smooth, leafy below only. *Leaves* flat, $\frac{1}{12}$ in. *Male spikelet* 1, short, suberect; *female* $\frac{1}{3}$ in., about equalling the capillary pedicels, 6–8-fld. *Glumes* pale brown, membranous, concolorous, midrib indistinct. *Perigynia* brown, 3-gonous; smooth, obscurely nerved. *Nut* oblong, 3-gonous, dotted.—DISTRIB. N. Scandinavia (Arctic), Kamtschatka, N. America.

31. **C. glau'ca,** *Murr.* ; stems wiry, leaves narrow flat glaucous, female spikelets 4–6 suberect cylindric many-fld., perigynia sessile turgid equalling the short dark glumes, beak short terete deflexed entire.

Rocks, woods, and pastures, N. to Shetland; ascends to 2,000 ft. in Yorkshire; Ireland; Channel Islands; fl. June–July.—*Rootstock* creeping, stoloniferous. *Stems* few, $\frac{1}{2}$–2 ft., trigonous, smooth. *Leaves* erect or recurved, $\frac{1}{8}$–$\frac{1}{4}$ in. broad. *Male spikelets* several; *female* $\frac{1}{4}$–2 in., shortly pedicelled; bracts with short sheaths or 0. *Glumes* close-set, acute or obtuse, dark brown, midrib greenish, margins green or not. *Perigynia* not ribbed, obovoid or orbicular, rough.—DISTRIB. Europe, N. Africa, Siberia.

C. *glau'ca* proper (*C. recur'va,* Huds.); spikelets cylindric, glumes acute.— VAR. C. *Michelia'na,* Sm.; spikelets cylindric, glumes obtuse. Alpine.—

VAR. *C. stictocar'pa,* Sm.; spikelets short, glumes acute, fruit dotted. Alpine.

32. **C. palles'cens,** *L.* ; stems wiry, leaves hairy erect flat, female spikelets 2–3 close oblong, perigynia elliptic 3-gonous obtuse smooth shining green larger than the ovate cuspidate glumes, beak obscure entire.

Marshy copses and meadows, from Caithness southd.; ascends to 1,900 ft. in the Highlands; Ireland; Channel Islands; fl. June–July. *Rootstock* tufted, stolons 0. *Stems* 1–2 ft., 3-quetrous, slender, often leafy, scaberulous. *Leaves* ⅛–⅓ in. broad, green; hairs scattered, soft. *Male spikelet* erect, pale yellow-red; *female* ½ in., shortly pedicelled; bracts foliaceous, sheaths short. *Glumes* few, very pale brown, margins whitish, midrib green. *Nut* obovoid, 3-quetrous, dotted and striate.—DISTRIB. Europe (Arctic) (excl. Spain, Greece), Siberia, N. America.

33. **C. panice'a,** *L.* ; stems curved, leaves glaucous flat, bract-sheaths long, spikelets oblong, perigynia 3-gonous smooth dotted exceeding the ovate dark glumes, beak very short terete decurved obliquely notched.

Wet meadows, &c., N. to Shetland; ascends to 2,300 ft. in the Highlands; Ireland; Channel Islands; fl. June–July.—*Rootstock* tufted, creeping, stoloniferous. *Stems* 1–2 ft., leafy, 3-gonous, smooth. *Leaves* ⅒–½ in. broad, margins rough. *Male spikelets* 1–2, narrow, clavate; *female* 2–4, ⅓–¾ in., rather distant, inclined, oblong or cylindric, dark, lower rarely long-pedicelled; bracts variable. *Glumes* obtuse acute or cuspidate, midrib broad green rarely concolorous, edges pale. *Perigynia* ovoid, pale brown, nerves obsolete. *Nut* 3-quetrous, brown, dotted.—DISTRIB. Europe (Arctic), N. Asia, N. America.

34. **C. vagina'ta,** *Tausch;* stems curved, leaves recurved keeled, bract long, sheath loose, spikelets short lax-fld., perigynia 3-gonous turgid smooth longer than the ovate brown obtuse glumes, beak cylindric de-curved obliquely 2-fid. *C. sali'na,* Don; *C. Mielichofe'ri,* and *C. phæo-sta'chya,* Sm. not Schk.; *C. scot'ica,* Spr.; *C. sparsiflo'ra,* Steud.

Rocky mts. of Dumbarton to Sutherland, alt. 2,000–3,800 ft.; fl. July.— Habit, &c., of *C. panice'a,* but leaves more radical, broader, recurved, never glaucous; bracts shorter, sheaths looser; spikelets laxer-fld.; glumes paler, more obtuse; perigynia longer obovoid and decurved, beak longer, cylindric. *Nut* elliptic, 3-quetrous, beaked.—DISTRIB. Arctic Europe and America, Germany, Siberia.—Stigmas 2–4 (*Boott*).

35. **C. capillar'is,** *L.* ; stems short, leaves recurved flat, bracts large, sheath long, female spikelets few-flowered pendulous, perigynia pedicelled elliptic-lanceolate 3-gonous smooth shining much exceeding the pale hyaline obtuse glumes, beak slender entire.

Grassy mountain banks and rocks, local, York to Shetland; ascends to 2,700 ft. in the Highlands; fl. June–July.—*Rootstock* tufted. *Stems* 4–8 in., 3-gonous, smooth. *Leaves* chiefly radical, soft. *Spikelets* small; *male* very slender; *female* 2–3, ¼–½ in., shorter than the capillary pedicel; upper bracts with hyaline tips, lower much larger. *Glumes* scattered, embracing the fruit, caducous, margins broad pale, nerves brown. *Perigynia* slightly decurved, not ribbed. *Nut* elliptic-ovoid, 3-quetrous.—DISTRIB. Arctic N. and Alpine Europe, N. Asia, N. America.

36. **C. pen'dula,** *Huds.* ; stems tall leafy, leaves broad flat, bract-sheaths long, spikelets many very long slender drooping, perigynia ovoid turgid 3-gonous smooth membranous green much exceeding the awned glumes, beak short terete decurved smooth notched.

Damp woods, &c., from Lanark and Elgin southd.; local in Ireland; fl. May-June.—*Rootstock* tufted, stolons 0. *Stems* 3–6 ft., 3-gonous, smooth or scaberulous. *Leaves* pale-green, not glaucous beneath, $\frac{1}{4}$–$\frac{1}{2}$ in. broad. *Spikelets* 3–5 in., *males* 1–3, inclined, sometimes interruptedly male and fem., or fem. at the base only ; *females* subsessile, graceful, curved, obtuse, dense-fld., pedicels wholly included in the very leafy bracts. *Glumes* spreading, small, oblong, obtuse or 2-fid, pale brown, awn hispid, midrib green, margins pale ragged. *Perigynia* narrow at both ends, gibbous above. *Nut* short, broad, pale, 3-quetrous.—DISTRIB. From Belgium southd., N. Africa.

†† *Perigynia hairy. Spikelets short in all but 42, 43.*

37. **C. præ'cox,** *Jacq.* ; leaves short flat curved, bract-sheaths short, spikelets few close, perigynia broadly ellipsoid turgid 3-gonous hispid brown equalling the brown obtuse or cuspidate glumes, beak very short.

Moors, heaths, and pastures, N. to Shetland ; ascends to 2,300 ft. in the Highlands ; Ireland ; Channel Islands ; fl. April–May.—*Rootstock* creeping, stoloniferous. *Stems* 6–12 in., 3-gonous, quite naked above, smooth or scaberulous. *Leaves* $\frac{1}{4}$–$\frac{1}{3}$ in. broad, densely tufted, keeled. *Spikelets* crowded, subsessile ; *male* slender, erect ; *female* 2–3, $\frac{1}{4}$–$\frac{1}{2}$ in., inclined, oblong, rather dense-fld. *Glumes* small, broad, edges brown, midrib green ; bracts variable. *Perigynia* subsessile, coriaceous, olive, opaque. *Nut* 3-gonous, short, brown, crowned with a minute ring.—DISTRIB. Europe, N. Asia ; introd. in N. America.

38. **C. tomento'sa,** *L.* ; slender, leaves glaucous beneath, bract-sheaths 0, spikelets few short close, perigynia obovoid 3-gonous pubescent green exceeding the acute glumes, beak very short notched.

Wet meadows, N. Wilts ; fl. June.—*Rootstock* creeping, stoloniferous. *Stems* 10–18 in., 3-quetrous, scaberulous above. *Leaves* $\frac{1}{8}$ in. broad, curved, glabrous, flat. *Spikelets* subsessile ; *male* solitary, erect ; *female*, 1–2, inclined, $\frac{1}{4}$–$\frac{1}{2}$ in. ; bracts long or short. *Glumes* very small, ovate, sometimes cuspidate, midrib green, edges brown. *Perigynia* coriaceous, turgid, beak distinct slightly decurved. *Nut* 3-gonous, obovoid, pale.—DISTRIB. From the Baltic to France, and E. to Italy and Turkey.

39. **C. pilulif'era,** *L.* ; slender, leaves recurved, bract-sheath 0, spikelets few short few-fld. close, perigynia stipitate subglobose pubescent equalling the ovate cuspidate brown glumes, beak very short notched.

Heaths, woods, and moors, N. to Shetland ; ascends to 3,300 ft. in the Highlands ; Ireland ; Channel Islands ; fl. June.—*Rootstock* tufted, stolons 0. *Stems* 6–20 in., 3-gonous, scaberulous. *Leaves* $\frac{1}{8}$–$\frac{1}{6}$ in., chiefly radical, broad, short, flat, keeled. *Spikelets* sessile ; *male* slender ; *female* ⅛-½ in., sub-globose ; bracts hardly leafy. *Glumes* rather spreading, brown, midrib green, edges brown or narrowly pale. *Perigynia* opaque, hardly 3-gonous, pedicel flat. *Nut* subglobose, brown.—DISTRIB. Europe (Arctic), W. Asia, Kamts-chatka, N. America.—VAR. *Lee'sii,* Ridley (*C. saxum'bra,* Lees), is a drawn-out form growing in shaded places in Yorks.

40. **C. monta'na,** *L.* ; slender, leaves narrow, bract glumaceous, spike-lets few small close, perigynia obovoid 3-gonous hirsute ribbed exceeding the obovate mucronate glumes, beak notched. *C. colli'na,* Willd.

Heaths, woods, and fields, local; Worcester, Hereford, Monmouth, Gloster, Devon, Hants, Sussex ; fl. April–May.—*Rootstock* very stout, creeping and tufted. *Stems* 6–18 in., 3-quetrous, strict, then curved, scaberulous above. *Leaves* ₁₂-⅛ in. broad, usually shorter than the stem, straight or flexuous ; young pubescent. *Spikelets* sessile, usually very close, bright red-brown, polished ; *male* ½ in., solitary, stout ; *female* 1–3, ½-½ in., more slender, few-fld. ; bract broad ; membranous, awn subulate or green or 0, sheath very short. *Glumes* broad, chestnut-brown, embracing the spreading substipitate perigynia, obtuse or retuse and mucronate, midrib pale, edges brown. *Nut* pedicelled, pale.—DISTRIB. Europe, W. Siberia.

41. **C. ericeto'rum,** *Poll.* ; stems short curved, leaves recurved keeled, bract glumaceous, spikelets few capitate, perigynia obovoid pubescent equalling the broadly ovate obtuse ciliate brown glumes, beak entire. *C. cilia'ta,* Willd.

Chalk banks, Norfolk, Suffolk, Cambridge; fl. May–June.—*Rootstock* branched, creeping and tufted. *Stems* 2–6 in., 3-gonous, stiff, smooth. *Spikelets* sessile, small ; *male* ½ in., solitary, fusiform ; *female* ½-½ in., 6–10-fld. ; bracts hardly sheathing. *Glumes* pale brown, midrib concolorous ; margins broadly scarious. *Perigynia* subsessile, opaque, short. *Nut* obovoid, sessile, pale, 3-quetrous.—DISTRIB. Europe (Arctic), N. Asia.

42. **C. digita'ta,** *L.* ; stems slender curved, leaves recurved flat, bract subulate, sheath long brown, spikelets slender curved, perigynia remote stipitate narrow-obovoid 3-gonous pubescent embraced by the mucronate shining glumes, beak very short straight entire.

Copses on limestone, York, Derby, Notts, Hereford to Devon and Wilts; fl. May.—*Rootstock* very stout, tufted. *Stems* 6–10 in., smooth, obtusely 3-gonous. *Leaves* ⅛-⅛ in. broad, soft, flat, linear. *Spikelets* and their slender pedicels enclosed in the bracts ; *male* solitary, ⅓-¾ in., *female* 2–3, ½-1 in., 6–8-fld. ; bracts membranous, brown, tip green subulate. *Glumes* con-volute, broad, obtuse, scarious, pale red-brown, midrib narrow green, margins hyaline. *Perigynia* concealed, curved, green, beak terete. *Nut* stipitate, 3-quetrous, brown. —DISTRIB. Europe, Siberia.

C. DIGITA'TA proper ; bracts subulate, female spikelets close, perigynia equal-ling the glumes.

Sub-sp. C. ORNITHOPO'DA, Willd. ; bract ovate awned; female spikelets distant, perigynia longer than the glumes, beak 0.—Derby, York.

43. **C. hu'milis,** *Leyss.* ; stems shorter than the stiff involute curved leaves, bracts glumaceous, spikelets very remote slender lax-fld., perigynia stipitate narrow-obovoid 3-gonous pubescent embraced by the very broad scarious glumes, beak short entire or 0.   *C. clandesti'na,* Good.

Dry grassy hills, Hereford, Gloster, Wilts, Somerset, Dorset, Hants; fl. May. *Rootstock* very stout, creeping, tufted. *Stems* 1–3 in., 3-gonous, smooth, concealed amongst the narrow leaves. *Spikelets* white, scattered up the whole stem; *male* solitary, narrow, $\frac{1}{4}$–$\frac{1}{2}$ in.; *female* 3–5, $\frac{1}{8}$–$\frac{1}{4}$ in., enveloped in the obtuse apiculate or truncate scarious silvery bracts. *Glumes* silvery, clouded with pink and green. *Perigynia* greenish, 1-ribbed on 2 faces. *Nut* 3-quetrous, pale brown, smooth, beaked.—DISTRIB. From Belgium southd., Siberia.

*** *Stigmas* 3 (2–3 in 59, *paludosa*). *Fruit* with a long, usually slender, often forked beak (beak short in 46 *extensa*, 55 *strigosa*, 56 *vesicaria* sub-sp. *saxatilis*, 59 *paludosa*, and 60 *riparia ;* see also 34 **vaginata** and 35 **capillaris**).

† *Perigynia hairy, nerved. Male spikelets several.*

44. **C. hir'ta,** *L.* ; leaves long flat hairy, bracts leafy, sheaths long, spikelets long-pedicelled erect cylindric, perigynia ovoid turgid ribbed much longer than the ovate awned glumes, beak 2-cuspidate.

Damp copses, &c., from Inverness southd.; ascends to 1,200 ft. in N. England; Ireland ; Channel Islands ; fl. May–June.—*Rootstock* long, creeping, jointed, scaly. *Stems* slender, 1–2 ft., leafy, glabrous, 3-gonous, shining. *Leaves* $\frac{1}{4}$–$\frac{1}{2}$ in. broad, sheaths split. *Spikelets : male* 2–3, close, $\frac{1}{2}$–1 in., pale, glistening ; *female* 2–5, 1–1$\frac{1}{2}$ in., green, distant ; pedicel exserted ; lower bract exceeding the stem. *Glumes* small, broad, scarious, midrib green ; awn rigid scabrid often spreading ; of male lanceolate, hyaline. *Perigynia* large, $\frac{1}{3}$ in., sessile, plano-convex, narrowed into the beak. *Nut* 3-quetrous, beaked, pale brown. —DISTRIB. Europe, N. Africa, N. Asia.—VAR. *hirtæfor'mis,* Pers., with glabrous leaves and glumes ; and VAR. *ebractea'ta,* Syme (found near Epsom), with glabrous obtuse glumes and leaves and bracts reduced to a subulate point, are abnormal forms.

45. **C. filifor'mis,** *L.* ; slender, leaves very long, margins involute, bracts long, sheaths short, spikelets erect, perigynia narrow-ovoid turgid pubescent equalling the lanceolate acuminate glumes, beak 2-cuspidate.

Bogs and marshes, local, Sutherland to Devon and Hants ; Ireland; fl. May. —*Rootstock* creeping, stoloniferous. *Stems* 2–3 ft., slightly scabrid above, 3-gonous, leafy. *Leaves* numerous, very slender, stiff ; sheaths red-brown, stout, lower 2–3 very long, edges filamentous, *without* a blade. *Male spikelets* 2–3, very slender, 1–2$\frac{1}{2}$ in., brown ; *female* 1–3, $\frac{3}{4}$–1 in., rather lax-fld.; lower bracts filiform, overtopping the stem. *Glumes* chestnut-brown, margins concolorous ; midrib green, of male fl. narrower and more membranous. *Perigynia* $\frac{1}{6}$ in., plano-convex, green, narrowed into the beak. *Nut* stipitate. 3-quetrous.—DISTRIB. Europe (Arctic), Siberia, N. America.

†† *Perigynia glabrous. Male spikelets solitary; female short, distant; bracts leafy, sheaths long (except* 46 extensa *and* 47 flava).

46. **C. exten'sa,** *Good.* ; slender, leaves setaceous, bracts very long, sheaths short, spikelets subsessile dark, perigynia rhombic or elliptic-obovate 3-gonous inflated ribbed much larger than the broad acute or mucronate glumes, beak short straight smooth 2-toothed.

Brackish marshes, from Orkney southd.; Ireland; Channel Islands; fl. June. —*Rootstock* tufted. *Stems* 10–18 in., 3-gonous, smooth, slender. *Leaves* mostly below the middle, $\frac{1}{12}$–$\frac{1}{12}$ in. broad, rigid, flexuous, shorter than the stem; margins involute. *Spikelets* 3–4, subsessile; *male* $\frac{1}{2}$–$\frac{3}{4}$ in., slender, brown; *female* 2–4, $\frac{1}{3}$–$\frac{3}{4}$ in., 2 upper or all contiguous; bracts at length horizontal, sheath variable. *Glumes* small, brown, midrib green, edges concolorous. *Perigynia* rather spreading, opaque, coriaceous, dull green, dotted, gradually narrowed into the beak. *Nut* shortly elliptic, olive-brown, 3-quetrous.—DISTRIB. Europe, N. Africa, W. Asia, N. and S. America.

47. **C. fla'va,** *L.* ; stems curved, leaves flat, bracts long, sheaths short, spikelets oblong green, perigynia spreading or deflexed ovoid 3-gonous inflated ribbed much exceeding the obtuse or subacute glumes, beak long slender scabrid 2-cuspidate.

Heaths, bogs, and marshes, from Orkney southd.; ascends to 2,900 ft. in the Highlands; Ireland; Channel Islands; fl. May–June.—*Rootstock* tufted, stolons 0. *Stems* 3–18 in., angles rather acute. *Leaves* $\frac{1}{8}$ to nearly $\frac{1}{4}$ in. broad, chiefly radical, often recurved, shorter than the stem, usually flat. *Spikelets* usually contiguous, sessile or pedicelled; *male* $\frac{3}{4}$–1 in., fusiform; *female* 2–4, $\frac{1}{4}$–$\frac{1}{2}$ in., subcylindric, squarrose; bracts spreading, sheaths variable. *Glumes* small, usually green clouded with brown. *Perigynia* sessile, variable in size, green, coriaceous, gradually narrowed into the beak. *Nut* short, 3-gonous, olive-brown, angles acute.—DISTRIB. Europe (Arctic), Madeira, W. Asia, India, N. America.—In the following sub-species the spikelets are so variable that I have not introduced them.

C. fla'va proper ; beak of perigynium decurved scabrid.—VAR. *C. lepidocar'pa,* Tausch.; perigynium small, beak short nearly straight or abruptly deflexed almost smooth.

Sub-sp. C. ŒDE'RI, *Ehrh.*; usually smaller, beak of smaller perigynium straighter scabrid.

48. **C. dis'tans,** *L.* ; slender, leaves flat, bracts leafy, sheaths long, spikelets distant lower pedicelled, perigynia suberect ovoid turgid ribbed exceeding the ovate glumes, beak slender flat scabrid 2-cuspidate.

Marshes and wet meadows, from Caithness southd.; ascends to 1,900 ft. in the Highlands; Ireland; Channel Islands; fl. June.—*Rootstock* tufted, creeping. *Stems* 10–18 in., 3-gonous, smooth or slightly scabrid above, leafy below the middle. *Leaves* $\frac{1}{10}$–$\frac{1}{6}$ in. broad, glaucous, often recurved. *Spikelets : male* $\frac{3}{4}$–1 in., long-pedicelled, slender; *female* 1–3, $\frac{1}{2}$–1 in., very distant, erect, cylindric-oblong; lower bract not equalling the stem, and often not its spikelet. *Glumes* broad, subacute, brown, midrib green, margins and tip white. *Perigynia* sessile, green, opaque, gradually narrowed into the

slender flattened short straight beak, mouth membranous. *Fruit* broad,
obovoid, 3-quetrous, brown.—DISTRIB. Europe, N. America?

C. DIS'TANS proper ; leaves darker, bracts usually longer, female spikelets 3–4
longer, glumes more obtuse mucronate or awned less hyaline at the tip,
perigynia broader black-dotted.—Brackish marshes, rare inland ; extends to
N. Africa.

Sub-sp. C. FUL'VA, *Good.* ; bracts rarely equalling the stem, female spikelets 2–3,
glumes obtuse or acute not mucronate, tips more or less hyaline, perigynia
broad at the base not dotted. *C. Hornschuchia'na,* Hoppe ; *C. speirosta'chya,*
Sm. Chiefly subalpine.—I have again gone into the question of *C. ful'va,*
Good., this time with Mr. Baker, and we do not doubt this being the plant
figured by Goodenough, and afterwards confounded with something else.—
C. xanthocar'pa, Degl., is a sterile form, or a hybrid with *C. dis'tans.*

49. C. biner'vis, *Sm.* ; slender, leaves rigid recurved flat, bracts leafy,
sheaths long, spikelets all pedicelled cylindric, perigynia ovoid 3-gonous
faintly ribbed green much exceeding the oblong-ovate glumes, beak flattened 2-cuspidate scabrid.

Heaths, moors, &c., N. to Shetland ; ascends to 3,200 ft. in the Highlands ;
Ireland ; fl. June–July.—Similar to *C. dis'tans,* and perhaps another sub-sp.,
but much coarser, often 3 ft. ; rootstock stout, tufted, creeping ; stems
3-quetrous, smooth, leafy ; leaves often short, more rigid, keeled and recurved ; female spikelets ½–1 in., browner, more numerous, stouter, longer ;
beak stouter and more 2-fid ; glumes obtuse or mucronulate, red-brown ;
perigynia and leaves beneath at times dotted ; I do not find the nut always
obovoid, but like *ful'va.*—DISTRIB. W. Europe, N. Africa, W. Asia.

50. C. læviga'ta, *Sm.* ; tall, leaves short broad, spikelets distant
pedicelled inclined or drooping cylindric, perigynia subsessile elliptic-
oblong or ovoid turgid ribbed green dotted longer than the acuminate
glumes, beak long slender 2-cuspidate.

Wet copses and marshes, from Mull and Aberdeen southd. ; ascends to 1,000
ft. in Northumbd. ; Ireland ; fl. June.—Closely allied to *C. biner'vis,* but
much larger, 1–3 ft. ; leaves almost ⅓ in. broad, dotted beneath ; sheath
auricled opposite the blade ; male spikelets 1–2¼ in., 3-gonous, rarely 2,
with obtuse mucronate glumes ; female 1½–2 in., drooping ; perigynia
nearly ¼ in., 3-gonous, opaque, purple-dotted, beak obscurely scabrid, cusps
long nearly straight. *Nut* stipitate, 3-quetrous, dotted, pale.—DISTRIB.
From Holland southd., N. Africa, N. America.

51. C. puncta'ta, *Gaud.* ; slender, leaves short flat subrecurved, bracts
leafy, sheaths long, perigynia spreading ovoid membranous turgid dotted
not ribbed shining longer than the ovate glumes, beak slender 2-fid
smooth.

Marshy places near the sea, Suffolk, Hants to Cornwall, Wales, Kirkcudbright ;
S. Ireland ; Guernsey ; fl. June.—Similar to *C. dis'tans,* but very distinct in
the tumid shining membranous hardly 3-gonous perigynia, ribbed at the
2 obscure angles only, and the more slender subterete beak. *Glumes* obtuse
or mucronate, mucro scabrid.—The nut does not differ materially from
ful'va—DISTRIB. Norway, Friesland, France, Italy, Switzerland, Azores.

52. **C. frig'ida,** *Allioni ;* slender, leaves long flat, bracts leafy, sheaths long, spikelets red-brown shortly cylindric long pedicelled, perigynia hardly stipitate narrow lanceolate exceeding the oblong acute glumes, beak long slender, margins scabrid or smooth.

Wet turf, Aberdeen, alt. 2,700 ft.; fl. Aug.—*Rootstock* tufted, shortly creeping. *Stems* 6–18 in., 3-gonous, slender, with 1–2 leaves. *Radical leaves* numerous, flat, ⅛–¼ in. broad, green. *Bracts* long, leafy, shorter or longer than the inflorescence, smooth. *Spikelets* inclined, rich dark brown, remote ; *male* solitary, slender, fusiform ; *female* 3–5, ¾–1½ in., long-pedicelled. *Glumes* numerous, erect, loosely appressed, dark purple with a green midrib, of the male subacute. *Perigynia* ⅛–¼ in., dark brown, not ribbed, smooth ; beak straight, half as long as the body. *Stigmas* 3.—DISTRIB. Alps of Central and S. Europe, Central Asia.—A rare instance of an alpine Scotch plant not being Scandinavian.

53. **C. depaupera'ta,** *Good.* ; slender, leaves long flat, bracts leafy, sheaths long, spikelets 3–4-fld. green, perigynia stipitate narrowed at both ends turgid ribbed green much larger than the lanceolate pale glumes, beak slender tip membranous.

Dry woods, Surrey, Somerset, and Kent, very rare ; fl. May–June.—*Rootstock* tufted, shortly creeping. *Stems* 10–24 in., 3-gonous, slender, smooth, leafy. *Leaves* ⅛–½ in. broad. *Bracts* scabrous, almost equalling the stem, flat. *Spikelets* erect, very short and distant ; *male* very slender, 1–1¼ in., many-fld. ; *female* 3–5, ½–⅔ in., pedicel of lowest sometimes 2–3 in. *Glumes* remote, membranous, acuminate, embracing the base of the perigynia, back green ; of the male numerous, obtuse, ciliate. *Perigynia* large, nearly ⅓ in., obscurely 3-gonous, polished, ribs many slender ; beak very long, straight, obscurely scabrid. *Fruit* obovoid, or obtusely 3-gonous, pale.—DISTRIB. From France and Spain eastd. to Russia and Turkey, N. Asia.

††† *Perigynia glabrous. Male spikelet solitary ; female long, curved or drooping. Bracts leafy, sheaths long.*

54. **C. sylvat'ica,** *Huds.* ; slender, leaves flat broad soft, spikelets long-pedicelled very slender drooping lax-fld., perigynia stipitate short turgid 3-gonous equalling the lanceolate thin pale glumes, beak membranous very long slender 2-fid.

Damp woods, from Argyll and Aberdeen southd. ; ascends to 2,000 ft. in the Highlands ; Ireland ; fl. May–June.—*Rootstock* tufted, shortly creeping. *Stems* 1–3 ft.. 3-quetrous, smooth. *Leaves* ⅓–½ in., membranous, bright green. *Bracts* leafy, sheaths long, *spikelets* 1–1½ in., distant or contiguous : *male* pale ; *female* sometimes branched at the base. *Glumes* lax or scattered, acuminate, membranous, with pale green back and keel ; of the males brownish, obtuse, mucronate or subacute. *Perigynia* subsessile, elliptic, rather small, brown, straight, rather shorter than the slightly compressed herbaceous smooth beak, angles ribbed, faces quite smooth. *Nut* broad 3-quetrous.—DISTRIB. Europe, N. Asia.

55. **C. strigo'sa,** *Huds.* ; leaves flat broad, spikelets very slender drooping, perigynia remote elliptic-lanceolate 3-gonous ribbed a little

longer than the oblong-lanceolate acute glumes, beak obscure obliquely truncate smooth hyaline.

Woods and copses, from York and Chester to Kent and Somerset; local in Ireland; fl. May–June.—*Rootstock* tufted, creeping. *Stems* 1–2 ft., smooth, 3-gonous, leafy. *Leaves* $\frac{1}{4}$–$\frac{1}{2}$ in. broad, usually short, pale green, flaccid; sheaths auricled opposite the blade. *Male spikelet* 1, $\frac{1}{2}$–2 in., slender, pale; *female* 1$\frac{1}{2}$–3 in., lowest long-pedicelled, often flexuous; rachis exposed between the glumes, many-fld.; bracts leafy, usually shorter than the stems, sheaths long. *Glumes* small, membranous, white, base enveloping the perigynia, centre pale green. *Perigynia* $\frac{1}{8}$ in., membranous, green, slightly decurved, narrowed at both ends. *Nut* narrow-elliptic, 3-gonous, pale.—DISTRIB. N. and Mid. Europe, Italy, W. Asia, N. Africa.

†††† *Perigynia glabrous. Male spikelets several (rarely one in 56, vesicaria); female stout, usually curved and drooping. Bracts leafy, sheaths 0.*

56. **C. vesica'ria,** *L.* ; stout, leaves flat, spikelets cylindric, perigynia large spreading conic-ovoid inflated ribbed pale exceeding the lanceolate subacute scarious-tipped glumes, beak stout 2-cuspidate smooth.

Bogs and marshes, from Inverness and Perth southd.; Ireland; fl. May–June. —*Rootstock* tufted and creeping. *Stems* 1–2 ft., scabrid, 3-quetrous. *Leaves* long, $\frac{1}{6}$–$\frac{1}{4}$ in. broad, soft, sheath-edges filamentous. *Spikelets* many, stout; *male* 1–3, 1$\frac{1}{2}$–2$\frac{1}{2}$ in., slender, pale brown, sometimes female at the top; *female* 1–3 in., shortly pedicelled, inclined or drooping, $\frac{1}{2}$ in. diam. when ripe; bracts overtopping the stem, sheath 0. *Glumes* much smaller than the perigynia, narrow, chestnut-brown, midrib pale, tip obtuse ; of the males linear-oblong. *Perigynia* nearly $\frac{1}{3}$ in., obscurely 3-gonous, nerves faint, dull yellow, shining; beak rigid, brown, pungent. *Nut* broadly elliptic, 3-gonous, pale, beak long.—DISTRIB. Europe (Arctic), N. Africa, N. and W. Asia, N.W. India, Greenland.

C. VESICA'RIA proper ; stems 1–2 ft., male spikelets 2–3, female large 1–2$\frac{1}{2}$ in. stout cylindric pale, perigynia elongate ribbed with a slender rigid 2-cuspidate beak, stigmas 3.—VAR. *C. involu'ta,* Bab., with margins of leaves involute, apiculate glumes and narrower perigynia, is intermediate between this and *C. ampulla'cea*.—Hale Moss, Manchester.

Sub-sp. C. GRAHA'MI, *Boott ;* perigynia brown less strongly nerved, beak shorter less strongly cuspidate more slender, stigmas 2.—Lofty mts. Perth, Forfar.

Sub-sp. C. SAXAT'ILIS, *L.*; stems 4–10 in., male spikelet 1 rarely 2, female small $\frac{1}{4}$–$\frac{1}{2}$ in. ovoid very dark, perigynia ovoid ribs faint or 0 with a short notched or 2-fid beak, stigmas usually 2. *C. pul'la,* Good., *C. vesica'ria* var. *alpiy'ena*, Fries.—Scotch alps from Ben'Lomond northd., alt. 2,500 to 3,300 ft.—DISTRIB. Scandinavia, N. Russia, N. America.—This and *C. dichroa*, Anderss., are alpine forms of *C. vesica'ria,* to which var. *Graha'mi* forms a passage. I accept Boott's authority for its being the true *saxat'ilis* of Linnæus, from the testimony of his Herb., and of his pupil Solander (in *Herb. Banks*). The Swedish authorities, however, refer *saxatilis* to *rig'ida.*

57. **C. ampulla'cea,** *Good.* ; stout, leaves glaucous, margins involute, female spikelets stout cylindric pale, perigynia spreading ovoid inflated ribbed

exceeding the obovate-lanceolate scarious-tipped glumes, beak very slender 2-cuspidate smooth. *C. rostra'ta,* Stokes (an earlier name).

Marshes and bogs, N. to Shetland; ascends to 2,700 ft. in the Highlands; Ireland; fl. June.—*Rootstock* tufted and creeping. *Stem* 1-2 ft., 3-gonous. smooth, angles obtuse. *Male spikelets* 2-3, slender, sometimes fem. at the base; *female* 2-4, 1-2¼ in., sessile or pedicelled, inclined, often squarrose. dense-fld.; bracts overtopping the stem, sheath 0. *Glumes* much as in *C. vesica'ria,* and perigynia similar and shining, but much smaller, yellow-brown, ⅛-⅙ in., as broad as long, 3-quetrous, horizontal or deflexed when ripe, abruptly contracted into the long beak. *Nut* obovoid, yellow.— DISTRIB. Europe (Arctic), Siberia, Himalaya, N. America.

58. **C. Pseu'docype'rus,** *L.*; stout, leaves broad, male spikelets solitary, female very long-pedicelled drooping cylindric, perigynia stipitate spreading elliptic-lanceolate ribbed green equalling the narrow awned glumes, beak very long smooth 2-cuspidate.

Banks of rivers, lakes, &c., from Elgin and Isla to Kent and Devon; rare in Ireland; fl. June.—*Rootstock* tufted. *Stems* 1-3 ft., 3-quetrous, scabrid. *Leaves* ¼-½ in., flat, scabrid. *Spikelets* towards the top of the stem; *male* slender, 2-3 in., pale, often female at the top; *female* 4-5, 2-3 in., subsquarrose, pale green; pedicels capillary, curved, sometimes 4 in.; bracts overtopping the stem, very broad, sheath 0. *Glumes* small, lanceolate or subulate, awn serrate. *Perigynia* ⅓ in., slender, horizontal or deflexed, 3-gonous, shining, ribs many close; beak rigid, pungent, deeply split. *Fruit* pale, 3-gonous; style persistent.—DISTRIB. Europe, temp. N. and S. Africa, Asia, America, Australia.

59. **C. paludo'sa,** *Good.*; tall, stout, leaves broad glaucous, male spikelets stout, anthers mucronate, female erect dark cylindric, perigynia ovoid ribbed exceeding the lanceolate awned glumes, beak short 2-toothed. *C. acu'ta,* Curt., not L.

River-banks and ditches, from Ross southd.; ascends to 1,200 ft. in Northumbd.; Ireland, local; fl. May-June.—*Rootstock* creeping, stoloniferous. *Stems* 2-3 ft., 3-quetrous, scabrid. *Leaves* erect, ¼-½ in., flat, sheath sometimes leafless, edges filamentous. *Bracts* erect, long, broad. *Spikelets* rather close; *male* 2-3, 1½-2 in., dark brown, base sometimes fem.; *female* shortly pedicelled. *Glumes* of male fl. obtuse, brown with hyaline tips, upper cuspidate; of fem. cuspidate. *Perigynia* 3-gonous, gradually narrowed into the beak, angles acute, upper part granulate. *Stigmas* 2 or 3. *Fruit* lenticular or 3-quetrous.—DISTRIB. Europe, N. Africa, W. Siberia, N.W. India, N. America.—VAR. *C. spadice'a,* Roth., has the female glumes with a serrulate awn.

60. **C. ripa'ria,** *Curtis;* tall, stout, leaves very broad, male spikelets stout, anthers cuspidate, female long stout cylindric, perigynia ovoid ribbed a little exceeding the ovate-oblong cuspidate glumes, beak short 2-toothed.

River-banks and ditches, from Banff and Dumbarton southd.; local in Ireland; Channel Islands; fl. May.—Much the largest British species, 3-5

H H

ft. *Rootstock* creeping and tufted. *Stem* 2–5 ft., 3-quetrous, scabrid. *Leaves* flat, ½ in. broad and upwards, sheath-edges filamentous. *Bract* broad, overtopping the stem. *Spikelets* very large; *male 3–6,* crowded, 1½–2 in. diam., dark brown, acute, sometimes fem, at the base, mucro of anthers longer than in *C. paludo'sa; female 4–6,* 2–3 in., pedicelled, inclined, sometimes compound at the base or male at the top. *Glumes* narrow, margins brown, midrib green, tip scabrid ; of male slender, acute. *Perigynia* ⅓ in., erecto-patent, dull green, narrowed into the beak; ribs many, close. *Fruit* elliptic, 3-quetrous, yellow.—DISTRIB. Europe, N. Africa, Siberia, N. and S. America.—This and the preceding are allied to *C. aquat'ilis.*

ORDER LXXXIX. **GRAMIN'EÆ.**

Herbs, usually tufted and slender. *Stem* cylindric or compressed, jointed, internodes usually hollow. *Leaves* alternate, narrow ; sheath split to the base, with often a transverse membrane (ligule) or ring of hairs at its mouth. *Spikelets* in terminal spikes, racemes, or panicles, usually composed of one pair of flowerless (empty) glumes enclosing or subtending one or more sessile or stalked normally flower-bearing (but sometimes also empty) glumes, which are distichously arranged on a slender rachis (*rachilla*). *Flowering glumes* boat-shaped, enclosing a 1–2-sexual flower, and a flat often 2-nerved scale (*palea*) with inflexed edges. *Perianth* of 2 (rarely 0 or 3 or more) minute scales, placed opposite the palea. *Stamens* 3 (rarely 1, 2, 6, or more), filaments capillary ; anthers 2-celled, versatile, pendulous. *Ovary* 1-celled, style long short or 0, stigmas usually 2 long or short feathery ; ovule 1, basal, erect, anatropous. *Fruit* a membranous utricle, often adherent to the palea, and sometimes to the flowering glume. *Seed* usually adnate to the pericarp, testa membranous, albumen hard floury ; embryo small, outside the base of the albumen, cotyledon reduced to a sheath enclosing the plumule, radicle conical below, obliquely dilated above into a broad scutellum which extends upwards and backwards beyond the cotyledon with its back against the albumen.—DISTRIB. All climates ; genera 300 ; species about 3,200.—AFFINITIES obscure.—PROPERTIES. Nutritious herbage, and farinaceous seeds ; stem and leaves used for textile and other purposes.

The tribes and genera of Grasses are most difficult of classification and definition. Many systems have been proposed. The primary divisions of Fries, adopted in earlier editions of this work, namely *Clisantheæ* (styles long, stigmas slender with simple hairs protruded at the top of the glume) and *Euryantheæ* (style short, stigmas feathery protruded from the sides of the glume), has broken down under Bentham's searching revision of the Order (see *Journ. Linn. Soc.* xix. 14, and *Gen. Plant.* iii. 1074). I have followed Bentham's classification.

SERIES A. PANICA'CEÆ. *Spikelets* jointed upon the pedicel below the lowest glumes; rachilla not jointed at the base nor produced beyond the uppermost glume, persistent. *Glumes* 4 or fewer, the terminal only bearing a 2-sexual flower, rarely a lower one bears an imperfect flower.

TRIBE I. **PANICE'Æ.** *Spikelets* dorsally compressed (laterally in *Spartina*). *Fl. glume* 3- or more-nerved, not awned, hardening round the fruit. (See 11, *Mibora* in Agrostideæ.)

Glumes 4. Pedicels of spikelets naked or hairy ....1. Panicum.
Glumes 4. Pedicel of spikelets with stiff bristles ....2. Setaria.
Glumes 3. Style long. Spikelets laterally compressed ......3. Spartina.

TRIBE II. **ORYZE'Æ.** *Spikelets* laterally compressed. *Empty glumes* minute or 0. *Fl. glume* and palea 1-nerved or keeled..........4. Leersia.

SERIES B. POA'CEÆ. *Spikelets* rarely jointed upon the pedicel below the lowest glume (except *Alopecurus, Polypogon,* and *Holcus*); rachilla jointed at the base above the (usually 2) lowest flowerless glumes, often produced beyond the uppermost glume, and deciduous. *Glumes* 3 or more, 2 lowest flowerless and often persistent (*Nardus* has no empty glumes; and all but the uppermost spikelet in *Lolium* have only one lowest empty glume).

TRIBE III. **PHALARIDE Æ.** *Spikelets* laterally compressed; rachilla not produced beyond the uppermost glume. *Glumes* 6, uppermost only with a 2-sexual flower. *Palea* 0 or in the perfect flowers 1-nerved.

Glumes 6, 3d and 4th imperfect, not awned. Stamens 3.......5. Phalaris.
Glumes 6, 3d and 4th empty, awned. Stamens 2........6. Anthoxanthum.
Glumes 6, 3d and 4th awned, triandrous; 5th diandrous....7. Hierochloe.
Glumes 3 or 4. Spikelets jointed on the pedicel..............8. Alopecurus.

TRIBE IV. **AGROSTIDE'Æ.** *Spikelets* terete or laterally compressed; rachilla produced or not beyond the fl. glume. *Glumes* 3, flower solitary 2-sexual; palea 2-nerved.

* Rachilla not produced beyond the fl. glume.

Panicle effuse. Fl. glume hardening round the fruit ....9. Milium.
Panicle dense, cylindric. Fl. glume enclosing the fruit.........10. Phleum.
Spikelets dorsally compressed in a simple subdistichous spike.
11. Mibora.
Panicle loose. Fl. glume small hyaline............................12. Agrostis.
Panicle contracted. Empty glumes awned ....13. Polypogon.
Rachilla with long silky hairs ....14. Calamagrostis.

** Rachilla produced beyond the fl. glume.

Empty glumes large, boat-shaped; fl. glume minute, 4-toothed
15. Gastridium.
Empty glumes large; fl. glume 2-fid, awned ....16. Apera.
Empty glumes large; fl. glumes awned, rachilla ciliate......17. Deyeuxia.
Spikelets large, rachilla long silky..............................18. Ammophila.
Empty glumes plumose; fl. glume 3-awned...... ....19. Lagurus.

TRIBE V. **AVENE'Æ.** *Spikelets* panicled, terete or laterally compressed, 2- rarely 3–4-fld.; rachilla produced beyond the fl. glume (except in some *Airæ*). *Glumes* 4 or more, 2 lowest empty usually larger than the others, 2 or more upper flowering with a dorsal bent and twisted (except some *Airæ* and *Deschampsiæ*) awn.

* *Rachilla not produced beyond the uppermost fl. glume* ....20. Aira.
** *Rachilla produced beyond the uppermost fl. glume; flowers all perfect or the upper imperfect.*

Flowers 2; awn bent in the middle, tip clubbed ....21. Corynephora.
Flowers 2; awn straight ....22. Deschampsia.
Flowers 2; lower not awned, upper imperfect awned ....23. Holcus.
Flowers 2–6; fl. glume deeply 2-fid and awned ....24. Trisetum.
Flowers 2–6; fl. glume entire or 2-toothed, long-awned ....25. Avena.

*** *Rachilla produced beyond the fl. glumes. Flowers 2, upper 2-sexual, lower male* ....26. Arrhenatherum.

TRIBE VI. **CHLORIDE'Æ.** *Spikelets* 1–2-seriate on a flattened rachis, laterally compressed, 1- or more-fld ....27. Cynodon.

TRIBE VII. **FESTUCE'Æ.** *Spikelets* panicled or subspicate, terete or laterally compressed; rachilla usually produced beyond the fl. glume. often bearing a rudimentary glume. *Glumes* 6 or more, 2 lowest empty, the others flowering; awn terminal or 0.

SUB-TRIBE 1. **Triodieæ.** *Spikelets* 2- or more-fld.; rachilla not bearded. *Fl. glumes* 1–3-nerved, broadly 3-toothed ....28. Triodia.

SUB-TRIBE 2. **Arundineæ.** *Spikelets* 2- or more-fld.; rachilla bearded with long silky hairs ....29. Phragmitis.

SUB-TRIBE 3. **Seslerieæ.** *Spikelets subspicate or capitate, with empty glumes (imperfect spikelets) on the pedicels below them.*
Pedicels with soft glumes below the spikelets ....30. Sesleria.
Pedicels with bristle-like glumes below the spikelets ....31. Cynosurus.

SUB-TRIBE 4. **Eragrosteæ.** *Spikelets* 2- or more-fld. *Fl. glumes* 1- or 3-nerved.
Spikelets in a subcylindric spike. Fl. glumes scarious ....32. Kœleria.
Spikelets in a contracted panicle, conical, terete ....33. Molinia.
Spikelets in an effuse panicle with whorled branches ....34. Catabrosa.

SUB-TRIBE 5. **Miliceæ.** *Spikelets* 2- or more-fld. *Fl. glumes* 3–5-nerved, upper empty, convolute and forming a club ....35. Melica.

SUB-TRIBE 6. **Eufestuceæ.** *Spikelets* 3–many-fld. *Fl. glumes* 5- or more-nerved.

Spikelets few-fld., clustered in a secund panicle ....36. Dactylis.
Spikelets panicled, pendulous. Glumes broad, scarious, obtuse.
                                                                              37. Briza.
Spikelets panicled. Fl. glumes compressed, keeled, tips nerved
awnless.   38. Poa.

Spikelets very many-fld. Fl. glumes convex, obtuse, tip
  nerveless, awnless ....39. Glyceria.
Spikelets panicled or spicate. Fl. glumes convex, tip nerved
  acute or awned. Ovary glabrous...40. Festuca.
Spikelets panicled, many-fld Fl. glumes convex. Ovary tip villous.
                                          41. Bromus.
Spikelets racemed or spicate. Fl. glumes convex. Ovary tip villous.
                                          42. Brachypodium.

TRIBE VIII. **HORDEE'Æ.** *Spikelets* 1- or more-fld., sessile in the
  notches of a simple rachis; rachilla produced beyond the uppermost fl.
  glume (except in *Nardus*). *Fl. glume* with a terminal awn or 0.

SUB-TRIBE 1. **Triticeæ.** *Spikelets solitary in the notches, 3- or more-fld.*
Spikelets many-fld., with their sides to the rachis ....43. Lolium.
Spikelets many-fld., with their faces to the rachis ....44. Agropyrum.

SUB-TRIBE 2. **Leptureæ.** *Spikes solitary in the notches, 1-2-fld.*
Empty glumes 1-3. Spikelets distichous ....45. Lepturus.
Empty glumes 0. Spikelets secund. Style long.................46. Nardus.

SUB-TRIBE 3. **Elymeæ.** *Spikelets 2 or more in each notch, collateral ;
  the central perfect ; the lateral perfect or reduced to bristles.*
Spikelets 1-fld.. ......47. Hordeum.
Spikelets 2-6-fld..48. Elymus.

1. PA'NICUM, *L.*

Spikelets without bristles at the base, or with slender hairs only, spiked
racemed or panicled, 1-fld., or if 2-fld. the lower male. *Empty glumes* 2
in the 2-fld. spikelets, 3 in the 1-fld. ; lowest small or minute, 2d larger,
strongly nerved. *Fl. glume* nerved, hardening and enclosing the palea
and fruit. *Scales* 2, fleshy, truncate. *Stamens* 3. *Ovary* glabrous ; stigmas
penicillate, shorter than the styles. *Fruit* compressed or plano-convex.
—DISTRIB. Chiefly tropical ; species 270.—ETYM. The Latin name.

SECTION 1. **Digita'ria**, Scop. (gen.). *Spikelets* unilateral, on digitate
spikes. *Flowering glume* not awned.

P. gla'brum, *Gaud.* ; spikes about 3, spikelets 1-fld. *Digita'ria
humifu'sa*, Pers. ; *D. filifor'mis*, Koch.
Sandy soil in the S.E. counties, from Norfolk to Hants, local ; fl. July-Aug.—
  Annual. *Root* fibrous. *Stems* 6-12 in., prostrate or decumbent. *Lèaves*
  narrow, flat ; sheaths flat, mouth hairy ; ligule short. *Spikes* 2-3 in.,
  flexuous, channelled on the face. *Spikelets* $\frac{1}{12}$ in., in pedicelled pairs, plano-
  convex, elliptic, purplish ; empty glumes hairy, lower minute appressed or
  0.—DISTRIB. Most warm climates.

SECTION 2. **Echinochlo'a**, *Beauv.* (gen.). *Spikelets* in racemes or
panicles. *Flowering glume* awned or pointed.

P. CRUS-GAL'LI, *L.* ; spikelets panicled 1-fld.

Fields and waste places in S.E. England; naturalized; fl. July.—Annual.
Stems 1–4 ft., stout, ascending. *Leaves* ½ in. diam., flat, glabrous, edges
rough, often waved; ligule 0. *Panicle* 3–6 in.; branches subunilateral;
rachis 3-quetrous, pubescent; pedicel hairy. *Spikelets* ⅓ in., plano-convex,
greenish; upper empty glume hispid, pointed or rigidly awned; fl. glume
polished.—DISTRIB. All temp. and trop. regions.

2. SETA'RIA, *Beauv.*

Spikelets in a dense cylindric spike-like panicle, as in *Panicum*, but awnless, and with stout rough bristles at the base on one side.—DISTRIB. All
warm and trop. regions; species 10.—ETYM. *seta*, a *bristle.*

S. vir'idis, *Beauv.* ; bristles clustered scabrid.

Sporadic, from Aberdeen to Kent and Devon; a colonist; fl. July–Aug.—
Annual. *Stems* suberect, scaberulous above. *Leaves* flat, smooth, edges
rough; sheaths smooth, edges ciliate; mouth with a ring of hairs. *Panicle*
1–3 in., green; branches whorled, hispid, 3-quetrous; bristles ¼–⅓ in., flexuous, purplish. *Spikelets* ₁₂ in., elliptic, obtuse; empty glumes membranous;
fl. glumes shining, punctulate, striate.—DISTRIB. All warm climates; introd.
in America.

S. VERTICILLA'TA, *Beauv.* ; bristles single or in pairs barbed.

Cultivated fields, Norwich, Surrey, and Middlesex; fl. July–Aug.—Habit of
S. vir'idis, but panicle usually narrower, and the bristles truly barbed, their
asperities pointing downwards.—DISTRIB. As of *S. vir'idis.*

3. SPARTI'NA, *Schreber.*

Spikelets long, laterally compressed, sessile in 2 ranks on one-sided
panicled erect 3-gonous appressed spikes, 1-fld., with rarely a rudimentary
2d flower. *Empty glumes* 2, narrow, unequal, pointed or awned; upper
5-nerved, exceeding the flowering; lower smaller. *Fl. glume* sessile,
coriaceous, edges membranous, awnless, 1-nerved, palea long 2-nerved.
Scales 0. *Stamens* 3. *Ovary* glabrous; style very long, stigmas long
hairy. *Fruit* terete or subcompressed, enveloped in the palea and fl.
glume.—DISTRIB. Chiefly warm climates; species 5 or 6.—ETYM.
σπαρτίνη, a *cord*, from the use of the leaves.

S. stric'ta, *Roth ;* point of rachis of spike subulate.

Muddy salt creeks and marshes, rare, from Lincoln to Devon and Kent;
Channel Islands; fl. July–Aug.—*Rootstock* long, branched, extensively
creeping. *Stems* 1–4 ft., erect, strict, stout, polished, leafy throughout.
Leaves ¼–½ in. broad, strict, erect, coriaceous, convolute, smooth, pungent,
glaucous above; ligule short, silky. *Panicle* 3–8 in., strict; spikes 2–8,
1–3 in.; rachis angled, smooth, point equalling or exceeding the upper
spikelet. *Spikelets* ½–⅔ in., erect, yellowish green; empty glumes acute,
lowest acuminate, upper 2-toothed and tipped with a stiff awn, a little
silky; keel somewhat scabrid.—DISTRIB. Belgium and Spain to Austria
and Italy, N. America.—Very variable.—*S. alterniflo'ra*, Loisel, from

Southampton, has the tip of the rachis exceeding the spike.—*S. Townsen'dii*, Groves, from Hythe (Hants), has shorter leaves and a flexuous tip to the rachis.—American specimens show similar variations.

4. LEER'SIA, *Soland.*

Spikelets jointed on the pedicel, panicled, much compressed, imbricate, gibbous, 1-fld. *Empty glumes* 0. *Fl. glume* hard, awnless, broad, 3-nerved, often ciliate. *Palea* as long, hard, 1-nerved. *Scales* 2, ovate, entire, short. *Stamens* 1–3, or 6. *Ovary* glabrous; styles short, stigmas feathery. *Fruit* laterally compressed, enclosed in the hardened glume and palea.—DISTRIB. Temp. and trop. regions; species 5.—ETYM. *J. D. Leers*, a German botanist.

L. oryzoi'des, *Swartz;* panicle effuse, stamens 3.

Wet meadows and watery places, rare; Surrey, Hants, Sussex; fl. Aug.–Oct. —Perennial. *Root* creeping. *Stems* 2–3 ft., decumbent below, smooth, shining, leafy, nodes villous. *Leaves* $\frac{1}{3}$–$\frac{1}{2}$ in. broad, flat, scabrid, glaucous; sheaths compressed, almost smooth; ligule truncate, torn. *Panicle* 3–7 in., very lax and few-fld., partially enclosed in the sheath when flowering, branches in $\frac{1}{2}$ whorls, capillary, flexuous; upper part open with imperfect ovaries; lower part included in the sheath, with fertile ovaries. *Spikelets* $\frac{1}{4}$ in., dimidiate-oblong, thin but rigid, translucent, smooth or scaberulous, pale green, keels with long rigid cilia.—DISTRIB. From Denmark southd., trop. Asia, Africa, and S. America.

5. PHAL'ARIS, *L.*

Spikelets much laterally compressed, in open or contracted or spike-like panicles, 1-fld., with 2 or more rudimentary glumes (scales or pedicels) beneath the fl. glume on one side. *Empty glumes* 2, enclosing the flowering, subequal, keeled. *Fl. glume* broad. *Palea* much smaller. *Scales* 2, minute. *Stamens* 3. *Ovary* glabrous; styles long, stigmas slender feathery. *Fruit* compressed, enclosed in the coriaceous glume and palea.—DISTRIB. Trop. and temp. regions; species 9 or 10.—ETYM. The old name.

SECTION 1. **Phal'aris** proper. *Panicle* spike-like. *Empty glumes* broadly-winged.

P. CANARIEN'SIS, *L.*; panicle ovoid. *Canary grass.*

Fields and waste places; an escape; fl. July.—Annual, glaucous. *Stems* 1–3 ft., erect, scaberulous. *Leaves* flat, upper sheaths inflated; ligule large. *Panicle* 1–1$\frac{1}{2}$ in., ovoid or subcylindric, compact, pale green. *Spikelets* nearly $\frac{1}{4}$ in. diam., orbicular; empty glumes membranous, acute, wings broad, keel green, nerves 2, stout; fl. glume much shorter, silky, obscurely nerved; scales 2, lanceolate, acute.—DISTRIB. Warm and temp. Europe, N. Africa, W. Asia; introd. in N. America.

SECTION 2. **Digra'phis,** *Trin.* (gen.). *Panicle* contracted, interrupted. *Empty glumes* hardly winged.

GRAMINEÆ.

P. arundina cea, *L.* ; panicle elongate, branches short.

Rivers, lakes, &c., N. to Shetland ; ascending to nearly 1,400 ft. in N. England ; Ireland ; Channel Islands ; fl. July–Aug.—Perennial, glabrous. *Rootstock* creeping. *Stems* 2–6 ft., stout, erect. *Leaves* ⅓–½ in. broad, flat ; sheaths smooth ; ligule large. *Panicle* 4–8 in., suberect ; branches scabrid, spreading only when flowering. *Spikelets* ⅓ in., ovate, often purplish ; empty glumes 3-nerved, acuminate, glabrous ; fl. glume rather shorter, ovate-lanceolate, nerveless, silky ; scales 2, narrow, silky.—DISTRIB. N. temp. and Arctic regions.

6. ANTHOXAN'THUM, *L.* VERNAL GRASS.

Sweet-scented. Spikelets cylindric, in a spike-like panicle, 1-fld., proterogynous. *Glumes* 6 (or 5 and a palea) ; 2 lowest persistent acute or mucronate, veŕy unequal, lowest 1-nerved ; 2d large, 3-nerved ; 3d and 4th short, keeled, hairy, 2-fid with a dorsal bent awn ; 5th fl. glume, and 6th minute hyaline obtuse awnless, outer broad 5–7-nerved, inner narrow (palea ?) 1-nerved. *Scales* 0. *Stamens* 2 ; anthers large, linear, purple or yellow. *Ovary* glabrous ; styles long, stigmas feathery. *Fruit* terete, acute, enclosed in the brown shining fl. glume and palea.—DISTRIB. N. temp. and cold regions of the Old World ; species 4 or 5.—ETYM. ἄνθος and ξανθός, from the *yellow anthers.*

A. odora'tum, *L.* ; perennial, panicle interrupted below, awn short scarcely exserted.

Meadows, &c., N. to Shetland ; ascends to 3,400 ft. in the Highlands ; Ireland ; Channel Islands ; fl. May–June.—*Stems* 6–18 in., shining, glabrous or scabrid. *Leaves* flat, hairy ; sheaths furrowed, often pubescent, mouth hairy. *Panicle* 1–5 in., pubescent or villous ; branches short. *Spikelets* ¼–⅓ in., fascicled, often squarrose, green, glabrous or hairy ; 2 lowest glumes ovate, acute, upper lanceolate, almost awned ; two succeeding awned glumes curved, with obliquely truncate tips ; fl. glume glabrous.—DISTRIB. Europe (Arctic), N. Africa, N. Asia, Greenland ; introd. in N. America.—Odour of Woodruff.

A. PUEL'II, *Lecoq.* and *Lamotte ;* annual, very slender, much branched, panicle lax, awn long slender much exserted.

Pastures and fields, Roxburgh to Devon and Hants, &c. ; a modern introduction.—*Stems* very numerous from the root, 6–10 in., bent at the nodes. *Leaves* narrow, glabrous or hairy ; sheaths smooth. *Panicle* 1–2 in., subacute. *Spikelets* pedicelled ; 2 lowest glumes very unequal, scaberulous, membranous, lower half the length of the acuminate upper ; awned glumes straight, with erose tips.—DISTRIB. S. and Central Europe, Mediterranean.—Odour faint.

7. HIEROCHLO'E, *Gmelin.* HOLY GRASS.

Spikelets laterally compressed, panicled, 3-fld. ; upper flower 2-sexual, 2-androus ; 2 lower male, 3-androus. *Empty glumes* subequal, about equalling the flowering, membranous, keeled, 3-nerved. *Fl. glumes*

pedicelled, 5-nerved, awned or not. *Palea* of the 2-sexual fl. keeled,
1-nerved ; of the male fl. 2-nerved. *Scales* lanceolate. *Ovary* glabrous ;
styles long, stigmas feathery. *Fruit* terete, free.—DISTRIB. Arctic and
cold regions ; species 10.—ETYM. ἱερός and χλόα, *sacred grass*, from its
being formerly strewed on church floors.

H. borea′lis, *R.* and *S.* ; fl. glumes shortly awned near the tip.

Wet banks, Caithness, (Forfar, extinct); fl. May–June.—*Rootstock* creeping.
Stems 10–18 in., tufted, smooth, glabrous. *Leaves* flat, acute, edges scabrid ;
sheaths smooth; ligule long. *Panicle* pyramidal, subunilateral, sparingly
branched, pedicels smooth. *Spikelets* ⅓ in. broad, ovate, fulvous or brown,
shining ; empty glumes membranous, translucent, acute, toothed, mucronate ; fl. glumes similar, of the male fl. hispid, ciliate ; of the 2-sexual fl.
hairy above.—DISTRIB. Arctic, Alpine, and N. Europe, N. and W. Asia,
N. America.

8. ALOPECU′RUS, *L.* FOX-TAIL GRASS.

Spikelets much laterally compressed, crowded in spike-like cylindric panicles, jointed on the pedicel, 1-fld. *Empty glumes* subequal, often connate below, a little exceeding the flowering. *Fl. glume* hyaline, convolute, edges
often connate at the base, 3-nerved ; awn bent, dorsal. *Palea* 0 (in British
species). *Scales* 0. *Stamens* 3. *Ovary* glabrous ; styles distinct or connate, stigmatic hairs short simple. *Fruit* laterally compressed, enclosed
in the palea.—DISTRIB. N. and S. temp. and cold regions ; species 20.—
ETYM. ἀλώπηξ and οὐρά, *fox-tail*.

> * *Empty glumes connate to or nearly to the middle.*

1. **A. agres′tis,** *L.* ; panicle acute, keel of subglabrous glumes narrowly
winged shortly ciliate, awn twice as long as the fl. glume.

Fields and roadsides, from Cumberland southd.; a casual N. of it; Channel
Islands; fl. May–Oct.—Annual, scaberulous. *Stems* 1–2 ft., lowest internode prostrate. *Leaves* flat, edges rough; sheath smooth; ligule large,
truncate. *Panicle* 2–3 in., slender, often purplish, flexuous; branches very
short, hairy, with 2 spikelets. *Empty glumes* ⅓ in., lanceolate, pale, connate to the middle, acute, incurved; fl. glume a little exserted, glabrous.—
DISTRIB. Europe, N. Africa, Siberia; introd. in N. America.—A troublesome
weed.

2. **A. alpi′nus,** *Sm.* ; panicle short ovoid obtuse, keel of acute silky
empty glumes silkily ciliate, awn very short.

Springs and edges of alpine streams, alt. 2,100 to 3,600 ft., Ross, Aberdeen,
Perth, Forfar, Inverness; fl. July–Aug.—Perennial. *Stems* creeping below,
then erect, 6–18 in., rather stout, smooth, contracted at the top. *Leaves*
short, broad, flat; sheaths inflated, upper much longer than its leaf; ligule
short, obtuse. *Panicle* ½–¾ in., ⅓–½ in. broad, dense; branches with 4–6
spikelets, short, silky. *Empty glumes* ⅛ in., ovate, very silky, shining;
fl. glume glabrous, obtuse; awn variable, dorsal, rarely 0; anthers
linear, yellow.—DISTRIB. Arctic regions, Fuegia.—VAR. *Watso′ni*, Syme,

from Aberdeenshire, has a laxer panicle, more acute purplish empty glumes, with an awn ½-¼ of their length.

** *Empty glumes distinct or connate towards the base.*

3. A. praten'sis, *L.* ; stem erect, panicle slender cylindric obtuse, keel of hairy acute empty glumes villously ciliate, awn twice as long as the fl. glume.

Meadows and pastures, N. to Shetland; ascends to 2,000 ft. in N. England; Ireland; Channel Islands; fl. May–June.—Perennial, stoloniferous. *Stems* 1–3 ft., erect or lowest internode inclined, smooth. *Leaves* scaberulous, flat; sheath smooth, upper inflated, longer than its leaf; ligule large, truncate. *Panicle* 1½–2½ in., dense, soft, pale green; branches very short, with 3–6 spikelets. *Empty glumes* ⅛ in., ovate-lanceolate, acute; fl. glume subacute, glabrous, ciliate, margins connate ¼-½ of its length.—DISTRIB. Europe, N. Africa, N. and W. Asia, N.W. India; introd. in America.—An excellent fodder.

4. A. genicula'tus, *L.* ; stem decumbent below, panicle cylindric, keel of hairy empty glumes ciliate. *A. palus'tris*, Syme.

Wet meadows and ditches, ascending to nearly 2,000 ft. in the N. of England; fl. May–Aug.—Perennial, rarely annual, glabrous. *Stems* 8–18 in., procumbent and rooting below. *Leaves* flat, scaberulous above and on the edges. *Panicle* 1–1½ in., slender, dense-fld.; branches with 1 spikelet. *Empty glumes* hairy and silky as in *A. praten'sis*, but smaller, ⅒ in.; anthers purplish, linear.—DISTRIB. Europe, N. Africa, N. and W. Asia, India; introd. in N. America.

A. GENICULA'TUS proper; not glaucous, sheaths cylindric, panicles cylindric, empty glumes connate at the base, fl. glume with a subbasal awn not twice its length. Common, N. to Shetland; Ireland; Channel Islands.— *A. pro'nus*, Mitt., is a prostrate form.

Sub-sp. A. FUL'VUS, *Sm.*; glaucous, sheaths rather inflated, panicle longer paler, empty glumes smaller ⅒ in. connate below obtuse villously ciliate, awn inserted near the middle of and a little longer than the fl. glume, anthers small yellow.—N. Wales, Chester and Norfolk to Devon and Sussex.

Sub-sp. A. BULBO'SUS, *Gouan ;* taller, more erect, not glaucous, lowest internodes tuberous, upper sheath inflated, panicle slender subacute, empty glumes acute, free to the base, keel shortly ciliate, flowering glume longer than the palea awn subbasal twice its length.—Salt marshes, N. Wales, York and Chester, to Kent and Cornwall.

9. MIL'IUM, *L.* MILLET-GRASS.

Spikelets minute, in an effuse panicle, 1-fld. ; rachilla produced. *Empty glumes* 2, equalling the flowering, broad, subequal, obtuse, awnless, 3-nerved. *Fl. glume* rigid, ovate, tumid, obscurely 3-nerved, awnless. *Palea* 2-nerved. *Scales* 2, fleshy, acute, toothed on one side. *Ovary* glabrous; styles short, stigmas feathery. *Fruit* terete, included in the hardened glumes and palea.—DISTRIB. Temp. and trop. ; species 5 or 6.—ETYM. Latin for *Millet*, misapplied.

M. effu'sum, *L.* ; leaves broad flat thin.

Damp woods from Argyll and Elgin southd.; ascends to 1,000 ft. in the Lake District; Ireland; fl. May–June.—Perennial, pale green. *Stems* 2–4 ft., tufted, erect, smooth, shining, leafy. *Leaves* ⅓–½ in., linear-oblong, acute, scabrid above; sheaths smooth; ligule long, truncate, torn. *Panicle* 5–10 in., very lax and slender; branches capillary, spreading or deflexed, in remote whorls, few-fld. *Empty glumes* 1⁄10 in., elliptic-ovoid, obtuse, scaberulous, edges hyaline; fl. glume quite smooth, white and polished when ripe.—Distrib. Europe (Arctic), N. and W. Asia, N. America.

10. PHLE'UM, *L.* Cat's-tail Grass.

Spikelets in crowded spike-like panicles, 1-fld., with rarely a rudimentary 2d. *Empty glumes* exceeding the flowering, equal, much laterally compressed, keeled, awned, or mucronate. *Fl. glume* hyaline, awned or not, 3–5-nerved, toothed. *Palea* small. *Scales* 2, hyaline, toothed on the outer margin. *Stamens* 3. *Ovary* glabrous ; styles long, stigmas slender feathery. *Fruit* compressed, enclosed in the fl. glume and palea.—Distrib. N. and S. temp. and Arctic regions ; species 10.—Etym. The old Greek name.

* *Empty glumes truncate, tip scarious. Fl. glume 3-nerved.*

1. **P. praten'se,** *L.* ; leaf-sheaths appressed, panicle cylindric, keel of empty glumes hispid, awn rigid ⅓ their length. *Timothy-grass.*

Pastures, &c., N. to Shetland ; ascends to 1,400 ft. in N. England ; (? native N. of the Caled. Canal, *Wats.*); Ireland; Channel Islands; fl. June–Aug.—Perennial. *Stems* 6–18 in., tufted, ascending, smooth. *Leaves* short, flat ; ligule long. *Panicle* 1–6 in., obtuse, green. *Spikelets* crowded, shortly pedicelled ; empty glumes with a stout green keel, ciliate with stiff setæ, sides pale ; awns ⅛ in., scabrid, rigid ; fl. glumes membranous, cuspidate, 5-nerved ; anthers oblong, yellow or purple.—Distrib. N. Africa, Siberia, W. Asia ; introd. in N. America.—An excellent fodder. Syme distinguishes two varieties.
P. praten'se proper ; stems erect rarely geniculate, spikes stout cylindric, glumes greenish, keel dark.—Var. *P. nodo'sum, L.* ; stem recumbent geniculate, lower internodes swollen, leaves narrower, spike slender, glumes pale, keel green.—Var. *stolonif'era,* Bab., has copious stolons.

2. **P. alpi'num,** *L.* ; upper sheaths inflated, panicle ovoid or oblong, empty glumes equalling their rigid awn, keel hispid.

By alpine springs and rills, Perth, Forfar, Aberdeen; alt. 2,100 to 3,600 ft.; fl. July.—Perennial. *Stems* 6–18 in., solitary, creeping below, smooth, ascending, rigid. *Leaves* short, flat, spreading; lower sheaths appressed; ligule short. *Panicle* ½–1¼ in., dull purple and green. *Empty glumes* as in *P. praten'se,* but larger, ⅛ in. including the awn.—Distrib. Arctic and Alpine Europe and Asia, Himalaya, N. America, Fuegia.

** *Empty glumes gradually pointed; fl. glume* 3-*nerved, with the pedicel of a* 2d *at its base.*

3. P. arena'rium, *L.* ; upper sheaths inflated, panicle cylindric-oblong narrowed at the base, empty glumes hardly awned, keel ciliate above.

Sandy dunes, &c., local, from Aberdeen (E. Scotland only) southd.; Ireland; Channel Islands ; fl. May–June.—Annual, glabrous. *Stems* 2–6 in., crowded, leafy. *Leaves* broad, flat; sheaths smooth; ligule long. *Panicle* ½–1 in., most contracted at the base, glaucous. *Spikelets* ⅟₁₀ in., crowded ; empty glumes lanceolate, acuminate, punctulate; fl. glume very small, hairy; anthers minute, short, yellow.—DISTRIB. Europe, W. Africa.

4. P. phalaroi'des, *Koel.* ; sheaths hardly inflated, panicle cylindric long, keel of shortly awned empty glumes almost smooth or ciliate. *P. Bœhmeri,* Schrad. (not Wibel).

Sandy and chalky fields, Norfolk and Bedford to Essex and Herts; fl. July.— Annual, glabrous. *Stems* 10–18 in., ascending, smooth. *Leaves* short, flat, scabrid, rather glaucous ; upper ligules long. *Panicle* 2–4 in., rather narrow, obtuse, green, slightly interrupted. *Spikelets* ⅟₁₂ in., fascicled; empty glumes linear-oblong, obliquely truncate below the short rigid subulate awn, very coriaceous, green ; margins white, punctulate ; *fl. glume* minute ; anthers linear-oblong.—DISTRIB. Europe, N. Asia.

11. MIBO'RA, *Adans.*

A minute annual. *Spikelets* minute, subsessile, laterally or distichously arranged on a simple slender flexuous rachis, dorsally compressed, 1-fld. *Empty glumes* 2, a little exceeding the flowering, broad, subequal, concave, not keeled, truncate, awnless, membranous, 1-nerved, upper next the rachis. *Fl. glume* hyaline, very hairy, truncate, 5-nerved. *Palea* 2-nerved or 0. *Scales* 2, very minute. *Stamens* 3 ; anthers short. *Ovary* glabrous ; styles 2, very long, stigmas slender hairy. *Fruit* ellipsoid, compressed, enclosed in the fl. glume and palea.—DISTRIB. W. Europe from Hanover to Spain and Italy, N. Africa.—ETYM. unknown.

M. ver'na, *Adans.* ; stems capillary. *Sturm'ia min'ima,* Hoppe ; *Knap'pia agrostide'a,* Sm. ; *Chamagros'tis min'ima,* Borkh.

Wet sands, especially near the sea, Anglesea ; Channel Islands ; formerly in Essex ; naturalized in Haddington ; fl. March–April.—*Stems* 1–3 in., tufted. *Leaves* short, strict, setaceous, scaberulous, obtuse, margins involute; sheaths white, inflated ; ligule short. *Spike* ½–1 in,

12. AGROS'TIS, *L.* BENT.

Spikelets very small, in an open or contracted panicle with whorled branches, 1-fld., rachilla not produced. *Empty glumes* 2, exceeding the flowering, unequal, membranous, awnless, convex, keeled. *Fl. glume* hyaline ; awn slender, dorsal, or 0. *Palea* 2-nerved or 0. *Scales* glabrous, entire. *Stamens* 3. *Styles* short, stigma feathery. *Fruit* enclosed in the glume, terete, glabrous.—DISTRIB. All temp. and cold climates ; species 100.—ETYM. An old Greek name.

SECTION 1. **Tricho'dium**, *Michx.* (gen.). *Upper empty glume smaller than the lower. Palea very minute or 0.*

1. **A. cani'na,** *L.* ; leaves narrow flat smooth.

Moors, heaths, &c., N. to Shetland; ascends to 1,500 ft. in Derby; Ireland ; Channel Islands; fl. July–Aug.—Perennial, glabrous, smooth. *Stems* 6–24 in., sometimes stoloniferous. *Leaves* 2–5 in., $\frac{1}{8}$–$\frac{1}{2}$ in. broad; ligule oblong. *Panicle* 2–4 in., slender, flexuous, flowering open, fruiting contracted, purplish or green; branches capillary, scabrid. *Fl. glume* $\frac{1}{3}$ shorter than the empty, truncate, nerves 5, excurrent; awn from above the base, variable in length, bent in the middle.—DISTRIB. Europe (Arctic), N. and W. Asia; Himalaya; N. and S. America, Australasia.

2. **A. seta'cea,** *Curt.* ; leaves setaceous margins involute scaberulous.

Dry downs, Glamorgan, Berks, Surrey, Cornwall to Sussex ; fl. June–July.— Perennial. *Stem* 8–12 in., strict, scaberulous. *Leaves* very many, erect, rigid ; ligule oblong. *Panicle* 1$\frac{1}{2}$–3$\frac{1}{2}$ in., and spikelets much as in *A. cani'na,* but empty glumes scabrid, almost awned; fl. glumes with a longer rather twisted awn, and 2 minute tufts of hairs at the base ; palea minute.— DISTRIB. Belgium, France, Spain.

SECTION 2. **Agros'tis** proper. *Empty glumes subequal, or upper smaller than the lower. Palea 2-nerved.*

3. **A. vulga'ris,** *With.* ; ligule short truncate, fruiting panicle spreading.

Meadows and marshy places, N. to Shetland; ascends to 3,200 ft. in the Highlands; Ireland; Channel Islands; fl. June–Sept.—Perennial. *Stems* 6–24 in., ascending, smooth. *Leaves* short, flat, scabrid; sheaths smooth. *Panicle* 1–3 in., narrowly ovate, but variable in form. *Empty glumes* $\frac{1}{12}$ in., subequal, ovate or lanceolate, acute, nearly smooth, dull red or purplish and green; fl. glume a little shorter, truncate, 3-nerved, 3 times longer than the palea; awn 0 or short.—DISTRIB. Europe, N. Africa, Himalaya, N. America.

A. vulga'ris proper; slender, panicle usually elongate, fl. glume rarely awned. —VAR. *A. pu'mila,* L.; short, stout, panicle shorter, branches stouter, empty glumes broader, fl. glume usually awned.—VAR. *A. ni'gra,* With.; taller more robust, panicle more scabrid, branches more rigid, spikelets larger. —Salop, Stafford, Worcester, Warwick.

4. **A. al'ba,** *L.* ; ligule long acute, fruiting panicle contracted. *Fiorin Grass.*

Pastures and waste places, N. to Shetland; ascends to 1,400 ft. in Yorkshire ; Ireland; Channel Islands; fl. July–Sept.—Perennial. *Stems* 6–24 in., more or less prostrate below, scaberulous above. *Leaves* flat, scabrid or not; sheaths smooth. *Panicle* 1–8 in., branched or lobed, green or yellowish branches scabrid. *Empty glumes* much as in *A. vulga'ris,* but rather larger more rigid, not so shining; fl. glume 5-nerved, rarely awned.—DISTRIB.

Europe (Arctic), N. Africa, N. Asia, N. America.—*A. stolonif'era*, L. (*β subre'pens*, Bab.), is a more stoloniferous state.

13. POLYPO'GON, *Desf.* BEARD-GRASS.

Spikelets in a contracted or spikelike panicle, jointed on the pedicel, 1-fld. *Empty glumes* 2, much exceeding the flowering, equal, concave, keeled, 2-fid or notched, sinus awned. *Fl. glume* sessile, hyaline, base naked, broad, truncate, toothed, awned near the top. *Palea* small 2-nerved. *Scales* 2, falcate, entire. *Stamens* 1–3; anthers small. *Ovary* glabrous; styles short, free, stigmas feathery. *Fruit* terete, wrapped in the hyaline glumes and palea.—DISTRIB. Warm regions; species 10.— ETYM. πολύς and πώγων, from the *many awns.*

1. **P. monspelien'sis**, *Desf.* ; annual, awns much exceeding the sca-brid deeply-notched glumes, fl. glume not awned.

Damp pastures on the S.E. coast, rare; from Norfolk, Essex, Kent, to Hants; Channel Islands; introd. elsewhere; fl. June–July.—*Stems* ½–4 ft., erect, stout, smooth. *Leaves* large, broad, flat, scabrid; sheaths smooth; ligule large, obtuse. *Panicle* 1–6 in., oblong-ovoid or fusiform, cylindric or lobed, obtuse, pale greenish-yellow, dense, soft. *Empty glumes* nearly ½ in., narrow linear; awn straight, 2–3 times as long; fl. glume not ¼ as long as the empty, silvery.—DISTRIB. From Holland southd., N. Africa, W. Asia, India.

2. **P. littora'lis**, *Sm.* ; perennial, awns as long as the nearly smooth acute obscurely notched glumes, fl. glume awned.

Muddy salt marshes, very rare; Norfolk, Essex, Kent, Hants; fl. July.— Perennial, variable in size. *Stems* 1–6 ft., erect or decumbent, usually more slender below than in *P. monspelien'sis;* panicle smaller, more lobed, purplish; empty glumes much smoother, more acute, awns shorter, keel scabrid; fl. glume about ⅓ shorter than the empty; awn exserted.—DISTRIB. W. Europe.—Supposed by Duval-Jouve to be a hybrid between *monspelien'sis* and *Agrostis al'ba.*

14. CALAMAGROS'TIS, *Adans.*

Tall, perennial grasses. *Rootstock* stoloniferous. *Spikelets* in a close or spreading subsecund panicle, 1-fld., branches whorled. *Empty glumes* much exceeding the flowering, subequal, concave, acuminate, upper 3-nerved. *Fl. glume* enveloped in silky hairs, hyaline, 3–5-nerved, truncate, 2-fid, toothed, awned at the tip or back. *Palea* small, 2-nerved. *Scales* entire. *Stamens* 3. *Ovary* glabrous; styles short, stigmas feathery. *Fruit* terete, grooved in front, enveloped in the fl. glume and palea.—DISTRIB. N. temp. regions; species 5.—ETYM. κάλαμος and ἀγρόστις, *reed-grass.*

1. **C. Epige'jos**, *Roth;* panicle open, empty glumes subulate-lanceo-late and hairs twice as long as the flowering, awn dorsal inserted above the middle of the fl. glume.

Damp woods, local, from Mull and Aberdeen to Devon and Kent; Aran Island and Derry in Ireland; Channel Islands; fl. July–Aug.—*Stem* 2–6 ft., simple, stout, scabrid above. *Leaves* very long, flat, scabrid, glaucous beneath, point slender; ligule acute, torn. *Panicle* 4–12 in., purplish-brown; branches suberect. *Empty glumes* ¼ in., very narrow, tip and keel scabrid; awn produced for ½ its length beyond the fl. glume.—DISTRIB. Europe (Arctic) (excl. Spain), Himalaya, N. Asia.

2. **C. lanceola'ta,** *Roth ;* panicle open, empty glumes lanceolate twice as long as the flowering, hairs rather longer than the fl. glume, awn terminal minute. *Arundo Calamagros'tis,* L.

Damp copses, hedges, &c., from the Border to Devon and Kent; fl. July–Aug.—*Stem* 2–4 ft., slender, erect, smooth. *Leaves* much narrower and smoother than *C. Epige'jos,* more convolute; ligule shorter. *Panicle* 2–6 in., purplish; branches spreading, tip drooping. *Empty glumes* nearly as large as in *Epigejos,* but broader; awn minute.—DISTRIB. Europe (Arctic), N. and W. Asia.—Andersson observes that the pedicel of an upper flower is sometimes present.

15. GASTRID'IUM, *Beauv.*

Annual. *Spikelets* small, in a contracted panicle, 1-fld. ; rachilla produced, glabrous. *Empty glumes* 2, subequal, gibbous at the base, acute or shortly awned. *Fl. glume* minute, hyaline, truncate, toothed ; awn dorsal or 0. *Palea* narrow. *Scales* oblong, entire. *Stamens* 3. *Styles* short, distinct, stigmas feathery. *Fruit* subglobose, shining, loosely enclosed in the fl. glume and palea.—DISTRIB. S. and W. Europe, trop. Africa, temp. S. America.—ETYM. γαστρ δεον, *a ventricle.*

G. lendig'erum, *Gaud.* ; glumes lanceolate, awn exserted. *Agrostis austra'lis,* L. *Nit-grass.*

Maritime sandy marshes, &c., from S. Wales, Warwick, and Norfolk southd. ; Channel Islands; (a colonist, *Wats.*); fl. June–Oct. *Stem* 6–10 in., densely tufted, erect or ascending, slender, leafy. *Leaves* short, flat, scaberulous; sheaths smooth; ligule oblong. *Panicle* large, 1–3 in., cylindric-fusiform, dense, pale green, glistening; branches scaberulous; pedicels swollen at the top. *Empty glumes* ⅛ in., erect, very acuminate, the swollen base polished, keel obscurely scabrid; fl. glume white, shining, 4-toothed; awn near the top, very slender.—DISTRIB. Europe, W. Asia, N. Africa.

16. AP'ERA, *Adans.*

Annual. *Spikelets* small, panicled, shining, 1-fld. ; rachilla produced, glabrous. *Empty glumes* 2, unequal, membranous, acute, keeled, upper 3-nerved. *Fl. glume* shorter than the empty, membranous, hardly 2-fid with a slender dorsal flexuous awn. *Palea* hyaline, 2-keeled. *Scales* ovate-lanceolate. *Stamens* 3. *Styles* short, distinct, stigmas feathery. *Fruit* narrow, free within the fl. glume and palea.—DISTRIB. Europe and W. Asia ; species 2.—ETYM. ἀπήρος, undivided, alluding to the entire fl. glume.

A. Spi′ca-ven′ti, *Beauv ;* panicle large pyramidal effuse or contracted, awn much exceeding the fl. glume. *Agros′tis Anemagros′tis,* Syme.

Sandy, often inundated fields, S.E. England, from York to Kent and Hants ; fl. June–July.—*Stems* 1–3 ft., densely tufted, stout or slender, erect from a decumbent base, smooth. *Leaves* flat, glabrous or hairy, scabrid beneath and on the edges ; ligules short, torn. *Panicle* 3 in., sometimes 3 in. broad, green or purple ; branches many in a whorl, capillary, scabrid. *Empty glumes* $\frac{1}{10}$ in., lanceolate, acuminate, membranous, closed in fruit, nearly smooth, shining ; fl. glume hairy, awn subterminal, 3 times its length, erect, very slender ; palea with 2 tufts of silky hairs at the base.—DISTRIB. Europe, N. Africa, Siberia.

A. SPI′CA-VEN′TI proper ; panicle large broad effuse, anthers linear. *Agrostis, L.*
A. INTERRUP′TA, *Beauv.* ; panicle contracted interrupted, anthers oblong.—E. counties. *Agrostis, L.*

17. DEYEUX′IA, *Clarion.*

Perennial grasses. *Spikelets* panicled, 1-fld. ; rachilla produced into a hairy or penicillate tip, bearded below the fl. glume. *Empty glumes* sub-equal, keeled. *Fl. glume* longer or shorter, 5-nerved, tip entire or 2-4-toothed ; awn twisted, dorsal, or 0. *Palea* thin, 2-nerved. *Scales* ovate, entire or lobed. *Styles* short, distinct, stigma plumose. *Fruit* often oblique enclosed within the fl. glume and pale, free or subadherent.—DISTRIB. All temp. regions ; species 120.—ETYM. *N. Deyeux,* an eminent French chemist.

D. neglec′ta, *Kunth ;* panicle close, empty glumes oblong-lanceolate slightly exceeding the flowering, awn from below the middle of the fl. glume straight. *Calamagros′tis lappon′ica,* Hook., not Hartm. ; *C. stric′ta,* Nutt.

Bogs and marshes, very rare ; Delamere Forest ; (Forfar extinct) ; Caithness ; Ireland ; fl. June–July.—*Stems* 1–3 ft., strict, slender, smooth, polished. *Leaves* short, all flat or lower filiform with convolute margins, almost smooth ; ligule short. *Panicle* 2–6 in., narrow, erect, pale purplish and green. *Empty glumes* $\frac{1}{10}-\frac{1}{8}$ in., acute, nearly smooth.—DISTRIB. Scandinavia (Arctic), Dahuria, N. America.—Leaves in cultivated (Irish) specimens quite flat.

D. neglec′ta proper ; upper ligule obtuse, lower empty glumes acuminate, upper acute twice as long as the hairs.—(Forfar extinct), Cheshire.—VAR. *Hooke′ri,* Syme ; upper ligule acute, lower empty glumes acute, upper broader one third longer than the hairs.—Lough Neagh.

18. AMMOPH′ILA, *Host.* MARRAM-GRASS.

Spikelets large in a contracted panicle, much laterally compressed, 1-fld. ; rachilla produced beyond the fl. glume. *Empty glumes* 2, scarcely exceeding the flowering, rigid, subequal, long, narrow, keeled, subacute. *Fl. glume* rigid, with an oblique callus and a short pencil of silky hairs at the base, 5-nerved ; awn minute, subterminal. *Palea* equalling the glume, rigid, 2-nerved. *Scales* very acuminate. *Stamens*, ovary, and fruit of *Calamagros′tis.*

—DISTRIB. Shores of Europe, N. Africa ; species 4.—ETYM. ἄμμος and φιλω, *sand lover.*

1. **A. arundina'cea,** *Host.* ; panicle subcylindric white, fl. glume linear-oblong acute 3 times as long as the hairs. *Psamma arenaria,* R. and S.

Sand-hills by the sea, N. to Shetland ; Ireland; Channel Islands ; fl. July.— *Rootstock* widely creeping, binding the sand. *Stems* 2–4 ft., smooth or scabrid above. *Leaves* long, rigid, convolute, polished without, scabrid and glaucous within ; sheaths long ; ligule very long, 2-fid, torn. *Panicle* 3–6 in., straight, broadest and sometimes lobed at the base, yellowish ; branches short. *Spikelets* erect, pedicels scabrid ; empty glumes ⅓–½ in., acute, keel scabrid ; fl. glume and palea quite like the empty glumes in colour and texture. *Anthers* ¼ in., linear, yellow.—DISTRIB. Of the genus.

2. **A. bal'tica,** *Link ;* panicle elongated interrupted, fl. glume lanceolate acuminate twice as long as the hairs. *P. samma baltica,* R. and S.

Ross Links and Holy Isle, Northumbd. ; fl. Aug.–Sept.—Habit of *A. arundinacea,* but distinguished by its laxer less cylindric panicles, more lanceolate and acuminate glumes, and by the length of the hairs. The flowering glumes are 5-nerved in both, but more faintly in *A. bal'tica.*—DISTRIB. Seashores from Holland to Gothland.

19. **LAGU'RUS,** *L.* HARE'S-TAIL GRASS.

Annual. *Spikelets* in a dense villous ovoid head, 1-fld., laterally compressed ; rachilla produced. *Empty glumes* 2, much exceeding the flowering, equal, long, with slender feathery points. *Fl. glume* membranous, shortly stipitate, with the pedicel of an upper flower at the back of the palea, narrow, terete, with 2 short awns and an intermediate long bent and twisted one. *Palea* 2-nerved. *Scales* 2, fleshy. *Stamens* 3. *Ovary* glabrous ; styles very short, stigmas long feathery. *Fruit* smooth, embraced by the glume and palea.—DISTRIB. W. and S. Europe, N. Africa, W. Asia.—ETYM. λαγώς and οὐρά, *hare's tail.*

L. ova'tus, *L.* ; leaves short flat, ligule short.

Sandy places, Guernsey ; naturalized near Saffron Walden ; fl. June.—*Stems* 6–10 in., very many, erect or decumbent below, stout, pubescent, leafy below. *Leaves* broad, and inflated sheaths pubescent or villous. *Heads* 1–1½ in., ¾ in. broad, white, obtuse. *Empty glumes* ¼ in., very slender ; awn twice as long as the nearly glabrous fl. glume.

20. **AI'RA,** *L.*

Spikelets laterally compressed, loosely panicled, 2-fld., rachilla not produced. *Empty glumes* 2, equalling the flowering, membranous, subequal, acute, 1-nerved. *Fl. glumes* pedicelled, membranous, subequal, convex, 3-nerved, toothed, awned at the back. *Palea* 2-nerved, 2-fid. *Scales* acuminate, entire or 2-fid. *Stamens* 3. *Ovary* glabrous ; stigmas

I I

subterminal, feathery to the base. *Fruit* grooved, free or adnate to the
fl. glume and palea.—DISTRIB. Temp. regions; species 4–5.—ETYM. A
Greek name for some grass.

1. **A. caryophyl'lea,** *L.*; sheaths scabrid, panicle spreading, branches
long trichotomous.

Sandy meadows, &c., N. to Shetland; ascends to 1,400 ft. in the Highlands; Ireland; Channel Islands; fl. June–July.—*Stems* 2–10 in., tufted,
leafless and scabrid above, bent below, often purplish. *Leaves* setaceous,
short, scabrid, obtuse; ligule long. *Panicle* 1–2 in. *Spikelets* $\frac{1}{12}$ in., ovate,
shining; lower empty glume ovate, acuminate, exceeding the flowering,
keel scabrid; awn twice as long as its glume, twisted.—DISTRIB. Europe,
N. Africa; introd. in N. America.

2. **A. præ'cox,** *L.*; sheaths glabrous, panicle contracted, branches very
short with 1–2 spikelets.

Dry pastures, &c., N. to Shetland; ascends to 1,800 ft. in Yorkshire; Ireland;
Channel Islands; fl. April–June.—Habit of *A. caryophyl'lea,* but panicle very
different; whole plant greener; spikelets narrower; 2-fid points of
flowering glume shorter; awn inserted higher up.—DISTRIB. Europe;
introd. in N. America.

21. CORYNEPH'ORUS, *Beauv.*

Annual grasses. *Spikelets* small, panicled, 2-fld.; rachilla produced,
penicillate. *Empty glumes* 2, subscarious, subequal, acute. *Fl. glumes*
shorter, hyaline; awn dorsal, bent, twisted below the bearded joint, tip
clavate. *Palea* narrow, 2-nerved. *Scales* 2-fid. *Stamens* 3. *Styles* short
distinct, stigmas feathery. *Fruit* grooved, adhering to the fl. glume and
palea.—DISTRIB. Europe, W. Asia; species 2.—ETYM. κορύνη and Φέρω,
from the *clubbed* awn.

C. canes'cens, *Beauv.*; tufted, leaves short rigid. *Aira canescens,* L.

Sandy coasts, Norfolk, Suffolk; Channel Islands; fl. July.—*Tufts* hard, rigid,
pungent. *Stems* 4–8 in., bent below, glabrous above. *Leaves* 1–2 in.,
glaucous, involute; upper sheaths long, scaberulous; ligule lanceolate.
Panicle 1–3 in., narrow-oblong, spreading in flower; branches thickened at
the forks, short. *Spikelets* $\frac{1}{8}$ in., narrow, pale silvery or purplish; empty
glumes narrow, acuminate, tips hyaline; fl. glumes shorter, villous at the
base; keel channelled; awn included or shortly exserted, purple below,
bearded at the middle where bent, above gradually thickened, white;
anthers purplish.—DISTRIB. Europe, Siberia.

22. DESCHAMP'SIA, *Beauv.*

Perennial grasses. *Spikelets* panicled, 2-fld.; rachilla produced, sometimes bearing a male flower. *Empty glumes* subequal, keeled, shining,
truncate, toothed, 3–5-nerved; awn dorsal, straight or twisted. *Palea*
narrow, 2-nerved. *Scales* entire, lanceolate. *Stamens* 3. *Styles* distinct,

stigmas feathery. *Fruit* grooved, included in the fl. glume and palea, free.—DISTRIB. Temp. and cold regions ; species 20.—ETYM. M. H. Deschamps, a French chemist.

1. **D. flexuo'sa,** *Trin.* ; leaves filiform terete solid, branches of panicle spreading capillary, spikelets subterete. *Ai'ra flexuo'sa,* L.

Dry woods, heaths, &c., N. to Shetland, ascends to 3,700 ft. in the Highlands ; Ireland ; Channel Islands ; fl. June–Aug.—*Stems* ½–2 ft., erect, slender, polished, naked above. *Leaves* short, curved, obtuse, grooved (not involute), sheath of upper long ; ligule short, obtuse. *Panicle* 2–5 in. ; branches 2–3-nate. *Spikelets* ⅙ in. ; purplish or yellow-brown, shining ; empty glumes acuminate ; awn about ½ as long as the spikelet, erect, then bent or twisted. —DISTRIB. Europe (Arctic), N. and W. Asia, N. America, Fuegia.—Often viviparous.

VAR. *Ai'ra monta'na,* Huds.; subalpine, glumes larger more purple.—VAR. *A. seta'cea,* Huds. (*uligino'sa,* Weihe) ; leaves capillary, upper fl. glume longer pedicelled.

2. **D. cæspito'sa,** *Beauv.* ; leaves flat scabrid, branches of nodding panicle flexuous, spikelets much compressed. *Ai'ra cæspito'sa,* L. ; *A. ma'jor,* Syme.

Wet meadows, woods, &c., N. to Shetland ; ascends to nearly 3,000 ft. in the Highlands ; Ireland ; fl. June–July.—*Stems* 2–4 ft., rather stout, leafy, shining, smooth. *Leaves* coriaceous ; sheaths shining, smooth or rough, upper very long ; ligule obtuse. *Panicle* 4–8 in., linear-oblong, spreading in flower. *Spikelets* ¼ in., very shining, fulvous or purplish ; empty glumes narrow, obtuse, shorter than the flowering, keel scabrid or smooth, upper obscurely 3-nerved ; fl. glumes 1–3 (3d always imperfect), silky at the base ; awn short, inserted below the middle, variable in length.—DISTRIB. N. and S. temp., Arctic, and mountain regions.—A variable grass.

D. CÆSPITO'SA proper ; tall, leaves longer broader scabrid, branches of panicle rough, awn inserted below the middle usually equalling the glume.

Sub-sp. D. ALPI'NA, *R.* and *S.;* short, leaves narrower channelled smooth, branches of panicle smooth, awn inserted at the middle shorter. *Ai'ra lævi-ga'ta,* Sm. Usually viviparous. Wet rocks, Scotland ; ascends to 4,100 ft.; W. Ireland.

23. HOL'CUS, *L.*

Spikelets much laterally compressed, in open panicles, jointed on the pedicel, 2-fld. ; lower flower 2-sexual, upper male ; rachilla shortly produced. *Empty glumes* 2, boat-shaped, subequal, keeled, lower 1-nerved ; upper larger, 3-nerved, notched, acute or shortly awned. *Fl. glumes* shorter, 5-nerved, lower sessile, 2-sexual, awnless ; upper pedicelled, male, with a dorsal twisted awn. *Palea* 2-nerved, 3-toothed. *Scales* oblique, acuminate. *Stamens* 3. *Ovary* glabrous ; stigmas sessile feathery. *Fruit* laterally compressed, enclosed in the fl. glume.—DISTRIB. Europe, temp. Asia, and N. and S. Africa ; species 8.—ETYM. obscure.

1. **H. lana'tus,** *L.* ; softly tomentose, root fibrous, awn of the flowering glume included, tip scabrous or smooth.

Meadows, copses, waysides, &c., N. to Shetland; ascends to 1,800 ft. in
Yorkshire; Ireland; Channel Islands; fl. June–Aug.—Perennial, densely
tufted. *Stems* 6–24 in., ascending, slender, leafy. *Leaves* flat, soft, upper
sheaths inflated; ligule short. *Panicle* 2–5 in., pale green or pinkish;
branches 2–3-nate. *Spikelets* ⅓ in., elliptic-oblong; empty glumes acute,
nerves strong.—DISTRIB. Europe, N. Africa, Siberia; introd. in N.
America.

2. **H. mol'lis**, *L.* ; villous at the nodes, rootstock creeping, awn of fl.
glume exserted at length inflexed scabrid throughout.

Sandy and waste places, woods, &c.; less common than *H. lana'tus;* ascends
to 1,500 ft. in`N. England; fl. June–Aug.—Similar to *H. lana'tus,* .but
usually more slender; glumes more scabrid and very acuminate.—DISTRIB.
Europe except N. Russia, N. Africa.

24. TRISE'TUM, *Pers.*

Perennial grasses. *Spikelets* panicled, compressed, 2- rarely 3–6-fld. ;
rachilla produced, with sometimes a male fl. glume. *Empty glumes* membranous, unequal, keeled, acute. *Fl. glumes* shorter, with 2 awned points
and an intermediate dorsal twisted awn. *Palea* hyaline, 2-nerved. *Scales*
membranous, lanceolate. *Styles* distinct, stigmas feathery. *Fruit* furrowed ventrally, glabrous or tip pubescent, enclosed in the fl. glume and
palea.—DISTRIB. All temp. regions ; species 50.—ETYM. *tri, three,* and
seta, from the 3-awned fl. glume.

T. flaves'cens, *Beauv.* ; panicle open, branches in ½-whorls, fl. glume
2-cuspidate 3-awned. *Ave'na flaves'cens,* L.

Dry pastures from Elgin and the Clyde southd.; ascends to 1,600 ft. in N.
England; Ireland; Channel Islands; fl. July–Aug.—*Root* fibrous, stoloniferous. *Stems* 1–2 ft., erect, smooth, glabrous. *Leaves* flat, and sheaths
hairy; ligule truncate, ciliate. *Panicle* 2–4 in.; branches many, capillary.
Spikelets ¼ in., many, 3–4-fld., shining, yellowish; empty glumes ovate,
acuminate; awns very divergent.—DISTRIB. Europe, N. Africa, N. Asia,
Himalaya.

25. AVE'NA, *L.*

Annual or perennial grasses. *Spikelets* large, terete, panicled, 2- or
more-fld., upper flower usually imperfect. *Empty glumes* 2, equalling the
flowering, usually subequal, rather membranous, 1–11-nerved. *Fl. glumes*
more rigid, rounded at the back or 2-cuspidate, 2-fid with a long bent and
twisted awn from the sinus. *Palea* 2-nerved. *Scales* 2-fid. *Stamens* 3.
Ovary hirsute at the top ; styles short, distant, stigmas feathery. *Fruit*
furrowed, adherent to the glume, top hairy.—DISTRIB. Temp. and cold
regions ; species 40.—ETYM. The old Latin name.

* *Annual. Spikelets at length drooping. Empty glumes 5–11-nerved.*

1. **A. fat'ua**, *L.* ; spikelets 2–3-fld., empty glumes 9-nerved, fl. glume
below and pedicels hairy. *Wild Oat.*

Cornfields, N. to Shetland, but not indigenous in Scotland; Ireland; Channel Islands; (a colonist, *Wats.*); fl. June–Aug.—*Stems* 1–3 ft., stout, smooth : nodes hairy. *Leaves* flat, scaberulous; sheaths smooth; ligule short, torn. *Panicle* 6–10 in., equally spreading; branches whorled, scabrid. *Spikelets* 1 in., green. *Fl. glumes* 2-fid, with long fulvous hairs below, half as long as the brown awn.—DISTRIB. Europe, N. Africa, Siberia, N.W. India.

A. STRIGO'SA, *Schreb.* ; spikelets 2-fld., empty glumes 7–9-nerved, fl. glumes and pedicels glabrous.⸲

Cornfields, rare, not indigenous; fl. June–July.—Smaller and more slender than *A. fat'ua ;* leaves sometimes hairy; panicle unilateral, branches fewer; fl. glume more deeply 2-fid, segments awned.—DISTRIB. Of *A. fat'ua.*

** *Perennial. Spikelets suberect. Empty glumes 1–3-nerved.*

2. A. praten'sis, *L.* ; leaves glaucous glabrous, lower branches of panicle 2-nate with 1–2 3–6-fld. spikelets.

Moors and dry.pastures, Ross and Skye to Devon and Kent; ascends to 2,200 ft. in the Highlands; Channel Islands; fl. June–July.—*Root* fibrous, stoloniferous. *Stems* ½–3 ft., erect, scaberulous, bent at the base. *Leaves* flat or involute ; lower sheaths more or less scabrid; ligule ovate, acute. *Panicle* racemose, contracted after flowering. *Spikelets* ½–⅜ in., pale, shining; fl. glume scabrid; pedicel silky; awn faintly bent, equalling the glume.— DISTRIB. From Norway southd. and eastd. to Thrace and S. Russia (excl. Spain).

A. praten'sis proper; leaves involute, sheaths terete nearly smooth.—VAR. *A. alpi'na,* Smith *(A. planicul'mis,* Sm.,not Schrad.); leaves flat, lower sheaths much compressed. Mountains.

3. A. pubes'cens, *Huds.* ; leaves pubescent, lower branches of panicle 5-nate with 2–4 2-fld. spikelets.

Dry pastures, N. to Shetland; ascends to 1,600 ft. in N. England; Ireland; fl. June–July.—Habit of *A. praten'sis,* but less densely tufted; leaves flatter ; sheaths very pubescent; awns more spreading.—DISTRIB. From Norway southd. and eastd. to Bosnia (excl. Spain), Siberia.

26. ARRHENATH'ERUM, *Beauv.*

Perennial grasses. *Spikelets* panicled, terete, 2-fld. ; lower fl. male, upper female or 2-sexual ; rachilla hairy between the fl. glumes, produced. *Empty glumes* scarious, very unequal. *Fl. glumes* rigid, 5–7-nerved, 2-toothed, lower with a long basal bent and twisted awn, upper with a short dorsal awn or 0. *Palea* 2-nerved. *Scales* lanceolate, toothed laterally. *Stamens* 3. *Styles* short, distinct, stigmas feathery. *Fruit* pubescent, loosely enclosed in the fl. glume and palea, pubescent.— DISTRIB. Europe, N. Africa, W. Asia ; species 3.—ETYM. ἄῤῥην, *masculine,* and ἀθήρ, *awn.*

A. avena'ceum, *Beauv.* ; panicle narrow long nodding, lower fl. glume long-awned, upper usually awnless. *A. elatius,* Presl. ; *Ave'na ela'tior,* L.

Fields, hedgerows, and pastures, N. to Shetland; ascends to 1,500 ft. in N. England; Ireland; Channel Islands; fl. June–July.—*Rootstock* widely creeping, nodes often tuberous (var. *bulbo'sa*, Lindl.). *Stems* 2–4 ft., erect, slender, smooth. *Leaves* flat, scabrid; sheaths smooth; ligule truncate. *Panicle* 6–12 in.; branches 2–3-nate, suberect, very scabrid. *Spikelets* ⅓ in., pale, shining; lower empty glume much smallest; upper oblong-lanceolate, acute; fl. glume subsessile, hairy below the middle, half as long as the dark twisted bent awn (the 2 fl. glumes are sometimes equally awned).— DISTRIB. Europe, N. Africa, W. Asia; introd. in N. America.—A pest.

27. CY'NODON, *Rich.* DOG'S-TOOTH GRASS.

Perennial grasses. *Spikelets* laterally compressed, secund or on radiating spikes ; rachilla 1-fld. *Empty glumes* 2, much smaller than the flowering, awnless, spreading, subequal. *Fl. glume* convex, 3-nerved, awnless, keel ciliate. *Palea* narrow, 2-nerved. *Scales* fleshy, truncate. *Stamens* 3. *Ovary* glabrous ; styles 2, rather long, stigmas feathery. *Fruit* laterally compressed, enveloped in the fl. glume and palea.—DISTRIB. Temp. and trop. regions ; species 4.—ETYM. κύων and ὀδούς, *dog's tooth.*

C. Dac'tylon, *Pers.* ; leaves short involute, tips obtuse.

Sandy shores of S.W. England, Dorset to Cornwall ; casual at Kew ; Channel Islands ; fl. July–Aug.—*Stems* 4–10 in., stout, woody, prostrate and extensively creeping, with short suberect leafy and flowering branches, smooth ; fl. branches clothed with strongly furrowed sheaths. *Leaves* subulate, stiff, glaucous, strongly nerved; sheaths pale, mouth hairy. *Spikes* 3–6, 1–2 in., radiating, purplish; rachis convex, grooved above. *Spikelets* $\frac{1}{10}$–$\frac{1}{12}$ in., imbricate; empty glumes ovate, acute; keel scabrid.—DISTRIB. From Holland southd., Asia, Africa ; introd. in N. America.—The chief pasture (*Doab* and *Bermuda grass*) of many dry climates.

28. TRIO'DIA, *Br.*

Perennial grasses. *Spikelets* terete, panicled, 3–5-fld., upper flower often imperfect ; rachilla jointed between the flowers. *Empty glumes* exceeding the flowering, subequal, herbaceous, acute ; upper 3-nerved. *Fl glumes* convex, 3-toothed, keeled, 7-nerved. *Palea* broad, ciliate. *Scales* broad, fleshy. *Stamens* 3. *Ovary* stipitate, glabrous ; styles short, terminal, stigmas feathery. *Fruit* free, plano-convex.—DISTRIB. Temp. and trop.; species 20.—ETYM. τρεῖς and ὀδούς, from the *three teeth.*

T. decum'bens, *Beauv.* ; spikelets 6-10 turgid. *Dantho'nia,* DC.

Dry pastures and moors, N. to Shetland ; ascends to 1,800 ft. in Yorkshire ; Ireland ; Channel Islands; fl. July.—Perennial, bright green. *Root* fibrous. *Stems* 6–12 in., densely tufted, rigid, glabrous, leafy. *Leaves* obtuse, coriaceous, slender, at length involute, hairy below ; sheaths grooved, lower hairy ; mouth with a row of hairs. *Panicle* 1–2 in., erect; rachis and branches flexuous. *Spikelets* ¼–⅓ in., obovoid, shining, pale green and purplish, rachilla very short ; empty glumes large, ovate, acute, keel scabrid, margins hyaline; fl. glumes ovoid, not keeled, coriaceous, imbricate, bearded at the base, ciliate on the lower margins.—Distrib. Europe, N. Africa.

29. PHRAGMI'TES, *Trin.* Reed.

Perennial stout water-reeds. *Spikelets* subterete, panicled, 3-6-fld. ; lower fl. male, the rest 2-sexual. *Empty glumes* 2, short, unequal, membranous, keeled. *Fl. glumes* distant ; lowest naked, 1-3-androus ; the rest 3-androus, enveloped in long silky basal hairs, very long acuminate, 3-nerved, entire. *Palea* very short. *Scales* large, obtuse. *Stamens* 3. *Ovary* glabrous ; styles short, stigmas feathery. *Fruit* terete, loosely wrapped in the fl. glume.—Distrib. Arctic and temp. zones ; species 1 or 2.—Etym. obscure.

P. commu'nis, *Trin.* ; panicle very large soft dull purple nodding, silky hairs equalling the fl. glumes. *Arun'do Phragmi'tes,* L.

Edges of lakes, &c., N. to Shetland ; Ireland ; Channel Islands ; fl. July–Aug.—*Rootstock* extensively creeping, jointed. *Stems* 6–10 ft., stout, terete, erect, smooth. *Leaves* ½–1 in. broad, flat, rigid, acuminate, glaucous beneath, edges hispid ; sheath smooth, mouth bearded. *Panicle* 10–18 in., ovoid, dense ; branches smooth, with long scattered hairs. *Spikelets* ½–¾ in., 3–5-fld. (or fewer, *P. nigricans,* Dumort.), shining ; empty glumes lanceolate, ½ the length of the very narrow subulate flowering.—Distrib. Europe (Arctic), Asia, Africa, America, Australia.

30. SESLE'RIA, *Scop.*

Spikelets compressed, in a contracted dense ovoid or subcapitate bracteate panicle, 2-6-fld. ; bracts small or large, entire toothed or multifid, sheathing the lower peduncles ; rachilla jointed above the lower glumes. *Empty glumes* 2, longer than the flowering, subequal, 1- rarely 3-nerved. *Fl. glumes* 2-3, the upper rudimentary, keeled, 3-cuspidate. *Palea* 2-keeled. *Scales* 2, 3-5-toothed. *Stamens* 3. *Ovary* hairy at the top ; styles terminal, connate below, stigmas very long barbellate. *Fruit* oblong enclosed in the fl. glume and palea.—Distrib. Europe, N. Africa, W. Asia ; species 12.—Etym. *L. Sesler,* an Italian botanist.

S. cæru'lea, *Scop.* ; panicle oblong subsecund silvery-grey.

Hilly pastures, especially in limestone districts, from Ross to York and Lancaster ; ascends to 2,500 ft. in the Highlands ; W. Ireland, local ; fl. April–June.—Perennial. *Root* fibrous. *Stems* 6–18 in., erect, smooth. *Leaves* narrow, flat, glaucous above, tip scabrid mucronate, glabrous or hairy, upper very short ; sheaths compressed, breaking up into fibres ; ligules short, ciliate. *Panicle* ½–1 in., blue-grey, glistening. *Spikelets* ⅒ in., in subsessile fascicles, the lower embraced at the base by a small convolute bract or glume ; empty glumes ciliate, acuminate ; fl. glumes pubescent, 3 central nerves confluent ; awn very short.—Distrib. From Belgium to Greece and S. Russia.

31. CYNOSU'RUS, *L.* DOG'S-TAIL GRASS.

Spikelets dimorphous, in dense spike-like panicle ; terminal spikelet of each fascicle terete 2–5-fld. with an upper flowerless glume, lower spikelets reduced to an involucre of pectinately arranged distichous rigid subulate empty glumes surmounting the terminal. *Empty glumes* 2, shorter than the lowest fl. glume, unequal, rigid. *Fl. glumes* terete, 3-keeled, nerved, mucronate, coriaceous, opaque. *Palea* with 2 ciliate nerves. *Scales* with a basal lobe. *Stamens* 3. *Ovary* glabrous ; styles short terminal, stigmas feathery. *Fruit* adherent to the fl. glume and palea.—DISTRIB. N. temp. regions ; species 3 or 4.—ETYM. κυών and οὐρά, *dog's tail.*

1. **C. crista'tus,** *L.* ; spike linear unilateral, fl. glumes shortly awned.
Dry pastures and banks, N. to Shetland ; ascends to 1,800 ft. in N. England ; Ireland ; Channel Islands ; fl. July-Aug.—Perennial, tufted, stoloniferous. *Stems* 1–2 ft , terete, strict, smooth, naked above. *Leaves* almost filiform, slightly hairy ; sheaths smooth ; ligule 2-fid. *Spike* 1–2 in., strict, rigid ; rachis flexuous ; spikelets ₁⁄₂ in., sessile between 6–10 rigid scabrid serrulate concave glumes, and as long as these ; empty glumes cuspidate ; fl. glumes scabrid above, obscurely 3-nerved.—DISTRIB. Europe, N. Africa.

2. **C. echina'tus,** *L.* ; panicle dense, awn equalling the fl. glume.
Sandy seashores, Channel Islands ; introduced on the British coasts ; casual in Ireland ; fl. July.—Annual ; larger and more robust than *C. crista'tus ;* leaves broad, flat ; panicle ¾–1½ in., ovoid, lobed, squarrose, shining, branches ⅓ in., pectinate ; segments of branches ¼–½ in., subulate, slender, scarious, scaberulous ; spikelets fewer on each branch, ¼ in. *Empty glumes* hyaline ; fl. glumes green.—DISTRIB. Mid. and S. Europe, N. Africa, W. Asia.

32. KOELE'RIA, *Pers.*

Spikelets oblong, compressed, in a contracted spike-like panicle, 2–5-fld. ; rachilla jointed between the fl. glumes, terminated by 2–3 empty glumes. *Empty glumes* 2, rather shorter than the flowering, unequal, narrow, compressed, keeled, membranous ; lower 1- upper 3-nerved. *Fl. glumes* secund, rather close, membranous, acuminate, obscurely keeled, 5-nerved at the base, lowest sessile. *Palea* 2-fid. *Scales* 2, oblique. *Stamens* 3. *Ovary* glabrous ; styles terminal, short, stigmas feathery. *Fruit* almost linear, plano-convex, free.—DISTRIB. Temp. regions ; species 12.— ETYM. *G. L. Koeler,* a German writer on grasses.

K. crista'ta, *Pers.* ; panicle silvery interrupted below.
Banks and pastures, N. to Caithness ; ascends to 1,800 ft. in Yorkshire ; Ireland ; Channel Islands ; fl. June–July.—Perennial, pubescent or silky, pale green. *Root* fibrous, stoloniferous. *Stems* 1–3 ft., slender. *Leaves* narrow, glabrous beneath, soon involute ; sheaths striate ; ligule 0. *Panicle* 1–4 in., linear-oblong or lanceolate ; branches 2–3-nate, very short. *Spikelets* ⅛ in., shining, pale green ; empty glumes oblong-lanceolate, acute, broadly hyaline, glabrous pubescent or scabrid ; keel minutely scaberulous ; rachis

pubescent; fl. glumes linear-lanceolate, scaberulous, mucronate.—DISTRIB. N. and S. temp. regions.

33. MOLIN'IA, *Schrank.*

A tall rigid wiry perennial grass. *Spikelets* subterete, in a slender panicle, 1-5-fld., upper fl. imperfect. *Empty glumes* 2, much shorter than the flowering, equal, convex, 1-nerved. *Fl. glumes* longer, distant, conical, acute, awnless, cartilaginous, nerves 3, very strong. *Palea* 2-nerved, obtuse. *Scales* membranous, 1-toothed. *Stamens* 3. *Ovary* glabrous; styles short, stigmas feathery. *Fruit* sub-4-gonous, furrowed, 2-beaked with the persistent style-bases, inclosed in the cartilaginous fl. glume.—DISTRIB. N. temp. regions.—ETYM. *G. J. Molina*, a writer on Chilian botany, &c.

M. cæru'lea, *Mœnch;* leaves flat, stem wiry. *Eno'dium,* Gaud.
Wet moors, &c., N. to Shetland; ascends to 3,000 ft. in the Highlands; Ireland; Channel Islands; fl. July–Aug.—*Stems* 1-4 ft., terete, striate, node solitary, near the base, naked above. *Leaves* smooth, rigid, hairy at the base, tips very slender; sheaths smooth; ligule 0. *Panicle* 1-12 in., usually stout, very contracted; rachis flexuous, compressed; branches erect. *Spikelets* few, ⅓ in., narrow, reddish or violet-purple or green; empty glumes subacute; fl. glumes deciduous, ovate-lanceolate, subacute; anthers violet-brown.—DISTRIB. Europe (Arctic), N. and W. Asia, N. Africa.
M. cæru'lea proper; spikelets blue-purple 2-3-fld., fl. glumes 3-nerved.—VAR. *M. depaupera'ta,* Lindl.; spikelets green 1-fld., fl. glumes 3-5-nerved.

34. CATABRO'SA, *Beauv.*

A perennial soft grass. *Spikelets* minute, subterete, obconic, in branched effuse panicles, 1- or 3-4-fld.; rachilla jointed between the fl. glumes, glabrous. *Empty glumes* 2, much shorter than the flowering, unequal, scarious, convex, awnless; upper truncate, crenulate. *Fl. glumes* coriaceous, cuneate, terete, 5-nerved, torn at the membranous tip, back smooth, awnless; lower 1-nerved sessile; upper 3-nerved pedicelled, often flowerless or male. *Palea* as long. *Scales* ovate-oblong. *Stamens* 3. *Ovary* ovoid, glabrous; stigmas subsessile, feathery. *Fruit* compressed, not or obscurely furrowed, enclosed in the fl. glume.—DISTRIB. Temp. and cold regions. —ETYM. καταβρώσις, from the *erose* top of the glumes.

C. aquat'ica, *Beauv.*; panicle pyramidal.
Watery places, N. to Shetland, common; Ireland; Channel Islands; fl. May–June.—Terrestrial or aquatic, soft, bright green, flaccid. *Rootstock* stout, branched, creeping and rooting. *Stems* 6-12 in., bent below, then erect, compressed, smooth, striate, leafy, sometimes much branched. *Leaves* flat, ⅛-¼ in. broad, linear, obtuse, upper short; sheaths inflated, smooth; ligule ovate, obtuse. *Panicle* 1-4 in., rachis stout, grooved; branches whorled, divided, slender. *Spikelets* ⅒ in., subsolitary, pedicelled, green and purplish; empty glumes, green, tips very broad, nerves very obscure; fl. glumes fulvous or purplish, smooth, nerves green; anthers white.— DISTRIB. Europe, N. Africa, N. Asia, Himalaya, N. America.—Var. *littora'lis,* Parn. (*mi'nor,* Bab.), has smaller 1-fld. spikelets. Sandy coasts.

35. MEL'ICA, *L.*

Perennial grasses. *Spikelets* terete, racemed or panicled, 1–2-fld. ;
rachilla elongate bearing a clavate head of fl. glumes. *Empty glumes* 2,
membranous, convex, subequal, awnless, 3–5- or upper 7-nerved. *Fl.
glumes* cartilaginous, convex, 5–9-nerved, awnless. *Palea* 2-nerved.
Scales fleshy, free or connate. *Ovary* glabrous ; stigmas broad, sessile,
feathery. *Fruit* oblong, subterete, loosely wrapped in the fl. glume.
—DISTRIB. Temp. and subtrop. regions ; species 20.—ETYM. An old
name.

1. **M. nu'tans,** *L.* ; spikelets drooping racemed secund ovoid 2-fld.,
upper flower 2-sexual.

Woods, from Argyll and Ross to Monmouth and Hereford, W. of England
only ; ascends to 1,400 ft. in the Highlands ; fl. May–June.—Stoloniferous.
Stems 10–18 in., filiform, inclined, 3-quetrous, scabrid above, with scaly
sheaths below. *Leaves* flat, slender, sparsely hairy, edge and keel scabrid ;
ligule 0. *Racemes* 1–2 in., drooping ; rachis flexuous. *Spikelets* 6–10, $\frac{1}{4}$–$\frac{1}{3}$
in., very shortly peduncled, broad ; empty glumes purple, oblong, edges and
tips broadly scarious ; fl. glumes greenish, strongly keeled, purple below the
scarious tip, glabrous.—DISTRIB. Europe (Arctic), N. Asia.

2. **M. uniflor'a,** *Retz.* ; spikelets very few erect panicled on long
capillary peduncles 2-fld., upper flower male.

Woods, from the Clyde and Elgin southd. ; ascends 1,500 ft. in N. England ;
Ireland ; fl. May–July.—Habit, foliage, spikelets, and colouring of *M.
nu'tans,* but inflorescence very different ; ligule long, from the auricles of
the sheath ; rachis of panicle capillary, lower branches 2-nate, 1–1½ in. ;
pedicels capillary, scaberulous.—DISTRIB. Europe.

36. DAC'TYLIS, *L.* COCK'S-FOOT GRASS.

A perennial grass. *Spikelets* laterally compressed, secund, sessile,
densely imbricate at the end of the branches of a one-sided panicle, 3–4-fld. ;
rachilla glabrous. *Empty glumes* 2, mucronate, keeled, membranous ; upper
larger, 3-nerved. *Fl. glumes* larger, cartilaginous, keeled, 5-nerved ; awn
subterminal, short, scabrid. *Palea* as long, 2-fid, nerves ciliate. *Scales*
2, acutely toothed. *Stamens* 3. *Ovary* glabrous ; styles terminal, stigmas
feathery. *Fruit* dorsally compressed, ventrally grooved, loosely enveloped
in the glume.—DISTRIB. Europe (Arctic), N. Africa, Siberia, N. India ;
introd. in N. America.—ETYM. δακτυλίς, a *finger's* breadth ; of obscure
application.

D. glomera'ta, *L.* ; rough, leaves broad flat.

Pastures, wast places, &c., N. to Shetland ; ascends to 1,600 ft. in N. England ;
Ireland ; Channel Islands : fl. June–July.—*Stems* 2–3 ft., creeping below,
erect, stout, smooth. *Leaves* long, flat, keeled, compressed ; sheaths scabrid ;
ligule long. *Panicle* strict, 1–6 in., green and violet ; lower branches few,
long, strict, scabrid, horizontal in flower, erect in fruit. *Spikelets* ⅓ in.,
oblong, scabrid.

37. BRI'ZA, *L.* Quaking Grass.

Spikelets large, ovate or cordate, turgid, pendulous, in effuse panicles, many-fld. ; branches in ½-whorls ; pedicels capillary ; rachilla jointed between the fl. glumes. *Empty glumes* 2, longer or shorter than the lowest flowering, subequal, broad, rounded at the back. *Fl. glumes* imbricate, boat-shaped or saccate, very obtuse, many-nerved, upper often flowerless. *Palea* small, nerves ciliate. *Scales* 2, ovate-lanceolate. *Stamens* 3. *Ovary* glabrous ; styles short, terminal, stigmas feathery. *Fruit* broadly ovoid, compressed, enclosed in the hardened glume.—Distrib. Temp. Europe, Africa, and Asia : species 10.—Etym. An old Greek name.

1. **B. me'dia,** *L.* ; perennial, ligule short, spikelets ovate, empty glumes shorter than the 1st fl. glume.

Meadows and heaths, &c., from Ross southd. ; ascends to 2,100 ft. in Yorkshire ; Ireland ; Channel Islands ; fl. June.—*Stems* solitary, creeping below, 6–18 in., very slender, smooth. *Leaves* flat, smooth or scabrid ; sheaths smooth, upper inflated. *Panicle* pyramidal ; branches very long, capillary. *Spikelets* ¼–⅓ in., green or purplish, shining ; fl. glumes 5–9, sheathing one another.—Distrib. Europe, N. and W. Asia ; introd. in N. America.

2. **B. mi'nor,** *L.* ; annual, ligule long, empty glumes longer than the 1st fl. glume.

Fields, Hants to Cornwall ; Cork ; Channel Islands ; fl. July.—*Stems* tufted, 4–10 in. *Panicle* almost as in *B. me'dia*, but spikelets more numerous, smaller, ⅛ in. diam., broader than long.—Distrib. From France and Spain to Turkey, N. Africa.

38. PO'A, *L.* Meadow-Grass.

Spikelets compressed, 2–many-fld., in branched usually effuse panicles ; branches 2-nate or in ½-whorls ; rachilla jointed between the fl. glumes. *Empty glumes* 2, shorter than the lowest flowering, unequal, acute or obtuse, keeled ; lower 1–3-nerved ; upper larger, 3-nerved, awnless. *Fl. glumes* often webbed below, keeled, acute, 5–7-nerved, tips hyaline. *Palea* 2-fid, nerves 2 ciliate. *Scales* tumid below. *Stamens* 3. *Ovary* glabrous ; styles 2, short, terminal, stigmas feathery. *Fruit* obtusely 3-gonous, grooved, enclosed in the glume, glabrous.—Distrib. Chiefly cold and temp. regions ; species 80.—Etym. πόα, fodder.

* *Annual. Branches of the panicle solitary or 2-nate.*

1. **P. an'nua,** *L.* ; stems compressed, leaves obtuse, ligule long acute, branches of panicle 2-nate at length deflexed, fl. glumes 3–7 5-nerved.

Waste places, &c. ; ascends to 3,200 ft. in the Highlands ; Ireland ; Channel Islands ; fl. April–Sept.—Flaccid, bright green, sometimes glaucous, quite glabrous and smooth. *Stems* 6–12 in., weak. *Leaves* linear, subacute, often waved. *Panicle* 1–3 in., subpyramidal, subsecund, green or purplish. *Spikelets* ⅛ in., subsessile ; empty glumes broadly hyaline, upper broadest in the middle; fl. glumes glabrous or hairy below the middle. - Distrib. N. temp. Europe (Arctic), Asia, and N. Africa; N. America, native (?).

** *Perennial; rootstock creeping, stoloniferous.*

2. P. praten'sis, *L.* ; stem smooth terete, upper leaf shorter than its sheath, ligule long, branches of diffuse panicle 3–5-nate scabrid, fl. glumes 3–5 acute webbed, edges and keel silky, nerves 5 distinct.

Meadows, banks, and pastures, N. to Shetland; ascends to 2,400 ft. in Yorkshire; in Ireland to 2,800 ft.; fl. June–July.—Glabrous, pale green. *Stems* 1–2 ft., rather stout, very stoloniferous. *Leaves* linear, flat, acute, tip often concave; sheaths subcompressed. *Panicle* 2–6 in., pyramidal, closed or open after flowering; branches long or short, rarely 2–3-nate. *Spikelets* ⅓–¼ in., green or purplish; empty glumes acuminate, keel scabrid; fl. glumes broadly hyaline.—Distrib. N. temp. and Arctic regions.—*P. subcæru'lea,* Sm., is a small glaucous state, and *P. angustifo'lia,* L., one with slender leaves.—*P. strigo'sa,* Gaud., is a small state, growing in dark places, with convolute leaves, and a narrow panicle that closes after flowering.

3. P. compres'sa, *L.* ; stem smooth compressed, upper leaf equalling or exceeding its sheath, ligule short, branches of effuse panicle 2–3-nate scabrid, fl. glumes 4–9 obtuse nearly glabrous, nerves obscure.

Dry banks, walls, &c., from Ross southd.; rare in Ireland; Channel Islands; fl. June–July.—Smooth or slightly rough, glabrous, more or less glaucous. *Stems* 1–2 ft., usually much bent towards the base. *Leaves* flat, rough or not. *Panicle* 1–3 in., usually more secund and contracted than in *P. praten'sis,* but sometimes effuse. *Spikelets* ⅛ in., green or bluish-purple; empty glumes subequal; fl. glumes broadly hyaline, 3-nerved (*P. subcompres'sa,* Parn.) or 5-nerved (*P. polyno'da,* Parn.) and then more usually webbed.—Distrib. Europe, N. and W. Asia; native (?) of N. America.

*** *Rootstock shortly creeping, stolons 0.*

4. P. trivia'lis, *L.* ; stems and sheaths usually rough, ligule oblong acute, branches of effuse panicle 5-nate scabrid, fl. glumes 3–5 acuminate glabrous or webbed, nerves 5 distinct.

Woods and meadows, N. to Shetland; ascends to 2,500 ft. in the Highlands; Ireland; Channel Islands; fl. June–July.—Very near indeed to *P. praten'sis,* and chiefly distinguishable by the roughness, absence of stolons, and erect panicle. VAR. *P. Koeleri,* DC., which grows in woods, has smooth sheaths. VAR. *parviflora,* Parn., is a weak state with 1–2 fld. spikelets.—Distrib. Europe, N. Africa, N. Asia; introd. in N. America.

5. P. nemora'lis, *L.* ; stems and sheaths smooth, ligule short or 0, branches of the subsecund panicle 2–5-nate scabrid, fl. glumes 1–5 subacute, edges and keel pubescent, nerves obsolete.

Copses and woods, from Skye and Elgin southd.; ascends to 1,600 ft. in the Highlands; Ireland, rare; Channel Islands; fl. June–July.—Smooth, bright green, glaucous or not. *Stems* 1–3 ft., very slender, terete or subcompressed. *Leaves* linear, very narrow, flaccid; sheaths smooth, striate. *Panicle* 2–5 in., slender, nodding. *Spikelets* small, ⅛ in., yellow-green and purplish ; empty glumes often equalling the flowering, acuminate ; fl. glumes 2–5, scarely hyaline at the tip or margins.—DISTRIB. Europe, N. Asia, Himalaya, N. America.—Very variable. I am quite unable to define the forms into which it has been divided, and about which no two authors are agreed. I recognise the following British varieties, but I cannot correlate them with exotic ones satisfactorily. The characters taken from the length of the upper leaf and its sheath and the position of the upper node are valueless. VAR. *P. Parnell'ii,* Bab. ; smaller, more slender, with smaller spikelets.— VAR. *P. Balfour'ii,* Parn. (*P. monta'na,* Parn. ; ? *P. cæ'sia* and *glau'ca,* Sm.) ; stouter, panicle more erect, spikelets larger.—Alpine cliffs ; ascends to 3,000 ft.

6. **P. lax'a,** *Hœnke ;* stems slightly compressed, leaves short, ligules long torn acute, branches of narrow lax subsecund panicle solitary or 2-nate smooth, fl. glumes 2–4, keel and margins villous, nerves 3–5 obscure.

Rocks, on the alps of Aberdeen and Inverness ; alt. 2,000 to 3,600 ft. ; fl. July–Aug.—Flaccid, smooth, pale glaucous green. *Stems* 4–10 in., compressed and prostrate below. *Leaves* linear, obliquely mucronate, upper longer than its sheath ; sheaths compressed. *Panicle* 1–3 in., drooping, lax or open, branches rarely scaberulous. *Spikelets* 1/10–⅙ in., often viviparous, green and purplish ; empty glumes acuminate, tips hyaline.—DISTRIB. N., Alpine, and Arctic Europe, Siberia, N. America.—Often with difficulty distinguished from *P. alpi'na.*
P. LAXA proper ; leaves channelled tip concave, panicle open in flower closed in fruit. *P. flexuo'sa,* Sm. ; *P. mi'nor,* Gaud.
Sub-sp. P. STRICTA, Lindb. ; leaves flat to the tip, panicle open in flower spreading in fruit.

7. **P. alpi'na,** *L.* ; stems glabrous terete, leaves broad firm tip rounded, upper ligules long acute, branches of erect spreading panicle 2-nate, fl. glumes 3–9, keel and margins pubescent, nerves 3–5 obscure.

Rocks, &c., on lofty mts., N. Wales, York to Sutherland ; alt. 3–4,000 ft. ; W. Ireland, very rare ; fl. June–Aug.—Smooth. *Rootstock* stout, creeping. *Stems* 10–18 in., stout. *Leaves* rather short, strict, rigid, keeled, mucronate, edges thickened scabrid, upper shorter than its sheath ; lower sheaths broad, white, membranous, persistent, leafless, upper compressed smooth. *Panicle* oblong or pyramidal, 1–3 in. broad. *Spikelets* ⅓ in., green and purplish, often viviparous ; empty glumes unequal, ovate, acute ; fl. glumes with hyaline tips, webbed, also broad.—DISTRIB. N., Alpine, and Arctic Europe, N. and W. Asia, Himalaya, N. America.

8. **P. bulbo'sa,** *L.* ; lower nodes tuberous, stem and terete sheaths smooth, leaves narrow, ligules all long acute, branches of panicle 2-nate scabrid, fl. glumes 3–6, margins and acute keel pubescent, nerves obsolete.

Sandy seashores, Norfolk to Kent and Devon ; fl. April–May.—Glabrous, rather rigid. *Stems* 6–10 in. ; tubers ovoid, ¼–⅓ in., covered with lax sheaths. *Leaves* 1–1½ in., very narrow, keeled, curved, upper very short ; lower sheaths short, upper long compressed. *Panicle* 1–1½ in., ovate, compressed.

Spikelets $\frac{1}{10}$ in. broad, green and purplish-brown ; empty glumes ovate, acute, keel scabrid ; fl. glumes also broad, acute.—DISTRIB. Europe, N. Africa, Siberia.—Very near *P. alpi'na*, but differing in locality, habit, and the close ovate panicle.

39. GLYCE'RIA, *Br.* MANNA GRASS.

Perennial tall grasses ; leaf sheaths entire. *Spikelets* linear, subterete, in effuse or contracted panicles, many-fld., branches in $\frac{1}{2}$-whorls ; rachilla jointed between the fl. glumes. *Empty glumes* 2, shorter than the lowest flowering, 1-5-nerved, unequal, membranous, convex, awnless ; upper larger. *Fl. glumes* caducous, cartilaginous, convex, not keeled, tip obtuse usually scarious ; nerves 3-9 evanescent upwards. *Palea* 2-fid, nerves ciliate. *Scales* fleshy, truncate. *Stamens* 3. *Ovary* glabrous ; styles terminal, short or 0, stigmas feathery. *Fruit* oblong, enveloped and sometimes adnate to the fl. glume, channelled.—DISTRIB. Temp. and cold climates ; species 30.—ETYM. γλυκερός, in allusion to the *sweet* grain.

SECTION 1. **Hydrochloa.** *Scales* fleshy truncate or 0. *Fruit* thick obtuse, hardly compressed, furrow very narrow or 0.

1. **G. aquat'ica,** *Sm.* ; panicle much branched, spikelets oblong, fl. glumes 5-9 entire. *G. spectab'ilis*, Koch.

Watery places, from Elgin and Mull southd. ; Ireland; Channel Islands; fl. July–Aug.—*Rootstock* stout, extensively creeping. *Stems* 2-6 ft., stout, smooth, striate. *Leaves* 1-2 ft., $\frac{3}{4}$-$\frac{1}{2}$ in. broad, flat, suberect, acute; sheaths smooth ; ligule short. *Panicle* 6-12 in. *Spikelets* $\frac{1}{4}$ in., yellow-green and purple ; glumes scabrid; empty short, shining ; flowering rigid, obtuse ; nerves strong.—DISTRIB. Europe, Siberia, Himalaya, N. America.

2. **G. flu'itans,** *Br.* ; panicle subsimple, spikelets linear, fl. glumes 7-20.

Watery places, N. to Shetland ; ascends to 1,600 ft. in Yorkshire; Ireland ; Channel Islands ; fl. July–Aug.—*Rootstock* stout, widely creeping. *Stem* 1-3 ft., stout, branched, terete, floating or creeping, smooth. *Leaves* $\frac{3}{4}$-$\frac{1}{2}$ in. broad, flat, acute, short, or the upper long and floating ; sheaths long, compressed ; ligule broad, acute. *Panicle* 1-2 ft., simple or branched ; rachis 3-gonous ; branches remote, smooth. *Spikelets* $\frac{1}{4}$-2 in., erect, green or tips purplish ; empty glumes unequal, tips torn ; fl. glumes scabrid, tips often ragged.— DISTRIB. Europe, N. Africa, W. Siberia, Himalaya, N. America.—A very variable plant.

VAR. *G. plica'ta*, Fr., has divaricate fruiting spikelets, and fl. glumes twice as long as broad.—VAR. *G. pedicella'ta*, Towns., has broader blunter leaves, with rough furrowed sheaths, 3-toothed fl. glumes, not exceeding the palea (it never fruits).—VAR. *G. declina'ta*, Brebiss, is a dwarf form with smooth sheaths, few-fld. spikelets, and the palea longer than the 3-toothed fl. glume. Hampshire and Scilly Islands.

SECTION 2. **A'tropis,** *Rupr.* (gen.). *Scales* slender, distinct. *Fruit* thick, dorsally compressed, ventrally hollow or broadly furrowed.

3, **G. marit'ima,** *Wahlb.* ; perennial, stolons long, leaves involute, branches of contracted panicle solitary or 2-3-nate very short, fl. glumes 4-12 obtuse or apiculate. *Scleroch'loa marit'ima,* Lindl.

Muddy, &c., sea-coasts, N. to Shetland; Ireland; Channel Islands; fl. July.— *Rootstock* widely creeping. *Stems* 8-16 in., terete, smooth. *Leaves* involute or channelled, acute or pungent; sheaths smooth; ligule rather long, oblong. *Panicle* 1-3 in., oblong, contracted; branches appressed or horizontal, rarely deflexed, or more than 3-nate; rachis subterete, grooved on one side. *Spikelets* ¼-½ in., ovate-oblong or linear-elongate, green or purplish; empty glumes subacute; fl. glumes obscurely 3-toothed, tip hyaline or not.— DISTRIB. Europe, N. Africa, Siberia, N. America.—VAR. *his'pida,* Parn.; panicle rough.—VAR. *ripa'ria,* Towns.; more slender, spikelets fewer, flowers later. Hants.

4. **G. Borre'ri,** *Bab.* ; perennial, stolons 0, leaves flat, branches of contracted panicle solitary or 2-3-nate very short, fl. glumes 3-6. *G. confer'ta,* Fr. ; *Scleroch'loa confer'ta,* Bab.

Salt marshes, E. and S. coasts; from York to Kent and W. to Devon ; Waterford; fl. June-Aug.—*Stems* densely tufted, 12-20 in., stout. *Leaves* short, sheaths long; ligule short. *Panicle* 3-6 in., strict, rachis stout, lower branches finally spreading. *Spikelets* smaller than in *G. marit'ima,* fl. glumes nearly all alike.—DISTRIB. Holland, N.W. Germany.

5. **G. dis'tans,** *Wahlb.* ; perennial, leaves flat, branches of effuse panicle 4-5-nate long slender at length deflexed, fl. glumes 3-6 obliquely truncate. *Scleroch'loa dis'tans,* Bab. ; *S. multicul'mis,* Syme.

Sandy chiefly maritime places, N. to Shetland Ireland, rare; fl. July-Aug.— Rarely stoloniferous; leaves broad, flat; stem in the typical states tall, slender; panicle with long horizontal and deflexed branches, and spikelets ⅛-¼ in. or less.—DISTRIB. Europe, N. Africa, Siberia, Himalaya, N. America.

6. **G. procum'bens,** *Dumort.* ; annual, leaves flat, sheaths inflated, branches of rigid panicle short solitary or 2-3-nate, fl. glumes 3-5 obtuse mucronate. *Scleroch'loa procum'bens,* Beauv.

Muddy seashores of England ; Ireland, rare; Channel Islands; fl. June-July. —*Root* fibrous. *Stems* 3-6 in., short, stout, rigid, spreading, erect or decumbent. *Leaves* short, subacute, glaucous; ligule short ; sheaths large, grooved. *Panicle* 1-2 in., green; rachis angular; branches stout, erecto-patent, subdistichous, smooth. *Spikelets* ⅛-¼ in., subsessile, crowded.—DISTRIB. W. Europe from Holland to Spain and Italy, N. Africa, Siberia, Himalaya, N. America.

40. FESTU'CA, *L.* FESCUE GRASS.

Spikelets subterete, racemed or panicled, 3- or more-fld. ; rachilla jointed between the fl. glumes. *Empty glumes* 2, rarely 1, shorter than the lowest fl. glume, unequal, membranous, acute ; upper larger, 3-nerved ; lower 1-nerved. *Fl. glumes* convex, 3-5-nerved, mucronate or awned at or near the tip, upper sometimes empty. *Palea* 2-fid, nerves hairy. *Scales* 2, notched. *Stamens* 1–3. *Ovary* glabrous ; styles short, terminal, stigmas feathery. *Fruit* free or adherent to the fl. glume and palea.—DIS-TRIB. Arctic, cold, and temp. regions ; species 80.—ETYM. An old Latin name.

SECTION 1. **Schedono'rus,** *Beauv.* (gen.). Perennial, tall. *Flowers* 3-androus. *Fruit* free within the fl. glume.

* *Leaves flat ; ligule of upper sheath short.*

1. **F. ela'tior,** *L.* ; panicle diffuse nodding, fl. glumes 3–7 glabrous acute or almost awned, ovary glabrous.

River-banks and wet places, N. to Shetland ; ascends to 1,300 ft. in N. England ; Ireland ; Channel Islands ; fl. June–July.—*Rootstock* creeping, stoloni-ferous. *Stems* 2–6 ft., nodding, smooth. *Leaves* ¼–⅓ in. broad, smooth, striate. *Panicle* 3–6 in., contracted after flowering ; rachis 3-quetrous, smooth ; branches 2-nate, scabrid. *Spikelets* many, ⅓–¼ in., linear-oblong, green and dull purple ; empty glumes broadly hyaline ; fl. glumes scabrid above, rarely awned ; margins hyaline.—DISTRIB. Europe (Arctic), N. Africa, W. Siberia, Himalaya, N. America.—A large maritime state with rougher sheaths and branches of panicle divaricate after flowering is the *F. arundina'cea,* Schreb.

2. **F. praten'sis,** *Huds.* ; panicle subsecund nodding close, fl. glumes 4–10 glabrous obtuse or mucronate, ovary glabrous. *F. ela'tior,* L., in part.

River-banks and wet places, from Caithness southd. ; ascends to 1,600 ft. in N. England ; Ireland ; fl. June–July.—Perhaps only a sub-species of *F. ela'tior,* but smaller, less stoloniferous ; panicle much narrower, more distichous and simple, sometimes 10 in., branches shorter, in pairs, one with 1 spikelet, the other with 2 or more ; fl. glumes more numerous, some-times shortly awned.—DISTRIB. Europe (Arctic), N. Asia ; introd. in N. America.

F. lolia'cea, Curt. (not Huds.), with inflorescence racemose or spiked, spikelets distichous, awn 0, fl. glumes more obtuse, is a hybrid with *Lol'ium peren'ne.*

3. **F. gigante'a,** *Vill.* ; panicle very open nodding, fl. glumes 3–8 scabrid, awn double their length, ovary glabrous. *Bromus,* L.

Damp woods and hedgebanks, from Argyll and Elgin southd. ; ascends to 1,000 ft. in Yorkshire ; Ireland ; fl. July–Aug.—Habit of *Bromus.* *Root* fibrous. *Stems* 2–4 ft., smooth, terete, nodding. *Leaves* flaccid, ¼–½ in. broad, bright green, striate, scaberulous above ; sheaths smooth. *Panicle* 8–12 in. ; branches in pairs, and 3-quetrous rachis long slender scabrid. *Spikelets* ½ in., linear-ovate or -oblong, membranous, pale green ; empty glumes lanceolate, broadly hyaline ; fl. glumes 2-fid, awn variable, very slender, flexuous ; styles subterminal.—DISTRIB. Europe, Siberia.—Flowers sometimes few (*F. triflo'ra,* Sm.).

** *Leaves involute ; ligule of upper sheath long, not auricled.*

4. **F. sylvat'ica,** *Vill.* ; panicle open subsecund erect, fl. glumes 3–5 acute scabrid, top of ovary hairy. *F. Calama'ria,* Sm.

Woods in hilly districts, from Banff and Inverness to Wilts and Sussex; absent in E. England and ? Wales; rare in Ireland; fl. July.—*Root* fibrous; stolons short. *Stems* 2–3 ft., stiff, erect, smooth, terete. *Leaves* ¼–½ in., broad, glaucous above, edges scabrid; sheaths almost smooth, lower lax leafless brown. *Panicle* 3–6 in., ovate, much branched; rachis and 2–4-nate branches very slender, slightly scabrid. *Spikelets* ¼ in., broadly ovate, small, flat, pale yellow-green; axis scabrid; empty glumes linear-subulate; fl. glumes spreading, slender, acuminate, shortly awned, scaberulous.— Distrib. W. Europe to Austria, Germany, and Italy.—*F. decid'ua*, Sm., is a narrower-leaved 2–3-fld. variety.

Section 2. **Festu'ca** proper. Perennial. *Lower* (or all) *leaves* setaceous; ligule auricled. *Spikelets* panicled. *Flowers* 3-androus; awn short.

5. **F. ovi'na,** *L.* ; glaucous, leaves setaceous or upper flat, ligule 2-lobed, panicle subunilateral, spikelets 3–12-fld. purplish, fl. glumes terete mucronate or shortly awned.

Dry hilly pastures, woods, &c., N. to Shetland, at all elevations; Ireland; Channel Islands; fl. June–July.—One of the most abundant grasses, 3–24 in., slender, variable in size, colour, and habit; the following sub-species express its principal modifications.—Distrib. Europe, N. Africa, Siberia, Himalaya, N. and S. America, Mts. of Australasia.

F. ovi'na proper; densely tufted, leaves all setaceous, sheaths glabrous, panicle contracted subsecund, spikelets small 3–5-fld., fl. glumes ⅛ in. mucronate or awned often viviparous.—Upland copses, moors and sandy places; ascends to 4,300 ft. (Arctic).—Syme has grouped the prevalent forms as follows : *F. ovi'na* proper; leaves setaceous flaccid green, radical short, fl. glumes awned.—*F. tenuifo'lia*, Sibth.; leaves setaceous flaccid green radical longer, fl. glume mucronate.—Var. *F. glau'ca*, Lamk.; leaves stouter, rigid, glaucous, radical short often recurved, fl. glume awned.—Var. *ma'jor*; taller, panicle larger, stem leaves broader than the radical, fl. glume usually awned.

Sub-sp. F. durius'cula, *L* ; less densely tufted, stoloniferous, stem-leaves flat, sheaths downy, panicle more open, spikelets usually many-fld., fl. glumes ¼ in. narrow. *F. cæ'sia*, Sm.?—Pastures and meadows; ascends to 2,700 ft. in the Highlands.

Sub-sp. F. ru'bra, *L.*; taller, laxly tufted, stoloniferous, leaves flat or involute, lower sheaths hairy, panicle effuse subsecund, spikelets pale red, fl. glumes ¼–⅓ in. broader awned. Shaded places in low grounds.—Var. *F. arena'ria*, Osb. (*F. sabulic'ola*, Duf., *F. oraria*, Dum., *F. rubra*, Sm. not L.); rigid, creeping, leaves all involute. Sandy shores.

Section 3. **Vul'pia,** *Gmel.* (gen.). Annual. *Leaves* setaceous. *Panicle* contracted. *Spikelets* secund, racemose or spiked. *Fl. glume* awned. *Stamens* 1–3.

6. **F. Myu'ros,** *L.* ; panicle branched at the base only very long and slender, lower empty glume small, fl. glumes 5–8 equalling their slender awns, stamens 1–3.

Walls, sandy and gravelly pastures, N. to Sutherland; Mid. and S. Ireland; Channel Islands; fl. June.—*Root* fibrous. *Stems* 6–18 in., very slender,

K K

leafy, glabrous, lower setaceous; upper sheaths long, terete, grooved, smooth; ligule very short. *Panicle* 4–10 in., strict or flexuous, very narrow, opaque or shining, glabrous or pubescent, rachis and branches angular smooth or scaberulous; lower branches appressed; upper very short. *Spikelets* with the awns ½–¾ in., cuneate when expanded, rachilla smooth; empty glumes subulate; fl. glumes distant, terete, almost subulate, narrowed into the slender awn, nerves obscure.—DISTRIB. Europe, N. Africa; introd. in N. America.

F. MYU′ROS proper; slender, upper sheath exceeding the stem and often the long slender panicle, lower empty glume ⅓ shorter than the upper. *F. pseudo-myu′ros,* Koch. From York (ascending 1,000 ft.) southd.—*F. ambig′ua,* Le Gall, is shorter, with upper empty glume 3–6 times as long as the lower, stamen 1. I. of Wight, Suffolk, Norfolk, Dorset.

Sub-sp. F. SCIUROI′DES, *Roth;* stem shorter naked above, panicle shorter more open, lower branches slender, lower empty glumes twice as long as the upper. *F. bromoi′des,* Sm.—N. to Caithness.

7. **F. uniglu′mis,** *Sol.* ; panicle or spike short, branches very short, lower empty glume minute or 0, fl. glumes 4–10 shorter than the awns, flowers 2–3-androus. *Vul′pia membrana′cea,* Link.

Sandy sea-coasts, from Lancaster and Norfolk to Devon and Kent; S. Wales; E. Ireland, local; Channel Islands; fl. June–July.—*Root* fibrous. *Stems* 4–10 in., bent below, slender, glabrous. *Leaves* setaceous, short; upper sheaths large, inflated; ligule short, auricled. *Panicle* 1½–2 in., oblong, unilateral, pale green, shining, rachis smooth; lower branches 2-nate. *Spikelets* with the awns ¾–1 in., almost subulate; fruiting cuneate; pedicels short, stout; rachilla smooth; empty glumes with subulate tips or awns, upper 3-nerved, broadly hyaline; fl. glumes distant, very narrow, strongly nerved, scabrid, terete below, compressed above, gradually narrowed into the subulate awn; ovary glabrous.—DISTRIB. W. Europe from Belgium S. and E. to Turkey, N. Africa.

SECTION 4. **Catapo′dium,** Link (gen.). Annual, rigid. *Spikelets* sessile or subsessile alternate on a simple or branched flat (not excavate) rachis. *Fl. glumes* subacute or mucronate.

8. **F. rig′ida,** *Kth.* ; branches of panicle distichous with 3–5 subsessile spikelets, fl. glumes 7–10 acute, nerves faint. *Glyce′ria,* Sm. ; *Sclerochlo′a,* Link.

Dry rocks, walls; &c., from Ross southd. ; Ireland; Channel Islands; fl. June. —Often purple, glabrous, smooth. *Root* fibrous. *Stem* 3–6 in. *Leaves* involute, subsetaceous; sheaths terete, grooved; ligule oblong. *Panicle* 1½–2½ in., strict; rachis broadly channelled, edges scabrid. *Spikelets* ¼–½ in., rarely solitary (then spiked); pedicels short, stout, ½-terete; fl. glumes quite terete, shining, smooth.—DISTRIB. W. and S. Europe from Holland southd., Canaries.

9. **F. lolia′cea,** *Huds.* ; spikelets spiked distichous, fl. glumes 8–12 obtuse mucronate, nerves faint. *Glyce′ria,* Wats. *Sclerochlo′a,* Woods. *Trit′icum,* Sm. *Demaze′ria,* Nym.

Sandy shores, from Fife and Wigton southd.; Ireland; Channel Islands; fl. July–Aug.—Green, smooth, glabrous. *Root* fibrous. *Stems* spreading and erect, leafy. *Leaves* small, flat or involute; sheaths smooth; ligule oblong. *Spikes* strict, 1–2½ in.; rachis stout, ½-terete, hardly flexuous, edges smooth. *Spikelets* ¼–⅓ in., sessile, erect, green; fl. glumes broad, obtusely-keeled, smooth; lateral nerves strongest.—Distrib. S. and W. Europe from France to Dalmatia, N. Africa.

41. BRO'MUS, *L.* Brome Grass.

Annual or perennial grasses. *Spikelets* subterete or laterally compressed, panicled or racemed, 5–many-fld.; rachilla jointed between the fl. glumes. *Empty glumes* 2, shorter than the lowest fl. glume, unequal, coriaceous, acute, awnless; lower smaller 1–5-nerved; upper 3–9-nerved. *Fl. glumes* convex or keeled, 5–9-nerved, tip entire or 2-fid, 1–3-awned, mid awn often bent or twisted. *Palea* 2-fid, nerves ciliate. *Scales* entire. *Stamens* 3, rarely 2. *Ovary* hairy at the top; styles short, inserted below the top, stigmas feathery. *Fruit* linear, grooved, oblong, adherent to the palea.— Distrib. Temp. and cold climates; species 40.—Etym. Greek name for *Oat*.

Section 1. **Festucoides,** *Coss.* and *Dur. Lower empty glume* 1-nerved, upper 3-nerved. *Fl. glumes* distant, narrow, convex below, keeled and compressed above, 5-nerved, awn terminal. *Styles* lateral on the ovary.

* *Keels of palea pubescent. Fl. glumes spreading in flower, erect in fruit.*

1. **B. as'per,** *Murr.*; leaves flat and sheaths hairy, panicle secund nodding, fl. glumes 5–8 twice as long as the awn. *B. ramo'sus,* Huds.

Damp woods, hedgebanks, &c., from Elgin and Mull southd.; ascends to 1,200 ft. in Yorkshire; Ireland; fl. June–July.—Annual or perennial. *Root* fibrous. *Stem* 2–6 ft., smooth. *Leaves* ¼–½ in. diam., green, long, hairs scattered, reflexed on the sheath; ligule short. *Panicle* 3–5 in., lower branches 2–6-nate, long, lax, capillary, and rachis scabrid. *Spikelets* 1 in., narrow, glaucous green, rachilla scabrid; empty glumes acuminate, hairy or glabrous; fl. glumes with the awn ⅓–⅔ in., diverging in flower, then erect, more or less hairy, lateral nerves strong, tip 2-toothed; awn variable in length.—Distrib. Europe, Siberia; introd. in N. America.
B. sero'tinus, Benek., has sheaths all with reflexed hairs, lower panicle branches 2-nate, empty glumes unequal, fl. glume glabrous next the midrib. —*B. Beneken'ii,* Syme, has upper sheaths glabrate, lower panicle branches 3–6-nate, with a semilunar thickening at the base, empty glumes subequal, fl. glume hairy all over. Near London, a doubtful native.

2. **B. erec'tus,** *Huds.*; leaves involute hairy, panicle erect narrow, fl. glumes twice as long as the awn.

Fields and waste places in dry soil, from Fife to Kent and Sussex; Ireland, rare; fl. June–July.—Perennial. *Rootstock* stout, creeping; stolons 0. *Stems* 1–3 ft., rigid, smooth, bent below, then erect. *Leaves* narrow, rigid, almost subulate, hairy; hairs scattered on the upper sheaths, erect; ligule short. *Panicle* 4–6 in., subsimple, strict, branches 2–3-nate and rachis scabrid. *Spikelets* ¾–1½ in., subsolitary, green or purplish; empty glumes longer than in *B. as′per;* fl. glumes hairy all over (var. *villo′sa,* Bab.), or on the nerves only, much the same in both.—DISTRIB. Europe, N. Africa.

** *Annual. Keels of palea pectinate-ciliate. Fl. glumes erect both in flower and fruit (except B. max′imus).*

3. **B. ster′ilis,** *L.* ; leaves flat hairy, sheaths compressed, panicle very lax, branches few very long in ½-whorls, spikelets subsolitary, fl. glumes 7–10 much shorter than their awns.

Fields and waste places, N. to Caithness; ascends to 1,200 ft. in Derby; Ireland; Channel Islands; fl. June–July.—*Root* fibrous. *Stems* 1–2 ft., erect, smooth, leafy. *Leaves* ¼–½ in. broad, flaccid, ribbed ; sheaths glabrous or pubescent. *Panicle* 6–16 in. broad, nodding, very lax and open ; branches horizontal and drooping. *Spikelets* with the awns 2 in., nodding, green ; empty glumes long, narrow, upper twice the longest; fl. glumes very narrow, gradually narrowed into the very slender straight awn, with 7 strong equidistant ribs, margins hyaline.—DISTRIB. Europe, W. Siberia, N. Africa ; introd. in N. America.

4. **B. madriten′sis,** *L.* ; leaves narrow flat hairy, sheaths terete, panicle erect, branches few strict erect, fl. glumes 6–8 half as long as the straight stout awn.

Sandy waste places, S. Wales to Oxford, and Kent to Devon ; Tipperary ; Channel Islands ; (native? *Wats.*) ; fl. June–July.—*Stems* 6–14 in., strict from an ascending base, smooth, leafy. *Leaves* ¹⁄₁₆–½ in., more or less hairy or tomentose ; hairs on sheaths reflexed ; ligule short, truncate. *Panicle* 2–4 in., oblong, compressed ; branches appressed, short, and slender rachis scabrid. *Spikelets* ½–¾ in., excluding the awns, dull green, rachilla smooth ; lower empty glume subulate, half as long as the narrow linear upper ; fl. glumes ½ in., very narrow, scabrid, nerves 7 lateral close together, margins and 2-fid tip broadly hyaline; awn slender ; stamens usually 2.—DISTRIB. From France and Spain to Turkey, N. Africa.

B. madriten′sis proper (*B. dian′drus,* Curt.) ; rachis pedicels and glumes scabrous. VAR. *Curtis′ii,* Bab.—VAR. *B. rig′idus,* Roth ; rachis pedicels and glumes pubescent. Channel Islands.

5. **B. max′imus,** *Desf.* ; leaves flat hairy, sheaths terete, panicle erect, branches few strict, fl. glumes 4–6 not ⅛ as long as the straight awn.

Sandy shores, Channel Islands ; fl. June–July.—*Root* fibrous. *Stems* 6–10 in., terete, pubescent. *Leaves* ¼–½ in. broad, bright green ; hairs on sheaths spreading or reflexed ; ligule short, truncate. *Panicle* 4–7 in., at length nodding ; branches short, appressed, and rachis pubescent. *Spikelets* with the awns 2–3 in., pale green or purplish ; rachilla glabrous ; empty glumes hyaline, very narrow, awned ; upper twice as long as the lower, almost equalling the lowest fl. glume ; fl. glumes ti in, narrow, scabrid, narrowed into the s abrid awn ; nerves faint ; stamens 2–3.—DISTRIB. France, Spain, Mediterranean.

SECTION 2. **Zeobro'wus**, *Griseb.* Annual or perennial. *Lower empty glume* 5- upper 7-9-nerved. *Fl. glumes* close, convex, 5-7-nerved, 2-fid, awned in the sinus. *Palea* with pectinate-ciliate nerves. *Styles* inserted below the top of the ovary. *Serrafal'cus*, Parl. (gen.).

* *Fl. glume longer than its palea.*

6. B. mol'lis, *L.* ; pubescent or tomentose, panicle ovoid strict, lower empty glume broadly ovate, fl. glumes 6–10 densely imbricate pubescent opaque equalling the slender awn.

Roadsides and waste places, N. to Shetland ; ascends to 1,800 ft. in N. England ; Ireland ; Channel Islands ; fl. May–July.—Glaucous green. *Stems* 4–24 in., terete. *Leaves* flat, ¼–½ in. broad, soft, edges scabrid ; sheaths terete, villous ; ligule short. *Panicle* 1–3 in., erect or nodding ; branches very short, subsimple. *Spikelets* ½–¾ in., compressed, oblong, tips conic ; empty glumes broadly ovate, acute, strongly nerved, upper much largest ; fl. glumes caducous, broadest and obtusely angled above the middle, 2-fid, pubescent.—DISTRIB. Europe, N. Africa, W. Siberia ; introd. in N. America.—Very variable ; Syme enumerates three forms, the typical with pubescent spikelets, and erect nearly straight awns ; a smaller glabrescent one (*glabres'cens*) ; and *Lloydia'nus* (*mollifor'mis*, Lloyd) with awns spreading outwards in fruit. *B. hordea'ceus*, Fr., is a prostrate maritime form.

7. B. racemo'sus, *L.* ; leaves and sheaths glabrate or hairy, panicle narrow, lower empty glume lanceolate, fl. glumes 6–10 imbricate scabrid shining equalling the slender awn.

Fields and waste places, N. to Shetland ; ascends to 1,200 ft. in the Highlands ; Ireland ; Channel Islands ; fl. June–July.—Very similar to *B. mol'lis*, but subglabrous, often 2–3-ft., rigid ; leaves rigid, more ciliate ; branches of panicle 3–5-nate, long and slender ; spikelets narrower, more acute, scabrid ; empty glumes narrower, especially the lower ; fl. glume broadest above the middle, margin obtusely angled.—DISTRIB. Europe (excl. Russia), N. Africa ; introd. in N. America.

B. commuta'tus, Schrad. (*B. praten'sis*, Ehrh., *B. arven'sis*, Sm.), is stouter, panicle more compound drooping, spikelets shorter, margins of caducous fl. glumes less rounded at the broadest part.

8. B. secali'nus, *L.* ; glabrate or sheaths hairy, panicle effuse hairy, fl. glumes 5–8 not imbricate terete scabrid longer than their awns.

Cornfields, from Isla and Ross southd. ; (a colonist, *Wats.*) ; fl. June–July.— *Root* of stout fibres. *Stems* 1–4 ft., strict, rigid, smooth, rarely pubescent (*B. veluti'nus*, Schrad., and *B. multiflo'rus*, Sm.) *Leaves* ⅛–¼ in. broad, glabrous or slightly hairy, scabrid above ; sheaths grooved ; ligule short. *Panicle* 3–5 in., oblong ; branches 3–5-nate and rachis flexuous, scabrid, subsimple. *Spikelets* ¼–¾ in., compressed ; empty glumes unequal, broadly oblong, scabrid, mucronate, upper larger ; fl. glumes spreading, linear-oblong, 7–9-nerved, margins incurved, straight, hyaline, notched, at length coriaceous ; awn variable.—DISTRIB. Europe, N. Africa, W. Siberia ; introd. in N. America.

*** Fl. glume not longer than its palea*

B. **ARVEN'SIS,** *L.* ; leaves and sheaths hairy, panicle pyramidal effuse, branches horizontal, fl. glumes imbricate equalling the straight awn.

Casual from Fife southd.; fl. July–Aug.—*Root* fibrous or creeping. *Stems* 1–2 ft., smooth. *Leaves* ½–¼ in. broad, flat, short, hairy beneath ; sheaths grooved, pubescent; ligule obtuse. *Panicle* 4–8 in., 3–5 in. broad ; branches 5–7-nate, widespreading, very long, capillary, scaberulous, with a few spikelets towards the tips. *Spikelets* ½ in., lanceolate, compressed, green or dull violet; rachilla smooth; empty glumes ovate-oblong, keel scabrid, upper much the largest, acute or shortly awned ; fl. glumes ⁷⁄₁₀ in., imbricate till quite mature, then rather distant, oblong, 2-fid, strongly nerved ; awn dark.—DISTRIB. Europe.

42. BRACHYPO'DIUM, *Beauv.*

Perennial grasses. *Spikelets* subsessile, distichous, terete, spiked, inserted broadside to the rachis, very many-fld. ; rachilla jointed between the fl. glumes. *Empty glumes* 2, rarely 1, much shorter than the flowering, straight. *Fl. glumes* densely imbricate ; awn terminal or 0, 7–9-nerved, nerves converging to the tip. *Palea* with ciliate nerves. *Scales* 2, ovate. *Stamens* 2–3. *Ovary* hairy at the top ; styles distant, stigmas feathery. *Fruit* adhering to the palea.—DISTRIB. Europe, temp. Asia, N. Africa ; species 5 or 6.—ETYM. βραχύς and ποδιόν, from the *subsessile* spikelets.

1. **B. sylvat'icum,** *R.* and *S.* ; root fibrous, leaves broad hirsute, spike drooping, awn equalling its fl. glume.

Copses, hedgerows, &c., N. to Orkney ; ascends to 1,000 ft. in N. England; Ireland; Channel Islands; fl. June–July.—*Stems* 1–3 ft., very slender, terete, inclined, leafy. *Leaves* ⅓–⅔ in. broad, flat, bright green, ciliate ; sheaths tereto ; ligule obtuse. *Spike* 2–6 in. ; rachis flattened, smooth, slender. *Spikelets* 8–18, 1–2 in., appressed, linear, very shortly pedicelled, hirsute or glabrate, green ; empty glumes cuspidate, strongly 3–5-nerved ; fl. glumes 8–10, ¼ in., linear-oblong, nerves strong, tip acuminate awned ; palea ciliate at the tip, equalling the glume.—DISTRIB. Europe, N. Africa, N.W. Himalaya.

2. **B. pinna'tum,** *Beauv.* ; rootstock creeping, leaves narrow involute, spike erect, awn shorter than its fl. glume.

Downs and hedgerows in chalky soil, from York to Devon and Kent; absent in Wales, Cambridge only in the E. counties; fl. July.—Glaucous. *Stems* 1–3 ft., very slender, terete, naked, smooth. *Leaves* rigid, almost glabrous, involute, rarely flat ; ligule ciliate. *Spike* 1–6 in. ; rachis flattened, smooth. *Spikelets* ¾–1½ in., erect, curved away from the rachis, glabrous or pubescent, green and purplish ; empty and fl. glumes much as in *B. sylvat'icum,* but glabrous or nearly so, and tips of the latter suddenly contracted into the short awn.—DISTRIB. Europe, N. Africa, Siberia.

43. LO'LIUM, *L.* Rye-grass.

Characters of *Agropy'rum*, but upper empty glume absent except in the terminal spikelet; lower persistent, facing the rachis.—Distrib. N. temp. regions; species 2 or 3.—Etym. An old Latin name.

1. **L. peren'ne,** *L.* ; perennial, empty glume shorter than the spikelet.

Waste places, N. to Shetland; ascends to 1,600 ft. in N. England; Ireland; Channel Islands; fl. May–June.—*Root* fibrous, stolons leafy. *Stems* 18 in., bent below, ascending, smooth, slightly compressed. *Leaves* flat, edges and upper surface scabrid; sheaths smooth, compressed; ligule short. *Spike* 4–10 in., strict, stout and 6-10-fld. or slender and 3–4-fld. (*L. ten'ue,* L.); rachis smooth, channelled on one side. *Spikelets* ⅓–½ in. (much longer in var. *ital'icum*), quite smooth, shining; empty glumes strongly ribbed, linear-lanceolate; fl. glume linear-oblong, terete, obtuse or cuspidate or awned, ribbed.—Distrib. Europe, N. Africa, W. Asia; introd. in N. America.— *L. ital'icum,* A. Br., *L. remo'tum,* Schrk., *L. multiflo'ra,* Lamk., and *L. peren'ne,* var. *arista'ta,* are cultivated annual or biennial forms, with many flowers, not known in a wild state.—*L. festuca'ceum,* Link (*Festuca lolia'cea,* Curt., not Huds.), is a hybrid with *F. ela'tior* (Nyman).

2. **L. temulen'tum,** *L.* ; annual, empty glume equalling or exceeding the spikelets. *Darnel.*

Cornfields, N. to Shetland; Ireland; Channel Islands; (a colonist, *Wats.*); fl. June–Aug.—Similar to *L. peren'ne,* but always annual, without stolons, empty glume longer, and fl. glumes more turgid, awn short or long or 0 (*L. arven'se,* With.).—Distrib. Europe, N. Africa, W. Siberia, India; introd. in N. America.—Fruit very poisonous.

44. AGROPY'RUM, *J. Gœrtn.*

Perennial grasses. Spikelets solitary, sessile, distichous, compressed, spiked, inserted broadside to the rachis, 3-many-fld. ; rachilla usually jointed between the fl. glumes. *Empty glumes* 2, shorter than the flowering, unequal. *Fl. glumes* rigid, awned or not ; nerves, 5–7, meeting in the tip. *Palea* with ciliate keels. *Scales* ovate, entire, ciliate. *Stamens* 3. *Ovary* hairy at the top ; stigmas distant, subsessile, feathery. *Fruit* grooved, usually adherent to the palea.—Distrib. All temp. climates ; species 20.—Etym. The classical name.

1. **A. cani'num,** *Beauv.* ; root fibrous, stolons 0, empty glumes 3–5-ribbed, fl. glumes 2-5-nerved. *Triticum,* Huds.

Woods, banks, and waste places, from Sutherland southd; ascends to 1,300 ft. in Yorkshire ; Ireland; fl. July.—Bright green. *Stems* 1–3 ft., slender. *Spike* 2–10 in., very slender. often flexuous and nodding ; rachis with scabrid edges. *Spikelets* ½–¾ in., green, rather slender ; pedicel very short, pubescent or glabrous ; empty glumes scabrid, cuspidate or shortly awned, nerves usually 3, very firm ; fl. glumes linear-lanceolate, smooth except at the 5-nerved tip ; awn scabrid, longer or shorter than the palea.—Distrib. Europe, Siberia, Himalaya, N. America.

2. **A. re'pens,** *Beauv.* ; rootstock creeping, rachis of spike not brittle, empty glumes 5-ribbed, fl. glumes 4–5 rigid cuspidate or acuminate rarely awned.   *Triticum,* L.   *Couch* or *Quitch Grass.*

Fields and waste places, N. to Shetland ; ascends to 1,300 ft. in N. England; Ireland; Channel Islands; fl. June–Aug.—Excessively variable.   *Rootstock* stout, long, creeping, jointed.   *Stems* 1–4 ft., bent and ascending, smooth, glabrous.   *Leaves* flat or involute, usually scabrid above and glabrous beneath, sometimes hairy ; sheaths terete ; ligule very short.   *Spike* 2–10 in., rigid, slender or robust, strict or curved, not nodding, rachis glabrous or pubescent.   *Spikelets* ¾–1 in., very rigid ; empty glumes acute obtuse or notched, rigid short points or awns of variable length; fl. glumes quite similar, but nerved only at the tip.—DISTRIB   Europe, N. Africa, N. Asia, Himalaya, N. America.—*Triticum cani'num,* var. *piflo'ra,* Mitt. (*T. alpi'num,* Don MSS.), found on Ben Lawers, and said to want the creeping rootstock, is (judging from the specimen) only *A. re'pens.*—There is no accordance amongst specialists as to the limits of the forms of this plant and their nomenclature.

A. RE'PENS proper ; stems solitary hollow, leaves flaccid usually hairy, nerves slender scabrid in one line, spikelets 3–7-fld., rachis slender.—VAR. *barba'ta* has empty glumes tapering subulate or awned, fl. glumes awned.—VAR. *obtu'sa* has empty glumes obliquely truncate, fl. glumes obtuse apiculate.

Sub-sp. A. PUN'GENS, *R.* and *S.* ; stems densely tufted solid above, leaves firm involute ribbed, ribs scabrid in one line, spikelets 5–12-fld., rachis with broad internodes.   Seashores and tidal rivers.—*A. littora'le,* Reichb. (*Triticum,* Host.), has glumes acuminate, fl. glume mucronate or awned.—*A. pycnan'-thum,* Gren. and Godr., is more glaucous, spike more compact, empty glumes rounded obtuse, fl. glumes obtuse mucronate.

Sub-sp. A. ACU'TUM, *R.* and *S.* ; stems loosely tufted solid geniculate at the base, leaves firm not so involute ribbed glabrous or hairy, ribs scabrid all over, spike arching lax long, spikelets 5–8-fld , rachis with very broad internodes.   *Triticum acu'tum,* DC.; *T. lax'um,* Fries.—Sandy shores.—Intermediate between *re'pens* and *jun'ceum.*

3. **A. jun'ceum,** *Beauv.* ; rootstock creeping, rachis of spike fragile, spikelets large shining, empty glumes 5-11-ribbed, fl. glumes 4–10 obtuse acute notched or truncate rarely mucronate.   *Triticum,* L.

Sandy seashores, local, Orkney to Devon and Kent; Ireland; Channel Islands; fl. July–Aug—Often glaucous, rigid, forming large masses. *Rootstock* stout; extensively creeping.   *Stems* bent below, ascending, smooth, sheathed at the base.   *Leaves* coriaceous, involute, pubescent above, glabrous beneath ; sheaths smooth, rather inflated ; ligule short.   *Spike* 2–4 in., stout, curved ; rachis very stout, smooth.   *Spikelets* ¾–1¼ in., distant, very stout, pale, rigidly coriaceous, smooth, shining ; empty glumes strongly or faintly nerved ; fl. glume obscurely so ; internodes of rachis much dilated upwards.—DISTRIB.   Europe, N. Africa, N. America?

45. LEPTU'RUS, *Br.*

Slender grasses.   *Spikelets* solitary, sessile, distichous, alternate in a small spike, placed broadside to and in excavations of the jointed rachis,

1-fld. ; rachilla produced, with sometimes a 2d fl. glume. *Empty glumes* 2 (rarely 1) enclosing the flowering, equal, hard, coriaceous, ribbed, placed in front of the spikelet, except in the terminal one. *Fl. glume* keeled. *Palea* 2-nerved. *Scales* glabrous, entire. *Stamens* 2-3. *Ovary* glabrous ; stigmas sessile, distant, terminal, feathery. *Fruit* enclosed in the fl. glumes.—DISTRIB. Europe, N. and S. Africa, Australia ; species 6.— ETYM. λεπτός and οὐρά from the *slender tail*-like spikes.

L. filifor'mis, *Trin.* ; annual, glabrous, spike slender.

Waste places by the sea, from Fife and Isla southd.; Ireland; Channel Islands; fl. July.—*Root* fibrous. *Stems* 4-10 in., bent or curved below, ascending, stout or slender, shining, terete, leafy. *Leaves* short, coriaceous, scaberulous, soon involute ; sheaths slightly compressed, smooth, upper inflated ; ligule very short, auricled. *Spike* 2-6 in., straight or curved, short or long ; rachis rigid, grooved, hollowed on one side. *Spikelets* ¼ in., green, appressed or spreading ; empty glumes rather oblique, linear-oblong, pointed ; fl. glumes with 1 green nerve ; palea with glabrous keels. —DISTRIB. From Gothland southd., excl. Russia ; N. Africa.

L. filifor'mis proper ; stem and spike slender, the latter nearly straight.— VAR. *L. incurva'tus,* Trin.; stem and spike stouter, the latter strongly curved. Ballast heaps, Fife.

46. NAR'DUS, *L.* MAT-WEED.

A small rigid perennial grass. *Spikelets* solitary, sessile, secund, in a simple unilateral spike, placed obliquely and in excavations of the slender rachis, 1-fld. *Empty glumes* 0. *Fl. glume* 1, slender, concave, keeled, shortly awned, persistent. *Palea* linear, entire, 2-keeled. *Scales* 0. *Stamens* 3. *Ovary* narrow, glabrous, contracted into a slender filiform hairy persistent stigma. *Fruit* adherent to the palea.—DISTRIB. Europe (Arctic), Azores, Greenland.—ETYM. obscure.

N. stric'ta, *L.* ; glabrous, leaves setaceous channelled scaberulous.

Heaths and dry pastures, N. to Shetland ; ascends to 3,300 ft. in the High-lands ; Ireland ; Channel Islands ; fl. June–July.—*Rootstock* stout, cree ing, densely tufted. *Stems* 2-8 in., erect, filiform, rigid, striate, angled ; base with long pale sheaths. *Leaves,* upper erect, lower almost horizontal; sheaths smooth ; ligule short. *Spike* 1-3 in., solitary ; rachis very slender, strict. *Spikelets* rather distant; fl. glume ⅓ in., slender, reddish or purplish ; divaricate after flowering, scabrid above, narrowed into the short awn.— Rejected by sheep, on account of the harsh foliage.

47. HOR'DEUM, *L.* BARLEY.

Spikelets 2-3-nate, subsessile, distichous, compressed, spiked, inserted broadside to the rachis, 1-fld., rachilla produced with a subulate rudi-mentary glume ; lateral spikelets, rarely the central, neuter male or 2-sexual. *Empty glumes* 2, exceeding the flowering or not, equal, col-laterally placed in front of the spikelet, awned. *Fl. glume* rounded at the back, awned. *Palea* as long, narrow, 2-keeled. *Scales* 2, ciliate. *Ovary* hirsute; stigmas 2, subsessile, feathery. *Fruit* free or adherent to the palea, grooved in front.—DISTRIB. N. temp. and warm regions, S. America; species 12.—ETYM. The old Latin name.

*   *Flowers of lateral spikelets 2-sexual, of middle male.*

1. **H. sylvat'icum,** *Huds.* ; perennial, spike subterete, empty glumes setaceous scabrid. *El'ymus europœ'us,* L.

Copses and woods in chalky soil, from Northumbd. to Hants and Kent; absent in Wales and the E. counties; Dublin (native?); fl. June–July.—Perennial, bright green. *Stems* 1-3 ft., strict, erect, smooth. *Leaves* ½–⅔ in., broad, flat, thin, scaberulous; sheaths hispid, hairs reflexed; ligule very short. *Spikes* 2–4 in., strict, erect, terete, green. *Spikelets* ¾ in., erect, subsessile; empty glumes awned, 3-nerved; fl. glumes linear-oblong, dorsally compressed, scabrid, shorter than the straight flexuous awn, nerved towards the tip; palea with smooth keels. *Fruit* very narrow.—DISTRIB. From Gothland to Spain, Italy, and Russia.

**   *Lateral spikelets flowerless or male, middle 2-sexual.*

2. **H. praten'se,** *Huds.* ; perennial, spike compressed, outer empty glume setaceous scabrid.

Wet meadows, &c., from Berwick southd.; Ireland, local; fl. June–July.— *Rootstock* creeping. *Stems* very slender, 1-2 ft., terete, scabrid above. *Leaves* ½–1 in., narrow, flat, at length involute, scabrid above, hairy beneath, as are the narrow sheaths; ligule very short. *Spike* 1-3 in., ¼–½ in. broad, inclined, linear, yellow-green. *Spikelets* with the awns ⅔ in., rather spreading, scabrid all over, not ciliate; fl. glumes terete, smooth, about equalling the awn, obscurely nerved; palea as long, acute, keels not ciliate.—DISTRIB. From Gothland to Spain and Russia, N. Asia, N.W. India, N. America.

3. **H. muri'num,** *L.* ; annual, spike compressed, outer empty glume of the mid. spikelet lanceolate ciliate, of the lateral setaceous scabrid. *Waybent, Barley-grass.*

Waste places, N. to Caithness; E. Scotland only; Ireland, very rare; Channel Islands; fl. June–July.—*Root* fibrous. *Stems* ascending, 6–18 in., smooth, glabrous. *Leaves* small, narrow, scabrid; sheaths inflated, glabrous; ligule very short. *Spikes* 1½-2 in., stout, inclined, green. *Spikelets* 1 in., densely imbricate; empty glumes filiform; fl. glumes lanceolate, flattened, much shorter than the straight awn; palea with distantly ciliate keels.—DISTRIB. From Gothland southd., N. Africa.—VAR. *arena'ria,* Bab., is a form from sandy places with the stem below branched and rooting.

4. **H. marit'imum,** *With.* ; annual, spike subterete, empty glume scabrid, upper of the mid. spikelet ½-lanceolate, the rest setaceous. *Squirrel-tail Grass.*

Waste maritime localities, Durham to Kent and Devon; absent in Wales; Channel Islands; fl. June.—*Stems* 6-12 in., bent below, then erect, terete, smooth, leafy. *Leaves* short, straight, narrow, flat, glaucous, scabrid; lower sheaths pubescent, upper inflated; ligule very short. *Spike* 1–2 in., stout, erect, subterete, at length yellow-brown. *Spikelets* ½ in. without the awns, rather spreading, rigid; larger empty glume of mid. spikelet green with scarious margins, shorter than the rigid awn; fl. glume lanceolate, flattened, about equalling the rigid awn.—DISTRIB. From Denmark southd., N. Africa.

48. EL'YMUS, *L.* LYME-GRASS.

Tall perennial grasses. *Spikelets* 2–3-nate, sessile, distichous, compressed, spiked, inserted broadside to the rachis, 2–7 fld. *Empty glumes* 2, equalling or exceeding the flowering, equal, placed in front of the spikelet. *Fl. glumes* 5-nerved, coriaceous, awned or not. *Palea* with 2 ciliate keels. *Scales* ovate, usually ciliate. *Stamens* 3. *Ovary* hirsute; stigmas sessile distant feathery. *Fruit* grooved, adnate to the fl. glume and palea.—DISTRIB. N. temp. regions; species 20.—ETYM. ἐλύω, the fruit being *rolled up* in the palea.

E. arena'rius, *L.* ; fl. glumes rigid acuminate, awn 0.

Sandy seashores, from Essex and N. Wales to Shetland; Ireland; Channel Islands; fl. July.—Glaucous. *Rootstock* stout, creeping, stoloniferous. *Stems* 3–6 ft., very stout, smooth, terete. *Leaves* rigid, strict, pungent, ¼–⅔ in. diam.; sheaths smooth, grooved; ligule very short. *Spike* 6–12 in., stout, strict; rachis flexuous, plano-convex, hirsute. *Spikelets* 1 in., imbricate, appressed; rachilla stout, pubescent; empty glumes linear-lanceolate, purplish; fl. glumes 1–3, lanceolate, ciliate and hirsute, keeled towards the cuspidate tip; palea as long as the glume.—DISTRIB. Europe, N. Asia, N. America.

CLASS III. ACOTYLE'DONES OR CRYPTOGAMS.

ORDER XC. **FIL'ICES.**

Perennial herbs (very rarely annual), sometimes shrubby or arborescent, with fibrous roots, or creeping rootstocks. *Leaves* (*fronds*) tufted or alternate on the rootstock, simple pinnatifid or 1–4-pinnate, usually circinate in vernation ; petiole (*stipes*) sometimes jointed at the base and rachis, grooved on the upper surface. *Fructification* of microscopic *spores*, contained in usually minute *capsules* that are collected in masses (*sori*) on the under surface or edge of the frond, or rarely on separate fronds or parts of the frond, and are naked or covered with an involucre formed of or upon the margin or back of the frond. *Capsules* membranous, sessile or stalked, often mixed with jointed club-shaped hairs (imperfect capsules). *Spores* usually obtusely 4-hedral.—DISTRIB. Chiefly humid temp. and trop. regions ; genera 75 ; species 2,500. —AFFINITIES. With *Lycopodia'ceæ.*

In germination the spore develops a flat cellular scale (*prothallus*), on the under surface of which are formed cavities some containing male and some female organs. The *male cavities* (*antheridia*) contain sperm-cells (*spermatozoids* or *antherozoids*), which enclose a spiral filament. The *female cavities* (*archegonia*) contain a solitary free *germ-cell*. The antherozoids find their way into the archegonia and fertilize the germ-cell, which thereupon develops into a plant, the prothallus withering away.

TRIBE I. **HYMENOPHYL'LEÆ.** *Frond* very membranous, translucent, reticulate. *Involucre* 2-valved, urceolate or 2-lipped. *Capsules* minute, membranous, reticulate, sessile on a clavate or filiform receptacle, girt by a complete horizontal or oblique ring. *Vernation* circinate.

Involucre 2-valved ....1. Hymenophyllum.
Involucre urceolate...2. Trichomanes.

TRIBE II. **POLYPODIE'Æ.** *Frond* more or less coriaceous, opaque. *Involucre* marginal or dorsal or 0. *Capsules* minute, membranous, reticulate, not raised on an elevated receptacle, stalked, partially girt by a vertical ring, bursting transversely. *Vernation* circinate.

* *Sori* marginal ; involucre continuous with the reflexed or recurved margin of the frond.

Sori oblong, short. Fronds all similar................................3. Adiantum.
Sori linear, continuous. Fronds all similar.............................4. Pteris.
Sori subglobose, on special fronds ....5. Cryptogramme.
Sori linear, on special fronds..6. Lomaria.

** *Sori* dorsal, linear ; involucre linear.

Involucre on a nerve, single ....7. Asplenium.
Involucre on a nerve, double......................................8. Scolopendrium.

*** *Sori* dorsal, globose ; involucre short.

Involucre lacerate, attached under the sorus...........................9. Woodsia.
Involucre hooded, on one side of the sorus.......................10. Cystopteris.
Involucre orbicular, peltate..11. Aspidium.
Involucre reniform...12. Nephrodium.

**** *Sori* dorsal ; involucre 0 (see *Ceterach* under *Asplenium*).

Sori globose or oblong ....13. Polypodium.
Sori linear ....14. Gymnogramme.

TRIBE III. **OSMUN'DEÆ.** *Frond* coriaceous or membranous. *Involucre* 0. *Capsules* sessile or shortly stalked, vertically 2-valved, with a short lateral or subterminal striate areola. *Vernation* circinate...15. Osmunda.

TRIBE IV. **OPHIOGLOS'SEÆ.** *Capsules* large, 2-valved, without a ring or areola, coriaceous, in spikes or panicles. *Vernation* straight.

Frond ovate, simple. Capsules spiked ....16. Ophioglossum.
Frond pinnate. Capsules panicled................................17. Botrychium.

1. HYMENOPHYL'LUM, *Sm.* FILMY-FERN.

Rootstock filiform, creeping. *Fronds* usually matted and 2-4-pinnatifid or -pinnate, pellucid, reticulate ; segments with a midrib ; veins 0. *Sori* marginal, axillary or terminal ; involucre free or sunk in the frond, 2-valved or 2-lipped, opening outwards ; capsules sessile on a columnar receptacle ; ring complete, oblique. — DISTRIB. Trop. and temp. regions ; species 70.—ETYM. ὑμήν and φύλλον, from the membranous fronds.

1. **H. tunbridgen'se,** *Sm.* ; frond ovate pinnate below, pinnatifid above, pinnæ spreading spinulose-serrate, involucre toothed.

Moist shaded rocks or copses, from Stirling, Mull, and Argyll southd. to W.
    York, and from Kent to Cornwall (ascending to 1,000 ft.) ; Ireland, rare ;
    Channel Islands ; frt. June–July.—*Rootstocks* capillary, interlaced. *Stipes*
    1–2 in., winged above. *Frond* 1–3 in., glabrous ; pinnæ distichous, flabel-
    lately pinnatifid, lobes linear ; involucre solitary, axillary, suborbicular, lips
    strongly irregularly toothed.—DISTRIB. Belgium, France, Germany, Italy,
    Canaries, S. temp. regions.

2. **H. unilatera'le,** *Willd.* ; frond oblong pinnate below or throughout, pinnules decurved spinulose-serrate, involucre entire. *H. Wilso'ni,*
Hook.

Moist shaded rocks or copses, Shetland to York ; Stafford, Salop, Wales,
    Devon, Cornwall ; ascends to 2,800 ft. in the Hebrides, and 2,400 in
    Ireland ; frt. June–July. -Perhaps only a sub-species of *H. tunbridgen'se,*
    but more rigid, darker green ; involucre more ovoid and turgid ; pinnæ
    pinnatifid chiefly on the upper side.—DISTRIB. As *H. tunbridgen'se.*

2. TRICHOM'ANES, *L.* BRISTLE-FERN.

Rootstock creeping or tufted, stout or slender. *Frond* erect or pendulous, simple pinnate or 1-4-pinnatifid, usually pellucid, reticulate ;
segments with a stout simple or forked midrib. *Sori* marginal, axillary
or terminal ; involucre elongate, free or sunk in the frond, tubular or
campanulate, mouth entire or 2-lipped, opening outwards ; capsules sessile
on a long often exserted receptacle.—DISTRIB. Chiefly trop. and damp
warm climates ; species 78.—ETYM. obscure.

T. radi'cans, *Sw.* ; rootstock creeping, frond 2-3-pinnatifid. *T.
specio'sum,* Willd.

Wet shaded rocks, Killarney, York, S. Wales, Argyll, Arran ; frt. July–Sept.—
    Rootstock slender, wiry, extensively creeping, tomentose. *Stipes* 2–6 in.,
    stout, wiry, ascending, naked below, winged above. *Frond* 4–12 in., mem-
    branous, firm ; rachis winged ; lower pinnæ 1–4 in., rhomboid-ovate ; pin-
    nules the same shape, pinnatifid, toothed, nerves 1 to each segment. *Sori*
    lateral, 1–4 to each pinnule ; tube of involucre short, lips small ; receptacle
    exserted.—DISTRIB. W. Europe, trop. Africa and America, Himalaya,
    Japan, Polynesia.

T. radi'cans proper (*T. specio'sum*, Willd., *T. brevise'tum*, Br., *Hymenophyl'lum ala'tum*, Sm.); frond deltoid, involucre scarcely winged.— VAR. *T. Andrew'sii*, Newm.; frond lanceolate, involucres many winged, receptacle larger. Kerry.

3. ADIAN'TUM, *L.* MAIDENHAIR.

Rootstock tufted or creeping. *Frond* compound, 2–4-pinnate, rarely simple; rachis and branchlets capillary; veins forked or netted. *Sori* rounded or. oblong, parallel with and on the margin; involucre formed of the reflexed often kidney-shaped coriaceous margin of the frond, opening inwards, surface veined.—DISTRIB. All temp. and hot climates; species 62.—ETYM. The old Greek name.

A. Capil'lus-Ven'eris, *L.* ; frond 3–4-pinnate, pinnules cuneate lobed crenate glabrous.

Damp rocks, walls, &c., especially near the sea, local, Dorset to Cornwall, I. of Man, Glamorgan; W. Ireland, local; frt. May–Sept.—*Rootstock* creeping, scaly. *Stipes* 4–9 in., slender, black, polished, naked. *Frond* 4–12 in., ovate, with a short terminal and many spreading capillary branches, the lower pinnate; pinnules ½–1 in., membranous, outer edge rounded; stalks ¼ in.; veins repeatedly forked. *Sori* in the crenatures of the pinnules; involucre subreniform.—DISTRIB. From France southd.; temp. and trop. Old and New World.

4. PTER'IS, *L.* BRAKE, or BRACKEN.

Rootstock usually creeping. *Frond* various; veins free, forked or netted. *Sori* continuous; involucre scarious or membranous, confluent with the recurved margin of the frond, not recurved in age.—DISTRIB. All regions; species 83.—ETYM. πτερόν, from the *wing*-like fronds.

P. aquili'na, *L.* ; frond coriaceous 3–4-pinnate, veins free.

Forests, heaths, moors, &c., N. to Shetland; ascends to 2,000 ft. in the Highlands; Ireland; Channel Islands; frt. July–Aug.—*Rootstock* stout, subterranean, extensively creeping. *Stipes* 1–6 ft., stout, erect, pale, dark at the base. *Frond* 2–3 ft.; rachis glabrous or pubescent; upper pinnæ simple, next cut into linear pinnules, lower stalked, 1 ft. or more, again pinnate; pinnules 1 in., sessile, auricled at the base; veins close, 1–2-forked. *Involucre* glabrous villous or ciliate, sometimes double, inner very narrow.— DISTRIB. Arctic Europe, and all temp. and many trop. regions.

5. CRYPTOGRAM'ME, *Br.* PARSLEY-FERN, ROCK-BRAKE.

Rootstock tufted, often elongate. *Fronds*, outer barren, inner fertile, 2–4-pinnatifid; veins forked, free. *Sori* terminal on the veins, subglobose, afterwards confluent along the margins of the fertile pinnules; involucre membranous, continuous with the recurved margin of the frond, spreading in age.—DISTRIB. N. temp. and Arctic regions; species 1.—ETYM. κρύπτος and γραμμή, from the *concealed* sori.

C. cris'pa, *Br.* ; fertile pinnules fusiform. *Alloso'rus*, Bernh.

Loose stony places in mt. districts, from Harris and Caithness to N. Devon;
absent in E. England; ascends to 3,500 ft. in the Highlands; Ireland, very
rare; frt. June–July.—*Rootstock* scaly, clothed with broken bases of fronds.
Stipes of barren fronds 1–2 in., of fertile 2–5 in. slender, naked, pale brown.
Fronds deltoid-ovate, submembranous; barren 2-pinnate, pinnules 2–3-
pinnatifid, cuneate or oblong, 2–3-toothed; fertile 2–3-pinnate; pinnules
fusiform or oblong-lanceolate, obtuse, entire, subpetioled.

6. LOMA'RIA, *Willd.* HARD-FERN.

Rootstock usually short or creeping. *Fronds* tufted, of 2 kinds; outer
barren or fertile below only; inner fertile; veins free, simple or forked.
Sori linear, close to the margin, continuous round the pinnule, often
covering its lower surface; involucre linear, close to and parallel with the
margin, opening inwards, scarious.—DISTRIB. Trop. and temp. chiefly
south regions; species 40.—ETYM. λῶμα, from the marginal sori.

L. Spi'cant, *Desv.*; barren fronds narrow-lanceolate pinnatifid above,
pinnate below. *Blech'num borea'le,* Sw.

Heaths, woods, banks, &c., N. to Shetland; ascends to 2,000 ft. in the Highlands; Ireland; Channel Islands; frt. July–Aug.—*Rootstock* stout, creeping,
scaly. *Stipes* of barren fronds 2–3 in., of fertile 6–9 in., polished, red-
brown. *Fronds* erect or spreading; barren 6–9 in., narrowed to the base,
coriaceous, green, glabrous; pinnules $\frac{1}{2}$–$\frac{3}{4}$ in., linear-oblong, sessile by a
broad base, obtuse, quite entire, sinus narrow, veins inconspicuous; fertile
pinnate, pinnæ distant, falcate, narrow, obtuse, dilated at the base, lower
minute very distant. *Involucre* marginal in a young state.—DISTRIB.
Europe (Arctic), Canaries, N.E. Asia, N.W. America.

7. ASPLE'NIUM, *L.* SPLEENWORT.

Rootstock usually short, tufted. *Fronds* various. *Sori* dorsal on the
veins, linear or oblong, oblique, distant from the midrib, except when the
frond is much divided; involucre oblong or linear, membranous, laterally
attached to the vein, opening towards the midrib.—DISTRIB. All climates
but very cold; species 280.—ETYM. α and σπλήν, having been a reputed
spleen medicine.

SUB-GEN. 1. **Asple'nium** proper. *Involucre* straight, narrow, margin
entire or erose. *Frond* not scaly beneath. *Veins* free.

* *Ultimate pinnules without a distinct midrib.*

1. **A. Ru'ta-mura'ria,** *L.*; frond oblong or ovate rigid irregularly 2-
pinnate, pinnæ 3–7 obovate-cuneate, tip rounded or truncate toothed.

Walls and rocks, N. to Orkney; ascends to 2,000 ft. in the Highlands;
Ireland; Channel Islands; frt. June–Oct.—*Rootstock* stout, shortly
creeping, without scales. *Stipes* tufted, 2–4 in., wiry, black below. *Frond*
1–2 in., recurved, often deltoid; pinnæ stalked, upper entire, lower again
pinnate; pinnules $\frac{1}{4}$–$\frac{1}{2}$ in., often rhomboid; midrib obsolete; veins flabel-
late, forked. *Sori* many, linear-oblong, 2–5 on each pinnule; involucre
entire or margins erose.—DISTRIB. Europe (Arctic), N. and S. Africa, N.
Asia, N.W. Himalaya, N. America.

2. **A. german'icum,** *Weiss ;* frond oblong-lanceolate pinnate, pinnæ few distant alternate cuneate-lanceolate simple-toothed or lobed. *A. alternifo'lium,* Wulf.

Rocks, very rare; N. Wales, N. England, Roxburgh, Perth, Fife; frt. June–Sept.—*Rootstock* densely tufted, creeping, without scales. *Stipes* 2–4 in., slender, erect, black below. *Frond* rather flaccid, 2–3 in.; pinnæ 7–9, ⅓–¾ in., very variable, sometimes fan-shaped, lower shortly stalked, irregularly 2-3-lobed; lobes crenate or toothed; midrib obsolete, veins forked. *Sori* 2–4 on each pinna or segment, parallel, linear-oblong; involucre entire, at length covering the breadth but not the length of the segments.—DISTRIB. Europe (excl. Greece and Turkey), Himalaya, China.

3. **A. septentriona'le,** *Hull ;* frond linear-lanceolate inciso-pinnatifid, pinnæ narrow erect, tips incised.

Walls and rocks, rare; Devon, Somerset, N. Wales (ascending to 3,000 ft.), northd. to Perth and Aberdeen; frt. June–Oct.—*Rootstock* densely tufted, hardly scaly. *Stipes* many, 3–4 in., erect, rigid, black below. *Frond* coriaceous, 1–2 in., lanceolate. simple or cleft into slender segments; midrib obsolete; veins forked. *Sori* 1–4 on each pinna, parallel, at length covering the pinna; involucre narrow.—DISTRIB. Europe, N. and W. Asia, Himalaya, N. America.

** Ultimate pinnules with a distinct midrib.

4. **A. Trichom'anes,** *L.* ; frond linear pinnate, rachis rigid, chestnut-brown, pinnæ ¼–⅔ in. many subsessile.

Walls and rocks, N. to Orkney; ascends to 2,000 ft. in Wales; Ireland; Channel Islands; frt. May–Oct.—*Rootstock* stout, shortly creeping; scales few, subulate, blackish. *Stipes* 1–4 in., crowded, naked, polished, red-brown, black below. *Frond* 6–12 in., rigid; pinnæ 15–40, horizontal, dark green lower smaller, base obliquely cuneate truncate rounded or auricled, sometimes incised; midrib subcentral; veins few, oblique, forked above the middle. *Sori* oblique, short; involucre pale brown, entire or erose.—DISTRIB. Europe, N. Africa, N. and W. Asia, N. America, S. temp. regions. —*A. an'ceps,* Sol., is simply a larger form.—*A. Clermont'æ,* Syme (*A. Petrar'chæ,* Newm., not DC.), from a garden wall at Newry, is considered by its author, with hardly a doubt, to be a hybrid with *A. Ru'ta-mura'ria,* from which latter it differs in its simply pinnate linear frond, more sessile pinnæ, and more divergent veins.

5. **A. vir'ide,** *Huds.* ; frond linear pinnate, rachis green slender, pinnæ ¼–½ in. many shortly stalked rhombic-ovate crenate.

Wet rocks in mt. districts, from Shetland to S. Wales and Derby; ascends to 2,800 ft. in the Highlands; W. Ireland; frt. June–Sept.— Perhaps an alpine sub-species of *A. Trichom'anes,* distinguished by its more flaccid habit, pale rachis, shorter paler and shortly stalked pinnæ.—DISTRIB. Europe (Arctic), N. and W. Asia, N. America.

6. **A. mari'num,** *L.* ; frond oblong or lanceolate coriaceous pinnate below, pinnæ 1–2 in. oblong-ovate crenate.

Sea-cliffs and caves, Shetland to York on the E. coast, and to Cornwall and Hants on the W. and S.; Ireland; Channel Islands; frt. June–Sept.—

Rootstock stout, clothed with purple-brown chaffy scales. *Stipes* 3–6 in., red-brown below, stout, polished. *Frond* 3–10 in.; rachis stout, winged, green; pinnæ acute or obtuse, base truncate cuneate or cordate auricled above sinuate-lobed or serrate, upper confluent; midrib and forked veins obscure. *Sori* large, oblique; involucre coriaceous.—DISTRIB. France, Spain, Italy, N. Africa, N. America.

7. A. lanceola'tum, *Huds.* ; frond broadly lanceolate membranous 2-pinnate, pinnules broad crowded acutely serrate.

Wet rocks; York, Wales, Gloster, Cornwall to Kent; Cork; Channel Islands ; frt. June–Sept.—*Rootstock* short, stout, clothed with subulate scales. *Stipes* 2–4 in., chestnut-brown, glossy. *Frond* 6–9 in., bright green ; pinnæ many, shortly petioled, lower smaller distant, cut to the rachis into ovate or obovate pinnules ; veins forked. *Sori* short, at length confluent.—DISTRIB. Europe from France, Spain, and Germany to Turkey, N. Africa.

8. A. Adian'tum-ni'grum, *L.* ; frond deltoid-ovate 2–3-pinnate, pinnules petioled inciso-pinnatifid and serrate.

Rocks and walls, N. to Shetland; ascends to 1,900 ft. in the Highlands ; Ireland; Channel Islands; frt. June–Oct.—*Rootstock* stout, oblique, scales subulate. *Stipes* 6–9 in., almost naked, polished, chestnut-brown. *Frond* 6–12 in., coriaceous ; rachis brown below, winged and compressed above, pinnæ polished ovate-lanceolate, pinnules ⅓–⅔ in. ; veins pinnate and forked. *Sori* copious, short, crowded, at length confluent; involucre free, pale brown, edges entire.—DISTRIB. Europe, N. Africa, W. Asia, Himalaya.

VAR. *A. acu'tum*, Bory; lower pinnæ triangular acuminate, segments narrow very acute ; S.W. Ireland.—VAR. *A. obtu'sum*, Willd. (*A. Serpenti'ni*, Tausch); pinnæ triangular, ultimate segments broad obtuse. Aberdeenshire, on Serpentine.

SUB-GEN. 2. **Athy'rium,** *Roth* (gen.). *Involucre* short, oblong or obliquely reniform, reflexed after dehiscence, margin laciniate. *Frond* not scaly beneath ; veins free.

9. A. Fi'lix-fœm'ina, *Bernh.* ; frond large membranous oblong-lanceolate 2–3-pinnate, pinnules very many close-set subsessile oblong serrate.

Moist woods, rocky places, &c., N. to Shetland; ascends to 2,200 ft. in N. England ; Ireland; Channel Islands; frt. July–Aug.—*Rootstock* stout, ascending, often 6–8 in., clothed with broad ferruginous scales. *Stipes* 6–12 in., stout, copiously scaly below, brittle, brown or pale yellow. *Frond* 1–5 ft., bright green, flaccid, waving ; pinnæ sessile, close-set, lanceolate, acuminate, spreading and ascending; pinnules ½–¾ in., sessile, spreading. obtuse, lower pinnatifid, upper coarsely serrate ; veins pinnate in the segments. *Sori* many, small ; involucre variously curved, membranous, very convex, margin fringed or erose.—DISTRIB. Europe (Arctic), Africa, N. and W. Asia, Himalaya, America.

A. Fi'lix-fœm'ina proper ; frond 2-pinnate, pinnules separate pinnatifid obtuse, basal shorter, sharply toothed at the sides and tip.—VAR. *A. rhœ'ticum*, Roth (var. *convex'a*, Newm.) ; frond 2-pinnate, pinnules narrow convex toothed, basal longest.—VAR. *A. mol'le*, Roth ; stipes short, frond small pinnate, pinnules oblong flat confluent below less toothed.—VAR. *A.*

inci'sum, Hoffm.; frond very large 3-pinnate, pinnæ broad, lower pinnules again pinnate flat toothed.—VAR. *Ath. latifo'lium,* Bab., is a form with very broad much imbricated sharply incised pinnules, once found near Keswick.

SUB-GEN. 3. **Cete'rach,** *Willd.* (gen.). *Involucre* almost obsolete. *Frond* covered with chaffy scales beneath. *Veins* anastomosing.

10. **A. Cete'rach,** *L.* ; frond pinnatifid. *Cete'rach officina'rum,* Desv. Rocks and walls, chiefly in W. counties, from Argyll and Perth southd.; Ireland; Channel Islands; frt. April–Sept.—*Rootstock* short, stout. *Stipes* 1–3 in., wiry, blackish, chaffy. *Frond* 4–6 in., erect or spreading, leathery, linear-lanceolate or oblong, bright opaque green above, beneath densely clothed with rusty ovate toothed scales; segments $\frac{1}{3}$–$\frac{1}{2}$ in., horizontal, broadly ovate or oblong, quite entire, lower segments free, sinus broad deep rounded. *Sori* linear, hidden under the scales; involucre a very narrow membrane, or a ridge on the swollen nerve.—DISTRIB. From Belgium southd., N. Africa, W. Asia, N.W. Himalaya.

8. SCOLOPEN'DRIUM, *Sm.* HART's-TONGUE.

Rootstock stout, short, inclined. *Fronds* tufted, simple, coriaceous ; veins free or anastomosing. *Sori* linear on opposite contiguous veins, almost confluent ; involucre linear, attached to the vein, those of the contiguous sori opening opposite one another.—DISTRIB. Temp. and trop. regions ; species 9.—ETYM. The old Greek name.

S. vul'gare, *Sm.*; frond oblong-ligulate, base cordate. Hedgebanks, rocks, copses, &c., N. to Shetland; Ireland ; Channel Islands ; frt. July–Aug.—*Rootstock,* stipes, and often midrib clothed with subulate scales. *Stipes* very stout, 4–8 in. *Fronds* 6–18 in., broadest in the middle, coriaceous, flaccid, bright green ; basal auricles converging ; margin undulate ; midrib stout ; veins in groups of 2–4, indistinct, free or casually anastomosing, horizontal. *Sori* parallel, at right angles to the midrib, very variable in length and number.—DISTRIB. From Gothland southd., N. Africa, W. Asia, Japan, N.W. America.—A multitude of varieties are cultivated, presenting wonderful departures from the normal state.

9. WOOD'SIA, *Br.*

Rootstock short, tufted. *Stipes* usually articulate above the base. *Fronds* pinnate. *Sori* globose ; involucre inferior, membranous, at first calyciform, then usually breaking up into capillary segments.—DISTRIB. Arctic and N. temp. regions, Andes, S. Africa ; species 14.—ETYM. *J. Woods,* an eminent English botanist.

W. hyperbo'rea, *Br.* ; frond lanceolate pinnate, pinnæ ovate-oblong or cordate. Wet alpine rocks, alt. 2,000–3,000 ft., N. Wales ; Durham to Dumfries, Forfar, and Perth ; frt. July–Aug.—*Rootstock* stout, subelongate. *Fronds* 3–6 in., densely tufted. *Stipes* shining, clothed with ferruginous scales. *Pinnæ* subdistant, $\frac{1}{4}$–$\frac{1}{2}$ in., pubescent and ciliate ; veins simple and forked. *Sori*

3-5 on each lobe.—DISTRIB. Arctic, N. and Alps of Mid. Europe, N. Asia, Himalaya, N. America.

W. HYPERBJ'REA proper; frond linear-lanceolate, pinnæ ovate-cordate, lobes few broad.—N. Wales, Forfar, Perth.

Sub-sp. W. ILVEN'SIS, *Br.*; frond broadly lanceolate, pinnæ deeply pinnatifid with oblong subcrenate lobes.—N. Wales to Forfar.

10. CYSTOP'TERIS, *Bernh.* BLADDER-FERN.

Delicate flaccid ferns. *Rootstock* short or creeping. *Fronds* tufted or scattered, 1-4-pinnate; veins pinnate or forked, venules free. *Sori* small, dorsal on the middle of a venule, globose; involucre membranous, attached by a broad base to the venule below the sorus, ovate, convex, acute, at length reflexed.—DISTRIB. Cool damp regions; species 5.—ETYM. κύστις and πτερίς, from the *bladder*-like involucre.

1. **C. frag'ilis,** *Bernh.*; rootstock tufted, frond ovate-lanceolate 1-2 pinnate, pinnæ deltoid-ovate.

Rocks and walls in mountain districts, Orkney to Cornwall and Sussex; absent in E. half of England, S. of York and N. of Middlesex; ascends to 4,000 ft. in the Highlands; Ireland; frt. July-Aug.—*Rootstock* densely clothed with pale brown lanceolate membranous scales. *Stipes* 2-4 in., brittle. *Frond* 4-8 in.; rachis slightly winged above, larger pinnæ 1-1½ in., lobes or teeth obtuse or acute. *Sori* 2-12 on each segment.—DISTRIB. Arctic, N. and S. temp regions.

C. FRAG'ILIS proper; frond tripinnatifid, pinnules generally incised half-way to the rachis, ultimate division contiguous.—VAR. *C. denta'ta,* Hook.; pinnæ ovate-lanceolate obtuse obtusely toothed, sori submarginal.—VAR. *C. Dickie'ana,* Sim; frond ovate-oblong obtuse membranous, pinnæ ovate obtuse subdeflexed segments broad crowded obtuse crenate.

Sub-sp. C. ALPI'NA, *Desv.*; frond quadripinnatifid, pinnules incised nearly to the rachis, ultimate divisions not quite contiguous.—Teesdale.

2. **C. monta'na,** *Link;* rootstock creeping, frond deltoid 3-4-pinnate, pinnæ and pinnules spreading.

Alpine wet rocks, alt. 2,300-3,600 ft., very rare, Perth, Forfar, and Aberdeen; frt. July-Aug.—*Rootstock* widely creeping, sparingly scaly. *Stipes* 6-9 in., very slender. *Frond* 4-6 in., as broad as long, lowest pinnæ 1-1½ in., segments cut to the rachis, deeply sharply toothed, especially towards the tip. *Sori* small, 18-24 on the lowest pinnules. *Involucre* cut at the edge.—DISTRIB. Arctic and alpine regions, Europe, Asia, America.

11. ASPID'IUM, *Sw.* SHIELD-FERN.

Habit various. *Sori* dorsal, globose; involucre superior, orbicular, peltate.—DISTRIB. All regions; species 55.—ETYM. ἀσπίς, a shield, from the form of the involucre.—The British species belong to the section *Polystichum,* having free veins.

1. **A. Lonchi'tis,** *Sw.*; frond linear-oblong pinnate

L L

Clefts of alpine rocks, &c., from Caithness to N. Wales and York; ascends to 3,200 ft. in the Highlands; Ireland; frt. June–Aug.—*Rootstock* densely tufted, oblique, scaly. *Fronds* densely tufted, 6–18 in., coriaceous, bright green, glabrous except the scaly short stout stipes and rachis and veins beneath. *Pinnæ* many, ½–1 in., narrow-ovate, base auricled and obliquely rhomboid, falcate, acuminate, spinulose-serrate. *Sori* in 2–3 rows on each side the midrib.—Distrib. Europe (Arctic), N. and W. Asia, Himalaya, N. America.

2. **A. aculea′tum,** *Sw.* ; frond ovate-lanceolate 2–3-pinnate.

Woods, shaded hedgebanks, &c., from Orkney southd.; ascends to 2,500 ft. in Yorkshire; Ireland; Channel Islands; frt. July–Aug.—*Rootstock* short, stout, and stipes and rachis densely clothed with ferruginous scales. *Frond* 6–12 in., scaly beneath; lower pinnæ 4–6 in., close-set, lanceolate; pinnules obliquely rhomboid-ovate auricled, teeth mucronate or awned. *Sori* 1-seriate on each side the midrib, dorsal on the veins.—Distrib. From Belgium southd., W. Asia, N. America, S. temp. regions.

A. aculea′tum proper; rather flaccid, pinnules sessile lower free, serratures spinulose.

Sub-sp. A. loba′tum, *Sw.* ; frond 2-pinnate, pinnules very rigid sessile decurrent confluent below, upper basal longest.—Var. *lonchitidoi′des ;* narrower. approaching *A. Lonchi′tis.*

Sub-sp. A. angula′re, *Willd.* ; submembranous, pinnules small petioled lax sometimes again pinnatifid, teeth large awned.—From the Clyde southd.

12. NEPHRO′DIUM, *Rich.*

Sori subglobose, dorsal or terminal on the venules ; involucre reniform, superior, attached by the sinus.—Distrib. All regions ; species 224. —Etym. νεφρός, from the *kidney*-shaped involucres.—The British species all belong to the sub-genus *Lastre′a,* Presl, with free veins.

1. **N. Fi′lix-mas,** *Rich.* ; rootstock tufted, stipes and rachis with lanceolate scales, frond 1–2-pinnate, pinnules deeply obtusely lobed contracted at the base, involucre convex eglandular. *Male Fern.*

Woods and shaded places, N. to Ross; ascends to 2,400 ft. in Yorkshire; Ireland; Channel Islands; frt. July–Aug.—*Rootstock* sometimes 6–10 in., solid and woody. *Fronds* 1–3 ft., and stipes more or less scaly beneath, oblong-lanceolate, rather rigid, rarely simply pinnate; segments entire or serrate at the tip; veins simple or forked. *Sori* large, 1-seriate. *Involucre* smooth, firm.—Distrib. N. temp. regions, India, Africa, Andes.

N. Fi′lix-mas proper; frond 2-pinnate, pinnæ long crowded acuminate, pinnules obtuse serrate, lower distinct.—Var. *affi′nis,* Fisch. (var. *inci′sa,* Newm.); pinnules oblong-lanceolate incised less crowded.—Var. *Borre′ri,* Newm.; rachis very scaly, frond bright golden yellow, pinnules very obtuse almost truncate less serrate.—Var. *L. abbrevia′ta,* DC.; frond pinnate, pinnæ pinnatifid or crenate oblong obtuse with one row of sori along the midrib.

2. **N. crista′tum,** *Rich.* ; rootstock shortly creeping, stipes with ovate or oblong scales, rachis naked, frond oblong-lanceolate sub-2-pin-

nate, pinnules deeply obtusely lobed toothed attached by a broad base, involucre flat eglandular.

Bogs and marshes, Notts, Hunts, Chester, Yorks; Renfrew; frt. Aug.—*Root-stock* and stout pale stipes clothed with large bullate acuminate pale scales. *Frond* 1–1½ ft., narrow, glabrous; pinnæ shortly petioled, oblong, base truncate obtuse apiculate, teeth short not awned, veins forked. *Involucre* quite glabrous, entire.—DISTRIB. Europe, W. Siberia, N. America. *Lastrea uligino'sa*, Newm., with pinnules more divided, teeth slightly spinulose, is intermediate between *N. spinulo'sum* and *crista'tum*, and occurs with the typical form.

3. **N. rig'idum**, *Desv.* ; rootstock tufted and stipes scaly below, frond oblong-lanceolate, lower pinnæ rhomboid lobed to the rachis with mucronate pinnules flat subglandular beneath, involucre gland-ciliate.

Mountains of Lancashire, York, and Westmoreland, rare; ascends to about 1,500 ft.; frt. July–Aug.—*Rootstock* and stout stipes densely clothed with long concolorous scales. *Frond* 12–18 in., subglandular beneath, narrow; pinnæ 2–3 in.; pinnules acutely toothed, lower with subpinnate venules. *Sori* close to the midrib. *Involucre* firm, convex.—DISTRIB. W. Europe to Greece, W. Asia, N. America.

4. **N. spinulo'sum**, *Desv.* ; rootstock tufted, stipes sparingly scaly, frond oblong-lanceolate, lower pinnæ subdeltoid, pinnules lobed to the rachis flat, teeth awned, involucre smooth gland-ciliate or not.

Woods and damp shaded places, from Aberdeen and Dumbarton southd.; Ireland; frt. Aug.-Sept.—*Rootstock* stout, suberect, and stout pale stipes 1 ft., sparingly clothed with ovate scales. *Fronds* 12–18 in., glabrous or glandular beneath; lower pinnæ 2–4 in., subdeltoid; pinnules 1 in., ovate-lanceolate, pinnatifid to the rachis, lobes oblong spinulose-toothed. *Sori* chiefly on the upper half of the frond.—DISTRIB. Europe (Arctic), excl. Turkey and Greece, S. Africa, N.E. Asia, N. America.
N. SPINULO'SUM proper; scales ovate concolorous, frond oblong-lanceolate eglandular beneath pale green, involucre not gland-ciliate.
Sub-sp. N. DILATA'TUM, *Desv.*; scales denser narrower centre dark brown, frond larger ovate-lanceolate or subdeltoid 2-3-pinnate more deeply cut darker and brighter green, pinnæ closer glandular beneath, involucre evanescent gland-ciliate.—VAR. *glandulo'sa*, Newm., is more glandular beneath, frond broader.—VAR. *na'na*, Newm., is smaller.—*Aspidium Boott'ii*, Tuck. (*Lastrea colli'na*, Newm.), has pinnules subentire.—*Aspidium dume-to'rum*, Sm., connects it with *dilata'tum.* — Ascends to 3,700 ft. in the Highlands; Channel Islands.
Sub-sp. ASPIDIUM REMO'TUM, *Braun ;* scales lanceolate concolorous extending up the rachis, frond oblong-lanceolate, pinnæ close lanceolate, pinnules ovate-oblong cut half away to the rachis eglandular beneath, lower only free, involucre eglandular.—Windermere.—Between *spinulo'sum* and *Fi'lix-mas.*

5. **N. æ'mulum**, *Baker ;* rootstock tufted, stipes densely scaly below, frond subdeltoid 3-pinnate, pinnules triangular-ovate concave above, glandular beneath, lowest largest spinous-serrate, edges of involucre

eglandular.   *N. fœnise'cii,* Lowe ;   *Lastrea recur'va,* Bree ;   *L. æ'mula,* Brack.

Hilly districts, from Orkney southd., local ; Ireland ; frt. July–Sept.—Hardly distinct from *N. spinulosum,* frond more triangular and divided, remarkably concave and curved upwards ; scales more fimbriate and undulate, glands of involucre sessile.—Distrib. Europe, Madeira, Azores.—Smells of hay.

6. **N. Thelyp'teris,** *Desv.* ; rootstock creeping, stipes naked, frond lanceolate pinnate, pinnæ deeply pinnatifid, margins entire recurved.

Bogs and marshes, Forfar to Kent and Somerset ; local in Ireland ; frt. July–Aug.—*Rootstock* long, black, hardly scaly. *Stipes* 1 ft., slender, straw-coloured. *Frond* 1–2 ft., membranous, glabrous or sparingly hairy beneath ; pinnæ 2–3 in., spreading, cut to the rachis into narrow oblong entire obtuse lobes ; upper venules simple, lower forked. *Sori* small, dorsal on the venule. *Involucre* gland-ciliate.—Distrib. Europe, Asia, Africa, N. America, N. Zealand.

7. **N. Oreop'teris,** *Desv.* ; rootstock short tufted, stipes naked above, frond pinnate, pinnæ pinnatifid glandular beneath, margins entire flat. *N. monta'num,* Baker.

Mountain heaths and pastures, N. to Shetland ; ascends to 3,000 ft. in the Highlands ; local in Ireland ; frt. July–Aug.—*Rootstock* erect or decumbent, and short stout stipes and rachis below scaly. *Frond* 1½–2 ft., broadly oblong-lanceolate ; pinnæ 3–4 in., spreading, sessile, lanceolate, lower smaller more obtuse and distant ; lobes flat, obtuse, entire, costa pubescent ; lower venules forked. *Sori* near the margins. *Involucre* membranous.—Distrib. Europe (excl. Sweden), W. Asia.—Fragrant.

13. POLYPO'DIUM, *L.*   Polypody.

Ferns of various habit. *Fronds* simple, lobed, pinnatifid or compound. *Sori* dorsal, globose ; involucre 0.—Distrib. All regions, but chiefly trop. ; species 390.—Etym. πολύς and πούς, from the many stipes of some.

Section 1. **Polypo'dium** proper. *Stipes* articulate with the rootstock.

1. **P. vulga're,** *L.* ; rootstock creeping densely scaly, fronds alternate pinnatifid, segments linear-oblong obtuse or acute entire crenate-serrate.

Walls, banks, trees, &c., N. to Shetland ; ascends to 3,400 ft. in Yorkshire ; Ireland ; Channel Islands ; frt. June–Sept. *-Rootstock* stout ; scales pale brown, lanceolate. *Stipes* stout, 3–4 in. *Frond* 6–12 in., linear-oblong or ovate-oblong, coriaceous, naked and glabrous ; segments ¼–⅓ in., broad ; venules pinnate, tips thickened. *Sori* large, 1-seriate, terminal on a lateral venule.—Distrib. Europe, N. and S. Africa, N. and W. Asia, N. America. *P. cam'bricum,* L., is a var. with pinnatifid segments.

Section 2. **Phegop'teris.** *Stipes* not articulate with the rootstock.

2. **P. Phegop′teris,** *L.* ; rootstock creeping scaly, fronds alternate pinnate, pinnæ pinnatifid, lowest pair deflexed, segments obtuse ciliate.

Damp shaded places, Shetland to Cornwall and Somerset; absent in England S. of Derby and E. of Gloster; ascends to 3,500 ft. in the Highlands; local in Ireland; frt. June–Aug.—*Rootstock* long, slender; scales scattered. *Stipes* 6–9 in., slender, base scaly. *Frond* 6–9 in., subdeltoid, slightly hairy beneath, rather membranous, pinnate below, pinnatifid above; pinnæ subopposite, elongate, sessile; lower 2–3 in.; segments subentire; venules 6–8 on each side, lower forked, tips thickened. *Sori* submarginal, dorsal on the venules.—DISTRIB. Europe, N. and W. Asia, N. America.

3. **P. Dryop′teris,** *L.* ; rootstock creeping scaly, fronds alternate deltoid 2-pinnate, pinnules deeply pinnatifid, segments obtuse subcrenate.

Dry shaded places, from Shetland southd. to Derby, Wales, Cornwall, and Devon; absent in England E. of Derby, Gloster, and Devon; ascends to 2,700 ft. in the Highlands; N. Ireland, rare; frt. July–Aug.—*Rootstock* long, slender; scales orange-brown. *Stipes* very slender, 6–12 in., scaly below. *Frond* 6–12 in., flaccid, glabrous; lower pinnæ largest; lowest segments sometimes free; venules forked, tips thickened. *Sori* submarginal, dorsal on the venules.—DISTRIB. Europe, N. and W. Asia, Himalaya, N. America.

P. DRYOP′TERIS proper; frond glabrous flaccid.

Sub-sp. P. ROBERTIA′NUM, *Hoffm.*; rootstock stouter, frond more coriaceous glandular-pubescent. *P. calca′reum,* Sm.—Limestone rocks from Perth to Wales and Derby; Salop and Stafford to Gloster; Wilts, Somerset, Bucks, Oxford. Extends to Tibet.

4. **P. alpes′tre,** *Hoppe;* rootstock short, frond oblong-lanceolate 2-pinnate, pinnules deeply pinnatifid, lobes toothed. *Pseudathyr′ium,* Newm.

Shaded rocks and streams, from Sutherland to Argyll and Perth, alt. 1,200–3,600 ft.; frt. July–Aug.—*Rootstock* stout, oblique, scaly. *Stipes* 4–6 in., tufted, stout, scaly below. *Frond* 1–2 ft., herbaceous, glabrous; pinnules 3–4 in., lanceolate, rachis nearly naked; venules pinnate in the lobes. *Sori* small, 1–4 on each lobe, marginal in the sinus, dorsal on the venule. —DISTRIB. W. Europe (Arctic) to Spain and Germany, W. Asia, Greenland, N.W. America.—Resembles *Asplenium Filix-fœmina.*

P. *alpes′tre* proper; stipes short, pinnæ spreading or ascending narrow-lanceolate broadest at the base, pinnules crowded.—VAR. *P. flex′ile,* Moore; stipes very short, pinnæ short spreading or deflexed, pinnules rather distant. Forfar.

14. GYMNOGRAM′ME, *Desv.*

Fronds of various habit, 1–3-pinnate; veins free or anastomosing. *Sori* dorsal, oblong or linear, usually spreading in irregular lines, branched and confluent on or between the veins; involucre 0.—DISTRIB. Chiefly warm regions; species 84.—ETYM. γυμνός and γραμμή, from the *naked sori.*

G. leptophyl′la, *Desv.* ; glabrous, annual, frond 2–3-pinnate.

Moist banks, Jersey; frt. March–May.—Annual. *Fronds* fragile, 1–3 in.,
shortly stipitate, broadly ovate-oblong; inner with longer stipes, narrower,
more fertile; pinnules obovate-cuneate, 2–3-lobed, lobes obtuse decurrent;
veins dichotomous. *Sori* oblong, simple or confluent.—DISTRIB. S. Europe,
Africa, Asia, America, Australasia.

15. OSMUN'DA, *L.* FERN-ROYAL.

Rootstocks often very large, tuberous or massive. *Fronds* coriaceous,
tufted, 1-2-pinnate, some of the pinnæ altered, contracted, and covered
with naked confluent sori; veins forked, free. *Capsules* globose, subsessile, with a short lateral or subterminal striate areola (an incomplete
contracted ring).—DISTRIB. Temp. and trop. regions; species 6.—ETYM.
After the god *Thor* (*Osmunder*).

O. rega'lis, *L.*; fronds 2-pinnate fertile at the top.

Bogs, marshy woods, &c., N. to Shetland; ascends to 1,000 ft. in N. England;
Ireland; Channel Islands; frt. June–Aug.—*Rootstock* large, densely clothed
with matted fibres, many-headed. *Stipes* 2–10 ft., stout, erect, naked,
brown. *Frond* glabrous; barren pinnæ 3–12 in., sessile or shortly petioled,
oblong, obtuse, truncate cordate or auricled at the often unequal base, serrulate; fertile pinnules subcylindric, lobed.—DISTRIB. Europe, Africa,
Asia, America.

16. OPHIOGLOS'SUM, *L.* ADDER'S-TONGUE.

Rootstock short, with fleshy fibrous roots. *Frond* consisting of a
barren oblong linear or lanceolate reticulately-veined blade, and a fertile
flattened distichous spike of opposite confluent globose capsules that burst
transversely and are obscurely striate at the top. *Spores* minute.—
DISTRIB. All climates; species 3 or 4.—ETYM. ὄφις and γλῶσσα, *snake's
tongue.*

O. vulga'tum, *L.*; blade ovate linear or elliptic-oblong.

Damp pastures, banks, woods, &c., N. to Shetland; ascends to 1,000 ft. in the
Lake District; Ireland; frt. May–July.—*Rootstock* not tuberous. *Frond*
6–9 in., stout or slender. *Blade* 2–4 in., obscurely petioled, coriaceous
midrib obsolete. *Spike* 1–2 in., peduncled; capsules 6–20.—DISTRIB.
Europe, W. Siberia, Himalaya, N. America, S. temp. regions.
O. VULGA'TUM proper; blade large ovate or oblong, epidermal cells flexuous,
spike 2–4 in., spores tubercled. — VAR. *ambig'ua*, Coss. and Germ.;
smaller, blade linear oblong, spike 1–2 in.—Orkney, Wales, Scilly, Donegal.
Sub-sp. O. LUSITAN'ICUM, *L.*; rootstock more tuberous, blade ½–1 in. oblong
or lanceolate, epidermal cells straight, spike ¼–½ in., spores smooth.–
Guernsey; frt. Jan.–Feb.—W. Europe, W. Africa.

17. BOTRYCH'IUM, *Sw.* MOONWORT.

Rootstock small, tuberous; roots of thick fleshy fibres. *Frond* consisting of an erect barren 1-4-pinnate flabellately-veined blade, and a

fertile branched receptacle, covered on the surface facing the blade with small globose coriaceous capsules which burst transversely. *Spores* minute. —DISTRIB. Temp. and trop. regions ; species 6.—ETYM. βότρυς, from the *clustered* sori.

B. Luna'ria, *Sw.* ; blade about the middle of the frond pinnate.

Pastures and grassy banks, N. to Shetland ; ascends to 2,700 ft. in the Highlands ; Ireland ; frt. June–Aug.—*Rootstock* tuberous, enclosing at its top the bud of the next year's frond. *Frond* 3–6 in., stout, terete, fleshy, glabrous ; blade ½–2 in., oblong ; pinnæ ½-circular or lunate, close-set, entire crenate toothed or subpinnatifid. *Receptacle* ½–3′in., erect, segments narrow, incurved. *Capsules* sub-2-seriate on the segments.—DISTRIB. Europe (Arctic), N. and S. temp. and cold regions.

A form with the frond deltoid, pinnules 3–4 pairs incised or pinnatifid, lobes linear or cuneate 1-nerved, found on the sands of Barry, has been doubtfully referred to *B. rutaceum,* Sw.

ORDER XCI.—EQUISETA'CEÆ.

Rootstock creeping. *Stem* erect, terete, jointed, grooved, hollow except at the joints, and with air-cells in their walls under the grooves, joints terminating in toothed sheaths ; teeth corresponding with the ridges ; branches if present arising from the sheath-bases, solid. *Capsules* 6–9, 1-celled, on the under surface of the peltate scales of a terminal cone. *Spores* of one kind, attached to 4 clubbed elastic threads (*elaters*), which are coiled round the spore when moist, and uncoil when dry.—DISTRIB. Chiefly temp. N. regions, a few are sub-trop. ; none are high southern ; genus 1 ; species 25.—AFFINITIES. None direct.—PROPERTIES. The cuticle abounds in siliceous cells ; whence the stems of some are used for polishing.

Germination and impregnation as in *Filices ;* but the prothallus is usually (functionally) 1-sexual.

1. EQUISE'TUM, *L.* HORSE-TAIL, PADDOCK-PIPES.

Characters of the Order.—ETYM. *equus, seta, horse bristle.*

  * *Fruiting stems simple or rarely branched, succulent ; barren appearing later, branched ; branches simple.*

1. **E. arven'se,** *L.* ; barren stems 6–19-grooved, branches spreading, sheaths of fruiting stems distant loose with teeth ribbed to the tip.

Roadsides, banks and fields, N. to Shetland ; ascends to 2,000 ft. in N. England ; Ireland ; Channel Islands ; frt. April.—*Barren stems* erect or decumbent, slightly scabrid, usually ending in a long naked point ; branches crowded, 4-gonous ; *fertile stems* (rarely with branches) stouter, shorter ; sheaths scarious.—DISTRIB. Europe (Arctic) N. Africa, N. Asia, Himalaya, N. America.

2. **E. praten'se,** *Ehrh.* ; barren stems scabrid 8–20-grooved, branches spreading, sheaths of fruiting stems close-set, ribs of teeth not reaching the tip.   *E. umbro'sum,* Willd. ; *E. Drummon'dii,* Hook.

Marshes, rare, from Caithness to York; ascends to 1,200 ft. in N. England; N. Ireland; frt. April.—Closely allied to *E. arven'se,* but greener, less glaucous, more scabrid, with more numerous ribs and branches; the barren stem terminates in an abrupt brush of branches as in *E. sylvat'icum. Barren stems* 1–2 ft., sometimes bearing a cone, slender, abrupt; branches simple, slender, usually spreading, 3–4-gonous, sheaths very short; *fertile* much stouter, sheaths very lax, funnel-shaped.   *Cones* ¾–1½ in.—DISTRIB. Europe (Arctic) N. of the Alps, Italy, Siberia, N. America.

3. **E. max'imum,** *Lamk.* ; barren stems 20–40-grooved, branches suberect, sheaths of short fruiting stems close large loose, teeth 2-ribbed.   *E. Telmetei'a,* Ehrh. ; *E. fluviat'ile,* Sm., not L.

Bogs, ditches, &c., from Skye, Lanark, and Edinburgh southd.; ascends to 1,200 ft. in Yorkshire; Ireland; Channel Islands; frt. April.—*Barren stems* 3–6 ft., ½ in. diam.; branches 4-gonous, slender, erecto-patent, sheaths very short; *fertile* 8–10 in., ½–¾ in. diam. including the large lax sheaths.   *Cone* 2–3 in., obtuse.—DISTRIB. From Denmark southd., N. Africa, N. and W. Asia, N. America.

** *Fruiting and barren stems subsimilar, simple or branched.*

4. **E. sylvat'icum,** *L.* ; stems 10–18-grooved, branches recurved or deflexed divided, stem-sheaths lax, teeth long obtuse, teeth of branch-sheaths 3-ribbed to the tip.

Copses and hedgebanks, from Shetland to Devon and Kent; ascends to 2,700 ft. in the Highlands; Ireland; frt. April–May.—Readily recognised by the elegant appearance of the whorls of compound recurved branches. *Stem* 1–2 ft., nearly smooth.   *Teeth* of branch-sheaths 3–5, large.   *Cones* ¾–1 in., short, ovoid-oblong, obtuse.—DISTRIB. Europe, N. Asia, N. America.

5. **E. palus'tre,** *L.* ; stems 5–12-grooved, branches simple, stem-sheaths short appressed, teeth acute, tips membranous.

Wet places, N. to Shetland; ascends to 2,500 ft. in the Highlands; Ireland; Channel Islands; frt. June–July.—Very variable.   *Stem* 6–18 in., deeply furrowed, branched throughout, slightly rough.   *Cones* short, blunt, those on the branches small.—DISTRIB. Europe, N. and W. Asia, N. America. —VAR. *polysta'chya* bears cones on the branches also.—VAR. *alpi'na* (or *subnu'da*) is a stunted state.

6. **E. limo'sum,** *L.* ; stems smooth faintly 10–30-striate, branches simple erect or 0, sheaths short appressed, teeth short rigid.

Sides of lakes and ditches, N. to Shetland; ascends to 2,500 ft. in the Highlands; Ireland; Channel Islands; frt. June–July.—Easily distinguished by the hardly furrowed stems, and close and short sheaths. *Stems* stout, 1–3 ft., slender, with short suberect branches or none. *Cones* short, oblong, obtuse —DISTRIB. Europe, N. Asia, N. America.

E. limo'sum proper; smooth, branches short rigid equalling the internodes, cones subsessile.—VAR. *E. fluviat'ile,* L.; scaberulous above, branches tapering longer than the internodes, cones peduncled.

7. **E. hyema'le,** *L.* ; stems scabrid 8–34-grooved, branches all sub-radical or 0, sheaths white with black tip and base, teeth black with deciduous tips. *Dutch Rush.*

Marshes in woods, local, from Ross and Moray to Kent and Hereford; ascends to 1,700 ft. in Forfar; rare in Ireland; frt. July–Aug.—Easily distinguished by its size, glaucous colour, scabridity, and stems simple or branched at the base only. *Stems* 1–3 ft.; branches simple, grooves shallow. *Cones* small, conoid, acute.—DISTRIB. Europe, N. Africa, N. Asia, N. America.

E. hyema'le proper; stems perennial or biennial, sheaths close, tip of teeth black.—VAR. *E. Moor'ei,* Newm. (*E. palea'ceum,* Schleich.); stems annual, sheaths loose, teeth truncate, tip white. Wicklow, near the sea.

8. **E. variega'tum,** *Schleich.* ; stems filiform more or less scabrid 4–14-grooved, branches basal, sheaths green below black above, teeth obtuse apiculate membranous.

Wet places and sandy shores, local, from Ross and the Clyde to York: Chester, Wales, Norfolk, Devon; Ireland; frt. July–Aug.—Usually small. *Stems* 4–12 in., often decumbent and branching dichotomously below. *Sheaths* short, rather distant, appressed; teeth membranous, white, or edges black. *Cone* small, ovoid, acute.—DISTRIB. Europe, N. Africa, Siberia, N. America.

E. variega'tum proper; erect, 1–2 ft., teeth of sheaths short acute. Wet banks, &c.—VAR. *E. arena'rium,* Newm.; stems decumbent more slender. teeth of sheaths 6–8 cuneate. Sandy shores, from Lancashire northd. —VAR. *E. Wilso'ni,* Newm.; tall, erect, 2–3 ft., teeth of sheaths short obtuse. Watery places.—VAR. *E. trachyo'don,* Braun (*E. Mackai'i,* Newm.); stem erect or almost decumbent stouter, branches longer flexuous, sheaths black, teeth at length white. Damp woods, N.E. Ireland, Scotland.

ORDER XCII.—**LYCOPODIA'CEÆ.**

Rootstock running, creeping, or a corm, or 0. *Stem* dichotomously branched, usually rigid, leafy throughout. *Leaves* imbricate all round or 2–6-fariously, small, simple, nerveless or 1-nerved. *Capsules (sporangia)* sessile in the axils of the leaves or of the scales of a terminal or axillary sessile or peduncled cone, 1–3-celled, compressed, often reniform, 2-valved. *Spores* marked with 3 radiating lines at the top.—DISTRIB. All climates; genera 4; species 100.—AFFINITIES. With *Filices.*

In germination the spore develops a prothallus upon which archegonia and antheridia are produced, as in *Filices.*

1. LYCOPO'DIUM, *L.* CLUB-MOSS.

Perennial. *Stem* erect prostrate or creeping. *Leaves* small. *Capsules* coriaceous, flattened, reniform, 1-celled, 2-valved.—DISTRIB. Of the Order ; species about 50.—ETYM. λύκος and πούς, from a fancied resemblance to a *wolf's foot.*

* Stem creeping.   Capsules in terminal cones.

1. **L. clava′tum,** *L.* ; leaves hair-pointed, cones peduncled.

Heaths and moors, Shetland to Cornwall, Hants, and Essex ; ascends to 2,500 ft. in Yorkshire ; Ireland ; frt. July–Aug.—*Stems* 1–3 ft., rigid, flexuous, much branched, densely leafy. *Leaves* imbricate all round, $\frac{1}{8}$–$\frac{1}{4}$ in., subsecund, incurved, linear-oblong or lanceolate, acuminate, hair-point variable in length. *Cones* 1–3 in., solitary or in pairs on a rigid erect peduncle covered with minute appressed subulate leaves, cylindric, obtuse ; scales appressed, broadly ovate or cordate, acuminate. *Capsules* orbicular-reniform.—DISTRIB. Arctic, and N. and S. temp. and cold regions.

2. **L. anno′tinum,** *L.* ; leaves acuminate entire or serrate, cones sessile, scales broadly ovate toothed.

Rocks and stony alpine moors, from Orkney to the Clyde and Perth ; N. Wales, Cumberland, Westmoreland, Lancashire, Leicester ; ascends to 2,700 ft. in the Highlands ; frt. June–Aug.—Habit of *L. clava′tum,* but less branched, branches constricted here and there, leaves more lax, obscurely 5-farious, sometimes spreading, linear-lanceolate ; scales of obtuse cone broad, abruptly acuminate.—DISTRIB. Europe (Arctic), N. and W. Asia, Himalaya, America.

3. **L. complana′tum,** *L.* ; leaves 2–4-farious lanceolate quite entire, cones peduncled or sessile, scales broadly ovate subentire.

Stony moors, heaths, &c., from Shetland to York, Derby, Wales, Somerset and Hants ; Ireland ; frt. July–Aug.—*Stem* 6–18 in., rigid, wiry, flexuous, sparingly leafy ; branches fastigiate, much forked, ascending or erect. *Leaves* $\frac{1}{8}$–$\frac{1}{4}$ in., dark green, appressed, of 2 sizes ; larger (lateral) adnate, subdecurrent, concave, obtuse ; smaller shorter, more subulate, free. *Cones* $\frac{1}{2}$–1 in., oblong, obtuse, terete.—DISTRIB. Temp. and cold regions of the N. hemisphere and mts. of the tropics.

L. COMPLANA′TUM proper ; leafy branches longer less crowded, leaves dimorphic, central ones on the flattened stem more erect and narrower than the lateral, spikes usually several peduncled. Gloster and Worcester. (Temp. regions and mts. of tropics.)

Sub-sp. L. ALPI′NUM, *L.* ; leafy branches shorter more crowded not flattened, leaves uniform, spikes solitary sessile.—Common in Wales and N. to Shetland, ascends to 4,000 ft. in the Highlands. (N. temp. and Arctic regions.)

4. **L. inunda′tum,** *L.* ; leaves secund on the sterile branches subulate-lanceolate quite entire, scales of cone subulate with much-dilated spinous-toothed bases.

Wet heaths and bogs, from Ross southd., local ; Ireland, very rare ; frt. June–Aug.—*Stems* short, 2–6 in., closely appressed to the ground. *Leave* secund, though inserted all round the stem, dark green, midrib indistinct. *Cones* 1–3 in., fusiform, on strict erect leafy branches, the leaves of which are erect and not secund; scales erect, narrow, much longer than the leaves, bases sometimes cordate.—Distrib. Europe, temp. and trop. N. and S. regions.

** *Stem decumbent at the base. Capsules axillary in the upper leaves.*

5. **L. Sela'go,** *L.* ; branches stout uniform in height, leaves subulate-lanceolate quite entire.

Moors and heaths, Shetland to Sussex and Cornwall ; ascends to 3,500 ft. in the Highlands ; Ireland ; frt. June–Aug.—*Stem* stout, rigid, shortly creeping at the base ; branches 2–8 in., $\frac{1}{4}$–$\frac{2}{3}$ in. diam., densely leafy, erect, strict, obtuse. *Leaves* erect, appressed, incurved, squarrose or spreading, acuminate, pungent or not; midrib 0; upper capsuliferous sometimes yellower.—Distrib. Temp. and cold N. and S. regions.

Order XCIII.—SELAGINELLA'CEÆ.

Land- or water-plants, stemless, or with branched, slender stems. *Leaves* small, imbricate all round the stem, or distichous and of 2 forms, long and slender in the stemless species. *Capsules* of two forms, the larger 2–4-valved, containing macrospores ; the smaller containing microspores.—Distrib. All temp. and warm climates ; genera 2 ; species about 100.—Affinities. Between *Marsilea'ceæ* and *Lycopodia'ceæ*.

In germination, the macrospores of *Selaginel'la* and *Isoë'tes* develop a cellular prothallus under the integuments, in the position of three radiating lines ; this is extruded, and upon its surface are developed many archegonia along the above lines, one only of which is fertilized. The microspores burst, and emit cells containing each an antherozoid, which, entering the archegonia, fertilize their germ-cell, as in *Filices.*

1. SELAGINEL'LA, *Beauv.*

Terrestrial plants. *Leaves* small, uniform and imbricate all round the stem, or of 2 forms, one large and distichous, and the other smaller unilateral on the stem. *Capsules* of 2 kinds, in terminal cones : 1, minute, oblong or globose, containing microspores ; 2, larger, 2–4-valved containing 1–6 macrospores.—Distrib. Chiefly trop. ; species about 150.—Etym. Diminutive of *Sela'go,* an old name for *Lycopo'dium.*

S. selaginoi'des, *Gray ;* decumbent, leaves lax lanceolate and ovate, scales of cone spinulose-ciliate. *S. spino'sa,* Beauv.

Bogs and marshes, from Shetland to Lincoln, Derby, Chester, and Wales ; Ireland ; ascends to 3,300 ft. in the Highlands ; frt. July–Aug.—*Stems* 2–6 in., slender, sparingly branched; branches ascending. *Leaves* inserted all round, $\frac{1}{8}$–$\frac{1}{4}$ in., incurved or squarrose, pale yellow-green, acuminate, midrib obscure. *Cones* on elongate erect branches, 1–3 in., terete, rather stouter than the branch; scales erect or spreading, broader and longer than the leaves, with long spinulose teeth. *Microspores* echinate, in 2-valved reniform capsules. *Macrospores* globose, in 3–4-valved and lobed capsules.— Distrib. N. and W. Europe, Siberia, Himalaya, N. America.

2. ISOE'TES, *L.* Quillwort.

Aquatic or terrestrial stemless plants. *Corm* depressed. *Leaves* long, subulate or filiform, often tubular and septate, base sheathing. *Capsules* sessile in the axils of the leaves, partially enclosed by and adnate to their sheathing bases, traversed by cellular threads ; those of the outer leaves contain globose macrospores, those of the inner contain oblong 3-gonous microspores. *Macrospores* with a crustaceous coat, marked on the upper hemisphere with 3 radiating lines, and bursting by 3 valves.—Distrib. Chiefly N. temp. and warm regions ; species 6 or 8.—Etym. ἴσος and ἔτος, *ever-green ;* of obscure application.

1. **I. lacus'tris,** *L.* ; aquatic, leaves subulate, macrospores covered with crested ridges or tubercled.

Bottoms of alpine and subalpine lakes, from Caithness to Salop and N. Wales ; ascends to 2,000 ft. in the Highlands ; Ireland ; frt. May–July.— *Corm* often as big as a hazel-nut. *Leaves* 10–20, 2–6 in., rigid, obscurely 4-gonous, dark green, of 4 septate tubes. *Capsule* ovoid or globose, partially covered by the inflexed edges of the sheath. *Macrospores* tubercled by the protrusion of the inner wall through perforations of the outer. *Microspores* granular.—Distrib. Europe (Arctic) N. of the Alps, W. Siberia, N. America.

I. lacus'tris proper ; leaves erect green, capsules ⅓ covered by the edges of the leaf-sheath, tubercles of macrospore short.—Var. *I. More'i,* Moore ; leaves 18 in. long. In deep water, Wicklow.

I. echinospo'ra, *Durieu ;* leaves spreading paler, capsules almost enclosed in the leaf-sheath, tubercles of macrospore longer more acute.—N. Wales, Aberdeen, Dumbarton, Kerry.

2. **I. Hys'trix,** *Durieu ;* terrestrial, leaves filiform, macrospores obtusely tubercled. *I. Duriœ'i,* Hook.

Sandy soil, inundated at times, Guernsey ; frt. May–June.—*Corm* short, stout, subglobose, 1 in. diam., clothed with the old spinescent dark horny leaf-bases, which consist of lateral subulate processes, and an intermediate tooth. *Leaves* 1–2 in., slender, plano-convex, obscurely tubular, sheath enveloping the capsule. *Macrospores* white.—Distrib. S. Europe, N. Africa.

Order XCIV.—MARSILEA'CEÆ.

Aquatic plants of various habit. *Rootstock* or stem creeping. *Leaves* filiform or bearing 4 obovate leaflets ; vernation circinate. *Fructification* of 2- or more-celled coriaceous oblong or globose capsules (formed of a metamorphosed leaf) placed near or on the rootstock, and containing on parietal

placentas many membranous sacs, enclosing macrospores and microspores.
—Distrib. Temp. and trop. regions ; genera 2 ; species 40.—Affinities.
With *Selaginellaceæ.*

1. PILULA′RIA, *L.* Pillwort.

Rootstock filiform, creeping. *Leaves* subsolitary, erect, setaceous. *Capsule* globose, 2–4-celled, 2–4-valved at the top , cells each with a longitudinal parietal placenta, on which are inserted many pyriform membranous sacs ; sacs in the upper part of the cell full of microspores immersed in mucilage ; those in the lower part contain each one macrospore. *Microspores* globular, full of antherozoids. *Macrospores* ovoid, with an outer coat of prismatic cells, pierced by a funnel-shaped opening, through which an inner glassy coat finally protrudes.—Distrib.   N. and S. temp. and cold regions ; species 3.—Etym. *pilula,* from the form of the capsule.

In germination a prothallus is developed at the top of the protruded portion of the inner coat of the macrospore, which bursts and frees it. After expulsion an archegonium is formed on the prothallus, and fertilization takes place by the contents of the microspore.

P. globulif′era, *L.* ; leaves setaceous, capsules pubescent.

Edges of lakes and ponds, from Skye and Sutherland to Cornwall and Hants ;
    N.E. and W. of Ireland, very rare ; frt. June–Aug.—*Rootstock* or stem 2–6
    in., glabrous, cylindric. *Leaves* 2–4 in., green. *Capsules* ¼ in. diam., ovoid
    or globose, shortly pedicelled, in the axils of the leaves or on the rootstock
    pubescent, brown, 4-celled.—Distrib. Europe N. of the Alps.

APPENDIX.

EXCLUDED SPECIES

DICOTYLEDONES.

RANUNCULACEÆ.
Anemone ranunculoides, L. In plantations only.
    „   *apennina*, L.
Thalictrum majus, Jacq. Confounded with *T. flexuosum*.
    „   *nutans*, Desf. A S. European species, not known as British.
Ranunculus alpestris, L. Clova Mts., G. Don ; never confirmed.
    „   *gramineus*, L. Said to have been found in Wales a century ago, but not confirmed.
Pæonia corallina, Retz. Steep holmes, introduced.
BERBERIDEÆ.
Epimedium alpinum, L. On rock-works, old castle gardens, &c.
PAPAVERACEÆ.
Papaver nudicaule, L. W. of Ireland, Giesecke ; never confirmed.
    „   *setigerum*, DC. A garden escape in the Fens.
Glaucium phœniceum, Crantz. Casually introduced into Norfolk.
FUMARIACEÆ.
Fumaria spicata, L. Authority unknown to me.
    „   *agraria*, Lag. Confounded with a form of *F. capreolata*.
CRUCIFERÆ.
Cardamine bellidifolia, L. Confounded with a form of *C. hirsuta*.
Malcolmia maritima, Br. Shores of Kent; not native.
Sisymbrium pannonicum, Jacq. An escape ; established at Crosby in Lancashire.
Erysimum virgatum, Roth. A garden escape.
    „   *orientale*, Br. An alien, casual in various localities.
Erucastrum Pollichii, Schimp. and Spenn. Almost naturalized in Essex.
Diplotaxis viminea, DC. Has been reported from Guernsey.
Vella annua, L. Reported from Salisbury Plain long ago ; never verified.
Alyssum incanum, L. A casual in several localities.
Lepidium hirtum, L. Confounded with *L. Smithii*.
    „   *sativum*, L. A garden escape.
Hutchinsia alpina, Br. Reported from Ingleborough ; never verified.
Clypeola Jonthlaspi, L. Authority unknown to me.
RESEDACEÆ.
Reseda Phyteuma, L. A casual on ballast heaps, Yorkshire.
CISTINEÆ.
Helianthemum ledifolium, L. Brean Downs ; never confirmed.

M M

VIOLACEÆ.
  Viola epipsila, Led. Confounded with *V. palustris.*
    ,,   *stricta,* Hornem. Confounded with *V. stagnina.*
FRANKENIACEÆ.
  Frankenia pulverulenta, L. Reported from Sussex ; never confirmed.
CARYOPHYLLEÆ.
  Silene annulata, Thore. Occurs in flax-fields.
    ,,   *alpestris,* Jacq. One of G. Don's discoveries ; unconfirmed.
    ,,   *Armeria,* L. A casual.
    ,,   *italica,* Pers. (*S. patens,* Engl. Bot.). Between Darenth and Dart-
        ford ; not wild.
  Saponaria Vaccaria, L. A casual in corn-fields.
  Cucubalus bacciferus, L. Isle of Dogs ; introduced.
  Arenaria fastigiata, Sm. Scotch Mts., G. Don ; never confirmed.
  Spergula pentandra, L. Reported to have been found in Ireland ; pro-
        bably confounded with a *Spergularia.*
    ,,   *pilifera,* DC. Authority unknown to me.
  Buffonia tenuifolia, Sm. Hounslow Heath, Dillenius ; never confirmed.
HYPERICINEÆ.
  Hypericum hircinum, L. A shrubbery plant.
    ,,   *elatum,* Ait.   ,,
    ,,   *barbatum,* Jacq. Perthshire, G. Don ; never confirmed.
MALVACEÆ.
  Malva borealis, Wallm. Reported from Kent ; but never confirmed.
    ,,   *verticillata,* L. Corn-fields in Wales.
    ,,   *parviflora,* L. Introduced with ballast.
  Lavatera sylvestris, Brot. Near Wareham, Scilly, Cornwall ; an escape.
GERANIACEÆ.
  Geranium nodosum, L. Garden stray in shrubberies, &c.
    ,,   *angulatum,* Curt.   ,,   ,,
    ,,   *striatum,* L.   ,,   ,,
CELASTRINEÆ.
  Staphylea pinnata, L. Shrubberies.
LEGUMINOSÆ.
  Coronilla varia, L. Confounded with *Hippocrepis,* and other plants.
  Medicago muricata, Willd. Confounded with *M. denticulata.*
  Melilotus parviflora, Lamk. A casual on ballast heaps, &c.
  Trifolium parviflorum, Ehr. A casual, near Dublin.
    ,,   *stellatum,* L. A casual, Shoreham.
    ,,   *resupinatum,* L. A casual, near Liverpool and elsewhere.
    ,,   *tomentosum,* L. Authority unknown to me.
    ,,   *agrarium,* L. Perth and Aberdeen, possibly native
  Lathyrus latifolius, L. A garden escape.
    ,,   *sphæricus,* Retz. Hertfordshire, an escape.
ROSACEÆ.
  Aremonia agrimonioides, DC. A garden escape.
  Potentilla alba, L. Reported by Hudson from Wales.
    ,,   *tridentata,* Sm. Clova Mts., G. Don ; never confirmed.
    ,,   *opaca,* Sm. Ditto ditto ; but the specimens are *P. intermedia.*
  Rosa Dicksoni, Lindl. Reported from the S. of Ireland ; not confirmed.
    ,,   *provincialis,* Ehr. A garden escape.

ROSACEÆ (*continued*)
Rosa cinnamomea, L. A garden escape.
  ,, *austriaca,* Crantz.   ,,
  ,, *lucida,* Ehr.   ,,
  ,, *rubella,* Sm. Reported from Shields; not confirmed.
Sanguisorba media, L. One of G. Don's discoveries; unconfirmed.
Rubus arcticus, L. Reported long ago from the Highlands; not confirmed.
  ,, *spectabilis,* Pursh. An American plant, naturalized in plantations.
Pyrus domestica, Sm. The Sorb or Service-tree. An introduced tree in
          Wyre forest.

SAXIFRAGEÆ.
Saxifraga Sibthorpii, Boiss. Argyllshire; a garden escape.
  ,, *Andrewsii,* Harv. (*S. Guthriana,* Hort.). A hybrid between
       S. umbrosa and one of the Aizoon group, is stated to have
       been found in Kerry, but never confirmed.
  ,, *Cotyledon,* L. Reported from the Lake District; never confirmed.
  ,, *rotundifolia,* L.   ,,   ,,   ,,
  ,, *leucanthemifolia,* Scop. Authority unknown to me.
  ,, *muscoides,* Wulf. One of Don's reputed discoveries; and re-
       ported from Westmoreland by Hudson.
  ,, *pygmæa,* Haw. (*moschata,* Engl. Bot. excl. syn.), a form of *S.
       muscoides,* Wulf., erroneously said by J. Donn to be a
       native of Scotland.
  ,, *pedatifida,* Sm. One of Don's reputed discoveries; also
       reported from Achil Island, but the specimens are *S. tri-
       furcata,* a garden plant.

CRASSULACEÆ.
Sedum Cepæa, L. A garden escape in Bucks.
  ,, *stellatum,* L.   ,,   in Sussex.

ONAGRARIEÆ.
Epilobium rosmarinifolium, Hænke. Said to have been found in Glen
        Tilt, but never confirmed.

UMBELLIFERÆ.
Echinophora spinosa, L. Reported from Dorset, &c.; not confirmed.
Bupleurum prostratum, Link. A corn-field casual.
Trinia Kitaibelii, Bieb. Confounded with *T. vulgaris.*
Ammi majus, L. An alien weed by the Severn.
Chærophyllum aureum, L. Scotland, G. Don; not confirmed.
      ,, *aromaticum,* L.   ,,
Siler trilobum, Scop. Naturalized at Cherry Hinton.
Angelica Archangelica, L. A garden relic.
Tordylium officinale, L. Recorded from near London; but confounded
       with *T. maximum.*

CAPRIFOLIACEÆ.
Diervilla canadensis, Willd. In shrubberies only.

RUBIACEÆ.
Asperula arvensis, L. A casual in corn-fields.
  ,, *taurina,* L. In shrubberies only.
Galium spurium, L. A casual flax-field plant.
  ,, *saccharatum,* All. One of G. Don's reputed discoveries.
  ,, *cinereum,* Sm.   ,,   ,,   ,,

RUBIACEÆ (*continued*).
   Crucianella stylosa, DC. A garden escape.
VALERIANEÆ.
   Centranthus Calcitrapa, Dufr. A garden escape. Walls at Eltham.
DIPSACEÆ.
   Dipsacus Fullonum, Mill. An escape from cultivation.
   Scabiosa maritima, L. A S. European species, has been gathered in Jersey.
COMPOSITÆ.
   Aster brumalis, Nees. A garden escape (American).
     „ *salignus*, Willd. An escape in Cambridgeshire.
     „ *longifolius*, Lamk. An American species, found near Perth.
   Xanthium strumarium, L. A casual weed.
     „ *spinosum*, L. „
   Solidago lanceolata, L. A garden escape (American).
   Anacyclus radiatus, Loisel. Ballast heaps, Cork.
   Anthemis tinctoria, L. Ballast heaps, &c.
   Achillea tanacetifolia, All. A garden escape near Sheffield.
     „ *decolorans*,Schrad. Known only in cultivation; of uncertain origin.
     „ *tomentosa*, L. A garden escape; reported from several localities.
   Artemisia cærulescens, L. Reported by Gerard, from Lincolnshire, Kent,&c.
   Petasites albus, Gærtn. In shrubberies.
     „ *fragrans*, Presl. „
   Tussilago alpina, L. One of G. Don's reputed discoveries.
   Senecio erraticus, Bert. A large state of *S. aquaticus.*
     „ *crassifolius*, Willd. Cork; an escape.
   Calendula officinalis, L. A garden escape.
     „ *arvensis*, L. „
   Carlina racemosa, L. Once found in Aran (Ireland), a single specimen.
   Centaurea Jacea, L. Found in Sussex, probably indigenous, being a very common Continental plant.
     „ *montana*, L. A garden escape.
     „ *intybacea*, L. One of G. Don's reputed discoveries.
     „ *salmantica*, L., and *C. leucophæa*, Jord., are said to have been found in Jersey.
   Cnicus oleraceus, L. Once found in Lincolnshire.
   Crepis pulchra, L. One of G. Don's reputed discoveries.
     „ *nicæensis*, Balb. A S. European species, is found sporadically where introduced with grass-seeds, &c.
   Hieracium amplexicaule, L. Old castle walls.
     „ *collinum*, Fries (*pratense*, Fries). Waste ground, fences, railway banks. Probably the *H. Auricula*. L., stated to be found in Westmoreland by Hudson (*H. dubium*, Fl. Dan., t. 1044).
     „ *præaltum*, Vill. A Southern species; occurs as an escape.
     „ *glomeratum*, Frol. An Eastern European plant; „
     „ *stoloniferum*, W. & K. A Southern species; occurs as an escape.
     „ *cerinthoides*, L. One of G. Don's reputed discoveries.
     „ *villosum*, L. Reported from the Highlands.
   Prenanthes purpurea, Lamk. Naturalized in Skye and near Edinburgh.
ERICACEÆ.
   Erica multiflora, L. Authority unknown.

ERICACEÆ (*continued*).
    Ledum palustre, L. N.W. Ireland; never confirmed.
    Ṿ*accinium macrocarpon*, Ait. Flintshire; introduced.
PRIMULACEÆ.
    Lysimachia ciliata, L. An American plant, introduced into Cumber-
        land.
       „   *punctata*, L. Introd. near Newcastle.
GENTIANEÆ.
    Gentiana acaulis, L. Reported from Wales; not confirmed.
    Swertia perennis, L.   „   „
BORAGINEÆ.
    Echinospermum Lappula, Lehm. An alien on ballast.
       „   *deflexum*, Lehm.   „
    Mertensia virginica, DC. A garden escape.
    Omphalodes verna, Mœnch. „   „
    Symphytum peregrinum, Ledeb. Derbyshire. Shrubberies only.
       „   *asperrimum*, Bieb. Shrubberies and cultivated fields.
       „   *tauricum*, Willd. A garden escape.
       „   *orientale*, L.   „   „
CONVOLVULACEÆ.
    Cuscuta approximata, Bab. A casual on Bokhara clover.
       „   *hassiaca*, Pfeiff. A casual on Lucerne.
SOLANEÆ.
    Nicandra physaloides, Gærtn. A garden escape.
    Datura Stramonium, L. A casual weed.
    Physalis Alkekengi, L. A casual.
    Lycium barbarum, L. A cottage ornament.
    Hyoscyamus albus, L. Ballast hills, Sunderland.
PLANTAGINEÆ.
    Plantago Psyllium, L. Ballast hills, Jersey.
       „   *argentea*, L. A casual, Ireland.
       „   *arenaria*, L. A casual, sandhills, Somerset and Jersey.
SCROPHULARINEÆ.
    Verbascum thapsiforme, Schrad. Reported by Hudson; not confirmed.
       „   *phlomoides*, L. An alien on Clapham Common.
       „   *phœniceum*, L. Reported in 1803 from Wales.
    Linaria supina, Desf. A casual on ballast, Devon and Cornwall.
       „   *purpurea*, L. A garden escape, old walls, &c.
       „   *spartea*, Hoffm. Walton Heath, a casual.
    Erinus alpinus, L. A S. European plant, has been found in Yorkshire,
        Cheshire, and elsewhere.
    Veronica peregrina, L. A casual weed of cultivation.
       „   *fruticulosa*, L. Reported from Scotland; never confirmed.
OROBANCHEÆ.
    Orobanche lucorum, Koch. Confounded with *O. elatior*.
       „   *ramosa*, L. A casual on hemp.
ACANTHACEÆ.
    Acanthus mollis, L. Scilly Is. and Cornwall, an escape.
LABIATÆ.
    Origanum Onites, L. Confounded with *O. vulgare*.
       „   *virens*, Link.   „   ,

LABIATÆ (*continued*).
    Teucrium regium, Schreb. Reported from Wales by mistake.
    Stachys annua, L. An alien in corn-fields in Kent.
AMARANTHACEÆ.
    Amaranthus retroflexus, L. A casual weed.
       ,,   *Blitum*, L.   ,,
CHENOPODIACEÆ.
    Atriplex hortensis, L. A garden escape.
       ,,   *nitens*, Reb.   ,,
    Chenopodium ambrosioides, L. A casual weed
       ,,   *multifidum*, L.   ,,
POLYGONEÆ.
    Polygonum Fagopyrum, L. An escape from cultivation.
    Rumex scutatus, L. Naturalized near Edinburgh.
EUPHORBIACEÆ.
    Euphorbia dulcis, L. An escape from cultivation.
       ,,   *salicifolia*, Host. An alien in Forfarshire.
       ,,   *Characias*, L. Mistaken for *E. amygdaloides*.
       ,,   *pilosa*, L. (*palustris*, Forster, not L.). Copses near Bath and
               at Westmerton, Sussex.
       ,,   *coralloides*, L. Slinfold, Sussex.
SANTALACEÆ.
    Thesium humile, Vahl. Devonshire; not indigenous.
       ,,   *intermedium*, Schrad. No authority for Britain.
CUPULIFERÆ.
    Castanea vulgaris, Lamk. Parks, plantations, &c., only.
SALICINEÆ.
    Salix dasyclados, Wimm. See *S. acuminata*, Sm.
       ,,   *grandifolia*, Ser. Authority doubtful.
       ,,   *hastata*, L. Reported from sands of Barrie; never confirmed.
       ,,   *petiolaris*, Sm. An American willow.
       ,,   *plicata*, Fries. Authority doubtful.
       ,,   *serpyllifolia*, Scop. Inserted by error.

MONOCOTYLEDONES.

ORCHIDEÆ.
    Gymnadenia odoratissima, Reichb. Authority doubtful.
IRIDEÆ.
    Crocus sativus, L. Cultivated at Saffron Walden.
       ,,   *aureus*, Sibth. Parks only.
       ,,   *biflorus*, Mill.   ,,
    Iris tuberosa, L. Penzance and Cork; introduced only.
      ,, *Xiphium*, L. Reported by error from Worcestershire.
      ,, *pumila*, L. Reported from Leicestershire; no doubt introduced.
      ,, *germanica*, L. Reported from Staffordshire;   ,,
AMARYLLIDEÆ.
    Narcissus poeticus, L. Shrubberies, parks, and sites of old gardens.
       ,,   *conspicuus*, Don.   ,,   ,,
       ,,   *minor*, L.   ,,   ,,
       ,,   *incomparabilis*, Curt.   ,,   ,,

LILIACEÆ.

Scilla bifolia, L. Reported a century ago from W. of England.
Lilium pyrenaicum, Gouan. A garden escape in Devonshire.
Allium roseum, L. A garden escape, Rochester and Suffolk.
    „   *paradoxum*, Don. Linlithgow.

JUNCEÆ.

Luzula nivea, Desv. A garden escape in Forfarshire, &c.
    „   *albida*, DC. A casual in Surrey.

TYPHACEÆ.

Typha minor, Sm. Reported by Dillenius from Hounslow Heath.

NAIADEÆ.

Potamogeton gracilis, Fries. Confounded with *pusillus*.
    „   *Kirkii*, Syme (accidentally omitted in the body of this work), from Maam (Galway), is an obscure plant, referred (in MSS.) by Tiselius to *heterophyllus*, but by Babington apparently correctly to *sparganifolius*, Læstd., of Sweden. It is probably a form of *P. polygonifolius*, with (when present) very long subcoriaceous floating leaves, and long linear submerged ones like those of *lanceolatus* without their square reticulations.

CYPERACEÆ.

Eriophorum capitatum, Host. One of G. Don's reputed discoveries.
Carex brizoides, L. Reported from Yorkshire; no doubt introduced.
    „   *ustulata*, Wahl. One of G. Don's reputed discoveries.
    „   *hordeiformis*, Wahl.   „   „
    „   *Mœnchiana*, Wendl. A form of *C. acuta.* „
    „   *vulpinoides*, Michx. Thames near Kew. A N. American species.

GRAMINEÆ.

Panicum sanguinale, Scop. A corn-field casual.
Setaria glauca, Beauv.   „   „
Phalaris paradoxa, L. Casual.
Stipa pennata, L. Reported last century; never confirmed.
Phleum asperum, Jacq. A casual, not found lately.
    „   *Michelii*, All. One of G. Don's reputed discoveries.
Avena planiculmis, Schrad. Arran Is. (Scotland); never confirmed.
Trisetum subspicatum, Link. No authority.
Briza maxima, L. Becoming naturalized in Jersey.
Poa sylvatica, Chaix (*sudetica*, Hænke). Found near Kelso.
Bromus patulus, Reich.
    „   *squarrosus*, L.   } Ballast or corn-field casuals.
    „   *tectorum*, L.   }
Agropyrum cristatum, Gærtn. One of G. Don's reputed discoveries.
Ægilops ovata, L. No authority.
Elymus geniculatus, Curt. Gravesend; never confirmed.

ACOTYLEDONES.

FILICES.

Asplenium fontanum, Presl. On old walls; only where planted.
Onoclea sensibilis, Willd. Near Warrington; no doubt an escape.

EQUISETACEÆ.

Equisetum ramosum, Schkuhr. Wales, Schkuhr; no authority given.

INDEX.

The names of varieties and synonyms are in italics.

Alsine, 63
  Cherleri, 65
  marina, 65
  stricta, 64
Althæa, 75
  hirsuta, 75
  officinalis, 75
Alyssum, 36
  calycinum, 36
  incanum, 529
  maritimum, 36
Amaranthaceæ, 510
Amaranthus Blitum, 534
  retroflexus, 534
Amaryllideæ, 398
Ammi majus, 531
Ammodenia, 65
Ammophila, 480
  arundinacea, 481
  baltica, 481
Anacamptis, 391
Anacharis Alsinastrum, 382
  canadensis, 382
  Nuttallii, 382
Anacyclus radiatus, 532
Anagallis, 265
  arvensis, 265
  cœrulea, 265
  phœnicea, 265
  tenella, 265
Anchusa, 277
  arvensis, 277
  officinalis, 278
  sempervirens, 278
Andromeda, 251
  polifolia, 252
Anemone, 3
  apennina, 529
  nemorosa, 3
  Pulsatilla, 3
  ranunculoides, 529
Angelica, 182
  Archangelica, 531
  sylvestris, 182
Antennaria, 208
  dioica, 209
  hyperborea, 209
  margaritacea, 209
Anthemis, 211
  anglica, 212
  arvensis, 211
  Cotula, 212
  maritima, 212
  nobilis, 212
  tinctoria, 532
Anthoxanthum, 472
  odoratum, 472
  Puelii, 472
Anthriscus, 176
  Cerefolium, 176

Anthriscus
  sylvestris, 176
  vulgaris, 176
Anthyllis, 102
  Dillenii, 102
  Vulneraria, 102
Antirrhinum, 295
  majus, 295
  Orontium, 295
Apargia autumnalis, 239
  hispida, 239
  Taraxaci, 239
Apera, 479
  interrupta (sub-sp.), 480
  Spica-venti, 480
Aphanes arvensis, 127
Apium, 169
  graveolens, 169
  inundatum, 170
  nodiflorum, 170
Apocynaceæ, 268
Aporanthus Trifoliastrum, 94
Apple, 136
  crab, 136
Aquifoliaceæ, 85
Aquilegia, 12
  vulgaris, 13
Arabidia, 139
Arabis, 26
  ciliata (sub-sp.), 27
  hirsuta, 27
  hispida, 26
  perfoliata, 27
  petræa, 26
  sagittata. 27
  stricta, 26
  Turrita, 27
Aracium, 231
Araliaceæ, 186
Arbutus, 250
  Unedo, 251
Archangel. yellow, 330
Arctium, 220
  intermedium, 221
  Lappa, 220
  majus, 220
  minus (sub-sp.), 220
  nemorosum, 220
  pubens, 220
  tomentosum, 220
Arctostaphylos, 251
  alpina, 251
  Uva-Ursi, 251
Aremonia agrimonoides, 530
Arenaria, 63
  Cherleri. 65
  ciliata, 65
  fastigiata, 530
  Gerardi, 63
  hirta, 63
  hybrida, 64
  laxa 64

Arenaria
  leptoclados, 64
  marginata, 68
  norvegica (sub-sp.), 65
  peploides, 65
  rubra, 68
  rubella, 63
  serpyllifolia, 64
  sphœrocarpa, 64
  tenuifolia, 64
  trinervia, 64
  uliginosa, 63
  verna, 63
  viscosa, 64
Aristolochia, 352
  Clematitis, 352
Aristolochiaceæ, 351
Armeria, 258
  duriuscula, 258
  maritima, 258
  plantaginea. 258
  pubescens, 258
  pubigera, 258
  vulgaris, 258
Armoracia, 25, 37
  amphibia, 25
Arnoseris, 229
  pusilla, 229
Aroideæ, 423
Arrhenatherum, 485
  avenaceum, 485
  elatius, 485
Arrow-head, 428
Artemisia, 215
  Absinthium, 216
  cærulescens, 532
  campestris, 215
  gallica, 216
  maritima, 216
  vulgaris, 215
Arthrocnemum fruticosum. 342
Arthrolobium ebracteatum, 105
Arum, 423
  italicum, 423
  maculatum. 423
Arundo Calamagrostis, 479
  Phragmites, 487
Asarabacca, 352
Asarum, 352
  europæum, 352
Ash, 267
  mountain, 136
Asparagus, 403
  officinalis, 403
Aspen, 369
Asperula, 195
  arvensis, 531
  cynanchica, 195
  odorata, 195
  taurina, 531

Carduus
  crispus, 225
  nutans, 224
  polyanthemos, 225
  pycnocephalus, 225
  tenuiflorus, 225
  Woodwardsii, 227
Carex, 448
  acuta, 454
  acuta, 465
  alpicola, 452
  alpina, 453
  ampullacea, 464
  aquatilis, 455
  arenaria, 450
  argyroglochin, 452
  atrata, 453
  axillaris, 452
  Bœnninghauseniana, 452
  binervis, 462
  brizoides, 535
  Buxbaumii. 453
  cæspitosa, 454, 455
  canescens, 452
  capillaris, 457
  ciliata, 459
  clandestina, 460
  collina, 459
  curta, 452
  Davalliana, 448
  depauperata, 463
  dichroa, 464
  digitata, 459
  dioica, 449
  distans, 461
  disticha, 449
  divisa, 449
  divulsa (sub-sp.), 451
  echinata, 451
  Ehrhartiana, 451
  elongata, 453
  ericetorum. 459
  extensa, 461
  filiformis, 460
  flava, 461
  frigida, 463
  fulva (sub-sp.), 462
  Gibsoni, 455
  glauca, 456
  Goodenovii, 455
  gracilis, 454
  Grahami (sub-sp.), 464
  Grypos, 452
  hirta, 460
  hordeiformis, 535
  Hornschuchiana, 462
  humilis, 460
  incurva, 449
  intermedia, 449
  involuta, 464
  irrigua (sub-sp.), 456
  lævigata, 462

Carex
  lagopina, 452
  lepidocarpa, 461
  leporina, 452
  ligerica, 450
  limosa, 456
  Micheliana, 456
  Mielichoferi, 457
  Mœnchiana, 535
  montana, 459
  muricata, 451
  Œderi (sub-sp.), 461
  ornithopoda, 459
  ovalis, 452
  pallescens, 457
  paludosa, 465
  panicea, 457
  paniculata, 450
  paradoxa (sub-sp.), 450
  pauciflora, 448
  pendula, 458
  Persoonii, 452
  phœostachya, 457
  pilulifera, 458
  polygama, 453
  præcox, 458
  Pseudo-cyperus, 465
  Pseudo-paradoxa, 451
  pulicaris, 448
  pulla, 464
  punctata, 462
  rariflora, 456
  recurva, 456
  remota, 452
  rigida, 454
  riparia, 465
  rupestris, 449
  salina, 457
  saxatilis (sub-sp.), 464
  saxumbra, 459
  scotica, 457
  spadicea, 465
  sparsiflora, 457
  speirostachya, 462
  spicata, 451
  stellulata, 457
  stictocarpa, 457
  stricta, 454
  strigosa, 463
  sylvatica, 463
  tenella. 452
  teretiuscula, 450
  tomentosa, 458
  tricostata, 454
  trinervis, 455
  ustulata, 535
  vaginata, 457
  Vahlii, 453
  vesicaria, 464
  vitilis, 452
  vulgaris, 455
  vulpina, 451

Carex
  vulpinoïdes, 535
  xanthocarpa, 462
Carlina, 221
  racemosa, 532
  vulgaris, 221
Carline-thistle, 221
Carnation, 54
Carpinus, 368
  Betulus, 368
Carrot, 184
Carum, 170
  Bulbocastanum, 171
  Carui, 171
  flexuosum, 174
  Petroselinum, 171
  segetum, 171
  verticillatum, 170
Caryolopha, 278
Caryophylleæ, 52
Castanea vulgaris, 534
Catabrosa, 489
  aquatica, 489
Catapodium. 498
Catch-fly, 54
  Nottingham, 56
Cat-mint, 322
Cat's-ear, 238
Cat's-foot, 209
Cat's-tail, 422
  grass, 475
Cauliflower, 32
Caucalis, 185
  Anthriscus, 185
  arvensis, 186
  daucoides, 185
  helvetica, 186
  infesta, 186
  latifolia, 185
  nodosa, 186
Celandine, 18
  lesser, 10
Celastrineæ, 87
Celery, 169
Centaurea, 222
  aspera, 223
  Calcitrapa, 223
  Cyanus, 222
  decipiens, 222
  intybacea, 532
  Isnardi, 223
  Jacea, 532
  leucophæa, 532
  montana, 532
  nigra, 222
  nigrescens, 222
  paniculata, 223
  salmantica, 532
  Scabiosa, 222
  solstitialis, 223
Centaury, 270
  yellow, 269

N N 2

THE END.

LONDON :
R. CLAY, SONS, AND TAYLOR,
BREAD STREET HILL.

THE

𝕾𝖙𝖚𝖉𝖊𝖓𝖙'𝖘 𝕮𝖆𝖙𝖆𝖑𝖔𝖌𝖚𝖊

OF

BRITISH PLANTS.

ARRANGED ACCORDING TO

THE STUDENT'S FLORA OF THE BRITISH ISLES,

BY

SIR J. D. HOOKER, K.C.S.I., C.B., &c.

COMPILED BY

REV. GEORGE HENSLOW, M.A., F.L.S., &c.

LONDON.

BATEMAN, High Street, Portland Town.

—

1879.

[Price 1s. 6d.   Post free.]

THE
STUDENT'S CATALOGUE OF BRITISH PLANTS.

Class I. DICOTYLEDONES.
Sub-class I. ANGIOSPERMÆ.
Division I. POLYPETALÆ.

Order I. **RANUNCULACEÆ.**

Tribe I. **CLEMATIDEÆ.**

1. **CLEMATIS,** *L.* Traveller's Joy.
    1. **Vitalba,** *L. Old Man's Beard.*

Tribe II. **ANEMONEÆ.**

2. **THALICTRUM,** *L.* Meadow-rue.
    1. alpinum, *L.*
    2. minus, *L.*

Sub-sp. minus proper.—Var. 1, *mariti-mum.*—Var. 2, *montanum,* Wallroth (sp.); *calcareum,* Jord.
Sub-sp. majus, *Jacq ; minus,* L. Herb.; *flexuosum,* Bernh.
Sub-sp. Kochii, *Fries* (sp.).
Sub-sp. saxatile, *Schleich* (sp.).

    3. flavum, *L.*—Var. 1, *sphærocarpum,* Bosw.—Var. 2, *riparium,* Jord. (sp.).—Var. 3, *Morisonii,* Gmel, (sp.).

3. **ANEMONE,** *L.*
    1. Pulsatilla, *L. Pasque-flower.*
    2. nemorosa, *L. Wood Anemone.*
    3. apennina, *L.*

3*. *ADONIS, L.* Pheasant's eye.
    1. autumnalis, *L.*

4. **MYOSURUS,** *L.* Mouse-tail.
    1. minimus, *L.*

Tribe III. **RANUNCULEÆ.**

5. **RANUNCULUS,** *L.* Buttercup, Crowfoot.

Section 1. Batrachium.
    1. aquatilis, *L.*

Sub-sp. heterophyllus, *Auct.* (sp.).—Var. I, *peltatus,* Fries (sp.).—Var. 2, *fissifolius,* Schrank.—Var. 3, *floribundus,* Bab. (sp.).—Var. 4, *confusus,* Godron (sp.).--Var. 5, *Baudotii,* Godron (sp.); *aquat :* var. *Symei,* Hook. and Arn.—Var. 6, *heterophyllus,* proper, Fries (sp.).—Var. 7, *pseudo-fluitans,* Bosw. (sp.); *penicillatus,* var. *rivulare,* Schur.
Sub-sp. pantothrix, *Broteri* (sp. in part). Var. 1, *Drouetii,* Schultz (sp.)—Var. 2, *trichophyllus,* Chaix (sp.).—Var. *Godronii,* Gren. (sp.)
Sub-sp. circinatus, *Sibthorp,* (sp.); *divaricatus,* Schrank.
Sub-sp. fluitans, *Lamk.* (sp.) ; *Bachii,* Wirtg.
Sub-sp. tripartitus, *DC.* (sp.).

    2. Lenormandi, *Schultz ; cænosus,* Gren and Godr.
    3. hederaceus, *L. ; cænosus,* Gussone. *Ivy-leaved Ranunculus.*

Section 2. Hecatonia.

    4. Lingua, *L. Spear-wort.*
    5. Flammula, *Lesser Spear-wort.*

Sub-sp. flammula proper.
Sub-sp. reptans, *L.* (sp.).

    6. ophioglossifolius, *Villars.*
    7. auricomus, *L. Goldielocks.* (Var. *depauperatus.*)
    8. sceleratus, *L.*
    9. acris, *L.*—Var. 1, *vulgatus,* Jord. (sp.)—Var. 2, *Boræanus,* Jord. (sp.).—Var. 3, *tomophyllus,* Jord. (sp.).
    10. repens, *L.*
    11. bulbosus, *L.*
    12. chærophyllus, *L*

Section 3. Echinella.

    13. hirsutus, *Curtis ; Philonotis,* Ehr.

14. arvensis, *L.*

15. parviflorus, *L.*

SECTION 4. Ficaria, *Dillen.* (gen.)

16. Ficaria, *L. ; Lesser Celandine .—*
VAR. 1, *divergens*, F.Schultz —VAR. 2,
incumbens, F. Schultz,

TRIBE IV. **HELTEVOREÆ.**

6. **CALTHA**, *L.* MARSH MARIGOLD.

1. palustris, *L.*

Sub-sp. PALUSTRIS proper; VAR. 1, *vul-
garis*, Schott (sp.),—VAR. 2, *Guerangerii*,
Boreau (sp.)—VAR. 3, *minor*, Bosw.
Sub-sp. RADICANS, *Forster* (sp.).

7. **TROLLIUS**, *L.* GLOBE-FLOWER.

1. europæus, *L.*

8. **HELLEBORUS**, *L.* HELLEBORE.

1. viridis, *L.*

2 foetidus, *L.*

8*. *ERANTHIS*, *Salisbury.* WINTER
ACONITE.
1. HYEMALIS, *Salisb.*

9. **AQUILEGIA**, *L.* COLUMBINE.

1. vulgaris, *L.*

9* *DELPHINIUM, L.* LARKSPUR.

1. AJACIS, *Reich.* D. *Consolida*, Brit.
Fl.

9** *ACONITUM, L.* MONKSHOOD,
WOLFSBANE.
1. NAPELLUS, *L.*

10. **ACTÆA**, *L.* BANE-BERRY.

1. spicata, *L.*

———

ORDER II. **BERBERIDEÆ.**

1. **BERBERIS**, *L.* BARBERRY.

1. vulgaris, *L.*

———

ORDER III. **NYMPHÆACEÆ.**

1. **NUPHAR**, *Smith.* YELLOW WATER-
LILY.

1. luteum, *L.*; VAR. 1, *majus*, Bosw.
—VAR. 2, *minus*, Bosw.

2. pumilum, *Smith.*

2. **NYMPHÆA**, *L.* WHITE WATER-
LILY.

1. alba, *L.*

ORDER IV. **PAPAVERACEÆ.**

1. **PAPAVER**, *L.* POPPY.

1. hybridum, *L.*

2. Argemone, *L.*

3. dubium, *L.*

Sub-sp. LAMOTTEI, Boreau (sp.).
Sub-sp. LECOQII, Lamotte (sp.).

4. Rhœas, *L.*

SOMNIFERUM, *L. Opium Poppy.*

2. **MECONOPSIS**, *Viguier.* WELSH
POPPY.

1. cambrica, *Vig.*

3. **CHELIDONIUM**, *L.* CELANDINE.

1. majus, *L.*

4. **GLAUCIUM**, *Juss.* HORNED POPPY.

1. luteum, *L.*

5. **RŒMERIA**, *DC.*

1. hybrida, *DC.; Gl. violaceum*, Juss.

ORDER V. **FUMARIACEÆ.**

1. **FUMARIA**, *L.* FUMITORY.

1. capreolata, *L.*

Sub-sp. PALLIDIFLORA, *Jord.* (sp.).
Sub-sp. BORÆI, *Jord.* (sp.).
Sub-sp. CONFUSA, *Jord.* (sp.) ; *agraria*,
Mitten.
Sub-sp. MURALIS, *Sonder.*

2. officinalis, *L.*

3. densiflora, *DC. ; calycina*, Bab. ;
micrantha, Lagasca.

4. parviflora, *Lamk.; tenuisecta*, Bosw.

Sub-sp. *parviflora* proper.
Sub-sp. VAILLANTII, *Loisel*, (sp.).

2. **CORYDALIS**, *DC.*

1. claviculata, *DC.*
LUTEA, *DC.*
SOLIDA, *Hook. ; bulbosa*, DC.

ORDER VI. **CRUCIFERÆ.**

TRIBE I. **ARABIDEÆ.**

1. **MATTHIOLA**, *Br.* STOCK.

1. incana, *Br.*

2. sinuata, *Br.*

1* *CHEIRANTHUS, L.* WALLFLOWER
CHEIRI, *L.*

2. **NASTURTIUM,** *Br*

1. officinale, *Br. ; Watercress.—*VAR.

1. *officinale* proper.—VAR. 2, *siifolium,* Reich, (sp.).

2. sylvestre, *Br.*

3. palustre, *DC. ; terrestre,* Sm.

4. amphibium, *Br.; Armoracia,* Koch.

3. **BARBAREA,** *Br.* WINTER-CRESS,

1. vulgaris, *Br.*

Sub-sp. VULGARIS proper.
Sub-sp. ARCUATA, *Reich.*
Sub-sp. STRICTA, *Andrz.* (sp.).
Sub-sp. INTERMEDIA, *Boreau* (sp.).
PRÆCOX, *Br. American Cress.*

4. **ARABIS,** *L.* ROCK-CRESS.

1. petræa, *Lamk. ; hispida, L. Cardamine hastulata,* Sm.

2. stricta, *Huds.*

3. ciliata, *Br.—*VAR. 1, *ciliata* proper.

VAR, 2, *hispida,* Bosw.

4. hirsuta, *Br.; sagittata, D.C. ; Turritis hirsuta, L.—*VAR. 1, *hirsuta* proper.—VAR. 2, *glabrata,* Bosw.

5. perfoliata, *Lamk ; Turritis glabra, L.*

TURRITA, *L.*

5. **CARDAMINE,** *L.* BITTER-CRESS.

1. hirsuta, *L.*
Sub-sp. HIRSUTA proper.
Sub-sp. FLEXUOSA, *Withering* (sp.) ; *sylvatica,* Link.

2. pratensis, *L. Lady's Smock,Cuckooflower.*

3. amara, *L.*

4. impatiens, *L.*

6. **DENTARIA,** *L.* CORAL-ROOT.

1. bulbifera, *L.*

TRIBE II. SISYMBRIEÆ.

7. **SISYMBRIUM,** *L.* HEDGE-MUSTARD.

1. Thaliana, *Hook. Thale-cress.*

2. Irio, *L. London Rocket.*

3. Sophia, *L. Flixweed.*

4. officinale, *L. Hedge-mustard.*

5. Alliaria, *Scopoli ; Garlic-mustard. Erysimum, L.; Alliaria,* Andrz.

POLYCERATIUM, *L.*

8. **ERYSIMUM,** *L.* TREACLE-MUSTARD.

1. cheiranthoides, *L.*

8*. *HESPERIS, L.* DAME'S VIOLET.

MATRONALIS, *L.*

TRIBE III. **BRASSICEÆ.**

9. **BRASSICA,** *L.* CABBAGE, &c.

SECTION 1. Brassica proper.

1. oleracea, *L. Wild Cabbage.*

2. campestris, *L. ; polymorpha,* Bosw.

Sub-sp. 1. CAMPESTRIS proper, (*Linn Herb.*)
—VAR. 1, *oleifera,* DC. (yields rape and colza).—VAR. 2, *Napo-brassica,* DC. *Rutabaga,* DC.—(Swedish Turnip).

Sub-sp. 2. RAPA, *L.* (sp.); VAR. 1, *rapifera,* Koch (turnip).—VAR. 2, *campestris,* Koch.
—VAR. 3, *sylvestris,* Lond. Cat. (navew).

3. monensis, *Huds.*
Sub-sp. MONENSIS proper.
Sub-sp. CHEIRANTHUS, *Villars* (sp.).

SECTION 2. Sinapis, *L* (Gen.).

4. nigra, Koch. *Black Mustard.*

5. adpressa, *Boiss ; Sinapis incana,* L.

6. Sinapistrum, *Boiss-; Sinapis arvensis, L. Charlock.*

7. alba, *Boiss ; Sinapis,* L. *White mustard.*

10. **DIPLOTAXIS,** *DC.* ROCKET.

1. muralis, *DC. ; Sisymbrium,* L. *Brassica brevipes,* Bosw.

Sub-sp. MURALIS proper.—VAR. 1, *muralis* proper.—VAR. 2, *Babingtonii,* Bosw.

2. tenuifolia, *DC. Sisymbrium,* L. ; *Sinapis,* Sm. ; *Brassica,* Boiss.

TRIBE IV. **ALYSSINEÆ.**

11. **DRABA,** *L.* WHITLOW-GRASS.

1. aizoides, *L.*

2. rupestris, *Br.*

3. incana, *L. ; confusa,* Ehrh.

4. muralis, *L.*

12. **EROPHILA.** *DC.* VERNAL WHITLOW-GRASS.

1. verna, *L.*

Sub-sp. VERNA proper.
Sub-sp. BRACHYCARPA, *Jord.* (sp.).
Sub-sp. INFLATA, *Watson* ; *Draba verna, b.* Hook.

12*. *ALYSSUM, L.*
CALYCINUM, *L.*
MARITIMUM, *L.*; *Königa*, Br.; *Lobularia*, Desv.; *Glyce*, Lindl.

13.COCHLEARIA, *L.* SCURVY-GRASS.
1. officinalis, *L.; polymorpha*, Bosw.
Sub-sp. OFFICINALIS proper; VAR. *littoralis*, Lond. Cat.
Sub-sp. ALPINA, *Watson; grœnlandica*, Sm.
Sub-sp. DANICA, *L.* (sp.).
2. anglica, *L.;*—VAR. 1. *gemina*, Hort; —VAR. 2, *Hortii*. Bosw.
ARMORACIA, *L. Horse-radish.*

TRIBE V. CAMELINEÆ.

13*. *CAMELINA, Crantz.* GOLD OF PLEASURE.
SATIVA, *L.*

14. SUBULARIA, *L.* AWL-WORT.
1. aquatica, *L.*

TRIBE VI. LEPIDINEÆ.

15. CAPSELLA, *Mœnch.* SHEPHERD'S PURSE.
1. Bursa-Pastoris, *DC.*

16. SENEBIERA, *DC.* WART-CRESS.
1. Coronopus, *Poiret; C. Ruellii*, Gaert.
2. DIDYMA, *Persoon; Coronopus*, Sm.

17. LEPIDIUM, *L.* CRESS.
SECTION 1. Nasturtiastrum, *Gren.* and *Godr.*
1. latifolium, *L. Dittander.*
2. ruderale, *L.*
SECTION 2. Lepia, *DC.*
3. campestre, *Br. Pepperwort.*
4. Smithii, *Hook; hirtum*, Sm. in part.
SECTION 3. Cardaria, *DC.*
DRABA, *L.*

TRIBE VII. THLASPIDEÆ.

8. THLASPI, *L.* PENNY CRESS.
1. arvense, *L.; Mithridate Mustard.*
2. perfoliatum, *L.*

3. alpestre, *L.*
Sub-sp. SYLVESTRE, *Jord.* (sp.).
Sub-sp. OCCITANUM, *Jord.* (sp.).
Sub-sp. VIRENS, *Jord.* (sp.).

19. IBERIS, *L.* CANDY-TUFT.
1. amara, *L.*

20. TEESDALIA, *Br.*
1. nudicaulis, *Br.*

21. HUTCHINSIA, *Br.*
1. petræa, *Br.*

TRIBE VIII. ISATIDEÆ.

21*. *ISATIS, L.* WOAD.
TINCTORIA, *L.*

TRIBE IX. CAKILINEÆ.

22. CRAMBE, *L.* SEA-KALE.
1. maritima, *L.*

23. CAKILE, *Gœrtn.* SEA ROCKET.
1. maritima, *L.*
TRIBE X. RAPHANEÆ.

24. RAPHANUS, *L.* RADISH.
1. Raphanistrum, *L. Wild Radish.*
2. maritimus, *L.*

ORDER VII. RESEDACEÆ.

1. RESEDA, *L.* MIGNONETTE.
1. Luteola, *L. Dyer's Weed, Weld.*
2. lutea, *L.*
ALBA, *L.; suffruticulosa, L.,* and *fruticulosa, L.*

ORDER VIII. CISTINEÆ.

1. HELIANTHEMUM, *Tourn.* ROCK-ROSE.
SECTION I. Helianthemum proper.
1. vulgare, *Gœrtn. Surrejanum*, Eng. Bot.; *Cistus tomentosus*, Sm.
2. polifolium, *Pers.*
SECTION 2. Tuberaria.
3. guttatum, *Miller.*
Sub-sp. GUTTATUM proper.
Sub-sp. BREWERI, *Planchon* (sp.).
SECTION 3. Pseudo-cistus.
4. canum, *Dunal; Cistus marifolius,* and *C. anglicus*, *L.* – VAR. 1, *canum* proper.—VAR. 2, *vineale*, Pers. (sp.)

ORDER IX. VIOLACEÆ.

1. **VIOLA,** *L.* VIOLET, PANSY, HEARTSEASE.

SECTION 1. Nominium.

1. palustris, *L.*
2. odorata, *L. Sweet Violet.*—VAR 1, *permixta*, Jord. (sp.).—VAR. 2, *sepincola*, Jord. (sp.).
3. hirta, *L.* ;—VAR. 1, *calcarea*, Bab.
4. canina, *L. Dog-violet.*

Sub-sp. CANINA proper ; *flavicornis*, Sm. ; *pumila*, Hook and Arn.
Sub-sp. LACTEA, *Sm.* (sp.'.
Sub-sp. STAGNINA, *Kitaib.* (sp.).
Sub-sp. SYLVATICA, *Fries* (sp.).—VAR. 1, *Riviniana*, Reich. (sp.). —VAR. 2, *Reichenbachiana*, Boreau (sp.).
Sub-sp. ARENARIA, *DC.* (sp.).

SECTION 2. Melanium.

5. tricolor, *L. Heartsease or Pansy.*

Sub-sp. TRICOLOR proper.
Sub-sp. ARVENSIS, *Murray* (sp).
Sub-sp. CURTISII, *Forster* (sp.). ; *sabulosa* Boreau.—VARS. *Mackaii*, *Symei* and *Forsteri*.
Sub-sp. LUTEA, *Huds*, (sp.). ;—VAR. *amœna*.

ORDER X. POLYGALEÆ.

1. **POLYGALA,** *L.* MILKWORT.

1. vulgaris, *L.*

Sub-sp. · VULGARIS proper.—VAR. 1 ;— VAR. 2, *grandiflora*, Bosw.
Sub-sp. OXYPTERA, *Reich.* (sp).:—VAR. *ciliata.*
Sub-sp. DEPRESSA. *Wenderoth* (sp.).

2. calcarea, *F. Schultz ; amara*, Doo.
3 amara, *L. ; uliginosa*, Fries. ; *austriaca*, Crantz.

ORDER XI. FRANKENIACEÆ.

1. **FRANKENIA,** *L.* SEA-HEATH.

1. lævis, *L.*

ORDER XII. CARYOPHYLLEÆ.

TRIBE I. SILENEÆ.

1. **DIANTHUS.** PINK and CARNATION.

1. Armeria, *L. Deptford Pink.*
2. prolifer, *L.*

3. deltoides, *L. Maiden Pink ;—* VAR. 1, *deltoides* proper.—VAR. 2, *glaucus*, L. (sp.).
4. cæsius, *Sm.* . *Cheddar Pink.*

CARYOPHYLLUS, *L. Wild Carnation,* *Clove Pink.*
PLUMARIUS, *L. Wild Pink.*

1*. *SAPONARIA, L.* SOAPWORT, FULLER'S HERB.

1. OFFICINALIS, *L.* VAR. *hybrida*, *L.*

2. **SILENE,** *L.* CATCHFLY.

1. inflata, *Sm. ; Cucubalus Behen,* L. VAR. *puberula.*
2. maritima, *L.*
3. conica, *L.*
4. anglica, *L. ; —* VAR. 1, *gallica,* Koch (sp.).—VAR. 2, *quinquevulnera,* L. (sp.).
5. acaulis, *L. Moss Campion.*
6. Otites, *L.*
7. nutans, *L. Nottingham Catchfly.*
8. noctiflora, *L.*

3. **LYCHNIS,** *L.* CAMPION.

1. Flos-cuculi, *L. Ragged Robin.*
2. Viscaria, *L.*
3. alpina, *L.*
4. diurna, *Sibth., dioica a, L. Red Campion.*
5. vespertina, *Sibth ; White Campion ; dioica b, L.*

4. **GITHAGO,** *Desfontaines.* CORNCOCKLE.

1. segetum, *Desf; Agrostemma Githago,* L.

TRIBE II. ALSINEÆ.

5. **HOLOSTEUM,** *L.*

1. umbellatum, *L.*

6. **CERASTIUM,** *L.* MOUSE-EAR CHICKWEED.

SECTION 1. Mœnchia, *Ehr.* (gen.).

1. quaternellum, *Fenzl ; Mœnchia erecta,* Ehr.

SECTION 2. Cerastium proper.

2. tetrandrum, *Curtis ; atrovirens* and *pedunculatum*, Bab.—VAR. *pumilum*, Curtis (sp.) ; *glutinosum*, Fries.
3. semidecandrum, *L.*
4. glomeratum, *Thuillier ; vulgatum,* L. and *viscosum*, L. in part.

c

5. triviale, *Link ; viscosum* L. of
Hook. and Arn.—VAR. 1, *triviale* proper.—VAR. 2, *holosteoides,* Fries (sp.).
—VAR. 3, *pentandrum.*—VAR. 4, *alpestre* Lond. Cat. (*alpinum,* Koch).

6. arvense, *L.* ; VAR. 1, *pubescens,*
Syme.—VAR. 2, *Andrewsii,* Syme.

7. alpinum, *L.*—VAR. 1, *lanatum,*
Lamk. (sp.).—VAR. 2, *pubescens.*

8. latifolium, *Sm.* VAR. 1, *Smithii ;
latifolium,* Sm. VAR. 2, *compactum,*
Syme.—VAR. 3, *Edmondstonei,* Watson (var. *nigrescens,* Syme).

9. trigynum, *Villars ; Stellaria ceras-
toides,* L.

7. STELLARIA, *L. Stitchwort.*
SECTION 1. **Malachium,** *Fries* (gen.).

1. aquatica, *Scopoli* ; *Cerastium aquaticum,* L.

SECTION 2. **Stellaria** proper.

2. nemorum, *L.*

3. media, *L. Chickweed.* VAR. 1,
media proper.—VAR. 2, *Borœana,*
Jord. (sp.).—VAR. 3, *neglecta,* Weihe
(sp.).—VAR. 4, *umbrosa,* Opitz (sp.).

4. Holostea, *L.*

SECTION 3. **Larbrea,** St. Hilaire (gen.).

5. glauca, *Withering.*

6. graminea, *L. ; scapigera,* Willd.

7. uliginosa, *L.*

8. ARENARIA, *L.* SANDWORT.
SECTION 1. **Alsine,** *Wahl.* (gen.).

1. verna, *L.* VAR. 1, *verna* proper.
—VAR. 2, *Gerardi,* Willd. (sp.?).

2. rubella, *Hook.*

3. uliginosa, *Schleich. ; Spergula
stricta,* Swartz ; *Alsine stricta,* Wahl.

4. tenuifolia, *L.* VAR. 1, *tenuifolia,*
proper.—VAR. 2, *laxa,* Jord. (sp.).—
VAR. 3, *hybrida ; b viscosa,* Bab.

SECTION 2, **Arenaria** proper.

5. trinervis, *L. Mœhringia,* Clairv.

6. serpyllifolia, *L.* VAR. 1, *sphœrocarpa,* Tenore (sp.).—VAR. 2, *Lloydii,*
Jord.—VAR. 3, *leptoclados,* Gussone
(sp.).

7. ciliata, *L.*

8. norvegica, *Gunner.*

SECTION 3. **Ammodenia,** *Gmel.* (gen.) ;
Honckenya, Ehr.

9. peploides, *L. Sea Purslane.*

SECTION 4. **Cherleria,** *L.* (gen.).

10. Cherleria, Fenzl (sub *Alsine*).

9. SAGINA, *L.* PEARL-WORT.

1. apetala, *L.*

Sub-sp. APETALA proper.
Sub-sp. CILIATA, Fries (sp.).
Sub-sp. MARITIMA, Don (sp.).—VAR. 1,
maritima proper.—VAR. 2, *debilis,* Jord.
(sp.).—VAR. 3, *densa,* Jord. (sp.).—VAR.
4. *alpina,* Syme.

2. procumbens, *L.*

3. saxatilis, *Wimmer ; Linnæi,* Presl ;
Spergula saginoides, Sm.

4. nivalis, *Fries.*

5. subulata, *Wimmer ; Spergula,*
Swartz.

6. nodosa, *L. ; Spergula, L. Knotted
Spurrey.*

10. SPERGULA, L. SPURREY.

1. arvensis, *L.*

11. SPERGULARIA, *Persoon.* SAND-
WORT-SPURREY.

1. rubra, *St. Hilaire.*

2. marina, *Cambessedes.*

Sub-sp. MARINA proper ; *Alsine marina,*
Wahlb ; *Arenaria media,* L. ; *A. margi-
nata,* DC.
Sub-sp. NEGLECTA, *Bosw.*—VAR. 1, *neg-
lecta* proper.—VAR. 2, *salina,* Presl (sp.).
—VAR. 3, *media,* Fries (sp.).
Sub-sp. RUPESTRIS, *Lond. Cat.*

TRIBE III. **POLYCARPEÆ.**

12. POLYCARPON, *L.*

1. tetraphyllum, *L.*

ORDER XIII. **PORTULACEÆ.**

1. MONTIA, *L.* BLINKS.

1. fontana, *L.* ; VAR. 1, *minor* Gmelin (sp.).—VAR. 2, *rivularis* Gmel.
(sp.).

1* CLAYTONIA, *L.*

1. PERFOLIATA, *Don.*

2. ALSINOIDES, *Sims.*

ORDER XIV. **PARONYCHIEÆ**.

1. **CORRIGIOLA**, *L.* STRAPWORT.
   1. littoralis, *L.*
2. **HERNIARIA**, *L.* RAPTUREWORT.
   1. glabra, *L.* VAR. *ciliata*, Bab. (sp.).
3. **ILLECEBRUM**, *L.*
   1. verticillatum, *L.*
4. **SCLERANTHUS**, *L.* KNAWEL.
   1. annuus, *L.* VAR. 1, *annuus*, proper. —VAR. 2, *biennis*, Reuter (sp.).
   2. perennis, *L.*

———

ORDER XIV*. *TAMARISCINEÆ.*

TAMARIX, L. TAMARISK. GALLICA, *L.*

———

ORDER XV. **ELATINEÆ**.

1. **ELATINE**, *L.* WATERWORT.
   1. hexandra, DC.; *tripetala*, Sm.
   2. Hydropiper, *L.*

———

ORDER XVI. **HYPERICINEÆ**.
1. **HYPERICUM**, *L.* ST JOHN'S WORT.

SECTION 1.
   1. Androsæmum, *L. Tutsan.*
      CALYCINUM, *L.*

SECTION 2.
   2. perforatum, *L.*
   3. quadrangulum, *L.*
Sub-sp. DUBIUM, *Leers* (sp.).
Sub-sp. TETRAPTERUM, *Fries* (sp.).
Sub-sp. UNDULATUM, *Schousb.* (sp.).; *bæticum*, Boiss.
   4. humifusum, *L.*
   5. linarifolium, *Vahl*, DC., &c.
   6. pulchrum, *L.*
   7. hirsutum, *L.*
   8 montanum, *L.*

SECTION 3.
   9. elodes, *L.*

ORDER XVII. **MALVACEÆ**.

1. **ALTHÆA**, *L.* MARSH-MALLOW.
   1. officinalis, *L. Guimauve.*
      HIRSUTA, *L.*
2. **MALVA**, *L.* MALLOW.
   1. sylvestris, *L. Common Mallow.*
   2. rotundifolia, *L.*
   3. moschata, *L. Musk Mallow.*
3. **LAVATERA**, *L.* TREE-MALLOW.
   1. arborea, *L.*

———

ORDER XVIII. **TILIACEÆ**.

1. **TILIA**, *L.* LIME-TREE OF LINDEN.
   1. parviflora, *L.*
      INTERMEDIA, DC. *Common Lime.*
      GRANDIFOLIA, *Ehrhart* ; VAR. *coralina. Large-leaved Lime.*

———

ORDER XIX. **LINEÆ**.

1. **LINUM**, *L.* FLAX.
   1. catharticum, *L. Purging Flax.*
   2. perenne, *L.*
   3. augustifolium, *L.*
      USITATISSIMUM, *L. Common Flax.*

2. **RADIOLA**, Gmelin. ALL-SEED.
   1. Millegrana, *Sm.*

———

ORDER XX. **GERANIACEÆ** (including **OXALIDEÆ** and **BALSAMINEÆ**.)

TRIBE 1. **GERANIEÆ**.

1. **GERANIUM**, *L.* CRANE'S-BILL.
   1. sanguineum, *L.*
   2. . sylvaticum, *L.*
   3. pratense, *L.*
   4. pyrenaicum, *L.*
      PHÆUM, *L.*
   5. molle, *L.*
   6. rotundifolium, *L.*
   7. pusillum, *L.*
   8. columbinum, *L.*
   9. dissectum. *L.*

10. **Robertianum,** *L.* *Herb-Robert.*—VAR. 1, *Robertianum* proper. —VAR. 2, *Raii,* Lindl (sp.); *modestum,* Jord.—VAR. 3, *purpureum* Jord. (sp.).

11. lucidum, *L.*

2. **ERODIUM,** *L'Héritier.* STORK'S-BILL.

1. cicutarium, *L.*—VAR. 1, *vulgatum,* Bosw.—VAR. 2, *chærophyllum,* DC. (sp.).

2. maritimum, *L.*

3. moschatum, *L.*

TRIBE II. **OXALIDEÆ.**

3. **OXALIS,** *L.* WOOD-SORREL.

1. Acetosella, *L.* *Wood-sorrel.* CORNICULATA, *L.* STRICTA. *L.*

TRIBE III. **BALSAMINEÆ.**

4. **IMPATIENS,** *L.* BALSAM.

1. Noli-me-tangere, *L.* *Yellow Balsam.* FULVA, *Nuttall.* PARVIFLORA, DC.

———

ORDER XXI. **ILICINEÆ** or **AQUI-FOLIACEÆ.**

1. **ILEX,** *L.* HOLLY.

1. Aquifolium, *L.*

———

ORDER XXII. **EMPETRACEÆ.**

1. **EMPETRUM,** *L.* CROWBERRY.

1. nigrum, *L.*

ORDER XXIII. **CELASTRINEÆ.**

1. **EUONYMUS,** *L.* SPINDLE-TREE.

1. europæus, *L.*

———

ORDER XXIV. **RHAMNEÆ.**

1. **RHAMNUS,** *L.* BUCKTHORN.

1. catharticus, *L.*

2. Frangula, *L.* *Berry-bearing Alder.*

———

ORDER XXV. **SAPINDACEÆ**; *Tribe* ACERINEÆ.

1. **ACER,** *L.* MAPLE.

1. campestre, *L.* *Common* or *Small-leaved Maple.* PSEUDOPLATANUS, *L.* *Great Maple. Sycamore, Plane* of Scotland.

———

ORDER XXVI. **LEGUMINOSÆ.** Sub-order PAPILIONACEÆ.

TRIBE I. **GENISTEÆ.**

1. **GENISTA,** *L.*

1. tinctoria, *L.* *Dyer's Greenweed.*

2. pilosa, *L.*

3. anglica, *L.* *Needle Furze.*

2. **ULEX,** *L.* FURZE, WHIN, GORSE.

1. europæus, *L.* VAR. 1, *europæus* proper.—VAR. 2, *strictus,* Mackay (sp.). *Irish Furze.*

2. nanus, *Forster.*

Sub-sp. NANUS proper.
Sub-sp. GALLII, *Planchon* (sp.).

3. **CYTISUS,** *L.* BROOM.

1. scoparius, *Link ; Spartium, L. ; Sarothamnus,* Koch.

———

TRIBE II. **TRIFOLIEÆ.**

4. **ONONIS,** *L.* REST-HARROW.

1. arvensis, *L.; Wild Liquorice.*

2. spinosa, *L.; campestris,* Koch. RECLINATA, *L.*

5. **TRIGONELLA,** *L.* FENUGREEK.

1. ornithopodioides, *DC. ; Trifolium,* L. *Aporanthus Trifoliastrum,* Bromfield.

6. **MEDICAGO,** *L.* MEDICK.

1. falcata, *L.*

Sub-sp. FALCATA proper.
Sub-sp. SYLVESTRIS, *Fries* (sp.).
SATIVA, *L. Purple Medick,* Lucern.

2. lupulina, *L. Black Medick.*

3. denticulata, *Willd.* VAR. 1, *denticulata* proper.—VAR. 2, *apiculata* Willd. (sp.).—VAR. 3, *lappacea,*Lamk. (sp.).

4. maculata, *Sibthorp.*

5. minima, *L.*

7. **MELILOTUS,** *Tournefort.* MELILOT.

1. officinalis, *L.*
2. alba, *Lamk ; vulgaris* Willd ; *leucantha,* Koch. *White Melilot.*

ARVENSIS, *Wallroth.*

8. **TRIFOLIUM**, *L.* TREFOIL, CLOVER.

　1. subterraneum, *L.*
　2. arvense, *L. Hare's-foot Trefoil.*
　3. Bocconi, *Savi.*
　4. incarnatum, *L. Crimson Clover.*
Sub-sp. *incarnatum* proper.
Sub-sp. MOLINERII, *Balb* is (sp.).
　5. ochroleucum, *L.*
　6. pratense, *L. Red Clover.*
　7. medium, *L. Meadow Clover.*
　8. maritimum, *Hudson.*
　9. striatum, *L.*
　10. scabrum, *L.*
　11. glomeratum, *L.*
　12. suffocatum, *L.*
　13. strictum, *L.*

　　HYBRIDUM, *L. Alsike Clover.* VAR. 1, *hybridum* proper.—VAR. 2, *elegans,* Savi (sp.).
　14. repens, *L. White* or *Dutch Clover.*
　15. fragiferum, *L.*
　16 procumbens, *L. agrarium,* Huds. *Hop Trefoil.*
　17. minus, *Sm. ; procumbens,* Huds ; *filiforme* of foreign authors.
　18. filiforme, *L.*

TRIBE III. **LOTEÆ.**

9. **ANTHYLLIS**, *L.* KIDNEY-VETCH.
　1. Vulneraria, *L.* VAR. 1, *Vulneraria* proper.—VAR. 2, *Dillenii,* Schultz. (sp.).

10. **LOTUS**, *L.* BIRD'S-FOOT TREFOIL.
1 corniculatus, *L.*
Sub-sp. CORNICULATUS, proper. VAR. 1, *vulgaris.*—VAR. 2, *crassifolius.*—VAR. 3, *villosus.*
Sub-sp. TENUIS, *Kitaibel* (sp.). ; *tenuifolius,* Reich. ; *decumbens,* Forst.
　2. major, *Scop.*
　3. angustissimus, *L.*—VAR. 1, *diffusus,* Smith (sp.).—VAR. 2, *hispidus,* Desf. (sp.).

TRIBE VI. **GALEGEÆ.**

11. **ASTRAGALUS**, *L.* MILK-VETCH
　1. glycyphyllos, *L.*
　2. hypoglottis, *L.*
　3. alpinus, *L ; Phaca astragalina,* DC.

12. **OXYTROPIS**, *DC.*
　1. uralensis, *DC.* ; *Halleri,* Bunge.
　2. campestris, *DC.*

TRIBE V. **HEDYSAREÆ.**

13. **ORNITHOPUS**, *L.* BIRD'S-FOOT.
　1. perpusillus, *L.*
　2. ebracteatus, *Brot. ; Arthrolobium ebracteatum,* DC.

14. **HIPPOCREPIS**, *L.* HORSESHOE-VETCH.
　1. comosa, *L.*

15. **ONOBRYCHIS**, *Tournefort.* SAINFOIN.
　1. sativa, *Lamk.*

TRIBE VI. **VICIEÆ.**

16. **VICIA**, *L.* VETCH, TARE.
SECTION 1. Ervum, *L.* (gen.).
　1. tetrasperma *Mœnch.*
Sub-sp. TETRASPERMA proper.
Sub-sp. GRACILIS, *Loisel.* (sp.).
　2. hirsuta, *Koch. Common Tare.*

SECTION 2. Cracca.
　3. Cracca, *L.*
　4. Orobus, *DC. ; Orobus Sylvaticus,* **L.** *Bitter Vetch.*
　5. sylvatica, *L. Wood-vetch.*

SECTION 3. Vicia proper.
　6. sepium, *L.*
　7. lutea, *L.*
Sub-sp. LÆVIGATA, *Sm.* (sp.).
　8. sativa, *L. Vetch.*
Sub-sp. SATIVA proper.
Sub-sp. ANGUSTIFOLIA, *Rcth* (sp.). VAR. 1, *segetalis,* Koch (p.).—VAR. 2, *Bobartii,* Forst. (sp.) ; *angustifolia,* Sm.
　9. lathyroides, *L.*
　10. bithynica, *L.*—VAR. 1, *latifolia,* Bosw.—VAR. 2, *angustifolia,* Bosw.

d

17. LATHYRUS, *L.* Everlasting Pea.

Section 1. **Aphaca.**

1. **Aphaca,** *L.*

Section 2. **Nissolia.**

2. **Nissolia,** *L.*

Section 3. **Lathyrus, proper.**

3. **hirsutus,** *L.*

4. **pratensis,** *L.*

    tuberosus, *L.*

5. **sylvestris,** *L.*

6. **palustris,** *L.*

7. **maritimus,** *Bigelow ; Pisum maritimum*, L.—Var. 1, *maritimus* proper. —Var. 2, *acutifolius*, Bab.

Section 4. **Orobus,** *L.*, (gen.).

8. **macrorrhizus,** *Wimmer ; Orobus,* L. Var. 1, *macrorrhizus* proper ; *Orobus tuberosus*, L.—Var. 2, *tenuifolius*, Roth. (sp.).

9. **niger,** *Wimmer ; Orobus,* L.

Order XXVII. **ROSACEÆ.**

Tribe I. **PRUNEÆ.**

1. PRUNUS, *L.* Plum and Cherry.

Section 1. **Prunus, proper.**

1. **communis,** *Huds.*

Sub-sp. spinosa, *L.* (sp.). *Sloe, Blackthorn.*

Sub-sp. insititia, *L.* (sp.). *Bullace.*

Sub-sp. domestica, *L.* (sp.). *Wild Plum.*

Section 2. **Cerasus.**

2. **Cerasus,** *L. Wild Cherry ; Dwarf Cherry.*

3. **Avium,** *L. Gean.*

Section 3. **Laurocerasus.**

3. **Padus,** *L. Bird Cherry.*

Tribe II. **SPIRÆEÆ.**

2. SPIRÆA, *L.*

1. **Ulmaria,** *L. Meadow-sweet.*

2. **Filipendula,** *L. Dropwort.*

    salicifolia, *L.*

Tribe III. **RUBEÆ.**

3. RUBUS, *L.* Bramble, Raspberry, &c.

1. **Chamæmorus,** *L. Cloudberry.*

2. **saxatilis,** *L.*

3. **Idæus,** *L. Raspberry.*

4. **fruticosus,** *L. Blackberry, Bramble.*

Sub-sp. suberectus, *Anderson* (sp.) ; *umbrosus*, Lees.—Var. *plicatus*, W. and N. (sp.).—Var. *fissus*, Lindl.—Var. *affinis*, W. and N. (sp.)., (*lentiginosus*, Lees, a form).

Sub-sp. rhamnifolius, *W.* and *N.* (sp.). —Var. *Lindleianus*, Lees ; *nitidus*, Bell Salter.—Var. *cordifolius*, W. and N.— Var *incurvatus*, Bab. (sp.).—Var. *imbricatus*, Hort. — Var. *Grabowskii*, Weihe. — Var. *Colemanni*, Bloxam.— Var. *macrophyllus*, var. *glabratus*, Bab. —Var. *ramosus*, Blox.

Sub-sp. corylifolius, *Smith* (sp.) ; *Sublustris*, Lees ; *purpureus*, Bab.—Var. *Balfourianus*, Blox. (sp.).—Var. *althæifolius*, Bab.—Var. *latifolius*, Bab.— Var. *Wahlbergii*, Arrh. (*conjungens*, Bab. a form).

Sub-sp. cæsius, *L.* (sp.). *Dewberry, tenuis*, Bell Salter.—Vars. *ulmifolius, intermedius, hispidus.* Hybrid (? *pseudo-idæus*, Lejeune.

Sub-sp. discolor. *W.* and *N.*—Var. *thyrsoideus*, Wimmer.

Sub-sp. leucostachys, *Smith* (sp.) ; *vestitus*, Weihe ; *Leightonianus*, Bab.

Sub-sp. villicaulis, *Weihe* (sp.) ; *carpinifolius*, Bab ; *pampinosus*, Lees ; *vulgaris*, W. and N. ; *adscitus*, Genev ; *devasus*, Muell.

Sub-sp. rubeolus, *Weihe* (sp.) ; *Salteri*, Bab ; *calvatus*, Blox.

Sub-sp. umbrosus, *Arrh.* (sp.) ; *macrophyllus*, var. *umbrosus*, Bab. ; *carpinifolius*.

Sub-sp. macrophyllus, *Weihe* (sp.); *Schlechtendahlii*, Weihe ; *amplificatus*, Lees.

Sub-sp. mucronulatus, *Boreau* (sp.) ; *mucronatus*, Blox.

Sub-sp. sprengelii, *Weihe* (sp.) ; *Borreri*, Salter ; *rubicolor*, Blox.

Sub-sp. dumetorum, *Weihe* (sp.) ; *nemorosus* of many ; Vars. *tuberculatus*, Bab ; *diversifolius*, Lindl., Bab. ; *concinnus*, Baker.

Sub-sp. radula, *Weihe* (sp.).—Vars. *rudis*, Weihe, Bab. ; *Leightoni*, Lees *denticulatus,* Bab., *mutabilis,* Genev., *obliquus,* Wirtg.

Sub-sp. BLOXAMI, *Lees* (sp.) ; VARS. *scaber,* Weihe *(Babingtonii,* Salter); *fusco-ater,* Weihe ; *Briggsii,* Blox.

Sub-sp. KOEHLERI, *Weihe* (sp.). —VAR. *infestus* Weihe.

Sub-sp. HYSTRIX, *Weihe* (sp.) ; Vars. *Lejeunii,* Weihe; *rosaceus,* Weihe.

Sub-sp. PALLIDUS, *Weihe* (sp.). VARS. *humifusus,* Weihe; *foliosus,* Weihe ; *hirtus,* Weihe *(fuscus,* Lees) ; *pygmæus,* Weihe and *Reuteri,* Merc.

Sub-sp. GLANDULOSUS, *Bell* (sp.) ; *Bellardi,* Weihe, *(dentatus,* Blox.) ; *rotundifolius,* Blox.

Sub-sp. GUNTHERI, *Weihe,* (sp.).

Sub-sp. PYRAMIDALIS, *Bab.* (sp.).

TRIBE IV. POTENTILLEÆ.

4. DRYAS, *L.*

1. octopetala, *L. depressa,* Bab.

5. GEUM, *L.* AVENS.

1. urbanum, *L.*

2. rivale, *L.* HYBRID, *intermedium,* Ehr.

6. FRAGARIA, *L.* STRAWBERRY.

1. vesca, *L. Wild Strawberry.*

ELATIOR, *Ehr. Haut-bois.*

7. POTENTILLA, *L.* CINQUEFOIL.

SECTION 1. Trichothalamus, *Lehm.* (gen.).

1. fruticosa, *L.*

SECTION 2. Comarum, *L.* (gen.).

2. Comarum, Nestl. *Comarum palustre,* L.

SECTION 3. Sibbaldia, *L.* (gen.).

3. procumbens, *Clairv. Sibbaldia procumbens,* L.

SECTION 4. Potentilla proper.

4. Tormentilla, *Sibthorp. Tormentilla officinalis,* Curtis. VAR. 1, *erecta,* L. (sp.) ; *sylvestris,* Necker. —VAR. 2, *procumbens,* Sibth. (sp.) ; *mixta,* Neck ; *Tormentilla reptans,* L.

5. reptans, *L. nemoralis,* Nestl.

6. verna, *L.*

7. salisburgensis, *Haenke ; alpestris,* Hall, f. ; *aurea,* Sm. ; *maculata,* Pour.

8. anserina, *L. Silver Weed.*

9. **Fragariastrum,** *Ehr. Fragaria sterilis,* L.

10. rupestris, *L.*

11. argentea, *L.*

TRIBE V. POTERIEÆ.

8. ALCHEMILLA, *L.* LADY'S MANTLE.

SECTION 1. Aphanes, *L.*

1. arvensis, *Lamk. Aphanes,* L.

SECTION 2. Alchemilla, proper.

2. vulgaris, *L.* VAR. *montana,* Willd. (sp.) ; *hybrida,* Pers.

3. alpina, *L.* VAR. *conjuncta,* Bab. (sp.) ; *argentea,* Don.

9. AGRIMONIA, *L.* AGRIMONY.

1 Eupatoria, *L.*

Sub-sp. EUPATORIA proper.

Sub-sp. ODORATA, *Mill.* (sp.).

10. POTERIUM, *L.*

1. Sanguisorba, *L. Salad Burnet.* MURICATUM, *Spach.*

2. officinale, *Hook. f. ; Sanguisorba, officinalis,* L *Great Burnet.*

TRIBE VI. ROSEÆ.

11. ROSA, *L.* ROSE.

1. spinosissima, *L. ; pimpinellifolia,* L. *Scotch Rose, Burnet Rose.*

2. villosa, *L.*

Sub-sp. POMIFERA, *Herm.* (sp.).

Sub-sp. MOLLIS, *Sm.* (sp.) ; *mollissima,* Willd, Sm.; *heterophylla,* Woods. —VAR. 1, *cœrulea,* Baker. —VAR. 2, *pseudo-rubiginosa,* Lejeune (sp.).

Sub-sp. TOMENTOSA, *Sm.* (sp.). —VAR 1, *subglobosa* Sm. (sp.) *(Sherardi,* Davies). VAR. 2, *farinosa,* Raw. (sp.). —VAR. 3, *scabriuscula,* Sm. (sp.). —VAR. 4, *sylvestris,* Woods *(Jundzilliana,* Baker ; *Britannica,* Deseg.). —VAR. 5, *obovata,* Baker.

3. involuta, *Smith.* —VAR. 1, *Sabini,* Woods (sp.) ; *gracilis,* Woods ; *nivalis,* Don ; *coronata,* Crep. —VAR. 2, *Doniana,* Woods (sp.). —VAR. 3, *gracilescens,* Baker. —VAR. 4, *Robertsoni,* Baker. —VAR. 5, *Smithii,* Baker ; *involuta,* Sm. —VAR. 6, *lævigata,* Baker. —VAR. 7, *Moorei,* Baker. —VAR. 8, *occidentalis,* Baker. —VAR. 9, *Wilsoni,* Borrer (sp.).

4. rubiginosa, *L.*

Sub-sp. RUBIGINOSA proper ; *Eglanteria,* Woods. *Sweet-briar.* —VAR. 1, *permixta,* Deseg. (sp.). —VAR. 2, *sylvicola,* Deseg. and Ripart (sp.).

Sub-sp. MICRANTHA, *Smith* (sp.).—VAR. 1, *Briggsii*, Baker.—VAR. 2, *Hystrix*, Leman (sp.).

Sub-sp. SEPIUM, *Thuill.* (sp.).—VAR. 1, *Billietii*, Puget (sp.); *sepium*, Borrer, E. B. S.—VAR. 2, *inodora*, Fries. (sp.); *pulverulenta*, Lindl.—VAR. 3, *crypto-poda*, Baker.

5. hibernica, *Smith.*—VAR. 1, *glabra*, Baker.—VAR. 2, *cordifolia*, Baker.

6. canina, *L.* Dog Rose.

SERIES 1. Ecristatæ.
VAR. 1, *lutetiana*, Leman (sp.).—VAR. 2, *surculosa*, Woods (sp).—VAR. 3, *sphœrica*, Gren. (sp.).—VAR. 4, *senticosa*, Ach. (sp.).—VAR. 5, *dumalis*, Bechst. (sp.); *sarmentacea*, Sm. ; *glaucophylla*, Winch. — VAR. 6, *biserrata* Merat (sp.) *vinacea*, Baker.—VAR. 7, *urbica*, Leman (sp.); *collina*, Woods; *Forsteri*, Smith ; *platyphylla*, Rau.—VAR. 8, *frondosa*, Steven (sp.); *dumetorum*, Woods.—VAR. 9, *arvatica*, Baker. —VAR. 10, *dumetorum*, Thuill. (sp.); *uncinella*, Besser.—VAR. 11, *pruinosa*, Baker ; *cæsia*, Borrer.—VAR. 12, *incana*, Woods (sp.); *canescens*, Baker.—VAR. 13, *tomentella*, Leman (sp.) ; *inodora*, Hook. *Fl. Lond* ; *obtusifolia*, Desv. —VAR. 14, *Andevagensis*, Bast. (sp.).— VAR. 15, *verticillacantha*, Merat.—VAR. 16, *collina*, Jacq. (sp.).—VAR. 17, *cœsia*, Smith (sp.).—VAR. 18, *concinna*, Baker. —VAR. 19, *decipiens*, Dumort.

SERIES 2. Subcristatæ. VAR. *sclerophyllea*, Scheutz. (sp.); *monticola*, Rep.— VAR. 20, *Reuteri*, Godet (sp.) ; *nuda*, Woods ; *Crepiniana*, Deseg.—VAR. 21, *subcristata*, Baker.—VAR. 22, *Hailstonei*, Baker.—VAR. 23, *implexa*, Gren. (sp.). —VAR. 24, *coriifolia*, Fries (sp.); *bractescens*, Woods.—VAR. 25, *Watsoni*, Baker. —VAR. 26, *celerata*, Baker.

SERIES 3. Subrubiginosæ.
VAR. 27, *Borreri*, Woods (sp.); *dumetorum*, Eng. Bot.—VAR. 28, *Bakeri*, Deseg. (sp.).—VAR. 29, *marginata ;* *Blondæana*, Ripart.

7. arvensis, *L.*

Sub-sp. ARVENSIS proper; *repens*, Scopoli. —VAR. 1, *bibracteata*, Bast. (sp.).

Sub-sp. STYLOSA, *Bast.* (sp.).—VAR. 1, *stylosa.* proper.—VAR. 2, *systyla*, Bast. ; *collina*, Engl. Bot.; *leucochroa*, Desv.—

VAR. 3, *opaca*, Baker. VAR. 4, *gallicoides*, Baker.—VAR. 5, *Monsoniæ*, Lindl. —VAR. 6, *fastigiata*, Bast. (sp.).

TRIBE VII. POMEÆ.

12. PYRUS, *L.* PEAR, APPLE SERVICE, &c.

SECTION 1. Pyrus proper.

1. communis, *L.* Wild Pear.—VAR. 1, *Pyraster*, L. — VAR. 2. *Achras*, Gærtn. (sp.).—VAR. 3, *cordata*, Desv. (sp.) *Briggsii*, Bosw.

2. Malus, *L.* Wild or Crab-apple. VAR. 1, *acerba*, DC. (sp.).—VAR. 2, *mitis.*

SECTION 2. Sorbus, *L.* (gen.).

3. torminalis, *Ehr.* Wild Service.

4. Aria, *L.* White Beam.

Sub-sp. ARIA proper.

Sub-sp. RUPICOLA, *Bosw.*

Sub-sp. LATIFOLIA, *Pers.*

Sub-sp. SCANDICA, *Fries* (sp.), *P. fennica*, Fries (*pinnatifida*, Sm. in part) is a HYB. acc. to Bosw. bet. *scandica* and *Aucuparia.*

5. Aucuparia, *Gærtn.* Mountain Ash, Rowan-tree.

SECTION 3. Mespilus, *L* (gen.).

GERMANICA, *L.* (*Mespilus*). Medlar.

13. CRATÆGUS, *L.* HAWTHORN, WHITETHORN, MAY.

1. Oxyacantha, *L.*

Sub-sp. OXYACANTHOIDES, *Thuill.* (sp.).

Sub-sp. MONOGYNA, *Jacquin* (sp.).

14. COTONEASTER, *Lindl.*

1. vulgaris, *Lindl.*

———

ORDER XXVIII. SAXIFRAGEÆ.

TRIBE 1. SAXIFRAGEÆ, proper.

1. Saxifraga, *L.* SAXIFRAGE.

SECTION 1. Porphyrion, *Tausch.*

1. oppositifolia, *L.*

SECTION 2. Micranthes, *Haw.* (gen.).

2. nivalis, *L.*

SECTION 3. Hydatica and Arabadea, *Haw.* (gen.).

3. stellaris, *L.* — VAR. *integrifolia,* Hook.

4.. umbrosa, *L.*

Sub-sp. UMBROSA proper. - VAR. 1, *umbrosa* proper. *London Pride, St. Patrick's Cabbage.* — VAR. 2, *punctata,* Haw. (sp.). — VAR. 3, *serratifolia,* Mackay (sp.). — var. *duplicato-serrata,* Lond. Cat. 1867, *serrata,* L. C. 1874.

Sub-sp. GEUM, *L.* (sp.). — VAR. 1, *serrata,* Bosw. — VAR. 2, *elegans* Mack. (sp.). — VAR. 3, *crenata,* Bosw. — VAR. 4, *gracilis,* Mack. (M. S. sp.).

Sub-sp. HIRSUTA, *L.* (sp.).

ANDREWSII, *Harvey; Guthriana,* Hort.

SECTION 4. Hirculus, *Haw.* (gen.).

5. Hirculus, *L.*

6. aizoides, *L.;* VAR. *autumnalis,* L.

SECTION 5. Nephrophyllum, *Gaud.*

7. tridactylites, *L.*

8. rivularis, *L.*

9. granulata, *L.*

10. cernua, *L.*

SECTION 6. Dactyloides, *Tausch.*

11. cæspitosa, *L.*

12. hypnoides, *L.*

Sub-sp. HYPNOIDES proper. VAR. 1, *platypetala,* Sm. (sp.) ; *elongella,* Sm. ; *condensata,* Gmel. — VAR. 2. *gemmifera,* Bosw ; *leptophylla,* Haw.
Sub-sp. DECIPIENS, *Ehr.* (sp.) ; *cæspitosa,* KOCH ; *palmata,* Sm.
Sub-sp. HIRTA, *Don* (sp.). — VAR. 1, *hirta* proper, Bosw ; *Sternbergii,* Willd. — VAR. 2, *affinis,* Don. (sp.) ; *sponhemica,* Gmel. — VAR, *incurvifolia,* Don. (sp.).

2. CHRYSOSPLENIUM, *L.* GOLDEN SAXIFRAGE.

1. alternifolium, *L.*

2. oppositifolium, *L.*

3. PARNASSIA, *L.* GRASS OF PARNASSUS.

1. palustris, *L.*

4. RIBES, *L.* CURRANT, GOOSEBERRY.

SECTION 1. Grossularia.
1. Grossularia, *L. Wild Gooseberry.* VAR 1, *Grossularia* proper. — VAR. 2, *Uva-crispa,* L. (sp.).

SECTION 2. Ribesia.

2. alpinum, *L.*

3. rubrum, *L. Wild Currant.* — VAR. 1. *sylvestre,* Reich. — VARS. *Smithianum* and *Bromfeldianum,* Bosw. *petræum,* Sm. — VAR. 2, *spicatum,* Robson (sp.). — VAR. 3, *sativum,* Reich.

4. nigrum, *L. Black Currant.*

———

ORDER XXIX. CRASSULACEÆ.

1. TILLÆA,
1. muscosa, *L.*

2. COTYLEDON, *L.* PENNYWORT.
1. Umbilicus, *L.*

3. SEDUM, *L.* ORPINE, STONECROP.

SECTION 1. Telephium.

1. Rhodiola, *DC.; Rhodiola rosea, L. Rose-root.*

2. Telephium, *L. Orpine.* — VAR. 1, *purpurascens,* Koch (sp.). — VAR. 2, *Fabaria,* Koch (sp.).

SECTION 2. Cepæa.

3. villosum, *L.*

SECTION 3. Sedum proper.

4. album, *L.* — VAR. 1, *teretifolium,* Haw. (sp.). — VAR. 2, *micranthum,* Bast.

5. anglicum, *L.*

DASYPHYLLUM, *L.*

6. acre, *L. Biting Stone-crop, Wall-pepper.*

7. rupestre, *Huds.*

Sub-sp. ELEGANS, *Lej.* — VAR. 1, *majus.* — VAR. 2, *minus.*
Sub-sp. FORSTERIANUM, *Sm.* (sp.) ; Vars. *glaucescens* and *virescens.*

SEXANGULARE, *L.*

REFLEXUM, *L.* — VAR. 1, *reflexum* proper. — VAR. 2, *albescens,* Haw. (sp.) ; *glaucum,* Sm.

3.* SEMPERVIVUM, *L.* HOUSE-LEEK TECTORUM, *L.*

———

ORDER XXX. DROSERACEÆ.

1. DROSERA, *L.* SUNDEW.

e

1. rotundifolia, *L.*

2. intermedia, *Hayne; longifolia,* L. in part.

3. anglica, *Huds.* ; *longifolia,* L. in part ; *obovata,* Mert. and Koch., a hyb ? between this and *rotundifolia.*

———

ORDER XXXI. **HALORAGEÆ.**

1. **HIPPURIS,** *L.* MARE'S-TAIL.

1. vulgaris, *L*

2. **MYRIOPHYLLUM,** *L.* WATER MILFOIL.

1. verticillatum, *L.*—VAR. *pectinatum,* DC. (sp.).

2. alterniflorum, *DC.*

3. spicatum, *L.*

3. **CALLITRICHE,** *L.* WATER STAR-WORT.

1. verna, *L.*

Sub-sp. VERNALIS, *Kuetzing* (sp.) ; *aquatica,* Sm.

Sub-sp. PLATYCARPA, *Kuetz.* (sp.) ; *stagnalis,* Kuetz.

Sub-sp. HAMULATA, *Kuetz.* (sp.).

Sub-sp. OBTUSANGULA, *Leg.* (sp.) ; *Lachii,* Warren, MSS.

Sub-sp. PEDUNCULATA, *DC.* (sp.).

2. autumnalis, *L.*—VAR. 1, *autumnalis* proper.—VAR. 2, *truncata,* Guss. (sp.).

———

ORDER XXXII. **ONAGRARIEÆ.**

1. **EPILOBIUM,** *L.* WILLOW-HERB.

SECTION 1. **Chamænerion.**

1. angustifolium, *L. Rose-bay* or *French Willow.* —VAR. *brachycarpum,* Leighton (sp.).

SECTION 2. **Lysimachion.**

2. hirsutum, *L. Codlins and Cream.*

3. parviflorum, *Schreb.* — VAR. 1, *rivulare,* Wahl. (sp.).—VAR. 2, *intermedium,* Mérat (sp.).

4. montanum, *L.*

5. lanceolatum, *Sebast* and *Maur.*

6. roseum, *Schreb.*

7. tetragonum, *L.*

Sub-sp. TETRAGONUM proper.

Sub-sp. OBSCURUM, *Schreb.* (sp.) ; *virgatum,* Gren. and Godr.

8. palustre, *L.* — VAR. *ligulatum,* Baker.

9. alsinifolium, *Villars.*

10. alpinum, *L.*—VAR. *anagallidifolium,* Lamk. (sp.).

2. **LUDWIGIA,** *L.*

1. palustris, *Elliot ; Isnardia,* L.

2* **ŒNOTHERA,** *L.* EVENING PRIMROSE.

BIENNIS, *L.*

ODORATA, *Jacq.*

3. **CIRCÆA,** *Tourn.* ENCHANTER'S NIGHTSHADE.

1. lutetiana, *L.*

2. alpina, *L.*

———

ORDER XXXIII. **LYTHRACEÆ.**

1. **LYTHRUM,** *L.* LOOSESTRIFE.

1. Salicaria, *L.*

2. hyssopifolia, *L.*

2. **PEPLIS,** *L.* WATER-PURSLANE.

1. Portula, *L.*

———

ORDER XXXIV. **CUCURBITACEÆ.**

Tribe CUCUMERINEÆ.

1. **BRYONIA,** *L.* BRYONY.

1. dioica, *L.*

———

ORDER XXXV. **UMBELLIFERÆ.**

SERIES 1. HETEROSCIADIEÆ.

TRIBE I. **HYDROCOTYLEÆ.**

1. **HYDROCOTYLE,** *L.* PENNY-WORT.

1. vulgaris, *L.*

TRIBE II. **SANICULEÆ.**

2. **ERYNGIUM,** *L.* ERYNGO.

1. maritimum, *L. Sea Holly.*

2. CAMPESTRE, *L.*

2* *ASTRANTIA, L.*
MAJOR, *L.*

3. **SANICULA,** *L.* SANICLE.
1. europæa, *L.*
SERIES 2. HAPLOZYGIEÆ.
TRIBE III. **AMMINEÆ.**
SECTION I. Smyrnieæ.

4. **PHYSOSPERMUM,** *Cusson.*
BLADDER-SEED.
1. cornubiense, *DC.*

5. **CONIUM,** *L.* HEMLOCK.
1. maculatum, *L.*

6. **SMYRNIUM,** *L.* ALEXANDERS.
1. Olusatrum, *L.*
SECTION 2. Amminæ proper.

7. **BUPLEURUM,** *L.* HARE'S-EAR.
1. rotundifolium, *L.*
2. falcatum, *L.*
3. tenuissimum, *L.*
4. aristatum, *Bartl; Odontites,* Sm.

8. **TRINIA,** *Hoffmann.* HONEWORT.
1. vulgaris, *DC.*

9. **APIUM,** *L.* (and *Helosciadium,*
Koch), CELERY.
SECTION 1. Apium proper.
1. graveolens, *L.* *Wild Celery.*
SECTION 2. Helosciadium, *Koch* (ger.).
2. nodiflorum, *Reich.*—VAR. 1, *nodi-florum* proper.—VAR. 2. *repens,* Koch
(sp.) ; *Sium repens* Sm.
3. inundatum, *Reich.*

10. **CARUM,** *L.* CARAWAY.
SECTION 1. Carum proper.
1. verticillatum, *L.*
CARUI, *L.* *Caraway.*
SECTION 2. Petroselinum, *Hoffm.*
(gen.).
2. segetum, *Benth.* *Corn Parsley.*
PETROSELINUM, *Benth.; P. sativum.*
Hoffm. *Common parsley.*
SECTION 3. Bunium, *L.* (gen.).
3. Bulbocastanum, *Koch.*

11. **SISON,** *L.*
1. Amomum, *L.*

12. **CICUTA,** *L.* WATER-HEMLOCK,
COWBANE.
1. virosa, *L.*

13. **SIUM,** *L.* WATER-PARSNIP.
1. latifolium, *L.*
2. angustifolium, *L.*

14. **ÆGOPODIUM,** *L.* GOAT-, GOUT-
or BISHOP'S-WEED.
1. Podagraria, *L.* *Herb Gerard.*

15. **PIMPINELLA,** *L.* BURNET-
SAXIFRAGE.
1. Saxifraga, *L.*
2. magna, *L.*
SECTION 3. Scandicineæ.

16. **CONOPODIUM,** *Koch.* EARTH-
NUT.
1. denudatum, *Koch ; Bunium flex-uosum,* With.

17. **MYRRHIS,** *Scop.* CICELY.
1. odorata, *Scopoli.*

18. **SCANDIX,** L. SHEPHERD'S
NEEDLE.
1. Pecten-Veneris, *L.*

19. **CHÆROPHYLLUM,** *L.* CHER-
VIL.
1. temulum, *L. ; temulentum,* Sm.

20. **ANTHRISCUS,** *Hoffm.* BEAKED-
PARSLEY.
1. vulgaris, *Pers. ; Scandix Anthris-cus,* L. ; *Chærophyllum Anthriscus,*
Lamk.
2. sylvestris, *Hoffm. Chærophyllum
sylvestre,* L.
CEREFOLIUM, *Hoffm. ; Scandix,* L. ;
chærophyllum, sativum, Gærtn. *Cher-vil.*

TRIBE IV. **SESELINEÆ.**
SUB-TRIBE 1. Seselineæ proper.
21. **SESELI,** *L.*
1. Libanotis, *Koch ; Athamanta,* L.
22. **FŒNICULUM,** *Adanson.* FEN-
NEL.
1. vulgare, Gærtn.
Sub-tribe 2. Coriandreæ.

22*. *CORIANDRUM, L.* Coriander.

Sub-tribe 3. Cachrydeæ.

23. CRITHMUM, *L.* Samphire.
   1. maritimum, *L.*

Sub-tribe 4. Œnantheæ.

24. ŒNANTHE, L. Water Dropwort.
   1. fistulosa, *L.*
   2. pimpinelloides, *L.*
   3. Lachenalii, *Gmelin; pimpinell-
      oides,* Sm.
   4. silaifolia, *Bieberstein; peucedani-
      folia,* Sm.; *Smithii,* Watson.
   5. crocata, *L.*
   6. Phellandrium, Lamk.; *Phellan-
      drium aquaticum,* L.

Sub-sp. Phellandrium proper.
Sub-sp. Fluviatile, *Coleman* (sp.).

25. ÆTHUSA, *L.* Fool's Parsley.
   1. Cynapium; *L.*

Sub-tribe 5. Schultzieæ.

26. SILAUS, *Besser.* Pepper Saxifrage.
   1. pratensis, *Besser.*

Sub-tribe 6. Selineæ.

27. MEUM, *Jacquin.* Meu, Bald-
money. Spignel.
   1. athamanticum, *Jacq.*

28. LIGUSTICUM, *L.* Lovage.
   1. scoticum, *L.; Haloscias,* Fries.

Sub-tribe 7. Angeliceæ.

29. ANGELICA, *L.* Angelica.
   1. sylvestris, L.

Tribe V. PEUCEDANEÆ.

30. PEUCEDANUM, *L.* Hog's-
fennel.

Section 1. Peucedanum proper.
   1. officinale, *L. Sulphur-wort.*
   2. palustre. *Mœnch. Milk Parsley.*

Section 2. Imperatoria, *L.* (gen.).
Ostruthium, *Koch, Master-wort.*

Section 3. Pastinaca, *L.* (gen.).

3. sativum, *Benth. Wild Parsnip.*

31. HERACLEUM, *L.* Cow-Pars-
nip, Hogweed.
   1. Spondylium, *L.*

32. TORDYLIUM, *L.*
   1. maximum, *L.*

Series 3. *DIPLOZYGIEÆ.*

33. DAUCUS, *L.* Carrot.
   1. Carota, L.—Var. 1, *Carota* pro-
      per.—Var. 2, *gummifer,* Lamk (sp.) ;
      maritimus, With.

34. CAUCALIS, L.

Section 1. Caucalis, proper. *Bur-
Parsley.*
   1. daucoides, *L.*

Section 2. Turgenia, *Hoffm.* (gen.).
   latifolia, *L.*

Section 3. Torilis, *L.* (gen.). *Hedge
Parsley.*
   2. Anthriscus, *Huds.*
   3. infesta, *Curtis.*
   4. nodosa, *Scop.*

Order XXXVI. ARALIACEÆ.

1. HEDERA, *L.* Ivy.
   1. Helix, *L.* Var. *Hodgensii.*

Order XXXVII. CORNACEÆ.

1. CORNUS, *L.* Cornel, Dogwood.
   1. sanguinea, *L. Dogwood, Dog-
      berry, Prickwood.*
   2. suecica, *L.*

Division II. MONOPETALÆ.

Ord. XXXVIII. CAPRIFOLIACEÆ.

Tribe 1. SAMBUCEÆ.

1. VIBURNUM, *L.*
   1. Lantana, *L. Wayfaring-tree.*
   2. Opulus, L. *Guelder-rose.*

2. SAMBUCUS, *L.* elder.
   1. Ebulus, *L. Dwarf Elder, Dane-
      wort.*

2. **nigra**, *L. Elder.*

3. **ADOXA**, *L.* MOSCHATEL.
   1. **Moschatellina**, *L.*

4. **LONICERA**, *L.* HONEYSUCKLE.
   1. **Periclymenum**, *L. Woodbine* or *Honeysuckle.*
   CAPRIFOLIUM, *L.*
   XYLOSTEUM, *L.*

5. **LINNÆA**, *Gronovius.*
   1. **borealis**, *Gronov.*

ORDER XXXIX. **RUBIACEÆ.**

Tribe. STELLATÆ.

1. **RUBIA**, *L.* MADDER.
   1. **peregrina**, *L.*

2. **GALIUM**, *L.* BEDSTRAW.
   1. **verum**, *L. Lady's Bedstraw.*—VAR. 1, *verum* proper.—VAR. *luteum*, Bosw.—VAR. 2, *ochroleucum*, Bosw.
   2. **Cruciata**, *Scopoli; cruciatum*, With. *Valantia Cruciata*, L. *Crosswort.*
   3. **palustre**, *L.*—VAR. 1, *palustre* proper.—VAR. 2, *elongatum*, Presl. (sp.).—VAR. 3, *Witheringii*, Sm. (sp.).
   4. **uliginosum**, *L.*
   5. **saxatile**, *L.*
   6. **sylvestre**, *Poll.*—VAR. 1, *montanum*, Vill. (sp.).—VAR. 2, *nitidulum*, Thuill. (sp.); *commutatum*, Bab.
   7. **Mollugo**, *L.*
   Sub-sp. ERECTUM, *Huds.* (sp.); *aristatum*, Sm.
   Sub-sp. SCABRUM, *With.* (sp.); *elatum*, Thuill.; *insubricum*, Gaud.—VAR. *Bakeri*, Bosw.
   8. **boreale**, *L.*
   9. **Aparine**, *L. Goose-grass, Cleavers.*
   Sub-sp. APARINE proper.
   Sub-sp. VAILLANTII, *DC.* (sp.).
   10. **tricorne**, *With.*
   11. **parisiense**, *L.;* Sub-sp. ANGLICUM, Huds. (sp.).

3. **ASPERULA**, *L.*
   1. **odorata**, *L. Wood-ruff.*
   2. **cynanchica**, *L. Squinancy-wort.*

4. **SHERARDIA**, *Dillen.* FIELD-MADDER.
   1. **arvensis**, *L.*

ORDER XL. **VALERIANEÆ.**

1. **VALERIANA**, *L.* VALERIAN.
   1. **dioica**, *L.*
   2. **officinalis**, *L. Cat's-Valerian, All-heal.*—VAR. 1, *officinalis* proper.—VAR. 2, *sambucifolia*, Mikan (sp.). PYRENAICA, *L.*

1*. *CENTRANTHUS, DC.* SPUR-VALERIAN.
   RUBER, *DC.*

2. **VALERIANELLA**, *Tournef; Fedia*, Vahl. CORN SALAD.
   1. **olitoria**, *Mœnch; Valeriana Locusta*, L. in pt. *Lamb's Lettuce.*
   CARINATA, *Loisel.*
   2. **auricula**, *DC.; F. tridentata*, Stev.
   3. **dentata**, *Poll.; Morisonii*, DC.—VAR. 1, *dentata* proper.—VAR. 2, *mixta*, Desv. (sp.); *F. eriocarpa*, R and S.—VAR. 3, *eriocarpa*, Desv.

ORDER XLI. **DIPSACEÆ.**

1. **DIPSACUS**, *Tournef.* TEASEL.
   1. **sylvestris**, *L. Wild Teasel.*—cult. form *Fullonum*, L. (Fuller's Teasel.
   2. **pilosus**, *L.*

2. **SCABIOSA**, *L.*
   SUB-GEN. I. Scabiosa proper.
   1. **succisa**, *L. Devil's-bit-Scabious.*
   2. **Columbaria**, *L.*
   SUB-GEN. II. Knautia, *Coulter* (gen.).
   3. **arvensis**, *L.*

ORDER XLII. **COMPOSITÆ.**
SERIES 1. TUBULIFLOREÆ.

f

TRIBE I. **CYNAREÆ.**

1. ARCTIUM, *L.* BURDOCK.

1. **Lappa,** *L.*

Sub-sp. LAPPA proper (sp.) ; *majus,* Schkuhr.

Sub-sp. MINUS, Schkuhr (sp.).—VAR. 1, *minus* proper.— VAR. 2, *intermedium,* Lange (sp.) ; *pubens,* Bab.—VAR. 3, *nemorosum,* Lej. (sp.).

2. CARLINA, *L.* CARLINE-THISTLE.

1. **vulgaris,** *L.*

3. SAUSSUREA, *DC.*

1. **alpina,** *L.*

4. CENTAUREA, *L.* KNAPWEED.

SECTION 1. **Jacea.**

1. **nigra** *L.* *Knapweed.*—VAR. 1, *nigra* proper. — VAR. 2, *decipiens,* Thuill ; *nigrescens,* Bab.

SECTION 2. **Cyanus.**

2. **Scabiosa,** *L.*

3. **Cyanus,** *L.* *Bluebottle, Cornflower.*

SECTION 3. **Seridia.**

4. **aspera,** *L. ; Isnardi,* **L.**

SECTION 4. **Calcitrapa.**

5. **Calcitrapa,** *L.* *Star-thistle.*

SOLSTITIALIS, *L.*

5. SERRATULA, *DC.* SAW-WORT.

1. **tinctoria,** *L.*

6. CARDUUS, *L.* THISTLE.

Sub-gen. 1. **Carduus proper.**

1. **nutans,** *L.* *Musk-thistle.*

2. **crispus,** *L. ; acanthoides,* Sm. VAR. 1, *crispus* proper.—VAR. 2, *polyanthemos,* Koch (sp.).—VAR. 3, *litigiosus,* Gr. and Godr.

3. **pycnocephalus,** *Jacq.; tenuiflorus,* Curtis.

SUB-GEN. 2. **Cirsium,** *Tourn.* (gen.) ; CNICUS, *L.*

4. **lanceolatus,** *L.* *Spear-thistle.*

5. **eriophorus,** *L.*

6. **acaulis,** *L.*

7. **arvensis,** *Curtis.*

Sub-sp. ARVENSIS proper.

Sub-sp. SETOSUS, *Bieb.* (sp.).

8. **palustris,** *L.*

9. **pratensis,** *Huds. ; Cirsium anglicum,* Lamk. **Woodwardi,** Hyb. bet. 6 and 9.

10. **tuberosus,** *L.*

11. **heterophyllus,** *L.* *Melancholy Thistle.*

SUB-GEN. 3. *Silybum, Gærtn.* (gen.).

MARIANUS, *Gærtn.*

7. ONOPORDON, *L.* COTTON THISTLE.

1. **Acanthium,** *L.*

TRIBE II. **EUPATORIEÆ.**

8 EUPATORIUM. HEMP AGRIMONY.

1. **cannabinum,** *L.*

TRIBE III. **TUSSILAGINEÆ.**

9. PETASITES, *Tournef* BUTTER-BUR.

1. **vulgaris,** Desf.; *Tussilago Petasites,* L. and *T. hybrida,* L.

10. TUSSILAGO, *Tournef.* COLT'S-FOOT.

1. **Farfara,** *L.*

TRIBE IV. **ATEROIDEÆ.**

11. ASTER, *L.*

1. **Tripolium,** *L.*

12. ERIGERON, *L.*

1. **alpinus,** *L ; uniflorum,* Sm.

2. **acris,** *L.*

CANADENSIS, *L.*

13. BELLIS, *L.* DAISY.

I. **perennis,** *L.*

14. SOLIDAGO, *L.* GOLDEN-ROD.

1. **Virgaurea,** *L.* VAR. 1, *virgaurea,* proper.—VAR. 2,*angustifolia,* Gaud.— VAR. 3, *cambrica,* Huds. (sp.).

15. LYNOSYRIS, *Cassini.* GOLDIE-LOCKS.

1. **vulgaris,** *Cass. ; Crysocoma,* **L.**

TRIBE V. **INULEÆ.**

16. INULA, *L.*

1. **Conyza** *DC. ; Conyza squarrosa,* L. *Ploughman's Spikenard.*

2. crithmoides, *L. Golden Samphire.*
3. salicina, *L.*
4. Helenium, *L. Elecampane.*
5. dysenterica, *L. Flea-bane.*
6. Pulicaria, *L.; P. vulgaris,* Gærtn.

Tribe VI. HELIANTHEÆ.

17. BIDENS, *L.* Bur-Marigold.
1. cernua, *L.* Var. 1, *discoidea,* — Var. 2, *radiata.*
2. tripartita, *L.*

17*. *GALINSOGA, Ruiz* and *Pavon.*
parviflora, *Cav.*

Tribe VII. ANTHEMIDEÆ.

18. ANTHEMIS, *L.*
1. arvensis, *L. Corn Chamomile.* Var. *anglica,* Spr. (sp.) ; *maritima,* Sm.
2. Cotula, *L. Stinking May-weed.*
3. nobilis, *L. Chamomile.*

19. ACHILLEA, *L.*
1. Ptarmica, *L. Sneeze-wort.*
2. Millefolium, *L. Yarrow, Milfoil.*

20. DIOTIS, *Desf.* Cotton-weed.
1. maritima, *Cass.*

21. MATRICARIA, *L.*
Sub-gen. 1. Matricaria proper.
1. Chamomilla, *L. Wild Chamomile.*
Sub-gen. 2. Pyrethrum, *Gærtn.* (gen.)
2. inodora, *L.* Var. 1, *inodora,* proper.—Var. 2, *maritima,* L. (sp.).— Var. 3, *salina.*

22. CHRYSANTHEMUM, *L.*
1. segetum, *L. Corn Marigold.*
2. Leucanthemum, *L.* Ox-eye Daisy.
3. Parthenium, *Pers. ; Matricaria,* L.; *Pyrethrum,* Sm. *Fever-few.*

23. TANACETUM, *L.* Tansy.
1. vulgare, *L.*

24. ARTEMISIA, *L.* Wormwood.
1. campestris, *L.*
2. vulgaris, *L. Mugwort.*

3. Absinthium, *L. Wormwood.*
4. maritima, *L.*

Tribe VIII. GNAPHALIEÆ.

25. GNAPHALIUM, Cud-weed.
1. luteo-album, *L.*
2. sylvaticum, *L.* -Var. 1, *rectum,* Sm. (sp.).—Var. 2, *norvegicum,* Gunn. (sp.).
3. uliginosum, *L.*
4. supinum, *L.*

26. ANTENNARIA, *Brown.*
1. dioica, *Br. Cat's-foot.*—Var. *hyperborea,* Don. (sp.).
margaritacea; *Br.*

27. FILAGO.
1. germanica, *L.*—Var. 1, *canescens,* Jord. (sp.).—Var. 2, *apiculata,* G. E. Sm. (sp.).—Var. 3, *spathulata,* Presl. (sp.).
2. minima, *Fries ; montana,* DC. *Gnaphalium arvense,* Willd.
3. gallica, *L.*

Tribe IX. SENECIONIDEÆ.

27*. *DORONICUM, L.* Leopard's-bane.
pardalianches, *L.*
plantagineum, *L.*

28. SENECIO, *L.*
Section 1. Senecio proper.
1. vulgaris, *L. Groundsel.* Var. 1, *radiatus,* Koch.—Var. 2, *hibernica,* Bosw.
2. sylvaticus, *L. ;*—Form *lividus,* Sm.
3. viscosus, *L.*
4. Jacobæa, *L. Ragwort.*
5. erucifolius, *L ; tenuifolius,* Jacq.
6. aquaticus, *Huds.*—Var. *barbaræifolius,* Reich.
squalidus, *L.*
saracenius, *L.*
7. paludosus, *L.*

Section 2. Cineraria, *L.* (gen.).
8. palustris, *DC.*

9. **campestris**, *DC.; Cineraria integ-rifolia*, With.

SERIES 2. **LIGULIFLOREÆ.**

TRIBE X. **CICHORACEÆ.**

29. **LAPSANA**, *L.* NIPPLEWORT.

1. communis, *L.*

30. **ARNOSERIS**, *Gærtn.* LAMB'S or SWINE'S SUCCORY.

1. pusilla, *Gærtn.*

31. **CICHORIUM**, *L.* CHICORY.

1. Intybus, *L.*

32. **HYPOCHÆRIS**, *L.* CAT'S-EAR.

1. glabra, *L.*—VAR. 1, *glabra* proper. —VAR. 2. *Balbisii*, Lois (sp.).

2. radicata, *L.*

3. maculata, *L.; Achyrophorus maculatus*, Scop.

33. **HELMINTHIA**, *Juss.* OX-TONGUE.

1. echioides, *Gærtn.*

34. **TRAGOPOGON**, *L.*

1. pratensis, *L. Goat's-beard.*—VAR. 1, *pratensis* proper.—VAR. 2, *minor*, Fries (sp.).—VAR. 3, *grandiflorus*, Bosw.

PORRIFOLIUS, *L. Salsify.*

35. **PICRIS**, *L.*

1. hieracioides, *L.*—VAR. 1, *hieracioides* proper.—VAR. 2, *arvalis*, Jord. (sp.).

36. **LEONTODON**, *L.* HAWKBIT.

SUB-GEN. 1. **Thrincia**, *Roth* (gen.).

1. hirtus, *L.; Hedypnois*, Sm.

SUB-GEN. 2. **Apargia**, *Willd.* (gen.).

2. hispidus, *L.; Hedypnois*, Sm.

SUB-GEN. 3. **Oporinia**, *Don* (gen.).

1. autumnalis, *L.; Apargia*, Willd. —VAR. 1, *autumnalis* proper.—VAR. 2, *pratensis*, Koch (sp.); *Hedypnois Taraxici*, Sm.

37. **LACTUCA**, *L.* LETTUCE.

1. virosa, *L.*

2. Scariola, *L.*

3. saligna, *L.*

4. muralis, *Frœsen; Prenanthes muralis*, L.

38. **TARAXICUM**, *Juss.* DANDELION.

1. officinale, *Wiggers; Leontodon Taraxicum*, L.—VAR. 1, *Dens-leonis*, Desf. (sp.).—VAR. 2, *erythrospermum*, Andrz. (sp.).—VAR. 3, *lævigatum*, DC. (sp.).—VAR. 4, *palustre*, DC. (sp.).

39. **CREPIS**, L. HAWK'S-BEARD.

SUB-GEN. 1. **Crepis** proper.

1. virens, *L.; tectorum*, Sm.

2. biennis, *L.*

3. hieracioides, *Waldst.* and *Kit.; succisæfolia*, Tausch.; *Hieracium molle*, Sm.

SUB-GEN. 2. **Barkhausia**, *Mœnch* (gen.).

4. foetida, *L.*

5. taraxifolia, *Thuill.*

SETOSA, *Haller fil.*

SUB-GEN. 3. **Aracium**, *Monn.* (gen.).

6. paludosa, *Mœnch; Hieracium paludosum*, L.

40. **SONCHUS**, *L.* SOWTHISTLE.

1. arvensis, *L.*

2. palustris, *L.*

3. oleraceus, *L.*

Sub-sp. OLERACEUS proper.
Sub-sp. ASPER, Hoffm. (sp.).

41. **MULGEDIUM**, *Cassini.* BLUE SOWTHISTLE.

1. alpinum, *Less.; Sonchus alpinus*, L.; *cœruleus*, Sm.

42. **HIERACIUM**, *L.* HAWKWEED.

SECTION 1. **Piloselloidia.**

1. Pilosella, *L. Mouse-ear Hawk-weed.*—VAR. *Peleterianum*, Mer. (sp.); *pilosissimum*, Fries.

AURANTIACUM, *L.*

SECTION 2. **Pulmonarea.**

2. alpinum, *L.*—VAR. 1, *melanocephalum*, Tausch. (sp.); *alpinum*, Backh.—VAR. 2, *holosericeum*, Backh. (sp.).—VAR. 3, *eximium*, Backh (sp.): *tenellum*, L. C.—VAR. 4, *calenduliflorum*, Backh. (sp.).

3. **nigrescens,** *Willd.*—VAR. 1, *pulmonarium,* Sm. (sp.); *nigrescens,* Backh.—VAR. 2, *gracilentum,* Backh. (sp.).—VAR. 3, *globosum,* Backh. (sp.). VAR. 4; *chrysanthum,* Backh. (sp.); *rupestre,* Bab.; *microcephalum,* L.C. VAR. 5, *senescens,* Backh. (sp.). - VAR. 6, *lingulatum,* Backh. (sp.); *saxifragum,* Bab.; *divaricatum,* Don.

4. **Lawsoni,** *Sm.*

Sub-sp. ANGLICUM, *Fries* (sp.); *cerinthoides,* Backh.; *b. amplexicaule* and *c. acutifolium* L.C.

Sub-sp. IRICUM, *Fries* (sp.); *Lapeyrousii,* Bab.

5. **murorum,** *L.*

Sub-sp. MURORUM proper.—VAR. 1, *nitidum,* Backh. (sp.).—VAR. 2, *aggregatum,* Backh (sp.).

Sub-sp. PALLIDUM, *Backh.* (sp.).—VAR. 1, *cinerascens,* Jord. (sp.); *lasiophyllum,* Backh.—VAR. 2, *argenteum,* Fries. (sp.).

Sub-sp. CÆSIUM, *Backh.* (sp.).—VAR. 1, *Gibsoni,* Backh. (sp.); *hypochœroides,* Gibson.—VAR. 2. *flocculosum,* Backh. (sp.); *stelligerum,* Backh.—VAR. 3, *obtusifolium,* Backh. (sp.); *cæsium,* var. *obtusifolium,* Bab.

6. **sylvaticum,** *Sm.*

Sub-sp. VULGATUM, *Fries* (sp.); *maculatum,* Sm.—VAR. 1, *cinereum,* Backh.—VAR. 2, *rubescens,* Backh.; *nemorosum; montanum; macrocephalum.*

Sub-sp. GOTHICUM, *Fries* (sp.); *latifolium,* L. C.

Sub-sp. TRIDENTATUM, *Fries* (sp.).

SECTION 3. **Accipitrina.**

7. **prenanthoides,** *Villars; denticulatum,* Sm.; *strictum,* Fries.

8. **umbellatum,** *L.; canadense,* Fries.—VAR. *filifolium,* Backh.

9. **crocatum,** *Fries; juranum,* Fries (*Borreri,* Bosw.)

Sub-sp. INULOIDES, *Tausch.* (sp.).

Sub-sp. CORYMBOSUM, *Fries* (sp.). *rigidum,* Backh.

10. **boreale,** *Fries; sabaudum,* Sm.; *heterophyllum,* Bladon.

ORDER XLIII. **CAMPANULACEÆ** (including **LOBELIACEÆ**).

SUB-ORDER I. **Lobelieæ.**

1. **LOBELIA.**
1. **Dortmanna,** *L.*
2. **urens,** *L.*

SUB-ORDER II. **Campanuleæ.**

2. **CAMPANULA,** *L.*
SUB-GEN. 1. campanula proper.
1. **rotundifolia,** *L.* Hare-bell. VAR. 1, *rotundifolia* proper.—VAR. 2, *montana,* Bosw.
2. **Rapunculus,** *L.*
3. **patula,** *L.*
    PERSICIFOLIA, *L.*
4. **latifolia,** *L.*
5. **rapunculoides,** *L.*
6. **Trachelium,** *L.* Nettle-leaved Campanula.
7. **glomerata,** *L.*
SUB-GEN. 2. **Wahlenbergia,** *Schrad.* (gen.).
8. **hederacea,** *L.*
SUB-GEN. 3. **Specularia,** *Heist.* (gen.).
9. **hybrida,** *L.*

3. **PHYTEUMA,** *L.* RAMPION.
1. **orbiculare,** *L.*
2. **spicatum,** *L.*

4. **JASIONE,** *L.* SHEEP'S-BIT.
1. **montana,** *L.*

ORDER XLIV. **ERICACEÆ.**

SUB-ORDER 1. **Vaccinieæ.**

1. **VACCINIUM,** *L.*
1. **Myrtillus,** *L.* Whortle-berry, Bilberry.
2. **uliginosum,** *L.*
3. **Vitis-Idæa,** *L.* Cowberry.

2. **OXYCOCCOS,** *Tournef.* CRANBERRY.
1. **palustris,** *Pers.; Vaccinium Oxycoccus,* L.

SUB-ORDER II. **Ericeæ.**

TRIBE I. **ARBUTEÆ.**

3. **ARBUTUS,** *L.*
1. **Unedo,** *L.* Strawberry-tree.

g

4. ARCTOSTAPHYLOS, *Adanson.*
BEARBERRY.
1. alpina, *Spreng.*
2. Uva-ursi, *Spreng.*

TRIBE II. **ANDROMEDEÆ.**

5. ANDROMEDA, *L.*
1. polifolia, *L.*

TRIBE III. **ERICINEÆ**

6. ERICA, *L.* HEATH.
1. Tetralix, *L. Cross-leaved Heath.*
Sub-sp. TETRALIX proper.
Sub-sp. MACKAYI, *Hook.* (sp.) ; *Mackaiana,* Bab.
2. cinerea, *L.*
3. ciliaris, *L.*
4. vagans, *L. Cornish Heath.*
5. mediterranea, *L. m,* var. *hibernica,* Hook ; *hibernica,* Syme.

7. CALLUNA, *Salisbury.* LING.
1. vulgaris, *Salisb.*

TRIBE IV. **RHODOREÆ.**

8. DABEOCIA, *Don.* ST. DABEOC'S HEATH.
1. polifolia, *Don.*

9. PHYLLODOCE, *Salisbury.*
1. cærulea, *Bab. ; taxifolia,* Salisb. ; *Menziesia,* Swartz.

10. LOISELEURIA, *Desvaux.*
1. procumbens, *Desv. ; Azalea,* L.

TRIBE V. **PYROLEÆ.**

11. PYROLA, *Tournef.* WINTER-GREEN.
Sub-gen. 1. **Pyrola** proper.
1. minor, *Sw.*
2. media, *L.*
3. rotundifolia, *L.*—VAR. 1, *rotundifolia* proper. — VAR. 2, *arenaria,* Koch ; *maritima,* Kenyon.
4. secunda, *L.*
Sub-gen. 2. **Moneses,** *Salisb.* (gen.).
5. uniflora, *L. M. grandiflora,* Salisb.

SUB-ORDER III. **Monotropeæ.**

12. MONOTROPA, *L.* BIRD'S-NEST.
1. Hypopithys, *L. ; H. glabra,* DC. VAR. *glabra,* Bernh.—VAR. *hirsuta,* Roth.

———

ORDER XLV. **OLEINEÆ.**

1. LIGUSTRUM, *Tournef.* PRIVET.
1. vulgare, *L.*

2. FRAXINUS, *Tournef.* ASH.
1. excelsior, *L.* ; VAR. *heterophylla,* Vahl.

———

ORDER XLVI. **APOCYNEÆ.**
VINCA, L. PERIWINKLE.
1. minor, *L.*
MAJOR, *L.*

———

ORDER XLVII. **GENTIANEÆ.**
SUB-ORDER I. **Gentianeæ.**

1. CHLORA, YELLOW-WORT, YELLOW CENTAURY.
1. perfoliata, *L.*

2. MICROCALA, *Link et Hoffm.*
1. filiformis, *Link ; Exacum,* Sm.

3. CICENDIA, *Adans.*
1. pusilla, *Griseb.*

4. ERYTHRÆA, *Pers.* CENTAURY.
1. Centaurium *Pers.*
Sub-sp. CENTAURIUM proper ; *latifolia,* E. B. ; *Chironia Centaurium,* Sm.—VAR. *littoralis,* Fries (sp.).
Sub-sp. LATIFOLIA, Sm. (sp.).
Sub-sp. PULCHELLA, *Fries* (sp.).

5. GENTIANA, *L.* GENTIAN.
1. campestris, *L.*
2. amarella, *L.*
Sub-sp. AMARELLA proper.
Sub-sp. GERMANICA, *Willd.* (sp.).
3. Pneumonanthe, *L.*
4. verna, *L.*
5. nivalis, *L.*
SUB-ORDER II. **Menyantheæ.**

6. **MENYANTHES,** *Tournef.* BUCK-
or BOG-BEAN.
  1. trifoliata, *L.*

7. **LIMNANTHEMUM,** *Link.*
  1. nymphæoides, *Link. ; Villarsia
nymphæoides,* Vent.

ORDER XLVIII. **POLEMONIACEÆ.**
1. **POLEMONIUM,** *L.* JACOB'S LAD-
DER.
  1. cæruleum, *L. Greek Valerian.*

ORDER XLIX. **CONVOLVULACEÆ.**
SUB-ORDER I. Convolvuleæ proper.
1. **CONVOLVULUS,** *L.* BINDWEED.
Sub-gen. 1. Convolvulus proper.
  1. arvensis, *L. Small Bindweed.*
SUB-GEN. 2. Calystegia, *Br.* (gen.).
  2. sepium, *L.*
  3. Soldanella, *L.*
SUB-ORDER II. Cuscuteæ(*Presl,* order).

2. **CUSCUTA,** *Tournef.* DODDER.
  1. europæa, *L.*
  2. **Epithymum,** *Murr.* ; VAR. *Tri-
folii,* Bab. (ꞏp).
EPILINUM, *Weihe.*

ORDER L. **BORAGINEÆ.**
TRIBE I. **ECHIEÆ.**
1. **ECHIUM,** *Tournef.* BUGLOSS.
  1. vulgare, *L.; italicum,* Huds. *Viper's Bugloss.*
  2. plantagineum, *L.; violaceum,*
Brit. *Fl.*

TRIBE II. **ANCHUSEÆ.**
1*. *BORAGO, Tournef.* BORAGE.
OFFICINALIS, *L.*

2. **SYMPHYTUM,** *Tournef.* COM-
FREY.
  1. officinale, *L.; VAR.* 1, *officinale
proper.* — VAR. 2, *patens,* Sibthorp
(sp.).
  2. tuberosum, *L.*

3. **ANCHUSA,** *L.* ALKANET.
SECTION 1. **Lycopsis,** *L.* (gen.).
  1. arvensis, *Bieb. Bugloss.*
SECTION 2. **Anchusa** proper.
OFFICINALIS, *L.*
SECTION 3. **Caryolopha,** *Fisch et Traut.*
(gen.).
SEMPERVIRENS.

TRIBE III. **LITHOSPERMEÆ.**
4. **LITHOSPERMUM,** *Tournef.*
GROMWELL.
  1. officinale, *L.*
  2. arvense, *L.*
  3. purpureo-cæruleum, *L.*

5. **MERTENSIA,** *Roth.*
  1. maritima, *Don.*

6. **PULMONARIA,** *Tournef.* LUNG-
WORT.
  1. augustifolia, *L.*
OFFICINALIS, *L.*

7. **MYOSOTIS,** *L.* SCORPION-GRASS.
SECTION 1. (Perennial).
  1. palustris, *With. Forget-me-not.*
VAR. 1, *palustris* proper.—VAR. 2,
strigulosa, Reich. (sp.).
  2. lingulata, *Lehm. ; cæspitosa,*
Schultz.
  3. repens, *Don.*
SECTION 2. (Annual or biennial).
  4. sylvatica, *Hoffm.*
Sub-sp. ALPESTRIS, *Schmidt* (sp.).
  5. arvensis, *Hoffm.* ; VAR. 1, *arvensis* proper.—VAR. 2, *umbrosa, Bab.*
  6. collina, *Hoffm.* VAR. *Mitteni.*
  7. versicolor, *Reich.*

TRIBE IV. **CYNOGLOSSEÆ.**
7*. *ASPERUGO, Tournef.* MADWORT
(MADDERWORT.)
PROCUMBENS, *L.*

8. **CYNOGLOSSUM,** *Tournef.*
HOUND'S-TONGUE.
  1. officinale, *L.*
  2. montanum, *Lamk.* ; *sylvaticum,*
Haenke.

Order LI. SOLANEÆ.

1. HYOSCYAMUS, *Tournef.* Hen-
bane.

1. niger, *L.*

2. SOLANUM, *Tournef.* Nightshade.

1. Dulcamara, *L.* *Bittersweet* or
Woody Nightshade. — Var. 1, *marinum*, Syme.

2. nigrum, *L.*—Var. 1, *nigrum* proper.—Var. 2, *miniatum*, Bernh. (sp.).

3. ATROPA, *L.* Dwale.

1. Belladonna, *L.* *Deadly Nightshade.*

——

Order LII. PLANTAGINEÆ.

1. PLANTAGO, *L.* Plantain, Rib-
grass.

1. major, *L.*

2. media, *L.*

3. lanceolata, *L.;* Var. 1, *lanceolata*
proper.—Var. 2, *Timbali*, Jord. (sp.).

4. maritima, *L.*

5. Coronopus, *L.*

2. LITTORELLA, *L.* Shore-weed.

1. lacustris, *L.*

——

Order LIII. SCROPHULARINEÆ.
Sub-order I. Antirrhinideæ.

Tribe I. VERBASCEÆ.

1. VERBASCUM, *L.* Mullein.

1. Thapsus, *L.*

2. Lychnitis, *L.*

3. pulverulentum, *Vill.*

4. nigrum, *L.*

5. Blattaria, *L.*

Sub-sp. Blattaria proper.

Sub-sp. virgatum, *With.* (sp.).

Tribe II. ANTIRRHINEÆ.

2. LINARIA, *Tournef.* Toad-flax.

Section 1. Cymbalaria, *Chav.*
Cymbalaria, *Mill.* *Ivy-leaved Toad-
flax.*

Section 2. Elatinoides, *Chav.*

1. spuria, *Mill.* *Male Fluellen.*

2. Elatine, *Mill.*

Section 3. Linariastrum, *Chav.*

3. vulgaris, *Mill.* Var. 1, *vulgaris*
proper.—Var. 2, *latifolia*, Bab. ; *speciosa*, Bromf.

4. Pelisseriana, *Mill.*

5. repens, *Ait.*

Section 4. Chænorrhinum, *DC.*

6. minor, *Desf.*

3. ANTIRRHINUM, *Tournef.* Snap-
dragon

1. Orontium, *L.*

majus, *L.*

Tribe III. CHELONEÆ.

4. SCROPHULARIA, *Tournef.* Fig-
wort.

Section 1. Scrophularia proper.

1. nodosa, *L.*

Sub-sp. alata, *Gilib.* (sp.) ; *umbrosa*,
Dumort ; *Ehrharti*, Stev.

2. aquatica, *L. ; Balbisii*, Hornem.

3. Scorodonia, *L.*

Section 2. Ceramanthe, *Reich.*

vernalis, *L.*

Tribe IV. GRATIOLEÆ.

4*. MIMULUS, *L.*

luteus. *L.*

Sub-order II. Rhinanthideæ.

Tribe V. SIBTHORPIEÆ.

5. LIMOSELLA, *L.* Mudwort.

1. aquatica, *L.*

6. SIBTHORPIA, *L.*

1. europæa, *L.*

Tribe VI. DIGITALEÆ.

7. DIGITALIS, *Tourn.* Foxglove.

1. purpurea, *L.*

Tribe VII. VERONICEÆ.

8. VERONICA, *Tourn.* Speedwell.

Section 1. Omphalospora.

1. agrestis, *L.*

Sub-sp. agrestis proper.

Sub-sp. POLITA, *Fries* (sp.).

2. Buxbaumii, *Ten.*

3. hederæfolia, *L.*

SECTION 2. **Veronicastrum.**

4. triphyllos, *L.*

5. arvensis, *L.*

6. verna, *L.*

7. serpyllifolia, *L.* VAR. *humi-fusa,* Dicks (sp.).

8. alpina, *L.*

9. saxatilis, *L.*

SECTION 3. **Chamædrys.**

9. officinalis, *L.* VAR. *hirsuta,* Hopkirk (sp.).

10. Chamædrys, *L.*

11. montana, *L.*

12. scutellata, *L.*

SECTION 4. **Beccabunga.**

13. Beccabunga, *L. Brooklime.*

14. Anagallis, *L.*

SECTION 5. **Pseudo-Lysimachia.**

15. spicata, *L.* VAR. 1, *spicata* proper. —VAR. 2, *hybrida, L.* (sp.).

TRIBE VIII. **EUPHRASIEÆ.**

9. **BARTSIA,** *L.*

SUB-GEN. 1. **Bartsia proper.**

1. alpina, *L.*

SUB-GEN. 2. **Eufragia,** *Griseb.* (gen.).

2. viscosa, *L.*

SUB-GEN. 3. **Odontites,** *Persoon* (gen.).

3. Odontites, *Huds.; Euphrasia Odontites,* L. VAR. 1, *verna,* Reich. (sp.). —VAR. 2, *serotina,* Reich. (sp.).—VAR. 3, *rotundata,* Ball (sp.).—VAR. 4, *divergens,* Jord. (sp.).

10. **EUPHRASIA,** *Tournef.* EYE-BRIGHT.

1. officinalis, *L.* VAR. 1, *officinalis* proper.—VAR. 2, *gracilis,* Fries (sp.). —VAR. 3, *maritima.*

11. **RHINANTHUS,** *L.* YELLOW-RATTLE.

1. Crista-galli, *L.*

Sub-sp. MAJOR, *Ehr.* (sp.). VAR. *apterus,* Fries.

Sub-sp. MINOR, *Ehr.* (sp.).

12. **PEDICULARIS,** *Tournef.* LOUSE-WORT.

1. palustris, *L.*

2. sylvatica, *L.*

13. **MELAMPYRUM,** *Tournef.* COW-WHEAT.

1. pratense, *L.* VAR. 1, *pratense* proper.—VAR. 2, *latifolium.*—VAR. 3, *montanum,* Johnst. (sp.).—VAR. 4, *ericetorum,* D. Oliv.

2. sylvaticum, *L.* VAR. *pallidiflora.*

3. arvense, *L.*

4. cristatum, *L.*

ORDER LIV. **OROBANCHEÆ.**

1. **OROBANCHE** *L.* BROOM-ROPE.

SUB-GEN. 1. **Orobanche proper.**

1. major, *L.; Rapum,* Thuill.

2. elatior, *Sutt.*

3. caryophyllacea, *Sm.; Galii,* Duby.

4. rubra, *Sm.; epithymum,* DC.

5. minor, *Sutt.*

Sub-sp. MINOR proper.

Sub-sp. AMETHYSTEA, *Thuill* (sp.); *Eryngii,* Duby.

Sub-sp. PICRIDIS, *F. Schultz* (sp.).

Sub-sp. HEDERÆ, *Duby* (sp.); *barbata,* E. B Suppl.

Sub-gen. 2. **Phelipæa,** *Tournef.* (gen.).

6. cærulea, *Vill.*

2. **LATHRÆA,** *L.* TOOTHWORT.

1. squamaria, *L.*

ORDER LV. **LABIATÆ.**

TRIBE I. **SATUREINEÆ.**

1. **MENTHA,** *L.* MINT.

1. sylvestris, *L.* VAR. 1, *sylvestris* proper. VAR. 2, *nemorosa,* Willd. (sp.).—VAR. 3, *mollissima,* Bork. (sp.). —VAR. 4, *alopecuroides,* Hull. (sp.).

2. rotundifolia, *L.*

VIRIDIS, *L. Spear-mint.*

h

3. **piperita,** *Huds. Peppermint.* VAR.
1, *officinalis,* Hull (sp.).—VAR. 2, *vulgaris,* Sole.

4. **aquatica,** *L.*

Sub-sp. PUBESCENS, *Willd.* (sp.).—VAR.
1, *palustris* Sole (sp.).—VAR. 2, *hircina,*
Hull (sp.).

Sub-sp. HIRSUTA, *L.* (sp.).—VAR. 1,
hirsuta proper.—VAR. 2, *subglabra,* Baker.—VAR. 3, *citrata,* Ehr. (sp.). *Bergamot Mint.*

5. **sativa,** *L.*

Sub-sp. SATIVA proper.—VAR. 1, (separate whorls and foliaceous bracts).—VAR.
2, *paludosa,* Sole(sp.).—VAR.3, *subglabra,*
Baker.

Sub-sp. RUBRA, *Sm.* (sp.).

Sub-sp. GRACILIS, *Sm.* (sp.).—VAR. 1,
gracilis proper. — VAR. 2, *cardiaca,*
Baker.

Sub-sp. PRATENSIS, *Sole* (sp.).

Sub-sp. GENTILIS, L. (sp.).—VAR. 1,
gentilis proper.—VAR. 2, *Wirtgeniana*
F. Schultz. (sp.).—VAR. 3, *Pauliana,*
F. Schultz. (sp.).

6. **arvensis,** *L.*—VAR. 1, *arvensis*
proper.—VAR 2, *nummularia,* Schreb.
(sp.).—VAR. 3, *agrestis,* Sole (sp.).—
VAR. 4, *præcox,* Sole (sp.).—VAR. 5,
Allionii, Boreau (sp.).—VAR. 6, *parietariæfolia,* Beck (sp.).

7. **Pulegium,** *L. Penny-royal.* VAR.
erecta, L. C.

2. **LYCOPUS,** *Tournef.* GIPSY-WORT.
1. **europæus,** *L.*

3. **ORIGANUM,** *Tournef.* MARJORAM.
1. **vulgare,** *L.*

4. **THYMUS,** *L.* THYME.
1. **Serpyllum,** *L.*

Sub-sp. SERPYLLUM proper.

Sub-sp. CHAMÆDRYS, *Fries* (sp.).

5. **CALAMINTHA,** *Moench.*

SUB-GEN. 1. **Calamintha** proper.
1. **officinalis,** *Moench. Calamint.*

Sub-sp. MENTHIFOLIA, *Host* (sp.).—VAR.
Briggsii, Syme.

Sub-sp. NEPETA, *Clairv.* (sp.).

Sub-sp. SYLVATICA, *Bromf.*

SUB-GEN 2. **Clinopodium,** *L.* (gen.).
2. **Clinopodium,** *Benth. ; Clinopodium vulgare,* L. *Wild Basil.*

SUB-GEN. 3. **Acinos,** *Moench* (gen.).
3. **Acinos,** *Clairv. ; Thymus Acinos,*
L. *Basil Thyme.*

5* *MELISSA, Tournef.* BALM.
OFFICINALIS, *L.*

TRIBE II. **MONARDEÆ.**

6. **SALVIA,** *L.* SAGE.
1. **Verbenaca,** *L. Clary.* — VAR.
clandestina. Lim. (sp.).
2. **pratensis,** *L.*

TRIBE III. **NEPETEÆ.**

7. **NEPETA,** *L.*
SUB-GEN. 1. **nepeta** proper.
1. **Cataria,** *L. Cat-Mint.*

SUB-GEN. 2. **Glechoma,** *L.* (gen.).
2. **Glechoma,** *Benth. ; Glechoma hederacea,* L. *Ground Ivy.* — VAR. 1,
Glechoma proper.—VAR. 2, *parviflora,*
Benth. — VAR 3, *hirsuta,* Waldst. and
Kit. (sp.),

TRIBE IV. **STACHYDEÆ.**

8. **PRUNELLA,** *L.* SELF-HEAL.
I. **vulgaris,** *L. ; Brunella,* Tourn.

9. **SCUTELLARIA,** *L. Skull-cap.*
1. **galericulata,** *L.*
2. **minor,** *L.*

10. **MELITTIS,** *L.* BASTARD-BALM,
1. **Melissophyllum,** *L. ; grandiflora,*
Sm.

11. **MARRUBIUM,** *L.* WHITE HOREHOUND.
1. **vulgare,** *L.*

12. **STACHYS,** *L.* WOUNDWORT.
SUB-GEN. 1. **Stachys** proper.
1. **sylvatica,** *L.*
2. **palustris,** *L.*
ambigua, Sm. (sp.), HYB. bet. 1 and 2.
3. **germanica,** *L.*
4. **arvensis,** *L.*

SUB-GEN. 2. **Betonica,** *L.* (gen.).

5. **Betonica,** *Benth. ; Betonica offici-nalis,* L. *Wood Betony.*

13. **GALEOPSIS,** *L.* HEMP-NETTLE.

1. **Ladanum,** *L.*

Sub-sp. LADANUM proper (*L. Herb.*) ;' *in-termedia,* Villars.

Sub-sp. ANGUSTIFOLIA, Ehrh. (sp.). — VAR. 1, (almost glabrous).—VAR. 2, *canescens,* Schultz (sp.).

2. **dubia,** *Leers* ; *villosa,* Huds; *och-roleuca,* Lamk.

3. **Tetrahit,** *L.*

Sub-sp. TETRAHIT proper.—VAR. 1, (cor. twice as long as cal. teeth).—VAR. 2, *bifida,* Bœnn. (sp.).

Sub-sp. SPECIOSA, *Miller* (sp.) ; *versicolor,* Curt.

13*. *LEONURUS,* L. MOTHERWORT.
CARDIACA, *L.*

14. **LAMIUM,** *L.* DEAD-NETTLE.

SECTION 1. **Lamiopsis.**

1. **purpureum,** *L.*

Sub-sp. PURPUREUM proper. — VAR. 1, (crenatures shallow). — VAR. 2, *deci-piens,* Sonder.

Sub-sp. HYBRIDUM, *Vill* (sp.) ; *dissec-tum,* With. ; *incisum,* Willd.

2. **intermedium,** *Fries*

3. **amplexicaule,** *L. Henbit Dead-nettle.*

SECTION 2. **Lamiotypus.**

4. **album,** *L.*

MACULATUM, *L.*

SECTION 3. **Galeobdolon,** *Huds.* (gen.).

5. **Galeobdolon,** *Crantz ; Galeobdo-lon luteum,* Huds. *Yellow Archangel.*

15. **BALLOTA,** *L.* BLACK HORE-HOUND.

1. **nigra.** *L.* VAR 1, *fœtida,* Lamk, (sp.) —VAR. 2, *ruderalis,* Swartz (sp).

TRIBE V. **AJUGOIDEÆ.**

16. **TEUCRIUM,** *L.* GERMANDER.

SECTION I. **Scorodonia.** *Moench* (gen.).

1 Scorodonia, *L. Wood Sage.*

SECTION 2 Scordium, *Benth.*

2. **Scordium,** *L. Water Germander.*

3. **Botrys,** *L.*

SECTION 3. **Chamædrys,** *Benth.*
CHAMÆDRYS, *L.*

17. **AJUGA,** *L.* BUGLE.

1. **reptans,** *L*

2. **pyramidalis,** *L.* ; *genevensis,* var. Benth.

3. **Chamæpitys,** *Schreber. Ground Pine.*

ORDER LVI **VERBENACEÆ.**

1. **VERBENA,** *L.* VERVAIN.

I. officinalis, *L.*

ORDER LVII. **LENTIBULARINEÆ**

1. **Pinguicula,** *Tournef.* BUTTER-WORT.

1. **vulgaris,** *L.*

Sub-sp, VULGARIS proper.

Sub-sp.' GRANDIFLORA, *Lamk.* (-p.).

2. **lusitanica,** *L.*

3. **alpina,** *L.*

2. **UTRICULARIA,** *L.*

1. **vulgaris,** *L.*

2. **neglecta,** *Lehm.*

3. **intermedia,** *Hayne.*

4. **minor,** *L.*

ORDER LVIII. **PRIMULACEÆ.**

TRIBE I. **PRIMULEÆ.**

1. **PRIMULA,** *L.*

1. **vulgaris,** *Huds. Primrose.* Vars. *acaulis* Jacq., *caulescens* and *intermedia,* (Hybs ?, Baker).

2. **elatior,** *Jacq. Oxlip.*

3. **veris,** *L.; officinalis,* Jacq. *Cow-slip, Paigle.*

Hybs. bet. 1 and 3. *elatior* of older Bots, probably *P. veris, b elatior,* L., and *P. variabilis,* Goupil.

4. **farinosa,** *Bird's-eye Primrose.*

5. **scotica,** *Hook.* ; VAR. *acaulis.*

1*. *CYCLAMEN, Tournef.* SOWBREAD.
HEDERÆFOLIUM, *Willd.; europæum,* Sm.

2. LYSIMACHIA, *L.* Loose-strife.

Section 1. Lysimastrum, *Duby.*
1. vulgaris, *L.*
2. nemorum, *L.* *Yellow Pimpernel.*
3. Nummularia, *L.* *Creeping Jenny, Money-wort.*

Section 2. Naumburgia, *Moench* (gen.).
4. thyrsiflora, *L.*

3. TRIENTALIS, *L.* Chickweed-Winter-green.
1. europæa, *L.*

4. GLAUX, *Tournef.* Sea Milkwort.
1. maritima, *L.*

Tribe II. ANAGALLIDEÆ.

5. CENTUNCULUS, *L.* Bastard Pimpernel.
1. minimus, *L.*

6. ANAGALLIS, *Tournef.* Pimpernel.
1. arvensis, *L.* *Poor Man's weatherglass.*—Var. 1, *phœnicea,* Lamk. (sp.). —Var. 2, *cærulea,* Lamk. (sp.).
2. tenella, *L.* *Bog Pimpernel.*

Tribe III. HOTTONIEÆ.

7. HOTTONIA, *L.* Water-violet.
1. palustris, *L.*

Tribe IV. SAMOLEÆ.

8. SAMOLUS, *Tournef.* Brookweed.
1. Valerandi, *L.*

———

Order LIX. PLUMBAGINEÆ.

1. ARMERIA, *Willd.* Thrift. Sea-pink.
1. vulgaris, *Willd.* ; *maritima* and *alpina,* Willd. ; *pubescens,* Link. ; *pubigera, b scotica,* Boissier ; *duriuscula,* Bab. ; *Statice Armeria,* L.—Var. *planifolia,* Bosw.
2. plantaginea, *Willd.*

2. STATICE, *L.* Sea-lavender.
1. Limonium, *L.*
Sub-sp. Limonium proper.—Var. 1, *Behen,* Drejer (sp.).—Var. 2, *serotina,* Gren. and Godr. (sp.).—Var. 3, *pyramidalis.*
Sub-sp. bahusiensis, *Fries* (sp.) ; *rariflora.* Drejer.
2. auriculæfolia, *Vahl.* ; *spathulata,* Hook. ; *binervosa,* G. E. Sm.—Var. 1, *occidentalis,* Lloyd (sp.).—Var. 2, *intermedia,* Syme.—Var. 3, *Dodartii,* Gir.
3. bellidifolia, *Gouan* ; *caspia,* Willd. ? *reticulata,* L.

———

Division III. APETALÆ.

Order LX. POLYGONEÆ.

1. POLYGONUM, *L.*
Section 1. Bistorta, *Tournef.*
1. Bistorta, *L.* *Bistort, Snake-root.*
2. viviparum, *L.*

Section 2. Persicaria, *Meissn.*
3. amphibium, *L.*—Var. 1, *aquaticum.*—Var. 2, *terrestre.*
4. lapathifolium, *L.*
Sub-sp. 1. lapathifolium proper. — Var. 1, *rubrum* Gray.
Sub-sp. 2. maculatum, *Dyer* and *Trimen ; nodosum,* Reich. ; *laxum* E. B. Sup. —Var. 1, *gracile.*—Var. 2, *densum.*
5. Persicaria, *L.* *Persicaria.*
Sub-sp. 1. Persicaria proper.
Sub-sp. 2, nodosum, *Pers.* (sp.).
6. mite, *Schrank.*
7. Hydropiper, *L.* *Water-pepper.*
8. minus, *Huds.*

Section 3. Avicularia, *Meissn.*
9. Aviculare, *L.* *Knotgrass.*—Var. 1, *vulgatum.*—Var. 2. *littorale,* Link. (sp.).—Vars. *agrestinum,* Jord. ; *arenastrum* Boreau ; *microspermum,* Jord ; *rurivagum* Jord.
10. maritimum, *L.*
Sub-sp. maritimum proper.
Sub-sp. Raii, Bab. (sp.) ; *Roberti,* Loisel.

SECTION 4. **Tiniaria,** *Meissn.*

11. **Convolvulus,** *L.* *Black Bindweed.*—VAR. 1, *Convolvulus* proper.—VAR. 2, *pseudo-dumetorum,* Wats.

12. **dumetorum,** *L.*

2. **RUMEX,** *L.* DOCK.

SECTION 1. **Lapathum,** *Meissn.*

1. **obtusifolius,** *L.*—VAR. 1, *Friesii,* Gren. and Gort. (sp.).—VAR. 2, *sylvestris,* Wallr. (sp.).

2. **acutus,** *L.; pratensis,* Mert. and Koch.; *conspersus,* Hartm.

3. **pulcher,** *L.* *Fiddle Dock.*

4. **maritimus,** *L.* *Golden Dock.*

Sub-sp. MARITIMUS proper.
Sub-sp. PALUSTRIS, *Sm.* (sp.); *Steini,* Becker.

5. **crispus,** *L.* VAR. *trigranulatus,* Bosw.—VAR. *subcordatus,* Warren.

6. **sanguineus,** *L.* VAR. *sanguineus* proper.—VAR. *viridis.* Sibth. (sp.); *nemorosus,* Schrader.

7. **conglomeratus,** *Murray; acutus,* Sm. and L. Herb. VAR. *rupestris,* Le Gall (sp.).

8. **Hydrolapathum,** *Huds.* VAR. *latifolius,* Borrer.

9. **aquaticus,** *L.; domesticus,* Hartman ; *longifolius,* DC.
ALPINUS, *L.* *Monk's Rhubarb.*

SECTION 2. **Acetosa,** *Tournef.*

10. **Acetosa,** *L.* *Sorrel.*

11. **Acetosella,** *L.* *Sheep's Sorrel.*

3. **OXYRIA,** *Hill.* MOUNTAIN SORREL.

1. **reniformis,** *Hook.*

ORDER LXI. **CHENOPODIACEÆ.**

TRIBE I. (Embryo annular.)

1. **BETA,** *L.* BEET.

1. **maritima,** *L.*

2. **CHENOPODIUM,** *Tournef.* GOOSEFOOT.

SECTION 1. (*Fls.* 5-merous. *Seed* horizontal. *Annual.*)

1. **Vulvaria,** *L.; olidum,* Curtis.

2. **polyspermum,** *L.;* VAR. 1. *acutifolium,* Sm. (sp.). — VAR. 2, *cymo-*

sum, Moq. Tand. ; Var. *obtusifolium,* Bosw.

3. **album,** *L.* ; VAR. 1, *candicans,* Lamk. (sp.).—VAR. 2, *viride,* L. (sp.).
— VAR. 3, *paganum,* Reich. (sp.). VAR. *virens.* Lond. Cat.

4. **ficifolium,** *Sm.*

5. **urbicum,** *L.*—VAR. 1, *deltoideum,* Lamk. (sp.).—VAR. 2, *intermedium,* Mert. and Koch (sp.).

6. **hybridum,** *L.*

7. **murale,** *L.*

SECTION 2. (*lat Fls.* 2-4-merous. *Seed* vertical. *Annual.*)

8. **rubrum,** *L.*

Sub-sp. RUBRUM proper. — VAR. 1, (Leaves deltoid, serrate, &c.). ; VAR. 2, *pseudo-botryoides,* Wats. (leaves rhomboid, almost entire, &c.)
Sub-sp. BOTRYOIDES, *Sm.* (sp.).

9. **glaucum,** *L.*

SECTION 3. (*Perennial.*)

10. **Bonus-Henricus,** *L.* *All-good.*

3. **ATRIPLEX,** *Tournef.* ORACHE.

SUB-GEN. 1. Atriplex proper.

1. **patula,** *L.*

Sub-sp. PATULA proper.—VAR. 1, *angustifolia,* Sm. (sp.).—VAR. 2, *erecta,* Huds. (sp.); VAR. *serrata,* Eng. Bot. Ed. 3.
Sub-sp. HASTATA, *L.* (sp.). VAR. 1, *hastata,* Huds. (sp.); *patula,* Sm. ; *Smithii,* Bosw.—VAR. 2, *triangularis,* Willd. (sp.) ; *prostrata,* Bab. ; *deltoidea,* Bab.
Sub-sp. BABINGTONII, *Woods* (sp.).

2. **littoralis,** *L.;* VAR. 1, *littoralis* proper.—VAR. 2, *marina,* L. (sp.) ; *serrata,* Huds.

3. **laciniata,** *L.* (Koch); *arenaria,* Woods ; *rosea,* L. ? *farinosa,* Dumort.
SUB-GEN. 2. **Obione,** *Gærtn.* (gen.) ; *Halimus,* Wallroth.

4. **portulacoides,** *L.*

5. **pedunculata,** *L.*

TRIBE II. (Embryo conduplicate.)

4. **SALICORNIA,** *L.*

1. **herbacea,** *L.* VAR. 1, *herbacea* proper.—VAR. 2, *procumbens,* Sm. (sp.).—VAR. 3, *ramosissima,* Woods. —VAR. 4, *pusilla,* Woods.

2. radicans, *Sm.*

TRIBE III. (Embryo spiral.)

5. SUÆDA, *Forsk.* SEABLITE.
 1. maritima, *Dumort ; Schoberia,*
     C. A. Meyer; *Chenopodina,* Moq.
     Tand.
 2. fruticosa, *Forsk. ; Schoberia,* C. A.
     Meyer.

6. SALSOLA, *L.* SALTWORT.
 1. Kali, *L.*

———

ORDER LXII. THYMELEÆ.

1. DAPHNE, *L.*
 1. Laureola, *L. Spurge Laurel.*
 2. Mezereum, *L. Mezereon.*

———

ORDER LXIII. ELEAGNEÆ.

1. HIPPOPHAE, *L.* SEA BUCK-
     THORN.
 1. rhamnoides, *L.*

———

ORDER LXIV. LORANTHACEÆ.

1. VISCUM, *L.* MISTLETOE.
 1. album, *L.*

———

ORDER LXV. SANTALACEÆ.

1. THESIUM, *L.* BASTARD TOAD-
     FLAX.
 1. linophyllum, *L. ; humifusum,*
     DC. ; *divaricatum,* var. *anglicum,* A.
     DC.

———

ORDER LXVI. ARISTOLOCHIEÆ.

1. ASARUM, *Tournef.* ASARABACCA.
 1. europæum, *L.*

1*. ARISTOLOCHIA, *Tournef.* BIRTH-
     WORT.
     CLEMATITIS, *L.*

———

ORDER LXVII. EUPHORBIACEÆ.

SUB-ORDER I. Euphorbieæ.

TRIBE I. EUPHORBIÆ.

1. EUPHORBIA, *L.* SPURGE.
 SECTION 1. (exstipulate).
 1. Helioscopia, *L. Sun Spurge.*
 2. platyphyllos, *L.*
 Sub-sp. PLATYPHYLLOS proper.
 Sub-sp. STRICTA, *L.* (sp.).
 3. hiberna, *L.*
     CORALLOIDES, *L.*
 4. amygdaloides, *L.*
 5. Peplus, *L.*
 6. exigua, *L.*
 7. portlandica, *L.*
 8. Paralias, *L.*
     ESULA, *L.*
     CYPARISSIAS, *L.*
     LATHYRIS, *L.*
 SECTION 2. (stipulate).
 9. Peplis, *L.*

TRIBE II. ACALYPHEÆ.

2. MERCURIALIS, *Tournef.* DOG'S
     MERCURY.
 1. perennis, *L.*
 2. annua, *L.*

SUB-ORDER II. Buxeæ.

3. BUXUS, *Tournef.* BOX.
 1. sempervirens, *L.*

———

ORDER LXVII. CERATOPHYLLEÆ

1. CERATOPHYLLUM, *L.* HORN-
     WORT.
 1. demersum, *L.*
 Sub-sp. DEMERSUM proper; *apiculatum,*
     Cham.
 Sub-sp. SUBMERSUM, *L.* (sp.).

ORDER LXIX. URTICEÆ.

1. URTICA, *Tournef.* NETTLE.
 1. urens, *L.*
 2. dioica, *L.*

3. **pilulifera**, *L.* VAR 1, *pilulifera* proper.—VAR. 2, *Dodartii*, L. (sp.).

2. **PARIETARIA**, *Tournef.* PELLITORY.

1. **officinalis**, *L.*; *diffusa*, Koch.

ORDER LXX. CANNABINEÆ.

1. **HUMULUS**, *L.* HOP.

1. **Lupulus**, *L.*

ORDER LXXI. ULMACEÆ

1. **ULMUS**, *L.* ELM.

1. **montana**, *Sm.*; *campestris*, L. herb₁; *stricta* and *glabra*, Lindl.; *major*, Sm. *Scotch*, *Wych*, or *Mountain Elm.*

CAMPESTRIS, *Sm.* ; *suberosa*, Ebr. ; *minor* and *glabra*, Miller ; *carpinifolia*, Lindl. *Common Elm.*

ORDER LXXII. SALICINEÆ.

1. **POPULUS**, *Tournef.* POPLAR.'

SECTION I. Leuce.

1. **alba**, *L.*

Sub-sp. ALBA proper. *White Poplar*, *Abele.*

Sub-sp. CANESCENS, *Sm.*; (sp.). *Grey Poplar.*

2. **tremula**, *L.* *Aspen.* VARS. *villosa* and *glabra.*

SECTION II. Aigeiros.

NIGRA, *L.* *Black Poplar.*

2. **SALIX**, *Tournef.* WILLOW.

SECTION 1. (filaments hairy, free).

1. **triandra**, *L.* *Almond-leaved* or *French Willow.* VAR. 1, *triandra* proper.—VAR. 2, *Hoffmanniana*, Sm. (sp.).—VAR. 3, *amygdalina*, L. (sp.); *contorta*, Crowe. LANCEOLATA, *Sm.*

2. **pentandra**, *L.* *Bay-leaved Willow.* *cuspidata*, Schultz (*Meyriana*, Willd.) HYB? bet. 2 and 3.

3. **fragilis**, *L.* *Crack Willow*, *Withy.* —VAR. 1, *fragilis* proper.—VAR. 2, *decipiens*, Hoffm. (sp.).

Russelliana, Sm. (*Bedford Willow*). HYB? bet. 3 and 4.

4. **alba**, *L.* *White Willow.* VAR. 1, *alba* proper.—VAR. 2, *cœrulea.* Sm. (sp.).—VAR. 3, *vitellina*, L. (sp.). *Golden Willow.*

SECTION 2. (filaments glabrous, free.)

5. **Caprea**, *L.* *Common Sallow, Goat Willow.* Sub-sp. CAPREA proper.—VAR. 1, (Leaves stipulate).—VAR. 2. (L. exstipulate). Sub-sp. CINEREA, *L.* (sp.).—VAR. 1, *cinerea* proper.—VAR. 2. *aquatica*, Sm. (sp.).—VAR. 3, *oleifolia*, Sm. (sp.).

6. **aurita**, *L.*

7. **repens**, *L.*; *fœtida*, Sm.—VAR. 1, *repens* proper.—VAR. 2, *fusca*, L. (sp.). VAR. 3, *prostrata*, Sm. (sp.).—VAR. 4, *ascendens*, Sm. (sp.).—VAR. 5, *parvifolia*, Sm. (sp.).—VAR. 6, *argentea*, Sm. (sp.).—VAR. 7, *rosmarinifolia*, L. (sp.).

HYB. *ambigua*, Ehr. bet. 6 and 7; *incubacea*, L. ; *repens* var. *incubacea*, Bosw. ; *spathulata*, Willd.

8. **nigricans**, *Sm.* VARS. true *nigricans*; *cotinifolia*, Sm. and *Forsteriana* Sm.; *ruprestris*, Sm. ; *Andersoniana*, Sm. ; *damascena*, Forbes, and *petrœa*, G. Anders. ; *hirta*, Sm. ; *floribunda*, Forbes (*tenuifolia*, Sm. ; *bicolor*, Hook) is a doubtful plant.

9. **phylicifolia**, *L.* *Tea-leaved Willow*, Brit. Forms (sp.) 1st, *Davalliana*, Sm. ; *Weigeliana*, Willd. (*Wulfeniana*, Sm.) ; *nitens*, G. Anders. ; *Croweana*, Sm. ; *Dicksoniana*, Sm. (*myrtilloides*, Sm. not L.); and *tenuior*, Borr.—2nd, *laxiflora*, G. Anders. ; *propinqua* Borr. ; *tetrapla*, Walker ; *Borreriana*, Sm. ; *phillyreæfolia*, Borr.; *tenuifolia*, Borr.—3rd, *radicans*, Sm. (*phylicifolia*, Sm.). Hyb. *laurina*, Sm. (*laxiflora*, Borr. ; *bicolor*, Sm.) bet. 5 and 9 acc. to Anders.

DAPHNOIDES, *Vill.* ; *acutifolia*, Willd. *Violet Willow.*

10. **viminalis**, *L.* *Osier.*

STIPULARIS, *Sm.* Hyb (?) bet. 5 or, CINEREA and 10.

SMITHIANA, *Willd.* (*mollissima*, Sm.) Hyb ? bet. 5 and 10.—VARS. *acuminata*, Sm. ; *ferruginea*, G. Anders. (*holosericea*, Borr. ; *rugosa*, Leefe.)

11. **lanata**, *L.*

12. **Lapponum,** *L.*—VARS. *arenaria,* Sm. (sp. and L. in pt.) ; *Stuartiana,* Sm. (sp.) ; *glauca,* Sm. (sp.).

13. **Myrsinites,** *L.; retusa,* Dicks.— VAR. 1, *procumbens,* Forbes (sp.), (*lævis,* Hook.).—VAR. 2, *arbutifolia,* Bosw.—VAR. 3, *serrata,* Bosw.

14. **Sadleri,** *Bosw.*

15. **Arbuscula,** *L.* Brit. forms *carinata,* Sm. ; — *prunifolia,* Sm. ;— *venulosa,* Sm. ;—*vacciniifolia,* Walker (*livida,* Sm.).

16. **herbacea,** *L.*

17. **reticulata,** *L.*

SECTION 3. **Synandræ,** *Anders.*

18. **purpurea,** *L. Purple Osier.*— VARS. 1, *Woolgariana,* Borr. (sp.), and *ramulosa,* Borr. 2, *Lambertiana,* Sm. (sp.) Hyb. *rubra,* Huds. bet. 10 and 18 ; 1, *rubra* proper ; 2, *Forbyana,* Sm. (sp.) ; 3, *Helix,* L. (sp.) *Rose Willow.*—*Doniana,* Sm. (sp.) intermed. bet. 7 and 18.—*Pontederana,* Schleich, hyb ? bet. 5 (CINEREA) and 18.

———

ORDER LXXIII. **CUPULIFERÆ.**
SUB-ORDER I. **Quercineæ.**

1. **QUERCUS,** *Tournef.* OAK.
    1. **Robur,** *L.* VAR. 1, *Sessiliflora,* Sm. (sp.).—VAR. 2, *pedunculata,* Ehr. (sp.).

2. **FAGUS,** *Tournef.* BEECH.
    1. **sylvatica,** *L.*
SUB-ORDER II. **Coryleæ.**

3. **CORYLUS,** *Tournef.* HAZEL.
    1. **Avellana,** *L.*

4. **CARPINUS,** *Tournef.* HORN-BEAM.
    1. **Betulus,** *L.*

———

ORDER LXXIV. **BETULACEÆ.**

1. **BETULA,** *Tournef.* BIRCH.
    1. **alba,** *L.*
Sub-sp. VERRUCOSA, *Ehr.* (sp.).
Sub-sp. GLUTINOSA, *Fries* (sp.).—VAR. 1, *denudata.*—VAR. 2, *pubescens.*
    2. **nana,** *L.*

2. **ALNUS,** *Tournef.* ALDER.
    1. **glutinosa,** *L.*

———

ORDER LXXV. **MYRICACEÆ.**

1. **MYRICA,** *L.* SWEET-GALE, BOG-MYRTLE.
    1. **Gale,** *L.*

———

ORDER LXXVI. **CONIFERÆ.**

TRIBE I. **ABIETINEÆ.**

1. **PINUS,** *L.* PINE, FIR.
    1. **sylvestris,** *L. Scotch Fir.*

TRIBE II. **CUPRESSINEÆ.**

2. **JUNIPERUS,** *L.* JUNIPER.
    1. **communis,** *L.* VAR. 1, *communis* proper.—VAR. 2, *nana,* Willd. (sp.).

TRIBE III. **TAXINEÆ.**

3. **TAXUS,** *Tournef.* YEW.
    1. **baccata,** *L.* VAR. *fastigiata,* Lindl. (*Irish* or *Florence-court Yew.*)

CLASS II. MONOCOTYLEDONS.

DIVISION I. PETALOIDEÆ.

ORDER I. **HYDROCHARIDEÆ.**

1. **HYDROCHARIS**, *L.* FROG-BIT.
   1. Morsus-Ranæ.
2. **STRATIOTES**, *L.* WATER SOLDIER.
   1. aloides, *L.*

2.* ANACHARIS, *Rich.* WATER-THYME. CANADENSIS, *Planch. ; Alsinastrum,* Bab. ; *Nuttallii,* Planch. ; *Udora,* Nutt.

ORDER II. **ORCHIDEÆ.**

TRIBE I. **OPHRYDEÆ.**

1. **ORCHIS**, *L.*
SUB-GEN. 1. Orchis proper.
   1. mascula, *L. Purple Orchis.*
   2. laxiflora, *Lamk.*
   3. latifolia, *L. ; palmata,* Bosw. *Marsh Orchis.*
   Sub-sp. LATIFOLIA proper.
   Sub-sp. INCARNATA, *L.* (sp.).
   4. maculata, *L. Spotted Orchis.*
   5. Morio, *L. Green-winged Orchis.*
   6. ustulata, *L. Dark-winged* or *Dwarf Orchis.*
   7. purpurea, *Huds. ; militaris,* Sm. ; *fusca,* Jacq.
   8. militaris, *L.*
   Sub-sp. MILITARIS proper.
   Sub-sp. SIMIA, *Lamk.* (sp.) ;— *tephrosanthos,* Vill.
SUB-GEN. 2. **Anacamptis,** *Rich.* (gen.).
   9. pyramidalis, *L.*
SUB-GEN. 3. **Himantoglossum,** *Rich.* (gen.).
   10. hircina, *L. Lizard Orchis.*

2. **GYMNADENIA**, *Br.*
   1. **Conopsea**, *Br. Fragrant Orchis.*

3. **HABENARIA**, *Br.*
   1. bifolia, *Br. Butterfly Orchis.*
   Sub-sp. BIFOLIA proper.
   Sub-sp. CHLORANTHA, *Bab.* (sp.).
   2. viridis, *Br. Frog-orchis.*
   3. albida, *Br. ; Gymnadenia,* Rich.
4. **NEOTINEA**, *Reich. fil.*
   1. intacta, *Reich. fil. ; Tinea,* Bivoni.
5. **ACERAS**, *Br.* MAN ORCHIS.
   1. anthropophora, *Br.*
6. **HERMINIUM**, *Ar.* MUSK ORCHIS.
   1. Monorchis, *Br.*
7. **OPHRYS**, *L.*
   1. apifera, *Huds. Bee Orchis.*
   Sub-sp. APIFERA proper ; VAR. *Trollii,* Heg. (sp.).
   Sub-sp. ARACHNITES, *Willd.* (sp.).
   2. aranifera, *Huds. Spider Orchis.*
   VAR. 1, aranifera proper.—VAR. 2, *fucifera,* Smith (sp.).
   3. muscifera, *Huds. Fly Orchis.*

TRIBE II. **ARETHUSEÆ.**

8. **EPIPOGIUM**, *Gmelin.*
   1. Gmelini, *Rich. ; aphyllum,* Sw.

TRIBE III. **NEOTTIDEÆ.**

9. **EPIPACTIS**, *Rich.* HELLEBORINE.
   1. latifolia, *Sw.*
   Sub-sp. LATIFOLIA proper (sp.) ; *viridiflora,* Hoffm. ; *purpurata,* Sm. ; *Helleborine,* var. *varians,* Crantz ; *media,* Bab.
   Sub-sp. VIRIDANS ; *Helleborine,* var. *viridans,* Crantz.
   Sub-sp. RUBIGINOSA, *Koch* (sp.) ; *ovalis,* Bab. ; *media,* Fries ; *Helleborine,* var. *rubiginosa,* Crantz.

k

2. palustris, *L.*

10. **CEPHALANTHERA,** *Richard.*
   1. grandiflora, *Bab.*   *White Helleborine.*
   2. ensifolia, *Rich.*
   3. rubra, *Rich.*, *Red Helleborine.*

11. **NEOTTIA,** *L.*   Bird's-nest Orchis.
   1. Nidus-avis, *L.*

12. **LISTERA,** *Br.*
   1. ovata, *Br.*   *Tway-blade.*
   2. cordata, *Br.*   *Lesser Tway-blade.*

13. **GOODYERA,** *Br.*
   1. repens, *Br.*

14. **SPIRANTHES,** *Rich.*   Lady's-tresses.
   1. autumnalis, *Rich.*
   2. æstivalis, *Rich.*
   3. Romanzoviana, *Cham.; cernua,* Hook.; *gemmipara,* Lindl.

Tribe IV. **MALAXIDEÆ.**

15. **MALAXIS,** *Sw.*
   1. paludosa, *Sw. Bog Orchis.*

16. **LIPARIS,** *Rich.*
   1. Loeselii, *Rich.*, *Sturmia,* Reich. *Fen Orchis.*

17. **CORALLORHIZA,** *Haller.* Coral-root.
   1. innata, *Br.*

Tribe V. **CYPRIPEDEÆ.**

18. **CYPRIPEDIUM,** *L.* Lady's Slipper.
   1. Calceolus, *L.*

———

Order III. **IRIDEÆ.**

1. **ROMULEA,** *Maratti.*
   1. Columnæ, *Seb.* and *Maur.; Trichonema,* Reich.

1*. **SISYRINCHIUM,** *L.* Blue-eyed Grass.

Bermudianum, *L. ; anceps,* Bab.

1**. **CROCUS,** *L.*
   nudiflorus, *L.*
   vernus, *L. Purple Crocus.*

2. **IRIS,** *L.*
   1. Pseud-acorus, *L. Yellow Flag.*
   2. fœtidissima, *L.   Fœtid Iris, Roast-beef plant.*

3. **GLADIOLUS,** *L.*
   1. communis, *L. ; var. illyricus,* Koch (sp.) ; *imbricatus,* Bab.

———

Order IV. **AMARYLLIDEÆ.**

1. **NARCISSUS,** *L.*
   1. Pseudo-narcissus, *L. Daffodil, Lent Lily.* Var. 1, *Pseudo-narcissus* proper.—Var. 2, *Bromfieldii,* Bosw.
   biflorus, *Curt.*

2. **GALANTHUS,** *L.* Snowdrop.
   1. nivalis, *L.*

3. **LEUCOJUM,** *L.* Snow-flake.
   1. æstivum, *L.*
   2. vernum, *L.*

———

Order V. **DIOSCOREÆ.**

1. **TAMUS,** *L.* Black Bryony.
   1. communis, *L.*

———

Order VI. **ALISMACEÆ.**

Tribe I. **ALISMEÆ.**

1. **ALISMA,** *L.*
   1. Plantago, *L. Water Plantain.—*Var. 1, *Plantago* proper.—Var. 2, *lanceolatum,* With. (sp).
   2. ranunculoides, *L.* Var. 1, *ranunculoides* proper.—Var. 2, *repens,* Daves (sp).
   3. natans, *L.*

2. **ACTINOCARPUS,** *Br.* Star-fruit.
   1. Damasonium, *Br.; Damasonium stellatum,* Pers.

3. **SAGITTARIA**, *L.*
   1. sagittifolia, *L. Arrow-head.*
   Tribe II. JUNCAGINEÆ.
4. **TRIGLOCHIN**, *L.*
   1. palustre, *L.*
   2. maritimum, *L.*
5. **SCHEUCHZERIA**, *L.*
   1. palustris, *L.*
   Tribe III. BUTOMEÆ.
6. **BUTOMUS**, *L.* Flowering Rush.
   1. umbellatus, *L.*

Order VII. NAIADEÆ.
Tribe 1. POTAMEÆ.
1. **POTAMOGETON**, *L.*
Section 1. (leaves broad, alternate; stipules free.)
   1. natans, *L.*
Sub-sp. natans proper.
Sub-sp. polygonifolius, *Pourret* (sp.); *oblongus*, Viviani. Var. *linearis*, Bosw.
Sub-sp. plantagineus, *Ducros.*
Sub-sp. kirkii, *Bosw.*; *sparganiifolius*,Bab.
   2. rufescens, *Schrad.; fluitans,* Sm.
   3. heterophyllus, *Schreb.*
Sub-sp. heterophyllus proper; *gramineus*, Fries and Koch.
Sub-sp. nitens, *Weber.*
Sub-sp.? lonchitis, *Tuckerman* (sp.); *salicifolius*, Wolfg. ?
   4. lanceolatus, *Sm.*
   5. lucens, *L.*—Var. 1, *lucens* proper.
   —Var. 2, *acuminatus*, Schum. (sp.); *longifolius*, Gay.—Var. 3, *decipiens*, Nolte (sp.).
   6. prælongus, *Wulf*
   7. perfoliatus, *L.*
   8. crispus, *L.*
Section 2. (leaves opposite, exstipulate.)
   9. densus, *L.*
Section 3. (leaves narrow, alternate.)
   10. compressus, *L.; zosteræfolius,* Schum.; *cuspidatus*, Sm.
Sub-sp. compressus proper.
Sub-sp. acutifolius, *Link* (sp.).
   11. obtusifolius, *Mert.* and *Koch.* *gramineus*, Sm.

   12. pusillus, *L.*—Var. 1, *pusillus* proper.—Var. 2, *tenuissimus,* Fries (sp.). *gracilis*, Bab.—Var. 3, *mucronatus,* Schrad.; *compressus*, Sm.
   13. trichoides, *Cham.*
Section 4. (stipules adnate to leaves).
   14. pectinatus, *L.*
Sub-sp. pectinatus proper.—Var. *marinus*, Huds.
Sub-sp. flabellatus, *Bab.* (sp.); *zosteraceus*, Bab.
Sub-sp. filiformis, *Pers.* (sp.); *marinus*, Linn. Herb.
2. **RUPPIA**, *L.*
   1. maritima, *L.*
Sub-sp. maritima proper; *spiralis*, Hartm.
Sub-sp. rostellata, *Koch.* (sp.).
3. **ZANNICHELLIA**, *L.* Horned Pond-weed.
   1. palustris, *L.*
Sub-sp. brachystemon, *Gay* (sp.); *palustris* E.B.; *major*, Bonn.
Sub-sp. pedunculata, *Reich.* (sp.); *pedicellata*, E.B.
Sub-sp. polycarpa, *Nolte* (sp.).
Sub-sp. macrostemon, *Gay* (sp.); *palustris*, Boreau.
   Tribe II. NAIADEÆ.
4. **ZOSTERA**, *L.* Grasswrack.
   1. marina, *L.*—Var. 1, *marina* proper.—Var. 2, *angustifolia*, Bosw.
   2. nana, *Roth.*
5. **NAIAS**, *L.*
   1. flexilis, *Rostkov.*

Order VIII. LILIACEÆ.
Series 1. (Rootstock creeping. Fruit a berry.)
   Tribe I. TRILLIDEÆ.
1. **PARIS**, *L.* Herb Paris.
   1. quadrifolia, *L.*
   Tribe II. ASPARAGINEÆ.
2. **ASPARAGUS**, *L.* Asparagus.
   1. officinalis, *L.*
3. **RUSCUS**, *L.* Butcher's Broom.

1. aculeatus, *L.*

4. **CONVALLARIA**, *L.* LILY OF THE VALLEY.

1. majalis, *L.*

5. **POLYGONATUM**, *Tournef.*

1. verticillatum, *All.*

2. multiflorum, *All. Solomon's Seal.*

3. officinale, *All.*

6. **SMILACINA**, *Desf.*

1. bifolia, *Desf.* ; *Maianthemum*, DC.

SERIES 2. (*Root* of thick fleshy fibres, *Fruit* capsular.)

TRIBE III. **ANTHERICEÆ.**

7. **SIMETHIS**, *Kunth.*

1. bicolor, *Kunth ; planifolia,* Woods.

SERIES 3. (*Root* bulbous, *Fruit* capsular.)

TRIBE IV. **LILIEÆ.**

8. **SCILLA**, *L.* SQUILL.

SECTION 1. (*anthers* purple, *seeds* angular).

1. verna, *Huds.*

2. autumnalis, *L.*

SECTION 2. (*anthers* yellow, *seeds* subglobose.)

3. nutans, Sm.; *Agraphis*, Link ; *Hyacinthus nonscriptus*, *L. Bluebell, Wild Hyacinth.*

9. **MUSCARI**, *Tourn.* GRAPE HYACINTH.

1. racemosum, *Miller.*

10. **ORNITHOGALUM**, *L.*

1. pyrenaicum, *L.*

UMBELLATUM, *L. Star of Bethlehem.*

NUTANS, *L.*

11. **ALLIUM**, *L.*

SECTION I. Porrum.

1. vineale, *L. Crow Garlic.* VAR. 1, *vineale* proper.—VAR. 2, *capsuliferum*, Bosw.—VAR. 3, *compactum*, Thuill. (sp.)

2. sphærocephalum, *L.*

3. Scorodoprasum, *L. ; arenarium,* Sm. *Sand Leek.*

4. Ampeloprasum, *L. Wild Leek.*

VAR. 1, *Ampeloprasum* proper.—VAR. 2, *bulbiferum*, Bosw.—VAR. 3, *Babingtonii*, Borrer (sp.) ; *Halleri*, Bab.

SECTION 2. Codonoprasum.

5. Schœnoprasum, *L. Chives.* — VAR. 1, *Schœnoprasum* proper.—VAR. 2, *sibiricum*, *L.* (sp.) ; *arenarium,* Sm.

6. oleraceum, *L. Field Garlic.* VAR. 1, *oleraceum* proper.—VAR. 2, *complanatum*, Fries (Boreau, sp.) ; *carinatum*, Sm.

7. carinatum, *L.*

8. triquetrum, *L.*

9. ursinum, *L. Ramsons.*

12. **GAGEA**, *Salisb.*

1. lutea, *Ker. Yellow Star of Bethlehem.*

13. **LLOYDIA**, *Salisb.*

1. serotina, *Reich.*

14. **FRITILLARIA**, *L.*

1. Meleagris, *L. Snake's Head.*

14*. **TULIPA**, *L.*

SYLVESTRIS, *L. Wild Tulip.*

14**. **LILIUM**, *L.*

MARTAGON, *L. Purple Martagon Lily.*

SERIES 4. (*Rootstock* creeping. *Fruit* capsular.)

TRIBE V. **COLCHICEÆ.**

15. **COLCHICUM**, *L.* MEADOW SAFFRON.

1. autumnale, *L.*

TRIBE VI. **VERATREÆ.**

16. **TOFIELDIA**, *Hudson.*

1. palustris, *Huds.* ; *borealis*, Wahl.

TRIBE VII. **NARTHECIEÆ.**

17. **NARTHECIUM**, *Hudson.* BOG ASPHODEL.

1. ossifragum, *Huds.*

ORDER IX. **JUNCEÆ.**

1. **JUNCUS**, *L.* RUSH.

SECTION 1. *Rootstock* perennial, *Stems* not septate.)

1. communis, *Meyer.*—VAR 1, *effusus*, *L.* (sp.).—VAR. 2, *conglomeratus*, *L.* (sp.).

2. glaucus, *Ehr.*

Sub-sp. ? DIFFUSUS, *Hoppe* (sp.).
3 balticus, *Willd. ; arcticus*, Hook.
4. filiformis, *L.*
5. acutus, *L.*
6. maritimus, *Sm.*
7. triglumis, *L.*
8. biglumis, *L.*
9. castaneus, *L.*
10. trifidus, *L.*
11. squarrosus, *L.*
12. compressus, *Jacq. ; bulbosus, L.*
Sub-sp. COMPRESSUS proper.
Sub-sp. GERARDI, *Loisel.* (sp.) ; *bottnicus,* Wahl. ; *cœnosus,* Bich.
SECTION 2. (Perennial. Stems septate.)
13. obtusiflorus, *Ehr.*
14. articulatus, *L.*
Sub-sp. ARTICULATUS proper ; *acutiflorus,* Ehr.
Sub-sp. SUPINUS, *Moench* (sp.).—VAR. 1, *uliginosus,* Sibth. (sp.).—VAR. 2, *subverticillatus,* Wulf. (sp.).—VAR. 3, *Kochii,* Bab. ; *nigritellus,* Koch.
Sub-sp. LAMPROCARPUS, *Ehr.*
SECTION 3. (Annual.)
15. bufonius, *L.*
16. capitatus, *Weigel.*
17. pygmæus, *Rich.*

2. LUZULA, *DC.* WOOD-RUSH.
1. sylvatica, *Bich.*
2. pilosa, *Willd.* — VAR. *Borreri,* Bromf.
3. Forsteri, *DC.*
4. campestris, *Willd.* — VAR. 1, *campestris* proper.—VAR. 2, *congesta,* Sm. (sp.) ; *multiflora,* Lej. ; *sudetica,* DC.
5. spicata, *DC.*
6. arcuata, *Hook.*

ORDER X. AROIDEÆ.
SUB-ORDER 1. AROIDEÆ proper.

1. ARUM, *L.* CUCKOO-PINT, LORDS AND LADIES.

1. maculatum, *L.*
2. italicum, *Miller.*
SUB-ORDER 2. ORONTIACEÆ.

2. ACORUS, *L.* SWEET-FLAG.
1 calamus, *L.*

———

ORDER XI. LEMNACEÆ.
1. LEMNA, *L.* DUCKWEED.
SUB-GEN. 1. Lemna proper.
1. minor, *L.*
2. trisulca, *L. Ivy-leaved Duckweed.*
SUB-GEN. 2. Telmatophace, *Schleiden* (gen.).
3. gibba, *L.*
SUB-GEN. 3. Spirodela, *Schleiden* (gen.)
4. polyrhiza, *L.*
2. WOLFFIA, *Horkel.*
1. arrhiza, *Wimm.*

———

ORDER XII. TYPHACEÆ.
1. SPARGANIUM, *L.* BUR-REED.
1. ramosum, *Hudson.*
2. simplex, *Huds.*
3. natans, *L.*—VAR. 1, *affine,* Schn. (sp.).—VAR. 2, *minimum,* Fries (sp.).
2. TYPHA, *L.* REED-MACE, CAT'S-TAIL, BULLRUSH.
1. latifolia, *L. ; media,* DC.
2. angustifolia, *L.*

ORDER XIII. ERIOCAULONEÆ.
1. ERIOCAULON, *L.* PIPE-WORT.
1. septangulare, *With.*

———

ORDER XIV. CYPERACEÆ.
TRIBE I. SCIRPEÆ
1. SCIRPUS, *L.*
1. lacustris, *L. Bullrush.*
Sub-sp. LACUSTRIS proper.
Sub-sp. TABERNÆMONTANI, *Gmel.* (sp.) ; *glaucus,* Sm.

i

Sub-sp. CARINATUS, *Sm.* (sp.).

2. triqueter, *L.*
3. Rothii, *Hoppe ; pungens,* Vahl.
4. maritimus, *L.*
5. sylvaticus, *L.*

2. ISOLEPIS, *Br.*

1. setacea, *Br.*
2. Savii, *Schultes.*—VAR. *b. monostachys ; pygmæa,* Kunth.
3. fluitans, *Br. ; Eleogiton,* Link.
4. Holoschœnus, *R* and *S.*

3. ELEOCHARIS, *Br.*

1. palustris, *Br.*—VAR. 1, *palustris proper.* — VAR. 2, *uniglumis,* Link (sp.).—VAR. 3, *Watsoni,* Bab. (sp.).
2. multicaulis, *Sm.*
3. acicularis, *Sm*
4. cæspitosa, *Link.*
5. pauciflora, *Link.*
6. parvula, *Hook.*

4. BLYSMUS, *Panzer.*

1. compressus, *Panz.*
2. rufus, *Link.*

5. ERIOPHORUM, *L.* COTTON-GRASS.

1. vaginatum, *L.*
2. alpinum, *L.*
3. polystachyon, *L.*

Sub-sp. ANGUSTIFOLIUM proper, *Roth* (sp.) ; *gracile,* Sm.

Sub-sp. LATIFOLIUM, *Hoppe* (sp.) ; *pubescens,* Sm.

4. gracile, *Koch.*

TRIBE II. **RHYNCHOSPOREÆ.**

6. CLADIUM, *P. Brown.*

1. Mariscus, *Br.*

7. RHYNCHOSPORA, *Vahl.*

1. alba, *Vahl.*
2. fusca, *R* and *S.*

TRIBE III. **CYPEREÆ.**

8. SCHŒNUS, *L.*

1. nigricans, *L.*

9. CYPERUS, *L.*

1. longus, *L   Galingale.*
2. fuscus, *L.*

TRIBE IV. **CARICINEÆ.**

10. KOBRESIA, *Willd.*

1. caricina, *Willd.*

11. CAREX, *L.* SEDGE.

SECTION 1. (*Spikelet* solitary, terminal).

1. pauciflora, *Light.*
2. pulicaris, *L.*
3. dioica, *L.*
4. rupestris, *All.*

SECTION 2. (*Spikelets* most or all 2-sexual, *Stigmas* 2.)

5. incurva, *Light.*
6. divisa, *Huds.*
7. disticha, *Huds. ; intermedia,* Good.
8. arenaria, *L.*
9. paniculata, *L.*

Sub-sp. PARADOXA. *Willd.* (sp.).

10. teretiuscula, *Good.*
11. muricata, *L.*

Sub-sp. MURICATA proper.

Sub-sp. DIVULSA, *Good.* (sp.).

12. vulpina, *L.*
13. stellulata, *Good. ; echinata,* Murr.
14. remota, *L. ; tenella,* Sm.
15. axillaris, *wood.*

Hyb ? *Bænninghauseniana,* Weihe, bet. 14 and 15.

16. leporina, *L. ; ovalis,* Good.
17. canescens, *L. ; curta,* Good. VAR. 1, *canescens* proper.—VAR. 2, *alpicola,* Wahl. ; *vitilis,* Fries.
18. lagopina, *Wahl. ; leporina,* L. in part.
19. elongata, *L.*

SECTION 3. (*Spikelets,* 2-4 lower wholly female, upper male at base only, *stigmas* 3.)

20. Buxbaumii, *Wahl. ; canescens,* Auct. ; *polygama,* Schkubr.
21. alpina, *Swartz. ; Vahlii,* Schk.

22. atrata, *L.*

SECTION 4. (*Spikelets* mainly 1-sexual, lower all or chiefly female.)

23. rigida, *Good.*

24. acuta, *L. ; gracilis,* Curtis.

25. stricta, *Good.; cæspitosa,* Gay.

26. aquatilis, *Wahl.*—VAR. 1, *aquatilis* proper.— VAR. 2, *minor,* Boott. Var. *Watsoni,* Bosw.

27. vulgaris, *Fries ; cæspitosa,* Sm.; *Goodenovii,* Gay.—VAR. ? *Gibsoni,* Bab. (sp.).

28. limosa, *L.*

Sub-sp. IRRIGUA, *Hoppe,* (sp.).

29. rariflora, *Sm.*

30. glauca, *Scop.*—VAR. 1, *glauca* proper ; *recurva,* Huds.—VAR. 2, *Micheliana,* Sm. (sp.).—VAR. 3, *stictocarpa,* Sm. (sp.).

31. pallescens, *L.*

32. panicea, *L.*

33. vaginata, *Tausch'; salina,* Don ; *Mielichoferi,* and *phæostachya,* Sm ; *scotica,* Spr. ; *sparsiflora,* Steud.

34. capillaris, *L.*

35. pendula, *Huds.*

36. præcox, *Jacq.*

37. tomentosa, *L.*

38. pilulifera, *L.*

39. montana, *L. ; collina,* Willd.

40. ericetorum, *Poll.*

41. digitata, *L.*

42. ornithopoda, *Willd.*

43. humilis, *Leyss ; clandestina,* Good.

44. hirta, *L.* Var. *hirtiformis,* Pers. var. *ebracteata.*

45. filiformis, *L.*

46. extensa, *Good.*

47. flava, *L.*—VAR. 1, *flava* proper.— VAR. 2, *Œderi,* Ehr. (sp.).—VAR. 3, *lepidocarpa,* Tausch (sp.).

48. distans, *L.*

Sub-sp. DISTANS proper.

Sub-sp. FULVA, *Good.* (sp.).—VAR. 1, (bracts long).—VAR. 2, *Hornschuchiana,* Hoppe (sp.) (bracts shorter) ; *speirostachya,* Sm.

49. binervis, *Sm.*

50. lævigata, *Sm.*

51. punctata, *Gaud.*

52. frigida, *Allioni.*

53. depauperata, *Good.*

54. sylvatica, *Huds.*

55. strigosa, *Huds.*

56. vesicaria, *L.*

Sub-sp. VESICARIA proper.—VAR. *Grahami,* Boott. (sp.).

Sub-sp. SAXATILIS, *L.* (sp.); *pulla,* Good.; *vesicaria* var. *alpigena,* Fries.

57. ampullacea, *Good.*

58. Pseudo-cyperus, *L.*

59. paludosa, *Good.*

60. riparia, *Curtis.*

———

ORDER XV. **GRAMINEÆ.**

SERIES I. CLISANTHEÆ, *Fries.* (*Spikelets* closed in flower.)

TRIBE I. **PANICEÆ.**

1. **PANICUM,** *L.*

SUB-GEN. 1. Digitaria, *Scop.* (gen.).

1. glabrum, *Gaud. ; Digitaria humifusa,* Pers.

SUB-GEN. 2. Echinochloa, *Beauv.* (gen.).

CRUS-GALLI, *L.*

2. **SETARIA,** *Beauv.*

1. viridis, *Beauv.*

VERTICILLATA, *Beauv.*

TRIBE II. **PHALARIDEÆ.**

3. **NARDUS,** *L.* MAT-WEED.

1. stricta, *L.*

4. **SPARTINA,** *Schreber.*

1. stricta, *Roth.*

Sub-sp. STRICTA proper.

Sub-sp. ALTERNIFLORA, *Loisel.* (sp.).

5. **CHAMAGROSTIS,** *Borkh.*

1. minima, *Borkh. ; Sturmi aminima,* Hoppe ; *Mibora verna,* Adans.; *Knappia agrostidea,* Sm.

6. **PHLEUM,** *L.* CAT'S-TAIL GRASS.

1. **pratense**, *L.*   *Timothy-grass.—* VAR. 1, *pratense* proper. — VAR. 2, *nodosum, L.* (sp.).

2. alpinum, *L.*

3. arenarium, *L.*

4. Boehmeri, *Wibel.*

7. **ALOPECURUS**, *L.*   FOX-TAIL GRASS.

1. agrestris, *L.*

2. alpinus, *Sm.*—Var. *Watsoni,* Bosw.

3. pratensis, *L.*

4. bulbosus, *L.*

5. geniculatus, *L.*

6. fulvus, *Sm.*

8. **PHALARIS**, *L.*

SUB-GEN. 1. Phalaris proper.
   CANARIENSIS. *Canary grass.*

SUB-GEN. 2. Digraphis, *Trin* (gen.).

1. arundinacea, *L.*

9. **ANTHOXANTHUM**, *L.* VERNAL GRASS.

1. odoratum, *L.*
PUELII, *Lecoq* and *Lamotte.*

10. **HIEROCHLOE**, *Gmelin,* HOLY GRASS.

1. borealis, *R.* and *S.*

11. **SESLERIA**, *Scop.*

1. cærulea, Scop.

SERIES II.   EURYANTHEÆ.   *Fries.*
   (*Spikelets* open in flower.)

   TRIBE III. **POACEÆ.**
   SUB-TRIBE Agrostideæ.

12. **AGROSTIS**, *L.* BENT.

SUB-GEN. 1. Trichodium, *Michx.* (gen.)

1. canina, *L.*

2. setacea, *Curt.*

SUB-GEN. 2. Agrostis proper.

3. vulgaris, *With.*—VAR. 1, *vulgaris* proper.—VAR. 2, *pumila,* L. (sp.).

4. alba, *L. Fiorin Grass.* Var. *stolnifera,* L. (*b* subrepens, Bab.)

SUB-GEN. 3. Apera, *Beauv.* (gen.).

5. Spica-venti, *L.; Anemagrostis,* Bosw.

Sub-sp. SPICA-VENTI proper.

Sub-sp. INTERRUPTA, *L.* (sp.).

SUB-GEN 4.   Gastridium, *Beauv.* (gen.)

6. australis, *L. Gastridium lendigerum,* Gaud. *Nit-grass.*

13. **MILIUM**, *L.* MILLET-GRASS.

1. effusum, *L.*

14. **POLYPOGON**, *Desf.* BEARD-GRASS.

1. monspeliensis, *Desf.*

2. littoralis, *Sm.*

15. **LAGURUS**, *L.* HARE'S-TAIL GRASS.

1. ovatus, *L.*

16. **PSAMMA**, *Beauv.* MARREM-GRASS.

1. arenaria, *R* and *S; Ammophila arundinacea,* Host.

2. baltica, *Roem.* and *Schultes.*

17. **CALAMAGROSTIS**, *Adans.*

SUB-GEN. 1. Calamagrostis proper.

1. Epigejos, *Roth.*

2. lanceolata, *Roth.; Arundo calamagrostis,* L.

SUB-GEN. 2. Deyeuxia, *Beauv.*

3. stricta, *Nutt ; lapponica,* Hook.— VAR. 1, *stricta* proper. — VAR. 2, *Hookeri,* Bosw.

SUB-TRIBE. Oryzeæ.

18. **LEERSIA**, *Soland.*

1. oryzoides, *Swartz.*

SUB-TRIBE. Chlorideæ.

19. **CYNODON**, *Rich.* DOG'S-TOOTH GRASS.

1. Dactylon, *Pers. Doab* and *Bermuda Grass.*

SUB-TRIBE. Avenaceæ.

20. **AIRA**, *L.*

SUB-GEN. 1. Deschampsia, *Beauv.* (gen.).

1. flexuosa, *L.*—VAR. 1, *montana,* L. (sp.). — VAR. 2, *uliginosa* Weihe, (sp.) ; *setacea,* Huds.

2. cæspitosa, *L. ; major,* Bosw.
Sub-sp. CÆSPITOSA proper.

Sub-sp. ALPINA, L. (sp.); *lævigata*, Sm.

SUB-GEN. 2. Airopsis, *Desv.* (gen.).

3. caryophyllea, *L.*
4. præcox, *L.*

UB-GEN. 3. Corynephorus, *Beauv.* (gen.).

5. canescens, *L.*

21. AVENA, *L.* OAT.

SUB-GEN. 1. Avena proper.

1. fatua, *L.*
   STRIGOSA, *Schreb.*
2. pratensis, *L.* — VAR. 1, *pratensis* proper.—VAR. 2, *alpina*, Kunth (sp.) ; *planiculmis*, Sm.
3. pubescens, *L.*

SUB-GEN. 2. Trisetum, *Pers.* (gen.).

4. flavescens, *L.*

SUB-GEN. 3, Arrhenatherum, *Beauv.* (gen.).

5. elatior, *L. ; Arrenatherum avenaceum*, Beauv.

22. HOLCUS, *L.*

1. lanatus, *L.*
2. mollis, *L.*

23. TRIODIA, *Br.*

1. decumbens, *Beauv, Danthonia*, DC.

SUB-TRIBE Festuceæ.

24. PHRAGMITES, *Trin.* REED.

1. communis, *Trin. ; Arundo Phragmites*, L.

25. MELICA, *L.*

1. nutans, *L.*
2. uniflora, *Retz.*

26. DACTYLIS, *L.* COCK'S-FOOT GRASS.

1. glomerata, *L.*

27. KOELERIA, *Pers.*

1. cristata, *Pers.*

28. MOLINIA, *Schrank.*

1. cærulea, *Moench ; Enodium*, Gaud. VAR. 1, *cærulea* proper.—VAR. 2, *depauperata*, Lindl. (sp.).

29. POA, *L.* MEADOW-GRASS.

1. annua, *L.*
2. pratensis, *L.; VARS. subcærulea*, Sm. ; *angustifolia*, L. (sp.) ; *strigosa*, Gaud.
3. compressa, *L.*— VARS. *subcompressa*, Parn. ; *polynoda*, Parn.
4. trivialis, *L.* Var. *b. Koeleri* DC. (sp.).
5. nemoralis, *L.*—VAR. 1, *nemoralis* proper.—VAR. 2, *Parnellii*, Bab. (sp.). —VAR. 3, *glauca*, Sm. ; *Balfourii*, Parn. ; *montana*, Parn.
6. laxa, *Haenke.*

Sub-sp. 1, LAXA proper ; *flexuosa*, Sm. ; *minor*, Bab.

Sub-sp. 2, STRICTA, Lind. (sp.) ; *laxa*, Bab. (ed. vi.)

7. alpina, *L.*
8. bulbosa, *L.*

30. CATABROSA, *Beauv.*

1. aquatica, *Beauv.*—VAR. *littoralis*, Parn.

31. GLYCERIA, *Br.* MANNA GRASS.

SECTION 1. Glyceria proper.

1. aquatica, *Sm. ; spectabilis*, Mert and Koch.
2. fluitans, *Br.*

Sub-sp. FLUITANS proper.

Sub-sp. PLICATA, *Fries.* (sp.) ; *pedicellata*, Towns.

SECTION 2. Sclerochloa, *Beauv.* (gen.)

3. maritima, *Wahlb.*
4. distans, *Wahlb ; Scl. multiculmis*, Bosw.

Sub-sp. DISTANS proper.

Sub-sp. *Borreri*, Bab. (sp.) ; CONFERTA, *Fries.*

5. procumbens, *Sm.*
6. rigida, *Sm.*
7. loliacea, *Watson ; Poa*, Huds. ; *Sclerochloa*, Woods.

32. BRIZA, *L.* QUAKING GRASS.

1. media, *L.*
2. minor, *L.*

m

33. FESTUCA, *L.* Fescue Grass.

Section 1. Schenodorus, *Beauv.*

1. elatior, *L.*

2. pratensis, *Huds.*—Var. 1, *pratensis* proper.—Var. 2, *loliacea,* Huds. (sp.).

3. gigantea, *Vill. ;* Gen. *Bromus,* L.

4. sylvatica, *Vill. ; Calamaria,* Sm.

Section 2. Festuca proper.

5. ovina, *L.*

Sub-sp. ovina proper.—Var. 1, *ovina* proper.—Var. 2, *tenuifolia,* Sibth. (sp.). Var. 3, *glauca,* (sp.).—Var. 4. *major,* Bosw.

Sub-sp. duriuscula, *L.* (sp.). *cæsia,* Sm.

Sub-sp. rubra, *L.* (sp.).—Var. *arenaria,* Osb. (sp.); *sabulicola,* Duf. ; *rubra,* Sm.

Section 3. Vulpia, *Gmel,* (gen.).

6. Myurus, *L.*

Sub-sp. Myurus proper ; *pseudo-myurus,* Koch.—Var. *ambigua,* Le Gall (sp.).

Sub-sp. sciuroides, *Roth.* (sp.) ; *bromoides,* Sm.

7. uniglumis, *L.*

34. BROMUS, *L.* Brome Grass.

Section 1. Bromus proper.

1. asper, *Murr. ; ramosus,* Huds. Var. 1, *serotinus.* Benek. (sp.).— Var. 2, *Benekenii,* Bosw.

2. erectus, *Huds.*

3. sterilis, *L.*

4. madritensis, *L.*—Var.1, *madritensis* proper. — Var. *Curtisii,* Bab., *diandrus,* Curt.—Var. 2, *rigidus,* Roth (sp.).

5. maximus, *Desf.*

Section 2. Serrafalcus, *Parl.*

6. mollis, *L.* Vars. *glabrescens* and *Lloydianus* Bosw. (*molliformis,* Lloyd.).

7. racemosus, *L.*—Var. 2, *commutatus,* Schrad. (sp.); *pratensis,* Ehr., *arvensis,* Sm.

8. secalinus, *L.* arvensis, *L.*

35. CYNOSURUS, *L.* Dog's-tail Grass.

1. cristatus, **L.**

2. echinatus, *L.*

Sub-Tribe Hordeaceæ.

36. BRACHYPODIUM, *Beauv.*

1. sylvaticum, *R.* and *S.*

2. pinnatum, *Beauv.*

37. TRITICUM, *L.* Wheat.

1. caninum, *Huds.*

2. repens, *L. Couch-grass.*

Sub-sp. 1. repens proper.—Var. 1, *repens* proper.—Var. 2, *barbatum.*—Var. 3, *obtusum.*

Sub-sp. 2. pungens, *Ræm.* and *Sch.* (sp.).—Var. 1, *pungens* proper. — Var. 2, *littorale,* Reich. (sp.).—Var. 3, *pycnanthum,* Gren. and Godon. (sp.).

Sub-sp. 3, acutum, *DC.* (sp.); *laxum,* Fries.

3. junceum, *L.*

38. LOLIUM, *L.* Rye-grass.

1. perenne, *L. ; tenue,* L. ; N.B. cult. forms *italicum,* A. Br. ; *multiflorum,* Lamk ; *perenne,* var. *aristatum.*

2. temulentum, *L. Darnel.* Var. *arvense,* With. (sp.).

39. LEPTURUS, *Br.*

1. filiformis, *Trin.*—Var. 1, *filiformis* proper.—Var. 2, *incurvatus,* Trin. (sp.).

40. ELYMUS, *L.* Lyme-grass.

1. arenarius, *L.*

41. HORDEUM, *L.* Barley.

1. sylvaticum, Huds. *Elymus europæus,* L.

2. pratense, *Huds.*

3. murinum, *L. Way-bent, Barley-grass.*

4. maritimum, *With.; Squirrel-tail grass.*

Class III. ACOTYLEDONES or CRYPTOGAMS.

Division I. VASCULARES.

Order I. FILICES.

Tribe 1. **HYMENOPHYLLEÆ.**

1. **HYMENOPHYLLUM,** Sm. Filmy-fern.
   1. tunbridgense, Sm.
   2. unilaterale, Willd.; Wilsoni, Hook.

2. **TRICHOMANES,** L. Bristle-fen.
   1. radicans, Sw.—Var. 1, speciosum, Willd. (sp.) ; brevisetum, Br. ; Hymenophyllum alatum, Sm.—Var. 2, Andrewsii, Newm. (sp.).

Tribe II. **POLYPODIEÆ.**

3. **ADIANTUM,** Maiden-hair.
   1. Capillus-Veneris, L.

4. **PTERIS,** Brake or Bracken.
   1. aquilina, L.

5. **CRYPTOGRAMME,** Br. Parsley-fern, Rock-brake.
   1. crispa, Br.; Gen. Allosorus, Bernh.

6. **LOMARIA,** Willd. Hard-fern.
   1. Spicant, Desv. ; Blechnum boreale, Sw.

7. **ASPLENIUM,** L. Spleenwort.
   Sub-gen. 1. Asplenium proper.
   1. Ruta-muraria, L.
   2. germanicum, Weiss ; alternifolium, Wulf.
   3. septentrionale, Hull.
   4. Trichomanes, L.—Var. anceps, Sol. (sp.).
   5. viride, Huds.
   6. marinum, L.

7. lanceolatum, Huds.
8. Adiantum-nigrum, L.—Var. acutum, Bory (sp.).—Var. obtusum.
Sub-gen. 2. Athyrium, Roth (gen.).
9. Filix-fœmina Bernh.—Var. 1 Filix-fœmina proper ; Ath. latifolium, Bab.—Var. 2, rhæticum, Roth (sp.). Var. convexum, Newm.—Var. 3, molle, Roth. (sp.).—Var. 4, incisum, Hoffm. (sp.).
Sub-gen. 3. Ceterach, Willd. (gen. Hemidictyum, Presl.
10. Ceterach, L. Cet. officinarum, Desv.

8. **SCOLOPENDRIUM,** Sw. Hart's-Tongue.
   1. vulgare, Sm.

9. **WOODSIA,** Br.
   1. hyperborea, Br.
   2. ilvensis, Br.

10. **CYSTOPTERIS,** Bernh. Bladder-fern.
    1. fragilis, Bernh.
    Sub-sp. 1, fragilis proper.—Var. 1, (sori distant from margin).—Var. 2, dentata, Hook. (sp.).—Var. 3, Dickieana, Sim. (sp.).
    Sub-sp. 2, alpina, Desv. (sp.).
    2. montana, Link.

11. **ASPIDIUM,** Sw. Shield-fern.
    1. Lonchitis, Sw.
    2. aculeatum, Sw.
    Sub-sp. lobatum, Sw.—Var. lonchitidoides.
    Sub-sp. aculeatum proper.
    Sub-sp. angulare, Willd.

12. NEPHRODIUM, *Rich.*

1. **Filix-mas,** *Rich. Male Fern.* VAR. 1, *Filix-mas* proper.—VAR. 2, *affine,* Fisch, (sp.) ; (var. *incisum.* Newm.)— VAR. 3, *Borreri,* Newm. — VAR. 4, *abbreviatum,* DC. (sp.).

2. **cristatum,** *Rich.*—VAR. *uliginosum,* Newm. (sp.).

3. **rigidum,** *Desv.*

4. **spinulosum,** *Desv.*
Sub-sp. SPINULOSUM proper.
Sub-sp. DILATATUM, *Desv.* (sp.).—VAR. 1, *glandulosum,* Newm.—VAR. 2, *nanum* Newm.—VAR. 3, *Boottii,* Tuck. (sp.). *Lastrea collina,* Newm.—VAR. 4, *dumetorum,* Sm. (sp.).
Sub-sp. REMOTUM, *Braun* (sp.).

5. **æmulum,** *Baker ; fænisecii,* Lowe ; *Lastrea recurva,* Bree ; L. *æmula,* Brack.

6. **Thelypteris,** *Desv.*

7. **Oreopteris,** *Desv.; montanum,* Baker.

13. POLYPODIUM, *L.* POLYPODY.
SECTION 1. Polypodium proper.

1. **vulgare,** *L.* VAR. *cambricum,* L. (sp.).
SECTION 2. Phegopteris.

2. **Phegopteris,** *L.*

3. **Dryopteris,** *L.*
Sub-sp. DRYOPTERIS proper.
Sub-sp. ROBERTIANUM, *Hoffm.* (sp.) ; *calcareum,* Sm.

4. **alpestre,** *Hoppe ;* (gen.) *Pseudathyrium,* Newm.—VAR. 1, *alpestre* proper. — VAR. 2, *humile* (*flexile,* Moore) (sp.).

14. GYMNOGRAMME, *Desv.*

1. **leptophylla,** *Desv.*

TRIBE III. **OSMUNDEÆ.**

15. OSMUNDA, *L.* FERN-ROYAL.

1. **regalis,** *L.*

TRIBE IV. **OPHIOGLOSSEÆ.**

16. OPHIOGLOSSUM, *L.* ADDER'S-TONGUE.

1. **vulgatum,** *L.*
Sub-sp. VULGATUM proper.—VAR. *ambiguum,* Coss. and Germ.
Sub-sp. LUSITANICUM, *L.* (sp.).

17. BOTRYCHIUM, *Sw.* MOON-WORT.

1. **Lunaria,** *Sw.*—VAR. *rutaceum,* Sw. (sp.).

ORDER II. **EQUISETACEÆ.**

1. EQUISETUM, *L.* HORSE-TAIL, PADDOCK-PIPES.

1. **arvense,** *L.*

2. **pratense,** *Ehr. ; umbrosum.* Willd. ; *Drummondii,* Hook.

3. **maximum,** *Lamk. ; Telmateia,* Ehr. ; *fluviatile,* Sm.

4. **sylvaticum,** *L.*

5. **palustre,** *L.*—VAR. 1, *palustre* proper. — VAR. 2, *polystachyon ;* VAR. *alpinum* (or *subnudum.*).

6. **limosum,** *L.*—VAR. 1, *limosum* proper.—VAR. 2, *fluviatile,* L. (sp.).

7. **hyemale,** *L. Dutch Rush.*—VAR. 1, *hyemale* proper.—VAR. 2, *Moorei,* Newm. (sp.) ; *paleaceum,* Schleich.

8. **variegatum** *Schleich.* — VAR. 1, *variegatum* proper.—VAR. 2, *arenarium,* Newm. (sp.).—VAR. 3, *Wilsoni,* Newm. VAR. 4, *trachyodon,* Braun (sp.) ; *Mackaii,* Newm.

ORDER III. **LYCOPODIACEÆ.**

1. LYCOPODIUM, *L.* CLUB-MOSS.

1. **clavatum,** *L.*

2. **annotinum,** *L.*

3. **alpinum,** *L.*

4. **inundatum,** *L.*

5. **Selago,** *L.*

ORDER IV. **SELAGINELLACEÆ.**

1. SELAGINELLA, *Beauv.*

1. **selaginoides,** *Gray ; spinosa,* Beauv.

2. ISOETES, *L.* QUILLWORT.

1. **lacustris,** *L.*
Sub-sp. LACUSTRIS proper.
Sub-sp. ECHINOSPORA, *Durieu* (sp.).

2. **Hystrix,** *Durieu ; Duriæi,* Hook.

ORDER V. **MARSILEACEÆ.**

1. PILULARIA, *L.* PILLWORT.

1. **globulifera,** *L.*

Printed in the United States
By Bookmasters